Methods in Enzymology

Volume 293

ION CHANNELS

PART B

METHODS IN ENZYMOLOGY

Methods in Enzymology

Volume 293

Ion Channels Part B

EDITED BY

P. Michael Conn

OREGON REGIONAL PRIMATE RESEARCH CENTER
BEAVERTON, OREGON

ACADEMIC PRESS
San Diego London Boston New York Sydney Tokyo Toronto

This book is printed on acid-free paper. ♾

Academic Press
15 East 26th Street, 15th Floor, New York, New York 10010, USA
http://www.academicpress.com

Academic Press
24-28 Oval Road, London NW1 7DX, UK
http://www.hbuk.co.uk/ap/

International Standard Book Number: 0-12-182194-3

PRINTED IN THE UNITED STATES OF AMERICA
98 99 00 01 02 03 MM 9 8 7 6 5 4 3 2 1

Table of Contents

Section III. Electrophysiology

Section IV. Expression Systems

Section V. Model Simulations

Contributors to Volume 293

Article numbers are in parentheses following the names of contributors.
Affiliations listed are current.

JOHN P. ADELMAN (4), *Vollum Institute, Oregon Health Sciences University, Portland, Oregon 97201*

MILES H. AKABAS (8), *Center for Molecular Recognition, Departments of Medicine and Cellular Biophysics, Columbia University College of Physicians and Surgeons, New York, New York 10032-3702*

AKINORI AKAIKE (18), *Department of Pharmacology, Kyoto University Graduate School of Pharmaceutical Sciences, Kyoto University, Kyoto, 606 8501, Japan*

DANIEL L. ALKON (12), *Laboratory of Adaptive Systems, National Institute of Neurological Disorders and Stroke, National Institutes of Health, Bethesda, Maryland 20892*

VLADIMIR AVDONIN (36), *Department of Physiology and Biophysics, University of Iowa, Iowa City, Iowa 52242*

TED BEGENISICH (22, 30), *Department of Physiology, University of Rochester Medical Center, Rochester, New York 14642-8642*

FRANCISCO BEZANILLA (17, 19), *Departments of Physiology and Anesthesiology, Brain Research Institute, University of California Los Angeles, Los Angeles, California 90095*

CHRIS T. BOND (4), *Vollum Institute, Oregon Health Sciences University, Portland, Oregon 97201*

J. GERARD G. BORST (20), *Abteilung Zellphysiologie, Max-Planck Institut für Medizinische Forschung, D-69120 Heidelberg, Germany*

EMMANUEL BOURINET (5), *Centre de Recherches de Biochimie Macromoleculaire, Montpellier, Cedex 5, France*

GARY BUELL (26), *Department of Molecular Biology, Serono Pharmaceutical Research Institute S.A., Geneva, Switzerland*

SHIN-HO CHUNG (24), *Department of Chemistry, John Curtin School of Medical Research, Australian National University, Canberra, ACT 2601, Australia*

NORMAN DAVIDSON (27), *Division of Biology, California Institute of Technology, Pasadena, CA 91125*

CAROL DEUTSCH (2), *Department of Physiology, University of Pennsylvania, Philadelphia, Pennsylvania 19104-6085*

DENNIS A. DOUGHERTY (28), *Division of Chemistry and Chemical Engineering, California Institute of Technology, Pasadena, California 91125*

ALISON L. EERTMOED (31), *Committee on Neurobiology, University of Chicago, Chicago, Illinois 60637*

MARKUS U. EHRENGRUBER (27), *Division of Biology, California Institute of Technology, Pasadena, California 91125*

RICHARD F. GABER (6), *Department of Biochemistry, Molecular Biology and Cell Biology, Northwestern University, Evanston, Illinois 60208*

PETER W. GAGE (24), *Department of Chemistry, John Curtin School of Medical Research, Australian National University, ACT 2601, Australia*

JUSTIN P. GALLIVAN (28), *Division of Chemistry and Chemical Engineering, California Institute of Technology, Pasadena, California 91125*

ALAN L. GOLDIN (29), *Department of Microbiology and Molecular Genetics, University of California Irvine, Irvine, California 92697-4025*

M. B. GOODMAN (13), *Department of Biological Sciences, Columbia University, New York, New York 10027*

MICHAEL E. GREEN (35), *Department of Chemistry, City College of the City Univer-*

sity of New York, New York, New York 10031-9137

WILLIAM N. GREEN (31), *Department of Pharmacological and Physiological Sciences, University of Chicago, Chicago, Illinois 60637*

PAVEL A. GUSEV (12), *Laboratory of Adaptive Systems, National Institute of Neurological Disorders and Stroke, National Institutes of Health, Bethesda, Maryland 20982*

JOHN W. HANRAHAN (11), *Department of Physiology, McGill University, Montreal, Quebec, Canada H3G 1Y6*

STEFAN H. HEINEMANN (23), *Max Planck Research Unit, Molecular and Cellular Biophysics, D-07747 Jena, Germany*

FRITJOF HELMCHEN (20), *Bell Laboratories, Lucent Technologies, Murray Hill, New Jersey 07974*

CHRISTIAN HENNESTHAL (23), *Max Planck Research Unit, Molecular and Cellular Biophysics, D-07747 Jena, Germany*

DONALD W. HILGEMANN (15), *Department of Physiology, University of Texas Southwestern Medical Center, Dallas, Texas 75235-9040*

RICHARD HORN (9), *Department of Physiology, Institute of Hyperexcitability, Thomas Jefferson University, Jefferson Medical College, Philadelphia, Pennsylvania 19107*

TOSHINORI HOSHI (36), *Department of Physiology and Biophysics, University of Iowa, Iowa City, Iowa 52242*

TAKAHIRO M. ISHII (4), *Vollum Institute, Oregon Health Sciences University, Portland, Oregon 97201*

MARK C. JASEK (27), *Division of Biology, California Institute of Technology, Pasadena, California 91125*

YANLIN JIA (11), *Department of Physiology, McGill University, Montreal, Quebec, Canada H3G 1Y6*

SHUJI KANEKO (18), *Laboratory of Neuropharmacology, Kyoto University Graduate School of Pharmaceutical Sciences, Kyoto University, Kyoto, 606 8501, Japan*

DAVID B. KANTOR (27), *Division of Biology, California Institute of Technology, Pasadena, California 91125*

ARTHUR KARLIN (8), *Center for Molecular Recognition, Departments of Biochemistry and Molecular Biophysics, Neurology, Physiology and Cellular Biophysics, Columbia University College of Physicians and Surgeons, New York, New York 10032-3702*

ZIE KONE (11), *Department of Physiology, McGill University, Montreal, Quebec, Canada H3G 1Y6*

CESAR G. LABARCA (28), *Division of Biology, California Institute of Technology, Pasadena, California 91125*

MARKUS LANZREIN (27), *Division of Biology, California Institute of Technology, Pasadena, California 91125*

HENRY A. LESTER (27, 28), *Division of Biology, California Institute of Technology, Pasadena, California 91125*

RICHARD A. LEVIS (14), *Department of Molecular Biophysics and Physiology, Rush University Medical Center, Chicago, Illinois 60612*

MIN LI (1), *Departments of Physiology and Neuroscience, Johns Hopkins University School of Medicine, Baltimore, Maryland 21205*

PAUL LINSDELL (11), *Department of Physiology, McGill University, Montreal, Quebec, Canada H3G 1Y6*

SHAWN R. LOCKERY (13), *Institute of Neuroscience, University of Oregon, Eugene, Oregon 97403*

PHILIPPE LORY (5), *Centre de Recherches de Biochimie Macromoleculaire, Montpellier Cedex 5, France*

CHIN-CHIH LU (15), *Department of Texas Southwestern Medical Center, Dallas, Texas 75235-9040*

JIEXIN LUO (11), *Department of Physiology, McGill University, Montreal, Quebec, Canada H3G 1Y6*

KARL L. MAGLEBY (25), *Department of Physiology and Biophysics, University of Miami School of Medicine, Miami, Florida 33101*

CERI J. MATHEWS (11), *Department of Physiology, McGill University, Montreal, Quebec, Canada H3G 1Y6*

JAMES MAYLIE (4), *Department of Obstetrics and Gynecology, Oregon Health Sciences University, Portland, Oregon 97201*

ROBERT L. NAKAMURA (6), *Department of Biochemistry, Molecular Biology and Cell Biology, Northwestern University, Evanston, Illinois 60208*

JOËL NARGEOT (5), *Centre de Recherches de Biochimie Macromoleculaire, Montpellier Cedex 5, France*

THOMAS J. NELSON (12), *Laboratory of Adaptive Systems, National Institute of Neurological Disorders and Stroke, National Institutes of Health, Bethesda, Maryland 20892*

MARTIN NIETHAMMER (7), *Department of Neurobiology, Massachusetts General Hospital and Harvard Medical School, Boston, Massachusetts 02114*

MARK W. NOWAK (28), *Departments of Psychiatry and Pharmacology, Medical University of South Carolina, Charleston, South Carolina 29425*

ROBERT E. OSWALD (32), *Department of Pharmacology, College of Veterinary Medicine, Cornell University, Ithaca, New York 14853*

L. DONALD PARTRIDGE (21), *Department of Neurosciences, University of New Mexico School of Medicine, Albuquerque, New Mexico 87131*

JAMES L. RAE (14), *Department of Physiology and Biophysics, Mayo Foundation, Rochester, Minnesota 55905*

KATHRYN RADFORD (26), *Biosignal, Montreal, Canada H3J 1R4*

SYLVAIN RICHARD (5), *Centre de Recherches de Biochimie Macromoleculaire, Montpellier Cedex 5, France*

BRAD S. ROTHBERG (25), *Department of Physiology and Biophysics, University of Miami School of Medicine, Miami, Florida 33101*

MARK S. P. SANSOM (34), *Laboratory of Molecular Biophysics, University of Oxford, Oxford, OX1 3QU, United Kingdom*

KOHJI SATO (10), *Department of Anatomy and Neuroscience, Ehime University School of Medicine, Ehime, 791-0204, Japan*

MASAMICHI SATOH (18), *Department of Molecular Pharmacology, Kyoto University Graduate School of Pharmaceutical Sciences, Kyoto University, Kyoto, 606 8501, Japan*

ERIN M. SCHUMAN (27), *Division of Biology, California Institute of Technology, Pasadena, California 91125*

SEAN-PATRICK SCOTT (33), *Department of Biochemistry and Biophysics, School of Medicine, University of Pennsylvania, Philadelphia, Pennsylvania 19104-6089*

MORGAN SHENG (7), *Howard Hughes Medical Institute and Department of Neurobiology, Massachusetts General Hospital and Harvard Medical School, Boston, Massachusetts 02114*

ZUFANG SHENG (2), *Department of Physiology, University of Pennsylvania, Philadelphia, Pennsylvania 19104-6085*

THEODORE M. SHIH (29), *Department of Microbiology and Molecular Genetics, University of California Irvine, Irvine, California 92697-4025*

SCOTT K. SILVERMAN (28), *Department of Chemistry and Biochemistry, University of Colorado at Boulder, Boulder, Colorado 80309*

ALLISTER H. SMEETON (32), *Department of Chemistry, University of Leicester, Leicester, LE1 7RH, United Kingdom*

RAYMOND D. SMITH (29), *Department of Microbiology and Molecular Genetics, University of California Irvine, Irvine, California 92697-4025*

PER STAMPE (30), *Department of Physiology, University of Rochester Medical Center, Rochester, New York 14642*

ENRICO STEFANI (17, 19), *Departments of Anesthesiology and Physiology, Brain Research Institute, University of California Los Angeles, Los Angeles, California 90095*

RÜDIGER STEFFAN (23), *Max Planck Research Unit, Molecular and Cellular Biophysics, D-07747 Jena, Germany*

WALTER STÜHMER (16), *Max-Planck-Institut für Experimentelle Medizin, D-37075 Göttingen, Germany*

MICHAEL J. SUTCLIFFE (32), *Department of Chemistry, University of Leicester, Leicester, LE1 7RH, United Kingdom*

DIETER SWANDULLA (21), *Department of Experimental and Clinical Pharmacology, University of Erlangen—Nürnberg, D-91054 Erlangen, Germany*

JACQUELINE C. TANAKA (33), *Department of Pathology, School of Dental Medicine, University of Pennsylvania, Philadelphia, Pennsylvania 19104-6002*

MASAYA TOHYAMA (10), *Department of Anatomy and Neuroscience, Osaka University Medical School, Osaka, Japan*

LIGIA TORO (29), *Departments of Anesthesiology and Molecular and Medical Pharmacology, and Brain Research Institute, University of California Los Angeles, Los Angeles, California 90095-1778*

JAMES S. TRIMMER (3), *Departments of Biochemistry and Cell Biology, Institute for Cell and Developmental Biology, State University of New York, Stony Brook, New York 11794-5215*

YOLANDA F. VALLEJO (31), *Department of Pharmacological and Physiological Sciences, University of Chicago, Chicago, Illinois 60637*

Z. GALEN WO (32), *Department of Pharmacology, College of Veterinary Medicine, Cornell University, Ithaca, New York 14853*

XIAO-MING XIA (4), *Vollum Institute, Oregon Health Sciences University, Portland, Oregon 97201*

JIA XU (1), *Department of Physiology, Johns Hopkins University School of Medicine, Baltimore, Maryland 21205*

YOUFENG XU (27), *Division of Biology, California Institute of Technology, Pasadena, California 91125*

HANNS ULRICH ZEILHOFER (21), *Department of Experimental and Clinical Pharmacology, University of Erlangen—Nürnberg, D-91054 Erlangen, Germany*

PATRICIA ZERR (4), *Vollum Institute, Oregon Health Sciences University, Portland, Oregon 97201*

Preface

The rapid growth of interest and research activity in ion channels is indicative of their fundamental importance in the maintenance of the living state. A brief examination of the table of contents of this volume of *Methods in Enzymology* and of its companion Volume 294 reveals methods that range from the molecular and biochemical level to the cellular, tissue, and whole cell level. These volumes supplement Volume 207 of this series.

Authors have been selected based on research contributions in the area about which they have written and their ability to describe their methodological contributions in a clear and reproducible way. They have been encouraged to make use of graphics, comparisons to other methods, and to provide tricks and approaches that make it possible to adapt methods to other systems.

I would like to express my appreciation to the contributors for providing their contributions in a timely fashion, to the Editorial Advisory Board for guidance, to the staff of Academic Press for helpful input, and to Mary Rarick and Patti Williams for assisting with the remarkable amount of clerical work needed to coordinate the chapters included.

P. Michael Conn

METHODS IN ENZYMOLOGY

VOLUME I. Preparation and Assay of Enzymes
Edited by SIDNEY P. COLOWICK AND NATHAN O. KAPLAN

VOLUME II. Preparation and Assay of Enzymes
Edited by SIDNEY P. COLOWICK AND NATHAN O. KAPLAN

VOLUME III. Preparation and Assay of Substrates
Edited by SIDNEY P. COLOWICK AND NATHAN O. KAPLAN

VOLUME IV. Special Techniques for the Enzymologist
Edited by SIDNEY P. COLOWICK AND NATHAN O. KAPLAN

VOLUME V. Preparation and Assay of Enzymes
Edited by SIDNEY P. COLOWICK AND NATHAN O. KAPLAN

VOLUME VI. Preparation and Assay of Enzymes (*Continued*)
Preparation and Assay of Substrates
Special Techniques
Edited by SIDNEY P. COLOWICK AND NATHAN O. KAPLAN

VOLUME VII. Cumulative Subject Index
Edited by SIDNEY P. COLOWICK AND NATHAN O. KAPLAN

VOLUME VIII. Complex Carbohydrates
Edited by ELIZABETH F. NEUFELD AND VICTOR GINSBURG

VOLUME IX. Carbohydrate Metabolism
Edited by WILLIS A. WOOD

VOLUME X. Oxidation and Phosphorylation
Edited by RONALD W. ESTABROOK AND MAYNARD E. PULLMAN

VOLUME XI. Enzyme Structure
Edited by C. H. W. HIRS

VOLUME XII. Nucleic Acids (Parts A and B)
Edited by LAWRENCE GROSSMAN AND KIVIE MOLDAVE

VOLUME XIII. Citric Acid Cycle
Edited by J. M. LOWENSTEIN

VOLUME XIV. Lipids
Edited by J. M. LOWENSTEIN

VOLUME XV. Steroids and Terpenoids
Edited by RAYMOND B. CLAYTON

VOLUME XVI. Fast Reactions
Edited by KENNETH KUSTIN

VOLUME XVII. Metabolism of Amino Acids and Amines (Parts A and B)
Edited by HERBERT TABOR AND CELIA WHITE TABOR

VOLUME XVIII. Vitamins and Coenzymes (Parts A, B, and C)
Edited by DONALD B. MCCORMICK AND LEMUEL D. WRIGHT

VOLUME XIX. Proteolytic Enzymes
Edited by GERTRUDE E. PERLMANN AND LASZLO LORAND

VOLUME XX. Nucleic Acids and Protein Synthesis (Part C)
Edited by KIVIE MOLDAVE AND LAWRENCE GROSSMAN

VOLUME XXI. Nucleic Acids (Part D)
Edited by LAWRENCE GROSSMAN AND KIVIE MOLDAVE

VOLUME XXII. Enzyme Purification and Related Techniques
Edited by WILLIAM B. JAKOBY

VOLUME XXIII. Photosynthesis (Part A)
Edited by ANTHONY SAN PIETRO

VOLUME XXIV. Photosynthesis and Nitrogen Fixation (Part B)
Edited by ANTHONY SAN PIETRO

VOLUME XXV. Enzyme Structure (Part B)
Edited by C. H. W. HIRS AND SERGE N. TIMASHEFF

VOLUME XXVI. Enzyme Structure (Part C)
Edited by C. H. W. HIRS AND SERGE N. TIMASHEFF

VOLUME XXVII. Enzyme Structure (Part D)
Edited by C. H. W. HIRS AND SERGE N. TIMASHEFF

VOLUME XXVIII. Complex Carbohydrates (Part B)
Edited by VICTOR GINSBURG

VOLUME XXIX. Nucleic Acids and Protein Synthesis (Part E)
Edited by LAWRENCE GROSSMAN AND KIVIE MOLDAVE

VOLUME XXX. Nucleic Acids and Protein Synthesis (Part F)
Edited by KIVIE MOLDAVE AND LAWRENCE GROSSMAN

VOLUME XXXI. Biomembranes (Part A)
Edited by SIDNEY FLEISCHER AND LESTER PACKER

VOLUME XXXII. Biomembranes (Part B)
Edited by SIDNEY FLEISCHER AND LESTER PACKER

VOLUME XXXIII. Cumulative Subject Index Volumes I–XXX
Edited by MARTHA G. DENNIS AND EDWARD A. DENNIS

VOLUME XXXIV. Affinity Techniques (Enzyme Purification: Part B)
Edited by WILLIAM B. JAKOBY AND MEIR WILCHEK

VOLUME XXXV. Lipids (Part B)
Edited by JOHN M. LOWENSTEIN

VOLUME XXXVI. Hormone Action (Part A: Steroid Hormones)
Edited by BERT W. O'MALLEY AND JOEL G. HARDMAN

VOLUME XXXVII. Hormone Action (Part B: Peptide Hormones)
Edited by BERT W. O'MALLEY AND JOEL G. HARDMAN

VOLUME XXXVIII. Hormone Action (Part C: Cyclic Nucleotides)
Edited by JOEL G. HARDMAN AND BERT W. O'MALLEY

VOLUME XXXIX. Hormone Action (Part D: Isolated Cells, Tissues, and Organ Systems)
Edited by JOEL G. HARDMAN AND BERT W. O'MALLEY

VOLUME XL. Hormone Action (Part E: Nuclear Structure and Function)
Edited by BERT W. O'MALLEY AND JOEL G. HARDMAN

VOLUME XLI. Carbohydrate Metabolism (Part B)
Edited by W. A. WOOD

VOLUME XLII. Carbohydrate Metabolism (Part C)
Edited by W. A. WOOD

VOLUME XLIII. Antibiotics
Edited by JOHN H. HASH

VOLUME XLIV. Immobilized Enzymes
Edited by KLAUS MOSBACH

VOLUME XLV. Proteolytic Enzymes (Part B)
Edited by LASZLO LORAND

VOLUME XLVI. Affinity Labeling
Edited by WILLIAM B. JAKOBY AND MEIR WILCHEK

VOLUME XLVII. Enzyme Structure (Part E)
Edited by C. H. W. HIRS AND SERGE N. TIMASHEFF

VOLUME XLVIII. Enzyme Structure (Part F)
Edited by C. H. W. HIRS AND SERGE N. TIMASHEFF

VOLUME XLIX. Enzyme Structure (Part G)
Edited by C. H. W. HIRS AND SERGE N. TIMASHEFF

VOLUME L. Complex Carbohydrates (Part C)
Edited by VICTOR GINSBURG

VOLUME LI. Purine and Pyrimidine Nucleotide Metabolism
Edited by PATRICIA A. HOFFEE AND MARY ELLEN JONES

VOLUME LII. Biomembranes (Part C: Biological Oxidations)
Edited by SIDNEY FLEISCHER AND LESTER PACKER

VOLUME LIII. Biomembranes (Part D: Biological Oxidations)
Edited by SIDNEY FLEISCHER AND LESTER PACKER

VOLUME LIV. Biomembranes (Part E: Biological Oxidations)
Edited by SIDNEY FLEISCHER AND LESTER PACKER

VOLUME LV. Biomembranes (Part F: Bioenergetics)
Edited by SIDNEY FLEISCHER AND LESTER PACKER

VOLUME LVI. Biomembranes (Part G: Bioenergetics)
Edited by SIDNEY FLEISCHER AND LESTER PACKER

VOLUME LVII. Bioluminescence and Chemiluminescence
Edited by MARLENE A. DELUCA

VOLUME LVIII. Cell Culture
Edited by WILLIAM B. JAKOBY AND IRA PASTAN

VOLUME LIX. Nucleic Acids and Protein Synthesis (Part G)
Edited by KIVIE MOLDAVE AND LAWRENCE GROSSMAN

VOLUME LX. Nucleic Acids and Protein Synthesis (Part H)
Edited by KIVIE MOLDAVE AND LAWRENCE GROSSMAN

VOLUME 61. Enzyme Structure (Part H)
Edited by C. H. W. HIRS AND SERGE N. TIMASHEFF

VOLUME 62. Vitamins and Coenzymes (Part D)
Edited by DONALD B. MCCORMICK AND LEMUEL D. WRIGHT

VOLUME 63. Enzyme Kinetics and Mechanism (Part A: Initial Rate and Inhibitor Methods)
Edited by DANIEL L. PURICH

VOLUME 64. Enzyme Kinetics and Mechanism (Part B: Isotopic Probes and Complex Enzyme Systems)
Edited by DANIEL L. PURICH

VOLUME 65. Nucleic Acids (Part I)
Edited by LAWRENCE GROSSMAN AND KIVIE MOLDAVE

VOLUME 66. Vitamins and Coenzymes (Part E)
Edited by DONALD B. MCCORMICK AND LEMUEL D. WRIGHT

VOLUME 67. Vitamins and Coenzymes (Part F)
Edited by DONALD B. MCCORMICK AND LEMUEL D. WRIGHT

VOLUME 68. Recombinant DNA
Edited by RAY WU

VOLUME 69. Photosynthesis and Nitrogen Fixation (Part C)
Edited by ANTHONY SAN PIETRO

VOLUME 70. Immunochemical Techniques (Part A)
Edited by HELEN VAN VUNAKIS AND JOHN J. LANGONE

VOLUME 71. Lipids (Part C)
Edited by JOHN M. LOWENSTEIN

VOLUME 72. Lipids (Part D)
Edited by JOHN M. LOWENSTEIN

Volume 73. Immunochemical Techniques (Part B)
Edited by John J. Langone and Helen Van Vunakis

Volume 74. Immunochemical Techniques (Part C)
Edited by John J. Langone and Helen Van Vunakis

Volume 75. Cumulative Subject Index Volumes XXXI, XXXII, XXXIV–LX
Edited by Edward A. Dennis and Martha G. Dennis

Volume 76. Hemoglobins
Edited by Eraldo Antonini, Luigi Rossi-Bernardi, and Emilia Chiancone

Volume 77. Detoxication and Drug Metabolism
Edited by William B. Jakoby

Volume 78. Interferons (Part A)
Edited by Sidney Pestka

Volume 79. Interferons (Part B)
Edited by Sidney Pestka

Volume 80. Proteolytic Enzymes (Part C)
Edited by Laszlo Lorand

Volume 81. Biomembranes (Part H: Visual Pigments and Purple Membranes, I)
Edited by Lester Packer

Volume 82. Structural and Contractile Proteins (Part A: Extracellular Matrix)
Edited by Leon W. Cunningham and Dixie W. Frederiksen

Volume 83. Complex Carbohydrates (Part D)
Edited by Victor Ginsburg

Volume 84. Immunochemical Techniques (Part D: Selected Immunoassays)
Edited by John J. Langone and Helen Van Vunakis

Volume 85. Structural and Contractile Proteins (Part B: The Contractile Apparatus and the Cytoskeleton)
Edited by Dixie W. Frederiksen and Leon W. Cunningham

Volume 86. Prostaglandins and Arachidonate Metabolites
Edited by William E. M. Lands and William L. Smith

Volume 87. Enzyme Kinetics and Mechanism (Part C: Intermediates, Stereochemistry, and Rate Studies)
Edited by Daniel L. Purich

Volume 88. Biomembranes (Part I: Visual Pigments and Purple Membranes, II)
Edited by Lester Packer

Volume 89. Carbohydrate Metabolism (Part D)
Edited by Willis A. Wood

Volume 90. Carbohydrate Metabolism (Part E)
Edited by Willis A. Wood

VOLUME 91. Enzyme Structure (Part I)
Edited by C. H. W. HIRS AND SERGE N. TIMASHEFF

VOLUME 92. Immunochemical Techniques (Part E: Monoclonal Antibodies and General Immunoassay Methods)
Edited by JOHN J. LANGONE AND HELEN VAN VUNAKIS

VOLUME 93. Immunochemical Techniques (Part F: Conventional Antibodies, Fc Receptors, and Cytotoxicity)
Edited by JOHN J. LANGONE AND HELEN VAN VUNAKIS

VOLUME 94. Polyamines
Edited by HERBERT TABOR AND CELIA WHITE TABOR

VOLUME 95. Cumulative Subject Index Volumes 61–74, 76–80
Edited by EDWARD A. DENNIS AND MARTHA G. DENNIS

VOLUME 96. Biomembranes [Part J: Membrane Biogenesis: Assembly and Targeting (General Methods; Eukaryotes)]
Edited by SIDNEY FLEISCHER AND BECCA FLEISCHER

VOLUME 97. Biomembranes [Part K: Membrane Biogenesis: Assembly and Targeting (Prokaryotes, Mitochondria, and Chloroplasts)]
Edited by SIDNEY FLEISCHER AND BECCA FLEISCHER

VOLUME 98. Biomembranes (Part L: Membrane Biogenesis: Processing and Recycling)
Edited by SIDNEY FLEISCHER AND BECCA FLEISCHER

VOLUME 99. Hormone Action (Part F: Protein Kinases)
Edited by JACKIE D. CORBIN AND JOEL G. HARDMAN

VOLUME 100. Recombinant DNA (Part B)
Edited by RAY WU, LAWRENCE GROSSMAN, AND KIVIE MOLDAVE

VOLUME 101. Recombinant DNA (Part C)
Edited by RAY WU, LAWRENCE GROSSMAN, AND KIVIE MOLDAVE

VOLUME 102. Hormone Action (Part G: Calmodulin and Calcium-Binding Proteins)
Edited by ANTHONY R. MEANS AND BERT W. O'MALLEY

VOLUME 103. Hormone Action (Part H: Neuroendocrine Peptides)
Edited by P. MICHAEL CONN

VOLUME 104. Enzyme Purification and Related Techniques (Part C)
Edited by WILLIAM B. JAKOBY

VOLUME 105. Oxygen Radicals in Biological Systems
Edited by LESTER PACKER

VOLUME 106. Posttranslational Modifications (Part A)
Edited by FINN WOLD AND KIVIE MOLDAVE

VOLUME 107. Posttranslational Modifications (Part B)
Edited by FINN WOLD AND KIVIE MOLDAVE

VOLUME 108. Immunochemical Techniques (Part G: Separation and Characterization of Lymphoid Cells)
Edited by GIOVANNI DI SABATO, JOHN J. LANGONE, AND HELEN VAN VUNAKIS

VOLUME 109. Hormone Action (Part I: Peptide Hormones)
Edited by LUTZ BIRNBAUMER AND BERT W. O'MALLEY

VOLUME 110. Steroids and Isoprenoids (Part A)
Edited by JOHN H. LAW AND HANS C. RILLING

VOLUME 111. Steroids and Isoprenoids (Part B)
Edited by JOHN H. LAW AND HANS C. RILLING

VOLUME 112. Drug and Enzyme Targeting (Part A)
Edited by KENNETH J. WIDDER AND RALPH GREEN

VOLUME 113. Glutamate, Glutamine, Glutathione, and Related Compounds
Edited by ALTON MEISTER

VOLUME 114. Diffraction Methods for Biological Macromolecules (Part A)
Edited by HAROLD W. WYCKOFF, C. H. W. HIRS, AND SERGE N. TIMASHEFF

VOLUME 115. Diffraction Methods for Biological Macromolecules (Part B)
Edited by HAROLD W. WYCKOFF, C. H. W. HIRS, AND SERGE N. TIMASHEFF

VOLUME 116. Immunochemical Techniques (Part H: Effectors and Mediators of Lymphoid Cell Functions)
Edited by GIOVANNI DI SABATO, JOHN J. LANGONE, AND HELEN VAN VUNAKIS

VOLUME 117. Enzyme Structure (Part J)
Edited by C. H. W. HIRS AND SERGE N. TIMASHEFF

VOLUME 118. Plant Molecular Biology
Edited by ARTHUR WEISSBACH AND HERBERT WEISSBACH

VOLUME 119. Interferons (Part C)
Edited by SIDNEY PESTKA

VOLUME 120. Cumulative Subject Index Volumes 81–94, 96–101

VOLUME 121. Immunochemical Techniques (Part I: Hybridoma Technology and Monoclonal Antibodies)
Edited by JOHN J. LANGONE AND HELEN VAN VUNAKIS

VOLUME 122. Vitamins and Coenzymes (Part G)
Edited by FRANK CHYTIL AND DONALD B. MCCORMICK

VOLUME 123. Vitamins and Coenzymes (Part H)
Edited by FRANK CHYTIL AND DONALD B. MCCORMICK

VOLUME 124. Hormone Action (Part J: Neuroendocrine Peptides)
Edited by P. MICHAEL CONN

VOLUME 125. Biomembranes (Part M: Transport in Bacteria, Mitochondria, and Chloroplasts: General Approaches and Transport Systems)
Edited by SIDNEY FLEISCHER AND BECCA FLEISCHER

VOLUME 126. Biomembranes (Part N: Transport in Bacteria, Mitochondria, and Chloroplasts: Protonmotive Force)
Edited by SIDNEY FLEISCHER AND BECCA FLEISCHER

VOLUME 127. Biomembranes (Part O: Protons and Water: Structure and Translocation)
Edited by LESTER PACKER

VOLUME 128. Plasma Lipoproteins (Part A: Preparation, Structure, and Molecular Biology)
Edited by JERE P. SEGREST AND JOHN J. ALBERS

VOLUME 129. Plasma Lipoproteins (Part B: Characterization, Cell Biology, and Metabolism)
Edited by JOHN J. ALBERS AND JERE P. SEGREST

VOLUME 130. Enzyme Structure (Part K)
Edited by C. H. W. HIRS AND SERGE N. TIMASHEFF

VOLUME 131. Enzyme Structure (Part L)
Edited by C. H. W. HIRS AND SERGE N. TIMASHEFF

VOLUME 132. Immunochemical Techniques (Part J: Phagocytosis and Cell-Mediated Cytotoxicity)
Edited by GIOVANNI DI SABATO AND JOHANNES EVERSE

VOLUME 133. Bioluminescence and Chemiluminescence (Part B)
Edited by MARLENE DELUCA AND WILLIAM D. MCELROY

VOLUME 134. Structural and Contractile Proteins (Part C: The Contractile Apparatus and the Cytoskeleton)
Edited by RICHARD B. VALLEE

VOLUME 135. Immobilized Enzymes and Cells (Part B)
Edited by KLAUS MOSBACH

VOLUME 136. Immobilized Enzymes and Cells (Part C)
Edited by KLAUS MOSBACH

VOLUME 137. Immobilized Enzymes and Cells (Part D)
Edited by KLAUS MOSBACH

VOLUME 138. Complex Carbohydrates (Part E)
Edited by VICTOR GINSBURG

VOLUME 139. Cellular Regulators (Part A: Calcium- and Calmodulin-Binding Proteins)
Edited by ANTHONY R. MEANS AND P. MICHAEL CONN

VOLUME 140. Cumulative Subject Index Volumes 102–119, 121–134

VOLUME 141. Cellular Regulators (Part B: Calcium and Lipids)
Edited by P. MICHAEL CONN AND ANTHONY R. MEANS

VOLUME 142. Metabolism of Aromatic Amino Acids and Amines
Edited by SEYMOUR KAUFMAN

VOLUME 143. Sulfur and Sulfur Amino Acids
Edited by WILLIAM B. JAKOBY AND OWEN GRIFFITH

VOLUME 144. Structural and Contractile Proteins (Part D: Extracellular Matrix)
Edited by LEON W. CUNNINGHAM

VOLUME 145. Structural and Contractile Proteins (Part E: Extracellular Matrix)
Edited by LEON W. CUNNINGHAM

VOLUME 146. Peptide Growth Factors (Part A)
Edited by DAVID BARNES AND DAVID A. SIRBASKU

VOLUME 147. Peptide Growth Factors (Part B)
Edited by DAVID BARNES AND DAVID A. SIRBASKU

VOLUME 148. Plant Cell Membranes
Edited by LESTER PACKER AND ROLAND DOUCE

VOLUME 149. Drug and Enzyme Targeting (Part B)
Edited by RALPH GREEN AND KENNETH J. WIDDER

VOLUME 150. Immunochemical Techniques (Part K: *In Vitro* Models of B and T Cell Functions and Lymphoid Cell Receptors)
Edited by GIOVANNI DI SABATO

VOLUME 151. Molecular Genetics of Mammalian Cells
Edited by MICHAEL M. GOTTESMAN

VOLUME 152. Guide to Molecular Cloning Techniques
Edited by SHELBY L. BERGER AND ALAN R. KIMMEL

VOLUME 153. Recombinant DNA (Part D)
Edited by RAY WU AND LAWRENCE GROSSMAN

VOLUME 154. Recombinant DNA (Part E)
Edited by RAY WU AND LAWRENCE GROSSMAN

VOLUME 155. Recombinant DNA (Part F)
Edited by RAY WU

VOLUME 156. Biomembranes (Part P: ATP-Driven Pumps and Related Transport: The Na,K-Pump)
Edited by SIDNEY FLEISCHER AND BECCA FLEISCHER

VOLUME 157. Biomembranes (Part Q: ATP-Driven Pumps and Related Transport: Calcium, Proton, and Potassium Pumps)
Edited by SIDNEY FLEISCHER AND BECCA FLEISCHER

VOLUME 158. Metalloproteins (Part A)
Edited by JAMES F. RIORDAN AND BERT L. VALLEE

VOLUME 159. Initiation and Termination of Cyclic Nucleotide Action
Edited by JACKIE D. CORBIN AND ROGER A. JOHNSON

VOLUME 160. Biomass (Part A: Cellulose and Hemicellulose)
Edited by WILLIS A. WOOD AND SCOTT T. KELLOGG

VOLUME 161. Biomass (Part B: Lignin, Pectin, and Chitin)
Edited by WILLIS A. WOOD AND SCOTT T. KELLOGG

VOLUME 162. Immunochemical Techniques (Part L: Chemotaxis and Inflammation)
Edited by GIOVANNI DI SABATO

VOLUME 163. Immunochemical Techniques (Part M: Chemotaxis and Inflammation)
Edited by GIOVANNI DI SABATO

VOLUME 164. Ribosomes
Edited by HARRY F. NOLLER, JR., AND KIVIE MOLDAVE

VOLUME 165. Microbial Toxins: Tools for Enzymology
Edited by SIDNEY HARSHMAN

VOLUME 166. Branched-Chain Amino Acids
Edited by ROBERT HARRIS AND JOHN R. SOKATCH

VOLUME 167. Cyanobacteria
Edited by LESTER PACKER AND ALEXANDER N. GLAZER

VOLUME 168. Hormone Action (Part K: Neuroendocrine Peptides)
Edited by P. MICHAEL CONN

VOLUME 169. Platelets: Receptors, Adhesion, Secretion (Part A)
Edited by JACEK HAWIGER

VOLUME 170. Nucleosomes
Edited by PAUL M. WASSARMAN AND ROGER D. KORNBERG

VOLUME 171. Biomembranes (Part R: Transport Theory: Cells and Model Membranes)
Edited by SIDNEY FLEISCHER AND BECCA FLEISCHER

VOLUME 172. Biomembranes (Part S: Transport: Membrane Isolation and Characterization)
Edited by SIDNEY FLEISCHER AND BECCA FLEISCHER

VOLUME 173. Biomembranes [Part T: Cellular and Subcellular Transport: Eukaryotic (Nonepithelial) Cells]
Edited by SIDNEY FLEISCHER AND BECCA FLEISCHER

VOLUME 174. Biomembranes [Part U: Cellular and Subcellular Transport: Eukaryotic (Nonepithelial) Cells]
Edited by SIDNEY FLEISCHER AND BECCA FLEISCHER

VOLUME 175. Cumulative Subject Index Volumes 135–139, 141–167

VOLUME 176. Nuclear Magnetic Resonance (Part A: Spectral Techniques and Dynamics)
Edited by NORMAN J. OPPENHEIMER AND THOMAS L. JAMES

VOLUME 177. Nuclear Magnetic Resonance (Part B: Structure and Mechanism)
Edited by NORMAN J. OPPENHEIMER AND THOMAS L. JAMES

VOLUME 178. Antibodies, Antigens, and Molecular Mimicry
Edited by JOHN J. LANGONE

VOLUME 179. Complex Carbohydrates (Part F)
Edited by VICTOR GINSBURG

VOLUME 180. RNA Processing (Part A: General Methods)
Edited by JAMES E. DAHLBERG AND JOHN N. ABELSON

VOLUME 181. RNA Processing (Part B: Specific Methods)
Edited by JAMES E. DAHLBERG AND JOHN N. ABELSON

VOLUME 182. Guide to Protein Purification
Edited by MURRAY P. DEUTSCHER

VOLUME 183. Molecular Evolution: Computer Analysis of Protein and Nucleic Acid Sequences
Edited by RUSSELL F. DOOLITTLE

VOLUME 184. Avidin–Biotin Technology
Edited by MEIR WILCHEK AND EDWARD A. BAYER

VOLUME 185. Gene Expression Technology
Edited by DAVID V. GOEDDEL

VOLUME 186. Oxygen Radicals in Biological Systems (Part B: Oxygen Radicals and Antioxidants)
Edited by LESTER PACKER AND ALEXANDER N. GLAZER

VOLUME 187. Arachidonate Related Lipid Mediators
Edited by ROBERT C. MURPHY AND FRANK A. FITZPATRICK

VOLUME 188. Hydrocarbons and Methylotrophy
Edited by MARY E. LIDSTROM

VOLUME 189. Retinoids (Part A: Molecular and Metabolic Aspects)
Edited by LESTER PACKER

VOLUME 190. Retinoids (Part B: Cell Differentiation and Clinical Applications)
Edited by LESTER PACKER

VOLUME 191. Biomembranes (Part V: Cellular and Subcellular Transport: Epithelial Cells)
Edited by SIDNEY FLEISCHER AND BECCA FLEISCHER

VOLUME 192. Biomembranes (Part W: Cellular and Subcellular Transport: Epithelial Cells)
Edited by SIDNEY FLEISCHER AND BECCA FLEISCHER

VOLUME 193. Mass Spectrometry
Edited by JAMES A. MCCLOSKEY

VOLUME 194. Guide to Yeast Genetics and Molecular Biology
Edited by CHRISTINE GUTHRIE AND GERALD R. FINK

VOLUME 195. Adenylyl Cyclase, G Proteins, and Guanylyl Cyclase
Edited by ROGER A. JOHNSON AND JACKIE D. CORBIN

VOLUME 196. Molecular Motors and the Cytoskeleton
Edited by RICHARD B. VALLEE

VOLUME 197. Phospholipases
Edited by EDWARD A. DENNIS

VOLUME 198. Peptide Growth Factors (Part C)
Edited by DAVID BARNES, J. P. MATHER, AND GORDON H. SATO

VOLUME 199. Cumulative Subject Index Volumes 168–174, 176–194

VOLUME 200. Protein Phosphorylation (Part A: Protein Kinases: Assays, Purification, Antibodies, Functional Analysis, Cloning, and Expression)
Edited by TONY HUNTER AND BARTHOLOMEW M. SEFTON

VOLUME 201. Protein Phosphorylation (Part B: Analysis of Protein Phosphorylation, Protein Kinase Inhibitors, and Protein Phosphatases)
Edited by TONY HUNTER AND BARTHOLOMEW M. SEFTON

VOLUME 202. Molecular Design and Modeling: Concepts and Applications (Part A: Proteins, Peptides, and Enzymes)
Edited by JOHN J. LANGONE

VOLUME 203. Molecular Design and Modeling: Concepts and Applications (Part B: Antibodies and Antigens, Nucleic Acids, Polysaccharides, and Drugs)
Edited by JOHN J. LANGONE

VOLUME 204. Bacterial Genetic Systems
Edited by JEFFREY H. MILLER

VOLUME 205. Metallobiochemistry (Part B: Metallothionein and Related Molecules)
Edited by JAMES F. RIORDAN AND BERT L. VALLEE

VOLUME 206. Cytochrome P450
Edited by MICHAEL R. WATERMAN AND ERIC F. JOHNSON

VOLUME 207. Ion Channels
Edited by BERNARDO RUDY AND LINDA E. IVERSON

VOLUME 208. Protein–DNA Interactions
Edited by ROBERT T. SAUER

VOLUME 209. Phospholipid Biosynthesis
Edited by EDWARD A. DENNIS AND DENNIS E. VANCE

VOLUME 210. Numerical Computer Methods
Edited by LUDWIG BRAND AND MICHAEL L. JOHNSON

VOLUME 211. DNA Structures (Part A: Synthesis and Physical Analysis of DNA)
Edited by DAVID M. J. LILLEY AND JAMES E. DAHLBERG

VOLUME 212. DNA Structures (Part B: Chemical and Electrophoretic Analysis of DNA)
Edited by DAVID M. J. LILLEY AND JAMES E. DAHLBERG

VOLUME 213. Carotenoids (Part A: Chemistry, Separation, Quantitation, and Antioxidation)
Edited by LESTER PACKER

VOLUME 214. Carotenoids (Part B: Metabolism, Genetics, and Biosynthesis)
Edited by LESTER PACKER

VOLUME 215. Platelets: Receptors, Adhesion, Secretion (Part B)
Edited by JACEK J. HAWIGER

VOLUME 216. Recombinant DNA (Part G)
Edited by RAY WU

VOLUME 217. Recombinant DNA (Part H)
Edited by RAY WU

VOLUME 218. Recombinant DNA (Part I)
Edited by RAY WU

VOLUME 219. Reconstitution of Intracellular Transport
Edited by JAMES E. ROTHMAN

VOLUME 220. Membrane Fusion Techniques (Part A)
Edited by NEJAT DÜZGÜNEŞ

VOLUME 221. Membrane Fusion Techniques (Part B)
Edited by NEJAT DÜZGÜNEŞ

VOLUME 222. Proteolytic Enzymes in Coagulation, Fibrinolysis, and Complement Activation (Part A: Mammalian Blood Coagulation Factors and Inhibitors)
Edited by LASZLO LORAND AND KENNETH G. MANN

VOLUME 223. Proteolytic Enzymes in Coagulation, Fibrinolysis, and Complement Activation (Part B: Complement Activation, Fibrinolysis, and Nonmammalian Blood Coagulation Factors)
Edited by LASZLO LORAND AND KENNETH G. MANN

VOLUME 224. Molecular Evolution: Producing the Biochemical Data
Edited by ELIZABETH ANNE ZIMMER, THOMAS J. WHITE, REBECCA L. CANN, AND ALLAN C. WILSON

VOLUME 225. Guide to Techniques in Mouse Development
Edited by PAUL M. WASSARMAN AND MELVIN L. DEPAMPHILIS

VOLUME 226. Metallobiochemistry (Part C: Spectroscopic and Physical Methods for Probing Metal Ion Environments in Metalloenzymes and Metalloproteins)
Edited by JAMES F. RIORDAN AND BERT L. VALLEE

VOLUME 227. Metallobiochemistry (Part D: Physical and Spectroscopic Methods for Probing Metal Ion Environments in Metalloproteins)
Edited by JAMES F. RIORDAN AND BERT L. VALLEE

VOLUME 228. Aqueous Two-Phase Systems
Edited by HARRY WALTER AND GÖTE JOHANSSON

VOLUME 229. Cumulative Subject Index Volumes 195–198, 200–227

VOLUME 230. Guide to Techniques in Glycobiology
Edited by WILLIAM J. LENNARZ AND GERALD W. HART

VOLUME 231. Hemoglobins (Part B: Biochemical and Analytical Methods)
Edited by JOHANNES EVERSE, KIM D. VANDEGRIFF, AND ROBERT M. WINSLOW

VOLUME 232. Hemoglobins (Part C: Biophysical Methods)
Edited by JOHANNES EVERSE, KIM D. VANDEGRIFF, AND ROBERT M. WINSLOW

VOLUME 233. Oxygen Radicals in Biological Systems (Part C)
Edited by LESTER PACKER

VOLUME 234. Oxygen Radicals in Biological Systems (Part D)
Edited by LESTER PACKER

VOLUME 235. Bacterial Pathogenesis (Part A: Identification and Regulation of Virulence Factors)
Edited by VIRGINIA L. CLARK AND PATRIK M. BAVOIL

VOLUME 236. Bacterial Pathogenesis (Part B: Integration of Pathogenic Bacteria with Host Cells)
Edited by VIRGINIA L. CLARK AND PATRIK M. BAVOIL

VOLUME 237. Heterotrimeric G Proteins
Edited by RAVI IYENGAR

VOLUME 238. Heterotrimeric G-Protein Effectors
Edited by RAVI IYENGAR

VOLUME 239. Nuclear Magnetic Resonance (Part C)
Edited by THOMAS L. JAMES AND NORMAN J. OPPENHEIMER

VOLUME 240. Numerical Computer Methods (Part B)
Edited by MICHAEL L. JOHNSON AND LUDWIG BRAND

VOLUME 241. Retroviral Proteases
Edited by LAWRENCE C. KUO AND JULES A. SHAFER

VOLUME 242. Neoglycoconjugates (Part A)
Edited by Y. C. LEE AND REIKO T. LEE

VOLUME 243. Inorganic Microbial Sulfur Metabolism
Edited by HARRY D. PECK, JR., AND JEAN LEGALL

VOLUME 244. Proteolytic Enzymes: Serine and Cysteine Peptidases
Edited by ALAN J. BARRETT

VOLUME 245. Extracellular Matrix Components
Edited by E. RUOSLAHTI AND E. ENGVALL

VOLUME 246. Biochemical Spectroscopy
Edited by KENNETH SAUER

Volume 265. Cumulative Subject Index Volumes 228, 230–262

Volume 266. Computer Methods for Macromolecular Sequence Analysis
Edited by Russell F. Doolittle

Volume 267. Combinatorial Chemistry
Edited by John N. Abelson

Volume 268. Nitric Oxide (Part A: Sources and Detection of NO; NO Synthase)
Edited by Lester Packer

Volume 269. Nitric Oxide (Part B: Physiological and Pathological Processes)
Edited by Lester Packer

Volume 270. High Resolution Separation and Analysis of Biological Macromolecules (Part A: Fundamentals)
Edited by Barry L. Karger and William S. Hancock

Volume 271. High Resolution Separation and Analysis of Biological Macromolecules (Part B: Applications)
Edited by Barry L. Karger and William S. Hancock

Volume 272. Cytochrome P450 (Part B)
Edited by Eric F. Johnson and Michael R. Waterman

Volume 273. RNA Polymerase and Associated Factors (Part A)
Edited by Sankar Adhya

Volume 274. RNA Polymerase and Associated Factors (Part B)
Edited by Sankar Adhya

Volume 275. Viral Polymerases and Related Proteins
Edited by Lawrence C. Kuo, David B. Olsen, and Steven S. Carroll

Volume 276. Macromolecular Crystallography (Part A)
Edited by Charles W. Carter, Jr., and Robert M. Sweet

Volume 277. Macromolecular Crystallography (Part B)
Edited by Charles W. Carter, Jr., and Robert M. Sweet

Volume 278. Fluorescence Spectroscopy
Edited by Ludwig Brand and Michael L. Johnson

Volume 279. Vitamins and Coenzymes, Part I
Edited by Donald B. McCormick, John W. Suttie, and Conrad Wagner

Volume 280. Vitamins and Coenzymes, Part J
Edited by Donald B. McCormick, John W. Suttie, and Conrad Wagner

Volume 281. Vitamins and Coenzymes, Part K
Edited by Donald B. McCormick, John W. Suttie, and Conrad Wagner

Volume 282. Vitamins and Coenzymes, Part L
Edited by Donald B. McCormick, John W. Suttie, and Conrad Wagner

Volume 283. Cell Cycle Control
Edited by William G. Dunphy

VOLUME 284. Lipases (Part A: Biotechnology)
Edited by BYRON RUBIN AND EDWARD A. DENNIS

VOLUME 285. Cumulative Subject Index Volumes 263, 264, 266–289

VOLUME 286. Lipases (Part B: Enzyme Characterization and Utilization)
Edited by BYRON RUBIN AND EDWARD A. DENNIS

VOLUME 287. Chemokines
Edited by RICHARD HORUK

VOLUME 288. Chemokine Receptors
Edited by RICHARD HORUK

VOLUME 289. Solid Phase Peptide Synthesis
Edited by GREGG B. FIELDS

VOLUME 290. Molecular Chaperones
Edited by GEORGE H. LORIMER AND THOMAS BALDWIN

VOLUME 291. Caged Compounds
Edited by GERARD MARRIOTT

VOLUME 292. ABC Transporters: Biochemical, Cellular, and Molecular Aspects
Edited by Suresh V. Ambudkar and Michael M. Gottsman

VOLUME 293. Ion Channels, Part B
Edited by P. MICHAEL CONN

VOLUME 294. Ion Channels, Part C
Edited by P. MICHAEL CONN

VOLUME 295. Energetics of Biological Macromolecules, Part B
Edited by GARY K. ACKERS AND MICHAEL L. JOHNSON

VOLUME 296. Neurotransmitter Transporters
Edited by SUSAN G. AMARA

VOLUME 297. Photosynthesis: Molecular Biology of Energy Capture
Edited by LEE MCINTOSH

VOLUME 298. Molecular Motors and the Cytoskeleton, Part B
Edited by RICHARD B. VALLEE

VOLUME 299. Oxidants and Antioxidants, Part A
Edited by LESTER PACKER

VOLUME 300. Oxidants and Antioxidants, Part B
Edited by LESTER PACKER

VOLUME 301. Nitric Oxide: Biological and Antioxidant Activities, Part C
Edited by LESTER PACKER

VOLUME 302. Green Fluorescent Protein (in preparation)
Edited by P. MICHAEL CONN

Section I

Assembly

[1] Methods Used to Study Subunit Assembly of Potassium Channels

By JIA XU and MIN LI

Introduction

Expression of ion channels *in vivo* requires the specific association of various subunits. The composition and stoichiometry of various subunits in the assembled channel complexes determine many aspects of channel properties including biophysical kinetics, subcellular localization, and, ultimately, the ion conduction properties of a given cell.

Potassium channels play important roles in the cellular functions of both excitable and nonexcitable cells. The diversity of potassium currents is thought to be greatly enhanced as a result of the heteromeric association of different subunits that are encoded by more than 100 known potassium channel genes. Because each individual subunit may have distinct kinetic and conducting properties, the combinatorial assembly of various subunits provides a means of creating vast biophysical heterogeneity in both excitable and nonexcitable cells.

Thus, studies of subunit interaction are essential for understanding the molecular steps of channel expression at the posttranslational level. In addition, detailed biochemical and biophysical analyses of subunit interaction will provide important insights into the molecular coupling of subunit assembly with ion conduction in the cellular membrane. In this article, we describe several methods that we have previously used to study the subunit assembly of the Shaker-type potassium channels.[1–3] Both the rationale and techniques should be applicable to other ion channels.

Technical Consideration and Rationale

One powerful experimental paradigm commonly used in the structure and function studies of ion channels is to introduce specific changes into a channel subunit, for example, by site-directed mutagenesis. Comparison of the kinetic properties of mutated subunits with those of wild-type subunits allows one to infer the specific function of the mutated residue. Using this approach, elegant studies have been undertaken to identify residues or regions involved in voltage-sensing, inactivation, ion permeation/selectivity,

[1] J. Xu, W. Yu, Y. Jan, L. Jan, and M. Li, *J. Biol. Chem.* **270,** 24761 (1995).
[2] J. Xu and M. Li, *J. Biol. Chem.* **272,** 11728 (1997).
[3] W. Yu, J. Xu, and M. Li, *Neuron* **16,** 441 (1996).

0076-6879/98 $25.00

and shape of the ion channel pore.[4–6] However, one prerequisite for the use of this approach is that the mutated channel has to possess measurable properties, for example, formation of conducting channels. If the same approach were applied to identification of residues that are important for mediating subunit assembly, critical amino acid substitutions may render the expression nonfunctional and thereby result in a null phenotype. Thus, site-directed mutagenesis alone would not be most informative as initial experiments to determine regions that are involved in subunit assembly.

To identify regions that are critical for subunit assembly, experiments are often designed on the basis of one assumption; that is, the region involved in subunit interaction is likely to play a role in subunit assembly. The first step is to identify a region(s) that is capable of binding to another subunit. After such regions are identified, the second step of an experiment can then be formulated to obtain additional evidence to test the hypothesis further. In this article, we describe three techniques: the yeast two-hybrid system, protein overlay binding, and chimeric construction/assay. The first two approaches are used to identify regions that are involved in subunit interaction. The third approach is used to probe the functional role of certain subunit associations.

Yeast Two-Hybrid System

The yeast two-hybrid system is an efficient genetic assay to test the protein–protein interaction of two potential interacting polypeptides.[7] The method takes advantage of the yeast GAL4 transcription factor, which is composed of two functionally essential protein modules: the DNA binding (DB) domain and the transcription activating (TA) domain. To test a potential interaction between two proteins, one first needs to express them as either TA or DB fusion protein. If the two testing proteins associate, their physical association in the yeast nuclei will link the TA and DB domains to form a functional transcription factor. As a result, it permits the expression of reporter genes. Two commonly used reporter genes are *HIS3* and *LacZ*. The *HIS3* facilitates a positive growth selection, whereas *LacZ* allows a convenient enzymatic β-galactosidase assay. Thus, the detection of growth and/or β-galactosidase activity allows a genetic readout of the fusion protein interaction. We have used the yeast two-hybrid system to study the subunit interactions of the Shaker-type voltage-gated potassium

[4] C. Miller, *Science* **252,** 1092 (1991).

[5] L. Y. Jan and Y. N. Jan, *Annu. Rev. Physiol.* **54,** 537 (1992).

[6] L. Salkoff, K. Baker, A. Butler, M. Covarrubias, M. D. Pak, and A. Wei, *Trends Neurosci.* **15,** 161 (1992).

[7] S. Fields and S.-K. Song, *Nature* **340,** 245 (1989).

channels. Because nuclear localization is critical for the yeast two-hybrid system to work, the full-length membrane-bound α subunits are not suitable for this assay. However, since a key protein domain (NAB) that is involved in subunit assembly is localized in the hydrophilic cytoplasmic N terminus, expression of truncated potassium channels without their transmembrane domains has enabled us to test various α–α, α–β, and β–β interactions.[1–3]

Strains and Reagents

Yeast Strain. HF7c [HF7c (*MATα ura3-52 his-200 ade 2-101 lys2-801 trp1-901 leu2-3,112 gal4-542 gal80-538 LYS2::GAL1*$_{UAS}$-*GAL1*$_{TATA}$-*HIS3 URA3::GAL4*$_{17mer(x3)}$-*Cycl*$_{TATA}$-*lacZ*), a gift from Dr. David Beach, Cold Spring Harbor Laboratory].[8]

Drop-Out Mix. Combine the following dry ingredients minus the appropriate supplements (e.g., leucine, tryptophan, and/or histidine) and mix thoroughly by shaking in a 250 ml flask for 1 hr:

Adenine	0.5 g	Lysine	2.0 g
Alanine	2.0 g	Methionine	2.0 g
Arginine	2.0 g	*myo*-Inositol	2.0 g
Asparagine	2.0 g	Phenylalanine	2.0 g
Aspartic Acid	2.0 g	Proline	2.0 g
Cysteine	2.0 g	Serine	2.0 g
Glutamine	2.0 g	Threonine	2.0 g
Glutamic Acid	2.0 g	*Tryptophan*	*2.0 g*
Glycine	2.0 g	Tyrosine	2.0 g
Histidine	*2.0 g*	Uracil	2.0 g
Isoleucine	2.0 g	Valine	2.0 g
Leucine	*4.0 g*		

Drop-Out Media (One Liter)

6.7 g Yeast nitrogen base without amino acids (DIFCO, Detroit, MI)
2 g Appropriate drop-out mixture (His−/Trp−/Leu− or His+/Trp−/Leu−)
20 g Glucose (2% final)
(add 20 g Bacto-agar for plates)

YPD Medium. See *Current Protocols.*[9]

[8] H. E. Feilotter, G. J. Hannon, C. J. Ruddell, and D. Beach, *Nucleic Acids Res.* **22,** 1502 (1994).

[9] F. M. Ausubel, R. Brent, R. E. Kingston, D. D. Moore, J. G. Seidman, J. A. Smith, and K. Struhl, "Current Protocol." Greens Publishing Associates, Inc., and John Wiley & Sons, New York, 1993.

Buffer Z (for One Liter)

100 m*M* $NaPO_4$	16.1 g $Na_2HPO_4 \cdot 7H_2O$; 5.5 g $NaH_2PO_4 \cdot H_2O$
10 m*M* KCl	0.75 g
1 m*M* $MgSO_4$	0.12 g
50 m*M* 2-Mercaptoethanol	2.7 ml
(pH to 7.0)	

X-Gal. Dissolve 5-bromo-4-chloro-3-indolyl-β-D-galactoside in *N,N*-dimethylformamide (DMF) at a concentration of 100 mg/ml. It can be stored at −20° for up to 1 month.

10× TE. 100 m*M* Tris (pH 7.5); 10 m*M* ethylenediaminetetraacetic acid (EDTA)

10× Li-acetate. 1.0 *M* Lithium acetate (adjust pH with dilute acetic acid to pH 7.5).

TE/Li-acetate. Mix 10× TE, 10× Li-acetate, and H_2O in a ratio of 1 : 1 : 8 (v/v).

TE/Li-acetate/8% glycerol. Mix 10× TE, 10× Li-acetate, glycerol, and H_2O in a ratio of 1 : 1 : 0.8 : 7.2 (v/v).

PLATE mix. Mix 10× TE, 10× Li-acetate, 50% polyethylene glycol (PEG 3350, Sigma, St. Louis, MO) in a ratio of 1 : 1 : 8 (v/v).

Construction of Fusion Protein Vectors

Vectors and Host Cells. We use a GAL4 centrometic vector system. For a detailed description of the DNA binding domain vector (pPC97) and the transcription activation domain vector (pPC86), please refer to Chevray and Nathans.[10] This vector system maintains one copy of plasmid per cell. The HF7c strain was chosen because it gives the more stringent growth selection and higher transformation efficiency.

Procedure for Subcloning N-Terminal Domains of Shaker Potassium Channels into Yeast Fusion Protein Vectors

1. To prepare linearized template DNA, add one unit of *Eco*RI to 20 μl restriction digestion buffer containing 2 μg of pSK.ShB plasmid DNA (a gift of Dr. Lily Jan, University of California at San Francisco). Incubate at 37° for 30 min. Transfer 5 μl of digestion mixture to 45 μl of a final polymerase chain reaction (PCR) mixture (Boehringer Mannheim, Germany), containing 1 unit of *Taq* DNA polymerase and 100 nmol each of ML111 and ML120 (ML111: 5′ GAC GCG GCC GCA CTA TCT GGC GGC TTG CGA AC 3′; ML120; 5′ GGC CCC GGG GGC CGC CGT TGC CGG C 3′). Denature the template DNA for 5 min at 94°. Amplify the DNA fragment with 15 cycles of PCR: 94° for 1 min, 55° for 1 min,

[10] P. M. Chevray and D. Nathans, *Proc. Natl. Acad. Sci. U.S.A.* **89,** 5789 (1992).

and 72° for 1 min. Separate the PCR products on 1% agarose gel. Cut a small gel slice containing a DNA fragment ~700 bp in size that corresponds to the coding region of amino acids 2 to 227 of the N-terminal ShB (NShB). Recover the DNA fragment using the standard Geneclean protocol (Bio 101, Vista, CA). Digest the recovered DNA in a 200-μl final volume overnight with *Sma*I at room temperature. After adjusting the buffer composition, digest the DNA with *Not*I enzyme. Precipitate the DNA fragment with ethanol, wash the pellet once with 80% cold ethanol, and resuspend the DNA pellet in 10 μl water. Estimate the DNA concentration by running 1 μl on a gel alongside 0.5 μg of *Hin*dIII marker of known DNA concentration.

2. To prepare vector DNA, digest 2 μg each of pPC97 and pPC86 individually in 200 μl containing 20 units of *Sma*I at room temperature for 3 hr. Add 11 μl of *Sal*I digestion buffer and 20 units of *Sal*I. After incubation at 37° overnight, add 5 units of calf intestine phosphatase (CIP) and incubate the mixture at 55° for exactly 30 min. Add ethylene glycol bis(β-aminoethyl ether) *N,N,N,N*-tetraacetic acid (EGTA) to a final concentration of 20 m*M*. Inactivate the enzyme by heating the mixture at 75° for 15 min. Extract the mixture once with an equal volume of chloroform. Precipitate the DNA with ethanol. Wash the pellet once with cold 80% (v/v) ethanol and allow the pellet to air dry. Resuspend the vector DNA pellet in 20 μl doubly distilled H_2O. Estimate the DNA concentration as described earlier.

3. Mix 1 μl (50 ng) of digested vector with 7 μl (100 ng) of the digested NShB insert, and perform a 10-min ligation with a rapid DNA ligation kit (Boehringer Mannheim, GmbH, Germany). Transform 100 μl of DH5α cells with 10 μl of the ligated product. Digest minipreparation DNA and determine insert sequence. The resultant constructs are pPC97-NShB and pPC86-NShB.

Preparation of Yeast Competent Cells

1. Inoculate HF7c cells into 20 ml liquid YPD medium and grow overnight to 1–4 × 10^7 cells/ml (OD_{600} 0.1–0.5) at 30° with a shaking speed of 240 rpm. Note that HF7c cells grow poorly at a higher shaking speed.
2. Dilute cells 1:10 in 200 ml of fresh, warm YPD medium. Growth to an OD_{600} of 0.2–0.4.
3. Harvest cells by centrifugation for 8 min at 1500*g* at 4°. Wash once with 200 ml cold sterile water, once with 100 ml cold TE/Li-acetate, and once with 50 ml cold TE/Li-acetate/8% glycerol solution (see previous page for composition).
4. Resuspend the cell pellet with TE/Li-acetate/glycerol to a final density of 2 × 10^9 cells/ml. You will find that the volume of yeast cell

pellet may contribute to more than 50% of the total volume. Note that the concentration of glycerol is critical: A higher percentage will significantly decrease the competency.

5. Divide the yeast cells into aliquots of 300–1000 μl and freeze immediately on dry ice. Note that we found that storage in smaller aliquots significantly reduces the competency. The competent cells can be stored at $-80°$ for several months at least without a significant decrease in transformation competency. We recommend that the cells always be frozen in TE/Li-acetate/8% glycerol even if transformation is going to be performed on the same day.

Yeast Transformation

1. Thaw yeast competent cells at room temperature. Vortex to make a homogeneous suspension. Thawing the competent cells on ice is not recommended, since these steps in the protocol are meant to further improve transformation efficiency.
2. Mix 50 μl of the yeast cell suspension with 0.5 μg of pPC97-NShB, 0.5 μg of pPC86-NShB plasmid DNA, and 50 μg of single-stranded salmon sperm carrier DNA in 1.5-ml microfuge tubes by vortexing at top speed for 2–5 sec. The best transformation efficiency is achieved with single-stranded carrier DNA of size range 7–9 kb.
3. Add 300 μl sterile PLATE mix, vortex thoroughly at top speed for 2 sec.
4. Incubate at 30° with a shaking speed of 240 rpm for 30 min.
5. Heat shock at 42° for 15 min.
6. The transformation mixture can be plated directly on selective medium or spun down and resuspended in an appropriate volume of sterile TE for plating.

Testing Domain Interactions by Growth

The growth assay of the yeast two-hybrid system uses the *HIS3* reporter gene as a selective marker. Interaction between the two fusion proteins allows for the expression of *HIS3* product. As a result, the yeast will be able to grow on media lacking histidine when the two testing proteins interact. If, however, the protein products of the two constructs cannot associate, the yeast will fail to grow on medium lacking histidine. In general, yeast transformant should be selected on Trp−/Leu− medium first to ensure that the transformants carry two plasmid constructs. To test the potential NShB interaction, experimental design should include the suggested positive and negative controls of following plasmid combinations: Control 1, pPC86 + pPC97.NShB; Control 2, pPC86.NShB + pPC97; Control 3, pPC86.NShB + pPC97.NShB and Control 4, pPC86.bzJun + pPC97.bzFos.[10]

To test the NShB–NShB interaction, transform yeast with plasmid combination control 3. The controls include two negative controls (Controls 1 and 2) in which the yeast is cotransformed with either one of the fusion constructs and the corresponding empty vector of the other domain of GAL4, and a positive control (control 4) to rule out potential problems in the overall transformation procedures. The following two protocols describe two slightly different techniques of growth selection. The plate assay is simple and fast, while the assay of growth on liquid medium may be more quantitative and have potentially higher sensitivity to resolve variations in affinity.

Growth Assay on Plates

1. Cotransform HF7c with pPC97-NShB and pPC86-NShB as described in the previous section. Plate the transformation mixture on His+/Trp−/Leu− plates, and allow yeast transformants to grow at 30° for 72 hr. Usually, yeast colonies become visible after 48 hr.
2. Inoculate a single colony from the plates into 3 ml His+/Trp−/Leu− liquid medium. Allow to grow overnight at 30° at a speed of 240 rpm. Note that yeast may settle on the bottom of the culture tube by the next day.
3. Vortex to resuspend the yeast culture. Dilute it 1:10 with sterile TE and dot 3 μl of diluted mixture on both His+/Trp−/Leu− and His−/Trp−/Leu− plates.
4. Allow to grow at 30° for 48 hr.

Procedure for Liquid Medium Growth Curve Assay

1. Follow steps 1 and 2 described in the section on Growth Assay on Plates.
2. After overnight culture, take an aliquot of the culture and measure the cell density at 600 nm. Stop the culture at an OD_{600} value of 0.1–0.5. For HF7c, 1 ml of yeast cells at an OD_{600} value of 0.25 has approximately 1×10^7 cells.
3. Inoculate 2×10^7 cells into two flasks containing 50 ml of either His−/Trp−/Leu− or His+/Trp−/Leu− medium. The starting OD_{600} value should be about 0.01.
4. Take the OD_{600} measurement and/or cell counts every 2–3 hr until 15 hr after cultures reach the stationary stage.
5. Plot logarithmic OD versus time.
6. Determine the doubling rate at log phase and the OD at stationary stage. (Note that HF7c cells with a plasmid combination of pPC97-NShB/pPC86-NShB have a doubling rate of 2.3 hr and the saturation OD_{600} value is about 1.6 in His−/Trp−/Leu− medium.)

Testing Domain Interactions by β-Gal Assay

Two protocols have been commonly used to test β-Gal activity. Here we describe the procedure for performing the β-Gal assay on a membrane support. Refer to *Current Protocols* for the liquid culture β-Gal assay.[9]

Screen Yeast Colonies with β-Gal Assay on a Nitrocellulose Filter

1. Wear gloves for all steps. Label the nitrocellulose filter with a fine marker. Carefully place the filter onto the plate to be assayed.
2. Slightly bend the filter in half, allowing first the center and then the edges to touch the plate. This will help to prevent air bubbles between the plate and the filter.
3. Mark with a syringe needle to orient the filter.
4. Lift the filter with a blunt-ended forceps. Submerge immediately in liquid nitrogen (for 5 sec to 5 min). The best way to do this is to place the filter into the liquid nitrogen vertically at a slight angle, so that the colonies face down and hit the liquid nitrogen first. This seems to keep the filter from cracking into pieces, and is particularly important when lifting large colonies.
5. Remove the filter from liquid nitrogen, and place the filter on Whatman 3 MM (Clifton, NJ) paper with colony side up. Allow to thaw for about 2 min, but do not allow the filter to dry out completely. They are sometimes difficult to rewet.
6. Place the filter (with the colony side up) on Whatman 3 MM papers that are presoaked with 1 mg/ml 5-bromo-4-chloro-3-indolyl-β-D-galactoside (X-Gal) in buffer Z (about 7–8 ml is needed for a 150-mm-diameter filter).
7. Incubate the filter in a sealed container at 37°. Visible blue color may develop in 30 min to 3 hr.

Protein Overlay Binding

The protein overlay assay tests the direct interaction between a soluble protein (domain) of interest and a target protein immobilized on membrane support. This method was first developed for studying DNA–protein interaction.[11] We have expressed the N-terminal domain of ShB (NShB) as a fusion protein containing two heart muscle kinase sites (RRASV) to allow radioactive labeling, and a Flag monoclonal tag sequence (DYKD) for affinity purification and immunodetection.[12,13] The binding partner in this

[11] B. Bowen, J. Steinberg, U. K. Laemmli, and H. Weintraub, *Nucleic Acids Res.* **8,** 1 (1980).
[12] M. Li, Y. Jan, and L. Jan, *Science* **257,** 1225 (1992).
[13] M. A. Blanar and W. J. Rutter, *Science* **256,** 1014 (1992).

case is either the full-length ShB protein or another N-terminal domain from a homologous K^+ channel subunit. Protein–protein interaction can be detected using a radioactively labeled NShB probe or by immunodetection using an antibody specific to the probe. The binding can then be visualized with HRP-conjugated secondary antibody followed by enhanced chemiluminescence (ECL) (Amersham, Arlington Heights, IL).

Radioactive Labeling of NShB

1. Express and purify the NShB in *Escherichia coli* as a fusion protein using a standard protocol. The purified protein is diluted to 100 μg per ml in PBS. Carry out the labeling reaction as follows:

6 μl	10× heart muscle kinase reaction buffer [200 m*M* Tris pH 7.8, 10 m*M* dithiothreitol (DTT), 1 *M* NaCl, 120 m*M* $MgCl_2$]
10 μl	[γ-^{32}P]ATP 10 μCi/μl (Amersham, Arlington Heights, IL)
15 μl	Purified NShB fusion protein
27 μl	Deionized H_2O
2 μl	10 units/μl heart muscle kinase (Sigma). The enzyme should be freshly made in 40 m*M* DTT
60 μl	

2. Incubate the mixture at 37° for 1.5 hr, then dilute to 500 μl with HKE buffer containing 10 m*M* HEPES (pH 7.5), 60 m*M* KCl, and 1 m*M* EDTA.
3. To remove the free radioactive material, dialyze the labeling mixture against 200 ml of HKE buffer for 5 hr with five buffer changes. Count an aliquot of labeled protein sample and estimate the specific activity. We use probes with specific activity above 5×10^4 cpm/pmol.

Procedure for Using Radioactive Protein Probes

1. Protein preparations [e.g., crude lysate of baculovirus infected *Spodoptera frugiperda* fall armyworm ovary (Sf9) cells] are fractionated by SDS–PAGE. The separated proteins are electrophoretically transferred to a nitrocellulose filter using a standard transfer procedure. All remaining steps should be carried out at 4° and all buffers should be chilled on ice prior to use.
2. Rinse the filter for 5 min with buffer A containing 10 m*M* HEPES–KOH, pH 7.5, 100 m*M* KCl; 1 m*M* EDTA, 1 m*M* 2-mercaptoethanol.
3. To initiate the denaturation and renaturation, treat the filter for 15 min with buffer A supplemented with 6 *M* guanidine hydrochloride.

4. To renature the protein, treat the filter stepwise with buffer A supplemented with a decreasing concentration of guanidine hydrochloride (3, 1.5, 0.75, 0.38, 0.19, and 0.09 *M*). Each wash lasts for 5 min.
5. Rinse the filter with buffer A twice, for 10 min each time.
6. To block nonspecific binding, incubate the filter with buffer A, 5% (w/v) Carnation nonfat milk, 0.05% (w/v) Nonidet P-40 (NP-40), 60 min at 4°. Transfer the filter to buffer A, 1% Carnation milk, 0.05% NP-40, and incubate for 30 min.
7. To carry out protein binding, transfer the filter to buffer A containing 0.05% NP-40 and supplement with ^{32}P- or ^{35}S-labeled protein probe. Incubate for 12 hr. It is necessary to remove free radioactive material by either dialysis or chromatography.
8. Wash the filter for 30 min with buffer A supplemented with 1% Carnation milk and 0.05% NP-40 with three buffer changes.
9. Wrap the filter with Saran wrap. Visualize the binding signal by autoradiography.

Procedure for Using Probe-Specific Antibodies

1. Follow the procedures for using radioactive protein probes from steps 1 to 6.
2. Incubate the filter with purified protein probe for 12 hr in buffer A containing 1% Carnation milk and 0.05% NP-40. The preferred probe concentration is 2 μg/ml.
3. Wash the filter twice with buffer A containing 1% Carnation milk and 0.05% NP-40.
4. Incubate with first antibody 1:5000 specific to the probe in buffer A containing 1% Carnation milk and 0.05% NP-40, for 30 min.
5. Wash three times with buffer A containing 1% Carnation milk and 0.05% NP-40.
6. Incubate with horseradish peroxidase (HRP)-conjugated secondary antibody 1:5000 specific to the first antibody in buffer A containing 1% Carnation milk and 0.05% NP-40, for 30 min.
7. Wash three times with buffer A containing 1% Carnation milk and 0.05% NP-40.
8. Incubate the filter with chemiluminescence reagents (Amersham).

Construction of Chimeric Channels

After identification of a protein–protein interaction region within a given subunit, it is essential to establish its functional role(s) in channel assembly or properties. In the Shaker-like potassium channels, different α

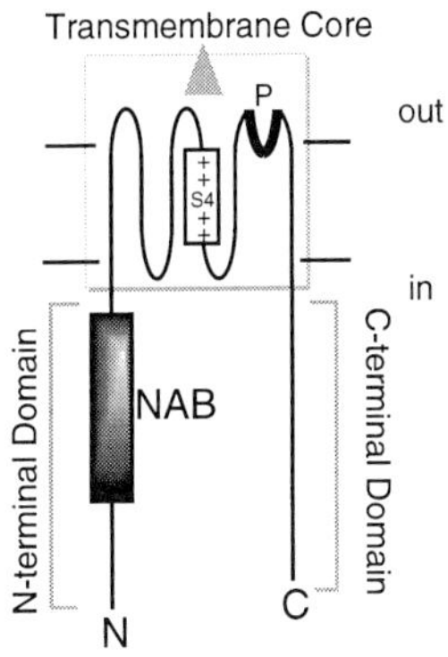

FIG. 1. Schematic diagram of the Shaker-type potassium channel.

subunits coassemble in a subfamily-specific manner, consistent with the binding specificity of the NAB–NAB interaction (Fig. 1). The chimeric channels are designed to test whether the NAB–NAB association plays a role in specifying the interacting subunit (see article in this volume dedicated to chimeric construction[14]).

Vector

The pShBΔ(6-46).2H3 plasmid is a pSK$^+$-based vector that contains ShBΔ(6-46) cDNA (a gift from Dr. Richard Aldrich, Stanford University). Two *Hin*dIII restriction sites were introduced by site-directed mutagenesis using the following oligonucleotides: ShBhind3s: 5′ GAA AGT TCG CAA GAT TCC AGA GTT GTA GCC 3′, and ShBch3s: 5′ CGT CGG TCT CGA TAC AAG CTT CCA GGG CAT TAT TGT G 3′. One *Hin*dIII site is located at amino acid 226 immediately prior to the S1 region; the other is located at amino acid 650 six amino acid residues prior to the stop codon. Introduction of these restriction sites produces a functional channel indistinguishable from the wild-type ShB.[12]

Procedure for Constructing Chimera

1. To prepare vector DNA, digest 10 μg of pShBΔ(6-46).2H3 in 100 μl containing 5 units of *Hin*dIII at 37° for 2 hr.
2. Add 5 units of CIP and incubate the mixture at 55° for exactly 30 min. Add EGTA to a final concentration of 20 m*M*. Inactivate the enzyme by heating the mixture at 75° for 30 min.

[14] T. M. Ishii, P. Zarr, X. Xia, C. T. Bond, J. Maylie, and J. P. Adelman, *Methods Enzymol.* **293,** [4], 1998 (this volume).

TABLE I
METHODS USED TO TEST SUBUNIT INTERACTIONS IN ION CHANNELS

Method	Reagent requirements	Advantages	Disadvantages	Additional comments
Yeast two-hybrid system	Hydrophilic protein/domains; no toxicity to nuclear transport; need to have cDNA of two interacting partners	No biochemical assays involved	Negative results are not informative	Amenable for large-scale genetic screen
Protein overlay	Soluble purified probe protein is preferable; probe can be obtained by *in vitro* translation	Simultaneous testing of multiple interacting proteins including unknown proteins from crude cell extracts	Not all protein–protein interactions can be detected presumably due to denaturation from filter immobilization	It tests direct protein–protein interaction
Coimmunoprecipitation	Antibodies against both testing proteins are required	Can be done with native preparation. May be used for studying multimeric complex	Interaction could be indirect	Solubilization step is critical and sometimes tricky

Gel-filtration chromatography	Purified soluble protein preparation	Can be used to determine stoichiometry in combination with sedimentation	Homogeneous biochemical behavior is necessary	Specific antibody is required if there is no adequate amount of proteins
Sucrose gradient sedimentation	Purified soluble protein preparation	(Same as gel filtration)	Homogeneous biochemical behavior is necessary	(Same as gel filtration)
Chemical cross-linking	Purified soluble protein preparation	Can provide information of stoichiometry	Cross-linking conditions varied with different protein interactions and amino acid composition	Careful titrations are needed to determine the type and concentration of chemical cross-linker
Column binding/pull down	Protein on column should be purified (not necessarily soluble)	Can use tissue extract (requires antibody)	One of the test proteins needs to be soluble	Antibody specific to testing protein may be necessary
Dominant suppression	Electrophysiologic or ligand binding assays	Direct functional test	Ability to suppress expression may not directly correlate its physical association with functional subunits	Conclusion is based on negative result

3. Extract the denatured protein once with an equal volume of chloroform. Precipitate the DNA with ethanol. Resuspend the vector DNA pellet with 20 μl water.
4. Estimate the vector DNA concentration by running 1 μl on a gel alongside 0.5 μg of λ *Hin*dIII marker.
5. To prepare linearized DNA for PCR amplification, add one unit of *Eco*RI to 20 μl restriction digestion buffer containing 2 μg plasmid DNA of pSK.DRK1 (a gift of Drs. Rolf Joho at Univ. of Texas Southwestern, and Arthur Brown at Case Western Reserve Univ.). Incubate at 37° for 30 min.
6. Transfer 5 μl of the digestion mixture to 95 μl of PCR reaction mixture containing 100 nmol of DrkTM1s and Cdrk primers (DrkTM1s: 5′ GAT GAA TTC CAA GCT TCG GTG GCC GCC AAG A 3′; Cdrk: 5′ GGA TAA GCT TCC CGG GGG GAG CTC AGA TAC TCT G 3′). Denature the template DNA for 5 min at 94°. Amplify the DNA fragment with 15 cycles of PCR: 94° for 1 min, 55° for 1 min, and 72° for 2 min.
7. Fractionate the PCR products on 1% agarose gel. Cut a small gel slice containing a fragment ~2.0 kb in size. Recover the DNA fragment using the standard Geneclean protocol as described earlier. Digest the recovered DNA overnight with *Hin*dIII enzyme in a 200-μl final volume. Precipitate the DNA fragment with ethanol and resuspend the DNA pellet in 10 μl water. Estimate the DNA concentration by running 1 μl on a gel alongside 0.5 μg of λ *Hin*dIII marker.
8. To ligate, mix 200 ng of linearized, phosphatase-treated pShBΔ(6-46).2H3 with a twofold molar excess of the insert DNA. Add ligase buffer, 1 unit of ligase (New England Biolabs, Beverly, MA), and water to a final volume of 10 μl. As a control, set up an identical reaction without the insert DNA. Incubate overnight at 14°.
9. Transform ligated DNA into *E. coli*. Prepare minipreparation DNA with the Promega (Madison, WI) Wizard system. Determine the insert and orientation by restriction enzyme digestion. Confirm the PCR fragment by DNA sequencing. This construct is now ready for preparation of cRNA for oocyte expression. The resultant chimeric channels can then be tested by electrophysiologic analyses and by toxin-binding studies.

Conclusion

The methods described in this article are part of a long list of methods that has been used to study subunit interaction in ion channels (see summary in Table I). Some of the listed methods have been discussed in other articles in this volume.[14–16]

The methods in our chapter are particularly suitable for testing the interaction between two hydrophilic polypeptides, such as the N-terminal interaction domains of the Shaker-type potassium channels. The yeast two-hybrid system is a powerful genetic assay and can be modified for a genetic screen to identify interesting mutations that modulate protein–protein interactions, such as temperature-sensitive mutants and genetic enhancers and suppressors. The overlay binding assay tests the direct interaction between two proteins. Although it may not work for every protein, positive results are usually very convincing since it requires relatively high affinity association to be detectable. Finally chimeric construction and testing assembly specificity is a functional approach applicable to both hydrophilic and hydrophobic regions by directly studying their roles in the subunit assembly of a functional channel.

[15] Z. Sheng and C. Deutsch, *Methods Enzymol.* **293,** [2], 1998 (this volume).
[16] J. S. Trimmer, *Methods Enzymol.* **293,** [3], 1998 (this volume).

[2] Assembly of Ion Channels

By ZuFang Sheng and Carol Deutsch

Introduction

Most ion channels are multisubunit conglomerates. Because synthesis and assembly of many different types of pore-forming subunits occur in a single cell, how do the right subunits find each other to give the correct stoichiometry and avoid scrambling to channel homogeneity? This problem is even more striking if we consider the vast number of nonchannel transmembrane proteins made simultaneously in a cell. Assembly is a multistep process that requires specific intersubunit recognition events. Each of these steps may include intermediate folded conformations of subunits and/or intermediate subunit stoichiometries. Such possibilities have not been explored for most types of ion channels, including K^+ channels, nor is it known which regions of the subunuits actually interact during each assembly step.

In some cases, the NH_2-terminal domains of ion channels can function

0076-6879/98 $25.00

as specific recognition motifs between subunits[1–4] (see also ref. 4a), but it is not clear that such elements contribute to stabilization of the mature multimeric protein or whether additional subunit–subunit interactions between transmembrane segments provide the energy to shift the equilibrium in a lipid bilayer toward multimerization and the final, mature channel that functions in the plasma membrane. Most voltage-gated K^+ channels are homotetrameric membrane proteins, each subunit containing six putative transmembrane segments, S1–S6. It is not clear what holds the tetramer together; intersubunit covalent linkages do not appear to be responsible.[5] In these channels the cytoplasmic NH_2 terminus contains a recognition domain, T1 ("first tetramerization"), that tetramerizes *in vitro* and confers subfamily specificity.[2,3,6,7] However, in the native channel there are also intramembrane association (IMA) sites in the central core of voltage-gated K^+ channels that provide sufficient recognition and stabilization interactions for channel assembly, and disruption of one or more of these interactions may suppress channel formation.[8,9] The relative contributions of different domain interactions (e.g., T1 and IMA) may vary from channel isoform to isoform. What are these T1 and IMA domains in the native full-length K^+ channel, and what are their relative contributions to channel formation?

Identification of the recognition and stabilization motifs in the primary sequence of channel proteins is a good beginning to understanding channel assembly; however, it still leaves many questions unanswered. How specific are these intersubunit interactions? How strong are they? At which stage in assembly are subunits integrated into the membrane? What are the spatial and temporal events involved in channel assembly? What is the subunit stoichiometry of the channel? What is the history of the subunits during assembly? Is recruitment of subunits a random event? What is the nature of the subunit pool? Where is it located? When are subunits recruited into multimeric channels, and where? We can address these issues both biochemically and biophysically, as described in the next section, using a variety of *in vitro* translation systems and *in vivo* expression systems.

The *in vitro* translation systems include rabbit reticulocyte (RRL) and

[1] S. Verrall and Z. W. Hall, *Cell* **68,** 23 (1992).
[2] M. Li, Y.-N. Jan, and L. Y. Jan, *Science* **257,** 1225 (1992).
[3] N. V. Shen, X. Chen, M. M. Boyer, and P. Pfaffinger, *Neuron* **11,** 67 (1993).
[4] T. Babila, A. Moscucci, H. Wang, F. E. Weaver, and G. Koren, *Neuron* **12,** 615 (1994).
[4a] J. Xu and M. Li, *Methods Enzymol.* **293,** [1], 1998 (this volume).
[5] L. M. Boland, M. E. Jurman, and G. Yellen, *Biophys. J.* **66,** 694 (1994).
[6] N. V. Shen and P. J. Pfaffinger, *Neuron* **14,** 625 (1995).
[7] J. Xu, W. Yu, J. N. Jan, L. Jan, and M. Li, *J. Biol. Chem.* **270,** 24761 (1995).
[8] L. Tu, V. Santarelli, Z. Sheng, W. Skach, D. Pain, and C. Deutsch, *J. Biol. Chem.* **271,** 18904 (1996).
[9] Z. Sheng, W. Skach, V. Santarelli, and C. Deutsch, *Biochem.* **36,** 15501–15513 (1997).

wheat germ agglutinin (WGA) systems, which contain cellular components necessary for protein synthesis (tRNA, ribosomes, amino acids, and initiation, elongation, and termination factors) and are capable of a variety of posttranslational processing activities (acetylation, isoprenylation, proteolysis, and some phosphorylation activity). Signal peptide cleavage and core glycosylation can be reconstituted and studied by adding canine pancreatic microsomal membranes to the translation reaction. These systems permit studies, for example, of transcriptional and translational control, association of proteins, and their membrane integration. However, the translation efficiency of high molecular weight proteins (>100,000) is relatively poor, and it is not clear that all aspects of *in vivo* processing have been reconstituted. Thus, caution must be used in extrapolating findings with the *in vitro* system to *in vivo* events.

The *in vivo* expression system most used for study of channel function and assembly has been *Xenopus* oocytes.[10] Mammalian cells are also used frequently and involve DNA transfection techniques.[10] Oocytes typically require injection of channel mRNA (typically 50 nl/oocyte; <0.1 ng to 100 ng mRNA/oocyte). This system is an intact cell system that expresses at high levels for both electrophysiological and biochemical measurements, which can be done simultaneously in parallel samples. Both the oocyte and a mammalian T-cell expression system are described later, as well as the methods used to study channel protein synthesis, integration into membranes, and oligomerization.

Broadly defined, assembly also involves trafficking, posttranslational modification, and localization of channel proteins in specific subcellular compartments, as well as the aforementioned processes of recognition and association (oligomerization). This article, however, focuses only on strategies and methods that can be used (1) to identify regions of a protein that are potentially involved in intersubunit interactions during assembly of the pore-forming unit of ion channels, (2) to determine the strength, kinetics, spatial, and temporal characteristics of the intersubunit interactions, and (3) to determine the subunit stoichiometry and history of subunits during assembly. For some cases we illustrate the approaches by describing experiments in our laboratory involving a voltage-gated K^+ channel, Kv1.3. However, these strategies and methods can be, and have been, used for other multimeric channels.

Strategies and Methods

The strategies used to address the issues just stated entail either direct or indirect determinations of various aspects of subunit association. The

[10] B. Rudy and L. E. Iverson, eds., *Methods Enzymol.* **207,** 225 (1992).

former category includes primarily biochemical approaches; the latter makes use of functional readouts. These strategies are protein based, yet each can have additional strategies at the DNA level. For example, strategies that entail constructing genes that link multiple channel domains in tandem, swapping channel domains to create chimeras, and/or deleting or mutating domains can be combined with the protein assays to elucidate mechanisms of channel assembly.

Identification of Putative Regions Involved in Intersubunit Interactions

Intersubunit association can be assessed by direct and indirect methods as described in the following subsections. To discover which regions of the channel interact across subunit boundaries, physical association between channel subunits or between peptide fragments of a channel and the full-length channel protein must be demonstrated. This can be done directly by (1) immunoprecipitation of one member of a complex by antibody against the other member, (2) cross-linking interacting proteins using bifunctional reagents, or (3) binding assays of interacting peptides. Such binding assays have been employed to show that K^+ channel subunits, or parts of these subunits, multimerize both *in vitro* and *in vivo*.[2-4,6] But these studies have been concerned primarily with cytoplasmic NH_2-terminal interactions. We describe one of these methods used in our laboratory, namely, immunoprecipitation. One important caveat concerning the association of peptide fragments of a channel with the channel protein is that it is not clear that such association faithfully reflects native associations between full-length subunits *in situ*. For instance, constraints imposed on a segment of the channel in the context of the full-length folded protein may lead to different interactions with another subunit compared with the isolated truncated channel peptide fragment. Therefore, for a transmembrane segment, it is ultimately important to determine not only whether these interactions occur in the native protein, but also the topology and orientation of the peptide fragment.

Immunoprecipitation. This method requires the use of antibodies (antisera) to a protein or a peptide construct. If the antibodies to native epitopes are not sufficiently good, an epitope tag may be used; c-*myc* (MEQKLISEEDL)[11] is excellent for this purpose. Such nonnative epitopes, however, should be inserted into a primary sequence at a nonperturbing distance (>15 amino acids) from putative topogenic determinants. The first step in this approach involves making the appropriate plasmid DNA either for use in transfections for subsequent *in vivo* expression, or for *in vitro* transcrip-

[11] G. Evans, G. Lewis, G. Ramsay, and M. J. Bishop, *Mol. Cell Biol.* **5,** 3610 (1985).

tion to produce mRNA for subsequent use in either *in vivo* or *in vitro* experiments. Standard methods of restriction enzyme analysis, agarose gel electrophoresis, and bacterial transformation are used for these studies. Plasmid DNA are purified using Qiagen columns (Valencia, CA), and capped mRNA is synthesized *in vitro* from linearized templates using Sp6 or T7 RNA polymerase (Promega, Madison, WI).

For *in vitro* immunoprecipitation experiments, proteins are translated *in vitro* with [^{35}S]methionine (2 μl/25 μl translation mixture; ~10 μCi/μl Dupont/NEN Research Products, Boston, MA) in rabbit reticulocyte lysate (commercial preparations are available from Promega, Madison, WI, and from MBI Fermentas, Amherst, NY; laboratory preparations can be made according to Jackson and Hunt[12] and Walter and Blobel[13]) in the presence (1.8 μl membrane suspension/25 μl translation mixture) or absence of canine pancreatic microsomal membranes (Promega or MBI Fermentas), according to the Promega *Protocol and Application Guide.* Two proteins that are proposed to interact are then cotranslated. Relative mRNA concentrations should be determined from the efficiencies of each construct to yield protein ratios that are desired. To maximize coimmunoprecipitation, microsomal membranes should be used in limiting concentration compared with the total mRNA concentration. The translation reaction can be visualized and quantitated using SDS–PAGE and phosphor imaging.

To perform immunoprecipitation from an *in vitro* translation system (RRL, microsomal membranes), 1–5 μl of cell-free translation products is mixed in 400 μl of buffer A [0.1 *M* NaCl, 0.1 *M* Tris (pH 8.0), 10 m*M* EDTA, and 1% (v/v) Triton X-100] containing 0.1 m*M* phenylmethylsulfonyl fluoride (PMSF). Ascites fluid (9E10 to *myc* epitope; 1 μl) or channel antisera (4 μl) are added and samples incubated at 4° for 30 min. Protein A Affi-Gel beads (10–20 μl; Bio-Rad, Richmond, CA) are added and the suspension mixed continuously at 4° for 6–15 hr with constant mixing. The beads are centrifuged and washed three times with buffer A and two times with 0.1 *M* NaCl, 0.1 *M* Tris (pH 8.0) prior to SDS–PAGE and fluorography. Where relevant, counts per minute in immunoprecipitated proteins should be corrected by the efficiency of recovering precipitated protein from the translation mixture. For example, for a *myc*-labeled peptide the correction factor can be calculated as the ratio of *myc*-peptide in the translation mixture to *myc*-peptide in the immunoprecipitate, as measured using the anti-*myc* antibody. The efficiency of immunoprecipitation will probably range from 10 to 25%. Each batch of membranes must be titrated for each protein or peptide construct to determine the maximal coimmunoprecipita-

[12] R. J. Jackson and T. Hunt, *Methods Enzymol.* **96,** 50 (1983).
[13] P. Walter and G. Blobel, *Methods Enzymol.* **96,** 84 (1983).

tion conditions, that is, the proper mRNA-to-membranes ratio. Also, a study of the time course of addition of membranes indicates that maximal coimmunoprecipitation occurs when membranes are added 5–10 min after translation has begun.[14,15]

To perform immunoprecipitation experiments from oocytes, the *in vitro* transcribed mRNA is mixed with [^{35}S]methionine/cysteine [10× concentration; Tran ^{35}S-label (ICN, Irvine, CA) 20 μCi/μl; 90% methionine, 10% cysteine] and injected directly into oocytes (50 nl/oocyte) as described later. Coinjections of mRNA encoding channel proteins that are proposed to interact are followed by incubation of the oocytes in 1.5-ml Eppendorf tubes at 18° for 4 hr in methionine/cysteine-free medium. The medium is then changed to a 1 m*M* methionine, 1 m*M* cysteine medium. For each time point, five to seven oocytes are frozen (−80°). Later, all samples are thawed and homogenized in 5× volumes (35 μl) of homogenization buffer (0.25 *M* sucrose, 50 m*M* Tris, pH 7.5, 50 m*M* potassium acetate, 5 m*M* $MgCl_2$, 1 m*M* dithiothreitol) while kept on ice. To the oocyte homogenate, 1.2 ml of buffer A (0.1 *M* NaCl, 0.1 *M* Tris (pH 8.0), 10 m*M* EDTA, and 1% Triton X-100) containing 0.1 m*M* PMSF is added. The solution is mixed continuously at 4° for 6 hr and centrifuged at 14,000*g* at 4° for 15 min and the supernatant removed. Antibody (the amount will be determined by the titer) is added to the supernatant and incubated for 30 min at 4°. Protein A Affi-Gel beads (15 μl) are added, and the suspension is mixed continuously at 4° overnight. The beads are centrifuged and washed three times with buffer A and twice with 0.1 *M* NaCl, 0.1 *M* Tris (pH 8.0) prior to SDS–PAGE and fluorography.

Yeast Two-Hybrid System. We can use another approach, the yeast two-hybrid method,[16] which relies on a functional readout, to learn whether two proteins interact *in vivo*. This method is a genetic assay based on the fact that eukaryotic transcriptional activators are bifunctional, containing discrete functional domains. One domain binds to DNA while the other activates transcription. Two fusion proteins are generated *in vivo* (in yeast), one from a plasmid containing the DNA-binding domain and protein X, and one from a plasmid containing the activation domain and protein Y. The two hybrid genes are expressed in a yeast host strain containing latent reporter genes, typically *lacZ* or *His3*. Expression of the reporter gene therefore indicates possible interaction between proteins X and Y. This technique is restricted to detecting interactions between cytoplasmic regions of channel proteins and cannot be used to detect interactions between

[14] Z. Sheng and C. Deutsch, unpublished data, 1997.
[15] D. Andrews, unpublished data, 1996.
[16] S. Fields and O. Song, *Nature* **340,** 245 (1989).

transmembrane segments of proteins. It has been used to identify and define the role of T1 recognition domains in voltage-gated K^+ channels.[7] A more detailed discussion of this technique is given by Xu and Li in this volume.[4a]

Dominant Negative Suppression. To learn which regions of channel subunits may associate, including transmembrane segments, a dominant negative suppression strategy can be used. Evidence for protein–protein association is obtained as follows. A full-length channel is coexpressed with a fragment of the channel that putatively interacts across subunit boundaries. If the fragment associates with the channel subunit, than it will result in scavenging of available monomers, competitively inhibiting association of a full-length subunit in the multimer, and/or associating with a multimer. In any of these cases, the result may be suppression of a fully functional channel. Thus suppression, measured by a variety of readouts, may be interpreted as evidence of protein–protein interaction, with the additional caveat that it must then be shown that such interactions also occur between the fragment region and channel protein in the full-length channel tetramer *in situ.* Some of these readouts are functional tests (e.g., current measurements), and others are biochemical (e.g., ligand binding, immunoprecipitation assays of channel formation; see earlier discussion). Ideally, several criteria should be experimentally demonstrated in order to use this strategy to infer putative sites. First, it should be shown that protein is being made and is stable *in vivo.* Second, if suppression occurs, it must be shown not to be a consequence of inhibition of transcription or translation, but rather due to physical interaction of the peptide fragment with channel protein. Third, it must be shown that if these peptide fragments suppress current, they are specific. Fourth, the orientation and topology of channel subunits and peptide fragments should be known. This strategy has been used in oocytes and mammalian cells to identify multimerization domains in hydrophilic NH_2-terminal segments[1,2,4,4a,8] and to probe for IMA sites within the hydrophobic core containing transmembrane segments.[8,9]

To determine experimentally which regions of channel subunits may be interacting, oocyte expression and electrophysiology are convenient and relatively simple tools for this purpose. While standard methods are used, some attention must be paid to conditions for suppression experiments. Specifically, the level of current expressed and the time after injection of the oocytes should be chosen to optimize detection of suppression (see later discussion). Oocytes can be isolated from *Xenopus laevis* females (Xenopus I, Ann Arbor, Michigan) as described previously.[17] Stage V–VI oocytes are selected and microinjected with mRNA (usually 0.1–10 ng)

[17] A. Goldin, *Methods Enzymol.* **207,** 266 (1992).

encoding for the channel of choice. The amount of mRNA should be adjusted to produce an appropriate current amplitude (see later discussion). Where applicable coinjections can be made in specified mole ratios, for example, for suppression experiments, we use 1:2 mole ratios of channel mRNA to truncated K^+ channel mRNA or transmembrane control mRNA, respectively. K^+ currents from mRNA-injected oocytes are measured with a two-microelectrode voltage clamp after 24–48 hr, at which time currents should be 2–10 μA. This level of expressed current is optimal for observing suppression because it avoids voltage-clamp artifacts that would mask true maximum current levels for control currents and therefore underestimate the extent of suppression. Electrodes (<1 MΩ) typically contain 3 *M* KCl, while the bath Ringer's solution contains (in m*M*): 116 NaCl, 2 KCl, 1.8 $CaCl_2$, 2 $MgCl_2$, 5 HEPES (pH 7.6). The appropriate holding potential for Kv1.3 is −100 mV. Data should be presented as box plots, which represent the central tendency of the measured current. This is a better means of displaying such data than bar graphs of the mean value because suppression data are quite variable and often non-Gaussian, and this method of analysis permits the entire range of data to be presented. The box and the bars indicate the 25–75 and 10–90 percentiles of the data, respectively. The horizontal line inside the box represents the median of the data. It may be necessary to carry out a statistical analysis of these results to determine whether experimental and control values are different. A nonparametric test such as a Mann–Whitney rank sum test is sufficient.

Characterization of Intersubunit Interactions

Avidity. In those cases in which the avidity (strength of association) between interacting channel proteins or peptides is to be determined by titrating the association with increasing amounts of detergent, sodium dodecyl sulfate (0–1.0% SDS in 0.1 *M* Tris, pH 7.8) or sodium *N*-dodecanoylsarcosinate (0–0.1% Sarkosyl in 50 m*M* NaCl, 0.1 *M* Tris, pH 7.5) should be added (100 μl) to the translation products (3.5 μl) and incubated for 30 min at 4° before diluting 10× with buffer A (+ PMSF) and continuing according to the procedures described earlier for immunoprecipitation.

Membrane Versus Nonmembrane Compartment. It is important to determine whether association requires membranes and, if so, whether proteins must first be integrated into the membranes for association to occur, and moreover, whether the proteins must be integrated into the same membrane. Finally, what is the temporal relationship between synthesis of associating proteins, membrane integration, and complex formation?

To demonstrate whether membranes are required for association of subunits and/or channel peptides, proteins are cotranslated in the presence and absence of microsomal membranes and then immunoprecipitated. The extent of membrane integration of each protein can be measured by extracting translation products made in the presence of microsomal membranes with either Tris buffer or carbonate buffer and comparing the pellet and supernatant content of ^{35}S-labeled protein. *In vitro* translation products (1–5 μl) are diluted in 750 μl of either sodium carbonate (0.1 *M* Na_2CO_3, pH 11.5) or Tris (0.25 *M* sucrose, 0.1 *M* Tris, pH 7.5) solution. Samples are incubated on ice for 30 min prior to centrifugation at 70,000 rpm (208 kG) TLA 100.3 rotor for 30 min. Supernatants are removed and 10% (v/v) trichloroacetic acid added to precipitate protein, which is resuspended in SDS–PAGE sample buffer. Membrane pellets are then dissolved directly in SDS–PAGE sample buffer. The pellet contains the membrane fraction of protein. As a measure of the physical and functional integrity of the microsomal membranes, control mRNA should be simultaneously translated in the assay, including one that encodes a known transmembrane protein and one that encodes a secreted protein. The former should remain in the pellet fraction at pH 7.5 and 11.5, whereas the latter will be in the pellet fraction (microsomal lumen) at pH 7.5 and in the supernatant at pH 11.5. The ratio of pellet to pellet plus supernatant for extractions done at pH 7.5 gives the fraction of pellet associated with microsomes, and for extractions done at pH 11.5, this ratio is the fraction of protein integrated into the microsomal membrane.

To determine whether two putatively interacting proteins can associate if they are integrated into different membranes, the two proteins are translated separately in the presence or absence of membranes, then mixed together, and immunoprecipitated. In this case, puromycin (1 m*M*, 15 min) is added to terminate translation prior to combining the two reaction mixtures. A control should be included to show that the channel proteins are actually being translated on the membranes. The ideal control is one in which one of the monitored proteins is itself glycosylated. Thus, synthesis and translocation (glycosylation) of this protein will verify whether the membranes are functioning properly.

To determine whether synthesis, membrane integration, and association occur sequentially or concurrently, and which of these steps is rate limiting, a study of the relative time course of the synthesis, membrane integration, and association of the interacting proteins can be done by measuring ^{35}S incorporation, carbonate extraction, and coimmunoprecipitation, respectively. In the case of Kv1.3, such a comparison has shown that synthesis and integration of channel protein are rapid and that the association step itself is the rate-determining, membrane-delimited step in complex formation.[9]

Complex Size. To determine the size of the associated channel complex, sucrose gradient experiments can be used. This method is based on the hydrodynamic properties of proteins and protein complexes. The fractional migration of a protein through a gradient is determined by the sedimentation characteristics of the protein or complex. Limitations of this method include experimental tailing of bands due to self-aggregation of proteins, and resolution that is often >30 kDa. However, the choice of detergent and salt concentration can have a significant impact on aggregation and resolution. *In vitro* translated protein (50 μl) is loaded on a 100-μl sucrose cushion (0.5 *M* sucrose, 100 m*M* KCl, 50 m*M* HEPES, pH 7.5, 5 m*M* $MgCl_2$, 1 m*M* DTT) and spun at 117 kG using a TLA rotor for 5 min to obtain pellets that contain the membrane fraction only. These pellets are solubilized for 30 min on ice in 100 μl of 0.05% dodecylmaltoside ($C_{12}M$), 50 m*M* NaCl, 50 m*M* Tris, 1 m*M* EDTA, pH 7.5, or in 1.5% CHAPS (Sigma, St. Louis, MO) buffer solution, 150–200 m*M* NaCl, 50 m*M* Tris-HCl (pH 8.0), 1 m*M* EDTA. Sometimes 0.015% phosphatidyl choline can be added to prevent aggregation. The solubilized proteins are then centrifuged for 1 hr at 4°, 139,000*g* (60,000 rpm) in a Beckman (Columbia, MD) TL-100 centrifuge to remove insoluble material. Twelve milliliters of 5–20% linear sucrose gradient (in 0.05% $C_{12}M$ or in 1% CHAPS, buffer as above) is poured into Nalgene (Fisher Scientific, Philadelphia, PA) ultratubes, using an HBI gradient maker. The solubilized protein is loaded on top of the gradient and sedimented at 164,000*g* (36,000 rpm) in a SW-40 Ti rotor on a Beckman L8-70M ultracentrifuge for 20 hr at 4°. Fractions (either 0.25 or 1 ml) are collected and either used directly in immunoprecipitation assays or precipitated with trichloroacetic acid (TCA, 10%) for 1 hr at 0°, and spun in an Eppendorf centrifuge (14,000 rpm) at 4° for 30 min to pellet the protein. The supernatant is removed and the pellet dried. SDS sample buffer is added to the pellet for separation on SDS–PAGE gels.

Determination of Subunit Stoichiometry and History during Assembly

Mass Tagging. To determine subunit stoichiometry, molecular weight markers can be engineered into subunits.[18,19] This is known as *mass tagging.*[19] Identical subunits can be labeled with additional peptide chains that will shift the molecular weights of monomer and multimer detected on SDS–PAGE. If assembly of subunits is random and stable, then the number of distinctly different protein bands observed on SDS–PAGE will indicate the subunit stoichiometry of the assembled channel. This approach has

[18] T. Sakaguchi, Q. Tu, L. H. Pinto, and R. A. Lamb, *Proc. Natl. Acad. Sci. U.S.A.* **94,** 5000 (1997).

[19] L. Heginbotham, E. Odessey, and C. Miller, *Biochem.* **36,** 10355–10342 (1997).

been used to biochemically demonstrate the tetrameric stoichiometry of the influenza virus M_2 channel[18] and of a prokaryotic K^+ channel.[19] In each case, two plasmids, each containing different engineered lengths of the C terminus of the channel were expressed, isolated, and run on SDS–PAGE. If two different species of monomers associate randomly to form a tetramer, then the relative proportion of the five resultant channel types will be described by a binomial distribution (see later section). Five distinct bands, representing each of the five possible channels subunit compositions, appeared on SDS–PAGE.[18,19]

This method can be adapted for use in low-yield expression systems by labeling the translated subunits with [^{35}S]methionine. However, this method does require that multimeric species be stable under conditions of SDS–PAGE, which is rarely the case. Moreover, quantitative analysis using binomial distributions may not be practical because the stability of each mass-tagged multimeric complex may not be identical, hence skewing the distribution for noncombinatoric reasons.[19] An alternative means to mass tagging a polypeptide is to engineer a glycosylation site into the subunit. In this case the cell itself labels the subunit *in vivo*. (Such strategies are used routinely to determine topologies of polytopic transmembrane proteins.[20,21]) However, the creation or deletion of consensus sites must be experimentally confirmed because consensus sites may not always be glycosylated. Sometimes specific lengths and amino acid sequences in the flanking regions are required for glycosylation.[20]

Functional Tagging. In addition to the direct methods already described, subunit stoichiometry can also be determined using a functional readout. A mutant is made that has altered conductance, pharmacology, and/or kinetics of gating. We refer to this approach as *functional tagging*. It has been used to determine not only subunit stoichiometry,[22] but also the contribution of individual subunits to specific K^+ channel functions.[23–26] Moreover, one can engineer a particular function of a channel to reveal its prior history, namely, to learn something about the assembly of individual subunits. In this regard the strategy and methods described later can be used to answer the following questions: Does synthesis and assembly of different channel subunits occur in the same shared compartment? Are

[20] Z. G. Galen and R. E. Ostwald, *J. Biol. Chem.* **270,** 2000 (1995).
[21] M. Holman, C. Maron, and S. Heinemann, *Neuron* **13,** 1331 (1994).
[22] R. MacKinnon, *Nature* **500,** 232 (1991).
[23] R. MacKinnon, R. W. Aldrich, and A. W. Lee, *Science* **262,** 757 (1993).
[24] M. P. Kavanaugh, R. S. Hurst, J. Yakel, J. P. Adelman, and R. A. North, *Neuron* **8,** 493 (1992).
[25] G. Panyi, Z. Sheng, L. Tu, and C. Deutsch, *Biophys. J.* **69,** 896 (1995).
[26] E. M. Ogielska, W. N. Zagotta, T. Hoshi, S. H. Heinemann, J. Haab, and R. W. Aldrich, *Biophys. J.* **69,** 2449 (1995).

subunits recruited randomly or preferentially? Does multimer formation occur in the plasma membrane? Are channel monomers and multimers in equilibrium in the plasma membrane? Is channel diversity temporally or spatially regulated? Any functional property of the channel can be used for this purpose. The major criterion that must be met, however, is that an order of magnitude difference in the chosen functional parameters for the wild-type and mutant subunits must exist, regardless of which parameter is being studied, whether it be time constants of gating, binding constants of some ligand, or single-channel conductances.

The functional tagging experiments are performed either by heterologously coexpressing a wild-type subunit with a mutant subunit, or by heterologously expressing a mutant subunit in a cell expressing endogenous wild-type channels. The appropriately modified function is measured and an analysis, as described later, is performed.

The method of analysis in functional tagging studies assumes a binomial distribution for the random formation of heteromultimeric channels. For a multimer of N subunits, the fraction of channels with exactly m mutant subunits will be

$$B(N, p, m) = \frac{N!}{m!(N-m)!} p^m (1-p)^{N-m}$$

where p is the fraction of mutant subunits in the membrane. The wild-type homomultimer is represented by $m = 0$, whereas $m = N$ represents the homomultimeric mutant channel. If the biophysical properties of each member of this population are known, it is possible to estimate both p and the validity of the underlying assumption, namely, that wild-type (WT) and mutant (MUT) subunits assemble randomly. Biophysical properties that can be quantified in this way include the kinetics of inactivation, the affinity of an open-channel blocker, and the single-channel conductance.

Three similar equations (Fig. 1) can be used to fit the data obtained from a cell expressing both endogenous and heterologous subunits, depending on the biophysical parameter to be measured. In the case of gating kinetics, $I(t)$ is the current at time t and $Y_m(t)$ is a function describing the gating kinetics for a channel with m mutant subunits (see later example). In the case of blocker affinity, $I([bk])$ is the current in the presence of blocking agent, $I(0)$ is the current in the absence of blocking agent, bk is the blocker molecule, $[bk]$ is the blocker concentration, and $F_{unbk,m}([bk])$ is the fraction of unblocked current for a channel with m mutant subunits. In the case of single-channel conductance, i_m is the single-channel current for a channel with m mutant subunits.

We introduce an approach that provides a general two-step test for possible sources of nonrandomness in channel assembly that can give us

Gating Kinetics: $I(t) = \sum_{m=0}^{N} B(N, p, m) Y_m(t)$

Affinity of Channel Blocker: $\frac{I([bk])}{I(0)} = \sum_{m=0}^{N} B(N, p, m) F_{unbk,m}([bk])$

Single Channel Conductance: $\frac{\text{number of openings to level } i_m}{\text{total number of openings}} = B(N, p, m)$

$$B(N, p, m) = \frac{N!}{m!(N-m)!} p^m (1-p)^{N-m}$$

FIG. 1. Readouts of functionally tagged subunits. Equations that describe gating kinetics, open-channel block, and single-channel conductance. In each case, $B(N, p, m)$ represents the binomial distribution, as described in the text, along with the functions and symbols used in these equations.

insights into the prior history of channel subunits. As outlined in Fig. 2, the first test determines the relative affinities of mixed subunits. WT and MUT subunits are heterologously coexpressed in a cell devoid of the channel in question, and the resulting channel population either conforms to or fails to conform to a binomial distribution. If it conforms, this means that WT and MUT subunits are recruited randomly and independently with the same probability. Failure means that subunits are not selected randomly, but since they are present at the same time in the same heterologous cell compartment, failure means there is some cooperativity, positive ("like prefers like") or negative ("like avoids like"). This will be manifest as an excess of homomultimers or an excess of heteromultimers, respectively, compared to a binomial distribution.

The second test determines whether subunits are recruited from integrated or segregated monomer pools and what must be the nature of the segregation. It should be used after verifying that the first test shows no cooperativity. MUT subunits are heterologously expressed in a cell that already has endogenous WT subunits. A binomial distribution can be interpreted as evidence that random recruitment of subunits occurred from an integrated pool of subunits. This test can only fail the binomial distribution if there is some segregation, assuming that these subunits did not fail the first test (i.e., did not show cooperativity). Segregation will be manifest either as an excess of endogenous WT channels or as an excess of both WT and MUT channels compared to the binomial distribution. The first case indicates temporal segregation of subunits (i.e., multimers were formed irreversibly at different times), while the second indicates spatial segregation (i.e., WT and MUT homomultimers were formed in spatially separate compartments).

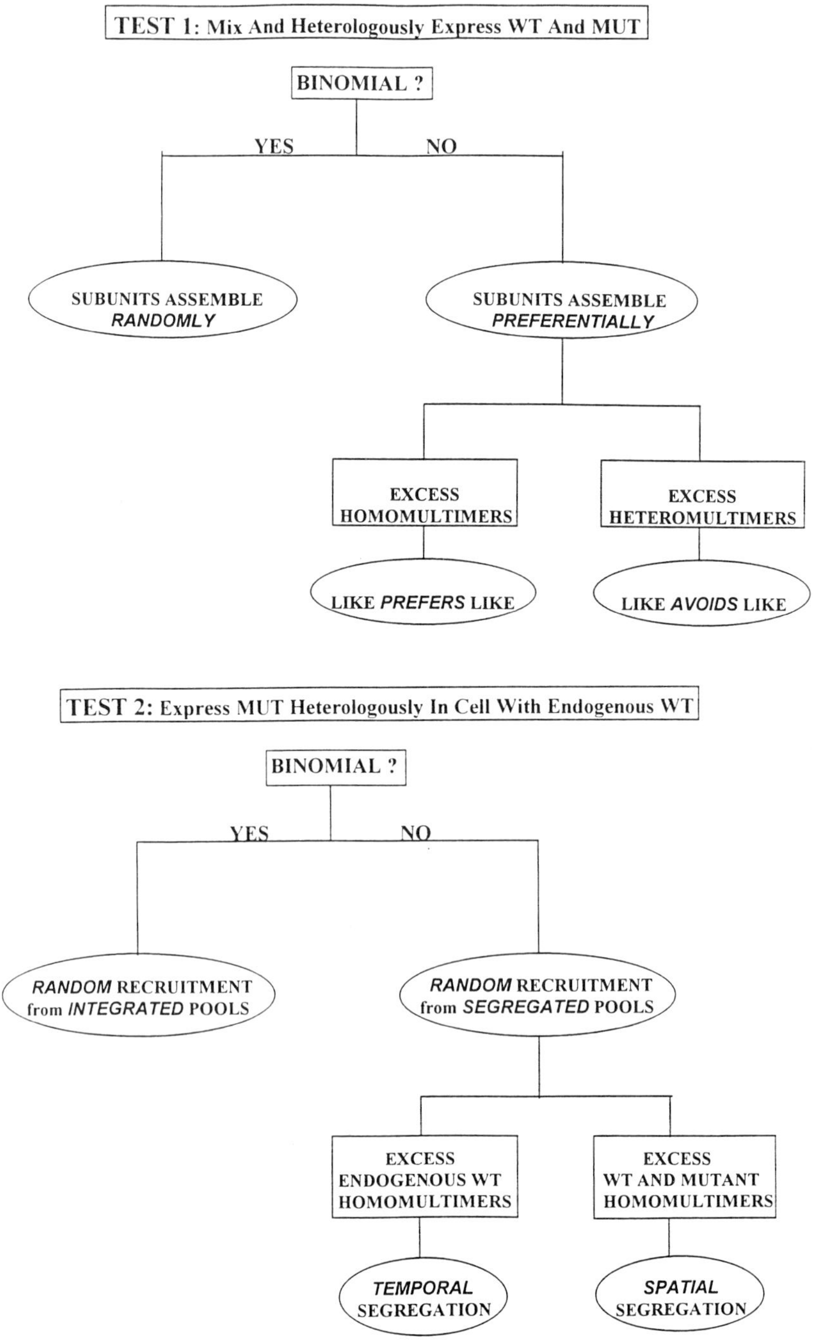
TEST 1: Mix And Heterologously Express WT And MUT
BINOMIAL ?
YES
NO
SUBUNITS ASSEMBLE RANDOMLY
SUBUNITS ASSEMBLE PREFERENTIALLY
EXCESS HOMOMULTIMERS
EXCESS HETEROMULTIMERS
LIKE PREFERS LIKE
LIKE AVOIDS LIKE
TEST 2: Express MUT Heterologously In Cell With Endogenous WT
BINOMIAL ?
YES
NO
RANDOM RECRUITMENT from INTEGRATED POOLS
RANDOM RECRUITMENT from SEGREGATED POOLS
EXCESS ENDOGENOUS WT HOMOMULTIMERS
EXCESS WT AND MUTANT HOMOMULTIMERS
TEMPORAL SEGREGATION
SPATIAL SEGREGATION

To illustrate a specific application of this approach, we describe our studies of Kv1.3. The wild-type homotetramer Kv1.3 inactivates with a time constant of 200 ms, and a point mutation in the S6 segment of this channel produces a mutant homotetramer that inactivates with a time constant of 4 ms, 50× faster than the wild-type channel.[25] When we expressed this functionally tagged subunit along with the wild-type subunit in a cytotoxic T cell (CTLL), which is devoid of endogenous channels, we found that the rate of inactivation depended exponentially on the number of mutant subunits (m). We could therefore account for the inactivation kinetics of a population of heterotetramers by a binomially weighted sum of inactivation rates when WT and MUT subunits were coexpressed in a cell line devoid of endogenous voltage-gated K^+ channels,[25] indicating a random association of mutant and WT subunits with no cooperative assembly. The equation for gating kinetics (Fig. 1) was used for this analysis. Here $Y_m(t)$ is a first-order process and equals $[(k_{i,m}/(k_{i,m} + k_{r,m})e^{-(k_{i,m}+k_{r,m})t} + k_{r,m}/(k_{i,m} + k_{r,m})]$ I_{peak}, where I_{peak} is the peak current. The inactivation and recovery rate constants for a channel with m mutant subunits are $k_{i,m}$ and $k_{r,m}$, respectively. The values for the rate constants were determined from the particular model invoked for the inactivation of Kv1.3 and from the measured parameters for the wild-type and mutant homomultimers. When we applied this method of analysis to the resulting current obtained from mutant-transfected Jurkat T cells, which also express endogenous WT channels, we found that the inactivation kinetics did not conform to a simple binomial distribution of channel types, but rather to a binomial distribution plus an additional term for preformed endogenous WT channels in the membrane, that is, excess WT channels compared to a binomial distribution of channel types, consistent with temporal, but not spatial segregation of the WT and mutant subunits[27] (Fig. 2).

In addition to the strategies and methods described, a variety of techniques have been used to define protein topology, that is, the location of a protein or region of protein with respect to a membrane-bound compartment. These methods include protease digestion, cell surface labeling with epitope-specific antibodies, cross-linking with sulfhydryl specific reagents,

[27] G. Panyi and C. Deutsch, *J. Gen. Physiol.* **107,** 409 (1996).

FIG. 2. Prior history of channel subunits. Flow diagram illustrating tests for subunit preferences in association (test 1) and for segregation of subunits (test 2). The results (rectangles) and interpretations (ovals) of heterologous coexpression of wild-type (WT) and mutant (MUT) subunits in a cell are shown in test 1. The results and interpretations of heterologous expression of MUT in a cell already expressing endogenous WT are shown in test 2, where it is known that WT and MUT subunits show no preferential association.

or cysteine accessibility. We refer the reader to a recent review by Skach[28] that describes assays of protein topology and the insights they provide regarding protein folding and general mechanisms of polytopic protein biogenesis. These approaches are designed to investigate the mechanisms of protein folding and assembly in the endoplasmic reticulum membrane, and are complementary to the approaches described herein.

[28] W. Skach, *Methods Enzymol.* **292,** 265–278 (in press).

[3] Analysis of K^+ Channel Biosynthesis and Assembly in Transfected Mammalian Cells

By JAMES S. TRIMMER

Introduction

Recent work in several laboratories has determined that K^+ channels are integral membrane, heterooligomeric glycoprotein complexes composed of four pore-forming α subunits and four cytoplasmic β-subunit polypeptides.[1,2] Although it is widely appreciated that the coassociation of functionally distinct α subunits contributes to the tremendous diversity of K^+ channels observed in native cells,[1] the discovery and characterization of a family of β subunits[2] indicates that these cytoplasmic polypeptides also make a significant contribution to channel diversity. In particular, the Kvβ3 β subunit, and the multiple isoforms arising from the alternative splicing of Kvβ1 transcripts, have dramatic effects on the inactivation of *Shaker*-related or Kv1 subfamily K^+ channel α subunits, while the Kvβ2 β subunit does not.[2] All of these β-subunit isoforms may also play a role in promoting the efficient surface expression of Kv1 subfamily α subunits.[3] It is clear from these data that the differential biosynthetic association and subsequent assembly of α and β subunits into the resultant $\alpha_4\beta_4$ complex[3,4] is a key mechanism used to generate diversity of K^+ channel function and ultimately neuronal excitability.

[1] L. Y. Jan and Y. N. Jan, *Trends Neurosci.* **13,** 415 (1990).
[2] O. Pongs, *Sem. Neurosci.* **7,** 137 (1995).
[3] G. Shi, K. Nakahira, S. Hammond, K. J. Rhodes, L. E. Schechter, and J. S. Trimmer, *Neuron* **16,** 843 (1996).
[4] D. N. Parcej, V. E. Scott, and J. O. Dolly, *Biochem.* **31,** 11084 (1992).

0076-6879/98 $25.00

The assembly of voltage-gated K$^+$ channels thus involves oligomerization of four integral membrane α subunits, synthesized on endoplasmic reticulum membrane-associated polyribosome complexes, and four cytoplasmic β subunits, translated from free cytoplasmic polyribosome complexes.[3,4] Subunit composition, subunit stoichiometry, and the relative position of each polypeptide subunit within the final complex are critical determinants of channel function; thus, studies characterizing the assembly of K$^+$ channels are crucial to understanding the molecular mechanisms important in governing electrical excitability. However, because K$^+$ channel complexes are formed from both integral membrane and cytoplasmic polypeptides, subunit assembly is quite complex. In addition, the intricate membrane topology of the integral membrane protein α subunits, with 24 membrane-spanning domains per tetramer, can only be achieved through the correct alignment of the cytoplasmic, transmembrane, and extracellular domains of each of the four component α-subunit polypeptides. Finally, each of the respective domains, including the 24 transmembrane domains, must not only assume the correct conformation, but also the proper orientation and position relative to each of the other segments in the channel complex. In addition to conformational changes and subunit assembly, the α- and β-subunit polypeptides may be modified posttranslationally through the enzymatic addition of oligosaccharide chains, phosphate and sulfate groups, and fatty acid chains, and by isomerization reactions involving disulfide bonds and proline residues. Additional posttranslational biosynthetic events can include the association of K$^+$ channel complexes with other cellular proteins involved in channel targeting and localization.[5]

To begin to address this complex series of dynamic events that together encompass the biosynthesis and assembly of voltage-gated K$^+$ channels, it is essential to have access to a cell culture system expressing the relevant channel polypeptides that is amenable to metabolic radiolabeling, and to have available subunit-specific antibodies to allow for the subsequent isolation of the radiolabeled channel subunits. A number of approaches have been used to express K$^+$ channel subunits in mammalian cells, including direct cytoplasmic injection of cRNA[6] and infection with recombinant vaccinia virus.[7] We have used cultured mammalian fibroblast cell lines expressing recombinant channel polypeptides via transient transfection as an expedient system for the expression of specific combinations of K$^+$ channel α-

[5] M. Sheng and E. Kim, *Curr. Opin. Neurobiol.* **6,** 602 (1996).

[6] S. R. Ikeda, F. Soler, R. D. Zuhlke, R. H. Joho, and D. L. Lewis, *Pflugers Arch.* **422,** 201 (1992).

[7] R. J. Leonard, A. Karschin, S. Jayashree-Aiyar, N. Davidson, M. A. Tanouye, L. Thomas, G. Thomas, and H. A. Lester, *Proc. Natl. Acad. Sci. U.S.A.* **86,** 7629 (1989).

and β-subunit polypeptides.[3,8–12] We have also generated subunit-specific polyclonal[9,10,13,14] and monoclonal antibodies[3,8,10] for the selective isolation of the respective subunits by immunoprecipitation to allow for their subsequent analysis. Here the culture and transfection procedures involved in expressing K^+ channel subunits, metabolic labeling, cell lysis and immunoprecipitation are described along with subsequent analysis of isolated channel subunits. Similar approaches have been applied to the study of the biosynthesis of a number of multisubunit membrane proteins. Here, the specific details pertaining to the study of voltage-gated K^+ channel subunits in particular are emphasized.

Cell Culture

African green monkey kidney fibroblast-like SV40 transformed COS-1 cells (American Type Culture Collection, Rockville, MD) are maintained in plastic tissue culture dishes in Dulbecco's modified Eagle's medium (DMEM, Life Technologies, Grand Island, NY) containing 10% newborn calf serum (Hyclone Labs, Logan, UT). Cells are seeded after harvesting in trypsin–EDTA solution (Life Technologies) at 10% confluence in 6-cm dishes ($\approx 2 \times 10^5$ cells in 6 ml media, dish surface area = 28.3 cm^2) for metabolic labeling, or 10-cm dishes ($\approx 5 \times 10^5$ cells in 16 ml media, surface area = 78.5 cm^2) for nonradioactive biochemical analysis. Cells can also be plated at 1% confluence onto 22 × 22-mm^2 glass coverslips in 3.5-cm dishes (6×10^4 cells in 2 ml media, surface area = 9.6 cm^2) for immunofluorescence analysis of the expressed channels.[8,10,12] Coverslips are sterilized by irradiation with an ultraviolet light for 20 min, precoated overnight in a sterile (0.22-μm filtered) solution of 25 μg/ml poly(L-lysine) (molecular weight >300,000, Sigma, St. Louis, MO) and washed three times with sterile water prior to cell plating. Cells are cultured at 37° in a humidified incubator in the presence of 5% CO_2 atmosphere for 8–18 hr, and not more than 24 hr, before transfection.

[8] Z. Bekele-Arcuri, M. F. Matos, L. Manganas, B. W. Strassle, M. M. Monaghan, K. J. Rhodes, and J. S. Trimmer, *Neuropharmacology* **35,** 851 (1996).

[9] K. Nakahira, G. Shi, K. J. Rhodes, and J. S. Trimmer, *J. Biol. Chem.* **271,** 7084 (1996).

[10] K. J. Rhodes, M. M. Monaghan, N. X. Barrezueta, S. Nawoschik, Z. Bekele-Arcuri, M. F. Matos, K. Nakahira, L. E. Schechter, and J. S. Trimmer, *J. Neurosci.* **16,** 4846 (1996).

[11] R. H. Scannevin, H. Murakoshi, K. J. Rhodes, and J. S. Trimmer, *J. Cell Biol.* **135,** 1619 (1996).

[12] G. Shi, A. K. Kleinklaus, N. V. Marrion, and J. S. Trimmer, *J. Biol. Chem.* **269,** 23204 (1994).

[13] J. S. Trimmer, *Proc. Natl. Acad. Sci. U.S.A.* **88,** 10764 (1991).

[14] K. J. Rhodes, S. A. Keilbaugh, N. X. Barrezueta, K. L. Lopez, and J. S. Trimmer, *J. Neurosci.* **15,** 5360 (1995).

Transfections

Plasmid DNA is introduced to the cultured cells in the form of an insoluble precipitate that adheres to the cell surface. We use a calcium phosphate precipitate[15] that is directly added to the culture media covering the cells. Transfection cocktails[16] are prepared from sterile (0.22-μm filtered) solutions of 2 *M* $CaCl_2$, and 2× HEPES-buffered saline (0.28 *M* NaCl, 50 m*M* HEPES (*N*-2-hydroxyethylpiperazine-*N'*-2-ethanesulfonic acid), 1.5 m*M* Na_2HPO_4, pH 7.05). The pH of the HEPES-buffered saline solution is critical and should be monitored closely. It is advisable to prepare fresh $CaCl_2$ and HEPES-buffered saline solutions approximately every 2 weeks. A 1-ml transfection cocktail is then prepared, and the entire 1-ml volume used to transfect a 10-cm dish. For a 6-cm tissue culture dish, use 375 μl, and for a 3.5-cm dish, 125 μl. The following example outlines the preparation of a 1-ml cocktail for one 10-cm dish.

To prepare the transfection cocktail, add 8 μg plasmid DNA to a 1.5-ml polypropylene microcentrifuge tube (the "C" tube). Larger tubes can be used for larger cocktails, however, always use polypropylene and not polystyrene tubes. Plasmid DNA should be CsCl-purified or of comparable purity (e.g., purified using Qiagen (Chatsworth, CA) or other commercially available plasmid purification systems) and should be at a minimum concentration of 0.2 mg/ml, although higher concentrations (1 mg/ml) are preferred. The volume is then adjusted to 439 μl total with H_2O. To this add 61 μl of 2 *M* $CaCl_2$, and mix by gentle vortexing. Prepare a second tube (the "H" tube) containing 500 μl of HEPES-buffered saline. To prepare the cocktail, use a P200 Pipetman (Rainin Instrument Co., Woburn, MA) to slowly add the contents of the C tube dropwise to the H tube, while constantly mixing the forming precipitate by continual gentle up-and-down pipetting of the contents of the H tube with a P1000 Pipetman (Fig. 1). For larger cocktails, a mechanical pipettor can be used to bubble the contents of the H tube[16]; however, for the microcentrifuge tubes normally used, mixing with the P1000 is easier to control and yields more consistent results. Once the entire contents of the C tube have been added to the H tube, the H tube is capped and mixed by gentle inversion three to four times, then allowed to sit at room temperature for 20 min for the calcium phosphate precipitate to form. The appropriate volume (see above) of this transfection cocktail is then distributed evenly into the culture media using a Pipetman. Gently swirl the dish to distribute the precipitate throughout the media, and incubate for 8–18 hr under standard growth conditions. Following

[15] F. L. Graham and A. J. van der Eb, *Virology* **52,** 456 (1973).

[16] F. M. Ausebel, R. Brent, R. E. Kingston, *et al.,* "Current Protocols in Molecular Biology." Wiley-Interscience, New York, 1990.

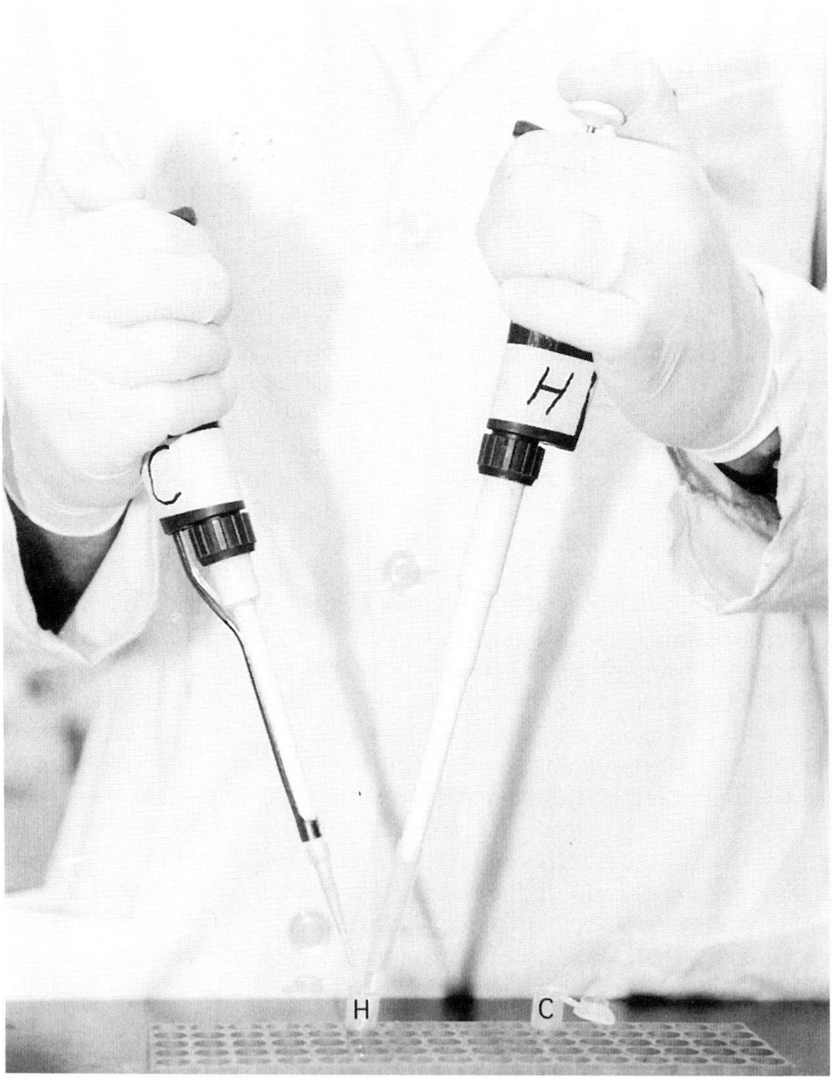

FIG. 1. Preparation of calcium phosphate transfection cocktail. Calcium chloride–DNA mixture is added dropwise via a P200 Pipetman (C) into tube containing HEPES-buffered saline, which is mixed by continual pipetting with a P1000 Pipetman (H).

this incubation, aspirate the media and replace with the appropriate volume of (see above) fresh media and incubate under standard growth conditions. For steady-state or pulse-chase metabolic labeling, begin labeling cells at 24 hr posttransfection. For other biochemical analyses, immunofluorescence, or for electrophysiology, process cells at 48 hr posttransfection.

Steady-State Metabolic Labeling of Transfected Cells

Steady-state labeling conditions result in the generation of large amounts of radiolabeled biosynthetic end product and are employed for biochemical analyses of extent of subunit association, subunit stoichiometry, and posttranslational modifications to K$^+$ channel α- and β-subunit polypeptides. The most commonly used metabolic label for these procedures is the radioactive amino acid [^{35}S]methionine, because of its high specific activity (1200 Ci/mmol) and the relatively high energy (167 keV beta) of [^{35}S], both of which contribute to the ease of detection of [^{35}S]methionine-labeled subunits. The main disadvantage of [^{35}S]methionine is the low abundance of this amino acid in K$^+$ channel α- and β-subunit polypeptides (1.5–3.5 mole percent) compared to other amino acids such as leucine (8.2–10.7 mole percent). However, the low specific activity of radiolabeled (^{3}H)leucine (25–200 Ci/mmol) combined with the relatively low energy (18 keV beta) of ^{3}H make this less useful as a metabolic label despite its high frequency in K$^+$ channel α- and β-subunit polypeptides.

Metabolic labeling using [^{35}S]methionine can be performed using either purified [^{35}S]methionine (DuPont-NEN, Wilmington, DE) or much less expensive protein hydrolyzates of *Escherichia coli* grown in the presence of [$^{35}SO_4$]$^{2-}$ (e.g., Expre^{35}S^{35}S; DuPont-NEN). Labeling with these ^{35}S-labeled protein hydrolyzates, which contain ≈73% of the label in [^{35}S]methionine, 22% in [^{35}S]cysteine, and 5% in other ^{35}S-labeled compounds (DuPont-NEN), can be performed in methionine-free media (Life Technologies) to restrict the labeling to methionine residues, or in methionine-free, cysteine free media (Life Technologies) to incorporate both [^{35}S]sulfur-containing amino acids, *Caution:* ^{35}S-Labeled compounds have been found to release volatile radioactive substances[17] and should be added to media under a chemical fume hood, because safety is more important than sterility in short-term labeling experiments such as these. All incubations should be performed in trays lined with activated carbon filter paper (β-safe) (Schleicher and Schuell, Keene, NH). Other appropriate precautions associated with the use of radiolabeled compounds should always be taken; consult your institutional environmental health and safety office for details.

[17] J. Meisenhelder and T. Hunter, *Nature* **335,** 120 (1988).

For optimal metabolic labeling of K^+ channel α- and β-subunit polypeptides expressed in transiently transfected cells,[9] cells are used 24 hr posttransfection. At this time the synthesis levels of the recombinant K^+ channel polypeptides are maximal,[12] although steady-state accumulation has not yet peaked (this is typically accomplished at ≈48 hr posttransfection). For [^{35}S]methionine labeling in a 6-cm tissue culture dish, cells are washed three times with 5 ml of prewarmed (37°) serum-free, methionine-free DMEM. Cells are then incubated for 15 min in 5 ml of prewarmed (37°) serum-free, methionine-free DMEM (Life Technologies) under standard growth conditions [37°, 5% CO_2 atmosphere] to deplete cellular methionine stores (the "starvation" step). After removal of the media, a labeling solution of 0.3 mCi/ml [^{35}S]methionine in prewarmed (37°) serum-free methionine-free DMEM is then added (1 ml for each 6-cm dish). Relatively short labeling times (4–6 hr) can be used for most applications; this will yield subunits labeled to high specific activity. Under these conditions it is not necessary to add unlabeled methionine or serum due to the short incubation period. Occasionally, it may be necessary to label for longer periods to ensure complete formation of the desired end product, as for the study of later posttranslational processing events such as oligosaccharide trimming (Fig. 2), interaction with plasma membrane-associated anchoring proteins, or for studies of turnover rates and stability.[3] In cases where it is necessary to label overnight (16–18 hr), media composed of nine volumes of serum-

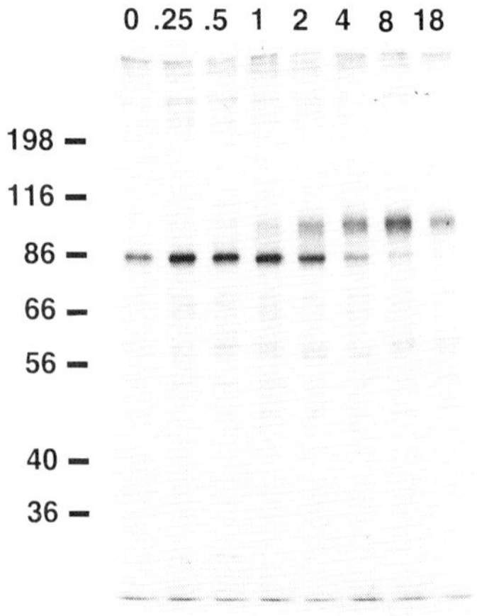

FIG. 2. Biosynthetic maturation of Kv1.4 in COS-1 cells. Kv1.4 transfected COS-1 cells were labeled with [^{35}S]methionine for 15 min, then subjected to chase with unlabeled methionine for the indicated times (in hours) at 37°. Cells were lysed and subjected to immunoprecipitation with anti-Kv1.4 antibody, followed by SDS–PAGE (7.5%) and fluorography. Numbers on left indicate mobility of prestained molecular weight markers (× 10^{-3}).

free, methionine-free DMEM and 1 volume of complete DMEM (which contains 10% v/v dialyzed newborn calf serum) (Sigma) is used to prepare the labeling solution. The starvation step prior to labeling can also be omitted. The use of this modified labeling solution helps to maintain cell viability and sustain protein synthesis during the longer incubation. After this labeling incubation is complete, remove the labeling solution; the radio-labeled cells are then washed and lysed as described later. For the steady-state condition used earlier, it is usually not advised to reuse the labeling solution, and it should be disposed of properly.

In vivo ^{32}P-labeling can be accomplished using a similar approach.[12,18] Cells are again used 24 hr posttransfection, washed three times with serum-free, phosphate-free DMEM and labeled for 16 hr in phosphate-free DMEM containing 1 mCi/ml ortho[^{32}P]phosphate (ICN, Irvine, CA). Labeling with these large amounts of ortho[^{32}P]phosphate requires caution, because the high energy of [^{32}P] (1.71 meV of beta) necessitates the use of proper shielding of all work areas and personnel.

Labeling can also be performed with other metabolic precursors, for instance with [^{3}H]glucosamine or [^{3}H]mannose; however, because these techniques have been described in great detail in a recent volume of this same series,[19] they are not addressed here.

Pulse-Chase Metabolic Labeling of Transfected Cells

Pulse-chase labeling is used to study time-dependent processes involved in K^+ channel biosynthesis, such as the α/α- and α/β-subunit associations leading to the assembly of $\alpha_4\beta_4$ K^+ channel complexes, the posttranslational modification of the α- and β-subunit polypeptides and other associated processing events, the transport of subunit polypeptides through the endomembrane secretory pathway, and the targeting of K^+ channel complexes to the cell surface and their subsequent incorporation into specialized plasma membrane domains via association with anchoring proteins. Pulse-chase labeling strategies are designed to selectively radiolabel only the pool of translation products generated during a discrete biosynthetic period. This cohort of radiolabeled core polypeptides can then be followed through the posttranslational events listed earlier. In the case of voltage-gated K^+ channel proteins, this has proven to be somewhat problematic given the early occurrence of a number of critical biosynthetic events such as α/α-[20] and α/β-[3] subunit association and constitutive biosynthetic phosphoryla-

[18] H. Murakoshi, G. Shi, R. H. Scannevin, and J. S. Trimmer, *Mol. Pharm.* **52,** 821 (1997).

[19] A. Varki, *Meth. Enzymol.* **230,** 16 (1994).

[20] K. K. Deal, D. M. Lovinger, and M. M. Tamkun, *J. Neurosci.* **14,** 1666 (1994).

tion.[12,18] Therefore, the standard pulse-chase protocols that have been used in studies of other multisubunit membrane proteins[21] have been modified to allow for analysis of such early and rapid posttranslational events.

Pulse-chase labeling protocols are performed as a variation of the steady-state labeling procedure described earlier. The most notable exception is that the labeling period, or "pulse" of radioactive [^{35}S]methionine, is limited in time, usually on the order of 15–30 min. These conditions work well for the analysis of the turnover rates of α- and β-subunit polypeptides.[3] In addition, certain K^+ channel biosynthetic events, such as the processing of aspargine-linked oligosaccharide chains on the rat Kv1.4 α subunit that occurs with a $t_{1/2}$ of $\approx$3 hr as the polypeptide reaches the Golgi apparatus (Fig. 2) can also be analyzed using these labeling conditions. Note that the biosynthetic maturation of glycoproteins containing asparagine-linked chains can be further analyzed in assays employing digestion of samples labeled in pulse-chase protocols with specific endoglycosidases, using changes in mobility on SDS gels to monitor the extent of modification. However, because this topic has been covered in detail in a recent volume,[22] it is not addressed here.

For these pulse-chase studies, 6-cm dishes of cells transfected 24 hr prior to labeling are prepared as described earlier for steady-state labeling and incubated in 0.5 ml of the [^{35}S]methionine labeling solution containing 0.4 mCi/ml for 15 min at 37° under 5% CO. Working quickly, the labeling solution is then removed, the cells washed two times with phosphate-buffered saline (PBS: 0.15 *M* NaCl, 10 m*M* sodium phosphate, pH 7.4), and the cells in 1 dish lysed as described later (this constitutes the 0-min chase sample). To the remaining dishes 5 ml of prewarmed chase medium, consisting of complete DMEM containing 10% v/v newborn calf serum and 5 m*M* additional unlabeled methionine (to swamp out for any residual [^{35}S]methionine), is added. The dishes are incubated in the case media for various times, typically 15 and 30 min, and 1, 2, 4, 8, and 18 hr. For studies of stability/turnover rates, chase periods can be extended for times up to 48 hr. Following the chase period, cells are washed two times with ice-cold PBS and lysed as described later.

These pulse-chase conditions are appropriate for studies of late events in K^+ channel maturation, but when applied to the study of α/α-[20] and α/β-subunit[3] interaction, or of constitutive biosynthetic phosphorylation of Kv2.1,[12,18] they failed. The precursor species (i.e., the free α- and β-subunit polypeptides,[3,20] or unphosphorylated Kv2.1 α subunit[12,18]) were not re-

[21] S. M. Hurtley and A. Helenius, *Annu. Rev. Cell Biol.* **5,** 277 (1989).

[22] S. R. Carlsson, *in* "Glycobiology: A Practical Approach" (M. Fukuda and A. Kobata, eds.), pp. 1–26. IRL Press, New York, 1993.

solved under these conditions, as both α/α and α/β-subunit interactions (Fig. 3), and phosphorylation of Kv2.1[12,18] appeared to be complete by the end of the 15-min pulse period. Because of these rapid kinetics, shorter pulse periods are necessary to capture these early and rapid events. Cells are labeled with [^{35}S]methionine as described earlier, but a shorter 5-min pulse is used.[3,12] It may be necessary to increase the concentration of [^{35}S]methionine in the media to 0.5 mCi/ml, or even 1 mCi/ml, to achieve adequate labeling, although studies focusing on the association of the Kv1.2 α subunit with the Kvβ2 β subunit have successfully used labeling solutions containing 0.3 mCi/ml of [^{35}S]methionine.[3] Shorter chase periods of 5 and 10 min are also added to the standard protocol to resolve these rapid events.[3,12] Cells are then harvested in detergent solution as described later.

Lysing COS Cells in Detergent Buffer

Following the radiolabeling procedures described earlier, it is necessary to separate the K$^+$ channel subunits from other radiolabeled cellular proteins. The first step in this procedure is to extract the detergent-soluble α-

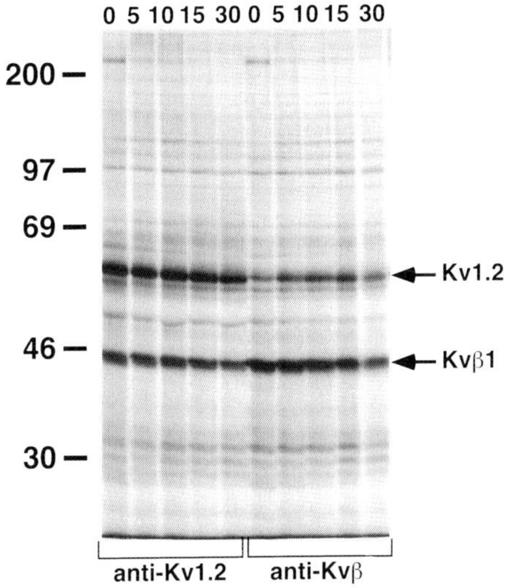

FIG. 3. Pulse-chase analysis of α/β-subunit interaction. Cells cotransfected with Kv1.2 and Kvβ1 were labeled with [^{35}S]methionine for 5 min, then subjected to chase with unlabeled methionine for the indicated times (in minutes) at 37°. Cells were lysed and subjected to immunoprecipitation with anti-Kv1.2 antibody (left-hand side), or anti-Kvβ-subunit antibody (right-hand side) followed by SDS–PAGE (9%) and fluorography. Numbers on left indicate mobility of prestained molecular weight markers ($\times 10^{-3}$); arrows on right denote mobility of Kv1.2 and Kvβ1.

and β-subunit polypeptides from detergent-insoluble cellular organelles and cytoskeleton. This can typically be accomplished by extracting the cells with buffered saline solutions containing nonionic detergents such as Triton X-100 or Nonidet P-40. Under these conditions, K^+ channel complexes are for the most part extracted intact (i.e., α/α- and α/β-subunit interactions) are maintained. Nuclei, mitochondria, and the bulk of the cytoskeleton, on the other hand, remain insoluble and can be easily removed by centrifugation. For this procedure, cultured adherent cells are rinsed two to three times with ice-cold buffered PBS. Ice-cold lysis buffer [1% Triton X-100, 20 m*M* Tris-HCl, pH 8.0, 10 m*M* EDTA, 0.15 *M* NaCl, 10 m*M* iodoacetamide, 10 m*M* azide, and 1 mg/ml bovine serum albumin (BSA)] is then added. The lysis buffer should contain protease inhibitors, because channel proteins are notoriously prone to proteolysis. We use a protease inhibitor cocktail containing 1 m*M* phenylmethylsulfonyl fluoride (PMSF), 2 μg/ml aprotinin, 10 μg/ml benzamidine (all from Sigma), 1 μg/ml leupeptin, and 2 μg/ml antipain (from Boehringr Mannheim, Indianapolis, IN). Stock solutions of iodoacetamide (1 *M*), BSA (50 mg/ml), and protease inhibitors (1000× in 0.15 *M* NaCl, 0.9% v/v benzyl alcohol) are prepared, stored at −20°, and added just before use. PMSF, which is quite toxic, can be prepared as a 500 m*M* stock in acetone. Cells are extracted using 1 ml lysis buffer per 6-cm dish by incubation on ice or on a rocker platform in a 4° cold room for 1 min. The resulting detergent lysate is then harvested to a 1.5-ml microcentrifuge tube with a Pasteur pipette, and the tube vortexed and incubated an additional 5 min on ice. The tube is then spun in a refrigerated microcentrifuge (Brinkmann Instruments, Westbury, NY) at 16,000*g* for 2 min at 4° to pellet the nuclei, and the supernatant saved as the detergent lysate. The lysate can be used immediately (which is best) or can be frozen in aliquots at −80°. If frozen, it will be necessary to centrifuge the thawed lysate for 5 min in a refrigerated microcentrifuge at 16,000*g* before use. In some instances (e.g., where background bands in immunoprecipitation products are a persistent problem) it may be necessary to centrifuge the lysate at 100,000*g* for 30 min to pellet additional insoluble material. In initial experiments it is also advisable to isolate the detergent-soluble lysate and detergent-insoluble pellet from unlabeled cells and assay for the presence of the specific subunit of interest by immunoblot analysis.

Immunoprecipitation of K^+ Channels

Immunoprecipitation reactions are performed to isolate selectively the K^+ channel subunit of choice. These can be performed under relatively mild conditions in order to maintain α/α- and α/β-subunit interactions,[3,9,20] or under harsher conditions when attempting to isolate a single subunit species in the absence of background protein bands.[12,18] The starting material for the immunoprecipitation reaction is the detergent extract of transfected cells in which the proteins have been radiolabeled using one of the metabolic labeling techniques described earlier. The basic strategy is to attach your subunit to an insoluble matrix through the specific interaction of a subunit-specific antibody. The matrix is then separated from the detergent-soluble lysate by centrifugation, and after repeated washing, the subunit eluted by heating in SDS sample buffer.

In the past we have made extensive use of whole, fixed *Staphylococcus aureus* bacteria (PANSORBIN, Calbiochem, La Jolla, CA) as the insoluble matrix. *Staphylococcus aureus* bacteria have on the surface a protein, "protein A," that binds to immunoglobulins from many species with high affinity.[23] This binding is mediated by the portion of the immunoglobulin (Ig) molecule, the FC region, that is at the opposite end of the molecule from the end involved in binding antigen; thus, binding to protein A does not interfere with antibody–antigen interaction. However, the binding of the antibody to antigen does induce a subtle allosteric change in the FC region, such that binding to protein A is enhanced on the binding to antigen.[23] Incubation of the cell lysate with antibody prior to the addition of the protein A-containing matrix can be used to isolate selectively preformed antibody–antigen complexes away from free antibody (and antigen) due to the increased affinity of antigen-bound immunoglobulins for protein A. Because of this, the incubations aimed at binding of antibody to antigen and that aimed at binding of antibody–antigen complexes to protein A are usually performed as sequential and not simultaneous steps.

We have also employed purified protein A covalently attached to agarose beads (Pierce Chemical, Rockford, IL) as a substitute for fixed whole *S. aureus* bacteria. Protein A-agarose is more expensive (2.5× the cost per mg IgG binding capacity) but typically yields lower backgrounds. In addition, because nonspecific binding to protein A-Sepharose is generally much lower than to whole fixed *S. aureus,* the preclearing step can usually be eliminated. One other consideration is whether your antibody preparation consists of antibodies that will bind to protein A.[24] Protein A binds strongly to IgGs from rabbit and guinea pig, and to human IgG_1 and IgG_2, and does not bind to IgGs from rat, sheep, horse, and goat, and to human IgG_3. Protein A binding to mouse IgGs is isotype specific, with strong binding to IgG_{2a}, IgG_{2b} and IgG_3, and weak binding to IgG_1. For antibodies that do not bind well to protein A, a sandwich technique can be employed. For this, fixed *S. aureus* or protein A-agarose can be precoated with commer-

[23] S. W. Kessler, *J. Immunol.* **115,** 1617 (1975).

[24] G. Kronvall and R. C. WIlliams, Jr., *J. Immunol.* **103,** 828 (1969).

cially available secondary antibody, which are affinity-purified rabbit antibodies raised against the non-protein A binding IgG used (e.g., rabbit anti-rat IgG, Pierce). Because the rabbit antibody will bind to the protein A on the fixed *S. aureus* or on the agarose bead, it will provide a "heterobifunctional" reagent to link the antibody to the matrix.

Agarose coated with protein G, a surface protein isolated from group G streptococci, can be used as an alternative to protein A-based matrices (Pierce). Protein G has a broader isotype binding specificity than does protein A, but is substantially more expensive. It is also possible, for some species and IgG isotypes, to obtain agarose with covalently attached secondary antibodies, such as goat anti-mouse IgG_1-agarose (Sigma). As a last resort, affinity matrices can be prepared by direct covalent attachment of the antibody of interest to agarose[25]; this should only be considered if using large amounts of antibody, and if it will be necessary to use it extensively for immunoprecipitation reactions. Note that if there is any uncertainty regarding the binding of the antibody preparation to protein A or any other affinity matrix, appropriate preliminary experiments should be performed to determine the efficiency of antibody recovery in the immunoprecipitation product before using the antibody in immunoprecipitation experiments.

When using either whole fixed *S. aureus* or protein A agarose as the insoluble matrix, it is necessary to wash the material before use. After thoroughly resuspending the entire contents of the stock bottle, an amount sufficient for the entire set of immunoprecipitation reactions is removed to a microcentrifuge tube. We typically use 50 μl of a 10% suspension of fixed *S. aureus* for the preclearing step, and an additional 50 μl of this suspension for the antibody binding step. For protein A agarose, 50 μl of a 50% suspension is used per immunoprecipitation reaction. In both cases, washing is accomplished by pelleting the matrices by microcentrifugation; the resulting pellet is washed three times by microcentrifugation/resuspension in lysis buffer containing 1 mg/ml BSA, and after the last wash resuspended in this buffer to the original volume. For fixed *S. aureus,* microcentrifugation is performed at 16,000*g* for 1 min at 4° to pellet, for protein A agarose, at 16,000*g* for 20 sec.

At this point, any necessary precoating steps should be performed. For precoating, 15 μg of commercially available affinity-purified rabbit polyclonal secondary antibody (e.g., rabbit anti-mouse IgG, Pierce) is added to the washed protein A-based affinity matrix to be used in each immunoprecipitation reaction (50 μl of a 10% suspension of fixed *S. aureus* or 50

[25] E. Harlow and D. Lane, "Antibodies. A Laboratory Manual," pp. 519–538. Cold Spring Harbor Laboratory Press, Cold Spring Harbor, New York, 1988.

μl of a 50% suspension of protein A agarose). After incubation for 45 min on a tube rotator (Scientific Equipment Products, Baltimore, MD) at 4°, the matrix is washed three times with lysis buffer containing 1 mg/ml BSA by centrifugation/resuspension as described earlier. After the last wash the matrix is resuspended in this buffer to the original volume.

If fixed *S. aureus* is being used, a preclearing step is performed to remove material that is present in your lysate that may bind nonspecifically to this matrix. If a single sample will be analyzed with a number of different antibodies, it is recommended that this preclearing step be performed in a single batch, to ensure uniformity of the cleared lysate used for the subsequent immunoprecipitation reactions. Add enough crude lysate to a glass or polypropylene centrifuge tube for all of the immunoprecipitation reactions (we typically use 100 μl of radiolabeled lysate, prepared as described earlier, for each immunoprecipitation reaction). For every 100 μl of lysate, add 50 μl of either a 10% suspension of fixed *S. aureus* and 875 μl of lysis buffer containing 1 mg/ml BSA. Incubate on a tube rotator in a 4° cold room for 45 min. Centrifuge at 16,000*g* for 30 min at 4° in a fixed-angle rotor (e.g., a Sorvall SS-34 rotor at 11,500 rpm) and retain the resultant supernatant as the "cleared lysate." For the subsequent immunoprecipitation reactions, 1 ml of the cleared lysate is used per reaction.

For immunoprecipitation reactions performed on multiple lysates with a single antibody (e.g., in an analysis of a set of samples from a pulse-chase labeling experiment), the preclearing step is performed in microcentrifuge tube, using 100 μl of lysate, 50 μl of fixed *S. aureus,* and 875 μl of lysis buffer per tube. Multiple microcentrifuge tubes can then be placed in a screw-cap 50-ml conical polypropylene tube and placed on the tube rotator. After the 45-min incubation at 4°, the tubes are spun in the microcentrifuge at 16,000*g* for 30 min at 4°, and the resultant supernatants used as the cleared lysate in the subsequent immunoprecipitation reactions.

To initiate the immunoprecipitation reaction itself, antibody is added to microcentrifuge tubes containing the cleared (if using fixed *S. aureus*) or crude (if using protein A agarose) lysates. If using fixed *S. aureus,* we recommend using flexible microcentrifuge tubes ("flex tubes", Brinkmann Instruments) because the vigorous technique used to resuspend the pellet in subsequent steps can shatter standard microcentrifuge tubes. The amount of antibody used in each reaction will vary depending on the concentration, purity, and avidity of each of the antibody preparations. We have found that 1–2 μl of serum in a 1-ml reaction is a good general starting point when using crude polyclonal antiserum. When using antibodies that have been affinity purified from polyclonal antiserum, start with 0.2–1 μg of immunoglobulin protein per tube, and for monoclonal antibodies, use 50–100 μl of hybridoma tissue culture supernatant, or 10–100 ng of purified monoclonal immunoglobulin protein per tube. However, it is important to stress that careful dilution series should be performed to optimize each new antibody before immunoprecipitation reactions on samples from involved experiments (e.g., pulse-chase experiments) are attempted. Following antibody addition, the samples are incubated on a tube rotator at 4° for 1 hr to overnight to allow for the formation of antibody–antigen complexes. The length of incubation varies among different antibodies, and is also dependent on the amount of antigen in the sample. Shorter incubation times result in lower backgrounds but in some cases also lower yields. Two hours is a reasonable starting point for most antibody–antigen combinations, but a simple time course will allow for a straightforward determination of the optimum conditions.

The insoluble affinity matrix is now added to capture the preformed antibody–antigen complexes. Because both protein A and commercially available secondary antibodies have a high affinity for binding to immunoglobulins, a short incubation period of 30–45 min is normally sufficient to capture the antibody–antigen complexes quantitatively. Following the incubation, the affinity matrix is pelleted by microcentrifugation at 16,000*g* at 4° for 1 min (for fixed *S. aureus*) or 20 secs (for agarose beads). The pellet from the fixed *S. aureus* cells is quite tight, and the wash buffer can be easily aspirated. However, this pellet is difficult to resuspend by vortexing, even at top speed. An alternative method is to scrape the bottom of the closed microcentrifuge tube containing the dry pellet back and forth across the bottom grid of a metal test tube rack, using a swiping motion similar to playing a washboard (Fig. 4). Two times across should be sufficient to effectively loosen the dry pellet, followed by addition of lysis buffer and vortexing to resuspend the pellet thoroughly. Simply vortexing the dry pellet also works to a certain extent, but extensive vortexing is required. It should be stressed that effective resuspension of the pellet is crucial to the specificity of the immunoprecipitation reaction. Resuspending an agarose bead pellet is quite easy and can be accomplished simply by adding lysis buffer; however, because this pellet is quite loose care should be taken during aspiration of the wash buffer. If aspirating with a Pasteur pipette, it is advisable to leave a small amount (e.g., 50 μl) of wash buffer over the pellet to prevent inadvertent aspiration of the agarose beads. An alternative approach is to fit the end of a Pasteur pipette with a 27-gauge needle.[25]

Overall, the pellets are washed six times in lysis buffer (without BSA) by resuspension/microcentrifugation, using 1 ml of lysis buffer per wash. After adding the buffer for the final wash, the resuspended pellet and buffer are transferred to a clean microcentrifuge tube, because material that may stick nonspecifically to the walls of the reaction tube will solubilize during boiling in sample buffer resulting in nonspecific background bands. The

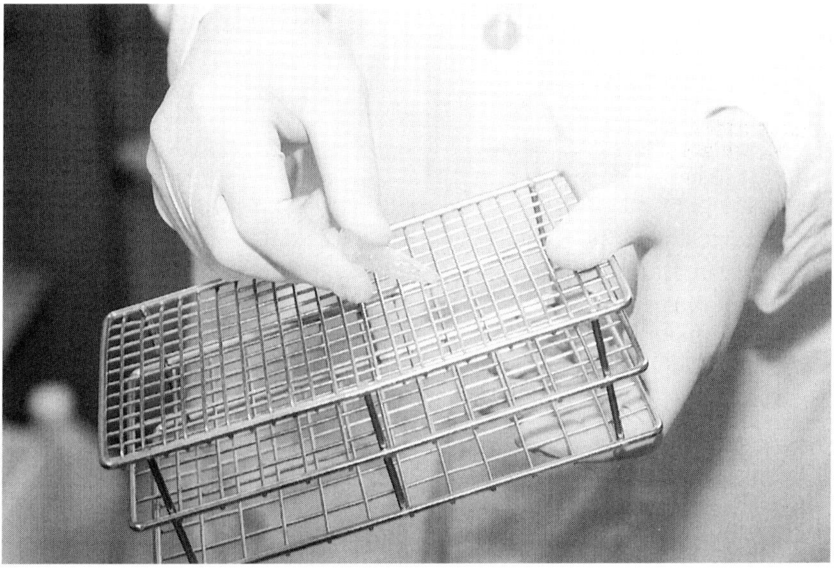

FIG. 4. Resuspending fixed *S. aureus* immunoprecipitation pellets. Flex tubes with pelleted *S. aureus* are drawn back and forth across the bottom of a wire test tube rack prior to wash buffer addition and vortexing.

microcentrifuge tube is then centrifuged, the supernatant aspirated, and the tube recentrifuged to collect any residual wash buffer sticking to the sides of the tubes. The remaining buffer can be removed with a P200 Pipetman, and the dry pellet loosened as described earlier. Sample buffer is then added, and the sample boiled for 3 min, or heated to 80° for 5 min, and then fractionated by SDS–PAGE. Radiolabeled ^{14}C-methylated (Amersham, Arlington Heights, IL) molecular weight standards can also be loaded onto the gel for subsequent determination of M_r. Radiolabeled sample proteins on the gel can then be imaged using either autoradiography (for ^{32}P-labeled samples), fluorography (for ^{35}S- or ^{3}H-labeled samples), or phosphorimaging (for any of the above).

For autoradiography, which is the direct exposure of film by beta particles or gamma rays, the SDS gel is first stained with Coomassie blue stain [0.6 g/liter Coomassie Brilliant Blue R-250 (Kodak, Rochester, NY) in 45.6% (v/v) H_2O/45.6% (v/v) methanol/8.8% (v/v) glacial acetic acid]. Coomassie blue staining of the gel is useful because it allows for the visualization of the intensity of staining of the immunoglobulins present in the immunoprecipitation reaction product. After equilibrium in the Coomassie blue staining solution (which takes ≈1 hr for a minigel, and 3 hr for a 1.2-mm-thick standard gel), the gel is destained for 3 hr (for a minigel) to 18 hr (for a 1.2 mm thick standard gel) in 83% (v/v) H_2O, 10% (v/v) methanol, and 7% (v/v) glacial acetic acid. Small bits of sponges (1 cm^3) can be added to assist in the destaining. This material, which can be any foam rubber packing material, should be thoroughly soaked in the destaining buffer for best results. Following destaining, the gel should be soaked for a short period (minigel—15 min, standard gel—45 min) in gel shrinking solution (49.5% v/v H_2O, 49.5% v/v methanol, and 1% v/v glycerol). This incubation will not only partially dehydrate the gel, but will also infuse the matrix with glycerol. Both of these considerations will help to prevent subsequent cracking of the gel upon drying. If Coomassie blue staining is not necessary, gels can be fixed in 45.6% v/v H_2O, 45.6% v/v methanol, 8.8% v/v glacial acetic acid, dehydrated in gel shrinking solution, and then dried. Gels are then dried on a vacuum gel dryer for 30 min (for minigels) or 2 hr (for large, 1.2-mm-thick gels) and then exposed to preflashed X-ray film. Preflashing can be accomplished with commercially available flash/filter sets, such as the Sensitize system (Amersham); film should be preflashed to an OD_{540} of 0.15. The gel is then placed face up in a cassette, followed by the X-ray film, flashed side down against the dried gel. An intensifying screen (e.g., Hyperscreen, Amersham; Reflection, DuPont-NEN; or Lightning Plus, Kodak) is then placed, face down, on the back side of the film, and the cassette closed and incubated at $-70°$.

For fluorography, which is the exposure of film by secondary light that was generated by the excitation of a fluor by a beta particle or a gamma ray, the gel can either be strained and destained, or fixed, as described in the previous section. Enhancement by fluorography is absolutely necessary for ^3H- and ^{35}S-labeled samples. Instead of incubation in the gel shrinking solution, gels are incubated in commercially available fluors such as Autofluor (National Diagnostics, Atlanta, GA), En^3Hance (DuPont-NEN), or Amplify (Amersham), and dried. The gel is then placed face up in a cassette, followed by the X-ray film, flashed side down against the dried gel. Intensifying screens will not work for ^3H- and ^{35}S-labeled samples; however, the cassette should still be incubated at $-70°$.

Recently, the advent of phosphorimaging technology has offered tremendous advantages over standard autoradiography and fluorography. Storage phosphor imaging plates are $\approx$250 times as sensitive as X-ray film and have a dynamic range of more than 300$\times$ that of X-ray film. Phosphorimagers, which process the image on these phosphor plates, have the capability to generate photographic quality figures from the information stored on the phosphor screen and to convert the stored information into digital form, which can then be analyzed and manipulated on lab computers with the aid of appropriate software. This has led to tremendous advances in the capabilities of autoradiography using phosphorimagers as a research and diagnostic tool. For phosphorimaging, the gels are prepared as for autoradiography (i.e., without fluorographic enhancement). The dried gel is then exposed to a storage phosphor screen (Kodak) at room temperature. The screen is then processed using a phosphorimager (Molecular Dynamics, Sunnyvale, CA).

Conclusion

The general methods for analysis of recombinant K^+ channel α- and β-subunit polypeptides expressed by transient transfection in mammalian cells have been reviewed. These methods have been successfully applied in the laboratory for analysis of the biosynthesis and subunit associations of a number of mammalian K^+ channel subunit and offer the advantages of a simple and reliable expression system that is amenable to metabolic radiolabeling and subsequent analyses.

Acknowledgments

I thank Drs. Robert Haltiwanger and Robert Scannevin for comments on the manuscript, and Drs. Gongyi Shi and Kensuke Nakahira for valuable contributions to this work. This work was supported by NIH grants NS34375 and NS 34383, and was done during the tenure of an Established Investigatorship from the American Heart Association.

Section II

Genetics

[4] Site-Directed Mutagenesis

By TAKAHIRO M. ISHII, PATRICIA ZERR, XIAO-MING XIA, CHRIS T. BOND, JAMES MAYLIE, and JOHN P. ADELMAN

Introduction

Ion channels are ubiquitous membrane proteins that are centrally important to the existence of all cells. Indeed, a sequence with remarkable homology to a mammalian K^+ channel resides in the *Escherichia coli* genome.[1] In the mammalian CNS, each cell may possess a unique complement of ion channels, a fingerprint of the personality of the cell. Prior to cloning, the unprecedented array of information from electrophysiologic recordings motivated detailed speculation concerning the underlying molecular architecture of ion channel proteins. Thus, Armstrong[2] used recordings coupled with biochemical approaches to infer the mechanism of channel inactivation, and proposed the model of a ball and chain inactivation particle for Na^+ channels. The concepts of a specialized segment residing in the membrane electric field that mediates the responses to voltage changes, a hydrophilic core for ion translocation surrounded by a hydrophobic interface with the lipid bilayer, and amino acids that determine the selectivity for different ions as well as the rate at which they flux across the membrane were incorporated into ever more detailed extrapolations of possible ion channel structures.[3]

The study of ion channels changed dramatically with the cloning and heterologous expression of the ionotropic acetylcholine receptor from the electric eel.[4,5] The unambiguous amino acid sequence was available for an entire, functional ion channel. This breakthrough was quickly followed by the cloning of voltage-gated sodium, calcium, and potassium channels; clearly, a new era had begun.[6–10] To date many ion channels have been

[1] R. Milkman, *Proc. Natl. Acad. Sci. U.S.A.* **91,** 3510 (1994).
[2] C. M. Armstrong and F. Bezanilla, *Nature* **242,** 459 (1973).
[3] B. Hille, "Ionic Channels of Excitable Membranes." Sinauer Associates Inc., Sunderland, MA, 1992).
[4] M. Noda, H. Takahashi, T. Tanabe, M. Toyosato, Y. Furutani, T. Hirose, M. Asai, S. Inayama, T. Miyata, and S. Numa, *Nature* **299,** 793 (1982).
[5] M. Mishina, T. Kurosaki, T. Tobimatsu, Y. Morimoto, M. Noda, T. Yamamoto, M. Terao, J. Lindstrom, T. Takahashi, M. Kuno, and S. Numa, *Nature* **307,** 604 (1984).
[6] M. Noda, S. Shimizu, T. Tanabe, T. Takai, T. Kayano, T. Ikeda, T. Takahashi, H. Nakayama, Y. Kanaoka, N. Minamino, K. Kangawa, H. Matsuo, A. Raftery, T. Hirose, S. Inayama, H. Hayashida, T. Miyata, and S. Numa, *Nature* **312,** 121 (1984).

0076-6879/98 $25.00

cloned; members of almost every class have been removed from their native environment and presented in public journals, each residue exposed for unmitigated viewing. Only a handful remain unknown at the molecular level and these will soon join the ranks of the cloned. This remarkable revolution has employed a diversity and integration of techniques worthy of the diversity of ion channel types.

Ion channel amino acid sequences are still being avidly scrutinized. Computer programs employing ever more complex algorithms labor to model the secondary and tertiary structures of these integral membrane proteins, producing detailed and aesthetically pleasing representations of the channels. They are captivating and provocative, compelling adult audiences to don 3D glasses and gaze into virtual deep pores, or experience the force of voltage moving the S4 voltage sensor. However, these representations may well be viewed as speculative.

How then might we gain *real* insight into the structure–function relationship of ion channels? Certainly, solving crystal structures would present a quantal leap. However, significant technical hurdles remain, and a crystal structure is a snapshot in time of a dynamic series of molecular motions. Most information concerning the molecular mechanisms underlying ion channel structure and function has been derived from site-directed mutagenesis studies. This strategy combines the ability to alter structures in a defined, predetermined manner and to express the channels in heterologous cell types at very high levels, presenting a virtually pure, noise-free ion channel population that can be studied electrophysiologically. Functional parameters are readily measured and compared to those of the parent channel, the differences in function being ascribed to the differences experimentally endowed on their structures. An iterative process results, as each mutation suggests insights into structure–function relationships that may be further tested with second-generation mutants.

This article attempts to document the most widely employed and useful techniques for site-directed mutagenesis of ion channels. We present some background on the evolution of these techniques and highlight major conceptual advances. Three basic protocols are described in detail, and a discus-

[7] M. Noda, T. Ikeda, H. Suzuki, H. Takeshima, T. Takahashi, M. Kuno, and S. Numa, *Nature* **322,** 826 (1986).

[8] T. Tanabe, H. Takeshima, A. Mikami, V. Flockerzi, H. Takahashi, K. Kangawa, M. Kojima, H. Matsuo, T. Hirose, and S. Numa, *Nature* **328,** 313 (1987).

[9] B. L. Tempel, D. M. Papazian, T. L. Schwarz, Y. N. Jan, and L. Y. Jan, *Science* **237,** 770 (1987).

[10] L. C. Timpe, T. L. Schwarz, B. L. Tempel, D. M. Papazian, Y. N. Jan, and L. Y. Jan, *Nature* **331,** 143 (1988).

sion of their relative merits and drawbacks is presented. An appendix is included with buffer compositions and some detailed supporting methods.

Oligonucleotides

Virtually every technique discussed in this article and almost all aspects of modern molecular biology rely on the availability of synthetic DNA chains of defined sequence. As the genetic code was being deciphered, chemical techniques were developed that permitted the efficient, random linkage of ribonucleotides into polymers using polynucleotide phosphorylase, and spurred on the development of nucleotide chemistry that would permit syntheses of chains with defined sequences.[11] Today a 20-mer may be efficiently synthesized in a few hours by machines that may make several chains at the same time; an experimenter may design an oligonucleotide in the morning and use it that evening.

Site-Directed Mutagenesis

The fundamental concept underlying almost all of the techniques discussed here is that an oligonucleotide of defined sequence may include one or more nucleotide differences from a sequence of interest, such as the coding sequence of an ion channel protein, and may be used to introduce those changes selectively and specifically into the channel coding sequence. Thus, once a structure–function model has been formulated, it may be tested by directly introducing single or multiple amino acid residue changes within the known primary amino acid sequence.

Early Strategies

The first site-directed mutagenesis strategies employed single-stranded M13 bacteriophage DNA.[12–15] An oligonucleotide containing a sequence complementary to the region of interest in the single-stranded template, and including one or more changes within the sequence, was combined with the template DNA, the mixture denatured, and slowly cooled. Nucleotide triphosphates and DNA polymerase Klenow fragment were then added

[11] H. G. Khorana, H. Buchi, H. Ghosh, N. Gupa, T. M. Jacob, H. Kossel, R. Morgan, S. A. Narang, E. Ohtsuka, and R. D. Wells, *Cold Spring Harb. Symp. Quant. Biol.* **31,** 39 (1966).

[12] C. A. Hutchison, S. Philips, M. H. Edgell, S. Gillam, P. Jahnke, and M. Smith, *J. Biol. Chem.* **253,** 6551 (1978).

[13] M. J. Zoller and M. Smith, *Nucleic Acids Res.* **10,** 6487 (1982).

[14] J. P. Adelman, J. Hayflick, M. Vasser, and P. H. Seeburg, *DNA* **2,** 175 (1983).

[15] J. Messing and J. Vieira, *Gene* **19,** 269 (1982).

and oligonucleotide-primed DNA synthesis used the parental strand as template. The partial heteroduplex was transformed into *E. coli,* using strains engineered for M13 production (JM101, JM109), and the reaction products were presented as phage plaques. Even though the synthesis of the mutant DNA strand resulted in only a partial heteroduplex, following transformation bacterial enzymes completed the replicative process and used the mutant strand as well as the wild-type strand for generating genomes for new phage particles. The shortcomings were that the *in vitro* reaction contained parental DNA strands that did not necessarily participate in mutagenesis reactions yet were transformed and replicated at high efficiency, and that there was no selection for mutant phage. Therefore, background parental phage dominated the population of plaques produced following transformation. The rare mutant plaques were identified by differential hybridization to the mutagenic oligonucleotide, not always an easy task, especially if the mutagenic chain contains only a single mismatch with the parental wild-type strand. Also, these early site-directed mutagenesis reactions were performed by first subcloning the target sequence into M13, and required that the mutant sequence be resubcloned into appropriate vectors for subsequent expression studies testing the function of the mutation. Although laborious by today's standards, the early mutagenesis techniques introduced the fundamental components that underlie almost all of today's widely employed strategies.

Mutant Selection Using ung, dut Bacterial Strains: An Overview

The ability to generate site-directed mutants rapidly has been greatly advanced by taking advantage of the powers of genetic selection. Many clever and efficient schemes effectively enrich for mutants over parental wild-type molecules. The method of Kunkel[16] was one of the first and remains in widespread use. The approach involves growing M13 single-stranded DNA, which will serve as substrate for mutagenesis on an *E. coli* strain harboring mutations in two genes involved in uracil metabolism, *dut*$^-$ and *ung*$^-$. This strain is deficient in dUTPase, the product of the *dut* gene, resulting in a buildup of dUTP, which competes with dTTP for incorporation into the DNA during replication. Incorporated uracils are not removed as the *ung*$^-$ mutation destroys uracil glycosylase activity. Single-stranded DNA prepared from a *dut*$^-$ *ung*$^-$ host contains ~20 uracil residues per genome and is employed in the mutagenesis reactions using dGATC as nucleotide substrates for the polymerase. In contrast to the parental strand, the *in vitro* synthesized mutant DNA strand contains no uracil residues.

[16] T. A. Kunkel, *Proc. Natl. Acad. Sci. U.S.A.* **82,** 488 (1985).

The heteroduplex is transformed into a wild-type host strains such as DH5α, where the parental strand is selectively destroyed due to the activity of uracil glycosylase. This results in an ~10^5-fold selective advantage for the mutant DNA strand (Fig. 1). Because there is no phenotypic selection, candidate clones are analyzed by DNA sequencing.

The need for subcloning steps has been obviated by the advent of phagemids, DNA molecules which may be manipulated between single-stranded and double-stranded DNAs.[17] Plasmid DNA that has been modified to contain the origin of replication from a phage such as M13 (f1 ori), a member of the f1 class of bacteriophage, goes through a double-stranded intermediate during DNA replication permitting isolation of either single-stranded or double-stranded forms. Therefore, when a host cell containing a double-stranded plasmid with an antibiotic resistance marker and the f1 ori (phagemid) is infected with an f1 helper phage that supplies the necessary transfactors for phage replication, single-stranded progeny of the phagemid are produced and packaged into phage particles, which may be isolated by standard techniques. These particles may be used to infect fresh cells, allowing the DNA to enter the cell but not resulting in a productive infection. Phagemid-infected cells may be plated to antibiotic-containing media, rescuing the plasmid form. Thus, the presence of the f1 ori permits manipulation of the DNA between single- and double-stranded forms, and has been introduced into many different vectors that also contain all of the elements necessary for expression studies. The single-stranded form is used for mutagenesis and the mutant is rescued as a double-stranded plasmid, eliminating the need for subcloning steps.

Detailed Method

Preparation of Uracil Containing DNA

1. Clone the sequence of interest into any common phagemid, and transfer the plasmid form to CJ236, or any other *dut*$^-$, *ung*$^-$ host strain also harboring a F′ episome; the episomal DNA carries the *tra* genes for pillus formation, which is used as a receptor for f1-like phage. Growth usually requires double drug selection, one for the plasmid (usually ampicillin) and one for the F′, depending on the strain; for CJ236 this is chloramphenicol.
2. From a freshly streaked plate, incubated overnight at 37°, inoculate 5 ml of LB broth with antibiotic (usually ampicillin) and chloramphenicol. Grow this culture to $OD_{600} < 1.0$.

[17] D. A. Mead and B. Kemper, *BioTechniques* **10,** 85 (1988).

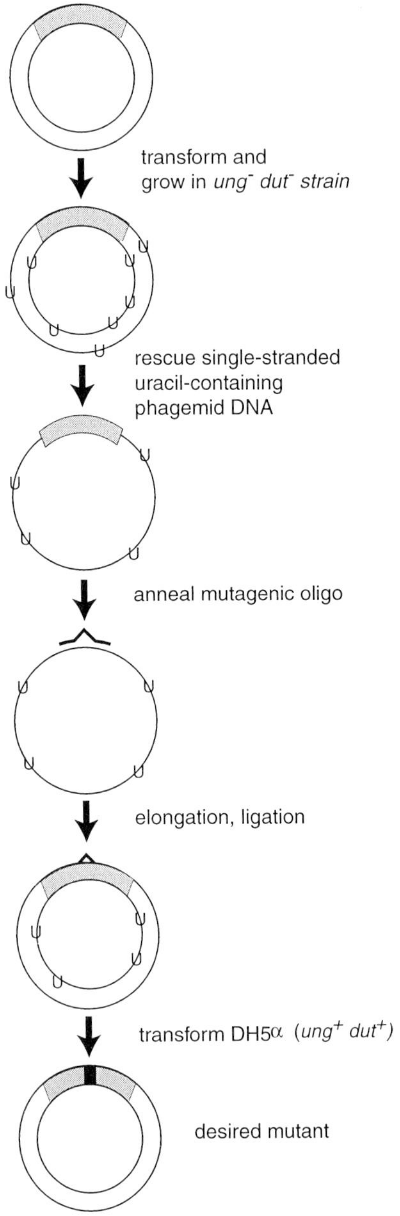

FIG. 1. Phagemid-based mutagenesis using *ung*$^-$ *dut*$^-$ strains for selection.

3. Use 0.5 ml of this culture to innoculate 25 ml of 2 × YT broth plus ampicillin; at this step chloramphenicol is not necessary and will interfere with high titer single-stranded phagemid rescue. Grow this culture to OD_{600} ~ 0.3–0.5 [~2 × 10^7 cfu (colony-forming units)/ml].
4. Infect the culture with MK801 helper phage, or an equivalent; several are commercially available. Optimal multiplicity of infection (MOI; number of infectious particles/bacteria cell) is ~20, and the phage stocks are usually 10^{10}–10^{11}/ml; several MOIs may be tested for best phagemid yields.
5. Incubate with shaking at 37° for 1 hr, then add kanamycin (for MK801; drug may differ with different helper phage) to 50 μg/ml. Continue incubation with shaking at 37° for 6 hr.
6. Centrifuge the culture at 15–20*g* for 10 min at 4°, rescue the supernatant, and respin. Incubate the supernatant containing the single-stranded phagemid particles with DNase-free RNase (1 μg/ml) for 15 min at 37°. Then add 0.2 volume of 20% (w/v) polyethylene glycol (PEG) 6000/2.5 *M* NaCl, mix, and incubate on ice for 15 min to 1 hr. Centrifuge at 15,000–20,000*g* for 10 min, remove the supernatant, and drain as much solution as possible. Resuspend the phagemid pellet by gently swirling in 250 μl of phage storage buffer. An aliquot (10 μl) may be removed for determining the phagemid titer. We routinely bypass the titer step unless the mutagenic reactions are unproductive.
7. Extract the uracil-containing phagemid DNA by extracting twice with phenol and chloroform, twice with chloroform, then ethanol precipitate the DNA. Generally, yields are 2–4 μg of single stranded uracil-containing DNA.

Mutagenesis. The mutagenic oligonucleotide must be phosphorylated so that the extension products may be ligated.

5 μl 10 μ*M* Oligonucleotide (50 pmol; C_f = 2.5 μ*M*)
2 μl 10× T4 kinase buffer
5U T4 DNA kinase
Adjust to a final volume of 20 μl with sterile distilled H_2O.
Incubate the reaction at 37° for 1 hr. Heat kill the kinase at 80° for 2 × 10 min. After the first 10 min, quench on ice (1 min), flash spin, and repeat.

Annealing Reaction

~250 ng Single-stranded uracil containing DNA (~0.25 pmol)
2 μl Kinase oligonucleotide (5 pmol, from kinase reaction)
2 μl 10× Annealing buffer
Adjust to a final volume of 20 μl with sterile distilled H_2O.

Heat the annealing reaction to 70° for 5 min and allow it to slowly cool to room temperature.

Elongation Reaction

20 μl Annealing reaction
3 μl 10× Synthesis buffer
10U T4 DNA polymerase
2U T4 DNA ligase
Adjust to a final volume of 30 μl with sterile distilled H_2O.

Incubate the reaction at 37° for 90 min. Transform *ung*$^+$ bacteria (we use DH5α) and plate to LBA plates.

Controls. Perform a parallel reaction without mutagenic oligonucleotide. Use the mutagenic oligonucleotide in a bidirectional polymerase chain reaction (PCR) with another, proven oligonucleotide, such as lacmer. Run the elongation products on a 1% (w/v) agarose gel along with a zero-time point aliquot. There should be a clear shift from single-stranded to closed circular DNA. If this is not the case, then perform several reactions varying the ratio of template to primer. Increasing primer concentration will help.

Titer the phagemid stock. Grow 10-ml cultures of CJ236 and DH5α in 2× YT broth to OD_{600} ~0.3. Add 1–10 μl of the phagemid stock to each culture and grow for an additional 3 hr. Spread the equivalent of 5, 1, 0.1, and 0.001 μl to LBA plates; incubate at 37° overnight. The phagemid titer on CJ236 should be ~10^9–10^{10} cfu/ml and at least 10^4 higher than on DH5α cells.

Plasmid-Based Mutagenesis

Many clever and elegant variations of plasmid-based site-directed mutagenesis techniques have been developed. One plasmid-based method that is highly efficient does not require subcloning, minimizes enzymatic steps, and from which mutants may be isolated in 1–2 days is used regularly in our laboratory.

Site-Directed Mutagenesis Using Antibiotic Resistance: An Overview

In this mutagenesis technique (see Fig. 2), the sequence of interest is cloned into a plasmid with two antibiotic resistance genes. One of the antibiotic resistance genes, tetracycline (tet), is wild type while the other, ampicillin resistance (amp), contains a nonsense mutation resulting in loss of function. Double-stranded plasmid DNA is prepared from cells grown in the presence of tetracycline. The DNA is denatured and annealed with three oligonucleotides, one including the mutation of interest, one correct-

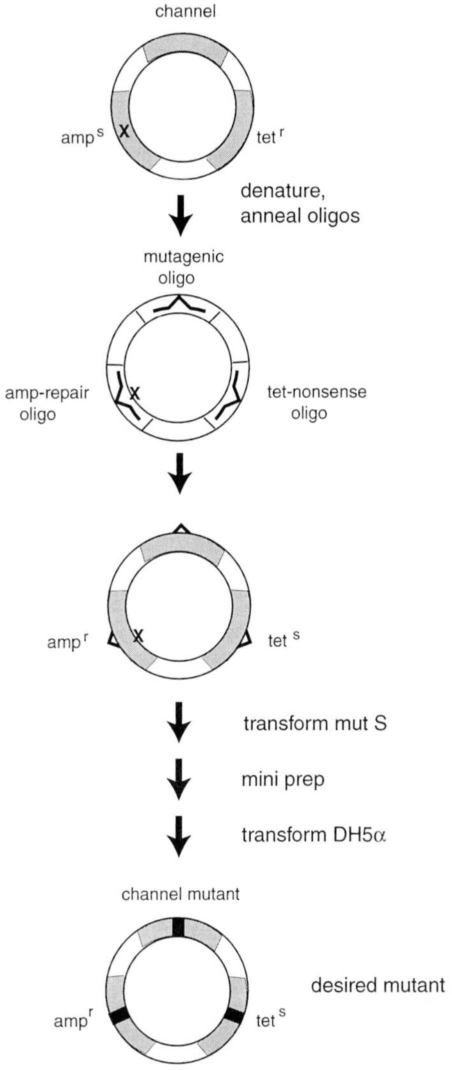

FIG. 2. Plasmid-based mutagenesis using antibiotic resistance selection.

ing the nonsense mutation in the amp gene, and one introducing a nonsense mutation in the tet gene; the mutagenic oligonucleotide is used in approximately 10-fold excess to the amp-repair and tet-knockout oligos. Deoxynucleotide triphosphates (dNTPs), T4 DNA polymerase (T4 pol), and T4 DNA ligase (ligase) are added, resulting in polymerization from the 3′ ends of each oligonucleotide, and ligation of the growing strands. The reaction products are transformed into an *Escherichia coli* strain that is deficient in mismatch repair (MutS) for segregation of the mutant and wild-type strands. The transformed cells are grown in liquid media containing ampicillin, and plasmid miniprep DNA is prepared. Minipreparation DNA is subsequently transformed into a standard strain of *E. coli,* such as DH5α, and plated to ampicillin-containing media. Because an excess of mutagenic oligonucleotide is used, most of the surviving clones contain the mutant, the repaired amp gene, and are sensitive to tetracycline. Mutations are verified by nucleotide sequence.

It is particularly useful if, in addition to the two antibiotic markers, the plasmid contains the necessary modules for expression studies, such as a promoter (SP6, T7, T3) for *in vitro* synthesis of mRNA, or sequences for expression in tissue culture cells [cytomegalovirus (CMV) or simian virus 40 (SV40) promoter, poly(A) signal]. Then, iterative mutagenesis cycles can be performed without the need for subcloning.[18] Thus, reaction progeny contain the desired mutation, are resistant to ampicillin, and are sensitive to tetracycline. In the next round of mutagenesis, three oligonucleotides are employed, one for the next desired mutant, one to repair the tet gene, and one to reintroduce the nonsense mutation in the amp gene; using this scheme, we routinely obtain about 40–60% efficiency.

Detailed Method

5′-Phosphorylation of Oligonucleotides. Mutagenic and antibiotic resistance oligonucleotides must be phosphorylated so that the extension products can be ligated, creating relaxed closed circular DNA.

5 μl 10 μM Oligonucleotide (50 pmol; $C_f = 2.5\ \mu M$)
2 μl 10× T4 kinase buffer
5U T4 DNA kinase
Adjust to a final volume of 20 μl with sterile distilled H_2O.

Incubate the reaction at 37° for 1 hr. Heat kill the kinase at 80° for 2 × 10 min. After the first 10 min, quench on ice (1 min), flash spin, and repeat.

Preparation of Double-Stranded DNA Template. Alkaline denaturation reaction:

1 μl Double-stranded miniprep DNA (adjusted to ~250 ng/μl)
2 μl 2 *M* NaOH; 2 m*M* EDTA, pH 8.0
Adjust to a final volume of 20 μl with sterile distilled H_2O.

Incubate 5 min at 55°. Add 2 μl of 2 *M* ammonium acetate, pH 4, and 75 μl 100% ethanol. Incubate in dry ice–ethanol bath for 15 min. Recover the

[18] R. N. Bohnsack, *Mol. Biotechnol.* **7,** 181 (1997).

DNA by centrifugation at top speed in a microcentrifuge for 10 min. Drain and wash the pellet two times with 500 μl 70% ethanol. Dry the pellet and dissolve the DNA in 10 μl of TE buffer (10 m*M* Tris, pH 7.5, 0.1 m*M* EDTA, pH 8.0) and proceed immediately to the annealing reaction.

Annealing Reaction

10 μl Alkaline denatured DNA
1 μl Amp repair oligonucleotide (0.25 pmol; diluted from the kinase reaction)
1 μl Tet knockout oligonucleotide (0.25 pmol; diluted from the kinase reaction)
1 μl Mutagenic oligonucleotide (2.5 pmol from kinase reaction)
2 μl 10× Annealing buffer
Adjust to a final volume of 20 μl with sterile distilled H_2O.

Heat the annealing reaction to 70° for 5 min and allow it to cool slowly to room temperature. Place the annealing reaction on ice.

Elongation Reaction

20 μl Annealing reaction
3 μl 10× Synthesis buffer
1 μl (10 U) T4 DNA polymerase
1 μl (2 U) T4 DNA ligase
Adjust to a final volume of 30 μl with sterile distilled H_2O.

Incubate the reaction at 37° for 90 min. Transform 100 μl of competent ES1301 *mutS* cells with 15 μl of the synthesis reaction. Following heat shock, plate 10 μl to LBA plates; this will give an indication of the complexity of the reaction (i.e., how many clones). Grow the remaining 90 μl in a 2-ml culture of LBA broth for 6 hr overnight. Next, prepare minipreparation DNA and transform competent DH5α cells. We usually transform 0.1, 1, 5, and 10 μl (of 25) from the minipreparation and plate to LBA.

Controls. Perform parallel reactions without antibiotic repair oligonucleotide. Use the mutagenic oligonucleotide in a bidirectional PCR with another, proven oligonucleotide.

PCR-Based Mutagenesis

The discovery of thermostable DNA polymerases with proofreading functions and high fidelity has made the PCR a practical tool for the rapid and efficient generation of site-directed mutants and chimeric molecules. We use two PCR-based approaches.

PCR for Site-Directed Mutants: An Overview

This method[19] is outlined in Fig. 3 and makes use of the selective restriction of *Dpn*I endonuclease sites at hemimethylated DNA, $^{5'}G^{m6}ATC^{3'}$; *Dpn*I sites occur quite commonly. The sequence of interest is cloned into any plasmid of choice and grown in a common strain of *E. coli,* requiring only that the *dam* modification system is intact as it is in almost all laboratory strains; we use DH5α. Two complementary oligonucleotides, each containing the desired mutation, are annealed to the plasmid, and the 3′ ends are extended by *Pfu* polymerase during a 16-cycle PCR. The elongation step is performed at a slightly suboptimal temperature, 68°, at which *Pfu* does not strand-displace, so the product of the PCR is nicked, double-stranded DNA: the progeny strands are unmethylated while the parental wild-type strands are *dam* methylated. The PCR product is digested with *Dpn*I, selectively cutting the parental wild-type DNA strands. After digestion, the nicked circular DNA is transformed into *E. coli* where the progeny strands are replicated while the nicked parental strands are not replicated; mutants are verified by nucleotide sequence analysis. Using this method, we routinely obtain 50–85% efficiencies.

Detailed Method

1 μl Plasmid minipreparation DNA (adjusted to ~100 ng/μl)
5 μl 10× *Pfu* buffer
5 μl 10 μM Oligonucleotide 1 (50 pmol; $C_f = 1\ \mu M$)
5 μl 10 μM Oligonucleotide 2 (50 pmol; $C_f = 1\ \mu M$)
1 μl 10 mM dNTPs ($C_f = 200\ \mu M$)
Adjust to a final volume of 50 μl with sterile distilled H_2O.
1 μl (2.5U) *Pfu* polymerase

Cycle conditions may vary with specific oligonucleotides, but this works most of the time; we recommend a Perkin-Elmer (Norwalk, CT) thermocycler (model 2400 or newer):

96°, 30 sec (one cycle)
96°, 30 sec/55°, 1 min/68°, 2 min for each kilobase of plasmid DNA (16 cycles)

Following the PCR, add 1 μl (10 U) *Dpn*I and incubate at 37° for 1 hr. Transform competent DH5α cells with 1 or 5 μl of the reaction.

Controls. Use the mutagenic oligonucleotides in bidirectional *Pfu* PCRs with another, proven oligonucleotide, such as lacmer.

[19] M. P. Weiner, G. L. Costa, W. Schoettlin, J. Cline, E. Mathur, and J. C. Bauer, *Gene* **151,** 119 (1994).

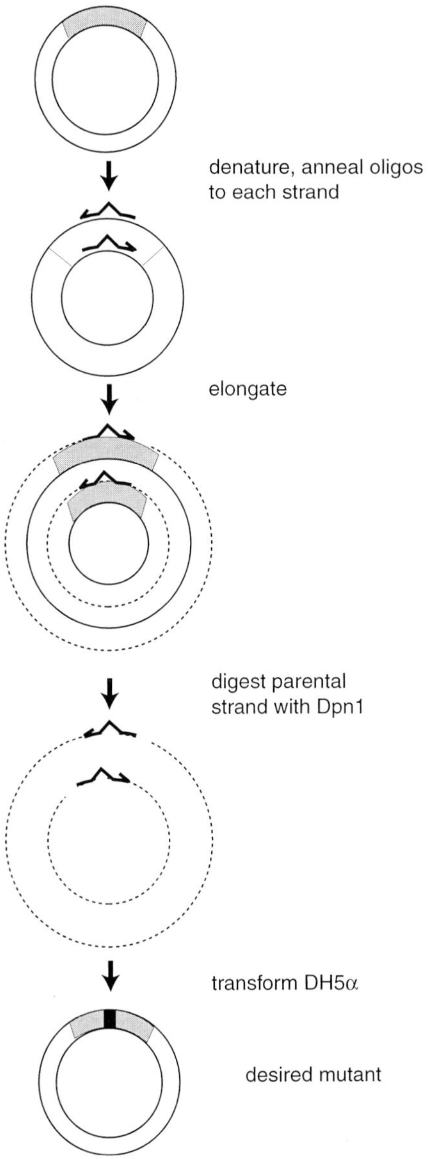

FIG. 3. PCR-based mutagenesis with *Dpn*I selection.

Overlap PCR for Chimera Construction: An Overview

The overlap PCR was described by Horton *et al.*[20] and is shown in Fig. 4. For chimera construction, two plasmids, each containing one of the donor sequences, are used in the PCR. A bipartite oligonucleotide, containing a 5′ sequence of ~12 bases, which is the site of overlap in the other sequence (A; nonhomologous to the substrate), and a 3′ sequence of 14 nucleotides (B), which is homologous to one of the strands in the substrate (B′), is used in the PCR with a downstream oligonucleotide complementary to the desired end of the donor sequence (C′). Similarly, the other donor plasmid is used in the PCR with a bipartite oligonucleotide containing the B′ sequence at the 5′ end and the A′ sequence at the 3′ end, and a downstream oligonucleotide (D) complementary to the desired end of the donor sequence (D′). The double-stranded PCR products are mixed, denatured, and reannealed, permitting duplex formation of the strands that have the only bipartite oligonucleotide sequences are complementary domains. The reannealed strands are used in the PCR with the original flanking oligonucleotides, C′ and D (oligonucleotides that are nested just inside the original outer chains). In the first reaction cycle, the 3′ ends are extended and the target sequences for the flanking oligonucleotides, C′ and D, are synthesized. In subsequent rounds, the flanking oligonucleotides serve as primers in the PCR, resulting in the chimeric sequence. Since the donor sequences are known, it is usually easy to design the positions of the flanking oligonucleotides such that there are convenient internal restriction sites for subcloning the PCR product. Indeed, we frequently make use of universal primers such as lacmer, and flanking sites in the multiple cloning site. Alternatively, sites may be built into the C′ and D oligonucleotides. Although we use this method for chimera construction, it may also be employed for point mutagenesis.

Detailed Method

First PCR, two reactions, one for each donor sequence:
1 μl (~100 ng) Minipreparation DNA
5 μl 10× Vent buffer
2 μl 10 m*M* dNTPs (C_f = 0.4 m*M*)
5 μl 10 μ*M* Oligonucleotide 1 (AB or B′A′; C_f = 1 μ*M*)
5 μl 10 μ*M* Oligonucleotide 2 (C′ or D; C_f = 1 μ*M*)
0.5 μl Vent polymerase (1U)
Adjust to a final volume of 50 μl with sterile distilled H_2O.

[20] R. M. Horton, H. D. Hunt, S. N. Ho, J. K. Pullen, and L. R. Pease, *Gene* **77,** 61 (1989).

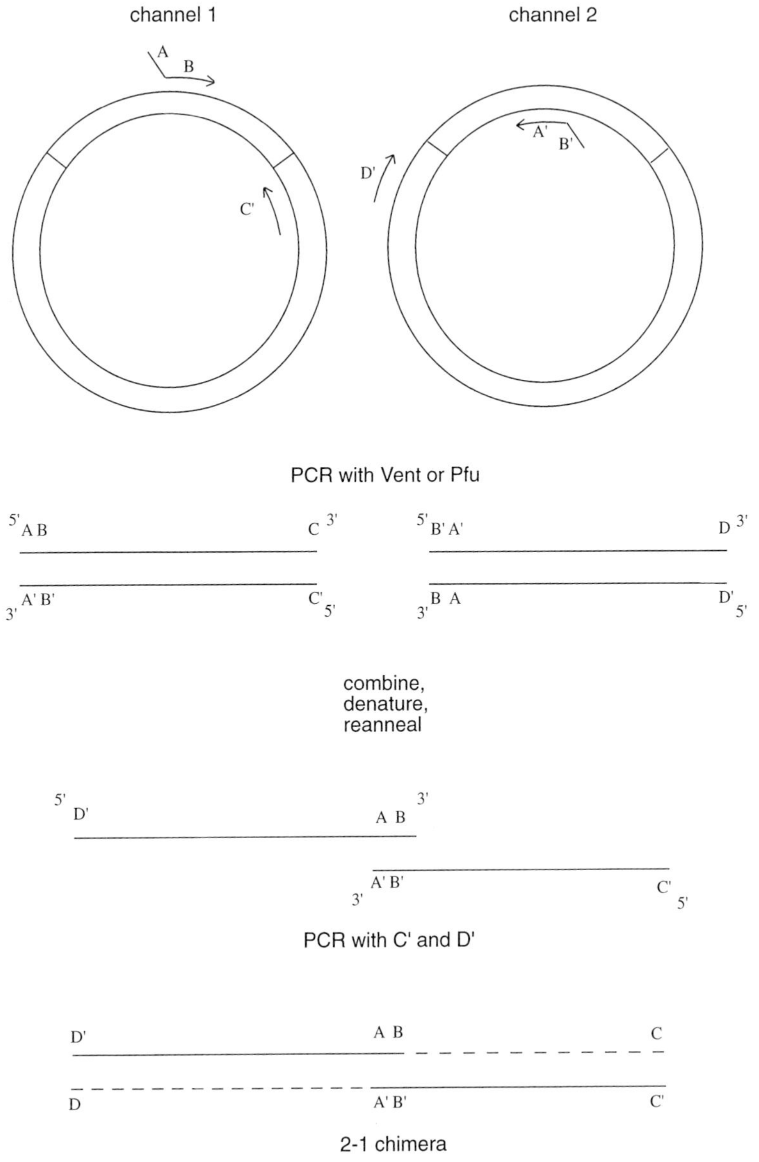

FIG. 4. Overlap PCR for chimera construction.

Cycle conditions may vary with specific oligonucleotides, but this works most of the time:

96°, 30 sec/55°, 20 sec/72°, 60 sec (20 cycles)

Although it may not always be necessary, we generally gel-purify the appropriate band from a small agarose gel, using the Qiaex II gel extraction kit.

Second PCR:

× μl (~50 ng) each fragment from the first PCR
5 μl 10× Vent buffer
2 μl 10 m*M* dNTPs (C_f = 0.4 m*M*)
5 μl 10 μM oligonucleotide C′ (or C″, a nested oligonucleotide)
5 μl 10 μM oligonucleotide D (or D″, a nested oligonucleotide)
0.5 μl Vent polymerase (1U)
Adjust to a final volume of 50 μl with sterile distilled H_2O.

Cycle conditions: 96°, 30 sec/55°, 20 sec/72°, 60 sec (20 cycles). The products are extracted with phenol and chloroform, and ethanol precipitated. The resuspended products are cut with the appropriate restriction endonuclease(s) and gel-purified prior to subcloning.

Conclusions

The remarkable progress during the past decade revealing the structure and function of ion channels has been powered by site-directed mutagenesis. However, the fly in the ointment is that one can never be sure that the structural point mutation has not interfered in some other subtle but significant manner with channel function. As ion channel crystal structures are solved, the integration of the two approaches will undoubtedly "turn up the gain," promising a larger and more detailed understanding of the dynamic processes underlying ion channel function.

Appendix

Composition of Buffers

10× T4 Kinase buffer:	500 m*M* Tris, pH 8.0
	100 m*M* $MgCl_2$
	200 m*M* Dithiothreitol (DTT)
10× Annealing buffer:	200 m*M* Tris, pH 7.5
	100 m*M* $MgCl_2$
	500 m*M* NaCl
10× Synthesis buffer:	100 m*M* Tris, pH 7.5
	5 m*M* dNTPs
	10 m*M* rATP
	20 m*M* DTT

Phage storage buffer: 10 m*M* Tris, pH 8.0
10 m*M* $MgSO_4$
100 m*M* NaCl

10× Pfu buffer: 100 m*M* KCl
100 m*M* $(NH_4)_2SO_4$
200 m*M* Tris, pH 8.8
20 m*M* $MgSO_4$
1% Triton X-100
1 mg/ml nuclease-free BSA

10× Vent buffer: 100 m*M* KCl
200 n*M* Tris, pH 8.8
100 m*M* $(NH_4)_2SO_4$
20 m*M* $MgSO_4$
1% Triton X-100

Recommended Supporting Methods

Plasmid Minipreparation. Grow 2-ml cultures in LB plus 50 μg/ml carbenicillin, an ampicillin analog with a longer shelf life. To prepare minipreparation plasmid DNA, resuspend the cell pellet in 200 μl of 50 m*M* Tris, pH 8.0, 10 m*M* EDTA, pH 8.0, with fresh RNase (100 μg/ml), then lyse the cells with 200 μl of 0.2 *N* NaOH, 1% sodium dodecyl sulfate (SDS), and precipitate the genomic DNA with 200 μl of 3 *M* potassium acetate, pH 5.5. After a 5-min incubation on ice and a 10-min spin in a microfuge, the 600 μl supernatant is rescued and the plasmid DNA is precipitated with 600 μl of 2-propanol, mixed well and centrifuge without chilling, the pellet washed twice with 70% (v/v) ethanol, lightly dried, and resuspended in 25 μl of TE. Generally, this yields ~500 ng/μl of plasmid DNA.

Gel Extraction. For rapid and efficient extraction of DNA from agarose gels, use the Qiagen gel extraction kit (Qiaex II), according to their instructions.

Primer Design. Although there are computer programs for optimizing design of oligonucleotides, and equations for calculating melting temperatures, they yield approximate results at best. There are, however, a few general guidelines for primer design that may help efficiencies. Nonetheless, certain mutants may require the synthesis of more than one oligonucleotide.

Length. For point mutants, the mutation should be placed in the center of the oligo. For either of the methods described earlier, we make oligonucleotides of ~25 bases, 12 bases on either side of the mutation mismatch. If the initial oligo fails, we generally will add a few bases at either end and may "slide" the sequence by a few bases relative to the target. We have used these same parameters to generate large deletions and insertions; these manipulations do not, generally, require a longer oligo.

G:::C content. Although oligonucleotides containing mostly A or T bases may be successful at priming and mutagenesis, a high A::T content generally will lower the overall efficiency, making for more sequencing to identify the correct mutant. In the best case, the oligo will have ~50% G:::C content, will have two or three of the 3′ most nucleotides, particularly the end position, as either a G or C.

Restriction sites in oligonucleotides. When including a restriction site at the 5′ end of the oligo, we precede the recognition sequence with GCAC as a foothold for the endonuclease; for a six-cutter this will add a total of 10 extra bases. This sequence works well with almost all restriction enzymes (a particular exception is *Xho*I).

Batch Preparation of Frozen Competent DH5α Cells

1. Grow a 5-ml culture of DH5α overnight in LB broth.
2. Make a 1:100 dilution into prewarmed LB broth; we generally grow 250 ml.
3. Follow the OD_{600} closely and when it reaches 0.3–0.4, chill the culture on ice for 10 min.
4. Spin at 650*g* (2K in a GSA rotor) for 10 min at 4°; drain and keep the cell pellet on ice.
5. Add 1/10 original volume of freshly mixed TSS:FSB (1:1).
6. Gently resuspend the cells by swirling (do not vortex) and add 5% (v/v) volume dimethyl sulfoxide (DMSO); incubate on ice for 5–10 min.
7. Precool Eppendorf tubes at 4° and aliquot the cells (200 μl is convenient). As cells are dispensed, immediately drop the tubes in liquid nitrogen; store at −80°.
8. For optimal transformation efficiency, the DNA solution volume should not exceed 5% of the volume of the competent cells. Fifty microliters of competent cells is usually saturated by approximately 1 ng of supercoiled plasmid DNA.

Note: The efficiency (~$5 \times 10^7/\mu g$ supercoiled DNA) has been optimized for DH5α and may not work as well for other strains.

Buffers

FSB, 100 ml

Potassium acetate (or KMES)	1.0 ml of 1.0 *M*
KCl	5.0 ml of 2.0 *M*
$MnCl_2 \cdot 4H_2O$	0.89 g

$CaCl_2 \cdot 2H_2O$	0.147 g
$Co(NH_3)_6Cl_3$ (hexamminecobalt chloride)	0.08 g
Glycerol	10.0 ml

Use 1 *N* NaOH to adjust pH to 6.3, filter to sterilize, store at 4°.

TSS, 100 ml

Bacto-tryptone	1.0 g
Bacto-yeast extract	0.5 g
NaCl	0.5 g
$MgSO_4$	0.48 g
PEG 8000	10.0 g

Use 1 *N* NaOH to adjust pH to 6.5, autoclave, store at 4°.

Acknowledgments

We thank Kevin Gibson, Stephan Tucker, and Zhao-Ping Liu for useful discussions. We thank Ms. Chris Fenner for artwork and endless patience.

[5] Molecular Physiology of Human Cardiovascular Ion Channels: From Electrophysiology to Molecular Genetics

By SYLVAIN RICHARD, PHILIPPE LORY, EMMANUEL BOURINET, and JOËL NARGEOT

Introduction

Ion channels are transmembrane proteins responsible for electrical and ionic signaling across lipid membranes in living cells. The current view of voltage-gated ion channels has evolved from a mathematical model describing changes in membrane conductances to a determination of macromolecular structures.[1,2] Identification of genes coding for ion channel proteins and discovery of naturally occurring mutations in these genes have influenced the concept of inherited human "ion channel disorders" or channelopathy.[3,4] Studying the functional properties of defective ion channels will increase our knowledge not only of various human diseases per

[1] B. Sakmann and E. Neher, "Single Channel Recordings." Plenum Press, New York, 1983.

[2] B. Hille, "Ion Channels of Excitable Membranes," 2nd Ed. Sinauer Associates, Sunderland, Massachusetts, 1992.

[3] E. P. Hoffman, *Annu. Rev. Med.* **46,** 431 (1995).

[4] V. Rojas, *News Physiol. Sci.* **11,** 36 (1996).

0076-6879/98 $25.00

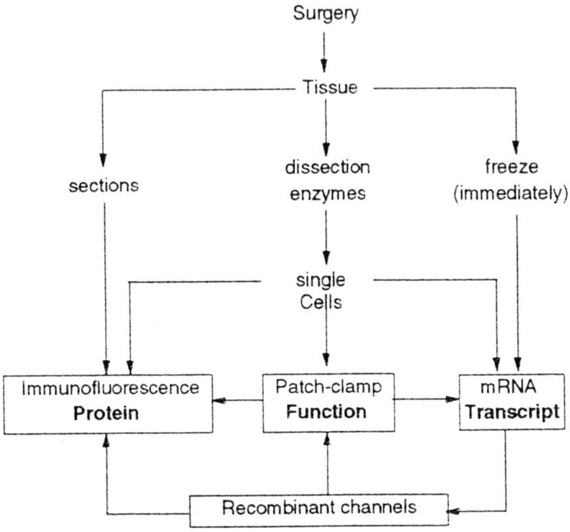

FIG. 1. Studies of human ion channels: Toward single-channel technologies.

se but also of the molecular physiology of normal ion channels.[5] Understanding the changes in expression, structure, function, and regulation of ion channel proteins that occur not only during development of normal tissues but also during pathology of mature tissues is a major challenge for the development of new therapies.

Ion channels are studied most commonly in animal tissues for practical reasons. However, studies of human ion channels are important because of species-dependent differences. In addition, no single animal model can adequately represent the wide variety of causes and manifestations of some clinical syndromes. The development of modern techniques such as patch-clamp, molecular biology, and genetics, together with the evolution of surgical techniques and single-cell methods, facilitates studies in human tissues. In this article, we focus on selected aspects of studies on the molecular physiology of human ion channels. The specific aims are to describe (1) the methods for isolation and culture of myocytes; (2) the techniques to combine both molecular and electrophysiological approaches; and (3) the advantages and limits inherent in the use of human cells and to draw perspectives from such studies. Most of our experience was obtained in cardiovascular myocytes but the techniques and methods presented here (Fig. 1) are also applicable to other human tissues.

[5] M. Ackerman and D. E. Clapham, *N. Engl. J. Med.* **336,** 1575 (1997).

Probing Ion Channel Properties and Function in Human Cardiovascular Myocytes

The patch-clamp techniques are appropriate for studying functional properties of ion channels in acutely isolated and cultured human cardiac and vascular cells. Investigation of ion channels in freshly isolated human cardiomyocytes, started a decade ago, has already provided valuable information concerning the nature, biophysics, pharmacology, and regulation of ionic currents in normal and diseased tissues.[6] Although less common, studies in cultured myocytes are also of potential interest. One advantage is to provide more time for longer term experiments. Tissue culture is adequate for studies combining electrophysiology, molecular biology, and gene manipulation techniques.

Single adult human cardiomyocytes are terminally differentiated but they can easily be primary cultured. Such a model *in vitro* can provide valuable information about the changes in electrical activity that occur during cell dedifferentiation,[7] and about the mechanisms by which the myocardium adapts to pathological conditions (e.g., hypertrophy). Similarly, cultures of vascular myocytes present several advantages. First, cells can be studied for weeks, which is of interest when tissue samples are rare. Second, the role and plasticity of ion channels during cell dedifferentiation and proliferation can be investigated *in vitro*. For example, cell culture promotes expression of T-type Ca^{2+} currents and of atypical Na^{+} currents in human coronary myocytes, suggesting phenotypic modulation of ion channel expression.[8,9]

The following paragraphs contain information on methods used to prepare single cardiac and vascular cells for immediate electrophysiologic recordings as well as in establishing cultures for long-term studies.

Obtaining Tissue Samples

Cardiac Samples. Single myocytes can be isolated from small pieces of atrial or ventricular tissues during corrective and transplant cardiac surgery[6] after approval of the procedure by the local ethical committee in agreement with national guidelines on this matter. Most studies are performed on samples (0.5–1 cm^3) of the right atrium excised during the routine cannulation procedure before cardiopulmonary bypass during open heart sur-

[6] E. Coraboeuf and J. Nargeot, *Cardiovasc. Res.* **27,** 1713 (1993).

[7] A. Bénardeau, S. Hatem, C. Rücker-Martin, S. Tessier, S. Dinanian, J. L. Samuel, E. Coraboeuf, and J. J. Mercadier, *J. Mol. Cell Cardiol.* **29,** 1307 (1997).

[8] J. F. Quignard, J. M. Frapier, M. C. Harricane, B. Albat, J. Nargeot, and S. Richard, *J. Clin. Invest.* **99,** 1 (1997).

[9] J. F. Quignard, F. Ryckwaert, B. Albat, J. Nargeot, and S. Richard, *Circ. Res.* **80,** 377 (1997).

TABLE I
SOLUTIONS FOR CARDIAC CELLS

Solution	Transport (T) (m*M*)	Enzymes 1 (E1) (m*M*)	Enzymes 2 (E2) (m*M*)	Storage (S) (m*M*)
$CaCl_2$	—	0.01	0.01	0.01
NaCl	136	136	136	119
KCl	10.8	10.8	10.8	—
$MgCl_2$	1.1	1.1	1.1	1.7
HEPES	25	25	25	25
Dextrose	22	22	22	—
Glutamate	10	10	10	—
Penicillin G	60 μg/ml	60 μg/ml	60 μg/ml	60 μg/ml
Streptomycin	100 μg/ml	100 μg/ml	100 μg/ml	—
Collagenase	—	0.6 mg/ml	0.3 mg/ml	—
Elastase	—	0.2 mg/ml	—	—
Pronase	—	—	0.5 mg/ml	
BSA	—	—	—	0.1%
KH_2PO_4	—	—	—	1 mg/ml
Succinate	—	—	—	5
Creatine	—	—	—	5
Glucose	—	—	—	6
pH	7.4 (NaOH)	7.4 (NaOH)	7.4 (NaOH)	7.4 (KOH)

gery.[10–12] Clinical diagnosis is generally aortic or mitral disease (stenosis or insufficiency), coronary artery disease, or congenital heart defects. Specimens of the left ventricular septum (0.1–0.8 cm^3) can also be excised.[13] Alternatively, larger fragments of both left and right ventricles can be obtained from heart transplants. Patients have end-stage heart failure (New York Heart Association, classes III and IV) caused by either dilated or ischemic heart disease with severe alteration of the left ventricular function (ejection fraction between 15 and 30%). Tissue samples are removed immediately after surgery and transferred to a sterile saline solution containing double-strength antibiotics (solution T; Table I) for transportation (at 4°) and also used later for dissociation. Surgery occurs overnight for reasons

[10] H. Ouadid, J. Seguin, S. Richard, P. A. Chaptal, and J. Nargeot, *J. Mol. Cell Cardiol.* **23,** 41 (1991).

[11] B. Le Grand, S. Hatem, E. Deroubaix, J. P. Couetil, and E. Coraboeuf, *Circ. Res.* **69,** 292 (1991).

[12] J. Feng, G. R. Li, B. Fermini, and S. Nattel, *Am. J. Physiol.* **270,** H1676 (1996).

[13] J. P. Benitah, P. Bailly, M. C. D'Agrosa, J. P. Da Ponte, C. Delgado, and P. Lorente, *Pflügers Arch.* **421,** 176 (1992).

inherent to the organ gift procedure. The dissection starts the next morning (i.e., between 1 and 6 hr after tissue excision). The transport solution can be saturated with 100% O_2 to optimize tissue sample conservation.

Of note is that the only possibility to get tissue from healthy hearts is when the transplantation is canceled at the last moment for technical reasons (this has never happened to us in 25 transplantations during the last 3 years) in agreement with the organ gift procedure.

Coronary Tissue. Human coronary tissue is available exclusively from heart transplants. The explanted diseased hearts have either ischemic cardiopathy (arteries are atheromatous) or, alternatively, dilated cardiopathy. In the latter case, arteries are not significantly atheromatous, which is determined from clinical investigation prior to the surgery and confirmed during dissection. The tissue is collected aseptically by dissection of the arteries (several centimeters in length) after cardiectomy, and transferred either to a saline transport solution (solution T, Table II) or into Hanks' solution. Samples are stored at 4° until the artery is dissected. Arterial tissues are not fragile and it is possible to store the tissue for more than 24 hr at 4° before dissection.

TABLE II
SOLUTIONS FOR CORONARY CELLS

Solution	Transport (T) (m*M*)	Freshly isolated cells, enzymes (F) (m*M*)	Primary cultured cells: Enzymes (C1) (m*M*)	Primary cultured cells: Enzymes (C2) (m*M*)
$CaCl_2$	0.01	0.01	0.01	0.01
NaCl	132	132	132	132
KCl	5	5	5	5
$MgCl_2$	1.5	1.5	1.5	1.5
HEPES	25	25	25	25
Penicillin G	60 μg/ml	60 μg/ml	60 μg/ml	60 μg/ml
Streptomycin	100 μg/ml	100 μg/ml	100 μg/ml	100 μg/ml
Papain	—	0.2 mg/ml		0.2 mg/ml
Collagenase	—	0.7 mg/ml	1 mg/ml	0.7 mg/ml
Elastase	—	0.2 mg/ml	1 mg/ml	0.7 mg/ml
DTT	—	0.1		0.1
KH_2PO_4	—	—	—	—
Succinate	—	—	—	—
Creatine	—	—	—	—
Glucose	—	—	—	—
pH	7.3 (NaOH)	7.3 (NaOH)	7.3 (NaOH)	7.3 (NaOH)

Dissection and Enzymatic Isolation

Cardiac Cells. Human cardiomyocytes are often more difficult to dissociate than animal cells. There is more variability (as described later) among tissue samples from different donors, which influences the time of exposure to enzymes. Reasons include age and disease of patients. Specimens from young patients are much more easy to dissociate. In contrast, samples from failing hearts are generally fibrous, which markedly influences the rate of success. Dissociation procedures used in several laboratories generally derive from the method originally described by the groups of Powell and Bustamante.[14,15] In our laboratory, we use two successive enzymatic cocktails (solutions E1 and E2; Table I). The cardiac tissue is chopped aseptically into 2- to 5-mm^2 pieces and incubated first with collagenase (type A or H; Boehringer Mannheim) and elastase (Boehringer Mannheim) under permanent mechanical agitation using a suction-force pump for 25 min at 37°. The digestion is arrested by adding 1% bovine serum albumin (BSA) to the solution. Then the tissue is placed into a second dissociation solution containing collagenase and pronase E (type XIV, *Streptomyces griseus,* Sigma, St. Louis, MO) for 5 min. Digestion is then stopped using the 1% BSA-containing solution and the tissue is stored at 4° (solution S; Table I). Prior to electrophysiological recording, dissociation is achieved by gentle mechanical trituration using a Pasteur pipette with a large fire-polished tip. Cells are plated onto the recording chamber and rinsed with the reference recording extracellular solution. A yield of approximately 60% of cells with clear cross-striation and without significant granulation is commonly obtained except when the tissue is too fibrotic. This percentage may decrease when extracellular Ca^{2+} is gradually reintroduced with the recording solution.

Coronary Myocytes

Dissection. After removal of the fat and connective tissues, vessels are cut into pieces of 1 cm in length under a dissecting microscope. The media is aseptically dissected in phosphate-buffered saline (PBS). The adventitia is mechanically removed using watchmaker's forceps. This separation is easy, except when atheromatous plaques are present. The two parts are identified based on their topological localization and visual aspect. The adventitia is fibrous and pearly, whereas the media is smooth and pinkish-gray colored. The media is cut longitudinally and the endothelial lining is

[14] T. Powell, M. F. Sturridge, S. K. Suvarna, D. A. Terrar, and V. W. Twist, *Br. Med. J.* **283,** 1013 (1981).

[15] D. J. Bustamante, T. Watanabe, D. A. Murphy, and T. F. McDonald, *Can. Med. Assoc. J.* **126,** 791 (1982).

destroyed by gentle rubbing of a sterile paintbrush along the transluminal surface. Then samples are placed in the incubator for at least 2 hr before isolation, which is achieved by a combination of enzymatic and mechanical methods. We use mainly two enzymatic methods depending on how the cells will be used.

Enzymatic dispersion. For patch-clamp recordings of *freshly isolated cells,* the media is chopped into pieces approximately 1–3 mm in diameter, stored for several hours at 4° in solution T with low Ca^{2+} (Table II), and supplemented with 0.2 mg/ml papain (Sigma). This favors tissue impregnation by the enzyme. After this step, 0.1 m*M* dithiothreitol (DTT), 0.3 mg/ml collagenase (type CLS2; Worthington Biochemical Incorporation), and 0.2 mg/ml elastase (Boehringer) are added and the tissue is subjected to enzymatic digestion (solution F; Table II). After a 25-min incubation at 37° in a shaking water bath, samples are rinsed using solution T and the dissociation is achieved by gentle mechanical trituration using a Pasteur pipette with a fire-polished tip. Alternatively, the dissociation may be started immediately after tissue dissection. Well-relaxed long spindle-shaped cells are obtained in a 70% yield. The cells can be used immediately for electrophysiological recordings or stored for several hours at 4°. The dedifferentiation of cells is slowed at this temperature and they keep their initial morphology.

For *primary cultures,* tissue samples are first incubated for 12 hr at 37° and then transferred in Hanks' solution (or solution C1; Table II) containing 0.6 mg/ml collagenase (Worthington) and 1 mg/ml elastase (Boehringer Mannheim) for 50 min at 37°. It is sometimes useful to increase the incubation time. A second enzymatic incubation (0.6 mg/ml collagenase, 0.6 mg/ml elastase, and 0.2 mg/ml papain with 0.1 m*M* DTT; solution F; Table II) for another 50 min may improve the yield of cells.

Cultures

Details regarding cell counts, tissue culture media, flasks, etc., are not covered here.

Cardiac Cells. There are only a few studies of cultured myocytes[7,12] but future studies will probably increase. After isolation, single myocytes are plated in a flask for 2 hr in order to eliminate nonmuscle cells. Then the myocytes that are not attached are resuspended in Dulbecco's modified Eagle's medium (DMEM) supplemented with 10% calf serum (human serum may be more appropriate), nonessential amino acids, insulin (1 n*M*), penicillin (1 IU/ml), and streptomycin (0.1 μg/ml).[7] The maintenance medium (with lower serum) is changed every 2–3 days (Table III).[7] Cells can be studied for up to 3 weeks and undergo marked changes in morphology

TABLE III
MAINTENANCE MEDIUM FOR CARDIAC CELLS

Composition	Start volume (%) (1 liter)	Maintenance volume (%) (1 liter)
Ham's F10	87	43.5
DMEM	—	43.5
Myoclone	—	5
Human serum	10	5
Glutamine (L)	2 mM	2 mM
HEPES	10 mM	10 mM
Penicillin	1 IU/ml	1 IU/ml
Streptomycin	1 μg/ml	1 μg/ml

and main ionic currents.[7] Adult cardiac myocytes are terminally differentiated and incapable of mitotic division.[16]

Coronary Myocytes. Extensive descriptions of cell culture techniques and their applications to studies of vascular smooth muscle can be found in *Cell Culture,* edited by M. Butler and M. Dawson,[17] and in the review by Campbell and Campbell.[18] We routinely use three methods for growing human arterial myocytes *in vitro:* explant culture, enzymatic dispersion, and subculture passaging. For primary cultures, the most common way to initiate cultures from vessels is enzymatic dispersion as described earlier. Enzymatically dissociated cells are centrifuged [100g, 7 min, room temperature (22–26°)], resuspended in the culture medium, and plated (10,000 cells per milliliter for electrophysiology) in 35-mm disposable petri dishes (Falcon) that are placed at 37° in an air–CO_2 incubator at 95% (v/v) CO_2. Cells attach within 24 hr. Unattached cells (rounded appearance) are generally nonviable. Cells spread out 2–3 days after plating and acquire a fibroblastic or an epithelial morphology. In the same time, they transform into dedifferentiated cells and start to proliferate. Their size increases four- to eightfold indicating a hyptertrophic process. Cultures are plated using Ham's F10 (Eurobio, France) supplemented with 10% human (v/v) serum (Institut Jacques Boy SA, Reims, France), 2 mM glutamine, and antibiotics (Eurobio; Table III). This medium is changed every day for 1 week. Thereafter, the maintenance medium (DMEM and Ham's F10; 1 : 1, v/v; Eurobio) (Table III), which contains lower human serum (5%) replaced by 5% myoclone super plus (fetal bovine; Life Technology, Cergy Pontoise, France),

[16] R. Zack, *Am. J. Physiol.* **31,** 211 (1973).

[17] M. Butler and M. Dawson, "Labfax." Bios Scientific Publishers. Academic Press, New York, 1992.

[18] J. H. Campbell and G. R. Campbell, *Clin. Sci.* **85,** 501 (1993).

is changed every 2–3 days. Confluence is obtained 15–20 days after plating. Note that cells do not grow when initial plating density is too low. The use of human serum is crucial to start the culture. We do not use any coating agents since attachment on the polystyrene substratum is normally good. The muscular origin of the cultured cells can be confirmed using α-SM actin labeling. Complete characterization (determination of cell growth, cell cycle time, etc.) can be achieved using a number of cell biology techniques detailed elsewhere.[17]

To avoid possible problems related to enzymatic procedures, the *explant technique* is a very easy alternate method that takes advantage of the migration and proliferative properties of arterial myocytes. Small pieces of media are transferred directly into the culture dishes with a minimal amount of solution to favor attachment within 48 hr to the bottom of the dish.

For *subcultures,* cells that have reached confluence are rinsed with Ca^{2+}/Mg^{2+}-free PBS, and 0.1% trypsin (Life Technology) is added for 2–3 min. After the monolayer detaches from the dish, enzymatic activity is arrested using a serum-containing culture solution. Then the cells are gently triturated for dispersion, plated, and grown exactly as described for primary cultures. It should be noted that cell lines derived from human arteries are commercially available (e.g., human aortic cells T/G HA-VSMC from ATCC, Rockville, MD) and can be used as an alternative.

Storage of Frozen Cells. Given the low frequency of heart transplants, we found it very convenient to freeze the coronary myocytes. To do this, mass cultures of cells are prepared using flasks rather than petri dishes. After confluence is reached, cells are trypsinized, rinsed, and dispersed as described for subcultures. They are then centrifuged at 100*g* for 5 min, resuspended in 70% maintenance medium plus 20% calf serum and 10% DMSO, aliquoted to 1 ml samples, stored at $-80°$ (temperature should be lowered very gradually: 1° per min), and then transferred in liquid nitrogen. This method allows storage of primary cultures originating from tissues of various patients with different diseases that could be used for long-term studies.

Electrophysiology

The use of single cells and the use of the patch-clamp techniques[19] apply quite well to studies of biophysical properties as well as to regulation by hormones, transmitters, intracellular messengers, and other regulators of ion channels in human cells. The patch-clamp techniques also provide experimental means for merging the tools of modern molecular and cellular

[19] O. P. Hamill, A. Marty, E. Neher, B. Sackman, and F. J. Sigworth, *Pfluegers Arch.* **391,** 85 (1980).

biology with those of electrophysiology.[20] The technical details of the patch-clamp techniques are not presented here, but various recording configurations and valuable experience drawn from many laboratories have been reviewed in detail in several reviews and books including *Single-Channel Recording,* edited by Sakmann and Neher and in this series in *Ion Channels,* edited by Rudy and Iverson.[21] The patch-clamp techniques allow identification and characterization of the ion channels present in human cell types and the expansion of the catalog of recognized ion channels when new channels are potentially identified.

During the course of our studies using various cell types such as aortic, coronary, pulmonary, and both atrial and ventricular myocytes, we have not encountered any particular difficulty related to the use of human tissue. When using human cells mandatory appropriate precautionary measures must be taken to ensure the safety of the experimenter and any assistants.

Probing Ion Channel Transcripts/Proteins in Correlation with Patch-Clamp Studies

Relevance of Protein Versus Transcript Detection to Electrophysiological Data

Combinations of methods have been used to study the molecular basis of ion channel expression. One way to provide a comprehensive analysis of gene expression in correlation with electrophysiological studies is to look at the transcript level, an area where most of the work has been performed to date. Highly resolved RNase protection assays or, to a lesser extent, quantitative reverse transcriptase–polymerase chain reaction (RT-PCR) can be developed to identify specific patterns of gene expression in given cardiac tissues or selected myocyte populations.[22] RNase protection provides a good idea of the expression pattern of an ion channel gene family[23] although relative abundance of single mRNA species is indicative of the tissue specificity of gene expression, but not ion channel protein expression or function. At the cellular level, single cardiomyocyte RT-PCR analysis

[20] M. Cahalan and E. Neher, *Methods Enzymol.* **207,** (1992).

[21] B. Rudy and L. E. Iverson, eds., *Methods Enzymol.* **207,** (1992).

[22] S. W. Kubalak, W. C. Miller-Hance, T. X. O'Brien, E. Dyson, and K. R. Chien, *J. Biol. Chem.* **269,** 16961 (1994).

[23] J. E. Dixon, W. Shi, H. S. Wang, C. McDonald, H. Yu, R. S. Wymore, I. S. Cohen, and D. McKinnon, *Circ. Res.* **79,** 659 (1996).

has been performed in only a few studies[24,25] when compared to neurons.[26] Transcript detection experiments contribute to identification of the expression pattern of ion channel genes. However, there are several limitations in the detection and analysis of ion channel transcripts, mainly due to a low level of expression of these genes. First, transcripts are detected in total RNA or mRNA samples prepared from a pool of cells, especially for RNase protection assays. Several cell subtypes are usually present that potentially introduce a bias in the validity of the results. Second, single-cell PCR assays combined with patch-clamp experiments allow correlations between functional data and gene expression at a cellular level. However, while using this technique, one must give greater attention and special care to both experimental design and setup in order to avoid artifacts. In addition, quantitation in single-cell PCR assays is not as efficient as regular transcript detection, that is, RNase protection. Single-cell PCR experiments are considered tedious if one expects to provide convincing evidence that expression patterns of ion channel genes indeed are reflecting the functional data. A further approach in this correlative analysis would be to combine both immunocytochemistry and electrophysiology.[27] Detection of the presence, the amount, and the distribution of channel proteins is a more challenging and reliable approach than transcript analysis.[28] Data obtained in our laboratory indicate at single-cell levels clear correlations between the pattern of protein expression and electrophysiologic data. The following section describes a series of experimental protocols that could be performed to correlate molecular and functional data from cardiac tissue to single myocytes.

Protocols for Detection of Ion Channel Transcripts and Proteins

RNA Extraction. Human heart tissue is obtained as described earlier. The amount of tissue can vary between 100 mg and several grams. In the case of a heart transplant, large samples of the four cavities are excised after cardiectomy and frozen immediately in dry ice. Specimens can also be stored (as described earlier) and dissected later to remove contaminating tissues (e.g., coronary arteries). This provides samples with equivalent prop-

[24] K. A. Krown, K. Yasui, M. J. Brooker, A. E. Dubin, C. Nguyen, G. L. Harris, P. M. McDonough, C. C. Glembotski, P. T. Palade, and R. A. Sabbadini, *FEBS Lett.* **376,** 24 (1995).

[25] V. K. Sharma, H. M. Colecraft, D. X. Wang, A. I. Levey, E. V. Grigorenko, H. H. Yeh, and S. S. Sheu, *Circ. Res.* **79,** 86 (1996).

[26] H. Monyer and B. Lambolez, *Curr. Opin. Neurobiol.* **5,** 382 (1995).

[27] C. Racca, M. V. Catania, H. Monyer, and B. Sakmann, *Eur. J. Neurosci.* **8,** 1580 (1996).

[28] D. M. Barry and J. M. Nerbonne, *Annu. Rev. Physiol.* **58,** 363 (1996).

erties both for cell dissociation (electrophysiology and immunocytochemistry) and RNA extraction (see Fig. 1). We currently obtain RNA of good quality using the methods described by Chomczynski and Sacchi[29] with necessary modifications. Briefly, the tissue is homogenized in a denaturing solution containing 4 *M* guanidinium thiocyanate (Fluka, Ronkonkoma, NY), 25 μM sodium citrate, 0.5% (v/v) *N*-laurylsarcosyl sodium salt, and 0.1 *M* 2-mercaptoethanol, using a Polytron (1 ml of solution/0.1 g of tissue). Homogenization is then completed (per 1 ml) sequentially with 0.1 ml of 2 *M* sodium acetate (pH 4.2), 1 ml of water-saturated phenol, and 0.2 ml of chloroform/isoamyl alcohol mixture (49:1). The homogenate is maintained on ice for 15 min. The suspension is then centrifuged (10,000*g*, for 20 min at 4°), and the upper phase, containing RNA, is subjected to several extractions with phenol/chloroform (1:1) in order to remove any contaminant proteins.

After a series of nucleic acid precipitations with 2-propanol and ethanol, RNA samples are washed twice with 75% (v/v) ethanol. For several assays, such as RT-PCR RNase protection, total RNA can be used successfully. However, for several other studies, it is often necessary to purify poly $(A)^+$ RNA. Affinity purified poly$(A)^+$ RNA (i.e., mRNA) is usually required for Northern blots, long-distance/RACE (rapid amplification of cDNA ends) PCR, or *Xenopus* oocyte expression.[30] Moreover, good mRNA samples that contain large-size transcripts/cDNA are the basis for cDNA library construction that allows cloning of cardiac cDNA isoforms of human ion channels.[31] Several cDNA libraries are now commercially available. However, new technologies in cDNA library construction allow preparation of specific and normalized cDNA libraries from small amounts of tissue, thus allowing cDNA library preparation from given cardiac areas (atrium versus ventricle) and given stages of human cardiac development and pathology.

Reverse Transcriptase–Polymerase Chain Reaction. To perform PCR, the RNA fractions are treated with DNase I (1 unit/50 μg RNA; GIBCO, Grand Island, NY) for 30 min at 37°. The DNase activity is then stopped using heat inactivation (5 min at 95°). First-strand cDNA synthesis is carried out using 5 μg of DNase treated RNA, 1 μl of oligo(dT) primers (500 μg/ml), and Superscript II (50 U, GIBCO) in a 50-μl reaction volume for 1 hr at 37°, and according to the manufacturer's conditions. The reverse transcriptase is then heat inactivated by incubating for 5 min at 95°.

[29] P. Chomczynski and N. Sacchi, *Anal. Biochem.* **162,** 156 (1987).

[30] P. Lory, F. A. Rassendren, S. Richard, F. Tiaho, and J. Nargeot, *J. Physiol.* (*Lond.*) **429,** 95 (1990).

[31] T. Collin, J. J. Wang, J. Nargeot, and A. Schwartz, *Circ. Res.* **72,** 1337 (1993).

Primers designated to amplify the transcript encoding are generally assayed on full-length cDNAs, when available. Each PCR amplification is performed in a reaction volume of 50 μl, using 2 μl of RT reaction, dNTP (200 μl of each dATP, dCTP, dTTP, dGTP), 2 mM $MgCl_2$, 1 pmol of each primer, and 1.5 U of *Taq* polymerase (Eurobio). The reaction samples are prepared on ice. After a first denaturation performed for 3 min at 94°, PCR amplification is performed using a "hot-start" protocol chosen for its highest performance with regard to specificity and efficiency of the amplification. The amplification protocol is performed as described: denaturation 1 min at 94°, annealing 1 min at variable temperatures (see later discussion), extension 1 min at 72°, for a total of 30 cycles. The "hot-start" protocol consists of an initial annealing temperature of 65°. Then, the annealing temperature is decreased at each cycle degree by degree to a maintained temperature between 58° and 54°. The maintained annealing temperature is optimized for each couple of primers in order to give the best yield of amplification. Multiplex PCR reactions for the simultaneous amplification of cDNA fragments encoding for close isoforms should be designed in the same way in order to allow maximal yield for each primer pair. Most of the time, one-fifth (10 μl) of each PCR reaction is electrophoretically separated on 2% agarose gels and stained with ethidium bromide for further analysis (i.e., densitometric analysis and sequencing).

RNase Protection Assays. PCR fragments, ranging from 250 to 600 bp, produced as described earlier are subcloned into the pCRII vector using the TA cloning kit (Invitrogen, San Diego, CA). Careful sequencing and restriction analyses are performed in order to optimize the probe design and to produce high-quality templates. The various steps in the RNase protection assays are accomplished according to the Ambion guides. *In vitro* transcription reactions are performed according to the Maxiscript kit (Ambion, Austin, TX) using [α-^{32}P]CTP. The riboprobes are gel-purified on a denaturing polyacrylamide/urea gel in order to remove unincorporated nucleotides as well as shorter transcription products. RNA samples (5 or 10 μg) are hybridized with approximately 50,000 cpm/tube (^{32}P) of the purified probe(s). Depending on the size of the protected fragments, up to three independent probes can be used in one reaction sample: one of them should be a loading control probe, such as GAPDH, cyclophilin (2) or the elongation factor EF1α.[32] Using the Ambion ribonuclease protection assay kit (RPA II) the unprotected RNA is digested with a mixture of RNase A and T1. RNase reactions are controlled using one tube without RNase and one tube without RNA sample. Then, the reactions are terminated

[32] W. C. Miller-Hance, M. LaCorbiere, S. J. Fuller, S. M. Evans, G. Lyons, C. Schmidt, J. Robbins, and K. R. Chien, *J. Biol. Chem.* **268,** 25244 (1993).

using sodium sarcosyl and proteinase K before precipitation with 2-propanol. Each pellet is manipulated carefully, resuspended in gel loading buffer, and heated for 3 min at 95° before gel loading. Gel electrophoresis is performed on a denaturing polyacrylamide/urea gel. RNA expression is then quantitated directly from dried gels using a phosphorimager (Molecular Dynamics, Sunnyvale, CA) as a ratio of the loading control signal. Alternatively, RNase protection assays are performed directly from tissues (without the RNA extraction/purification step) using the Direct Protect lysate RPA kit from Ambion.[33] Tissue samples are placed in 50 μl of the lysis solution, then vortexed and stored at $-20°$ until the RNase protection assays are to be performed. The advantage here is that RNA preparation is shunted. However, the sensitivity seemed lower than with the RPA II procedure.

Single-Cell PCR. To date, very little information has been reported regarding successful detection of mRNA transcripts encoding ion channels or receptors using single-cell RT-PCR from freshly isolated adult cardiac myocytes. Sharma and co-workers[23] have described the characteristics of muscarinic m1 receptors in single adult rat ventricular myocytes. An important confirmation from their work, however, related to the use of subtype specific antibodies in immunofluorescence experiments. Single-cell RT-PCR protocols can be found in the Materials and Methods section of their paper.[23] Additional protocols can be obtained from the references.[22,24,25]

Indirect Immunofluorescence. Freshly isolated myocytes are harvested into the isolation solution after enzymatic dissociation and rinsed twice under gentle cycles of centrifugation/resuspension. Centrifugation (~100 g) is performed at 22° and the supernatant carefully removed. The myocytes are then fixed in suspension for 5–10 min using a 4% (w/v) paraformaldehyde solution. Fixed cells are washed under gentle cycles of centrifugation/resuspension. The cells either can be maintained in suspension or can be forced to stick onto a cover glass, which is usually achieved after several (~12) hours at 22°. The following procedures are derived from conventional indirect immunofluorescence methods. Cells are permeabilized using a low concentration of Triton X-100 prepared in PBS (0.03% for 5–10 min). For ion channel proteins, optimization of permeabilization requires comparison between Triton X-100 and saponin treatments, as well as different working concentrations. Saturation can be performed using PBS–BSA (3%) or PBS–FCS solutions for 30 min at 37°. Myocytes are then incubated for 1 hr at 37° or overnight at 4° with the first antibody at the required dilution in PBS–BSA solution. Then, the cells are washed three times with PBS and incubated for 45 min at 37° with a goat anti-rabbit or a goat anti-mouse

[33] P. J. Gruber, S. W. Kubalak, T. Pexieder, H. M. Sucov, R. M. Evans, and K. R. Chien, *J. Clin. Invest.* **98,** 1332 (1996).

second antibody fluorescein isothiocyanate (FITC)-conjugated (Organon Teknika, Durham, NC). After three washes with PBS followed with distilled water, treatment with Hoescht 33258 (Sigma, St. Louis, MO) for 1 min is required to label nuclei. The cells adhering to the cover glass should then be mounted in 15% (w/v) Airvol 205, 33% glycerol dissolved in Tris buffer, or using Biomedia gel mount. In our laboratory, digitized images are obtained using a microscope (Leica) interfaced with a Kodak camera. Digitized images can be analyzed on a Silicon Graphics workstation using the Imgswork and Iris Showcase 3.2 packages (Silicon Graphics). High-quality prints are obtained on a Colorease thermal sublimation printer (Kodak).

Immunofluorescence on Sections. Small pieces of tissue obtained from human heart samples (see earlier section) are prepared for immunochemistry. Sections are incubated for several hours (overnight) with a buffer containing 4% paraformaldehyde in order to obtain reasonable fixation. The tissues are then submersed in a 15% (w/v) sucrose solution, prepared in PBS at 4° for 24 hr. Floating sections are then transferred to a sliding microtome and processed for immunocytochemistry or immunofluorescence. Little information is available on the immunocytochemical detection of ion channel proteins from heart sections,[34] especially from human heart. However, for processing the sections, several detailed protocols can be found.[35]

Single-Cell Immunofluorescence. The patch-clamp technique allows recording on single cells. Based on the principle of the single cell RT-PCR methodology described earlier, the whole-cell configuration is used to combine electrophysiologic recordings and delivery of a cell tracer. Lucifer Yellow of Cascade Blue (Molecular Probes, Eugene, OR) is added to the pipette solution in order to further identify the recorded cell. After analyzing the electrophysiologic properties of a given cell, the petri dish is filled with a paraformaldehyde solution (4% final) while the cell is maintained under voltage-clamp conditions. The cell is then carefully released from the pipette and forced to stick on a glass coverslip. Immunofluorescence labeling procedures can be performed on this recorded cell as described earlier.

Studying Recombinant Ion Channels in Mammalian Cells: Some Optimized Procedures

Interest of Recombinant Ion Channel Studies

The cloning of a variety of human ion channel cDNAs has allowed the development of genetic studies, and several human genes encoding ion

[34] S. A. Cohen, *Circulation* **94,** 3083 (1996).
[35] W. Schulze and M. L. Fu, *Mol. Cell Biochem.* **163,** 159 (1996).

channel subunits have been identified and cytogenetically mapped, thus opening the way to the identification of human genetic diseases linked to mutations within these genes.[3,36] Using the candidate gene approach, particular attention is given to regions of the genome harboring known ion channel genes. As a consequence, several complementary developments have allowed the possibility of functional screening of these cloned channels and their mutants, as well as the identification of complex protein associations that give rise to functional channels.[37,38] Moreover, infection strategies now allow expression of recombinant ion channels for functional studies in native cardiac myocytes.[39]

Functional Studies of Recombinant Ion Channels in Mammalian Cells

Although many expression studies have been performed in *Xenopus* oocytes,[40] several laboratories have now focused their interests on mammalian cells using electrophysiologic screening. We briefly comment on the recent developments that allow functional screening of recombinant ion channels as expressed in mammalian cells. Advantages are summarized: (1) a large choice of recipient cell lines exists (see ATCC or ECACC catalogs) and properties of recombinant channels explored in a given cell line are often closer to native situations, compared to *Xenopus* oocytes; and (2) stable cell lines can be produced in order to combine several approaches such as electrophysiology, biochemistry, molecular biology, or immunofluorescence. A variety of transfection procedures have now been developed in many laboratories. The cells to be transfected are usually grown in standard culture medium, DMEM, supplemented with 10% fetal calf serum, 1.5 m*M* glutamine, and antibiotics (0.1 μg/ml streptomycin and 100 UI/ml penicillin). The day before transfection, cells should be plated at low density (40–50% confluence) onto glass coverslips. A standard transfection can be performed using Lipofectamine (GIBCO, Grand Island, NY), according to the manufacturer's instructions. Plasmid DNA for mammalian transfection can be purified by adsorption to macroporous silica gel anion-exchange columns (Qiagen, Chatsworth, CA). The expression of the corresponding mRNA in transfected cells can be verified using RT-PCR. Several transfection agents are now available from several suppliers and can be

[36] E. P. Hoffman, *Annu. Rev. Med.* **46,** 431 (1995).

[37] F. Lehmann-Horn and R. Rudel, *Rev. Physiol. Biochem. Pharmacol.* **128,** 195 (1996).

[38] D. W. Wang, K. Yazawa, A. L. George, Jr., and P. B. Bennett, *Proc. Natl. Acad. Sci. U.S.A.* **93,** 13200 (1996).

[39] J. Barhanin, F. Lesage, E. Guillemare, M. Fink, M. Lazdunski, and G. Romey, *Nature* **384,** 78 (1996).

[40] D. C. Johns, H. B. Nuss, N. Chiamvimonvat, B. M. Ramza, E. Marban, and J. H. Lawrence, *J. Clin. Invest.* **96,** 1152 (1995).

tested for their specific interest to a given cell line. The next easy objective is to identify the transfected cells in order to optimize electrophysiologic screening. Several procedures can be followed.

Cell Selection Procedure Based on Antibiotic Resistance

To yield transfected cells using antibiotics, cDNA should be subcloned into an expression vector containing a selectable marker, such as pCEP4 from Invitrogen.[41] Transfection with such a construct confers resistance to cells cultured in the presence of hygromycin B. The day after transfection, culture medium should be supplemented with 200 μg/ml of hygromycin B (Sigma), and maintained for about 4–5 days in this culture medium prior to electrophysiology. Surviving cells are presumed to express the recombinant channels. This strategy is also used to yield a stable cell line.[42]

Cell Selection Procedure Based on Green Fluorescent Protein Expression

Green fluorescent protein (GFP) can be used to visualize the transfected cells. Informations regar this procedure is available[43,44] and is not repeated here. The use of GFP is well adapted for the identification of transfected cells for electrophysiology, even though the GFP expression and the recombinant channel expression are obtained with two independent expression plasmids. In addition, GFP expression can be detected on the same day after transfection and does not seem to affect the properties of recombinant channels.

Selection Procedures Based on CD8 Expression

This procedure also allows visualization of transfected cells. Cells are transfected with a plasmid that permits expression of the cell surface marker CD8. $CD8^+$ cells can then be identified using the Dynabeads M450 CD8 (Dynal), which are coated with a monoclonal antibody specific for the CD8 antigen. Compared to GFP this procedure does not require special optical equipment (i.e., fluorescence) to visualize the $CD8^+$ cells.[45]

Taken together, the above-described procedures allow identification of living transfected cells as required for a variety of functional studies of recombinant channels, especially electrophysiologic studies.

[41] J. Nargeot, N. Dascal, and H. A. Lester, *J. Membr. Biol.* **126,** 97 (1992).

[42] P. Lapie, C. Goudet, J. Nargeot, B. Fontaine, and P. Lory, *FEBS Lett.* **382,** 244 (1996).

[43] P. Lory, G. Varadi, D. F. Slish, and M. Varadi, *FEBS Lett.* **315,** 167 (1993).

[44] S. R. Kain and P. Kitts, *Methods Mol. Biol.* **63,** 305 (1997).

[45] R. Rizzuto, M. Brini, F. De Giorgi, R. Rossi, R. Heim, R. Y. Tsien, and T. D. Pozzan, *Curr. Biol.* **6,** 183 (1996).

Conclusion and Perspectives

Studies of human cardiovascular ion channels are essential and exciting as emphasized in the Introduction. However, these studies can be hampered by several difficulties relating mainly to regular obtention of homogeneous and abundant sources of tissues. In addition, cardiac tissues are not available routinely and some are not available at all (e.g., sinoatrial node kept in place to attach the new heart during transplantation) and samples are most often taken from diseased tissues. Samples from diseased tissues are of potential interest for studying pathology; however, the interpretations can be complicated when there is a lack of normal tissue. Most patients also receive medication prior to and during surgery. Therefore, precise clinical data including history of the pathology and drug treatment received are clearly required.

Although less obvious than with animal models, working with human tissue offers a wealth of diversity of pathological situations that can be interpreted directly at the molecular level. However, this approach cannot be exclusive and therefore should be used in parallel with animal models. In this context, the experience gained with human cell studies can help to select animal models more closely related to the human pathology studied. Interest in such approaches is also not limited to the cardiovascular system. Although more difficult to obtain and work with, it is possible to obtain brain tissue samples from routine neurosurgery (e.g., ablation of epileptic center). This could be of crucial interest since ion channel activity plays a central role in most neurophysiologic functions. Reinforcing this remark, mutations of genes are implicated in major neurological disorders such as epilepsy, ataxia, or migraine.[46] Comparing a given pathological channel in its native cell will certainly be a bonus to the functional studies of the same cloned mutant channel expressed in a host mammalian cell. Future investigation may tend to evolve toward such approaches.

Acknowledgments

We thank Dr. G. Dayanithi and Dr. J. Lemos for suggestions regarding the manuscript. We warmly thank our collaborators S. Lemaire, J. F. Quignard, C. Ménard, C. Choby, M. Mangoni, and V. Leuranguer for contributions to some of the experiments described in this article. This work was supported by grants from the Foundation de France, the Association Française Contre les Myopathies, and the French MENESR (ACCSV9).

[46] M. E. Jurman, L. M. Bolan, Y. Liu, and G. Yellen, *BioTechniques* **17,** 876 (1994).

[6] Studying Ion Channels Using Yeast Genetics

By ROBERT L. NAKAMURA and RICHARD F. GABER

Introduction

A number of recent ion channel investigations have involved heterologous expression in *Saccharomyces cerevisiae.* With this system, the advantages of yeast genetic analysis are now applicable toward the study of ion channels. In addition, the microbial aspect of the system allows experiments to be designed on a scale much larger than can be accommodated by traditional electrophysiologic analysis. For instance, the ability to perform genetic screens and selections can be a powerful method for sorting through libraries of channel mutations. Similarly, screening a cDNA library for novel ion channels can be as straightforward as performing a yeast transformation and plating for growth on the appropriate test medium. The ability to generate specific host cell genotypes and the availability of the entire genome sequence are also important components of the system.

The analysis of heterologous ion channels expressed in *S. cerevisiae* can produce information from a new perspective: simple growth phenotypes of *S. cerevisiae* are used to characterize channel function. Although such output is less quantitative than standard electrophysiologic analysis, it can nevertheless be compelling. Indeed, for some channel characteristics the yeast system can provide greater sensitivity than conventional electrophysiology.

The purpose of this chapter is to describe the yeast microbial genetic system for analysis of heterologous ion channels. Although ion channels expressed in *S. cerevisiae* cells have been analyzed by conventional electrophysiologic protocols or by microphysiometry, these techniques will not be extensively discussed. Instead, the emphasis is on using genetic techniques to alter the function of K^+ channels and the reliance on growth rates to measure the output of the altered channels.

Potassium Transport in *Saccharomyces cerevisiae*

A brief introduction into the biology of potassium transport of *S. cerevisiae* is useful for understanding the genetic backgrounds that are utilized in the study of potassium channels. Nutritional uptake of potassium in *S. cerevisiae* is dependent on the Trk1 and Trk2 high-affinity potassium trans-

0076-6879/98 $25.00

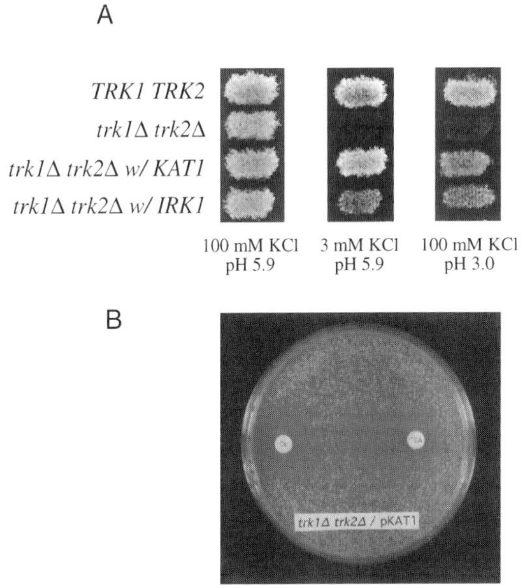

FIG. 1. (A) Growth phenotypes of wild-type (*TRK1 TRK2*), K^+ uptake-deficient (*trk1Δtrk2Δ*), and *trk1Δtrk2Δ* cells expressing *KAT1* and *IRK1* on K^+-permissive medium (100 m*M* KCl), K^+-limiting media (3 m*M* KCl, pH 5.9), and low-pH medium containing permissive concentrations of potassium (100 m*M* KCl, pH 3.0). (B) Filter disk assay showing that Kat1-dependent growth of *trk1Δtrk2Δ* cells is sensitive to the presence of external TEA and cesium. Similar concentrations of inhibitors had no effect on *S. cerevisiae* cells expressing endogenous K^+ transporters (not shown).

porters.[1–3] The electrogenic activity of the plasma membrane H^+-ATPase generates a hyperpolarized membrane potential,[4] allowing cells to maintain high intracellular concentrations of potassium[5] (~150 m*M* K^+) under conditions in which the potassium concentration of the growth medium is very low (<50 μ*M* K^+).[6] Deletion of *TRK1* and *TRK2* abolishes high-affinity uptake, severely impairing growth on potassium-limiting media (<7 m*M* K^+) (Fig. 1A).[7] This provides a conditional negative phenotype to *trk1Δtrk2Δ* cells that is the basis for detecting function of heterologous

[1] J. Ramos, P. Contreras, and A. Rodriguez-Navarro, *Arch. Microbiol.* **143,** 88 (1985).
[2] R. F. Gaber, C. A. Styles, and G. R. Fink, *Mol. Cell. Biol.* **8,** 2848 (1988).
[3] C. H. Ko, A. M. Buckley, and R. F. Gaber, *Genetics* **125,** 305 (1990).
[4] G. W. Borst-Pauwels, *Biochim. Biophys. Acta* **650,** 88 (1981).
[5] T. Ogino, J. A. den Hollander, and R. G. Shulman, *Proc. Natl. Acad. Sci. U.S.A.* **80,** 5185 (1983).
[6] A. Rodriguez-Navarro and J. Ramos, *J. Bacteriol.* **159,** 940 (1984).
[7] C. H. Ko and R. F. Gaber, *Mol. Cell. Biol.* **11,** 4266 (1991).

potassium channels by restoration of growth. In addition, *trk1Δtrk2Δ* cells are sensitive to low pH (pH 3.0) even in the presence of otherwise permissive concentrations of K^+ (100 m*M* KCl) (Fig. 1A).[7] Suppression of this low-pH hypersensitivity by a heterologous K^+ channel provides a second, and in some cases more sensitive, growth phenotype by which function of the channel can be assessed.

Some investigations of ion channels have exploited the phenotype of K^+ uptake-deficient cells that also contain a disruption of the *ENA1* locus.[8] This strain is hypersensitive to sodium due to the deletion of the Na^+-ATPase, Ena1, which is responsible for the extrusion of that ion.[9] In this genetic background, a highly selective K^+ channel can suppress the *ena1*-dependent Na^+ sensitive phenotype, while K^+ channel mutations that increase permeability to sodium fail to do so.

Not all phenotypes conferred by the expression of heterologous K^+ channels require the use K^+-uptake deficient cells (see later discussion). However, by demonstrating that a particular channel can rescue the growth phenotype of *trk1Δtrk2Δ* cells, one can establish that the channel is functional in yeast.

Plasmid-Mediated Expression of Heterologous Channels

A number of standard yeast genetic and molecular biological methods are necessary to study the expression of heterologous ion channels.[10,11] In general, a heterologous channel is expressed from a typical *S. cerevisiae* plasmid. Plasmids contain a genetic marker (*URA3, TRP1, LYS2,* etc.) used to select for the presence of the plasmid on the appropriate medium. Depending on the desired level of expression, vectors are low copy (centromeric) or high copy (2μ). Regulation of expression is also controlled transcriptionally by a yeast promoter such as *GAL1, PGK1,* or *ADH1.*

Potassium-Permissive Media

Maintenance of *trk1Δtrk2Δ* strains requires growth on standard media that is supplemented with potassium. This prevents the K^+ uptake-deficient strains from being exposed to nonpermissive conditions, which can result

[8] N. Uozumi, W. Gassmann, Y. Cao, and J. I. Schroeder, *J. Biol. Chem.* **270,** 24276 (1995).

[9] R. Haro, B. Garciadeblas, and A. Rodriguez-Navarro, *FEBS Lett.* **291,** 189 (1991).

[10] C. Guthrie and G. R. Fink, *Methods Enzymol.* **194** (1991).

[11] F. M. Ausubel, R. Brent, R. E. Kingston, D. D. Moore, J. G. Seidman, J. A. Smith and K. Struhl, eds., "Current Protocols in Molecular Biology," pp. 13.0.1–13.13.9. John Wiley and Sons, New York, 1989.

in the accumulation of extragenic suppressors.[12] These suppressors are usually the result of mutations in genes encoding endogenous membrane proteins.[13,14]

Potassium-Limiting Media

The growth media used to detect function of a heterologous K^+ channel contain limiting concentrations of K^+. Yeast Nitrogen Base (YNB) media (DIFCO, Detroit, MI) normally contains approximately 7 m*M* K^+, which is sufficiently low for detecting heterologous K^+ channel-dependent growth in *trk1Δtrk2Δ* cells. However, some circumstances and some genetic backgrounds require even lower concentrations of K^+, which can be achieved using low-salt (LS) media.[2,3,6] LS media can be prepared to contain micromolar concentrations of potassium and is essentially free of other cations that may interfere with K^+-dependent growth assays. To prepare agar with very low K^+ concentrations and to remove impurities, powdered agar is washed with 50 m*M* Tris, pH 5.9.[1]

Yeast Nitrogen Base Media (7 mM K^+, pH 5.9)

0.16% Bacto-yeast nitrogen base (without amino acids or ammonium sulfate)
3.8 m*M* Ammonium sulfate
0.2% Amino Acid mix (dependent on selection markers[10])
Distilled water

Low-Salt Media

8 m*M* H_3PO_4
2 m*M* $MgSO_4$
0.002 m*M* $CaCl_2$
0.1% Vitamin mix (contains 0.00002% biotin, 0.0004% calcium pantothenate, 0.00002% folic acid, 0.0004% niacin, 0.0002% *p*-aminobenzoic acid, 0.0004% pyridoxine hydrochloride, 0.0002% riboflavin, 0.0004% thiamin hydrochloride) (all values w/v)
0.1% Trace elements (contains 0.00005% boric acid, 0.000004% copper sulfate, 0.00001% potassium iodide, 0.00002% ferrous chloride, 0.00004% manganese sulfate, 0.00002% sodium molybdate, 0.00004% zinc sulfate) (all values w/v)

[12] M. Vidal, A. M. Buckley, F. Hilger, and R. F. Gaber, *Genetics* **125,** 313 (1990).
[13] M. B. Wright, E. A. Howell, and R. F. Gaber, *J. Bacteriol.* **178,** 7197 (1996).
[14] H. Liang, C. H. Ko, T. Herman, and R. F. Gaber, *Mol. Cell. Biol.* **18,** 926 (1998).

0.2% Amino acid mix (dependent on selection markers[10])
Distilled water

For plates add 2% agar.

pH of Media

Because *trk1*Δ*trk2*Δ cells are sensitive to low pH, growth media containing supplementary amino acids are normally adjusted to pH 5.9 with NaOH or NH_4OH. For some purposes such as studying channel selectivity, it may be necessary to avoid extraneous cations in the growth media. In these cases, the pH can be adjusted using arginine base or Tris–base.[1]

The external pH can be important to the activity of an ion channel. Some channels expressed in yeast are only functional over a narrow range of pH, necessitating that media be prepared and buffered to alternate pH levels.[15] Media prepared to pH 5.9 will, over a period of several days of yeast growth, become more acidic due to the extrusion of protons. Buffering the media with MES (2-[N-Morpholino]ethanesulfonic acid) and Tris–base stabilizes the pH over extended periods of growth.

Media Prepared for Analysis of Ion Selectivity

The yeast system is exquisitely suited for testing the selectivity and characterizing the pharmacology of some K^+ channels. Channel-dependent function in the presence of competing ions or channel blockers can be assessed by the addition of such compounds (Na^+, Li^+, NH_4^+, Ca^{++}, Cs^+, Ba^{2+}, TEA^+, etc.) to the growth media as chloride salts. The concentration required to inhibit channel-dependent growth or to confer cytotoxicity (see later section) will vary depending on the ratio of K^+ to competitor ion in the medium. Alternatively, a rapid method for determining the effect of a competing ion on K^+ uptake by a heterologous channel is to spread a lawn of cells expressing the channel on K^+-limiting medium and apply the inhibitor to a sterile dish of filter paper placed on the plate. The compound will diffuse to form a concentration gradient around the disc. *trk1*Δ*trk2*Δ cells expressing a channel that is sensitive to the inhibitor fail to grow where the concentration of the inhibitor is sufficiently high, resulting in a zone of inhibition. This technique is shown in Fig. 1B where Kat1-dependent growth of *trk1*Δ*trk2*Δ cells is inhibited by Cs^+ and TEA.

[15] R. L. Nakamura, J. A. Anderson, and R. F. Gaber, *J. Biol. Chem.* **272,** 1011 (1997).

Cloning and Expression of K^+ Channels in *Saccharomyces cerevisiae*

The microbial aspect of *S. cerevisiae* allows for rapid screening of cDNA libraries. In addition, the conditional phenotype of *trk1Δtrk2Δ* cells makes them ideal hosts for cloning some K^+ channels by complementation. Although several inward-rectifying K^+ channels have been cloned based on their ability to fulfill the host cell requirement for nutritional uptake of K^+, these are not the only types of proteins that have been cloned by this strategy. K^+ transporters and other types of K^+ channels have also been identified using K^+ transport-deficient strains of *S. cerevisiae.*[16–18] Presumably, the magnitude of the yeast membrane potential provides a sufficiently strong driving force that K^+ uptake is conferred even under conditions when other types of K^+ channels are rarely open.

KAT1 and *AKT1* from *Arabidopsis thaliana* were the first K^+ channels to be cloned using K^+ uptake-deficient strains of *S. cerevisiae.*[19,20] cDNA libraries constructed in yeast expression vectors were screened by testing for the ability of these cells to grow on media containing low concentrations of K^+. Approximately 40,000 colonies were screened in the cloning of *KAT1,* and 60,000 for the cloning of *AKT1.* The cDNA clones encoding Kat1 and Akt1 complemented the growth deficiency on low K^+ media, and were subsequently shown by electrophysiologic analysis to encode inward-rectifying voltage-gated K^+ channels.[21,22] Both channels are of the six transmembrane class similar in structure to the Shaker superfamily of channels.

Inward rectifying K^+ channels of the IRK (two transmembrane) family also function in *S. cerevisiae* cells. Both the guinea pig *IRK1* channel and the mouse *IRK1* channels have been expressed in *trk1Δtrk2Δ* cells and shown to confer growth on low K^+ (this work).[23] Recently, *trk1Δtrk2Δ* cells were used to clone the *ORK1* K^+ channel from *Drosophila.*[17] This is a member of the two-pore family of channels which includes the *S. cerevisiae*

[16] D. P. Schachtman and J. I. Schroeder, *Nature* **370,** 655 (1994).

[17] S. A. Goldstein, L. A. Price, D. N. Rosenthal, and M. H. Pausch, *Proc. Natl. Acad. Sci. U.S.A.* **93,** 13256 (1996).

[18] M. A. Banuelos, R. D. Klein, S. J. Alexander-Bowman, and A. Rodriguez-Navarro, *EMBO J.* **14,** 3021 (1995).

[19] J. A. Anderson, S. S. Huprikar, L. V. Kochian, W. J. Lucas, and R. F. Gaber, *Proc. Natl. Acad. Sci. U.S.A.* **89,** 3736 (1992).

[20] H. Sentenac, N. Bonneaud, M. Minet, F. Lacroute, J. M. Salmon, F. Gaymard, and C. Grignon, *Science* **256,** 663 (1992).

[21] D. P. Schachtman, J. I. Schroeder, W. J. Lucas, J. A. Anderson, and R. F. Gaber, *Science* **258,** 1654 (1992).

[22] F. Gaymard, M. Cerutti, C. Horeau, G. Lemaillet, S. Urbach, M. Ravallec, G. Devauchelle, H. Sentenac, and J. B. Thibaud, *J. Biol. Chem.* **271,** 22863 (1996).

[23] W. Tang, A. Ruknudin, W. P. Yang, S. Y. Shaw, A. Knickerbocker, and S. Kurtz, *Mol. Biol. Cell* **6,** 1231 (1995).

outward-rectifying K^+ channel *TOK1* (*DUK1, YKC1*).[24] Even the single transmembrane ion channel M2 from influenza virus has been shown to function in *S. cerevisiae*.[25]

Structure–Function Studies

The yeast system provides a unique perspective for structure–function analyses by its ability to generate experimental data rapidly based on mutational analysis. Large-scale site-directed and random mutageneses can be performed, and their effects on channels determined by analysis of growth phenotypes. Two properties of the *S. cerevisiae* system can be exploited: (1) the ability to screen large numbers of channels and/or (2) the ability to perform genetic selections. With an extensive genetic screen, structural features can be readily identified even if revealed through the occurrence of extremely rare mutations. In addition, the ability to select for mutations that direct a specific effect on the activity of the channel provides an important tool for identifying structural features. For example, to identify channel sites involved with TEA binding, channel-dependent growth can be assessed in the presence of TEA. Channel mutations that alter resistance to this blocker are likely to contribute to its binding site.

Site-Directed Mutagenesis

In the yeast system, large numbers of mutant channels can be analyzed in a single experiment. This allows the opportunity to analyze comprehensively not only one residue, but several residues simultaneously. For example, we have examined the effect of all 8000 possible mutant sequence combinations at the glycine–tyrosine–glycine sequence in the pore of Kat1 (unpublished data, 1994). A library of the channel mutants was expressed in *trk1Δtrk2Δ* cells and each mutant was screened for function by testing for the ability to confer growth on millimolar concentrations of K^+. Several hundred mutants were found to be capable of conferring growth under these conditions. Of these, fewer than 10 different mutants retained the ability to confer growth on micromolar concentrations of K^+. Using similar approaches we have determined that several positions in the pore of Kat1 never form functional channels when mutated. However, in combination with a second or third site mutation, the multiply mutant channel can regain function.[15] These types of compensatory effects reveal structural

[24] K. A. Ketchum, W. J. Joiner, A. J. Sellers, L. K. Kaczmarek, and S. A. Goldstein, *Nature* **376,** 690 (1995).

[25] S. Kurtz, G. Luo, K. M. Hahnenberger, C. Brooks, O. Gecha, K. Ingalls, K. Numata, and M. Krystal, *Antimicrob. Agents Chemother.* **39,** 2204 (1995).

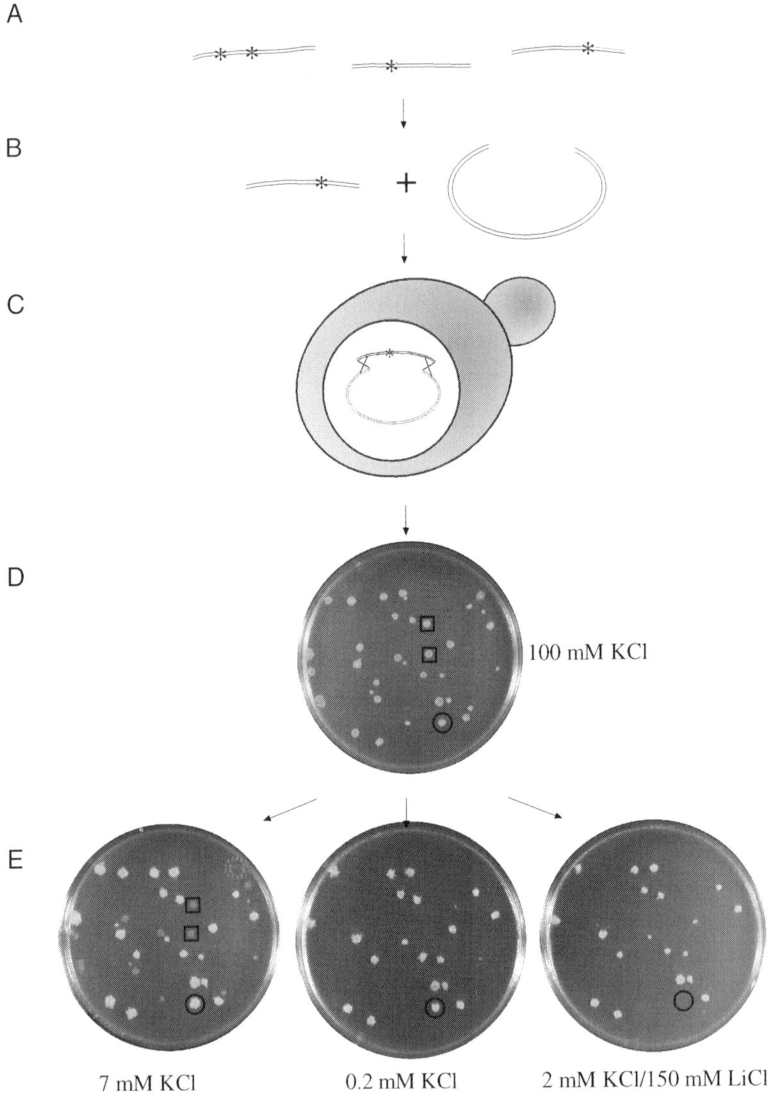

FIG. 2. Random mutagenesis of defined regions of ion channels. (A) PCR performed under mutagenic conditions generates potentially mutant fragments of DNA. (B) The fragment and linearized vector are cotransformed into *S. cerevisiae* and (C) homologous recombination occurs *in vivo,* forming a circular plasmid. (D) Transformants are allowed to develop under permissive conditions and (E) channel-dependent growth is assessed on K^+-limiting media in the presence or absence of competitor ions. Two colonies that express channels that are only weakly functional on 7 m*M* KCl are boxed. A colony that expresses a channel that is sensitive

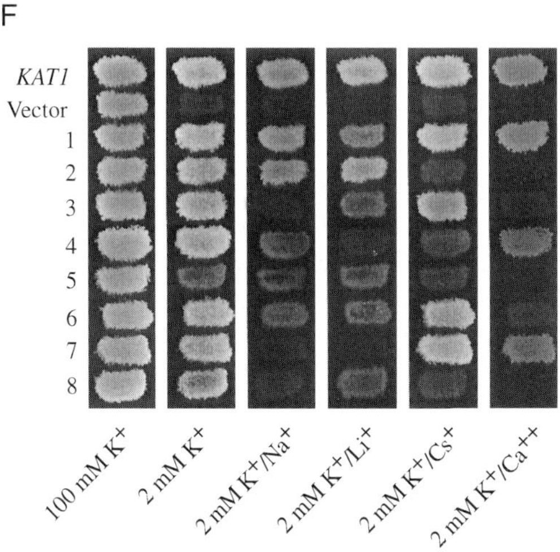

to lithium is circled. (F) Mutant Kat1 channels (numbered 1 through 8) identified in a screen for decreased ion selectivity. Growth of *trk1Δtrk2Δ* cells assessed on permissive medium (100 m*M* KCl), K^+-limiting conditions (2 m*M* KCl), and K^+-limiting conditions supplemented with 500 m*M* NaCl, 125 m*M* LiCl, 1 m*M* CsCl, and 75 m*M* $CaCl_2$.

relationships that would be nearly impossible to identify without a facile genetic screen.

Random Mutagenesis

Random mutagenesis is another important approach for structure–function studies of ion channels. Standard protocols including hydroxylamine mutagenesis can be utilized to mutate a channel expressed from a plasmid.[26] Alternatively, polymerase chain reaction (PCR) can be performed under mutagenic conditions to introduce random mutations to a defined region of a channel[27,28] (Fig. 2A). Transfer of the PCR-generated mutant fragment into an expression vector can be achieved by standard cloning procedures or the amplified fragments can be conveniently integrated into a plasmid by cotransformation into *S. cerevisiae* (Fig. 2B). In the latter case, a vector harboring part of wild-type channel gene is digested with restriction enzymes to leave overlapping ends homologous to the ends

[26] J. Sambrook, E. F. Fritsch, and T. Maniatis, *in* "Molecular Cloning: A Laboratory Manual," pp. 1–15.105. Cold Spring Harbor Laboratory Press, New York, 1989.

[27] D. W. Leung, E. Chen, and D. V. Goeddel, *Technique,* **1,** 11 (1989).

[28] D. Muhlrad, R. Hunter, and R. Parker, *Yeast* **8,** 79 (1992).

of the amplified fragment. The fragment and linearized plasmid undergo homologous recombination *in vivo* resulting in a circular plasmid capable of transforming recipient cells (Fig. 2C). The colonies that develop are suitable for genetic screening by replica-plating to various test media (Figs. 2D and E).

This technique was used to identify amino acids in the pore of Kat1 that play critical roles in ion selectivity. A 500-base-pair fragment encoding the S5-pore-S6 region of Kat1 was amplified by PCR under mutagenic conditions and cotransformed into a *trk1*Δ*trk2*Δ host with a gapped plasmid containing the overlapping region of wild-type *KAT1*. Ura^+ transformants were selected on K^+-permissive medium (100 m*M* KCl) and replica plated to K^+-limiting media supplemented with potential competitor ions. Several examples of mutants that were identified in this screen are shown in Fig. 2F. Note the independence of sensitivity to the test ions; for example, mutant 3 is sensitive to sodium and resistant to lithium, whereas mutant 4 has the opposite phenotype.

Analysis of Channel-Dependent Yeast Growth Phenotypes

The basic analyses of ion channels expressed in *S. cerevisiae* involve growth phenotypes. Initially, a channel is tested for function in K^+ uptake-deficient cells by growth on limiting concentrations of K^+. Controls include cells containing an empty vector (negative control) and cells expressing a functional K^+ channel (positive control). A convenient method for detecting functional channels is by replica plating. Patches of cells are grown on K^+-permissive medium and blotted to K^+-limiting media using sterile velvet. Blotting off the surface of the selective media with fresh velvet after the initial replica plating ensures a sparse inoculant of evenly distributed cells on all test plates. When carefully executed, this technique can be used to distinguish between subtle growth differences (see Figs. 1A, 1B, and 2F for examples of this technique). In addition, differences in growth rates can be detected on solid media by streaking for single colonies (see Fig. 5A for an example of this technique).

K^+ channel-dependent growth rates can be quantified by measuring the growth of liquid cultures. The media are essentially the same as solid media but prepared without agar. Cultures are inoculated with cells expressing a K^+ channel and cell density is determined by spectroscopy. At a wavelength of 600 nm, an optical density (OD) of 1 is roughly equal to 5×10^7 cells/ml. Inoculation at approximately 5×10^4 cells/ml typically provides a large enough number of cells to result in a detectable OD after 24–48 hr, yet a small enough number to avoid spontaneous mutants in the inoculant that could take over in liquid culture.

Differential growth rates conferred by heterologous channels can also be visualized by serial dilution of liquid cultures onto solid media. This technique can provide a compelling visual representation of differences in growth rates. Cells are precultured in liquid medium under K^+-permissive conditions and cell density is quantified by spectroscopy. Equal numbers of cells are then serially diluted and spotted onto solid K^+-limiting media and formation and growth of colonies is assessed (Fig. 3). Both serial dilution and liquid culture experiments provide quantitative methods for detecting subtle channel phenotypes. Nevertheless, the qualitative examination of patches of K^+ channel-expressing yeast cells on solid media is sensitive, rapid, easy, and a recommended first step.

Assessing Channel Selectivity

Of significant interest in ion channel biology is the issue of selectivity. Selectivity of a heterologous channel expressed in K^+ uptake-deficient cells can be assessed by examining growth on K^+-limiting media supplemented with potentially inhibitory concentrations of a test ion. For example, growth conferred by a highly selective K^+ channel such as Kat1 is virtually unaffected by the presence of high concentrations of NaCl in the medium. In contrast, the presence of competitor ions can abolish growth of cells expressing mutant *KAT1* channels that have decreased selectivity (Figs. 3 and 4A).

Several factors should be considered when interpreting a result in which a mutant channel is unable to rescue growth of *trk1*Δ*trk2*Δ cells in the

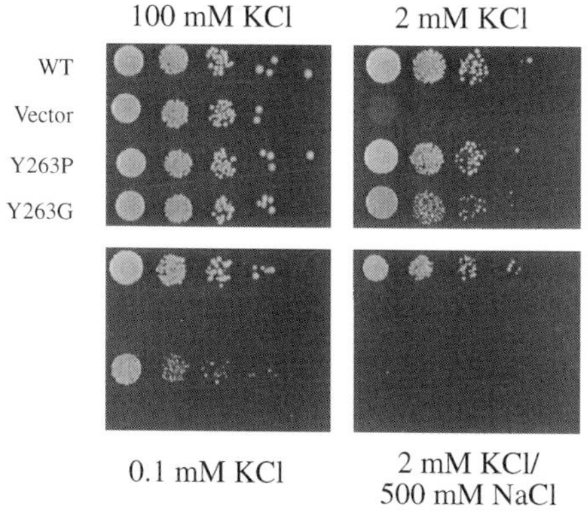

FIG. 3. Serial dilution of mutant *trk1*Δ*trk2*Δ cells expressing *KAT1* channels. Growth rates of *trk1*Δ*trk2*Δ cells expressing wild-type *KAT1*, vector, and two mutants of *KAT1* at permissive conditions (100 m*M* KCl), K^+-limiting conditions (2 m*M* KCl and 0.1 m*M* KCl), and K^+-limiting conditions supplemented with sodium (2 m*M* KCl/500 m*M* NaCl).

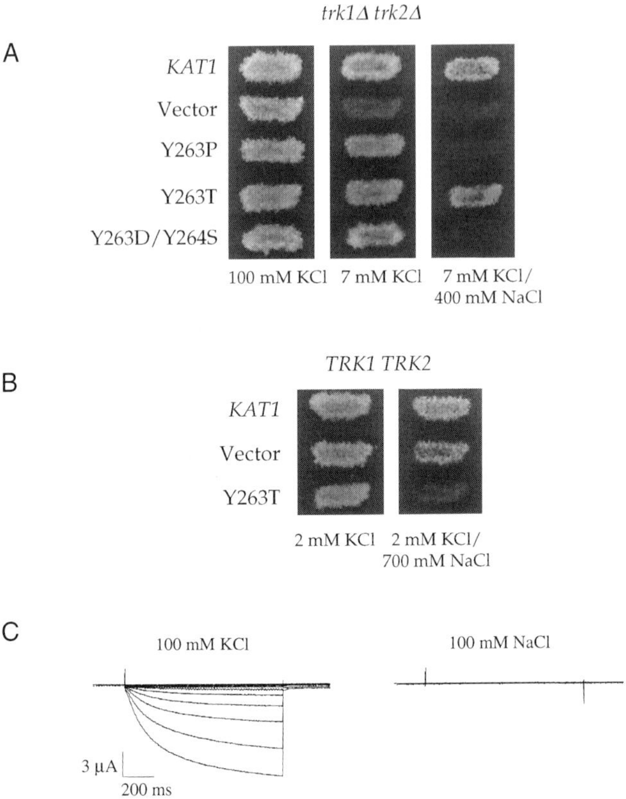

FIG. 4. (A) Wild-type *KAT1*, vector, and three mutants of *KAT1* expressed in *trk1*Δ*trk2*Δ yeast cells on permissive medium (100 m*M* KCl), K^+-limiting medium (7 m*M* KCl), and high sodium medium (7 m*M* KCl/400 m*M* NaCl). (B) Wild-type *KAT1*, vector, and $KAT1^{Y263T}$ expressed in wild-type yeast cells on permissive medium (2 m*M* KCl) and high sodium medium (2 m*M* KCl/700 m*M* NaCl). (C) Two-electrode voltage-clamp measurement of $KAT1^{Y263T}$ in bath solutions containing 100 m*M* KCl or 100 m*M* NaCl. Cells were held at −40 mV and hyperpolarized to −160 mV in 10-mV steps. [Reprinted by permission from R. L. Nakamura, J. A. Anderson, and R. F. Gaber, *J. Biol. Chem.* **272,** 1013 (1997). Copyright © 1997 by The American Society for Biochemistry and Molecular Biology, Inc.]

presence of a competing ion. One possibility is that a mutation confers increased blockage by the competing ion, preventing sufficient K^+ uptake through the channel to sustain growth of the cells. Mutations that alter selectivity of a channel might simultaneously decrease K^+ uptake due to lower expression, decreased conductance, altered gating, or other factors. In such cases, if the test ion can normally compete with K^+ for uptake, a further decrease in K^+ uptake can result in the inability of the mutant channel to rescue the *trk1*Δ*trk2*Δ phenotype.

A mutant may also be unable to confer growth in the presence of a competitor ion due to increased permeation of the ion. At sufficiently high intracellular concentrations many cations, with the notable exception of K^+, are toxic. The yeast system can be used to identify mutations that confer even slight increases in permeation of some test ions. For this assay, mutant channels are expressed in wild-type *S. cerevisiae* cells. In this background, cells expressing a wild-type K^+ channel or harboring an empty vector grow equally well under K^+-limiting conditions due to the function of the endogenous Trk1 and Trk2 transporters (Fig. 4B). Similarly, both strains are capable of equally strong growth in the presence of supplementary cations due to the K^+-selective uptake of the wild-type K^+ channel and/or Trk1 and Trk2. However, mutant K^+ channels that allow permeation of the test ion will confer toxicity to the cells (Fig. 4B, $KAT1^{Y263T}$). As a method for detecting permeation of a toxic ion, this test can be extremely sensitive because the growth medium can be prepared with high concentrations of the toxic ion and low concentrations of K^+, for example, Na^+ to K^+ ratios of 700:2. Such extreme concentration gradients create a large driving force for uptake of the test ion, which can then be detected due to its cytotoxic effects even if its permeation relative to K^+ is only slightly increased.

Sensitivity of Yeast System

The yeast system can be extremely sensitive for detecting subtle modifications of channel activity. We have found that many mutant Kat1 channels, particularly those that confer only weak suppression of the *trk1*Δ*trk2*Δ phenotype are unable to elicit detectable currents when expressed in *Xenopus* oocytes. The ability to detect function of weakly conducting K^+ channels in *S. cerevisiae* may be due to the difference in time scale of the experiments; current recordings from a typical two-electrode voltage-clamp experiment last but a few seconds. In contrast, growth of *S. cerevisiae* cells occurs over a period of hours to days during which the membrane potential may confer essentially constant activation of a voltage-gated K^+ channel like Kat1.

The difference time scale may also account for a discrepancy between the two systems of analysis with regard to their abilities to detect changes in ion selectivity. For example, we have identified a mutant *KAT1* channel that elicits highly K^+-selective currents when assayed by two-electrode voltage clamp in *Xenopus* oocytes[15] (Fig. 4C). Sodium permeation by the mutant channel was undetectable, and reversal potential measurements failed to indicate a change in Na^+ permeation compared to the wild-type channel. However, using the toxic ion sensitivity assay described earlier, expression of this mutant channel in wild-type *S. cerevisiae* cells resulted in an increased sensitivity to Na^+ detected as impaired growth, indicating that the mutation does mediate permeation to this ion (Fig. 4B). Thus, expression of K^+ channels in *S. cerevisiae* can reveal changes in selectivity that are undetectable by conventional electrophysiology. Evidently this results from the ability of the yeast cells to integrate the output signal from a channel over a period of days. Because the output is measured by growth and is cumulative, the sensitivity of the yeast system can be exceptional.

Comparing Yeast and Electrophysiologic Systems

Several investigations have examined the selectivity of plant K^+ channels expressed in both the yeast and *Xenopus* oocyte systems.[8,15,29,30] In these studies, selectivity of mutant K^+ channels was assessed in the yeast system by testing for growth in the presence of competing ions. In the oocytes, selectivity was determined by measuring reversal potentials or by conductance ratios. Channel block was assessed by measuring K^+ currents in the presence of a test ion. As indicated previously, differences can arise in the characterization of some mutants in the yeast and oocyte systems. However, in most cases, the inferences regarding channel selectivity reached in both systems were in agreement.

We have also compared the *S. cerevisiae* and *Xenopus* oocyte systems for the ability to assess the effects of external pH on channel activity. pH sensitivity of a channel can suggest structural features such as a titratable residue located in proximity to the ion permeation pathway. A pH-sensitive channel, $KAT1^{Y253R}$, was expressed in oocytes where two-electrode voltage-clamp experiments revealed that K^+ currents were much smaller in magnitude at pH 5.0 than at pH 7.4 (Fig. 5A). When expressed in *trk1Δtrk2Δ* cells on low potassium medium, the channel was unable to confer growth at pH 5.0, but able to confer growth above pH 7.0 (Fig. 5B). Thus, both systems of analysis are capable of revealing pH-dependent channel activity.

Conclusions

Although the emphasis of this chapter is on the analysis of ion channels in yeast using growth phenotypes, standard patch-clamp techniques have also been used to investigate K^+ channel expressed in *S. cerevisiae* cells.

[29] J. A. Anderson, R. L. Nakamura, and R. F. Gaber, *Soc. Exp. Biol. Symp.* **48,** 85 (1994).

[30] D. Becker, I. Dreyer, S. Hoth, J. D. Reid, H. Busch, M. Lehnen, K. Palme, and R. Hedrich, *Proc. Natl. Acad. Sci. U.S.A.* **93,** 8123 (1996).

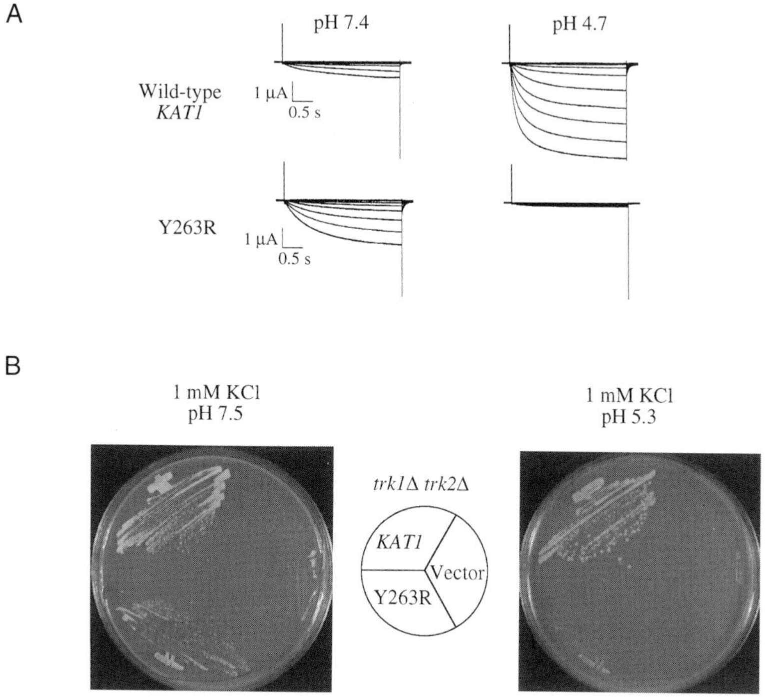

FIG. 5. (A) Two-electrode voltage-clamp measurement of $KAT1^{Y263R}$ in bath solutions containing 100 m*M* KCl prepared at pH 7.4 or pH 4.7. Cells were held at −40 mV and hyperpolarized to −150 mV in 10-mV steps. (B) Wild-type *KAT1,* vector, and $KAT1^{Y263R}$ expressed in *trk1Δtrk2Δ* cells on K^+-limiting media (1 m*M* KCl) prepared at pH 7.5 or pH 5.3. [Reprinted by permission from R. L. Nakamura, J. A. Anderson, and R. F. Gaber, *J. Biol. Chem.* **272,** 1016 (1997). Copyright © 1997 by The American Society for Biochemistry and Molecular Biology, Inc.]

Kat1, Akt1, and the endogenous Tok1 channel have each been directly characterized by electrophysiologic experiments in yeast.[31–34] The ability to perform such experiments in *S. cerevisiae* has several advantages. Using the yeast genetic system, high-throughput screens can first be carried out after which potentially interesting mutant channels can be analyzed by electrophysiology without switching to an alternate system. Furthermore, by deleting the genes encoding the endogenous K^+ transporters and the

[31] M. C. Gustin, B. Martinac, Y. Saimi, M. R. Culbertson, and C. Kung, *Science* **233,** 1195 (1986).
[32] A. Bertl and C. L. Slayman, *J. Exp. Biol.* **172,** 271 (1992).
[33] A. Bertl, C. L. Slayman, and D. Gradmann, *J. Membr. Biol.* **132,** 183 (1993).
[34] A. Bertl, J. A. Anderson, C. L. Slayman, and R. F. Gaber, *Proc. Natl. Acad. Sci. U.S.A.* **92,** 2701 (1995).

Tok1 channel, K^+ flux can be virtually eliminated when assayed by patch clamping (A. Bertl, personal communication, 1997). This provides a clean background in which K^+ channels can be expressed.

Microphysiometry has also been used to the study the influenza M2 ion channel in the yeast system.[25] This technology relies on microelectrodes that can detect the extrusion of protons from *S. cerevisiae* cells expressing M2 channels. Compounds that inhibit channel activity can be applied and their effect detected as a change in external pH. Microphysiometry provides the means by which large numbers of compounds can be screened rapidly for effects on ion channel activity.

The yeast genetic system provides a novel approach by which ion channels can be studied. It allows large-scale experiments to be performed using standard yeast genetic and microbial techniques. Data are in the form of simple growth rates that can be visually assessed on solid media. In this regard, cumulative effects on yeast growth have proven particularly beneficial for the study of ion selectivity. Slight changes in ion permeation are "amplified" in the yeast system by assessing growth on media prepared with extremely high (but normally permissive) concentrations of toxic cations. The ability to perform genetic manipulations with yeast is another important property of the system. Deletion of the endogenous K^+ transporter genes has produced strains that are severely impaired for a potassium uptake providing useful backgrounds for the identification of K^+ channels and assessment of their function.

Traditional electrophysiologic methods are powerful techniques that cannot be supplanted by the study of channel-dependent growth phenotypes in *S. cerevisiae*. However, the yeast system has proven to be a powerful and exquisitely sensitive addition to the array of tools with which channels can be discovered and new insights into ion channel biology revealed.

[7] Identification of Ion Channel-Associated Proteins Using the Yeast Two-Hybrid System

By MARTIN NIETHAMMER and MORGAN SHENG

Introduction

A wide variety of ion channels are involved in generating and regulating the electrical behavior of a neuronal cell. Typically, these ion channels are not randomly distributed on the cell surface but rather localized at specific subcellular sites. Good examples include ligand-gated ion channels (iono-

0076-6879/98 $25.00

tropic neurotransmitter receptors) at postsynaptic sites and voltage-gated sodium channels at nodes of Ranvier. Presumably the targeted localization of these ion channels depends on protein–protein interactions between ion channel subunits and the cytoskeleton. In addition, it is likely that ion channels interact with intracellular proteins that are involved in their modulation (e.g., protein kinases such as the tyrosine kinase Src[1,2] or protein kinase C[3]) or that are involved in downstream signaling mechanisms activated by the ion channel.[3,4] Thus, knowledge about intracellular proteins associated with ion channels is critical to our understanding of signal transduction by ion channels and of mechanisms of ion channels clustering at specific subcellular sites. Currently, few such proteins have been identified.

At the neuromuscular junction, 43K/rapsyn is involved in ACh receptor clustering and postsynaptic specialization.[5,6] In the spinal chord, glycine receptors are clustered due to association with gephyrin, which may link receptor subunits to tubulin.[7] Both rapsyn and gephyrin were initially identified biochemically as proteins that copurified with their respective receptors.

Biochemical copurification of associated proteins may not be feasible for many ion channels, especially those of low abundance and low solubility (such as NMDA receptors). An alternative approach is the yeast two-hybrid system. The two-hybrid system is a widely used yeast-based genetic approach for detecting protein–protein interactions and provides a powerful way of isolating ion channel-associated proteins in an *in vivo* setting. It can be used both to assay known interactor proteins and for screening of cDNA libraries for novel proteins, giving immediate access to the primary DNA sequence of the interacting protein, as well as allowing for relatively easy determination of the binding domains involved. Since ion channels are transmembrane proteins, using them in a two-hybrid screen is not trivial, even though it has been done recently with some success.

In this article, we describe the yeast two-hybrid assay we have used to identify proteins binding to Shaker-type channels and NMDA receptor subunits. We also discuss some general and specific considerations about using ion channel proteins in the two-hybrid system.

[1] T. C. Holmes, D. A. Fadool, R. Ren, and I. B. Levitan, *Science* **274,** 2089 (1996).
[2] X. M. Yu, R. Askalan, G. J. Keil, and M. W. Salter, *Science* **275,** 674 (1997).
[3] S. Tsunoda, J. Sierralta, Y. Sun, R. Bodner, E. Suzuki, A. Becker, M. Socolich, and C. S. Zuker, *Nature (London)* **388,** 243 (1997).
[4] J. E. Brenman, D. S. Chao, S. H. Gee, A. W. McGee, S. E. Craven, D. R. Santillano, Z. Wu, F. Huang, H. Xia, M. F. Peters, S. C. Froehner, and D. S. Bredt, *Cell* **84,** 757 (1996).
[5] S. C. Froehner, *Ann. Rev. Neurosci.* **16,** 347 (1993).
[6] M. Gautam, P. G. Noakes, J. Mudd, M. Nichol, G. C. Chu, J. R. Sanes, and J. P. Merlie, *Nature (London)* **377,** 232 (1995).
[7] J. Kirsch, I. Wolters, A. Triller, and H. Betz, *Nature (London)* **366,** 745 (1993).

Principles of Two-Hybrid System

The yeast two-hybrid system[8,9] makes use of the modular nature of eukaryotic transcription factors. Transcription activators typically consist of a DNA binding domain, which recognizes specific sequence motifs in the regulatory region of genes, and a distinct activator domain.[10] Typically, the DNA binding domain and the transcription activating domain function independently; that is, hybrids between different DNA binding domains and activation domains form fully functional transcription factors.[11] Furthermore the two domains need not be covalently linked.[12,13] This latter property is exploited in the yeast two-hybrid system.

A plasmid construct encoding a fusion protein of a known protein (the "bait") is made with a DNA binding domain, and cotransfected into reporter yeast strains with a plasmid encoding a second fusion protein, consisting of a known protein or a random cDNA fragment derived from a library, fused to a transcriptional activation domain. Thus, by using an entire library of cDNA fragments for the second construct, a screen for unknown interactors can be performed. Neither of the two fusions alone can activate the reporter genes; only when the expressed fusion proteins bind each other within the yeast nucleus is the transcription activator reconstituted. This is detected by activation of one or more reporter genes whose upstream regulatory region carries the appropriate DNA sequence elements for binding of the bait-DNA binding domain.

Since its inception, the two-hybrid system has been used to identify numerous protein–protein interactions in a variety of systems. More recently, several two-hybrid screens have successfully identified proteins that interact with ion channels (Table I).[14–18] It is noteworthy that all of these proteins, for reasons discussed later, interact with the intracellular carboxy-terminal region of the channel proteins. A number of other ion channels

[8] S. Fields and O.-k. Song, *Nature (London)* **340,** 245 (1989).

[9] C. T. Chien, P. L. Bartel, R. Sternglanz, and S. Fields, *Proc. Natl. Acad. Sci. U.S.A.* **88,** 9578 (1991).

[10] L. Keegan, G. Gill, and M. Ptashne, *Science* **14,** 699 (1986).

[11] R. Brent and M. Ptashne, *Cell* **43,** 729 (1985).

[12] J. Ma and M. Ptashne, *Cell* **55,** 443 (1988).

[13] J. L. McKnight, T. M. Kristie, and B. Roizman, *Proc. Natl. Acad. Sci. U.S.A.* **84,** 7061 (1987).

[14] E. Kim, M. Niethammer, A. Rothschild, Y. N. Jan, and M. Sheng, *Nature (London)* **378,** 85 (1995).

[15] M. Wyszynski, J. Lin, A. Rao, E. Nigh, A. H. Beggs, A. M. Craig, and M. Sheng, *Nature (London)* **385,** 439 (1997).

[16] H.-C. Kornau, L. T. Schenker, M. B. Kennedy, and P. H. Seeburg, *Science* **269,** 1737 (1995).

[17] M. Niethammer, E. Kim, and M. Sheng, *J. Neurosci.* **16,** 2157 (1996).

[18] H. Dong, R. J. O'Brien, E. T. Fung, A. A. Lanahan, P. F. Worley, and R. L. Huganir, *Nature (London)* **386,** 279 (1997).

TABLE I
ION-CHANNEL BAITS USED IN TWO-HYBRID SCREENS

Ion Channel	Residues[a]	Vector	Results	Ref.
Kv1.4	568-655*	pBHA	PSD-95 SAP97 Chapsyn-110	14
NMDA-R1	834-938*	pBHA	α-actinin-2	15
NMDA-R2A	834-1464*	pGBT9	PSD-95	16
	1349-1464*	pBHA	PSD-95 SAP97	17
NMDA-R2B	1361-1482*	pBHA	PSD-95 SAP97	17
GluR2	834-883*	pPC97	GRIP	18

[a] * denotes C-terminal residue of the ion channel.

have also been shown to interact with known proteins in the two-hybrid system (such as Kv1.1-1.3)[14]; these channel parts have not, however, been tested as bait in library screens.

Materials and Methods

A variety of vectors and reporter strains have been designed for use in the two-hybrid system. The most commonly used DNA binding domains are derived from the *Saccharomyces cerevisiae* transcription factor Gal4p and the *Escherichia coli* repressor *lexA,* while the most commonly used activation domains stem from Gal4p and the herpes simplex virus protein VP16.[19] Reporter genes usually encode for yeast nutritional marker genes or the *E. coli* gene β-galactosidase. These vectors have been described in detail elsewhere,[19] and we thus restrict this article to the system used in our laboratory, which utilizes the *lexA* DNA binding domain and the Gal4p activation domain.

Formulations for solutions and media needed for two-hybrid screens are given in the Appendix.

Vectors and Reporter Yeast Strains

Most two-hybrid vectors share some common features such as bacterial *ori* (for prokaryotic replication) and *bla* (for ampicillin resistance) sequences, as well as yeast origin sequences such as 2μ or ARS (for replication in yeast). Expression of the selectable nutritional markers (e.g., *TRP1* or

[19] P. L. Bartel and S. Fields, *Methods Enzymol.* **254,** 241 (1995).

LEU2) from these vectors is usually controlled by an ADH1-derived yeast promoter. We use two vectors in our system. The bait protein is inserted into vector pBHA [a derivative of pBTM116, with a hemagglutinin (HA) tag inserted between the LexA sequence and the multiple cloning site]; and the library is fused into vector pGAD10. The pGAD10 vector is part of the commercially available matchmaker two-hybrid system (Clontech Laboratories, Palo Alto, CA). We use the yeast host strain L40 (genotype *MAT***a** *trp1-901 leu2-3,112 his3Δ200 ade2 lys2-801am LYS2::(lexAop)*$_4$-*HIS3 URA3::(lexAop)*$_8$-*lacZ*), which is mutant for the chromosomal copies of *LEU2, TRP1,* and *HIS3*. Two independent reporter genes *HIS3* and *lacZ*, with upstream *lexA* binding sites are stably integrated into the genome.[20] Activation of both of these reporter genes confers histidine-independent growth and β-galactosidase activity, thereby allowing nutritional selection and blue-white X-Gal screening for yeast harboring interacting bait and activation domain fusions. pBHA contains the *TRP1* gene, and pGAD10 the *LEU2* gene as selectable markers. We can thus select for cotransformed yeast by growing the yeast on leucine/tryptophan-deficient media. Interaction of the two fusion proteins can be selected for by additionally making the plates histidine deficient. Some leaky His3p expression usually occurs even in the absence of interaction of the fusion proteins. This can be overcome by adding the competitive inhibitor 3-amino-1,2,4-triazole (3-AT) to the selection plates (we usually use a concentration of 2.5–5 m*M*). 3-AT can also help to suppress any intrinsic low-level transcriptional activating activity that a bait may exhibit.

Making Bait Constructs

Generally, we find that the following types of baits tend to be more successful in the two-hybrid system: (1) defined modular domains of proteins, (2) C-terminal ends, and (3) whole proteins. Presumably, modular domains and C termini will more likely fold in a native confirmation as fusion proteins and thus make the screen more "realistic." Using domains requires prior knowledge about the bait protein, such as the presence of recognized domains that can fold in varying contexts, such as SH3 or PDZ domains. Using the whole protein is often impractical for a number of reasons—the proteins may be large, and thus hard to subclone and transform, and, for reasons that are not understood, large baits often display transcriptional activation by themselves. However, if these problems do not occur, whole proteins are advantageous because they make no presumption about domain structure and may be more likely to fold correctly. Even

[20] S. M. Hollenberg, R. Sternglanz, P. F. Cheng, and H. Weintraub, *Mol. Cell Biol.* **15,** 3813 (1995).

more worrisome for screening with ion channels or other transmembrane proteins is that these proteins will not fold correctly when not membrane inserted and will likely never even reach the yeast nucleus due to hydrophobic transmembrane regions. The whole-protein approach thus usually is only feasible for small, intracellular proteins and not for transmembrane proteins.

Generally, an intracellular C terminus seems to be the best suited part of ion channels for two-hybrid screens. It is probably no coincidence that all successful two-hybrid screens with ion channels have used the intracellular C-terminal region as a bait (Table I), and to date we know of no instance of successful yeast two-hybrid screens with N termini or extra- or intracellular loops of ion channels. Extracellular regions are often glycosylated *in vivo.* If this glycosylation is involved in the binding of other proteins or the tertiary structure of the channel, the interaction is not likely to show up in the two-hybrid system. While intracellular N termini should behave similarly to C termini in the two-hybrid system (with respect to folding) they at present have an additional problem. All vectors commercially available to this point fuse the bait to the C-terminal end of *lexA* or Gal4 DNA binding proteins, thus likely masking interactions that might take place with a free N terminus. An N-terminal fusion vector has been reported[21]; however, to our knowledge it has not yet been used in any ion channel screens.

Construction of the bait is done by standard molecular techniques.[22] The DNA encoding the channel part of interest is subcloned into the multicloning site of pBHA (Fig. 1). pBHA has stop codons in all three frames downstream of the multicloning site. Nevertheless, when making a C-terminal bait it is important to include the stop codon at the C terminus. The easiest way to generate the insert (especially of shorter fragments) is by PCR (polymerase chain reaction) using specific primers, adding the appropriate restriction sites at the ends. Typically we insert the bait in frame beginning at the *Eco*RI site. Because frameshifts that result in faulty bait proteins are a major concern, it is essential to sequence the final bait plasmid before using it in the two-hybrid system. For sequencing of the *lexA*–bait fusion junction we use the primer 5′-CTT CGT CAG CAG AGC TTC ACC ATT G-3′ which corresponds to amino acids 181–188 of the lexA protein (~100 nucleotides upstream of the *Eco*RI site). Once the sequence is confirmed, it is of equal importance to test any bait for self-activation, by cotransforming the bait plasmid with an insert-free pGAD10 and testing reporter gene activation.

[21] F. Béranger, S. Aresta, J. de Gunzburg, and J. Camonis, *Nucleic Acids Res.* **25,** 2035 (1997).

[22] J. Sambrook, E. F. Fritsch, and T. Maniatis, eds., "Molecular Cloning: A Laboratory Manual." Cold Spring Harbor Laboratory Press, New York, 1989.

```
                    HA tag
        Tyr Pro Tyr Asp Val Pro Asp Tyr Ala
GAA TTG TAC CCA TAC GAC GTC CCA GAC TAC GCT

GAA TTC CCG GGG ATC CGT CGA CCT GCA GCC AAG
 EcoRI   SmaI   BamHI   SalI    PstI

CTA ATT CCG GGC GAA TTT CTT ATG ATT TAT GAT
  *                          *         *

TTT TAT TAT TAA ATA AGT TAT AAA AAT AAG TGT
             *    *      *      *

ATA CAA ATT TTA AAG TGA CTC
                     *
```

FIG. 1. Multicloning site of pBHA. The hemagglutinin tag (HA) and unique restriction sites are indicated. Downstream stop codons in all three frames are marked by asterisks (*). The LexA sequence ends directly 5′ of the sequence shown here.

What to Do When Bait is Self-Activating

Self-activation is a poorly understood process. Many baits, probably for different reasons, will activate the reporter genes to variable degrees even when cotransfected with an empty pGAD10 vector. For instance, this may be due to the bait behaving like a true transcription activator, binding to a transcription factor, or being nonspecifically able to recruit transcriptional machinery. Self-activation is a common problem with two-hybrid baits, and will make a successful screen next to impossible. Since it is impossible to predict if a bait is activating, all bait constructs must be tested by cotransfection with empty library vector. Basically, two steps can be taken if a bait does activate transcription. If the self-activation is weak (i.e., β-Gal activity and growth on His^- plates are weak), it may still be possible to detect strong interaction above that background (especially if His^- growth is suppressed with 3-AT). The other option is to alter the bait. Often subtracting or adding a few amino acids on either side of the bait can abolish self-activation. This may be because these modifications allow for proper folding of the bait, thus avoiding nonspecific binding of an incorrectly folded bait to proteins of the transcriptional machinery, for instance. In other cases, larger bits may have to be added or deleted from the bait. Unfortunately, this is not a predictable process. We routinely make a series of alternative overlapping baits to minimize this problem. Generally, we find that longer baits are more likely to be self-activating. For instance, we originally attempted to use the C-terminal 350 amino acids of NR2A (residues 1115–

1464) for a two-hybrid screen. That construct proved to be self-activating. (Note that self-activation is system dependent, and a bait may not be self-activating in the context of one vector, but strongly in the context of another.) By contrast, a shorter construct restricted to the C-terminal 116 amino acids of NR2A[17] displays virtually no self-activation of reporter genes.

If the two-hybrid system is simply used to test for interaction between specific known proteins (rather than for a library screen), a third option should be considered. Simply reversing the two proteins between DNA binding and activation domain vectors can, and most of the time will, abolish the problem. That is to say, it is unlikely that both interacting proteins will self-activate reporter gene transcription.

Another approach has been reported by Cormack and Somssich,[23] adding a specific repressor sequence to the bait construct. With this method they were able to repress self-activation that even high levels of 3-AT could not suppress. Although this approach remains to be tested in a screen situation, it may provide a last ditch solution when all else fails.

Sequence of Events for Library Screen

All the protocols needed for a library screen are given below. The following summary of the events will help to plan the experiments. It takes about 2–3 weeks to complete an entire screen. First, yeast are cotransformed with a bait plasmid and cDNA library and plated on $trp^-leu^-his^-$ plates. Three days later, the yeast colonies that grow independently of histidine are further assayed for expression of the second reporter gene (β-galactosidase activity). Positive colonies need to be restreaked again on $trp^-leu^-his^-$ plates. We streak the colonies with a sterile loop onto sectors of 150-mm plates. This helps with the isolation of pure clones and also confirms that colonies picked were truly showing histidine-independent growth and β-galactosidase activity. This step is analogous to purification of a positive plaque from a conventional hybridization screen. After 2 days, these re-streaked colonies are again assayed for β-galactosidase activity. If it is still not possible to isolate single colonies, another restreaking is necessary. Plasmid DNA is isolated from the yeast and transformed into bacteria by electroporation. The isolated library plasmids are then transformed one more time into yeast together with the original bait plasmid in order to check if this reconfers the interaction phenotype. This step is needed to confirm isolation of truly interacting plasmids from the library.

[23] R. S. Cormack and I. E. Somssich, *Anal. Biochem.* **248,** 184 (1997).

Yeast Transformation

To transform yeast, we use a method described by Bartel and Fields.[19] This method is a modified version of the lithium acetate method developed by Ito *et al.*[24] and improved by Schiestl and Gietz[25] and Hill *et al.*[26] Dimethyl sulfoxide (DMSO) is added to improve transformation efficiency.[27] All the volumes and amounts given here are for testing interactions between two known proteins. For library screens, these must be adjusted as described at the end of the protocol.

1. Grow overnight culture of yeast in 20–40 ml YPD (yeast extract, peptone, dextrose) medium (inoculated from a recently streaked YPD plate grown for 2–3 days at 30°) in a sterile 250-ml flask, in shaking water bath at 30°.

2. Dilute overnight culture into fresh YPD (~50- to 100-fold) in a sterile flask (the YPD volume should never exceed more than 20% of total flask volume; ideally it should be less than 10%). The total final volume should be about 10 ml culture per intended transformation. Grow for another 3–4 hr with shaking at 30° (the final OD_{600} should be about 0.20–0.25). Some protocols such as that described by Bartel and Fields[19] grow the yeast until an OD_{600} of 0.5 is reached. Generally, we find reduced efficiency with ODs under 0.2; for most transformations any OD between 0.2 and 0.8 seems to work well.

Time saver: The procedure outlined in steps 1 and 2 will give optimal transformation efficiency. This is really only needed for library screens. For standard two-hybrid assays testing interaction between two known proteins, transformation efficiency is less important. In this case the yeast can be diluted the night before and grown overnight to an OD of 0.2–0.8 (assume a 2-hr doubling time). We usually try to err on the side of overgrowth, since OD values higher than 0.25 are easily tolerated, whereas lower OD values will diminish the transformation efficiency. For this procedure it is useful to grow the 20 ml culture an extra day to saturation, before the final dilution.

3. Harvest the cells by centrifuging 5–10 min at 1500–2000*g* at 4°, in 50-ml sterile conical tubes. Decant supernatant and wash the cells by resuspending the pellet in 10 ml sterile water per 50-ml tube and pelleting again at 1500–2000*g* for 5–10 min. At this stage, if several 50-ml tubes have been used, pool the cells after resuspending each pellet in water.

[24] H. Ito, Y. Fukuda, K. Murata, and A. Kimura, *J. Bacteriol.* **153,** 163 (1983).

[25] R. H. Schiestl and R. D. Gietz, *Curr. Genet.* **16,** 339 (1989).

[26] J. E. Hill, A. M. Myers, T. J. Koerner, and A. Tzagoloff, *Yeast* **2,** 163 (1986).

[27] J. Hill, K. A. Donald, and D. E. Griffiths, *Nucleic Acids Res.* **19,** 5791 (1991).

4. Resuspend the washed yeast cell pellet in freshly made 1× LiAc/1× TE, pH 7.5, buffer at a density of 100 μl per transformation (i.e., per 10 ml of the original culture).

5. Preload transforming DNAs into 1.5-ml microfuge tubes. This can be done toward the end of the yeast growth phase, and the DNA kept on ice. Use 200 ng to 1 μg of each plasmid for the cotransformation, with ~100 μg of carrier single-stranded salmon sperm DNA (10 mg/ml stock). Whenever feasible, use master mixes to minimize tube-to-tube variability. If the total volume of DNA is greater than 10 μl, it is worthwhile to add the appropriate amounts of 10× LiAc and 10× TE to bring contents to 1× buffer concentrations. However, for most applications, it is sufficient to mix 2 μl of each plasmid (assuming typical minipreparation concentrations of 100–300 ng/μl) and 6 μl (60 μg) of salmon sperm DNA. Add 100 μl of yeast cells (resuspended in step 4) to each tube of DNA.

6. Add 0.6 ml of freshly made 40% polyethylene glycol (PEG) solution in 1× LiAc/TE (i.e., 0.8 volume of 50% PEG, 0.1 volume 10× LiAc, 0.1 volume 10× TE) to each transformation. Invert the tube several times and vortex *briefly* to mix contents. Because 40% PEG is quite viscous, take care to mix well but without excess agitation of cells.

7. Incubate 30 min in 30° water bath, shaking moderately. Mix occasionally.

8. Add 70 μl of DMSO. Vortex briefly or invert several times.

9. Heat shock for 15 min in a 42° water bath. No mixing is required during this stage.

10. Spin 5–10 sec in microfuge (high speed) and carefully aspirate supernatant.

11. Resuspend pellets well in 1 ml YPD medium (the pellet will be rather clumpy because of the PEG), and incubate 2–3 hr in 30° shaking water bath, to allow for *HIS3* expression. This rather long outgrowth is due to the slow doubling time of yeast, as well as the fact that for successful *HIS3* expression both the bait and the library plasmid need to be expressed before activation of the *HIS3* gene can occur.

12. Spin cells 30 sec in microfuge. With a quick shake of the wrist, decant most of the YPD from the microtube. Resuspend the pellet in the remaining YPD (about 50 μl). Plate half of the resuspension on each of the selection plates ($trp^-leu^-his^-$ and trp^-leu^-). If only plating on trp^-leu^- plates, no outgrowth is needed. Just directly resuspend the heat shocked cells in 50–100 μl YPD and plate. The yeast can either be plated on 60-mm plates or on sectors of 150-mm plates. Grow 2–3 days at 30° and perform X-Gal assay. Library screens are always grown 3 days.

For the screening of a cDNA library, the following adjustments need to be made: if a library is screened, the preceding protocol needs to be scaled up. A single transformation (to be plated on a 150-mm plate) is done with 5–10 μg of each plasmid (bait and library), 200 μg of carrier DNA, and 200 μl of yeast (still diluted at 100 μl per 10 ml of original culture). Double the volumes of PEG and DMSO as well. We normally perform a library screen with the equivalent of 5–10 such 200-μl transformations. In this case, the transformation can be pooled in 50-ml conical tubes or Falcon 2059 tubes rather than doing separate microtubes. In the final step, the outgrowth is then resuspended into 400 μl YPD per 150-mm petri dish ($trp^-leu^-his^-$ selection, with 2.5–5 m*M* 3-aminotriazole). In addition, do not forget to make a 1 : 200 to 1 : 400 dilution (i.e., 2–4 μl of the resuspension into 100 μl YPD) and plate these on 100-ml leu^-trp^- plates (containing His!) to test for transformation efficiency. This step is critical to evaluate the efficiency of the screen. We get routinely between 50,000 and 150,000 transformants per 150-mm plate using the amount of yeast and DNA described earlier (i.e., for a 10-plate screen 500,000 to 1,500,000 colonies total).

Reporter Gene Activation Assays

Usually, only a small fraction of transformed yeast grows on his^- plates. Once the yeast colonies grow up (usually after 2 or 3 days), they can be tested for β-galactosidase activity. The His selection even in the presence of 3-AT is not always complete (and sometimes spontaneous mutations can make the yeast His^+ again), so β-galactosidase activation serves as an independent secondary screen to ensure that colonies growing on the $trp^-leu^-his^-$ plates contain interacting fusion proteins. When testing for interaction of already isolated proteins, we usually plate the yeast on $trp^-leu^-his^-$ and trp^-leu^- plates. The number of colonies growing without histidine compared to the growth with histidine correlates well with β-galactosidase activity, and can serve as an additional measure of strength of interaction.

There are a variety of ways to assay for β-galactosidase activity. We use a filter assay, both for isolation of colonies in a screen and to test interactions. The relative times it takes for colonies to turn blue in our experience correlates well with the more quantitative colorimetric assays[14] and has the advantage of convenience (protocols of quantitative liquid colorimetric assays are described elsewhere[19,28]).

[28] P. L. Bartel, C.-T. Chien, R. Sternglanz, and S. Fields, *in* "Cellular Interactions in Development: A Practical Approach" (D. A. Hartley, ed.), p. 153. Oxford University Press, Oxford, 1993.

1. X-Gal is kept as a stock solution of 20 mg/ml in *N*,*N*-dimethylformamide at −20°. When performing the assay, add 1.67 ml X-Gal stock per 100 ml Z buffer (see Appendix). Some protocols will also add 0.27 ml 2-mercaptoethanol to this, but this is usually unnecessary. Put a filter paper of appropriate size (e.g., Whatman 1, Clifton, NJ) into a petri dish (one per yeast plate), and add X-Gal solution. Be careful not to add too much (about 3.5–4 ml for a 150-mm plate); there should be no excess pooling of solution.

2. For library screens we use nitrocellulose membranes (e.g., Protran, Schleicher & Schuell, Keene, NH, pore size 0.45 μm). Colonies stick better to nitrocellulose and stay more discretely separate from each other. Since nitrocellulose is more expensive, filter paper (e.g., Whatman 1) can be substituted when testing yeast two-hybrid interactions between known proteins.

3. Place the filter onto the agar plate (make sure to avoid air bubbles) until completely moist. Mark the filter in some way that will allow for orientation of filter to agar. Grab with forceps, and pull off in a smooth, rapid motion. Be careful not to drag the filter/nitrocellulose over the plate, as colonies may get smeared. Submerge in liquid nitrogen for about 10–20 sec (be careful when using nitrocellulose as it gets very brittle during freezing; supported nitrocellulose circles are more sturdy, e.g., Optitran, Schleicher & Schuell). This freeze–thaw step permeabilizes the yeast.

4. Place the filter onto the X-Gal soaked filter, *yeast colony side up.* Again make sure that no air gets trapped underneath. It is best to hold the filter with forceps and lower it gently from one side as it thaws.

5. Start a timer and wait for the yeast colonies to turn blue. Some protocols place the yeast at 30° during this time. We keep it at room temperature, which seems to work well in our experience. Since X-Gal time can be somewhat variable depending on room or bench temperature and colony density, it is important to have a positive control included in every assay (including library screens). Weak interactors can take quite a while; it is thus advisable to add a negative control as well because many weakly self-activating baits will cause the yeast to turn blue after many hours (we routinely watch for at least 4 hr for a library screen).

6. Whether performing a library screen or restreaking colonies, one can just pick the positive colonies with a sterile toothpick directly from the filters (there is sufficient viable yeast left after freezing for subsequent growth).

The filter X-Gal assay is very reliable. It is, in addition, semiquantitative, in that it assays the relative time it takes the different colonies to turn blue. Unlike colorimetric assays, the threshold of "blueness" is set by the observer—this may cause some variability between different people. Usually, the differences seem to be not large, and the results are consistent and reproducible, especially when compared with a positive control as internal reference. It is also not linear, in that it sets a lower limit for the speed at which the assay can work (usually around 12–15 min at room temperature). On the other hand, it does have the advantage of great sensitivity.

Recovery of Plasmids from Yeast

There are several protocols for the isolation of plasmid DNA from yeast. Generally, it is not a trivial procedure, because of the tough yeast cell wall. Also, DNA yields are low, and the DNA is contaminated with genomic DNA and cellular proteins. We use glass beads to disrupt the cell walls mechanically. A variation of this method described in the Clontech yeast protocol handbook (PT3024-1, Clontech Laboratories) additionally uses an enzyme preparation (lyticase) to weaken the cell walls. We do not find this to be necessary, but it may increase the yield for some low copy number plasmids. The resulting DNA is very impure (it is not useful for sequencing or restriction digests), but sufficient for electroporation into bacteria. A cleaner, higher yield DNA preparation is then obtained from bacteria by standard plasmid preparation methods.

1. Inoculate single yeast colonies in 2 ml of the appropriate selection medium (leu^-) and incubate at 30° overnight.
2. Transfer cultures to microtubes and pellet in microfuge (5 sec). Decant the supernatant and resuspend the pellet in the residual liquid by vortexing.
3. Add 0.2 ml of yeast lysis buffer, 0.2 ml of neutralized phenol/chloroform/isoamyl alcohol (25:24:1) and about 0.3 g (it is not critical to be very exact here) acid-washed glass beads[29] (Sigma, St. Louis, MO).
4. Vortex for 2 min at room temperature. This step is critical to disrupting the yeast cell walls.
5. Spin 5 min at room temperature in microfuge. Transfer the supernatant to a clean tube and ethanol precipitate the DNA. Wash with 70% (v/v) ethanol and dry under vacuum.
6. Resuspend the pellet in a small volume of TE (25 μl).
7. Transform *E. coli* with electroporation.

Bartel and Fields[19] describe essentially the same method, except they directly lyse a single yeast colony without overnight growth by adding 50 μl of lysis solution and phenol/chloroform, and 0.1 g of glass beads.

[29] J. E. Celis, ed., "Cell Biology." Academic Press, San Diego, 1994.

The final DNA is rather crude, but works well for electroporation. We usually get 20–200 colonies of *E. coli* with 1.5 μl of the yeast plasmid prep. Of course, this preparation will result in isolation of both the bait and the library plasmid. In a library screen it is essential to recover the library plasmid vector (pGAD10) rather than the known bait construct. Fortunately, the *LEU2* marker carried by pGAD10 complements the bacterial *leuB* marker. This can be exploited by using a bacterial strain with a *leuB* mutation, such as HB101[30,31] or MH4.[32] After electroporation, the cells are plated on M9 minimal plates (with 50 μg/ml ampicillin) containing all amino acids except leucine. As a backup, we usually also plate some of the cells on LB (50 μg/ml Amp) plates. This will not select for pGAD10 library plasmids but the cells usually grow better, which sometimes helps when transformation efficiencies are very low. The disadvantage is that they will then need to be rescreened on M9 plates to distinguish between library plasmids and bait plasmid.

Library plasmid DNA is then prepared from the transformed bacteria by standard methods. It is advisable at this stage to perform some crude restriction mapping, especially if a large number of clones have been isolated. This may give hints about multiple hits, an important factor in judging the veracity of the results (see below).

Since it is possible for multiple copies of pGAD10 to get transformed into the same yeast cells, clones may be isolated during this procedure that were not responsible for the original reporter gene activation. We thus usually pick at least two colonies from each bacterial transformation. Also, it is absolutely essential to retransform the isolated library plasmid into yeast together with the original baits. These retransformants are then assayed for histidine-independent growth and β-galactosidase activity. Failure to reconfer such activity eliminates clones that falsely showed up as positive in the original screen, a common occurrence. It is not unusual for half or more of the clones to drop out at this point. At this stage, it is also sensible to retransform the isolated clones with an entirely unrelated bait, as well as empty bait vector, which can eliminate pGAD10 constructs that can activate transcription by themselves or nonspecifically interact with different baits.

Judging Quality of Library Screen

The number of transformants (as calculated from leu$^-$trp$^-$ plating efficiency) is a measure of completeness of the library screen. Ideally, the total

[30] H. W. Boyer and D. Roulland-Dussoix, *J. Mol. Biol.* **41,** 459 (1969).

[31] F. Bolivar and K. Backman, *Methods Enzymol.* **68,** 245 (1979).

[32] M. N. Hall, L. Hereford, and I. Herskowitz, *Cell* **4,** 1057 (1984).

number of screened colonies should exceed the number of independent clones of the library. Note that in a random fragment library, only one-sixth of clones will generate true in-frame fusions with the GAL4 activation domain.

If positive clones are isolated, the main concerns are to eliminate false positives, that is, clones not binding to the bait, or clones where the binding has no *in vivo* biological relevance. The first type of false positives, clones not binding (or nonspecifically binding) to the bait but spuriously isolated during the screen, is eliminated by retransformation with the original bait vector. This class of false positives will not induce *HIS3* or β-galactosidase activity when coexpressed with the original bait in fresh yeast, or they will do so by interacting indiscriminately with various unrelated baits.

The other type of false positives is much more difficult to exclude—clones that indeed bind specifically to the bait in yeast, but whose interaction is of no biological significance *in vivo*. For example, the isolated "interacting" protein might never be in the same cellular or subcellular localization as the protein used for the bait.

So how is a screen assessed for the potential validity of the isolated clones? Generally, outcomes of screens can be grouped into three categories: (1) no positive clones are isolated, that is, even if there were blue colonies in the original screen, the recovered library plasmids do not reconfer *HIS3* and β-galactosidase activity with the original bait; (2) several identical or overlapping fragments of the same gene, or fragments of related genes are isolated, with very little else; or (3) many different, unrelated fragments are isolated.

In the case of category 1, if a large enough number of colonies was screened, the bait may either not "work" in the two-hybrid system or there may be no interacting clones of high enough affinity represented in the library. Either way, not much time is wasted. Category 2 represents the ideal scenario—not only were the fragments isolated several times (which at least shows a robust interaction), but it may even be possible to map the region of interaction based on common regions shared by overlapping cDNA fragments of the same gene. This scenario generally predicts that the isolated clones will be interesting to study and are true interactors *in vivo*. Situation 3, on the other hand, can be difficult. Unless there is a compelling reason to follow up some of the clones (such as previously suspected interactors), it becomes a game of chance. It may be advisable to modify the bait and perform another screen. We often perform more than one screen in parallel with slightly differing baits or different libraries, thus possibly isolating the same protein from more than screen. Using more than one library also increases the complexity of the screen.

Some clones such as heat shock proteins or ribosomal proteins seem to pop up frequently in two-hybrid screens.[33] The laboratory of Erica Golemis at the Fox Chase Cancer Center (Philadelphia, PA) maintains a web page that lists some of the most common false-positive clones isolated in two-hybrid screens; this may be useful in excluding some of the screen results (http://www.fccc.edu/research/labs/golemis/InteractionTrapInWork.html).

If any of the clones are judged to be useful, that is, the interaction in the two-hybrid system is specific, an independent assay should be performed to confirm the interaction in an independent manner, such as an *in vitro* filter overlay assay using fusion proteins.[34] It may also be possible to confirm the interaction in recently developed mammalian and bacterial two-hybrid systems.[35,36]

The problem of proving the validity of the interaction *in vivo* is the major challenge, and is discussed in Ref. 37.

Limitations, Perspectives, and Outlooks

When doing a two-hybrid screen, it is important to keep in mind that many protein–protein interactions, even if they do occur *in vivo,* are simply undetectable by this method. Examples are baits or interactions that are toxic when expressed in yeast, or proteins that do not express, do not fold or modify properly, are degraded, or never enter the nucleus. Expression can be verified by methods described elsewhere,[38,39] but some of the other problems may be unsolvable, even by altering the bait. For reasons outlined earlier, ion channels are likely to be susceptible to the folding problem, especially with loops flanked by transmembrane domains. It may thus be fundamentally impossible to detect some type of interactions between ion channels and other proteins using the two-hybrid system. Nevertheless, we feel that the yeast two-hybrid system is a powerful approach for identification of novel interactions. Particularly, the speed of the screen makes it a very viable option because it allows for the testing of many different parts of ion channels. Our understanding of ion channel-associated protein complexes has rapidly expanded in recent years thanks to the two-hybrid system, and this will be a major factor in dissecting functional aspects of synapses in the future.

[33] P. N. Hengen, *Trends Biochem. Sci.* **22,** 33 (1997).
[34] M. Li, Y. N. Jan, and L. Y. Jan, *Science* **257,** 1225 (1992).
[35] Y. Luo, A. Batalao, H. Zhou, and L. Zhu, *BioTechniques* **22,** 350 (1997).
[36] S. L. Dove, J. K. Joung, and A. Hochschild, *Nature* (*London*) **386,** 627 (1997).
[37] M. Wyszynski and M. Sheng, *Methods Enzymol.* **293,** [20], 1998 (this volume).
[38] J. A. Printen and G. F. Sprague, Jr., *Genetics* **138,** 609 (1994).
[39] K. Langlands and E. V. Prochownik, *Anal. Biochem.* **249,** 250 (1997).

Appendix: Solutions

For convenience, we provide recipes for most of the solutions and media needed for the two-hybrid screen.

YPD Medium

Per liter, to 950 ml of deionized water, add:

Bacto-peptone	20 g
Bacto-yeast extract	10 g
Glucose	20 g

Shake until dissolved, adjust to 1 liter, autoclave for 20 min at 15 lb/in^2 on liquid cycle. (*Note:* Some of the glucose will caramelize during autoclaving, giving the medium a tea-colored appearance. This is not a problem—just do not leave the medium in the autoclave for a long time after the cycle is finished.) For YPD plates add 15 g/liter Bacto-agar before autoclaving.

Minimal Plates and Media

Yeast Minimal Plates. Per liter, to 800 ml of deionized water, add:

Yeast nitrogen base without amino acids	6.7 g
Bacto-agar	15 g

Adjust volume to 860 ml, autoclave for 20 min at 15 lb/in^2 on liquid cycle. Let cool to about 45° (water bath). Add 40 ml 50% glucose (filter sterilized) and 100 ml 10× dropout (HLT$^-$, see below). For LT$^-$ plates, 10 ml 100× His (see below). (*Note:* It is advantageous to add a stir-bar into the flask before autoclaving. It will allow you to mix in the drop-out and glucose without creating bubbles, which are a nuisance during plate pouring.)

If background depression on HLT$^-$ plates is desired, add an appropriate amount of 1 *M* 3-aminotriazole (usually to a final concentration of 2.5 or 5 m*M*).

Selection medium is made the same way, except without Bacto-agar.

10× Drop-Out HLT$^-$. This solution is deficient for His/Leu/Trp. Per liter, to 950 ml of deionized water, add:

Adenine	200 mg
L-Arginine hydrochloride	200 mg
L-Aspartic acid	1000 mg
L-Glutamic acid	1000 mg
L-Isoleucine	300 mg
L-Lysine hydrochloride	300 mg
L-Methionine	200 mg
L-Phenylalanine	500 mg
L-Serine	1350 mg

L-Threonine	2000 mg
L-Tyrosine	300 mg
Uracil	200 mg
L-Valine	1500 mg

Adjust to 1 liter, filter sterilize (0.2 μm), store at 4°.

100× Histidine/100× Tryptophan. Per 100 ml, dissolve 200 mg L-histidine (monohydrate) or 200 mg L-tryptophan, filter sterilize (0.2 μm), store at 4°.

1 M 3-Aminotriazole. Dissolve 3-amino-1,2,4-triazole in deionized water (8.4 g/100 ml). 3-AT takes a long time to dissolve and may in fact not go completely into solution. Filter sterilize the final solution (0.2 μm); this will also remove any remaining particles. Store at 4°. (*Note:* At temperatures even slightly below 4° 3-AT may fall out of solution. If that occurs put it at room temperature until it is redissolved. Protect from light.)

M9 Minimal Medium. Per liter, to 650 ml of autoclaved deionized water (less than 50°), add:

5× M9 salts	200 ml
1 *M* $MgSO_4$ (autoclaved or filter sterilized)	2 ml
50% Glucose (filter sterilized)	8 ml
1 *M* $CaCl_2$ (filter sterilized)	0.1 ml
10 mg/ml Proline (filter sterilized)	4 ml
1 *M* Thiamine hydrochloride (filter sterilized)	1 ml
100 mg/ml Ampicillin	0.5 ml

If necessary, supplement with appropriate drop-outs, for example, for L^- medium:

10× HLT^- dropout	100 ml
100× Histidine	10 ml
100× Tryptophan	10 ml

Adjust to 1 liter with sterile deionized water. (*Note:* Make sure to add the $MgSO_4$ and $CaCl_2$ after dilution of the M9 salts. Ideally, add $CaCl_2$ at the very last, i.e., after dropouts have been added as well.) M9 is not a very rich medium, so the final concentration of ampicillin is only 50 μg/ml. For M9 plates, add 15 g Bacto-agar to the 650 ml water before autoclaving.

The 5× M9 salts is made by dissolving the following salts in deionized water to a final volume of 1 liter:

$Na_2HPO_4 \cdot 7H_2O$	64 g (or 33.9 g of anhydrous Na_2HPO_4)
KH_2PO_4	15 g
NaCl	2.5 g
NH_4Cl	5 g

Autoclave the salt solution or filter sterilize.

50% Glucose. Dissolve glucose to 50% (w/v). (*Note:* Add glucose slowly to water, and not the other way around, otherwise the glucose will solidify.

Slight heating may be necessary to get the glucose into solution.) Filter sterilize (0.2 μm). Store at 4°.

Yeast Transformation Solutions

The 10× LiAc is 1 *M* lithium acetate dihydrate (102 g/liter). Adjust pH with acetic acid to 7.5. The 10× TE is 100 m*M* Tris, 10 m*M* EDTA. Adjust pH to 7.5. The 50% PEG is 50 g/100 ml polyethylene glycol 3350 in deionized water. Stir overnight to dissolve. Filter sterilize (0.2 μm) all three solutions, store at room temperature.

Yeast Lysis Buffer

Triton X-100	2% (v/v)
SDS	1% (w/v)
NaCl	100 m*M*
Tris, pH 8.0	10 m*M*
EDTA, pH 8.0	1 m*M*

After preparing the solution, filter sterilize (0.2 μm) and store at room temperature.

X-Gal Assay

Z-Buffer for X-Gal Staining. Per liter,

$Na_2HPO_4 \cdot 7H_2O$	16.1 g (8.53 g anhydrous Na_2HPO_4)
$NaH_2PO_4 \cdot H_2O$	5.5 g
KCl	0.75 g
$MgSO_4 \cdot 7H_2O$	0.25 g

Adjust to pH 7.0 with NaOH. Autoclave and store at room temperature.

X-Gal Stock Solution. X-Gal (5-bromo-4-chloro-3-indolyl-β-D-galactopyranoside) is kept as a stock solution of 20 mg/ml in *N,N*-dimethylformamide at −20° in the dark.

Acknowledgments

We thank Elaine Aidonidis for assistance with the manuscript, Roland Roberts for helpful comments, and Cheng-Ting Chien for help in teaching us the yeast two-hybrid system. M.S. is an Assistant Investigator of the Howard Hughes Medical Institute.

[8] Substituted-Cysteine Accessibility Method

By ARTHUR KARLIN and MYLES H. AKABAS

Introduction

The functional properties of ion channels, such as gating, ion selectivity, single-channel conductance, multi-ion occupancy, transient blocking, desensitization, and inactivation, have been extensively studied by electrophysiologic techniques. The molecular bases for these functions are being gradually revealed by protein chemical methods, by genetic manipulation, and by direct structural determinations. We describe an approach to channels that combines chemical and genetic approaches. The substituted-cysteine accessibility method (SCAM)[1,2] can be used to identify all of the residues that line a channel, to size the channel, to determine differences in the structures of the channel in different functional states, to locate the gates and selectivity filters, and to map the electrostatic potential profile in the channel.

SCAM is based on certain assumptions. In membrane-embedded, channel proteins, we assume that the sulfhydryl group of a native or engineered cysteine residue (Cys) is in one of three environments: in the water-accessible surface, in the lipid-accessible surface, or in the protein interior. We assume that the channel lining is part of the water-accessible surface and further that in the membrane-spanning domain of the protein, the channel lining is the only water-accessible surface. We assume that hydrophilic, charged reagents will react much faster with sulfhydryls in the water-accessible surface than in the lipid-accessible surface or in the interior. Additionally, electrophilic reagents react much faster, in the case of thiosulfonates at least 5×10^9 faster,[3] with ionized $-S^-$ than with un-ionized –SH. In the lipid-accessible surface and in the protein interior, ionization of –SH is suppressed because of the low dielectric constant of the environment.

Thus, we mutate one at a time every residue in the membrane-embedded segments of a channel protein to Cys, express the mutants in heterologous cells, and determine whether these substituted Cys react with small, charged, sulfhydryl-specific reagents. In many cases, the reaction of a reagent with a Cys results in an irreversible alteration in the function of the channel. In

[1] M. H. Akabas, D. A. Stauffer, M. Xu, and A. Karlin, *Science* **258,** 307 (1992).
[2] M. H. Akabas, C. Kaufmann, P. Archdeacon, and A. Karlin, *Neuron* **13,** 919 (1994).
[3] D. D. Roberts, S. D. Lewis, D. P. Ballou, S. T. Olson, and J. A. Shafer, *Biochemistry* **25,** 5595 (1986).

0076-6879/98 $25.00

these cases, the reaction of the small quantity of mutant channel expressed at the surface of a cell can be readily detected electrophysiologically. If the function of a channel protein with a substituted Cys is irreversibly altered by exposure to the reagents, we assume that this effect is due to reaction with the substituted Cys. We infer that such a reactive Cys is exposed in the water-filled lumen of the channel.

As a background we require that the function of the wild-type channel not be irreversibly altered by the reagent. If it is, then a pseudo wild type must be engineered with the reactive native Cys suitably mutated to retain channel function. A susceptible, endogenous Cys need not be in the membrane-spanning domain of the channel protein, but if it is, then it provides a promising starting point for Cys substitution and the application of SCAM.

In probing a series of Cys mutants with a reagent, some will be affected and others will not. The lack of an effect could indicate the lack of a reaction. Alternatively, a reaction could have no detectable functional effect. Negatives have to be interpreted cautiously.

Only Cys-substitution mutants with near wild-type function are used, and these are assumed to have near-wild-type three-dimensional structure. No previously buried wild-type Cys are assumed to be revealed by the mutation. In the near wild-type structure of the mutant, we assume that the position of the substituted Cys side chain is nearly the same as that of the wild-type side chain, so that SCAM reports on the exposure of the wild-type side chain. In our combined experience of substituting Cys in the acetylcholine receptor (AChR), the $GABA_A$ receptor, and the cystic fibrosis transmembrane conductance regulator (CFTR), only 2 of more than 200 Cys-substitution mutants failed to yield functional channels on the cell surface. Mutation to cysteine is very well tolerated.[4] In a subsequent chemical modification, an accessible Cys side chain is "mutated" to an unnatural, longer and charged side chain, which is not well tolerated, and function is often significantly altered.

As a first approximation, we have classified a residue as either exposed or buried, based on the presence or absence of significant irreversible effects of relatively high concentrations of reagents. This approach, however, neglects fluctuations in protein structure and the possibility that even buried residues could react at appreciable rates. No Cys is likely to be completely unreactive. Thus, the rate constants for the reaction at consecutive residues should be determined and should be the basis for the classification of residues as exposed or buried. Reactivity, however, depends on factors in addition to simple accessibility to reagent; these include the acid dissociation of the Cys–SH, the steric constraints on the formation of an activated

[4] H. R. Kaback, *Proc. Natl. Acad. Sci. U.S.A.* **94,** 5539 (1997).

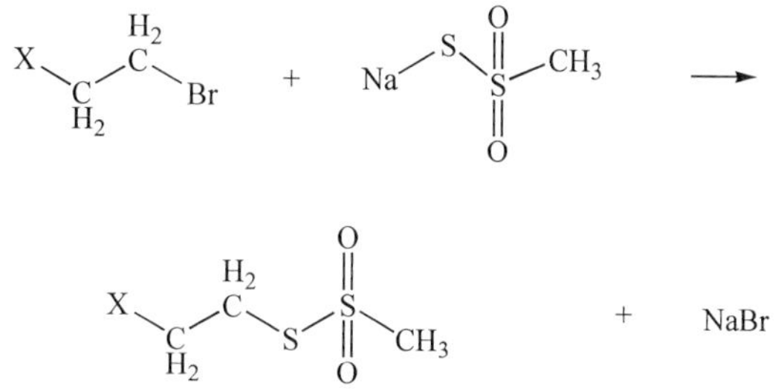

FIG. 1. Reaction scheme for the synthesis of MTS reagents. For the different reagents, X and the abbreviations are NH_3^+ (MTSEA), $(CH_3)_3N^+$ (MTSET), SO_3^- (MTSES), OH (MTSEH).

complex, and, for charged reagents, the electric field along the path to the residue. Some of these factors can be dissected out by comparing the rates of reaction of reagents with the same reaction mechanism that differ mainly in charge or that differ mainly in size. The thiosulfonate derivatives offer a set of reagents that can be so compared.

We describe the synthesis and properties of some thiosulfonate derivatives, and we also describe applications of SCAM to ion channels, to 1chtransport proteins, and to the membrane-embedded binding sites of G-protein-coupled receptors. The use of SCAM to study the voltage-induced conformational changes in the S4 region of voltage-dependent Na^+ channels is described in another chapter in this volume.[4a]

Sulfhydryl-Specific Reagents

The positively charged 2-aminoethylmethane thiosulfonate (MTSEA)[5] and 2-trimethylammonioethylmethane thiosulfonate (MTSET),[6] the negatively charged 2-sulfonatoethylmethane thiosulfonate (MTSES),[6] and the neutral 2-hydroxyethylmethane thiosulfonate (MTSEH)[6a] are a set of rapidly reacting, sulfhydryl-specific reagents that differ only in their headgroups and whose rates of reaction with a Cys-substitution mutant are readily compared. These reagents were synthesized according to the scheme shown in Fig. 1.

Synthesis of MTSEH

MTSEH, for example, is synthesized by dissolving 40 g of sodium methane thiosulfonate and 41 g of 2-bromoethanol (95%) in acetonitrile. The

[4a] R. Horn, *Methods Enzymol.* **293,** [9], 1998 (this volume).
[5] T. W. Bruice and G. L. Kenyon, *J. Prot. Chem.* **1,** 47 (1982).
[6] D. A. Stauffer and A. Karlin, *Biochemistry* **33,** 6840 (1994).
[6a] J. M. Pascual and A. Karlin, *J. Gen. Physiol.* (1998), in press.

stirred mixture under argon is heated overnight at reflux temperature. The mixture is cooled and filtered; the filtrate is concentrated, mixed with dichloromethane, filtered, and concentrated again to yield a yellow oil. One milliliter of the yellow oil is mixed with 9 ml of chloroform, and the small amount of precipitate that forms is removed by centrifugation in a clinical centrifuge. The supernatant is layered on a silica gel (Merck grade 9385, 230–400 mesh) column (32-cm length, 1.9-cm diameter), preequilibrated with chloroform (the commercial product stabilized with 0.75% ethanol). The column is eluted under mild pressure at 3–4 ml/min with 90 ml of chloroform and with 260 ml of 98% chloroform–2% methanol; the next 100 ml contains the pure product, as determined initially by thin-layer chromatography (TLC). Fractions are spotted on strips of silica on aluminum foil with fluorescent indicator and developed in 98% chloroform–2% methanol. The components are visualized first under short UV and then by spraying with a mixture containing 1 m*M* 5,5′-dithiobis-2-nitrobenzoate (DTNB), 0.5 m*M* dithiothreitol (DTT), and trimethylamine in methanol, in which the product gives a white spot against a yellow background. The 100 ml containing the product is reduced to about 10 ml on a rotary evaporator; 20 ml of dichloromethane is added, and the volume is reduced again, first by rotary evaporation and then on a high vaccum line with liquid nitrogen traps. About 0.6 g of liquid is recovered. By 5-thio-2-nitrobenzoate (TNB) assay (see later section), the average purity of three preparations is 98%. The NMR spectrum is consistent with the structure of MTSEH. Mass spectrometry, +FAB ionization, gave MH^+ 157.

MTSEA, MTSET, MTSES, and MTSEH are now commercially available (Toronto Research Chemicals, Inc., North York, Ontario, Canada).

Synthesis of AEAETS

A doubly charged thiosulfonate, 2-aminoethyl-2-aminoethane thiosulfonate (AEAETS)[6a,7,8] is synthesized as follows: to 10 g (0.044 mol) of cystamine dihydrochloride in 20 ml water in a 250-ml round-bottom flask, immersed in ice–water, is added dropwise over about 10 min with continuous stirring, 9.6 ml of 30% hydrogen peroxide (0.1 mol) diluted with 10 ml water. The stirred mixture is kept at 45° overnight and 2 days at room temperature. The volume is reduced on a rotary evaporator connected to a high vaccum pump through liquid nitrogen traps, with the reaction mixture at 47°. The resulting thick, yellow oil is dispersed in 250 ml of methanol. A white precipitate is removed by filtration. The volume of the filtrate is reduced by rotary evaporation to about 100 ml, and this solution is stored

[7] L. Field, T. C. Owen, R. R. Crenshaw, and A. W. Bryan, *J. Am. Chem. Soc.* **83,** 4414 (1961).
[8] L. Field, H. Harle, T. C. Owen, and A. Ferretti, *J. Organic Chem.* **29,** 1632 (1964).

overnight in the cold room. About 6 g of white precipitate is collected and is dried. The crude product is mixed with 600 ml of methanol at 50°, and the small amount of undissolved solid is removed by filtration. To the filtrate is added 240 ml of diethyl ether, and the mixture is kept at 4° for 2 days. A precipitate (about 3 g), which sticks to the flask, is collected and dried. This is dissolved in 300 ml of methanol at 50° and filtered. To the filtrate is added 120 ml of diethyl ether, and the mixture is kept at 4° for 2 days. Crystals (1.4 g) are collected and dried. By TLC on cellulose with fluorescent indicator, developed in 60% ethanol–30% 0.1 *N* HCl–10% *tert*-butanol, the product in methanol gives one ninhydrin-positive spot with an R_f of 0.26. TNB assay gives 95% purity based on a molecular weight of 256.9. The m.p. is 168°–170°. Mass spectrometry by direct probe electron impact with no solvent gives peaks of 256 and 258, corresponding to the compound with two $^{35}Cl^-$ and to the compound with one $^{35}Cl^-$ and one $^{37}Cl^-$. The NMR spectrum is consistent with the structure of AEAETS.

Assay of Thiosulfonates and Other Reagents That React with Sulfhydryls

The TNB assay for reagents that react with thiols is conveniently carried out as follows[6]: in a 1-ml cuvette is mixed 700 μl 100 m*M* Tris buffer (pH 8.0), 100 μl 0.3 m*M* DTT in water, 100 μl 0.6 m*M* DTNB in 100 m*M* Tris buffer (pH 8.0), and 100 μl of the reagent to be assayed, at a concentration of not more than approximately 0.5 m*M*. (DTNB is conveniently kept at 4° as a 10 m*M* stock solution in methanol.) The absorbance at 412 nm is read both before adding the reagent (A) to be assayed and after (A'), and the concentration of reagent (m) in the solution from which the aliquot of volume V_M is taken is calculated as

$$m = [AV_{\text{TNB}} - A'(V_{\text{TNB}} + V_{\text{M}})]/(\varepsilon V_{\text{M}}) \quad (1)$$

where, additionally, V_{TNB} is the volume in the cuvette before the addition of sample, and $\varepsilon = \varepsilon_{\text{TNB}} - \varepsilon_{\text{DTNB}}/2 = 13{,}700$. Thiosulfonate reagents should be assayed when received and approximately every 6 months thereafter. Note that disulfides do not give a net change in absorbance in the TNB assay.

Hydrolysis of Thiosulfonates

The MTS derivatives and AEAETS hydrolyze to a sulfenic acid (RSOH) and a sulfinic acid ($R'SO_2H$) (Fig. 2). The latter is a relatively strong acid, with a pK_a of 2.3 for R′ = CH_3.[5] The sulfenic acid disproportionates to form a thiol and another sulfinic acid.[9] Further reactions can take place

[9] J. P. Hendrickson, D. J. Cram, and G. S. Hammond, "Organic Chemistry." McGraw-Hill, New York, 1970.

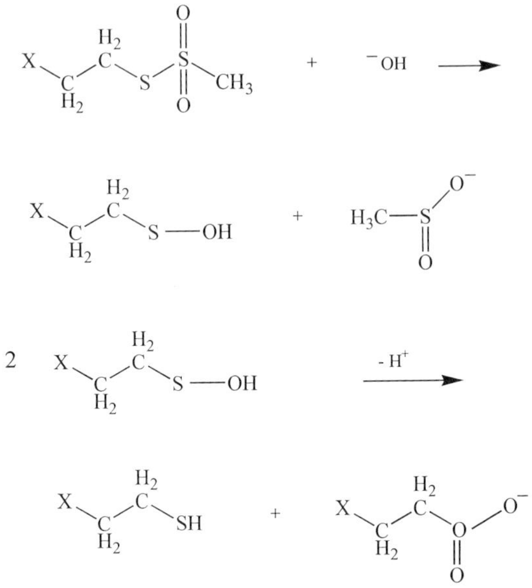

FIG. 2. Hydrolysis of methane thiosulfonates.

because both the parent thiosulfonate and the sulfenic acid (RSOH) can react with a thiol to form a disulfide. The reaction of the sulfenic acids with TNB led to underestimates of the rates of hydrolysis of the MTS reagents.[6] During the hydrolysis and subsequent reactions of the MTS reagents, the absorbance at 245 nm (A_{245}) first rises and then falls (Karlin, unpublished work). These data can be fit with a kinetic equation [Eq. (2)] describing the simplified sequential reactions, A → B → C, where A, B, and C are the components that have appreciable extinction coefficients, the thiosulfonate, the sulfenic acid, and a subsequent product.

$$\text{Absorbance} = a_0[\varepsilon_C + r\exp(-k_A t) + q\exp(-k_B t)] \tag{2}$$

where a_0 is the initial concentration of the thiosulfonate, ε_A, ε_B, and ε_C are the molar extinction coefficients of A, B, and C, and k_A and k_B are the rate constants for the first and second reactions, respectively,

$$r = \varepsilon_A + (\varepsilon_B k_A - \varepsilon_C k_B)/(k_B - k_A)$$

and

$$q = (\varepsilon_C - \varepsilon_B)k_A/(k_B - k_A)$$

TABLE I
RATE CONSTANTS FOR HYDROLYSIS OF THIOSULFONATE DERIVATIVES[a]

Reagent	pH	Temperature	$(10^4)k_A$ (1/sec)	$t_{1/2}$ (min)	$(10^4)k_B$ (1/sec)	$t_{1/2}$ (min)
MTSEA	7.0	20.0	9.9 ± 0.4	12.0	2.3 ± 0.4	50.0
MTSEA	6.0	20.0	1.25 ± 0.03	92.0	0.49 ± 0.20	236.0
MTSEA	7.0	4.0	1.00 ± 0.01	116.0	0.13 ± 0.003	890.0
MTSET	7.0	20.0	10.3 ± 1.7	11.2	12.3 ± 1.2	9.4
MTSET	6.0	20.0	2.1 ± 0.11	55.0	2.1 ± 0.11	55.0
MTSES	7.0	20.0	0.31 ± 0.04	370.0	0.31 ± 0.04	370.0
AEAETS	7.0	20.0	18 ± 0.5	6.4	25 ± 2	4.6

[a] Determined in 190 m*M* $NaPO_4$ buffer (Karlin, unpublished).

When k_A equals k_B,

$$\text{Absorbance} = a_0[\varepsilon_C + (\varepsilon_A - \varepsilon_C)\exp(-kt) + (\varepsilon_B - \varepsilon_C)kt\exp(-kt)] \quad (3)$$

These equations are fit to the data by estimating ε_A and ε_C from the initial and final absorbance and letting ε_B, k_A and k_B be free parameters (Table I).

The rates of hydrolysis at neutral pH dictate that these reagents should be dissolved in neutral buffer only immediately before use. A convenient procedure is to make a concentrated solution in unbuffered water, to store this on ice (up to a few hours), and to dilute to the desired concentration in buffer just before use.

Rate Constants for Reactions with 2-Mercaptoethanol

The rate constants for the reactions of the thiosulfonates with small thiols (Fig. 3) in solution are used to normalize the rate constants for the reactions of these reagents with proteins in solution and in membranes in an approach to the electrostatic properties around the sites of reaction.[6,6a,10–12] It is the ionized $-S^-$ form of 2-mercaptoethanol (2-ME) that reacts with thiosulfonates,[3] but otherwise 2-ME is uncharged and serves as a model small molecule. Rate constants were determined in a Hi-Tech Scientific PQ/SF-53 (Salisbury, England) rapid-kinetics spectrophotometer, as previously described.[6] The rate constants shown in Table II were deter-

[10] M. Cheung and M. K. Akabas, *J. Gen. Physiol.* **109,** 289 (1997).
[11] M. Holmgren, P. L. Smith, and G. Yellen, *J. Gen. Physiol.* **109,** 527 (1997).
[12] N. Yang, A. L. George, and R. Horn, *Biophys. J.* **73,** 2260 (1997).

TABLE II

RATE CONSTANTS FOR REACTIONS OF THIOSULFONATES WITH 2-MERCAPTOETHANOL[a]

Constant	MTSEH	MTSES	MTSEA	MTSET	AEAETS
$(10^{-4})k_X$ (1/*M*/sec)	0.95 ± 0.02	1.7 ± 0.1	7.6 ± 0.4	21.2 ± 0.8	26.1 ± 0.6
k_X/k_{MTSEH}	1	1.8	8	22	27

[a] pH 7.0, ionic strength 129 m*M*, 20°.

mined at 20° in 58 m*M* $NaPO_4$–0.1 m*M* EDTA, pH 7.0, calculated to have the same ionic strength (129 m*M*) as oocyte bath solution.[6a]

Characteristics of Thiosulfonates Relevant to Their Reactions in Channels

In the all-*trans* configuration, MTSEA, MTSET, and MTSES, each fits into a cylinder, 6 Å in diameter and 10 Å in length. The amine head of MTSEA, however, is smaller than the trimethylammonium head of MTSET and the sulfonate head of MTSES; and in the AChR channel, MTSEA reacted with a larger set of substituted Cys than the other two.[2] MTSEH is the same size as MTSEA. In the all-*trans* configuration AEAETS fits into a cylinder 6 Å in diameter and 13 Å in length, and the nitrogens are about 10 Å apart. Reversal-potential measurements with MTSEA or MTSET replacing nearly all external Na^+ indicated that these reagents have significant permeability relative to Na^+ through the open AChR channel expressed in oocytes.[2] Although the conductances of these reagents through the open channel have not been measured, the amplitudes of the macroscopic current indicated that the conductances are not much less than that of Na^+. Other channels that were largely accessible to the MTS reagents include the $GABA_A$ receptor channel,[13] the cyclic nucleotide-gated channel,[14] the N-methyl-D-aspartate (NMDA) receptor channel,[15] and the CFTR channel.[16] Access of MTS reagents to the lumen of voltage-gated K^+ channels is more limited.[17–19]

MTSEA, but not MTSET or MTSES, readily permeated into phospholipid vesicles and crossed cell membranes, both in intact cells and in excised

[13] M. Xu and M. H. Akabas, *J. Gen. Physiol.* **107,** 195 (1996).

[14] Z. P. Sun, M. H. Akabas, E. H. Goulding, A. Karlin, and S. A. Siegelbaum, *Neuron* **16,** 141 (1996).

[15] T. Kuner, L. P. Wollmuth, A. Karlin, P. H. Seeburg, and B. Sakmann, *Neuron* **17,** 343 (1996).

[16] M. Cheung and M. H. Akabas, *Biophys. J.* **70,** 2688 (1996).

[17] J. M. Pascual, C. C. Shieh, G. E. Kirsch, and A. M. Brown, *Neuron* **14,** 1055 (1995).

[18] L. L. Kurz, R. D. Zuhlke, H. J. Zhang, and R. H. Joho, *Biophys. J.* **68,** 900 (1995).

[19] A. Gross and R. MacKinnon, *Neuron* **16,** 399 (1996).

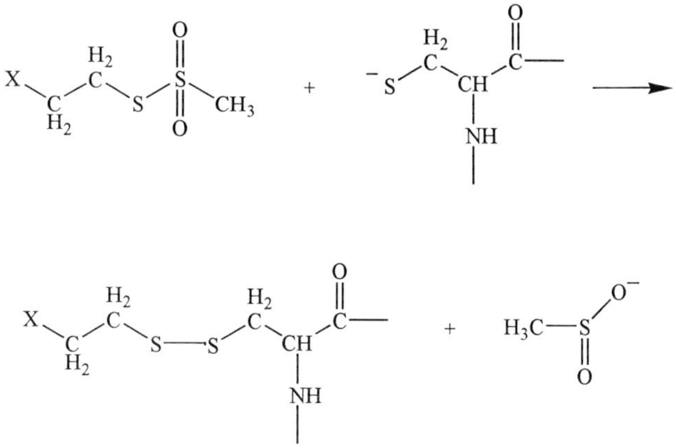

FIG. 3. Reaction of methane thiosulfonates with a cysteinyl residue.

patches, to react with cysteines on the trans side.[20] Two related characteristics of the thiosulfonates are relevant to their permeation through membranes, their charge, and their partition coefficient between water and nonpolar solvent. MTSET is permanently positively charged. The charge on MTSEA depends on the pH of the medium and the pK_a of the amine. The pK_a is difficult to determine because of the rapid hydrolysis of MTSEA at alkaline pH, but it appears to be 8.5 or greater (Karlin, unpublished). No doubt, it is the deprotonated amine that is responsible for the rapid permeation of MTSEA through the bilayer. The protonated form of MTSEA is as lipophobic as the permanently charged MTSET: the partition coefficients of MTSEA hydrobromide and of MTSET bromide between unbuffered water and *n*-octanol are similar (Table III). Also, MTSES is completely deprotonated and negatively charged at neutral pH and strongly prefers water. The uncharged MTSEH and methylmethane thiosulfonate prefer water only slightly more than *n*-octanol (dielectric constant 10.3), although considerably more than hexane (dielectric constant 1.89), and both reagents would be expected to permeate through membranes rapidly.

This lipid permeability of MTSEA and of completely uncharged MTS reagents limits their use as a sidedness reagent. The reaction due to reagent that has reached the bath on the *trans* side, however, can bequenched by the addition of an impermeant thiol, like cysteine itself, to the trans side.[20] This can be done in excised patches or in whole-cell patches by inclusion of 10 mM cysteine or glutathione in the patch pipette.

[20] M. Holmgren, Y. Liu, Y. Xu, and G. Yellen, *Neuropharmacol.* **35,** 797 (1996).

TABLE III
PARTITION COEFFICIENTS OF METHANE THIOSULFONATES[a]

Solvent	MMTS	MTSEH	MTSEA(H^+ Br^{2-})	MTSET$^+$(Br^-)	MTSES$^+$ (Na^+)
Water/*n*-octanol	1.3	3.2	400	690	2500
Water/hexanes	14	720	43,000	Not done	Not done

[a] At 22° to 25° (Karlin, unpublished).

Other Reagents

p-Chloromercuribenzene sulfonate (PCMBS),[13,21,22] Ag^+,[23] Hg^{2+},[24] and *N*-(6-phosphonyl-*n*-hexyl)maleimide[25] have been used to probe the accessibility and sidedness of Cys in membrane-spanning segments and in extramembranous domains of channels and other intrinsic membrane proteins. Cysteinyl residues have been tagged with spin labels[26] or with fluorescent labels,[25] and the ability of polar molecules to quench the ESR (electron spin resonance) signal or the fluorescence has been used to determine the accessibility of the attached labels.

Screening for Accessibility

SCAM involves several steps. Probe reagents must be selected. A functional assay that is likely to be sensitive to the chemical modifications of exposed Cys must be developed. The susceptibility of the wild-type protein to the reagents must be determined, and if endogenous Cys are reactive, they must be mutated to tolerated substitutes, generating a pseudo wild-type protein for further mutation. Consecutive residues in membrane-spanning segments of the target protein (pseudo wild type if necessary) must be mutated to Cys. These mutants must be expressed. The function of the mutants must be assayed. The reactivity of each mutant toward the probe reagents must be assayed, if possible, in the different functional states of the protein.

Expression

Wild-type and mutant target proteins have been expressed in a variety of cell types. The requirement is that the cell does not express the target

[21] R. T. Yan and P. C. Maloney, *Cell* **75,** 37 (1993).
[22] M. Xu, D. F. Covey, and M. H. Akabas, *Biophys. J.* **69,** 1858 (1995).
[23] Q. Lu and C. Miller, *Science* **268,** 304 (1995).
[24] G. M. Preston, J. S. Jung, W. B. Guggino, and P. Agre, *J. Biol. Chem.* **268,** 17 (1993).
[25] J. J. Falke, A. F. Dernburg, D. A. Sternberg, N. Zalkin, D. L. Milligan, and D. E. Koshland, *J. Biol. Chem.* **263,** 14850 (1988).
[26] W. L. Hubbell, H. S. McHaourab, C. Altenbach, and M. A. Lietzow, *Structure* **4,** 779 (1996).

protein endogenously. *Xenopus* oocytes are convenient both for expression and for electrophysiologic assay by two-electrode voltage-clamp techniques.[27] HEK 293 cells have proven to be a convenient expression system for patch-clamp assay of function[28] and for binding assays.[29] The cells are transfected with the mutant genes in an appropriate vector.

Assays for Reaction

For ion channels and transporters, changes in current or in transport rate offer convenient and sensitive assays.[1,4,14,15,17,18,30–32] For G-protein-coupled receptors, reactions with Cys exposed in the ligand-binding crevice have been detected by the irreversible effects on ligand binding.[29]

Initially, we anticipated that attaching charged groups to Cys in conducting pathways would result in a decrease in conductance and inhibition of voltage-clamp current. The reality is more complicated. We observed inhibition at some positions and potentiation at others. We analyzed at the macroscopic level one reaction in the AChR α-subunit second membrane-spanning (αM2) segment that resulted in potentiation. The potentiation is due to a shift in the EC_{50} for ACh to lower concentrations, but the maximum current also decreased.[2] The reaction of MTSEA with two other Cys-substitution mutants in βM2 is studied at the single-channel level: the reaction of MTSEA with βG255C results in a large decrease in the single-channel conductance with only small changes in gating kinetics, while the reaction with βV266C results in a large increase in the mean open time and the appearance of subconductance states (Zhang and Karlin, 1998, *Biochemistry,* in press). As a rule, every reaction is likely to result in some change in gating kinetics, and in some cases this is the entire effect. The result can be inhibition or potentiation of the macroscopic response. In order for an assay to be sensitive to changes in gating kinetics, as well as changes in conductance, the test stimulus should be less than saturating. In the case of ligand-gated ion channels, the optimal test concentration of the agonist is the EC_{50}.[33]

A direct chemical assay for the reaction of the probe reagent would eliminate the ambiguity inherent in a functional assay. A chemical assay would require the isolation of the channel protein after the reaction, the

[27] B. Rudy and L. E. Iverson, *Methods Enzymol.* **207,** 225 (1992).

[28] J. Chen, Y. Zhang, G. Akk, S. Sine, and A. Auerbach, *Biophys. J.* **69,** 849 (1995).

[29] J. A. Javitch, D. Fu, J. Chen, and A. Karlin, *Neuron* **14,** 825 (1995).

[30] M. Xu and M. H. Akabas, *J. Biol. Chem.* **268,** 21505 (1993).

[31] J. A. Mindell, H. Zhan, P. D. Huynh, R. J. Collier, and A. Finkelstein, *Proc. Natl. Acad. Sci. U.S.A.* **91,** 5272 (1994).

[32] S. Chen, H. A. Hartmann, and G. E. Kirsch, *J. Membr. Biol.* **155,** 11 (1997).

[33] H. Zhang and A. Karlin, *Biochemistry* **36,** 15856 (1997).

determination of the extent of reaction by some means, and the subtraction of the extent of (functionally silent) reaction with wild type. An oocyte might express a femtomole of channel protein on its surface. Changes in the functional properties of this quantity of channel protein are easily assayed, but the extent of reaction with a femtomole of a specific Cys is difficult to assay chemically, even if the channel protein could be isolated efficiently. (A tritiated reagent might yield 40 cpm per fmol; a ^{125}I-iodinated reagent might yield 1000 cpm per fmol.) Practically, a biochemical assay would require many oocytes or plates of cells.

Endogenous Cysteines

If application of the sulfhydryl reagents to wild type does not have a detectable effect, then either the endogenous cysteines are inaccessible to the reagents or their modification has no functional effect. If the number of endogenous cysteine residues is small and if mutating them does not alter the function of the protein, then substituting them by Ser or Ala is the safest course. This obviates the possibility that the introduction of a Cys might cause an initially inaccessible endogenous Cys to be susceptible to modification.

If application of the sulfhydryl reagents does cause a functional change in the wild-type protein then the accessible endogenous cysteine(s) residues must be identified and a suitable residue must be substituted into that position.[21,34,35] It is important to keep in mind that more than one endogenous cysteine might be accessible and that the reactions of two Cys could, in the worst case, have cancelling effects on the assayed function. In the dopamine transporter, removal of one endogenous Cys uncovered the functional effect of the reaction of a second endogenous Cys with MTSEA (Javitch, unpublished).

Mutagenesis

In a suitable background construct, one must generate a large number of Cys-substitution mutants (20–25 per membrane-spanning segment). We have used two mutagenesis procedures, altered-sites mutagenesis (Promega, Madison, WI) and unique-site elimination mutagenesis[36] (Clontech, Palo Alto, CA). The altered-sites procedure has been successfully used

[34] J. A. Javitch, X. Li, J. Kaback, and A. Karlin, *Proc. Natl. Acad. Sci. U.S.A.* **91,** 10355 (1994).
[35] Y. Olami, A. Rimon, Y. Gerchman, A. Rothman, and E. Padan, *J. Biol. Chem.* **272,** 1761 (1997).
[36] W. P. Deng and J. A. Nickoloff, *Anal. Biochem.* **200,** 81 (1992).

with a variety of proteins, AChR, $GABA_A$ receptor, CFTR, and the dopamine receptor. The success rate for obtaining mutants is generally about 90% on the first try; however, it requires an extra subcloning step in order to transfer the mutant from the pAlter plasmid to a suitable expression plasmid. The unique-site elimination procedure was used successfully with some cDNAs, but with others it resulted in a large number of deletion products.

Screening Conditions

First, one should determine which Cys are reactive enough to be considered accessible. We recommend that MTSEA and MTSES be used in the initial screen. We found that MTSEA had the widest reactivity, related no doubt to its high rate of reaction with thiols (Table II) and to its relatively small size. In anion-conducting pathways, residues that react with the cationic MTSEA and not with MTSES should be tested also with the anionic mercurial, PCMBS.[13] We have screened by applying 2.5 m*M* MTSEA for 1 min and for 5 min: for a reaction with a rate constant of 1 $M^{-1}sec^{-1}$, 5 min is the half-time for reaction of 2.5 m*M* reagent. MTSES, which reacts with 2-ME about four times slower than MTSEA, should be applied at 10 m*M*. A reasonable test concentration for PCMBS is 0.5 m*M*.

At some positions, the maximum effect of complete reaction may be small, and many repetitions may be required to obtain reliable results. For each reagent, the effects on all of the mutants should be compared to the effect on wild type by an analysis of variance. The number of residues that are found to be significantly affected depends on the statistical test chosen; small effects fall on one side or the other of statistical significance, depending on the strictness of the test.[37]

Channel Lining

The side chains of the residues that are exposed in the channel contact water and conducted ions. In the wider parts of the channel, the ions are completely hydrated, but in the narrower parts of the channel, where the ions are partially dehydrated,[38] the side chains contact the ion itself. These side chains are likely to play key roles in the magnitude and the selectivity of ion conduction. The MTS reagents are small enough to permeate into regions of the channel that are at about 6 Å in diameter, too narrow for a fully hydrated ion, and therefore potentially MTS reagents can react with

[37] M. H. Akabas and A. Karlin, *Biochemistry* **34,** 12496 (1995).

[38] B. Hille, "Ionic Channels of Excitable Membranes." Sinauer Associates, Sunderland, Massachusetts, 1992.

the Cys substituted for these key residues. For example, MTSEA reacts with αT244C and αE241C,[2] residues in the narrowest part of the AChR channel that play key roles in selectivity and conductance of this receptor.[39–41] In the wider part of the AChR channel, three of six residues identified by SCAM as exposed in the open state are two Leu and a Val; the hydrophobicity of these residues might serve to weaken interaction between conducted ions and this part of the channel lining.

The reaction of a substituted Cys in the channel with any of the reagents, MTSEA, MTSET, MTSES, or PCMBS, sets a lower limit of about 6 Å for the diameter of the channel in the vicinity of that Cys.

Secondary Structure

If the channel lumen is surrounded by membrane-spanning segments, each with a regular secondary structure, then in each segment a stripe of residues would be exposed in the channel. If the segments were ideal α helices, with 3.6 residues per turn, then every third or fourth residue should be exposed in the channel, with adjacent residues exposed depending on the width of the accessible stripe. Multiple α helices usually associate as coiled coils. In the more common, left-handed coiled coils, the helices cross at an angle of $-20°$, and the pitch is about 3.5 residues per turn. In right-handed coiled coils, the crossing angle is about 40° and the pitch is 3.9–4.0 residues per turn. If the segments were β strands, then every other residue might be exposed. The accessibility of residues in the entire second membrane-spanning segment (M2) of the AChR α subunit was consistent with an α helix in the open state, but an interrupted α helix in the closed state.[2] The accessible stripes in the $GABA_A$ receptor,[13] CFTR,[42] and dopamine D2 receptor[29] were also consistent with an α-helical conformation. By contrast, the accessibility of residues in the M1 segment of the AChR,[37] of the P-loop of cyclic nucleotide gated channels,[14] voltage-dependent K^+ channels,[17,18,23] Na^+ channels,[32,43–45] and the M2 segment of the

[39] K. Imoto, C. Busch, B. Sakmann, M. Mishina, T. Konno, J. Nakai, H. Bujo, Y. Mori, K. Fukuda, and S. Numa, *Nature* **335,** 645 (1988).

[40] A. Villarroel, S. Herlitze, M. Koenen, and B. Sakmann, *Proc. R. Soc. Lond.* (*B*) **243,** 69 (1991).

[41] B. N. Cohen, C. Labarca, L. Czyzyk, N. Davidson, and H. A. Lester, *J. Gen. Physiol.* **99,** 545 (1992).

[42] M. H. Akabas, C. Kaufmann, T. A. Cook, and P. Archdeacon, *J. Biol. Chem.* **269,** 14865 (1994).

[43] M. T. Perez-Garcia, N. Chiamvimonvat, E. Marban, and G. F. Tomaselli, *Proc. Natl. Acad. Sci. U.S.A.* **93,** 300 (1996).

[44] N. Chiamvimonvat, M. T. Perez-Garcia, R. Ranjan, E. Marban, and G. F. Tomaselli, *Neuron* **16,** 1037 (1996).

[45] J. P. Benitah, G. F. Tomaselli, and E. Marban, *Proc. Natl. Acad. Sci. U.S.A.* **93,** 7392 (1996).

NMDA receptor[15] were not consistent with a single, regular secondary structure.

The pattern of exposure of Cys to reagents may correctly report the secondary structure; however, this pattern is subject to distortions due to the Cys substitution itself, to the confounding effects of multiple subunits with possibly nonidentical conformations, and to structural fluctuations of the structure during the reaction times that are long relative to the frequency of fluctuations.

Reaction Rates

The initial screen identifies positions that are susceptible to the probing reagents. The Cys at these positions should be further characterized by the rate constants of their reaction with the reagents.[6a] These rate constants can be used to extract additional information about the channel. The rate constants depend on a number of factors including (1) the permeability of the pathway to the substituted Cys, (2) for charged reagents, the electrostatic potential along the pathway and at the residue, (3) the acid dissociation of the Cys–SH, and (4) the local steric constraints on the formation of an activated complex between the reagent and the Cys. Some of these factors can be dissected out by comparing the rates of reaction of reagents that differ mainly in charge or that differ mainly in size but that react with the same reaction mechanism. The influence of the electric field on the reaction rates can be examined by carrying out the reaction at different transmembrane electrostatic potentials. At zero transmembrane potential, the intrinsic electrostatic potential at a substituted Cys can be estimated by comparing the rate constants for the reactions of reagents that differ in charge. Insights into the structural differences between different functional states of a channel can be obtained by the comparison of the rate constants in these different states.

State Dependence of Accessibility

In screening Cys mutants in different functional states, we found evidence for large differences in rates of reaction at certain positions in the AChR[1,2,37] and in the $GABA_A$ receptor.[13,30] Some of these differences have been quantitated by determination of the rate constants.[6a,33,46,47] In the absence of ligand the channel is predominantly in the closed state, notwithstanding that mutation of some residues changes the spontaneous open

[46] J. M. Pascual and A. Karlin, *Soc. Neurosci. Abstracts* **23,** 388 (1997).
[47] G. G. Wilson and A. Karlin, *Neuron* (1998), in press.

probability of the channel.[28] In the presence of ligand, the channel undergoes transitions between the open, desensitized, and closed states; once again, the proportion of time spent in a particular state may be affected by the Cys mutation. In addition, rate constants for the reactions of MTS reagents with Cys in different functional states have been measured in the K^+ channel[11,48] and in the Na^+ channel.[49–51] As stated earlier, the rate constants can depend on a number of factors other than accessibility.

Location of Channel Gates

The strategy to locate a gate is to identify, as near as possible to the closed gate, the channel-lining residues on its two sides.[47] In the closed state, only residues on the extracellular side of the gate should be accessible to charged reagents added to the extracellular medium, and only residues on the intracellular side of the gate should be accessible to charged reagents added to the intracellular medium. Thus, the region of the closed gate is bracketed between two sets of residues. Furthermore, a reagent that is conducted by the channel should react also with residues on the far side of the gate when it is open. In the AChR, we observed that MTS reagents, applied extracellularly in the absence of agonist, were able to react with residues as close to the cytoplasmic end of the channel as αT244C[2] and as the aligned βG255C.[52] We inferred that the gate was at least as close to the cytoplasmic end of the channel as these residues. Similarly in the homologous $GABA_A$ receptor, the residue aligned with αThr-244, α_1Val-257, was accessible to $PCMBS^-$ in the closed state of this channel, implying that the gate was located at a similar cytoplasmic position.[13,22] The results of the application of the MTS reagents from both the extracellular and the intracellular sides of the membrane in whole-cell, patch-clamped cells expressing AChR were consistent with the location of the activation gate in the region of the channel between αE241 and αT244.[47] A related approach has also been used to locate a gate in the voltage-dependent K^+ channel.[11]

Two complications of this analysis are the difficulty in defining the functional state of the channel. For example, the AChR opens spontaneously in the absence of ligand, and mutation can increase this tendency. Nevertheless, the open probability is orders of magnitude smaller in the

[48] H. P. Larsson, O. S. Baker, D. S. Dhillon, and E. Y. Isacoff, *Neuron* **16,** 387 (1996).

[49] N. Yang and R. Horn, *Neuron* **15,** 213 (1995).

[50] N. Yang, A. L. George, Jr., and R. Horn, *Neuron* **16,** 113 (1996).

[51] S. Kellenberger, T. Scheuer, and W. A. Catterall, *J. Biol. Chem.* **271,** 30971 (1996).

[52] H. Zhang and A. Karlin, *Biochemistry* **37,** 1998, in press.

absence of ACh than in its presence. Furthermore, in the presence of ACh, the receptor enters another closed state, the desensitized state. The rates of reaction should be determined in the predominantly closed states of the channel (no ligand, and long-time exposure to ligand) and compared with the rates in the predominantly conducting state in the presence of ACh. By comparing the rates, the accessibility in the different states can be clarified. Also, MTSEA (but not MTSET or MTSES) readily crosses lipid membranes.[20] To use MTSEA as a sidedness reagent, as required to locate the closed gate, a quencher such as free cysteine or glutathione should be added to the side opposite the side of addition of MTSEA.[20,47] These thiols will rapidly react with free MTS reagent.

Location of Channel Blocker Binding Sites

The protection of accessible Cys in the channel by putative channel blockers can be used to identify the binding site for the blocker in the channel. In a channel, protection from modification implies that the protected cysteine residue either is at, or is distal to, the blocker binding site. (Allosteric effects of blockers are hard to exclude definitively, but if the blocker acts as a voltage-dependent open channel blocker and the Cys is in the channel, it is likely that the protection by the blocker is direct.) The picrotoxin binding site was located near the cytoplasmic end of the $GABA_A$ receptor channel,[22] and the binding site for agitoxin was mapped in K^+ channels.[19] In AChR, the binding site for voltage-dependent, open channel blocker QX-314 was identified as lying in the region of the channel lining between αSer-248 and αVal-255,[53] similar to the position inferred from the effects of mutations.[54] Closer to the intracellular end of the channel, αT244C was also protected against extracellular MTSEA by QX-314.

Electrostatics and Rate Constants

For charged reagents, the rates of reactions may depend both on the transmembrane electrostatic potential and intrinsic electrostatic potential due to the charges of the protein itself. A theoretical framework[6a,33] for analyzing these rate constants is briefly described next.

We consider the reaction of a reagent, X, maintained at a fixed concentration outside the membrane, with a single site or atom, S, in a membrane-

[53] J. M. Pascual and A. Karlin, *Biophys. J.* **72,** A151 (1997).

[54] R. J. Leonard, C. G. Labarca, P. Charnet, N. Davidson, and H. A. Lester, *Science* **242,** 1578 (1988).

spanning channel. We take the site to be in a free-energy well with a barrier on either side. The kinetic steps are

$$\begin{array}{c} X_{EX} \underset{k_{-1}}{\overset{k_1}{\rightleftharpoons}} X_S \underset{k_{-2}}{\overset{k_2}{\rightleftharpoons}} X_{IN} \\ + \\ S \\ \downarrow k_S \\ SX \end{array} \tag{4}$$

where X_{EX} is the reagent in the extracellular medium, X_{IN}, in the intracellular medium, and X_S, in the channel at the site of reaction; k_1, k_{-1}, k_2, and k_{-2} are the first-order rate constants for the movements of X, from the extracellular medium to S, from S to the extracellular medium, from S to the intracellular medium, and from the intracellular medium to S, respectively; k_S is the second-order rate constant for the reaction of X_S and S. In the following, the concentrations of X and S are represented by x and s.

The solutions of the kinetic equations yield for the concentration of X at the site of reaction,

$$x_S = (a/b)[1 - \exp(-bt)] \tag{5}$$

where $a = k_1 x_{EX} + k_{-2} x_{IN}$, and $b = k_{-1} + k_2$, and the fraction of unreacted S,

$$s/s_0 = \exp\{-(ak_S/b)[t - (1/b)(1 - \exp(-bt))]\} \tag{6}$$

(Note that the concentration of X_S in any single channel after S in that channel has reacted is not relevant to the reaction in any other channel.)

Applying absolute reaction rate theory to the movement of X to and from the site, we obtain expressions for the rate constants that depend on free-energy differences between ground states and transition states at the peaks of the barriers.[55,56] The advantages and shortcomings of this approach to ion permeation in channels has been discussed.[57] For X with charge z, these free energies contain long-range electrostatic terms that depend on z. We explicitly state the long-range electrostatic energy terms and lump together all other free-energy terms with the frequency factors. Thus,

$$k_1 = g_1 \exp[-z\beta(\delta\psi_M/2 + \psi_1)] \tag{7}$$
$$k_{-1} = g_{-1} \exp[-z\beta(-\delta\psi_M/2 + \psi_1 - \psi_S)] \tag{8}$$
$$k_2 = g_2 \exp\{-z\beta[(1 - \delta)\psi_M/2 + \psi_2 - \psi_S)]\} \tag{9}$$
$$k_{-2} = g_{-2} \exp\{-z\beta[-(1 - \delta)\psi_M/2 + \psi_2]\} \tag{10}$$

[55] A. N. Woodhull, *J. Gen. Physiol.* **61,** 687 (1973).
[56] B. Hille, *J. Gen. Physiol.* **66,** 535 (1975).
[57] J. A. Dani and D. G. Levitt, *J. Theor. Biol.* **146,** 289 (1990).

The g_1, g_{-1}, g_2, and g_{-2} are each the product of a constant and $\exp(-\Delta G^{\ddagger}_{NE}/RT)$. The $\Delta G^{\ddagger}_{NE}$ are the non-long-range electrostatic standard free-energy differences, $G^{\ddagger}_{1,NE} - G^{\ddagger}_{EX,NE}$, $G^{\ddagger}_{1,NE} - G^{\ddagger}_{S,NE}$, $G^{\ddagger}_{2,NE} - G^{\ddagger}_{S,NE}$, and $G^{\ddagger}_{2,NE} - G^{\ddagger}_{IN,NE}$, respectively, where the subscripts EX, IN, 1, and 2 refer to the extracellular medium, the intracellular medium, barrier 1, and barrier 2, respectively, and NE indicates non-long-range electrostatic; z is the algebraic charge of X, $\beta = F/RT$, δ is the electrical distance to the site S, ψ_M is the membrane electrostatic potential difference, intracellular minus extracellular, and ψ_1, ψ_2, and ψ_S, are the intrinsic electrostatic potentials due to charges in the channel protein, at barrier 1, barrier 2, and S, respectively, relative to the potential (due to the protein charges) in the extracellular or intracellular medium at a distance from the membrane of many protein lengths. Following Woodhull,[55] the electrical distances from the extracellular medium to the barriers are $\delta/2$ for barrier 1 and $(1 + \delta)/2$ for barrier 2. We assume that k_S is independent of ψ_M and ψ_S; that is, that any separation of charge that might occur in the activated complex between X and S is over too short a distance to be influenced by the gradients in ψ_M and ψ_S at S.

Hereafter, we consider only conditions in which $x_{IN} = 0$ and, therefore, in which $a = k_1x_{EX}$, and k_{-2} does not enter the equations.

Experimentally, s/s_0 is estimated from the current, I, elicited by ACh, before and after the reaction with X as

$$s/s_0 = (I_{inf} - I)/(I_{inf} - I_0) \tag{11}$$

where I_{inf} is the current after reaction of all channels and I_0 is the current before any reaction. Given the number and precision of the data, it is not practical to fit them with Eq. (6); rather we fit them with the simpler exponential decay equation

$$s/s_0 = \exp(-k^*x_{EX}t) \tag{12}$$

where k^* is the effective second-order rate constant and x_{EX} is the fixed extracellular concentration of reagent.

We approximate Eq. (6) by Eq. (12) with

$$k^* = k_Sk_1/(k_{-1} + k_2) \tag{13}$$

The relative error in this approximation can be estimated as

$$\varepsilon = (\tau - \tau^*)/\tau^* \tag{14}$$

where τ is the time at which $s/s_0 = 1/e$ according to exact Eq. (6) and τ^* is the time at which $s/s_0 = 1/e$ according to the approximate Eq. (12).

$$\varepsilon = \leq k_Sk_1x_{EX}/(k_{-1} + k_2)^2 \tag{15}$$

The smaller x_{EX}, the smaller the error. The approximation is good when x_S reaches a steady state before much reaction of S has occurred.

To examine the dependence of k^* on ψ_M, we rewrite Eqs. (7)–(9):

$$k_1 = k_1{}^0 \exp(-z\beta\delta\psi_M/2) \tag{16}$$
$$k_{-1} = k_{-1}{}^0 \exp(z\beta\delta\psi_M/2) \tag{17}$$
$$k_2 = k_2{}^0 \exp[-z\beta(1-\delta)\psi_M/2] \tag{18}$$

where $k_1{}^0$, $k_{-1}{}^0$, and $k_2{}^0$ are the rate constants at $\psi_M = 0$. Substituting Eqs. (16)–(18) into Eq. (13), we obtain

$$k^* = k_S k_1{}^0 \exp(-z\beta\delta\psi_M)/[k_{-1}{}^0 + k_2{}^0 \exp(-z\beta\psi_M/2)] \tag{19}$$

We obtain δ from the slope of $\ln(k^*)$ as a function of ψ_M.

$$d\ln(k^*)/d\psi_M = -z\beta\{\delta - (1/2)k_2{}^0/[k_2{}^0 + k_{-1}{}^0 \exp(z\beta\psi_M/2)]\} \tag{20}$$

The derivative is equal to $-z\beta\delta$ only when $k_2{}^0 = 0$; that is, when X can jump from the extracellular medium to S but not from S to the intracellular medium—when X is impermeant. For a permeant, positively charged X, which, for example, leaves S at the same rate to either side of the membrane ($k_2{}^0 = k_{-1}{}^0$), the slope would be small and negative or even positive (apparent $\delta < 0$) for large, negative ψ_M and larger and negative (apparent $\delta >$ 0) for large, positive ψ_M. A constant slope is characteristic either of an impermeant reagent ($k_2{}^0 = 0$) or of the condition, $k_2{}^0 \gg k_{-1}{}^0$ (in which case the apparent $\delta = \delta - 0.5$). Experimentally, the slope of $(1/\beta)\ln(k^*)$ as a function of ψ_M is taken as the apparent $z\delta$.

Under certain conditions, the intrinsic electrostatic potential at S, ψ_S, can be estimated from the rate constants. Following Stauffer and Karlin,[6] we take the ratio of the rate constants, ${}^1k^*$ and ${}^2k^*$, for two reagents, ^{1}X and ^{2}X, with unequal charges z_1 and z_2. Equations (19) and (7) and (8) and (9) give

$$\begin{aligned}{}^1k^* = {}^1k_S\,{}^1g_1 \exp[-z_1\beta(\delta\psi_M/2 + \psi_1)]/\{{}^1g_{-1} \exp[-z_1\beta(-\delta\psi_M/2 + \psi_1 - \psi_S)] \\ + {}^1g_2 \exp[-z_1\beta((1-\delta)\psi_M/2 + \psi_2 - \psi_S)]\}\end{aligned} \tag{21}$$

and a similar equation for ${}^2k^*$. The ratio of the observed rate constants is

$$\begin{aligned}{}^1k^*/{}^2k^* = ({}^1k_S/{}^2k_S)({}^1g_1/{}^2g_1)\{\exp[-(z_1 - z_2)\beta(\delta\psi_M + \psi_S)]\}\{{}^2g_{-1} \\ + {}^2g_2 \exp[-z_2\beta(\psi_M/2 + \psi_2 - \psi_1]\}\{{}^1g_{-1} \\ + {}^1g_2 \exp[-z_1\beta(\psi_M/2 + \psi_2 - \psi_1]\}\end{aligned} \tag{22}$$

The ratio ${}^1k_S/{}^2k_S$, we assume, is close to the ratio of the rate constants for the reaction of the two reagents with a small thiol, such as 2-mercaptoethanol (2-ME)[6]; thus, we assume ${}^1k_S/{}^2k_S = {}^1k_{ME}/{}^2k_{ME}$. Let ρ be the ratio of the effective rate constants for the reactions of the two reagents with the substituted Cys divided by the ratios of the rate constants for the reactions of the two reagents with 2-ME:

$$\rho = ({}^1k^*/{}^2k^*)/({}^1k_{\mathrm{ME}}/{}^2k_{\mathrm{ME}}) \tag{23}$$

Let $\rho_0 = \rho$, calculated with ${}^1k^*$ and ${}^2k^*$ determined or extrapolated to $\psi_{\mathrm{M}} = 0$.

$$\rho_0 = \exp[-(z_1 - z_2)\beta\psi_{\mathrm{S}}]({}^1g_1/{}^2g_1)\{{}^2g_{-1} + {}^2g_2\exp[-z_2\beta(\psi_2 - \psi_1)]\} \div \{{}^1g_{-1} + {}^1g_2\exp[-z_1\beta(\psi_2 - \psi_1)]\} \tag{24}$$

This expression can be simplified under these conditions: (1) that because the reagents are similar except for charge, the nonelectrostatic contributions to the rate constants are similar, and ${}^1g_1 \cong {}^2g_1$, and ${}^1g_{-1} \cong {}^2g_{-1}$; and (2) for both reagents, $g_2 \ll g_{-1}$; that is, the jump of reagent from the site of reaction to the extracellular side is much faster than the jump from the reaction site to the intracellular side. This holds for impermeant reagents, which reach the site of reaction from the extracellular side but do not permeate further (i.e., $g_2 = 0$), and may hold for permeant reagents as well, because the narrowest part of the channel, presumably its highest barrier, is distal to the site of reaction. This is the case for most of the channel in the AChR, where the narrowest part is at the intracellular end.[40,41] In the closed state of a channel, $g_2 = 0$ for all reagents and for all positions more extracellular than the gate. Under the above two conditions,

$$\rho_0 \cong \exp[-(z_1 - z_2)\beta\psi_{\mathrm{S}}] \tag{25}$$

Dependence of Rate Constants on Membrane Potential

In the sixth membrane-spanning segment of CFTR, the rate constants for the reactions of $\mathrm{MTSES^-}$ and $\mathrm{MTSET^+}$ with eight channel-lining residues were measured at several membrane potentials.[10] For each residue and reagent, the apparent $z\delta$ was calculated. Opposite and equal slopes were obtained for the negatively and positively charged reagents, indicating that the effect of membrane potential was on the reagent and not on the structure of the protein or on the ionization state of the cysteine residue.

In the AChR channel, the rate constant for the reaction fo AEAETS at αT244C, near to the cytoplasmic end of the channel was dependent on membrane potential in the open state of the channel but not in the closed state, and the slope of the dependence, $z\delta$, was about 0.4.[46] The rate constants for the reactions of MTSEA and of MTSEH with αT244C were not dependent on membrane potential. MTSEA is significantly permeant through the open channel[2] and MTSEH is neutral. The absence of voltage dependence of the reactions of MTSEA and of MTSEH indicate that the voltage dependence of the reaction of AEAETS was not due to effects of membrane potential on gating kinetics or on the acid dissociation of the Cys. Also, in the M1 segment of the AChR β subunit, the reaction rate of βV229C with AEAETS was voltage dependent in the open state of the channel ($z\delta$ about 0.26), but the reaction rates of MTSEA and MTSEH were not voltage dependent.[33] A complication in the interpretation of such $z\delta$ values in terms of structure is that in each of the MTS reagents the reactive divalent sulfur is about 5 Å from the charged headgroup atom, and in AEAETS, the reactive sulfur is in the middle of the molecule, roughly 5 Å from each of the two charged nitrogens.

Intrinsic Electrostatic Potential

The rate constants for the reactions at zero membrane potential of MTSEA, AEAETS, and MTSEH, with substituted Cys in the M2 segment of the AChR α subunit, were compared according to Eq. (25).[6a] [Note that it is difficult to determine in each case whether the assumptions made in deriving Eq. (25) are satisfied.] The ratios formed with the rate constants for AEAETS (assuming only one of the two charges enters the field) and MTSEH and the ratios formed with the rate constants for MTSEA and MTSEH gave nearly the same values for ψ_S. At αT244C, αL251C, and αL258C, in the open state, the apparent intrinsic potentials (ψ_S) were about -230, -80, and -20 mV. In the closed state, these potentials were 70 to 100 mV more positive. The intrinsic electrostatic potential profile of the AChR channel could affect its cation selectivity and conductance.

In the voltage-gated Na^+ channel, Cys substituted for S4 Arg residues are accessible to MTSET and MTSES on the intracellular side of the hyperpolarized membrane and on the extracellular side of the depolarized membrane.[50] One of these Arg moves outside the transmembrane field on the extracellular side but into a region of an intrinsic negative potential of -46 mV.[12]

Location of Charge-Selectivity Filter

Most ion channels are selectively permeable to either anions or cations, but the location and mechanism of charge selectivity are largely unknown. In the cation-selective AChR, the negatively charged MTSES reacted with residues in the M2 segment (αV255C and βV266C) about one-third of the way in from the extracellular end of the channel,[2,52] indicating that the exclusion of ions coming from the extracellular side occurred closer to the cytoplasm than these residues. In the $GABA_A$ receptor, both anionic and cationic reagents reacted with Cys substituted for residues approximately two-thirds of the distance from the extracellular end, indicating that charge selectivity occurs near the intracellular end of the channel.[13]

A comparison of the rates of reaction of oppositely charged MTS reagents with channel-lining Cys in the CFTR chloride channel indicated that the charge-selectivity filter is located near the cytoplasmic end of the channel.[10] The rates of reaction of $MTSES^-$ and $MTSET^+$ were determined at zero membrane potential. For channel-lining cysteine residues in the extracellular half of the M6 segment the ratio of the rates of reaction of $MTSES^-$ to $MTSET^+$ were similar to the ratio of the rates of reaction with 2-ME in solution, suggesting that both anions and cations can enter the extracellular end of the CFTR channel. Only near the cytoplasmic end of the channel, in the region of an Arg residue, did the ratio of the rates indicate selectivity for anions over cations. This is the region of charge selectivity determination, because mutation of the Arg to Cys reduced the Cl^- to Na^+ permeability ratio from 37 : 1 in wild-type CFTR to 6 : 1 in the R352C mutant (Guinamard and Akabas, submitted).

[9] Explorations of Voltage-Dependent Conformational Changes Using Cysteine Scanning

By RICHARD HORN

Introduction

Sodium, calcium, and potassium channels belong to a superfamily of membrane proteins that are exquisitely sensitive to changes of membrane potential. Depolarization causes a movement of positive charge outward across the membrane electric field, and tends to open and subsequently inactivate these ion channels (for recent reviews, see Refs. 1 and 2). Both the charge movement and the consequent effects on the "gates" of the channel involve conformational rearrangements of the protein. The movement of charge has been studied using gating current measurements,[3] and the effect of voltage on the opening and closing of channels has been examined extensively since the development of the voltage-clamp technique in the 1950s. However, the effect of voltage on the secondary or tertiary structure of these ion channel proteins was largely unknown until very

[1] F. J. Sigworth, *Quarterly Rev. Biophys.* **27,** 1 (1994).
[2] R. D. Keynes, *Quarterly Rev. Biophys.* **27,** 339 (1994).
[3] E. Stefani and F. Bezanilla, *Methods Enzymol.* **293,** [19], 1998 (this volume).

0076-6879/98 $25.00

recently, when conformational changes of proteins were explored using the method of cysteine scanning mutagenesis.[4–6] This technique has the potential to reveal changes in the surface accessibility of specific residues mutated to cysteine and then reacted with hydrophilic cysteine reagents. If (1) the accessibility of a specific residue is affected by membrane potential, and (2) the reaction of the introduced cysteine with a reagent produces a measurable biophysical effect, then the rate of cysteine modification can be used to determine features of the voltage-dependent exposure of this residue, including both the steady-state voltage dependence and the kinetics of the underlying conformational changes. It may also be possible to determine whether the residue of interest lies in the membrane electric field and the local electrostatic potential in the vicinity of the thiol group.[7–9]

Reagents

The reagents used most extensively in cysteine scanning studies are the methane thiosulfonates (MTS), many varieties of which are available from Toronto Research Chemicals (North York, Ontario, Canada). The thiolate anion of a cysteine residue performs a nucleophilic attack on a MTS reagent, producing a mixed disulfide product. This allows the attachment of a diversity of adducts to cysteine residues. The vagaries of MTS reagents are described by Akabas and Karlin.[6] Some cations (e.g., Ag^{+}, Hg^{2+}, and Cd^{2+}) also react avidly with thiols, and may be used as hydrophilic cysteine reagents in electrophysiologic experiments. The main prerequisites for a useful thiol reagent in this context are that (1) it must be specific for the thiol group, (2) it must react rapidly on the time scale of a voltage-clamp experiment (a few minutes or less), (3) it must react under mild "physiological" conditions, and (4) for most purposes it should be hydrophilic to restrict it to the protein surface. Many of the MTS reagents satisfy these criteria. Commonly used hydrophilic MTS reagents, such as MTS ethyltrimethylammonium (MTSET) and MTS ethyl sulfonate (MTSES), are monovalent ions.

[4] M. H. Akabas, D. A. Stauffer, M. Xu, and A. Karlin, *Science* **258,** 307 (1992).
[5] J. J. Falke, A. F. Dernburg, D. A. Sternberg, N. Zalkin, D. L. Milligan, and D. E. Koshland, *J. Biol. Chem.* **263,** 14850 (1988).
[6] A. Karlin and M. H. Akabas, *Methods Enzymol.* **293,** [8], 1998 (this volume).
[7] D. A. Stauffer and A. Karlin, *Biochemistry* **33,** 6840 (1994).
[8] M. Cheung and M. H. Akabas, *J. Gen. Physiol.* **109,** 289 (1997).
[9] N. Yang, A. L. George, and R. Horn, *Biophys. J.* **73,** 2260 (1997).

Measuring Modification Rate

Characterization of the reaction of a reagent with a specific cysteine residue requires that a measurable biophysical effect be observed. This may be either a change in the amplitude of a current or a change in kinetics of the current. If, for example, the reaction of a cysteine residue with a particular reagent abolishes the current of a channel, either by blocking the pore or by an effect on gating, the reaction can be followed by the time course of the reduction of the current.

In our studies of sodium channels, on the other hand, the modification of introduced cysteine residues causes changes in the kinetics of inactivation without large changes in the amplitudes of the peak sodium current.[9–11] The analysis of these data is somewhat more complicated. Our approach has been to measure an inactivation time constant ($\tau_{control}$) before exposure to an MTS reagent and again after all channels have been modified by the reagent (τ_{MTS}). Because each channel has a single cysteine substitution, we assume that the modification is all-or-none for each channel. Therefore, during modification an increasing fraction of channels will be modified (F_{mod}), beginning at zero and ending at one. For simplicity we use a reagent concentration low enough that the modification rate is much slower than inactivation kinetics. We then estimate F_{mod} at any time during modification by fitting the kinetics of inactivation as a weighted sum of 2 exponentially decaying components, where the time constants are fixed as $\tau_{control}$ and τ_{MTS}, and the fractional weighting factor of the latter component is the free parameter F_{mod}. The fitting employs the program CLAMPFIT in the pCLAMP suite of programs (version 6.01, Axon Instruments, Burlingame, CA).

For reasons discussed later, if the reagent concentration and the exposure of the cysteine residue are both constant during the modification, the time course of modification plotted, for example, as F_{mod} versus time, is usually a single exponential function. Therefore, the time course can be fit statistically by an exponentially decaying function with a modification rate of ρ_{mod}, the inverse of the time constant of modification. We typically use the plotting program ORIGIN (MicroCal, Northampton, MA) to estimate ρ_{mod}.

Measuring Voltage-Dependent Accessibility

If a change in transmembrane voltage causes a corresponding change in the rate of modification, ρ_{mod}, of a cysteine residue by a hydrophilic

[10] N. Yang and R. Horn, *Neuron* **15,** 213 (1995).
[11] N. Yang, A. L. George, Jr., and R. Horn, *Neuron* **16,** 113 (1996).

cysteine reagent, this will be referred to as a change in accessibility. "Accessibility" is defined here strictly in an operational sense in that it is observed through its effect on ρ_{mod}. By this definition a change in accessibility could be due either to an actual change in the position of the cysteine residue with respect to the protein surface or to a change in the environment of the thiol group that affects its reactivity to the reagent. Usually, for example, the ionized thiolate moiety is the active form of the cysteine thiol. Therefore, a conformational change of the protein that changes the ionization of the thiol group will have a direct effect on ρ_{mod}, and for purposes of discussion here this will be called a change in accessibility. In principle, the ionization of the thiol can be explored, for example, by varying pH.

In the simplest case a cysteine residue can have one of two accessibilities for extremes of voltage. This can be depicted by kinetic Scheme I. The cysteine residue is in state Cys_{V-} at hyperpolarized voltages and Cys_{V+} at depolarized voltages. The unmodified channel conformations are shown in the upper row of Scheme I, and the irreversibly modified channels are shown below. The modification rate is ρ_{mod+} from the depolarized conformation and ρ_{mod-} from the hyperpolarized conformation. If ρ_{mod+} differs from ρ_{mod-}, accessibility, as defined earlier, is voltage dependent. The voltage dependence of accessibility is denoted in this simple model by the voltage-dependent rate constants $\alpha(V)$ and $\beta(V)$. For generality $\alpha'(V)$ and $\beta'(V)$, the conformational rate constants for the modified cysteine, are also considered to be voltage dependent and may differ from $\alpha(V)$ and $\beta(V)$.

By controlling the experimental conditions it may be possible to estimate both ρ_{mod+} and ρ_{mod-}. For example, large depolarizations or hyperpolarizations can be used to put all channels into either the Cys_{V+} or the Cys_{V-} state, respectively, and either ρ_{mod+} or ρ_{mod-} can then be determined directly. When these two rates are known, the overall rate of modification (ρ_{mod}) at different membrane potentials can be used to infer the voltage-dependent accessibility of a specific cysteine residue. Although the kinetics of irreversible modification for Scheme I have two exponentially decaying components when the reagent concentration is fixed, the gating kinetics are typically

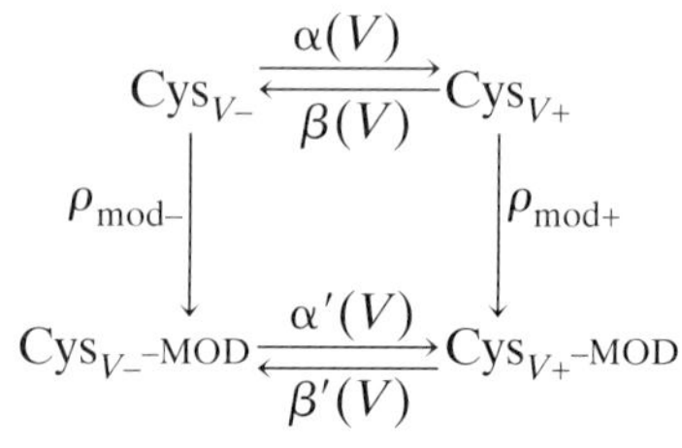

SCHEME I

rapid for voltage-gated ion channels, and it is possible to reduce ρ_{mod+} and ρ_{mod-} by decreasing reagent concentration. For a simple bimolecular reaction between the thiol and the reagent, ρ_{mod+} and ρ_{mod-} will each be linear functions of reagent concentration. If modification is rate limiting for a fixed reagent concentration [i.e., $\rho_{mod+}, \rho_{mod-} \ll \alpha(V) + \beta(V)$], the time course of modification at any voltage will be first order with an overall modification rate equal to a weighted sum of ρ_{mod+} and ρ_{mod-}, expressed as follows:

$$\rho_{mod} = \Pr\{Cys_{V+}\}\rho_{mod+} + \Pr\{Cys_{V-}\}\rho_{mod-}$$

The weighting factor for each rate is the steady-state probability of the channel being in either of its two conformations. If the channel can be driven with high probability into state Cys_{V-} by hyperpolarization, for example, the modification rate ρ_{mod} will equal ρ_{mod-}.

By measuring ρ_{mod} as a function of membrane potential, it is possible to estimate the voltage dependence of the steady-state probability of accessibility of the cysteine residue. Figure 1, for example, shows the rate of modification of the cysteine mutant R1454C of the human skeletal muscle sodium channel as a function of membrane potential (data from Yang *et al.*[9]). The reagent used was MTSET at a concentration of 100 μM. These data show that ρ_{mod} approaches 0 at hyperpolarized voltages, indicating that $\rho_{mod-} = 0$ and therefore that this cysteine residue is inaccessible to extracellular MTSET at sufficiently hyperpolarized potentials, in this case more negative than −140 mV. If the absolute rates (i.e., ρ_{mod}) at different voltages are normalized to a maximum of unity, they will represent the voltage dependence of the steady-state probability of accessibility of the cysteine residue. For the example in Fig. 1 accessibility can be fit reasonably well by a Boltzmann function (solid curve), consistent with the simple two-

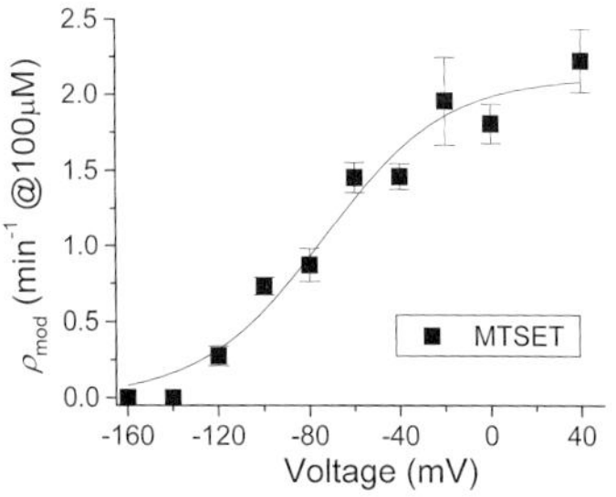

FIG. 1. The rate of modification of the sodium channel mutant R1454C by 100 μM MTSET as a function of membrane potential. Data from Yang *et al.*[9] The Boltzmann function (smooth curve) has a midpoint of −73 mV and a slope equivalent to a movement of 0.9 elementary charges.

state gating model of Scheme I. This interpretation has qualifications. First, the conformational transitions underlying exposure of the cysteine residue may be more complicated (e.g., more kinetic states), but the steady-state measurement obscures these subtleties. Second, the conformational transitions responsible for opening and closing of the conductive pore of the channel are likely to be considerably more complicated than those responsible for changing the accessibility of this single cysteine residue.

Measuring Kinetics of Voltage-Dependent Cysteine Accessibility

Although the rate of cysteine modification can be set to a low level, by varying either membrane potential or reagent concentration, the modification of an individual thiol group occurs very rapidly, on the time scale of nanoseconds or less, during a successful encounter with a reagent molecule. A low macroscopic rate of modification is therefore due, in general, to a low rate of successful encounters between reagent molecules and thiol groups. The latter rate, in turn, depends on (1) the concentration of reagent in the vicinity of the thiol group, (2) the probability that a random encounter between an "accessible" thiol and the reagent leads to modification, and (3) the probability that the thiol group is physically exposed on the protein surface. Membrane potential has the possibility of affecting any or all of these three factors: (1) If the site of modification is in the membrane electric field, the concentration of a charged reagent near the thiol will be voltage dependent.[8,9] (2) Membrane potential could influence the intrinsic reactivity of a thiol group, for example, by changing its ionization state. (3) Voltage can influence the access of reagent in solution to the thiol group by affecting the conformation of the channel, for example by exposing or burying the cysteine residue.

The methods used for Fig. 1 (described more fully in Yang and Horn[10]) permit the estimation of steady-state accessibility of any residue substituted by a cysteine. It is also possible to estimate the kinetics of the change of cysteine accessibility after a change of voltage, using a related method. Imagine the time course of accessibility during a pulse train of equal-duration voltage steps (Fig. 2A). Without loss of generality we can assume that cysteine accessibility changes from unmeasurably low to a finite level on depolarization. In the hypothetical example of Fig. 2A the exposure of a cysteine at a depolarized potential has a time constant of 500 μs, and its burial at a hyperpolarized voltage has a time constant of 50 μs. We ask for this example whether the total time that the cysteine residue is exposed is affected by the durations (Δt) of individual depolarizations. For simplicity we can assume a 50% duty cycle, namely, that the channel is depolarized 50% of the time in all cases.

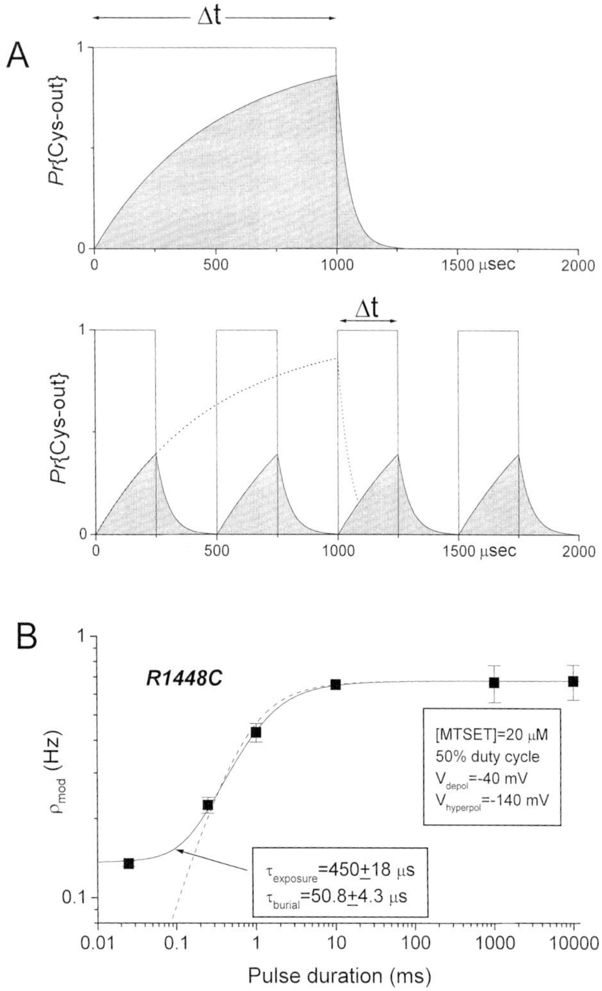

FIG. 2. Effect of individual pulse durations on voltage-dependent exposure of a cysteine residue. (A) One (above) versus 4 (below) cycles of depolarizations. The cysteine exposure and burial are assumed to be exponential at the depolarized and hyperpolarized voltages. In this hypothetical simulation, the exposure has a time constant of 500 μs and the burial 50 μs, as depicted by the curves. The integral of cysteine exposure over this 2-ms period is shown by the shaded region below the theoretical curves of exposure probability. Although the channel is depolarized for the same total time (1 ms) in both cases, the total exposure of the cysteine residue is greater for a single 1-ms depolarization than for four 250-μs depolarizations. (B) The rate of modification of R1448C plotted as a function of individual pulse duration (Δt). Data from Yang and Horn.[10] The solid curve is the best fit of Eq. (5), normalized and divided by Δt, with the indicated parameters. The dashed line is the best fit in which τ_{burial} is fixed at 0.

If the duration of individual depolarizations is much longer than the time required for cysteine accessibility to reach steady state, then the rate of modification by a fixed concentration of cysteine reagent will not depend on the duration of individual pulses. This follows from the fact that the total exposure time (i.e., the integral of the probability of cysteine exposure) will be independent of pulse duration for sufficiently long pulses, namely, for pulse durations much longer than the time constants for exposure and burial of the cysteine residue at each voltage. Therefore for such long-duration pulses the total exposure time will be a function only of the duty cycle and the exposure probabilities at each voltage, but not of the durations of individual pulses. Figure 2A shows, however, that when the pulse duration approaches the voltage-dependent kinetics of cysteine exposure, the integral of the exposure, depicted as the total shaded area under the curves, is affected by pulse duration. In this example a single depolarization of 1-ms duration produces an approximately two-fold greater total exposure of a cysteine residue than is obtained with four 250-μs pulses, although the channel is depolarized an equal duration for each voltage-clamp protocol. This is due to the fact that for the shorter depolarizations, the cysteine residue has less chance of being exposed.

We systematically varied pulse durations as in Fig. 2A to estimate the voltage-dependent rate constants, $\alpha(V)$ and $\beta(V)$, for a specific cysteine mutant (R1448C of the hSkM1 sodium channel) at a depolarized (-40-mV) and a hyperpolarized (-140-mV) voltage.[10] The data for this experiment are replotted in Fig. 2B (filled squares), which shows ρ_{mod} as a function of individual pulse duration. Consistent with the expectations of Scheme I and Fig. 2A, when pulse durations are sufficiently long (>10 ms), ρ_{mod} is independent of pulse duration over a range of 3 orders of magnitude. Furthermore, when individual pulse durations approach the kinetics of cysteine exposure (Fig. 2A), ρ_{mod} decreases, presumably because this cysteine residue does not have enough time to get fully exposed during short depolarizations.

These data may be used to estimate the time constants of cysteine exposure and burial at the two voltages used for pulsing, as follows. If the voltage dependence of cysteine accessibility has the simple form shown in Scheme I, the transition between less accessible and more accessible states will have exponential kinetics after a step of voltage. The integral of exposure probability (i.e., accessibility) can be determined analytically if the time constants of exposure and burial are known, as shown later. Therefore, these time constants can be estimated numerically from the effects of pulse duration on ρ_{mod}, because ρ_{mod} is directly proportional to accessibility.

Let V_1 and V_2 be defined as the depolarized and hyperpolarized voltages, respectively. The steady-state exposure probabilities at these two voltages are correspondingly defined as $p_{\infty,1}$ and $p_{\infty,2}$. The initial conditions at the moment of changing the voltage will depend on pulse duration, as shown in Fig. 2A, and are defined as $p_{\text{init},1}$ and $p_{\text{init},2}$ for the depolarizing and hyperpolarizing pulses, respectively. The time course of exposure probability at each voltage is exponential according to:

$$P_{V_1}(t) = p_{\infty,1} + (p_{\text{init},1} - p_{\infty,1})e^{-\rho_1 t} \quad (1)$$

$$P_{V_2}(t) = p_{\infty,2} + (p_{\text{init},2} - p_{\infty,2})e^{-\rho_2 t} \quad (2)$$

The rates ρ_1 and ρ_2 are the inverses of the time constants for changes in cysteine accessibility at V_1 and V_2. In the example of Fig. 2, ρ_1 and ρ_2 depict the rates of exposure and burial, respectively, of the cysteine residue. The steady-state exposure probabilities, $p_{\infty,1}$ and $p_{\infty,2}$, can be estimated from experiments like that of Fig. 1 by considering the Boltzmann fit as a scaled representation of exposure probability. The initial conditions, $p_{\text{init},1}$ and $p_{\text{init},2}$, for a pulse duration Δt are specified uniquely from the fact that there is no discontinuity of exposure probability when the membrane potential changes, an observation pointed out by Fred Sigworth (Yale University). Making use of this fact leads to the following expressions for initial conditions:

$$p_{\text{init},2} = \frac{p_{\infty,1} + (p_{\infty,2} - p_{\infty,1})e^{-\rho_1 \Delta t} - p_{\infty,2}e^{-(\rho_1+\rho_2)\Delta t}}{1 - e^{-(\rho_1+\rho_2)\Delta t}} \quad (3)$$

$$p_{\text{init},1} = p_{\infty,2} + (p_{\text{init},2} - p_{\infty,2})e^{-\rho_2 \Delta t} \quad (4)$$

These calculated initial conditions can be substituted into Eqs. (1) and (2), which can then be integrated to calculate the exposure probability for a pulse train of arbitrary Δt. The integral for one cycle of depolarization and hyperpolarization is given by:

$$\int_0^{\Delta t} \{P_{V_1}(t) + P_{V_2}(t)\}\, dt = (p_{\infty,1} + p_{\infty,2})\Delta t + \frac{p_{\text{init},1} - p_{\infty,1}}{\rho_1}(1 - e^{-\rho_1 \Delta t}) + \frac{p_{\text{init},2} - p_{\infty,2}}{\rho_2}(1 - e^{-\rho_2 \Delta t}) \quad (5)$$

This integral is normalized by dividing it by Δt, which allows the direct comparison of exposure probability as a function of Δt. The above set of equations can be used to estimate the exposure and burial rates, ρ_1 and ρ_2, from the effect of Δt on ρ_{mod} (Fig. 2B). We estimate these rates by nonlinear least squares minimization employing a variable metric algorithm, although other optimization algorithms (e.g., simplex) would also work. The solid curve in Fig. 2B plots the scaled and normalized integral of Eq. (5), using the best-fit time constants for exposure and burial of the cysteine in this experiment. The dashed line shows the expectation if the burial time constant is fixed at 0. The deviation between this theoretical curve and the data indicates that the burial was not infinitely fast at −140 mV for this S4 mutant.

Although a specific example from the author's laboratory has been used to illustrate a method for estimating the voltage-dependent kinetics of cysteine exposure, the theory is general for ion chanels obeying Scheme I. The assumption of first-order kinetics may not, however, be generally applicable, and should be evaluated on a case-by-case basis.

Measuring State-Dependent Accessibility of Cysteine Residue

A change in membrane potential typically produces a cascade of conformational changes in an ion channel, beginning with the movement of a voltage sensor through the electric field across the membrane. The downstream consequences of the movement of the voltage sensor are likely to include changes in the surface accessibilities of various amino acid residues. Therefore changes in accessibility of specific residues may be more tightly coupled to some gating process, like the opening or inactivation of a channel, than to the membrane potential itself. This can be explored by applying a cysteine reagent selectively during specific gating states.[12–14] Because the kinetics of gating may be rapid, solution application has to be even more rapid. For example, a sodium or potassium channel may open for only a few milliseconds before inactivating. To distinguish residues exposed only when a channel is open, and not when inactivated, or vice versa, solution application has to be on a submillisecond timescale. One technique involves the use of a solenoid to switch between two converging solution inputs applied to an excised patch.[15]

State-Dependent Movement of Fluorescently Tagged Cysteine Residues

Voltage dependence in a membrane protein requires that certain residues change their position or orientation with respect to the electric field when the membrane potential changes. It is also likely that some amino acid residues reorient with respect to one another and to the hydrophilic surface of the protein when the membrane potential changes. This movement has recently been tracked by tagging individual cysteine residues

[12] G. Yellen, D. Sodickson, T.-Y. Chen, and M. E. Jurman, *Biophys. J.* **66,** 1068 (1994).
[13] Y. Liu, M. E. Jurman, and G. Yellen, *Neuron* **16,** 859 (1996).
[14] M. Holmgren, M. E. Jurman, and G. Yellen, *J. Gen. Physiol.* **108,** 195 (1996).
[15] Y. Liu and J. P. Dilger, *Biophys. J.* **60,** 424 (1991).

with fluorophores and measuring the change of fluorescence intensity as a function of membrane potential.[16] Because the emission spectrum of a fluorophore may be sensitive to its environment,[17] these data suggest that membrane potential alters the environment of specific residues. A number of fluorophores are available as cysteine reagents from Molecular Probes (Eugene, OR).

This method has the potential of measuring the kinetics of the conformational changes of a protein in a rapid fashion over a wide range of voltages, and is therefore inherently more appealing than the tedious method described in Fig. 2. The main drawback, however, in the interpretation of such experiments is that fluorophores are typically large compounds and are likely to perturb the voltage-dependent conformational changes of the protein. With respect to Scheme I, the methods of Fig. 2 allow estimation of $\alpha(V)$ and $\beta(V)$ of an unmodified channel altered only by the introduced cysteine residue, whereas fluorescently tagged cysteines allow estimates of $\alpha'(V)$ and $\beta'(V)$ of a modified channel.

[16] L. M. Mannuzzu, M. M. Moronne, and E. Y. Isacoff, *Science* **271,** 213 (1996).
[17] J. R. Lakowicz, "Principles of Fluorescence Spectroscopy." Plenum Press, New York, 1983.

[10] Assessment of Distribution of Cloned Ion Channels in Neuronal Tissues

By KOHJI SATO and MASAYA TOHYAMA

Introduction

In neuronal tissues, many ion channels are differentially expressed with spatial and time diversity.[1–4] Investigating where and when an ion channel is expressed is of great importance to estimate the physiologic aspects of the ion channel. Two important methods are used for this purpose: immunohistochemistry and *in situ* hybridization histochemistry. Although immunohistochemistry is a powerful tool to detect the exact localization of a protein, it is sometimes time consuming to produce a good antibody.

[1] K. Sato, J. Mick, H. Kiyama, and M. Tohyama, *Neuroscience* **64,** 459 (1995).
[2] T. Furuyama, H. Kiyama, K. Sato, H. T. Park, H. Maeno, H. Takagi, and M. Tohyama, *Mol. Brain Res.* **18,** 141 (1993).
[3] K. Sato, H. Kiyama, and M. Tohyama, *Neuroscience* **52,** 515 (1993).
[4] K. Sato, H. Kiyama, and M. Tohyama, *Brain Res.* **590,** 95 (1992).

METHODS IN ENZYMOLOGY, VOL. 293
0076-6879/98 $25.00

2. After incubation, add 2 μl of tRNA (25 mg/ml), 180 μl of doubly distilled water, 100 μl of buffered phenol, and 100 μl of CIAA (chloroform : isoamyl alcohol, 24 : 1).
3. Vortex vigorously and centrifuge at 15,000 cpm at 4° for 5 min. Pour the supernatant (about 200 μl) into new tubes; add 200 μl of 4 *M* ammonium acetate and 800 μl of 100% ethanol.
4. Freeze for 30 min at −80° and centrifuge at 15,000 cpm at 4° for 10 min.
5. Dissolve the DNA precipitate in doubly distilled water and measure using a liquid scintillation counter. The count should be above 10^7 dpm/pmol.

Prehybridization

1. After warming to room temperature, slide-mounted sections are fixed in 4% paraformaldehyde in 0.1 *M* phosphate buffer (pH 7.4) for 15 min. (All steps are performed at room temperature unless otherwise indicated.)
2. Rinse sections three times (5 min each) in 4× SSC (pH 7.0). (1× SSC contains 0.15 *M* sodium chloride and 0.015 *M* sodium citrate.)
3. Dehydrate sections through a graded ethanol series (70–100%). Defat the sections with chloroform for 3 min, and immerse in 100% ethanol (twice for 5 min each time) before subjecting to hybridization.

Hybridization and Wash

1. Hybridization buffer: In a 50-ml sterilized tube, mix the following reagents (total 10 ml):

Deionized formamide	5 ml
20× SSC	2 ml
1.2 M Phosphate buffer (pH 7.4)	1 ml
100× Denhardt's solution	100 μl
tRNA (25 mg/ml)	100 μl
Doubly distilled water	1.8 ml
Dextran sulfate	1 g

Note: First, dissolve dextran sulfate completely in doubly distilled water, then add other reagents.

2. Labeled oligonucleotide probes are added to hybridization buffer to give concentrations of about 1× 10^7 dpm/ml. Add also 20 μl of 1 *M* DTT into 1 ml of hybridization buffer. Apply hybridization buffer

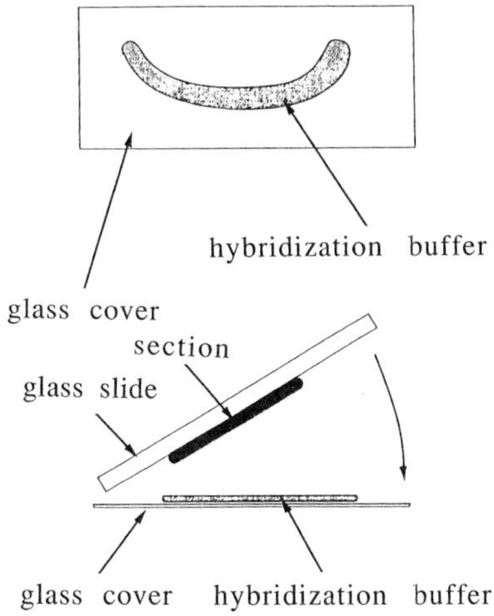

FIG. 1. Application of hybridization buffer on section. *Top:* Apply about 200 μl of hybridization buffer in a curved shape on a glass cover. *Bottom:* Slowly place a glass slide onto the glass cover, being careful not to create air bubbles. When hybridization buffer covers all sections, turn it over rapidly.

on tissues (0.2 ml/slides) as shown in Fig. 1. *Note:* The concentration of labeled oligonucleotide probes in hybridization buffer should be optimized.

3. Hybridization is performed in humid chambers for 24 hr at 41°.
4. The sections are rinsed in 1× SSC (pH 7.0) for 10 min. Coverglasses should lift off at this step or else they should be removed slowly.
5. Wash three times in 1× SSC at 55° for 20 min each time.

Film Autoradiography

The sections are then dehydrated through a graded ethanol series (70%–100%) and autoradiographed with X-ray film for a few days (Kodak, Rochester, NY, X-Omat) at room temperature. Film autoradiography gives a general view of the expression pattern and information about the intensity of the signal.

Emulsion Autoradiography

For detailed observation, coat the sections with an emulsion. The following procedure must be performed in a dark room:

1. In a 50-ml tube, melt 10 ml of Kodak NTB-2 emulsion (Kodak, Rochester, NY) with 10 ml of doubly distilled water at 42°.

2. Pour emulsion into a clean dipping chamber kept at 42°.
3. Dip slides slowly and smoothly into dipping chamber. Withdraw slowly and wipe the back side.
4. Place slides on a table with the section side up. Dry at least 1 hr.
5. Expose sections to the emulsion at 4° for 2–5 weeks (depending on the intensity) in a tightly sealed dark box containing desiccant.
6. Develop the sections with D-19 developer (Kodak, Rochester, NY), fix with photographic fixer, and wash with tap water.

Thionin Staining

The sections are counterstained with thionin solution to allow morphologic identification. However, strong staining sometimes weakens hybridization signals, so it is necessary to find the optimum condition.

1. Dissolve 1.25 g of thionin, 10 g of NaOH, and 50 ml of acetic acid in 500 ml water. Stir well and filter.
2. Place developed microslides in a slide rack and immerse for about 1 min in a glass staining dish filled with thionin solution.
3. Rinse slides in water for about 1 min.
4. Dehydrate sections through a graded ethanol series (70–100%), 3 min each.
5. Rinse slides in xylene three times for 3 min each time.
6. In a hood, mount slides with Entellan (Merck, Frankfurt, Germany).

Control Studies

To certify the specificity of the hybridization signal, control studies are recommended. One control system involves control competition experiments using 100-fold excess of the unlabeled probes together with the labeled probes, and another control uses RNase A pretreatment [1 μg/ml RNase A in RNase buffer (0.5 *M* NaCl, 10 m*M* Tris-HCl, 5 m*M* EDTA, pH 8.0)] for 30 min at 37° just before hybridization. If the signal is specific, these experiments should show no positive signals.

In situ Hybridization Histochemistry with cRNA Probe

Design of cRNA Probe

If an adequate length of the corresponding cDNA fragment (300–2000 bp) is already in a plasmid that contains T7, T3, or Sp6 promoters, such as pBluescript (Stratagene, La Jolla, CA) or pGEM (Promega, Madison, WI), this plasmid is able to be used directly. Otherwise the fragment must be subcloned into one of the above-mentioned plasmids. If there is no corresponding cDNA, produce a PCR product with specific primers and subclone it into one of the above-mentioned plasmids.

Construction of Template

1. Plasmid (10 μg) is linearized by cutting with a restriction enzyme (for the antisense probe) or another restriction enzyme (for the sense probe). Be careful to select adequate enzyme sites depending on the orientation of the cDNA and the location of promoters. *Note:* Restriction enzymes that make 5′-protruding terminals or blunt ends are preferable.
2. Check carefully with agarose gel electrophoresis to determine if linearization is complete.
3. The linearized plasmid is purified by phenol–chloroform and precipitated by adding 0.1 volume 3 *M* sodium acetate and 2.5 volume 100% ethanol, mixing well, and freezing at least for 30 min at −80° and centrifuging at 15,000 cpm at 4° for 10 min.
4. The DNA precipitate is dissolved with doubly distilled water to give a concentration of 1 μg/μl.

Synthesis of ^{35}S-Labeled cRNA Probe

1. Prepare the following reaction mix (10 μl total)

5× transcription buffer (200 m*M* Tris-Cl, pH 8.3, 30 m*M* $MgCl_2$, 10 m*M* spermidine chloride, 0.1% Triton X-100)	2 μl
100 m*M* DTT	0.5 μl
RNase inhibitor	0.5 μl
10 m*M* ATP, CTP, GTP	0.5 μl each
[α-^{35}S]UTP (NEG-039H, NEN)	5 μl
Linearized template solution	0.5 μl
T3, T7, or Sp6 RNA polymerase	1 μl

2. Incubate for 1 hr at 36°, add 1 μl of RNase-free DNase I, and incubate again for 15 min.
3. Add 40 μl of diethyl pyrocarbonate (DEPC)-treated doubly distilled water, 25 μl of 7.5 *M* ammonium acetate, 150 μl of cold 100% ethanol, and 1 μl of 10 mg/ml yeast tRNA to the reaction mix. Freeze for at least 30 min at −80° and centrifuge at 15,000 cpm at 4° for 10 min.

4. Dissolve the pellet with 50 μl of DEPC-treated doubly distilled water and perform ethanol precipitation again. Dissolve the final pellet with 50 μl of DEPC-treated doubly distilled water and measure using a liquid scintillation counter. The count should be greater than 5×10^5 dpm/μl.

Prehybridization

1. After being warmed to room temperature, slide-mounted sections are fixed in 4% paraformaldehyde in 0.1 *M* phosphate buffer (pH 7.2) for 15 min.
2. After washing with 0.1 *M* phosphate buffer two times, the sections are treated with 10 μg/ml proteinase K in 50 m*M* Tris-HCl buffer and 5 m*M* EDTA (pH 8.0) for 5 min at room temperature.
3. Slides are postfixed in the same fixative described earlier and washed once with DEPC-treated doubly distilled water.
4. Slides are acetylated with 0.25% acetic anhydride and 0.1 *M* triethanolamine for 10 min, washed with 0.1 *M* phosphate buffer once, and dehydrated in an ascending alcohol series (70, 80, 90, 100, and 100%), and air-dried.

Hybridization and Wash

1. In a 50-ml Falcon tube, the following reagents are mixed (hybridization buffer, 10 ml total):

Deionized formamide	5 ml
100 m*M* Tris-Cl (pH 8.0)	2 ml
5 *M* NaCl	600 μl
DEPC-treated doubly distilled water	1.8 ml
100× Denhardt's solution	200 μl
tRNA (25 mg/ml)	200 μl
Salmon sperm DNA (10 mg/ml)	200 μl
Dextran sulfate	1 g

Note: First, dissolve dextran sulfate completely with DEPC-treated doubly distilled water, and then add other reagents. Salmon sperm DNA should be added after denaturation.

2. Labeled cRNA probes are added to hybridization buffer to give concentrations of about 5×10^6 dpm/ml. Add also 20 μl of 1 *M*

DTT into 1 ml of hybridization buffer. Apply hybridization buffer onto tissues (0.2 ml/slides) as shown in Fig. 1.

3. Perform hybridization in humid chambers for 24 hr at 55°.
4. Rinse slides in 5× SSC, 1% mercaptoethanol (pH 7.0) at 55° for 15 min. Cover glasses should release spontaneously at this step; if not, remove them slowly.
5. Rinse slides in high stringency wash solution (50% deionized formamide, 2× SSC, 10% mercaptoethanol) at 65° for 30 min.

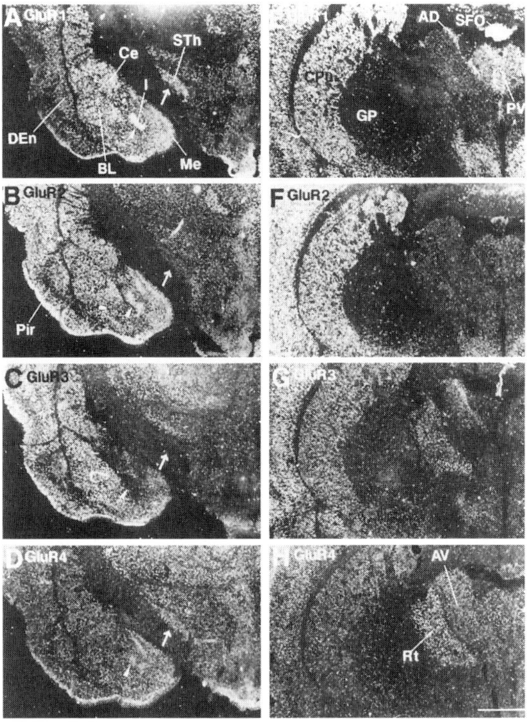

Fig. 2. Expression of GluR1–4 mRNAs at the level of the amygdala (A–D) and anterior thalamic nuclei (E–H). (A–D) GluR3 mRNA is absent in the subthalamic nucleus (the arrow in C), and in the intercalated nuclei of the amygdala (the arrowhead in C). (E–H) The globus pallidus shows low hybridization signals of GluR1–3 mRNAs in contrast with strongly labeled caudate putamen. GluR4 mRNA shows strong expression in the reticular thalamic nucleus (H). GluR1 mRNA is very strongly expressed in the subfornical organ (E). AD, anterodorsal thalamic nucleus; CPu, caudate putamen; DEn, dorsal endopiriform nucleus; GP, globus pallidus; I, intercalated nuclei of the amygdala; Me, medial amygdaloid nucleus; Pir, piriform cortex; Rt, reticular thalamic nucleus; SFO, subfornical organ; STh, subthalamic nucleus. Bar: 1 mm. [Reprinted from K. Sato, H. Kiyama, and M. Tohyama, *Neuroscience* **52,** 515–539 (1993), with kind permission from Elsevier Science Ltd., The Boulevard, Langford Lane, Kidlington OX5 1GB, UK.]

6. Rinse slides with RNase buffer (0.5 *M* NaCl, 10 m*M* Tris-HCl, 5 m*M* EDTA, pH 8.0) three times for 10 min each at 37°.
7. Incubate slides with 1 μg/ml RNase A in RNase buffer for 30 min at 37° and rinse again in RNase buffer for 10 min.
8. Rinse slides in high stringency wash solution (the same solution used in step 5) at 65° for 30 min.
9. Rinse slides with 2× SSC and 0.1× SSC for 15 min each at room temperature.

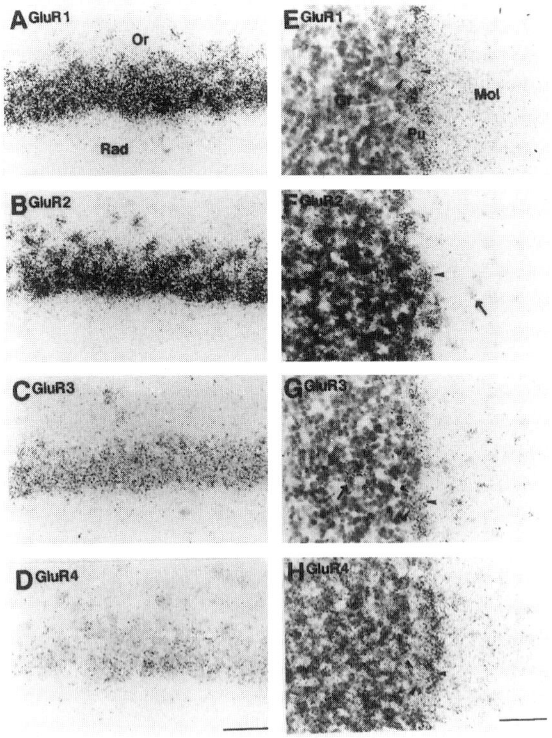

FIG. 3. Bright-field photomicrographs showing cells containing GluR1–4 mRNAs in the CA1 pyramidal cell layer of the hippocampus (A–D) and cerebellar cortex (E–H). (A–D) All four subunit mRNAs are expressed in most pyramidal cells in the hippocampus. (E–H) In the Purkinje cell layer, hybridization signals of GluR2 and 3 mRNAs accumulate on Purkinje cell bodies (arrowheads in F and G), while GluR1 and 4 mRNAs are strongly expressed around Purkinje cells, maybe on Bergman glias (arrowheads in E and H). The arrow in F indicates a putative stellae basket cell, and the arrow in G indicates a putative Golgi-type cell. Gr., granular cell layer; Mol, molecular layer; Or, stratum oriens; Pu, Purkinje cell layer; Py, stratum pyramidale; Rad, stratum radiatum. Bar: 50 μm.

10. Dehydrate slides in an ascending alcohol series (70, 80, 90, 100, and 100%), and air dry.

Autoradiography and Staining

These procedures are the same as those described for *in situ* hybridization with an oligonucleotide probe.

Control Studies

To certify the specificity of the hybridization signal, some control studies are recommended. One is hybridization with radiolabeled sense RNA probes, and another control is RNase A pretreatment (1 μg/ml RNase A in RNase buffer for 30 min at 37°) just before hybridization. If the signal is specific, these experiments should show no positive signals.

Observation of Developed Sections

Developed sections can be observed as two different images: a dark-field image and a bright-field image. The dark-field image is convenient to determine general distribution. Figure 2 shows examples of dark-field photomicrographs. Expression of glutamate receptor 1–4 (GluR1–4) mRNAs at the level of the amygdala (A–D) and anterior thalamic nuclei (E–H) was investigated with *in situ* hybridization histochemistry using specific oligonucleotide DNA probes against GluR1–4 mRNAs.[3] White grains indicate cells expressing glutamate receptor mRNAs. GluR1–4 mRNAs are abundantly expressed in the cerebral cortex and caudate putamen.

The bright-field image gives single-cell resolution. Figure 3 gives examples of bright-field photomicrographs. GluR1–4 mRNA expression was investigated in the hippocampus and cerebellum. Many black dots, which are exposed silver grains, are accumulated on cells expressing glutamate receptor mRNAs. In the hippocampus, pyramidal cells in CA1 express GluR1–4 mRNAs with different intensity (Figs. 3A–D). Purkinje cells strongly express GluR2 and 3 mRNAs (Figs. 3F and 3G, arrowheads).

Section III

Electrophysiology

[11] Patch-Clamp Studies of Cystic Fibrosis Transmembrane Conductance Regulator Chloride Channel

By JOHN W. HANRAHAN, ZIE KONE, CERI J. MATHEWS, JIEXIN LUO, YANLIN JIA, and PAUL LINSDELL

Introduction

The cystic fibrosis transmembrane conductance regulator (CFTR) has two sets of six membrane-spanning regions (TM1–TM12), two nucleotide-binding folds (NBF1, NBF2), and a regulatory domain containing numerous potential sites for phosphorylation by protein kinases.[1] It belongs to an important superfamily of ATP-binding cassette (ABC) transport proteins that has members in bacteria, yeast, and higher eukaryotes. Although it may have multiple functions, it is generally accepted that CFTR functions as an ATP-dependent, phosphorylation-activated Cl^- channel.[2–6] In this article we describe methods that we have found useful for studying the channel activity of CFTR. The reader is referred to Vol. 207 of this series for more general information on patch-clamp techniques.[6a]

Choice of Expression Systems

Functional studies of endogenous CFTR have been carried out using epithelial[7–11] and cardiac cells.[12–14] However, because it is often difficult to

[1] J. R. Riordan, J. M. Rommens, B.-S. Kerem, N. Alon, R. Rozmahel, Z. Grzelczak, J. Zielenski, S. Lock, N. Plavsic, J.-L. Chou, M. L. Drumm, M. C. Iannuzzi, F. C. Collins, and L.-C. Tsui, *Science* **245,** 1066 (1989).

[2] M. P. Anderson, D. P. Rich, R. J. Gregory, A. E. Smith, and M. J. Welsh, *Science* **251,** 679 (1991).

[3] N. Kartner, J. W. Hanrahan, T. J. Jensen, A. L. Naismith, S. Sun, C. A. Ackerley, E. F. Reyes, L.-C. Tsui, J. M. Rommens, C. E. Bear, and J. R. Riordan, *Cell* **64,** 681 (1991).

[4] M. P. Anderson, R. J. Gregory, S. Thompson, D. W. Souza, S. Paul, R. C. Mulligan, A. E. Smith, and M. J. Welsh, *Science* **253,** 202 (1991).

[5] C. E. Bear, C. Li, N. Kartner, R. J. Bridges, T. J. Jensen, M. Ramjeesingh, and J. R. Riordan, *Cell* **68,** 809 (1992).

[6] D. C. Gadsby, G. Nagel, and T.-C. Hwang, *Annu. Rev. Physiol.* **57,** 387 (1995).

[6a] B. Rudy and L. E. Iverson, eds., *Methods Enzymol.* **207** (1992).

[7] M. A. Gray, J. R. Greenwell, and B. E. Argent, *J. Membr. Biol.* **105,** 131 (1988).

[8] M. A. Gray, C. E. Pollard, A. Harris, L. Coleman, J. R. Greenwell, and B. E. Argent, *Am. J. Physiol.* (*Cell Physiol.*) **259,** C752 (1990).

[9] G. Champigny, B. Verrier, C. Gérard, J. Mauchamp, and M. Lazdunski, *FEBS Lett.* **259,** 263 (1990).

0076-6879/98 $25.00

obtain gigaohm seals on native epithelial cells, heterologous expression systems are generally preferred for biophysical studies. Many cell lines have been used successfully for transient and stable transfection, including Chinese hamster ovary (CHO) cells, fibroblasts, baby hamster kidney cells (BHK), and human enbryonic kidney (HEK 293) cells. The choice of preparations, of course, depends on the purpose of the experiment. We find transient expression of CFTR in *Spodoptera frugiperda* fall armyworm ovary (Sf9)[3] and *Xenopus* oocytes[15] to be inconvenient for detailed patch-clamp studies mainly due to the difficulty of controlling the level of expression. In our experience, the additional work involved in the preparation of stable lines is worthwhile (see Ref. 16). Hi-5 insect cells are reportedly[17] superior to Sf9 cells because they do not have other channel types that could potentially be confused with CFTR.[18] Single CFTR channels can be recorded on *Xenopus* oocytes, but this requires careful titration of the amount of RNA injected into the oocytes (or DNA for nuclear injections) to achieve low expression.

The ideal expression system for investigation of the unitary properties of CFTR would have a low but consistent level of CFTR expression so that, on average, one channel is observed per patch. This dream is seldom realized but patches containing single channels are sometimes obtained on CHO cells that have been stably transfected with pNUT–CFTR. We find 3–10 channels to be more typical.[19] The main disadvantage of CHO cells is the presence of endogenous channels, which have higher conductance and become active once membrane patches are excised. This problem is not serious when only short recordings are needed, but it becomes a severe limitation when long (10- to 20-min) recordings from excised patches are

[10] J. A. Tabcharani, W. Low, D. Elie, and J. W. Hanrahan, *FEBS Lett.* **270,** 157 (1990).

[11] C. Haws, W. E. Finkbeiner, J. H. Widdicombe, and J. J. Wine, *Am. J. Physiol.* (*Lung Cell. Mol. Physiol.*) **266,** L502 (1994).

[12] G. Nagel, T.-C. Hwang, K. L. Nastiuk, A. C. Nairn, and D. C. Gadsby, *Nature* **360,** 81 (1992).

[13] T. Baukrowitz, T.-C. Hwang, A. C. Nairn, and D. C. Gadsby, *Neuron* **12,** 473 (1994).

[14] P. Hart, J. D. Warth, P. C. Levesque, M. L. Collier, Y. Geary, B. Horowitz, and J. R. Hume, *Proc. Natl. Acad. Sci. U.S.A.* **93,** 6343 (1996).

[15] C. E. Bear, F. Duguay, A. L. Naismith, N. Kartner, J. W. Hanrahan, and J. R. Riordan, *J. Biol. Chem.* **266,** 19142 (1991).

[16] X.-B. Chang, J. A. Tabcharani, Y.-X. Hou, T. J. Jensen, N. Kartner, N. Alon, J. W. Hanrahan, and J. R. Riordan, *J. Biol. Chem.* **268,** 11304 (1993).

[17] I. C. H. Yang, T.-H. Cheng, F. Wang, E. M. Price, and T.-C. Hwang, *Am. J. Physiol.* **272,** C142 (1996).

[18] S. E. Gabriel, E. M. Price, R. C. Boucher, and M. J. Stutts, *Am. J. Physiol.* (*Cell Physiol.*) **263,** C708 (1992).

[19] J. A. Tabcharani, X.-B. Chang, J. R. Riordan, and J. W. Hanrahan, *Nature* **352,** 628 (1991).

desired, such as when studying modulation.[20] CFTR channel activity runs down rapidly when membrane patches are excised from CHO or epithelial cells into bath solution lacking protein kinase A (PKA; <100 sec at 20°, and <15 sec at 37°). This rundown can be advantageous when studying the pharmacology of the membrane-associated phosphatase, which presumably dephosphorylates and inactivates CFTR after patch excision.[21] We have found that the membrane-associated phosphatase activity of CHO cells can be used to remove constitutive phosphorylation on CFTR simply by holding excised, inside-out patches for at least 10 min before testing exogenous kinases (see below). However, because stable activity is not achieved, rundown prevents studies of CFTR regulation by exogenous phosphatases. Examining the direct effects of low ATP concentrations on channel gating is also problematic, because MgATP must be present as a substrate for PKA in order to maintain channel phosphorylation.

Fortunately, CFTR channels do not usually run down in patches excised from BHK cells, perhaps because CFTR expression levels are extremely high compared to those of endogenous phosphatases. As discussed later, BHK cells express up to 7000 channels per patch when transfectants are selected with 500 μM methotrexate.[22] This high density may enable most channels to escape dephosphorylation by endogenous phosphatases. These patches have very large currents and little rundown, therefore they are useful for macroscopic selectivity measurements[23] and for examining regulation by exogenous phosphatases.[24]

General Patch-Clamp Methods

Pipettes are prepared from borosilicate glass using a two-stage vertical puller (PP-83, Narishige Instrument Laboratory, Tokyo). For single-channel recording we use pipettes that have resistances of 4–6 MΩ when filled with 150 mM NaCl solution. For macroscopic current recording, a lower heat setting is used during the second pull so that pipettes have 1- to 2-MΩ resistance. We do not routinely coat pipettes (e.g., with Sylgard; Dow Corning, Midland, MI) because our experiments on the voltage-independent CFTR channel do not require rapid changes in membrane potential.

[20] Y. Jia, C. J. Mathews, and J. W. Hanrahan, *J. Biol. Chem.* **272,** 4978 (1997).

[21] F. Becq, T. J. Jensen, X.-B. Chang, A. Savoia, J. M. Rommens, L.-C. Tsui, M. Buchwald, J. R. Riordan, and J. W. Hanrahan, *Proc. Natl. Acad. Sci. U.S.A.* **91,** 9160 (1994).

[22] F. S. Seibert, J. A. Tabcharani, X.-B. Chang, A. M. Dulhanty, C. J. Mathews, J. W. Hanrahan, J. R. Riordan, *J. Biol. Chem.* **270,** 2158 (1995).

[23] P. Linsdell and J. W. Hanrahan, *J. Physiol.* (*Lond.*) **496,** 687 (1996).

[24] J. Luo, M. D. Pato, J. R. Riordan, and J. W. Hanrahan, *Am. J. Physiol.* **274** (*Cell Physiol.* 43) in press (1998).

Furthermore, because our CFTR-expressing cell lines have very good sealing properties, we do not find it necessary to firepolish the pipettes. The bath solution is always grounded through an agar bridge having the same ionic composition as the pipette solution. Single-channel currents are amplified (Axopatch 1B-C or 200 A-B, Axon Instruments, Inc., Foster City, CA), recorded on video cassette tape by pulse-coded modulation-type recording adapters (e.g., DR384, Neurodata Instrument Co., NY) and low-pass filtered during playback using eight-pole Bessel-type filters (900 LPF, Frequency Devices, Haverhill, MA). Alternatively, data are filtered and then sampled by a standard Digidata 2000 interface at frequencies that are 2- to 5-fold higher than the filter setting.

CFTR Channel Permeation

Selectivity

CFTR is highly selective for anions over monovalent cations and has anion permeability ratios consistent with a lyotropic sequence.[25–27] Low iodide permeability is commonly used as a diagnostic property of CFTR, but this criterion must be used with care since I^- permeability depends on the experimental protocol used. Single-channel studies indicate that P_I/P_{Cl} is initially high when first exposed to iodide ($P_I/P_{Cl} \sim 1.9$) but falls within tens of seconds to a very low value (Fig. 1). This switch leads to a hysteresis in the current–voltage relationship and is accelerated by holding the membrane patch at potentials that would drive Cl^- into the pore. The precise mechanisms of this switch are not presently known.[27] The switch to low iodide permeability does not occur under biionic conditions if the ionic strength is elevated to 400 m*M*.[28] Also, when only a fraction of the Cl^- is replaced by iodide, the reversal potential shifts in the direction expected when $P_I > P_{Cl}$. However, in most preparations, exposure to iodide eventually leads to a reduction in macroscopic conductance and $P_I < P_{Cl}$.

Because I^- exposure alters the permeability of the channel to I^- itself, the steady-state value of P_I/P_{Cl} after the switch should not be compared with the permeability ratios obtained for other ions in the absence of iodide.

[25] P. Linsdell, J. A. Tabcharani, and J. W. Hanrahan, *J. Gen. Physiol.* **110,** 365 (1997).

[26] P. Linsdell, J. A. Tabcharani, J. M. Rommens, X.-Y. Hou, X.-B. Chang, L.-C. Tsui, J. R. Riordan, and J. W. Hanrahan, *J. Gen. Physiol.* **110,** 355 (1997).

[27] J. A. Tabcharani, P. Linsdell, and J. W. Hanrahan, *J. Gen. Physiol.* **110,** 341 (1997).

[28] J. A. Tabcharani, X.-B. Chang, J. R. Riordan, and J. W. Hanrahan, *Biophys. J.* **62,** 1 (1992).

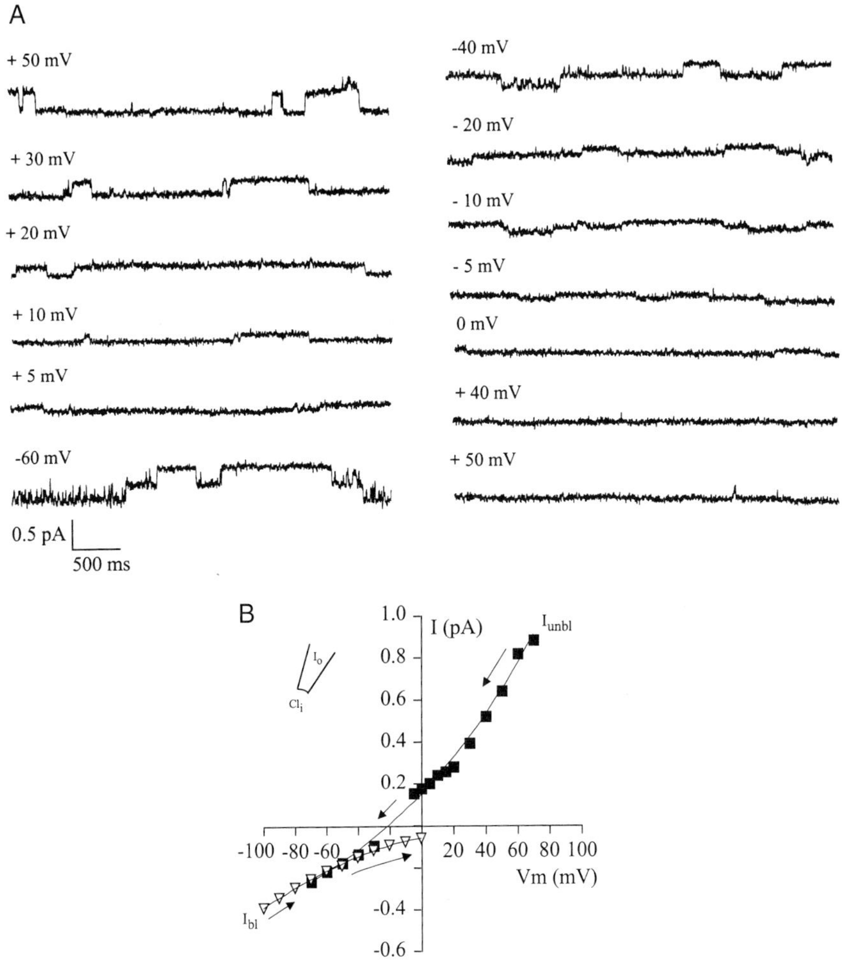

FIG. 1. Single CFTR channel currents recorded under biionic conditions showing both high (solid squares) and low (inverted triangles) permeability to iodide. The criterion of low iodide permeability needs to be used with caution when identifying CFTR's contribution to macroscopic Cl^- conductance. [With permission from J. A. Tabcharani *et al.*, *J. Gen. Physiol.* **110,** 341 (1997).]

CFTR shows a lyotropic or weak field strength selectivity sequence,[26,27] meaning that ions that are relatively easily dehydrated tend to be more permeant than those that retain their waters of hydration more strongly. P_I/P_{Cl} estimated before the switch is exactly the value expected based on the ratios for other halides.[27]

Selectivity is usually determined by exposing channels to different ions and measuring the resulting change in the zero current (reversal) potential. The simplest situation is to have equal concentrations of different ions present on either side. Under these biionic conditions, with Cl^- in the extracellular solution and a test anion X^- in the intracellular solution, the permeability of X^- relative to that of Cl^- (assuming zero cation permeability) is given by the Goldman–Hodgkin–Katz voltage equation:

$$P_X/P_{Cl} = \exp(E_{rev}F/RT) \quad (1)$$

where E_{rev} is the reversal potential, F is Faraday's constant, R is the gas constant, and T is the temperature in Kelvins. Because recordings made under biionic conditions involve different pipette and bath solutions, a liquid junction potential will occur between these two dissimilar solutions that must be corrected. Liquid junction potentials are determined for different anions relative to Cl^- as described previously[29] and are given in Table I.

Pore Size

CFTR also discriminates among anions on the basis of their size. The dimensions of the largest permeant anions presumably give some indication of the dimensions of the most constricted part of the pore. To estimate the pore diameter of CFTR we determine the unhydrated diameter of permeant ions, which can be related to permeability ratios according to an *excluded volume effect*.[30] Slight permeability to the external organic anions formate, bicarbonate, and acetate, and the apparent impermeability of the larger external anions propanoate, pyruvate, methane sulfonate, ethane sulfonate, and gluconate, suggested that the narrowest region of the CFTR pore has a functional diameter of ~5.3 Å.[26] In that study the functional diameter of an unhydrated ion was taken to be the geometric mean of its two smallest dimensions, based on the hypothesis that the ability of elongated, cylindrical shaped ions to pass through the channel would be relatively insensitive to their length.[26] Table I lists the minimum unhydrated dimensions of a large number of different anions, which we have estimated using Molecular Modeling Pro computer software (WindowChem Software Inc., Fairfield, CA).

Barrier Models

Permeating ions are thought to interact briefly with sites in the open pore. According to rate theory models, the free energies and positions

[29] E. Neher, *Methods Enzymol.* **27,** 123 (1992).

[30] T. M. Dwyer, D. J. Adams, and B. Hille, *J. Gen. Physiol.* **75,** 469 (1980).

TABLE I
CALCULATED MINIMUM DIMENSIONS AND MEASURED LIQUID JUNCTION POTENTIALS FOR DIFFERENT ANIONS[a]

Anion	Ion dimensions (Å)	Liquid junction potential (mV)
Chloride	3.62 × 3.62 × 3.62	0
Fluoride	2.72 × 2.72 × 2.72	−3
Bromide	3.90 × 3.90 × 3.90	0
Iodide	4.32 × 4.32 × 4.32	0
Cyanate (OCN^-)	3.56 × 3.56 × 5.68	ND[b]
$Au(CN)_4^-$	3.56 × 3.56 × 8.49	ND
Thiocyanate (SCN^-)	3.60 × 3.60 × 3.60	−1
$SeCN^-$	3.80 × 3.80 × 6.61	ND
Nitrate (NO_3^-)	3.10 × 4.76 × 5.17	−1
Formate	3.40 × 4.62 × 4.82	−3
$N(CN)_2^-$	3.65 × 4.47 × 7.71	ND
Bicarbonate	3.40 × 4.95 × 5.96	−7
Acetate	3.99 × 5.18 × 5.47	−5
Perchlorate (ClO_4^-)	4.63 × 4.64 × 4.91	−3
Propanoate	4.12 × 5.23 × 7.05	−6
Benzoate	3.55 × 6.30 × 8.57	−7
Pyruvate	4.09 × 5.73 × 6.82	−6
Hexafluorophosphate (PF_6^-)	4.92 × 4.98 × 5.16	−2
$Ni(CN)_4^-$	3.56 × 7.39 × 7.39	ND
Ethane sulfonate	4.94 × 5.37 × 7.44	−6
Methane sulfonate	5.08 × 5.43 × 5.54	−5
$C(CN)_3^-$	3.85 × 7.22 × 7.85	ND
Glutamate	4.67 × 6.52 × 10.78	−8
Isethionate	5.35 × 5.79 × 7.60	−8
Gluconate	4.91 × 6.86 × 12.09	−10
Glucoheptonate	5.50 × 6.51 × 13.54	−11
Glucuronate	5.23 × 7.73 × 9.43	−12
MES	6.00 × 6.99 × 9.65	−9
Galacturonate	6.51 × 6.66 × 8.10	−10
HEPES	6.64 × 6.64 × 12.86	−11
TES	6.63 × 6.75 × 11.35	−10
Lactobionate	7.57 × 9.32 × 13.11	−12

[a] Non-halide anions are given in order of increasing size (mean diameter, calculated as described in the text). Liquid junction potentials were measured as described (Neher, 1992) with 154 m*M* of the test ion in the intracellular (bath) solution and 154 m*M* Cl^- in the extracellular (pipette) solution.

[b] ND, not determined.

within the transmembrane electric field of binding sites and energy barriers control ion permeability and conductance. We have previously generated such models for Cl^- and gluconate permeation in CFTR using the AJUSTE computer program developed by Dr. Osvaldo Alvarez and colleagues.[31] The design and use of the program have been described in detail.[25,31] Models for CFTR permeation obtained using this program when the channel was assumed to have either two or three ion-binding sites are shown in Fig. 2. Both models can reproduce the permeation properties of CFTR under a range of ionic conditions.[26] The three-site model also explains the anomalous conductance of CFTR in Cl^-/SCN^- mixtures, and the loss of this behavior on mutation of a positively charged, pore lining amino acid. These models require many assumptions that are probably invalid. For example, anion-binding sites are assumed to be at fixed positions regardless of the presence of other anions in the pore. Nevertheless, such models provide a useful framework for thinking about anion permeation and the effects of mutations.

Macroscopic Current Recording from Excised Patches

When single-channel currents are too small to resolve (e.g., when studying relatively impermeant ions, fast channel blockers, or CFTR mutants having low conductance), we exploit the high expression of CFTR in BHK cells by recording macroscopic currents as described earlier (Fig. 3). This approach has been used to study CFTR channel block[23,25] and selectivity.[32,33] The macroscopic selectivity sequence for small anions is similar to that measured for single channels.[26,27] Reversal potentials are also obtained for large intracellular organic anions such as propanoate, which are not measurable as unitary currents (Fig. 4; see also Ref. 26). Large anions carry small unitary currents and also act as open channel blockers of Cl^- permeation.[34]

After obtaining a gigohm seal on the BHK cell, the patch is normally excised into bath solution containing MgATP but lacking protein kinase A. CFTR channels are inactive under these conditions, which allows the background or leak current to be measured. Current–voltage relationships are then obtained by applying a slow (37.5–100 mV/s) depolarizing voltage

[31] O. Alvarez, A. Villarroel, and G. Eisenman, *Methods Enzymol.* **207,** 816 (1992).

[32] P. Linsdell and J. W. Hanrahan, *Ped. Pulmonol.* **Suppl. 14,** 215 (Abstract) (1997).

[33] P. Linsdell, S.-X. Zheng, and J. W. Hanrahan, *Ped. Pulmonol.* **Suppl. 14,** 214 (Abstract) (1997).

[34] P. Linsdell and J. W. Hanrahan, *Am. J. Physiol.* (*Cell Physiol.*) **271,** C628 (1996).

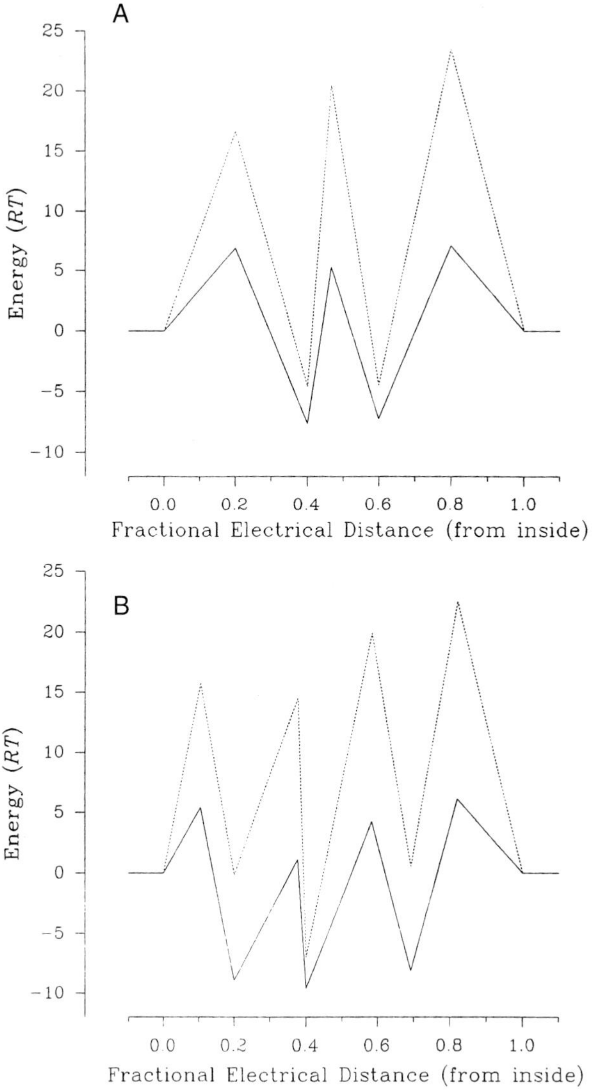

FIG. 2. Best fit free-energy profiles for Cl^- (solid line) and gluconate (dashed line) movement in the CFTR Cl^- channel pore, for both (A) the three-barrier, two-site model and (B) the four-barrier, three-site model. See Ref. 25 for parameters.

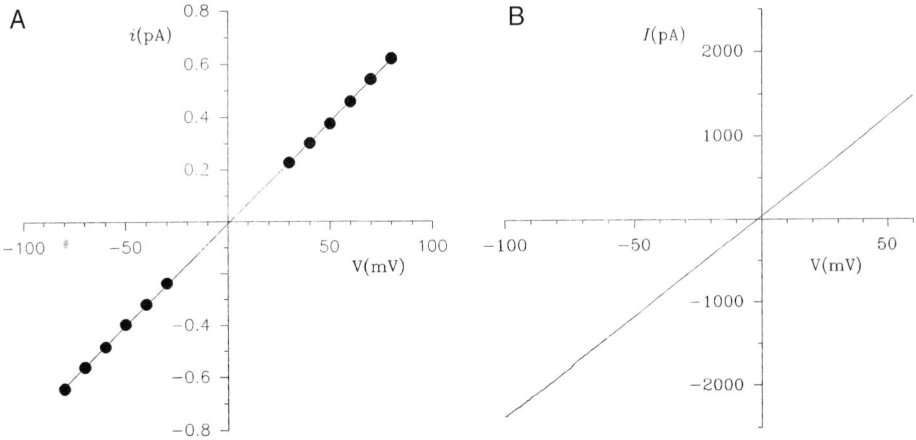

FIG. 3. Comparison of current–voltage relationship for (A) single-channel CFTR channel and (B) for the leak-subtracted macroscopic conductance of a BHK membrane patch. The slopes indicate conductances of 7.8 pS and 24.1 nS, respectively. Assuming similar P_0 values (0.4), these data indicate that the macroscopic current is carried by ~7700 channels.

ramp to the patch.[23] Adding PKA to the bath rapidly activates a large current, which is mediated exclusively by CFTR[23] (Linsdell and Hanrahan, unpublished observations, 1998). The leak current can be subtracted digitally to give a macroscopic current–voltage relationship for CFTR as in Figs. 3 and 4. This *I–V* curve usually has the same shape as the single-channel *I–V* relationship; however, this needs to be confirmed for mutants because there are instances in which mutations in the membrane-spanning segments can induce voltage-dependent gating.

Whole-Cell Currents

For whole-cell studies, pipette capacitance is canceled using the internal circuitry of the patch clamp while still in the cell-attached configuration. Excess suction is applied to break the patch, whole-cell capacitance is nulled, and currents recorded as described earlier. Current is measured while holding V_m at 0 mV, and during 1-sec alternating pulses to ±60 mV (1-sec duration). Under these ionic conditions, an increase in the outward current at 0 mV indicates elevation of Cl^- conductance. Current–voltage (*I–V*) relationships are generated by voltage steps from −100 to +100 mV in 10-mV increments (500-ms duration). Currents are plotted using Clampfit and the average current values between 400–500 ms are plotted against

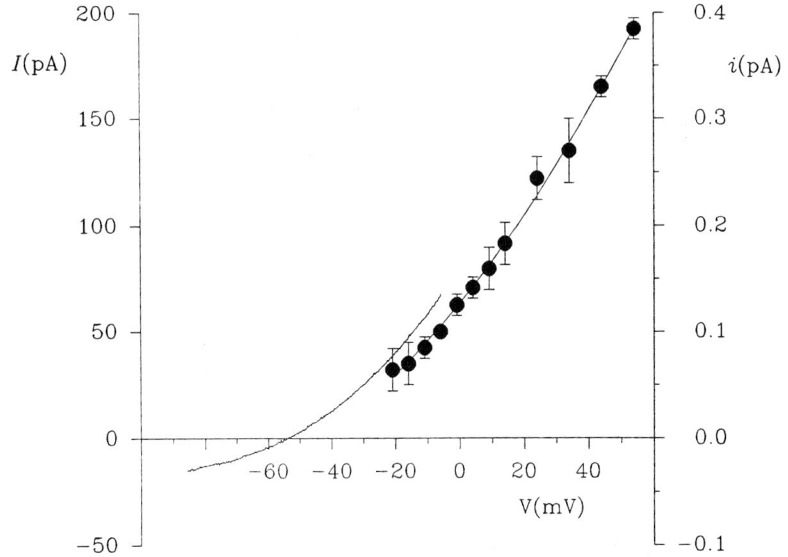

FIG. 4. Illustration of the usefulness of macroscopic selectivity measurements. The solid points indicate the current–voltage relationship obtained with external Cl^- and cytoplasmic propanoate. The single-channel currents were too small to allow calculation of a reversal potential (note scale on right). The solid curve without symbols shows a macroscopic *I–V* relationship (scale on left) that demonstrates finite propanoate permeability.

membrane potential. Currents are normalized to cell capacitance (pA/pF) to allow comparison of data from different cells.

Standard whole-cell patch-clamp methods are used when recording CFTR currents in transfected BHK cells. The standard bath (extracellular) solution contains (m*M*): 150 NaCl, 10 N-tris[Hydroxymethyl]methyl-2-aminoethanesulfonic acid, 2 $MgCl_2$ (pH 7.40), and is supplemented with 50 m*M* sucrose to prevent development of swelling-activated Cl current. The pipette (intracellular) solution contains (m*M*): 110 N-methyl-D-glucamine (NMDG)-aspartate, 30 NMDG-Cl, 1 $MgCl_2$, 10 TES, and 0.1 BAPTA [1,2-bis-(*o*-aminophenoxy)ethane-*N,N,N′,N′*-tetraacetic acid; tetrasodium salt] (pH 7.20) supplemented with 1 m*M* MgATP (from a 100 m*M* stock in buffer). Symmetrical 150 m*M* Cl^- solutions can also be used so that whole-cell *I–V* relationships can be compared with those for single channels; however, cell viability is compromised when intracellular chloride concentration is elevated. NMDG solutions have a short shelf-life and should be refrigerated, filtered (0.22 μm) frequently, and replaced every few days. Whole-cell CFTR currents are activated by adding cpt-cAMP or

8-bromo-cAMP (100 μM final concentration) to the bath solution from 50 mM stock solutions in water. The stocks are stable for several months when stored in small aliquots at −70°.

Regulation of CFTR Chloride Channel

Studying PKA Activation and Nucleotide Dependence of CFTR

Protein kinase A (PKA) catalyzes phosphorylation of CFTR at multiple sites[35–37] and activates the channel.[3,15,19,38,39] The major PKA sites phosphorylated *in vivo* are serines at positions 660, 700, 737, 795, and 813; however, studies of full-length CFTR and R domain peptides indicate that additional sites (namely, serine-422, -712, -753, and -768) can also be phosphorylated *in vitro*. Minor discrepancies between the sites phosphorylated in various studies probably reflect the different techniques and reagents used; e.g., serine at position 422 was phosphorylated in an NBF1-R domain peptide[40] but not in full-length CFTR[35]; whereas serine-753 was phosphorylated in full-length CFTR[22] but not in the peptide.[40] PKA phosphorylation sites on CFTR have distinct functions. Sites with distinct stimulatory and inhibitory effects have been identified.[41] We have recently suggested that one site, or set of sites, regulates burst duration (PKA_b) while another controls interburst duration (PKA_i).[42] A distinction has also been made between modulatory and activating PKA sites.[43] Some PKA sites capable of activating CFTR remain to be identified (the "cryptic" sites) because 10–15% responsiveness persists in a mutant (15SA) lacking serine-422 and all monobasic and dibasic PKA sites on the R domain (F. Seibert, unpubl. obs.; 1997). In defense of mutagenesis, this loss of activation by PKA is probably

[35] S. H. Cheng, D. P. Rich, J. Marshall, R. J. Gregory, M. J. Welsh, and A. E. Smith, *Cell* **66,** 1027 (1991).

[36] M. R. Picciotto, J. A. Cohn, G. Bertuzzi, P. Greengard, and A. C. Nairn, *J. Biol. Chem.* **267,** 12742 (1992).

[37] J. A. Cohn, A. C. Nairn, C. R. Marino, O. Melhus, and J. Kole, *Proc. Natl. Acad. Sci. U.S.A.* **89,** 2340 (1992).

[38] H. A. Berger, M. P. Anderson, R. J. Gregory, S. Thomson, P. W. Howard, R. A. Maurer, R. Mulligan, A. E. Smith, and M. J. Welsh, *J. Clin. Invest.* **88,** 1422 (1991).

[39] J. M. Rommens, S. Dho, C. E. Bear, N. Kartner, D. Kennedy, J. R. Riordan, L.-C. Tsui, and J. K. Foskett, *Proc. Natl. Acad. Sci. U.S.A.* **88,** 7500 (1991).

[40] R. R. Townsend, P. H. Lipniunas, B. M. Tulk, and A. S. Verkman, *Protein Sci.* **5,** 1865 (1996).

[41] D. J. Wilkinson, T. V. Strong, M. E. Mansoura, D. L. Wood, S. S. Smith, F. S. Collins, and D. C. Dawson, *Am. J. Physiol.* (*Lung Cell. Mol. Physiol.*) **273,** L127 (1997).

[42] J. Luo, M. D. Pato, F. S. Seibert, X.-B. Chang, J. R. Riordan, and J. W. Hanrahan, *Ped. Pulmonol.* **Suppl. 14,** 217 (Abstract) (1997).

[43] D. C. Gadsby and A. C. Nairn, *TIBS* **19,** 513 (1994).

not due to a general disruption caused by mutating the serines, because tyrosine phosphorylation causes very robust activation of the mutant.[44]

The phosphorylation state of CFTR must be controlled when studying ATP dependence of channel gating.[45,46] Phosphorylation of the dibasic PKA sites on CFTR reduces the EC_{50} for ATP-dependent gating and increases the maximum P_0 that can be achieved at high ATP concentrations.[46] Variations in the level of phosphorylation may help explain the wide range of EC_{50} values reported for ATP dependence in the literature. Also, demonstrating the strict dependence of CFTR gating on ATP may require exposure of patches to flowing ATP-free solution. Channel activity can persist for many minutes when patches are excised into stagnant bath solution that is nominally ATP free, even if the chamber is rinsed intermittently.

We use the catalytic subunit of type II bovine cardiac protein kinase A (PKA; ~0.3 mg/ml) to activate CFTR channels.[19] PKA is stable for more than 2 years at $-70°$ when stored in buffer containing (m*M*): 150 KCl (pH 7.0), 30 KH_2PO_4, 1 dithiothreitol (DTT), and 1 ethylenediaminetetraacetic acid (EDTA; disodium salt), however it should not be thawed and refrozen. Its potency declines after approximately 1 hr at room temperature (~23°); therefore, small aliquots are thawed and kept at 4° until used, typically within 2–3 weeks. In North America, Promega (Madison, WI) or Upstate Biotechnology Inc. (UBI, Lake Placid, NY) are reliable, though expensive, sources. Lyophilized PKA is avoided because of its low activity. MgATP and other nucleotides are stored at $-70°$ as 100 m*M* stock solutions in buffer (pH 7.4), and fresh aliquots are thawed each day. Note that all commercially available nucleotide preparations contain contaminants (nucleoside mono- and diphosphates).

Burst Analysis in Patches with Multiple CFTR Channels

The open probability of CFTR depends primarily on alterations in its burst kinetics. Open burst duration may reflect the NBF–NBF interactions that control ATP turnover at NBF1[13,45,47,48] (discussed in Ref. 46). Unfortunately, most patches contain multiple channels, which precludes use of the

[44] Y. Jia, F. Seibert, X.-B. Chang, J. R. Riordan, and J. W. Hanrahan, *Ped. Pulmonol.* **Suppl. 14,** 214 (Abstract) (1997).

[45] C. Li, M. Ramjeesingh, W. Wang, E. Garami, M. Hewryk, D. Lee, J. M. Rommens, K. Galley, and C. E. Bear, *J. Biol. Chem.* **271,** 28463 (1996).

[46] C. J. Mathews, J. A. Tabcharani, X.-B. Chang, J. R. Riordan, and J. W. Hanrahan, *J. Physiol. (Cambr.)* **508,** 365 (1998).

[47] T.-C. Hwang, G. Nagel, A. C. Nairn, and D. C. Gadsby, *Proc. Natl. Acad. Sci. U.S.A.* **91,** 4698 (1994).

[48] K. L. Gunderson and R. R. Kopito, *J. Biol. Chem.* **269,** 19349 (1994).

threshold crossing method for estimating open and closed times. To evaluate kinetics in multichannel patches, the mean number of channels open is determined by measuring the fraction of time spent at each multiple of the unitary current. The single-channel open probability (P_0) is calculated from:

$$P_0 = \sum_{i=1}^{N} t_i/TN \tag{2}$$

where t_i is the time spent above a threshold i set at 0.5, 1.5, 2.5, . . . , times the single channel amplitude, N is the number of channels locked open at the end of the experiment using AMP–PNP (see below), and T is the duration of the segment (typically >120 sec). The number of opening transitions during each segment is counted (see below) and used to estimate the mean burst (τ_{open}) and interburst (τ_{closed}) durations from the "cycle time" according to:

$$\tau_{open} = [(NP_0)T]/(n) \tag{3}$$

$$\tau_{closed} = [(N - NP_0)T]/(n - 1) \tag{4}$$

where N is the number of channels locked open by AMP–PNP (see below), T is the duration of the segment, and n is the number of identifiable bursts.

This method assumes there is one open and one closed state, although it is clear by inspection of recordings that CFTR has at least two closed states (interburst closures and flickery closures within bursts; see e.g., Refs. 49 and 50). Fortunately, brief closures within bursts have little impact on P_0 and do not depend on nucleotides. When they are excluded from the analysis, the kinetics of CFTR are well described by long openings (actually bursts) and closings (interbursts). To estimate the number of functional channels in patches (N) we add the nonhydrolyzable nucleotide adenylyl imidodiphosphate (AMP–PNP; tetralithium salt) to the bath from a 100 mM stock solution at the end of each experiment (final concentration of 1 mM). When P_0 is low (e.g., at low [ATP] or when channels are partially phosphorylated), the value of N obtained by locking channels with AMP–PNP exceeds that estimated from the maximum number that open simultaneously during long recordings. Although AMP–PNP does not cause locking at physiologic temperatures, it still increases P_0 and therefore improves the estimate of N.

Studying Regulation by PKC

In addition to the well-known activation of CFTR by PKA, it has recently become clear that PKC phosphorylation also has profound effects

[49] C. Haws, M. E. Krouse, Y. Xia, D. C. Gruenert, and J. J. Wine, *Am. J. Physiol.* (*Lung Cell. Mol. Physiol.*) **263,** L692 (1992).

[50] H. Fischer and T. E. Machen, *J. Gen. Physiol.* **104,** 541 (1994).

on CFTR and is required for activation by PKA.[20] Patch-clamp studies indicate that CFTR is constitutively phosphorylated by PKC (and PKA) in several mammalian cell types. This dependence must therefore be considered when designing experiments to study activation by other protein kinases. CFTR channel activity was weakly stimulated (10%) when freshly excised, multichannel patches were exposed to PKC.[19] However more recent work has revealed that PKC phosphorylation alone has no effect after prolonged rundown,[20] when membrane-associated phosphatases have removed constitutive phosphorylation (Fig. 5). We now interpret the stimulation by PKC observed previously as resulting from phosphorylation of permissive PKC sites, and reactivation of a few channels that still had residual PKA phosphorylation. Since the level of constitutive phosphorylation by PKC undoubtedly varies among cell lines and under different experimental conditions, PKC regulation must be considered when assessing the

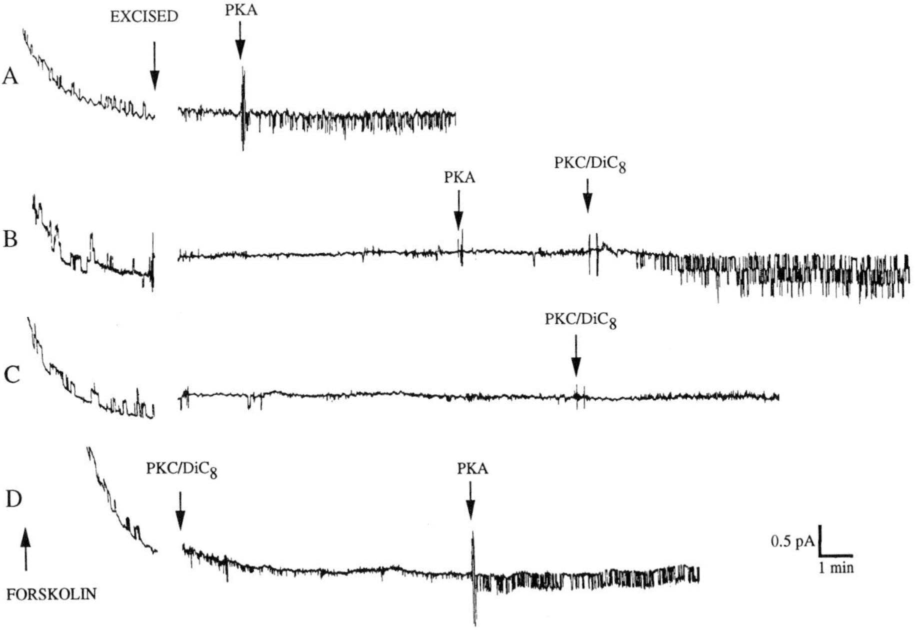

FIG. 5. The responsiveness of CFTR channels to PKA catalytic subunit under various conditions reveals the importance of constitutive phosphorylation. PKA activates CFTR channels when added soon after excision (A), but not after 10 min, although subsequent addition of PKC and the lipid activator DiC_8 does allow restimulation (B). PKC does not stimulate CFTR channels in the absence of PKA phosphorylation (C), but when present preserves responsiveness to PKA (D). [With permission from Y. Jia *et al.*, *J. Biol. Chem.* **272,** 4978 (1997).]

effects of other kinases or when comparing the responsiveness of CFTR in different preparations, particularly if negative results are obtained.

To study the role of PKC, the inhibitors chelerythrine chloride and Gö6976 are prepared in dimethyl sulfoxide (DMSO, 10 and 5 m*M* stocks, respectively) and stored in tightly sealed containers under nitrogen gas at −20°. Fresh aliquots are thawed each day and diluted immediately before use to yield final concentrations of 1 μM for chelerythrine, and 500 n*M* for Gö6976. Preincubating cells with chelerythrine in culture medium in a 5% CO_2 incubator at 37° for 30–120 min prior to patch-clamp recording abrogates PKA stimulation of CFTR channels in 80% of membrane patches. Gö6976 is routinely added to both the bath and pipette solutions, and several minutes are allowed to elapse after entering the whole-cell configuration to ensure adequate diffusion of Gö6976 into the cell before currents are recorded (see Ref. 51). For studies of exogenous PKC we use calcium and phospholipid-dependent PKC II from rat brain,[19] although other isoforms give similar results.[52] After thawing an aliquot, the PKC is kept at 4° and can be used over a period of 1–2 weeks. We also add 1,2-dioctanoylglycerol (8:0) DiC_8 to the bath solution. The DiC_8 is dissolved in chloroform at 10 m*M*, stored under nitrogen gas in a tightly sealed container at −20°, and diluted in bath solution immediately before being added at 5 μM final concentration.

Studying Regulation by Src

Tyrosine phosphorylation is not detectable on CFTR under control conditions; however, the tyrosine kinase $p60^{c\text{-}Src}$ is a potent channel activator when added to excised patches and phosphorylates immunoprecipitated CFTR protein *in vitro*.[53] Moreover, tyrosines on CFTR become phosphorylated *in vivo* when v-*Src* is coexpressed with CFTR. These results have implications for both the specificity and mechanism of channel regulation by phosphorylation. Although the physiologic relevance of tyrosine phosphorylation remains to be determined, at the very least, the results indicate that many kinases other than PKA and PKC may participate in regulating CFTR. To study CFTR regulation in excised patches, we add 30 U/mL of $p60^{c\text{-}Src}$ from UBI (Lake Placid, NY) to the cytoplasmic side. Control experiments are carried out using $p60^{c\text{-}Src}$, which has been heated to 100° for 10 min to destroy its enzymatic activity. When cells are transiently

[51] A. Marty and E. Neher, *in* "Single Channel Recording" (B. Sakmann and E. Neher, eds.) Vol. 2, pp. 31–52. Plenum Press, New York, 1995.

[52] H. A. Berger, S. M. Travis and M. J. Welsh, *J. Biol. Chem.* **268,** 2037 (1993).

[53] Y. Jia, M. A. Loo, F. Seibert, T. J. Jensen, Y. X. Hou, L. Cui, X.-B. Chang, D. M. Clarke, J. R. Riordan, and J. W. Hanrahan, *J. Biol. Chem.* submitted (1998).

transfected with an expression plasmid containing v-*Src*, CFTR channels can be activated in excised patches simply by exposing the patches to 1 m*M* MgATP. This activity is greatly enhanced by the protein tyrosine phosphatase (PTP) inhibitor dephostatin; therefore, there is an endogenous, membrane-associated PTP, which can dephosphorylate tyrosine residues on CFTR.

Studying Regulation by Protein Phosphatases

CFTR channel activity declines rapidly when patches are excised from CHO or human airway cells. This rundown is mediated by a membrane-associated phosphatase activity that is not sensitive to okadaic acid or calyculin A nor is it dependent on calcium and calmodulin,[19,21] but it does require magnesium consistent with PP2C.[24] As mentioned earlier, this rundown does not usually occur in patches from BHK cells expressing very high levels of CFTR. Adding PP2A, PP2C, or alkaline phosphatase individually to BHK patches with stable CFTR activity reduces P_0 by more than 90% but does not abolish it. PP1 and PP2B have no effect on channel activity.[24,52] Deactivation by PP2C is more rapid than by PP2A or alkaline phosphatase and has a time course resembling the spontaneous rundown, which is sometimes observed even in membrane patches excised from BHK cells. Deactivation by exogenous PP2A is associated with dramatic shortening of the mean burst duration. However, burst duration does not change after addition of PP2C or during spontaneous rundown (when it occurs). Thus, functionally distinct PKA sites may differ in their sensitivities to protein phosphatases; sites controling burst duration (PKA_b) are susceptible to PP2A but not PP2C, whereas those regulating interburst duration (PKA_i) can be dephosphorylated by either phosphatase.

PP1, PP2A, and PP2B are available from Promega and UBI. UBI also sells antibodies to PP2A, recombinant PP2Cα protein, and a polyclonal antibody to PP2Cα, although they are expensive. We have used PP2Cα, which is isolated from chicken gizzard.[54,55] Commercially available PP2A is avoided because of its low activity. Regardless of the source, the phosphatase activity of any enzyme preparation should be assayed prior to use, preferably under conditions that approximate those present during patch-clamp experiments. Full-length phospho-CFTR would be the most appropriate substrate for such studies but is difficult to purify in sufficient quantities for measuring release of radiolabeled phosphate. R domain peptide can be expressed in bacteria as a glutathione *S*-transferase fusion protein or with a histidine tag, purified, and then phosphorylated *in vitro*. Alterna-

[54] M. D. Pato and R. S. Adelstein, *J. Biol. Chem.* **258,** 7055 (1983).
[55] M. D. Pato and R. S. Adelstein, *J. Biol. Chem.* **258,** 7047 (1983).

tively, phosphorylated myosin light chains or other phosphatase substrates can be used, although the relative efficiency of dephosphorylation by the phosphatases will depend on substrate used; for example, PP2C dephosphorylates myosin light chains much more effectively than phosphocasein.

The protein phosphatases are stable for at least 6 months when stored as small aliquots at $-20°$; however, PP1 activity declines 2- to 3-fold once thawed and stored for 2–3 hr on ice. Protein phosphatases can be diluted in 50% glycerol and stored without freezing at $-20°$, although the glycerol may cause problems later when added during patch-clamp experiments. Patch-clamp solutions usually contain 150 m*M* NaCl; however, we found that this salt partially inhibits protein phosphatases relative to the same buffer lacking NaCl; that is, PP1 activity is reduced by 72%, PP2A by 66%, PP2B by 38%, and PP2C by 20%.[42] We elevate the Mg^{2+} concentration of the bath to 12 m*M* when studying regulation by PP2C, which requires ~2 m*M* Mg^{2+} for half maximal activity. Phosphatase inhibitors such as vanadate and fluoride must be used with care because they may act directly on CFTR. The PP1 and PP2A inhibitor calyculin A is more effective than okadaic acid when added to intact cells, although both have IC_{50} values in the 1- to 10-n*M* range *in vitro*. We typically use 10–100 n*M* of calyculin A or 1–10 μM of okadaic acid. Calyculin A concentrations higher than 100 n*M* are toxic to the human colonic epithelial cell line T_{84}. Neither inhibitor affects PP2C (at least up to 1 m*M* okadaic acid). Microcystin permeates poorly through cell membranes but is useful for whole-cell and inside-out patches.[56]

DRSCAN: A pCLAMP-Compatible Program for Analyzing Long Records

The pCLAMP suite of programs (Axon Instruments Inc.) copes adequately with well-behaved data. However, in our studies of CFTR (<10 pS) we found that setting the optimum threshold was sometimes difficult because the cursor position could not be set for individual traces. Moreover, experiments often last more than 20 min and involve several interventions; which require repeated switching to the General Parameters menu to enter start and end times of segments for each analysis. More serious problems arise when patches contain more than one channel, which is usually the case. To overcome these limitations while retaining the many useful features of pClamp, we have developed a companion program with a user-friendly graphical interface. The structure of this program, which we call DRSCAN,

[56] T.-C. Hwang, M. Horie, and D. C. Gadsby, *J. Gen. Physiol.* **101,** 629 (1993).

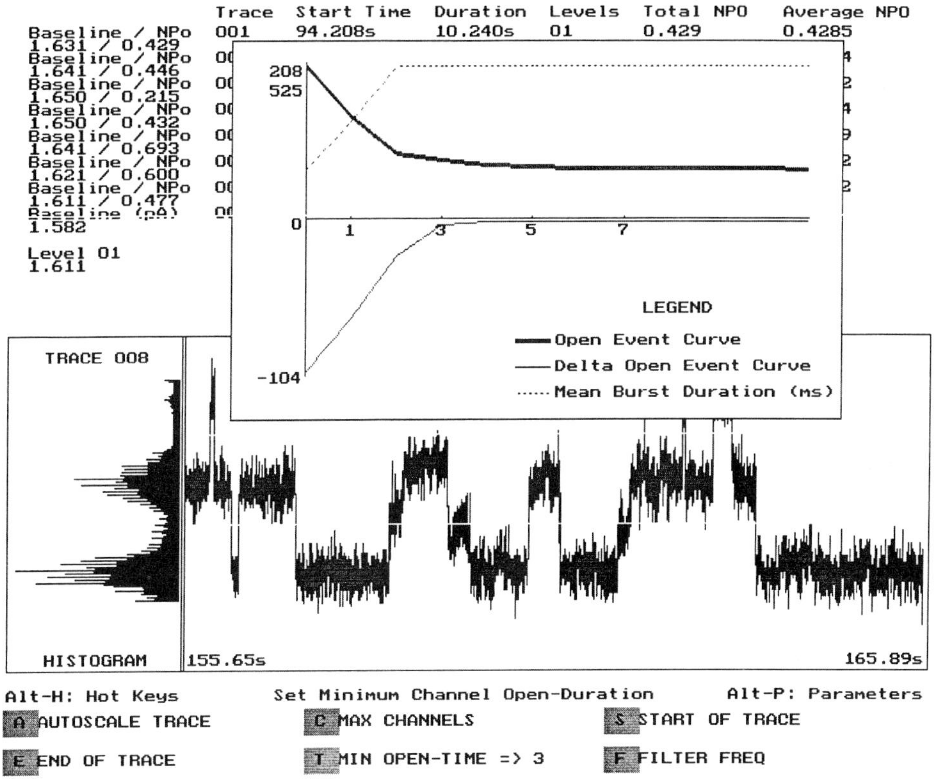

FIG. 6. The "event analysis" display provided by DRSCAN. The thin white lines correspond to event levels set by the user. Curves in the pop-up window guide the selection of durations for accepting or rejecting events.

is shown in Fig. 6. It reads pClamp Axon Binary Format (.ABF) data files that are acquired using FETCHEX in gap-free mode. The gain and duration of the displayed data are easily altered, and threshold level(s) can be set for individual screens using cursors. To aid threshold positioning, an all-points histogram is automatically drawn adjacent to the current trace (Fig. 6). DRSCAN performs threshold-crossing analysis for up to 99 cursors and calculates a running value for NP_0, mean burst duration, and mean interburst duration as described earlier. The results of DRSCAN analyses are output in Axon.EVL format and also in text format, which can be imported into graphing and statistical programs such as Origin 4.10 (Microcal, Northampton, MA). The program also calculates current amplitudes for selected segments of data using Gaussian fits to the all-point histogram,

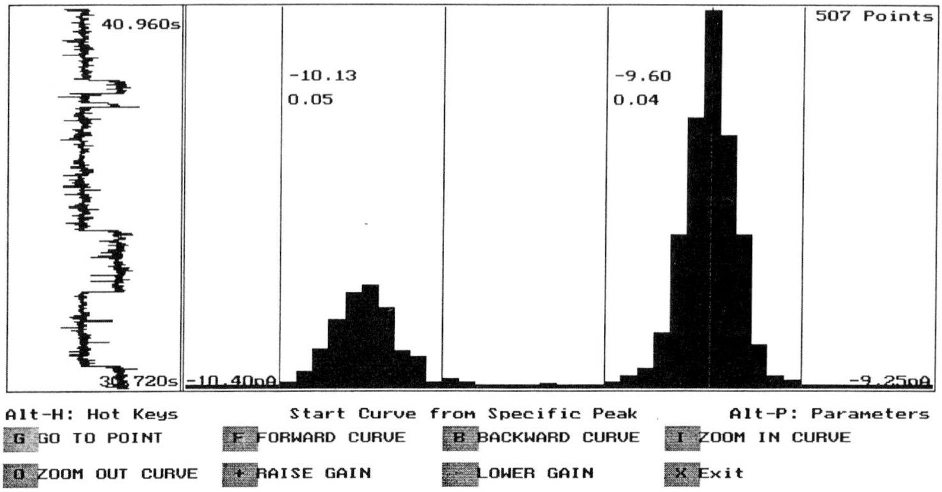

FIG. 7. Current–voltage analysis using Gaussian fits of all-points histograms. One-sided fits to the peaks or manual cursor positioning can be used when the histograms are skewed due to filtering.

allowing the generation of *I–V* relationships (Fig. 7). A brief overview of DRSCAN is given in the Appendix. Similar features could be incorporated into other programs intended for analyzing long continuous recordings of CFTR.

Output of Single-Channel Current Records for Production of Figures

Whether DRSCAN or pCLAMP is used for data analysis, routine plotting of current records is easily achieved using the Export Plot commands of the FETCHAN program. Although the output is not easily edited or of high quality, we find that Origin 4.10 (e.g., 16-bit version with the optional pCLAMP plug-in; Microcal) is a convenient and versatile way to plot and edit Axon binary data files, and yields publication-quality figures with multiple panels. A digital (Gaussian) filter can be applied to the data in FETCHAN by specifying a low-pass cutoff frequency from the General Parameters menu. The built-in fast Fourier transform (FFT) filter of Origin 4.1 also works well and improves the clarity of recordings.

Appendix

Hardware and Software Requirements

DRSCAN runs on stand-alone PCs with minimal hardware requirements. It requires a 486 or higher processor with at least 8 Mb of RAM and 1 Mb of free hard disk space. It was written using the Borland TURBO C++ for DOS compiler, and uses the TURBO C++ graphics library; therefore, it also requires at least an enhanced graphics adapter. An Axon Instruments compatible hardware interface is required for data acquisition. To prevent aliasing, data are digitized at a rate which is at least twice the analog filter cutoff frequency (−3 dB; eight-pole Bessel filter). DRSCAN requires MS DOS 3.31 or higher, although it can also be run as a DOS command from within any version of Microsoft Windows. It is a companion to the pCLAMP suite of programs, which is also required for data acquisition.

Program Description

DRSCAN is compiled as a project (TURBO C++ .PRJ) file consisting of both source code modules and object modules. The object modules are precompiled TURBO C++ graphics files (.OBJ files) linked with a set of specialized modular source code files (Fig. 8). These handle application file I/O routines (FIOROUT), DOS file system interface (DROUTINE), graphical user interface (GINTFACE), Axon binary file interface (ABFROUT), histogram generation algorithms and plotting routines (HISTROUT), and event list analysis algorithms (ANALYSIS, EVNTANAL). Finally, there are two display formatting modules, one for the text mode interface (TINTFACE) and another for the graphical mode interface (GRAFROUT).

Text Mode Interface (TINTFACE) Module

TINTFACE is a navigation system between submenus. It takes user requests in the form of menu item selections and communicates with the DROUTINE module, which specializes in reading and writing to and from the DOS file system. TINTFACE allows users to view the application output files in text format. On receiving a user request for an *I–V* analysis or an event list analysis, TINTFACE builds the necessary user application files and passes control to the ANALYSIS module. On running DRSCAN, the program parameters are validated by this module. The minimal parameter for starting DRSCAN are the program name and a user name, as in *"DRSCAN User_Name"*. The following files are automatically built:

1. A \DRSCAN\Data\User_Name directory is created (if none exists). All user files are kept here, separate from the application files located in the \DRSCAN root directory.
2. A user profile parameter file is created if none exists, to store the user's last data and analysis files source directories, display preferences, etc.

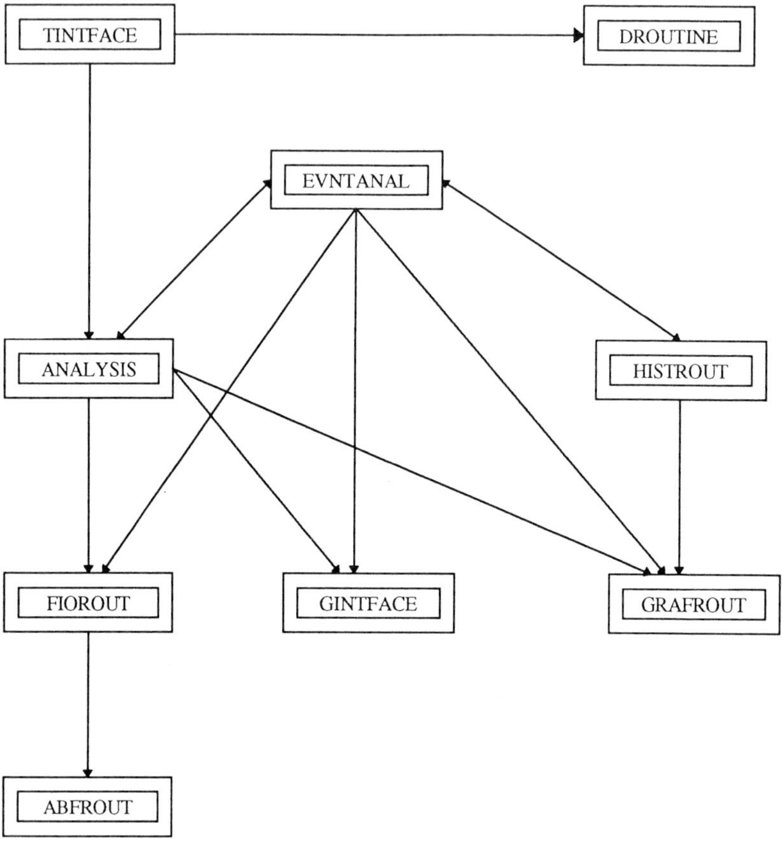

FIG. 8. Scheme showing the interrelationships betwen DRSCAN modules. An arrow from module M_1 to M_2 means that module M_2's computations are triggered by a service request issued in module M_1.

3. A Trace_Log file is created, which records chronological actions, as the user analyzes records in the open data file.
4. A circular event list file is created consisting of the FETCHEX data file name with extension ".EV?" where "?" can be either "L" or a number between 1 and 9. This allows users to have up to 10 different analyses on the same data file with results stored in a meaningful, easy-to-retrieve file-naming convention. The same holds for the *I–V* analysis, although with different extension conventions. Event list files are Axon ".EVL" file format compatible and therefore can be read by other programs such as Axon's READEVNT or PSTAT.

DOS File System Interface (DROUTINE) Module

This module specializes in handling the DOS file system. It maintains and displays the application directory entries, services formatting requests issued by TINTFACE, and organizes users' working spaces by keeping them apart from one another and separate from the application-wide system files. It adds flexibility to file naming and buffers DOS I/O calls to improve system performance.

ANALYSIS Module

This module is the driver for both the *I–V* and event list analyses. It is also the repository for application utilities (such as sorting and help routines). It is activated by TINTFACE, and it determines whether the user has requested *I–V* or event list analysis. It then performs a system initialization and loads the right table (help, menu, hot keys, short-cut keys, etc). It orders the correct screen resolution for both the histogram and trace curves. Requests are made to EVNTANAL for loading corresponding algorithms, to GINTFACE for setting the user graphical display resolution, to HISTROUT for processing and displaying the trace histogram, to GRAFROUT for displaying the trace, and to FIOROUT for reading Axon FETCHEX data files.

EVNTANAL Module

This module handles the algorithmic implementation of the DRSCAN application. It allows the user to select records in the data file to be processed, to set baselines, to correct baseline drift, and to set event level cursors at the desired place on the trace window. EVNTANAL also computes the NP_0 for the segment of data selected. This quantity is weighted by the duration of the given interval. The mean NP_0, total running NP_0, and "NP_0 so far" values are also computed (Fig. 6). Event files created by EVNTANAL reflect the true behavior of the data, as recorded by FETCHEX. This means that a typical event file created by EVNTANAL may contain events of zero length when there is a transition across two or more cursor levels.

The weighted NP_0 computation is based on Eq. (2), except that T is the duration of the trace in the window. At the end of an analysis, EVNTANAL can build a "reprocessed" event list file, which excludes events of duration less than some user specified parameter. The default is two data points. To help the user choose an optimal minimum duration to be excluded, a *"Delta Open Event Curve"* curve is also generated showing the number of open events computed during the segment of data as a function of event durations excluded (Fig. 6). The inflection point of this curve is taken as the optimal "minimum event length" that eliminates the most false events without excluding true ones. Analytically this point corresponds to the null value of the first derivative of the function f, such that $f(x) = y_i - y_{i-1}$ $(i \geq 2)$ and y_i is the number of events whose duration expands over i data points. In addition to ignoring spurious brief events, the event list reprocessing routine can calculate mean burst (τ_{open}) and interburst durations (τ_{closed}) according to Eqs. (3) and (4).

EVNTANAL also outputs the following files:

1. An event list file, whose format matches the Axon ".EVL" binary file. Event list files created by EVNTANAL record all events as they appear in the experiment file. All events are detected and written to the file whenever the current level crosses the threshold set by the user at the beginning of each trace analysis.
2. One or more reprocessed event list files. Reprocessed event list files rebuild the event list file created above by eliminating false events due to noise in the experiment file. This noise elimination process is carried out empirically using the first derivative of the *"Delta Open Event"* curve as described earlier.
3. An analysis log file. The analysis log file saves on disk the contents of the user graphical display screen. In the case of an event analysis this would contain information concerning the portion of the experiment records being processed. The analysis log file in this case would specify a trace starting time and a duration, the number of event levels detected in that window, and the NP_0 values for the trace, that is, "Trace NP_0", "Total Analysis NP_0 so far" and "Mean NP_0 so far". In the case of an I–V analysis, the log file would give the number of current intervals for each trace (also called "Peaks"), the range of current values in each peak, and the mean current value and its standard deviation.

HISTROUT Module

This module constructs the all-points amplitude histogram for the displayed current trace. The user adjusts the length and start time of the current trace to be analyzed according to the number of transitions and severity of baseline drift. HISTROUT allows the split screen to be toggled between "active" mode, which allows interactive cursor setting, and "passive" mode, in which the all-points amplitude histogram is simply displayed to the left of the trace (Fig. 6). Histogram bin width is calculated automatically based on the horizontal resolution of the screen and the current amplitude range of the selected data trace. The mean and variance of each peak is obtained by fitting data between two cursors with a Gaussian curve and these are displayed on the histogram. For example, Fig. 7 shows two peaks and user-defined vertical cursors. The mean currents and variances are −10.13 and −9.60 pA, and 0.05 and 0.04 pA, respectively.

HISTROUT is very precise in computing these parameters because it processes the file records point by point without unnecessary estimation. As a result most DRSCAN algorithms run in quadratic complexity time $O(n^2)$, with the exception of the event level detection, which runs in cubic complexity time $O(n^3)$.

File I/O (FIOROUT) Module

This module serves as an interface to the Axon binary file system, the ABFROUT. FIOROUT receives calls from EVNTANAL and ANALYSIS for reading Axon FETCHEX acquisition data files. It routes the same call with the appropriate Axon ".ABF" parameters. Selected channel reading and channel multiplexing reading routines in the Axon ABFFILES.C program are used to receive and service function calls from FIOROUT. ABFFILES.C is a public domain ".ABF" I/O utility program that comes with the distribution diskettes of pClamp6. We have modified some functions in that program to handle the reading and formatting of acquisition file records, which are then returned to the ANALYSIS or EVNTANAL modules.

ABFROUT Module

This is the collection of Axon ABFFILES.C I/O routines that deal with reading data records from data files.

Graphical Interface (GINTFACE) Module

This module handles the graphical user interface, receiving user keystrokes and translating them into application procedure calls. GINTFACE contains display resolution computation. It loads font files and interacts with the screen display aspect by detecting the hardware in use and managing the display area of the screen. All graphical menu item selections are serviced by this module. As such, it interacts with ANALYSIS (for displaying application global help topics and hot keys), and manages navigation among the various modules that generate user interface processes. GINTFACE also interacts with EVNTANAL for displaying the cursor levels set by the user (in Fig. 6, one such level cursor is shown as a thin clear line in the middle of the trace. It also displays the trace records read from the acquisition file and processed by GRAFROUT, and the histogram generated by HISTROUT and passed to EVNTANAL.

GRAFROUT Module

This module prepares the trace to be displayed by selecting meaningful data points that can fit in the display area. Data selection is based on a heuristic that filters the outliers in a cloud of points. The cloud of points has a size proportional to the (number of data points in the trace)/(horizontal display resolution). When the user sets a cursor level, this module changes pixels in the range of the cursor and performs textual display of the amplitudes and times specified by cursors.

[12] Identification of Ion Channel Regulating Proteins by Patch-Clamp Analysis

By Thomas J. Nelson, Pavel A. Gusev, and Daniel L. Alkon

Introduction

Microinjection of proteins into living cells permits measurement of physiologic activity under conditions in which the normal electrical activity of the cell is preserved. Alternatively, sections of membrane can be analyzed by the patch-clamp technique[1] to measure the effects on individual channels in the presence of or, after removal of the remainder of the cell, the absence of other cellular components. Because soluble cellular components are no longer present in the cell-unattached membrane patch configuration, any observed effect implies a direct effect on the channel or its closely associated components. Fibroblasts are particularly useful in patch-clamp studies. Like neurons, fibroblasts have a wide variety of ion channels, including voltage-dependent calcium channels,[2,3] potassium channels,[4] calcium-dependent po-

[1] B. Sakmann and E. Neher, eds., "Single Channel Recording." Plenum, New York, 1983.
[2] N. M. Soldatov, *Proc. Natl. Acad. Sci. U.S.A.* **89,** 4628 (1992).
[3] A. Peres, E. Sturani, and R. Zippel, *J. Physiol. (Lond.)* **401,** 639 (1988).
[4] A. S. French and L. L. Stockbridge, *Proc. R. Soc. Lond. B Biol. Sci.* **232,** 395 (1988).

0076-6879/98 $25.00

tassium channels,[5,6] and even action potentials.[7] However, fibroblasts are much easier to culture than neurons, grow more rapidly, and can be obtained in viable condition from autopsy samples. Thus, patch clamps of fibroblast membranes are invaluable in screening for proteins that interact with ion channels and examining their changes in various disease states, such as Alzheimer's disease.

In the patch-clamp technique, the membrane is maintained at a constant potential by electronic feedback. The patch clamp controller continuously supplies a precise current through an electrode, sufficient to maintain the potential constant.[8] A high degree of precision in voltage control can be obtained by placing the patch electrode in contact with a small region of the cell membrane. The remainder of the cell can be removed, allowing free access of the bath solution to the patch. Because the current is measured directly from the signal generator output of the controller, rather than from the ion channel itself, it is possible to measure single-channel currents with a high signal-to-noise ratio. Thus, patch clamping is one of the most sensitive techniques for studying the real-time function of single molecules and the proteins with which they interact.

One such protein that directly interacts with ion channels is calexcitin. Calexcitin is a low molecular weight calcium- and GTP-binding protein, originally found in the sea snail *Hermissenda,* and is a potent inhibitor of the i_A (rapidly inactivating) and i_{Ca-K^+} (calcium-dependent) potassium channels.[9,10] These channels were also found to be blocked after associative conditioning.[11] Associative learning causes activation of protein kinase C (PKC) and translocation of PKC from cytosol to membrane.[12] Therefore, because the phosphorylation state of calexcitin, which is a substrate of PKC, increases 2- to 3-fold after conditioning, we developed a procedure for purifying the native protein from squid (*Loligo pealei*) and analyzing

[5] P. T. Gray, S. Y. Chiu, S. Bevan, and J. M. Ritchie, *Proc. R. Soc. Lond. B Biol. Sci.* **227,** 1 (1986).
[6] A. Bakhramov, Y. S. Boriskin, J. C. Booth, and T. B. Bolton, *Biochim. Biophys. Acta* **1265,** 143 (1985).
[7] D. Lovisolo, G. Alloatti, G. Bonelli, L. Tessitore, and F. M. Baccino, *Pflugers Arch.* **412,** 530 (1988).
[8] T. G. Smith, J. Barker, B. Smith, and T. R. Colburn, *J. Neurosci. Meth.* **4,** 87 (1980).
[9] T. Nelson, C. Collin, and D. L. Alkon, *Science* **247,** 1479 (1990).
[10] T. Nelson, S. Cavallaro, C. Yi, D. McPhie, B. Schreurs, P. Gusev, A. Favit, O. Zohar, J. Kim, S. Beushausen, G. Ascoli, J. Olds, R. Neve, and D. Alkon, *Proc. Natl. Acad. Sci. U.S.A.* **93,** 13808 (1996).
[11] D. L. Alkon, J. Farley, M. Sakakibara, and B. Hay, *Biophys. J.* **46,** 1037 (1984).
[12] B. Bank, A. DeWeer, A. Kuzirian, A. H. Rasmussen, and D. L. Alkon, *Proc. Natl. Acad. Sci. U.S.A.* **85,** 1988.

its ability to inhibit potassium channels in human fibroblasts using the patch-clamp procedure.

The procedure outlined next has been used in our laboratory to purify calexcitin and characterize its effects on ion channels.

Isolation of Low Molecular Weight Ion Channel Regulating Proteins from Squid Optic Lobe by HPLC

Squid optic lobes (approximately 20–30) are homogenized in a high-speed shearing homogenizer with 50 ml HC buffer [10 m*M* Tris-HCl, pH 7.4, 20 μg/ml leupeptin, 50 μg/ml pepstatin, 50 m*M* NaF, 1 m*M* EDTA, 1 m*M* EGTA, 0.2–1 *M* dithiothreitol, and 0.1 m*M* phenylmethylsulfonyl fluoride (PMSF)]. The mixture is further sonicated for 20 sec with a 20-W probe sonicator at the maximal setting and centrifuged at 40,000 rpm in a Beckman (Palo Alto, CA) Ti70 rotor (100,000*g*) for 20 min. The pellet is resuspended in buffer with brief sonication, recentrifuged at 40,000 rpm for 20 min, and the combined supernatants are filtered through a 0.22-μm filter (Corning, Corning, NY, 25932-200) to remove any particulate matter. The filtrate is then passed through a YM30 ultrafiltration membrane (Amicon, Danvers, MA) to remove proteins above 25–30 kDa. The nonretained fraction is placed in a 50-ml Amicon ultrafiltration device equipped with a YM3 membrane and concentrated to a final volume of approximately 1–2 ml, and the retained fraction is saved for high-performance liquid chromatography (HPLC).

Purification of Calexcitin by High-Performance Liquid Chromatography

A 10× 250-mm AX-300 HPLC column (Synchrom, Inc., Lafayette, IN) is equilibrated with 10 m*M* NaF in HPLC-grade water. Because NaF is extremely corrosive, it is essential to minimize the duration of contact of NaF with the steel column by thoroughly washing the system with water after each run. Solvents must also be thoroughly degassed. A 2-ft length of 1/4-in (o.d.) copper tubing tightly coiled around the column, and connected to a circulating water cooling bath, is used to maintain the column at 4°. The stainless steel column has sufficient thermal conductivity to maintain the entire column assembly at an even temperature. A 100-μl aliquot of the concentrated optic lobe extract is injected and the column is washed for 20 min with a gradient of 0–1 *M* potassium acetate (adjust to pH 7.4 with HPLC-grade acetic acid), containing 10 m*M* NaF, at a flow rate of 2.0 ml/min. After 20 min, the proteins are eluted isocratically with 1 *M* potassium acetate, and 0.5-min fractions are collected. Experiments with tissues incubated with $[^{32}P]P_i$ to label all phosphoproteins indicated that the majority of proteins eluting after 25 min are phosphorylated. A large number of GTP-binding proteins are also retained under these conditions, and elute in several impure peaks between 20 and 30 min. Several HPLC runs may be required to obtain sufficient sample for biochemical and patch-clamp analysis. Each fraction is analyzed by sodium dodecyl sulfate–polyacrylamide gel electrophoresis (SDS–PAGE) to determine its purity. The phosphorylated form of calexcitin can be identified as the major 22-kDa Ca^{2+}-binding protein that elutes at approximately 29–32 min under these conditions, using a ^{45}Ca-overlay method to verify its binding to calcium.[10] Centricon-3 ultrafiltration units are coated by incubating at 23° for 1 hr with 1% bovine serum albumin, then thoroughly rinsed. The calexcitin is placed in the coated Centricon units and concentrated by centrifuging overnight at 5000 rpm in a SS-34 fixed-angle rotor (3000*g*) at 4° and stored in small aliquots at −80°.

Preparation of Proteins for Injection

The primary concern in preparing proteins for injection, apart from protein purity, is the presence of trace proteases copurifying with the protein or introduced from fingertips. The latter can be prevented by use of latex gloves. Protease inhibitors and preservatives such as glycerol must be removed before injection. PMSF, in particular, is extremely toxic to cells. Other inhibitors, such as diisopropyl fluorophosphate (DFP), are potent inhibitors of acetylcholinesterase. Protease inhibitors may be rapidly removed by the following procedure.

A polypropylene column (6-mm i.d.) containing 0.4 g (hydrated weight) of Sephadex G-25 is washed with 5 ml water. Commercial STE G-25 columns (5 Prime–3 Prime, Boulder, CO) designed for DNA purification may also be used provided that the buffer is first washed out. Care must be taken to avoid touching the top of the column with bare hands. The column is then placed in a 1.5-ml centrifuge tube and centrifuged at 3000 rpm in a Savant microcentrifuge (1000*g*) for 2 min at 4° to partially dry the column. The column is placed in a new 1.5-ml centrifuge tube, 40–45 μl protein is added, and the column is centrifuged at 3000 rpm for 3 min at 4°. The protein elutes in the centrifuge tube in an approximately 40-μl final volume. Potassium acetate (pH adjusted to 7.4 with HPLC-grade acetic acid) is added to 0.2 *M*, and the sample is frozen and lyophilized in a Speed-Vac to reduce the volume to 8 μl. Finally, the sample is centrifuged at 10,000*g* for 30 sec in a microcentrifuge to remove any particles that would clog the microelectrode.

Proteins that adsorb strongly to glass are also problematic. Unfortunately, treatment of microelectrodes with dichlorodimethylsilane/toluene or Sigmacote increases the electrode impedance to unacceptably high levels. For this reason, protein solutions are kept at least 0.1 mg/ml and only the tip of the microelectrode is filled. A Hamilton syringe with a 34-gauge needle is used to insert 1 μl of the protein solution (in 1 M potassium acetate, pH adjusted to 7.4 with acetic acid) into the microelectrode tip. The solution is allowed to fill the tip by capillary action. This requires approximately 30–60 min. Then the remainder of the electrode is filled with 9 μl 1 M potassium acetate. Several electrodes are typically made simultaneously. Once filled, the electrodes should be kept cold to minimize degradation of the protein.

Fibroblast Culture

Cultured skin fibroblasts are obtained from the Coriel Cell Repositories (Camden, NJ) and grown under standard conditions.[13] Cells are seeded (about five cells per square millimeter) in 35-mm Nunc petri dishes in Dulbecco's modified Eagle's medium (DMEM) (GIBCO, Grand Island, NY), supplemented with 10% (v/v) fetal calf serum, and used for patch-clamp experiments. Gigohm seals are most reliably obtained from cells 3–6 days after seeding.

Preparation of Inside-Out Patches of Fibroblast Membrane

Patch-clamp experiments are performed at room temperature (21–23°). Before recording, culture medium is replaced with 2 ml of the following solution (mM): 150 NaCl, 5 KCl, 0–2 $CaCl_2$, 1 $MgCl_2$, and 10 HEPES (NaOH), pH 7.4. Unpolished glass pipettes are made from Blue Tip capillary tubes (1.1- to 1.2-mm i.d.) using a BB-CH Mecanex (Geneva) puller and then filled with a high-potassium solution (mM): 140 KCl, 0–2 $CaCl_2$, 1 $MgCl_2$, and 10 HEPES (NaOH), pH 7.4. Pipette resistance should be 6–8 $M\Omega$.[14] Some investigators recommend coating the electrodes with Sylgard (Dow Corning, Midland, MI) to reduce the noise resulting from the saline meniscus[14]; however, we have found this to be unnecessary. In some cases, 0.01 mM EGTA is added to the culture medium and electrode solution. Tetrodotoxin (10^{-6} M) is added to the pipette solution to block Na^+ unitary currents. To exclude signals from Cl^- channels, gluconate salts of K^+, Na^+, Ca^{2+}, and Mg^{2+} may be used instead of chloride salts. With these bath and pipette solutions, it is possible to separate K^+ and Cl^- unitary currents.

Petri dishes are placed on the stage of an inverted microscope (e.g.,

[13] R. Etcheberrigaray, E. Ito, K. Oka, B. Tofel-Grehl, G. Gibson, and D. Alkon, *Proc. Natl. Acad. Sci. U.S.A.* **90,** 8209 (1993).

[14] A. Auerbach and F. Sachs, *in* "Voltage and Patch Clamping with Microelectrodes" (T. G. Smith, H. Lecar, S. Redman, and P. Gage, eds.) pp. 121–149. American Physiological Society, Bethesda, MD, 1985.

Axiovert 405M, Zeiss, Germany). The patch micropipettes, containing a Teflon-coated Ag wire and an Ag/AgCl pellet on one end for greater recording stability, are mounted on a suction pipette holder connected to the BNC connector of the amplifier head stage CV-4 1/100. The stage is mounted on a suitable patch-clamp micromanipulator, such as the Newport 421 (Newport Corp., Fountain Valley, CA) controlled with a piezoelectric positioning actuator (e.g., PCS-250, Burleigh Instruments, Fishers, NY). A slight positive pressure is applied to the electrode as it is lowered into the bath, to prevent debris from attaching to the tip. A seal is formed by pressing the electrode against the cell and applying a slight suction. To test the formation of gigaseals, −1-mV square pulses (100 Hz) are applied, and the impedance is calculated. After development of a 1- to 10-GΩ seal it is possible to obtain inside-out membrane patches whose cytoplasmic face is exposed to the bath solution. The patch is created by mechanically removing the patch micropipette from the fibroblast and briefly (1–2 sec) passing the tip through the air–water interface.[15]

Materials for Patch-Clamp Analysis

Records of single-channel activity can be obtained using an Axopatch-1D amplifier in voltage-clamp mode. Transients are filtered at a 10-kHz (−3-dB) four-pole Bessel filter and stored on tape. Because of the large volume of data, the most convenient storage medium is a PCM video recorder, such as the Toshiba (Japan) Model DX-900. Data may be subsequently filtered at 1 kHz (−3 dB) with an analog RC-type filter (e.g., a Frequency Devices (Haverhill, MA) 902LPF1 filter), and transferred to computer using one of several commercially available serial A/D interface boards (e.g., Digidata-1200).

Software should be standardized within a laboratory to facilitate equipment sharing in case of a malfunction. Two excellent software packages for data acquisition and single-channel off-line analysis are pClamp and Microcal Origin software (Microcal Origin Inc., Northampton, MA). The amplifier is the most critical component, and must have a wide unity-gain bandwidth (>5 MHz), low noise, and, most importantly, high input impedance (>10 GΩ). Amplifier, interface, and software in our laboratory were obtained from Axon Instruments (Foster City, CA).

Additional Considerations in Performing Patch-Clamp Experiments

In some experiments it is possible to record K^+ currents in the cell-attached configuration, but sometimes channel activity appears only after

[15] O. Hamill, A. Marty, E. Neher, B. Sakmann, and F. Sigworth, *Pflugers Arch.* **391,** 85 (1991).

excision of the membrane patch. The mean magnitude of unitary currents can be determined by Gaussian fitting and $\bar{x} \pm$ SE current values of the peaks are used to construct *I–V* curves. For each V_h (holding potential), duration of recording should be 1–2 min. The current–voltage relationships of the K^+ channels for inside-out patches from fibroblasts over the range ± 80 mV of pipette potential (V_h) are best fitted by a second-order polynomial using the standard error of current magnitude as a weight. Conductance is determined as a slope of first derivative of the polynomial function at $V_h = 0$ mV. Control recordings of channel activity are performed at 0–40 mV V_h for 10 min to establish the baseline for current magnitude (pA), mean open time (ms), open probability (P), and frequency of openings (Hz). The control and experimental recording should be performed at the same V_h because channel kinetics may depend on V_h level. For example, the mean open time of K^+ channels in our experiments with fibroblasts decreased on hyperpolarization of the patches.

Addition of Protein

Three-microliter aliquots of stock protein solution are kept at $-80°$ and applied to the experimental petri dish. Addition of protein is made using a 20-μl Pipetman at a distance of 5–8 mm from the tip of the patch micropipette. The latent period of protein effects is typically 2–4 min.

Data Analysis

Records of 45–60 sec in duration of single-channel activity are used to construct all-points amplitude histograms and open time distributions, and to obtain estimates of mean open probability and frequency of openings of K^+ channels.[16,17] More sophisticated statistical analyses, such as stochastic analysis of channel openings as Markov processes, can also be carried out.[18] The channel activity before and after protein application is statistically treated by the Student's *t* test, excluding from analysis patches with significant trends in single-channel activity during control recording.

In some cases, failure to get a protein effect on channel activity may be due to restricted access to the patch caused by the edges of the patch partially or completely sealing over the inner surface of the membrane. To prevent this, it is useful to check the effect on channel activity of some

[16] D. Colquhoun and F. Sigworth, *in* "Single-Channel Recording" (B. Sakmann and E. Neher, eds.) pp. 191–263. Plenum, New York, 1983.

[17] A. Auerbach and F. Sachs, *Biophys. J.* **45,** 187 (1984).

[18] D. Colquhoun and A. Hawkes, *Proc. R. Soc. Lond. Ser. B* **211,** 205 (1981).

other compounds that are known to affect channels under investigation. In our experiments we used different concentrations of intracellular Ca^{2+}, tetraethylammonium chloride, or substituted Cs^+ for Na^+. It is also possible to wash out the effect simply by changing the bath solution using a 1-ml pipette.

[13] Tight-Seal Whole-Cell Patch Clamping of *Caenorhabditis elegans* Neurons

By SHAWN R. LOCKERY and M. B. GOODMAN

Introduction

The nematode *Caenorhabditis elegans* is widely used to study the relationship between genes, neurons, and behavior. The adult hermaphrodite has a compact nervous system of only 302 neurons and the synaptic connections between these cells have been described completely.[1] The neural circuits for many of its behaviors have been delineated.[2] In addition, more than 350 genes affecting behavior have been identified.[3]

Caenorhabditis elegans presents a formidable challenge for electrophysiology, however. The animals are only 0.25–1.2 mm long and the cell bodies of *C. elegans* neurons are typically 2 μm in diameter.[1] In addition, the body is protected by a tough, pressurized cuticle that explodes when dissected.[4] The electrical properties of individual neurons in *C. elegans,* therefore, are largely unknown.

Here we present a reliable method for making tight-seal, whole-cell patch-clamp recordings from intact neurons in *C. elegans* at all larval stages. By combining this technique with cell-specific expression of green fluorescent protein (GFP),[5] whole-cell recordings can be made from identified neurons. Thus, it is now possible to describe the electrical properties of particular neurons and to find out how these properties are altered by mutations affecting neuronal development and behavior.

[1] J. G. White, E. Southgate, J. N. Thomson, and S. Brenner, *Phil. Trans. Roy. Soc. B* **314,** 1 (1986).
[2] C. I. Bargmann, *Ann. Rev. Neurosci.* **16,** 47 (1993).
[3] J. Hodgkin, R. H. Plasterk, and R. H. Waterston, *Science* **270,** 410 (1995).
[4] T. A. Tattar, J. P. Stack, and B. M. Zuckerman, *Nematologica* **23,** 267 (1977).
[5] M. Chalfie, Y. Tu, G. Euskirchen, W. W. Ward, and D. C. Prasher, *Science* **263,** 802 (1994).

METHODS IN ENZYMOLOGY, VOL. 293
0076-6879/98 $25.00

Overview

Recordings are made using simple modifications of existing techniques and equipment. Standard methods to immobilize worms during DNA injection,[6] which dehydrate the animal, are modified for electrophysiology. This involves immobilizing the animal with a waterproof glue (cyanoacrylate) under moist conditions. Because the worms are quite small, dissections are done with fine glass dissecting needles, based on the procedure for dissecting chromosomes.[7] Rupturing the membrane patch to obtain the whole-cell configuration (breaking in) appears to be more difficult in *C. elegans* neurons than in many other types of cells. To solve this problem, we developed a way to make extremely blunt pipettes with submicron openings. In addition, we stabilized the pipette mechanically to reduce movements produced by drift or suction pulses to a fraction of a micron. A commercially available patch-clamp amplifier was modified to accommodate the small capacitance and high input resistance of *C. elegans* neurons. With these modifications, the tight-seal, whole-cell patch-clamp recording configuration can be achieved with a success rate of 56%.

Preparing *Caenorhabditis elegans* for *in Situ* Electrophysiology

Animals

We use approximately synchronous cultures of larval worms. These are obtained by isolating eggs from gravid adults and allowing them to hatch on either sterile agar plates (NGM[8]) or plates seeded with bacteria (op50). Larvae are collected by washing them off with distilled water (1.5 ml) and pelleting them in a table-top microfuge (3 min at 3000 rpm). Animals are maintained and recorded at 23°C.

Gluing

An agarose pad (25 μm thick) is formed by pressing 10 μl of molten agarose (1.5%, medium EEO, Fisher Scientific, Pittsburgh, PA BP161) between two No. 0 coverslips (24 × 60 mm) and removing the top coverslip. The pH of the agarose from which the pad is made is adjusted to 9.0 (30 m*M* Na N-tris[Hydroxymethy]methyl-3-amino-propanesulfonic acid)

[6] A. Fire, *EMBO J.* **5,** 2673 (1986).

[7] S. D. Brown and A. H. Carey, *Methods Molec. Biol.* **29,** 425 (1994).

[8] J. E. Sulston and J. A. Hodgkin, *in* "The Nematode *Caenorhabditis elegans*" (W. B. Wood *et al.,* eds.), p. 587. Cold Spring Harbor Laboratory Press, Plainview, New York, 1988.

to accelerate polymerization of the glue. Worms are transferred immediately to the surface of the pad in a 0.35-μl sample from the pellet. The coverslip is sealed with beeswax over a hole in a glass plate to form a recording chamber.

Because water interferes with polymerization of the glue, the fluid in which the worms are transferred must either absorb into the agarose or evaporate. Absorption and evaporation are accelerated by spreading the fluid with a stream of air from a mouth tube. This also helps disperse worms across the agarose pad. When ready for gluing, worms are anesthetized by placing the chamber on the surface of a water-filled tissue culture flask (50 ml, Falcon 3014) stored at 4°.

Anesthetized worms are immobilized with cyanoacrylate glue (Nexaband Quick Seal, Veterinary Products Laboratory, Phoenix, AZ) applied with a small pipette held in a micromanipulator and viewed under a dissecting microscope. Glue pipettes (tip diameter of 17–19 μm) are pulled from thin-walled borosilicate capillary tubes (0.94-mm i.d., 1.20-mm o.d., Corning 8250, Garner Glass, Claremont, CA). The diameter of the glue drop ($\approx$50 μm) is set by adjusting the height of the column of glue at the time the pipette is filled (the taller the column, the larger the glue drop). To compensate for variations in worm size and posture, additional adjustments are made by sucking or blowing very gently on a mouth tube connected to the glue pipette. Worms are glued along one side, leaving the other side clear for dissection and recording. Gluing is done as quickly as possible, because the agarose pad tends to dry out, and worms glued to a dry pad become dehydrated and unsuitable for recording. With practice, about 25 worms can be glued in 5–10 min.

The chamber is rinsed with distilled water to remove unglued worms, then filled with extracellular saline. Air pockets appear on the glue after rinsing. Because air pockets make the preparation hard to see, we remove them with a stream of air bubbles from a mouth tube with a 0.4-mm opening. The recording chamber is then transferred to the stage of an inverted microscope (Zeiss Axiovert 135, Thornwood, NY).

Electrical recordings are made during the first hour after gluing. Pharyngeal pumping and movements of the nose and tail of glued worms can be observed for several hours after gluing, but the vigor of these movements declines steadily. At the same time, glued worms develop a Clr phenotype[9] and vacuoles sometimes appear in the head. Unglued worms in the same external saline are unaffected, suggesting these effects are due to the glue.

[9] S. G. Clark, M. J. Stern, and H. R. Horvitz, *Nature* **356,** 340 (1992).

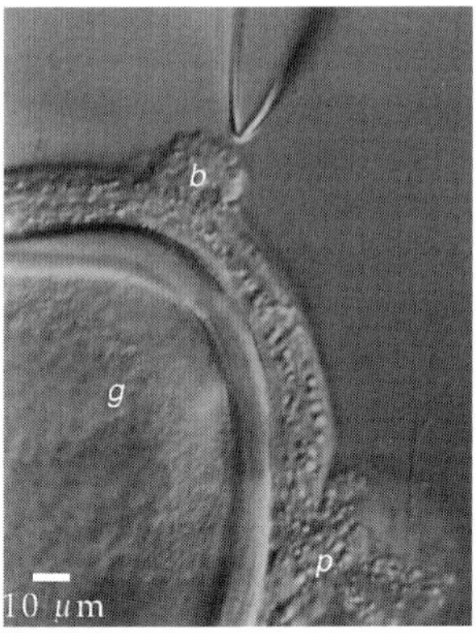

FIG. 1. DIC micrograph of the slit-worm preparation. The animal is an L1 larva. The pipette is visible entering from the top of the figure. The worm's nose is out of view to the left. Symbols: p, point of pressure release; g, glue; b, bouquet of neuronal cell bodies.

Dissection

Dissection proceeds in two steps. First, internal pressure is relieved by puncturing the cuticle at the level of the gonad primordium with a glass dissecting needle (cutter) held in a micromanipulator mounted on the microscope stage. Cutters are made from soda-glass rods (1.2 mm diam., LECO Co., St Joseph, MI) on a microelectrode puller. Second, cell bodies of neurons in the head are exposed using the cutter to open a slit in the cuticle. This is done by orienting the cutter tangent to the cuticle, working the tip under the cuticle surface, then moving the cutter along the tangent line, like opening an envelope with a letter opener. Neurons emerge through the slit, forming a hemispherical bouquet of 10–20 cell bodies (Fig. 1). This is called the *slit-worm preparation.*[10] Cell bodies are easier to see and record from in younger animals (L1–L3). Inspection of GFP-labeled neurons in the bouquet shows that exposed cell bodies remain attached to their processes. Moreover, measured capacitance in an identified neuron matches capacitance predicted from its surface area in undissected animals,[11] suggesting that exposed neurons remain intact.

Fabricating Patch Pipettes with Submicron Openings

We use blunt pipettes with submicron openings. This shape makes it easier to break in and minimizes access resistance. To make such pipettes,

[10] L. Avery, D. Raizen, and S. R. Lockery, *in* "*C. elegans:* Modern Biological Analysis of an Organism" (E. H. F. Epstein and D. C. Shakes, eds.), p. 251. Academic Press, Orlando, Florida, 1995.

[11] M. B. Goodman, D. H. Hall, L. Avery, and S. R. Lockery, *Neuron,* in press (1998).

we equip the inverted microscope of a fire-polishing setup (ALA Scientific Instruments, Westbury, NY) with a long-working distance, high-power, metallurgical objective (Nikon 100×/0.75, MPLAN SL WD). In addition, we fit the microscope for polarized light microscopy, using two inexpensive linear polarizers, one fixed and one rotatable. The fixed polarizer is made by trimming a sheet of polarizing plastic (M38,490, Edmund Scientific, Barrington, NJ) to fit below the nose piece and oriented at 45 deg with respect to the pipette. The rotatable polarizer is photographic quality (49 mm in diameter, Tiffen, Hauppauge, NY) and placed between the iris diaphragm and specimen stage in the transmitted light path. The rotatable polarizer is turned to maximize extinction.

Recording pipettes were made from soda-lime glass capillary tubes (0.8-mm i.d., 1.2-mm o.d., R6, Garner Glass, Claremont, CA) with a low softening temperature (700°). Capillaries are pulled to an internal tip diameter of 2–4 μm on a laser-based micropipette puller (P-2000, Sutter Instruments, Novaro, CA), coated with Sylgard 184, and fire-polished with a resistively heated platinum wire.

Figure 2 shows bright-field and polarized-light images of unpolished

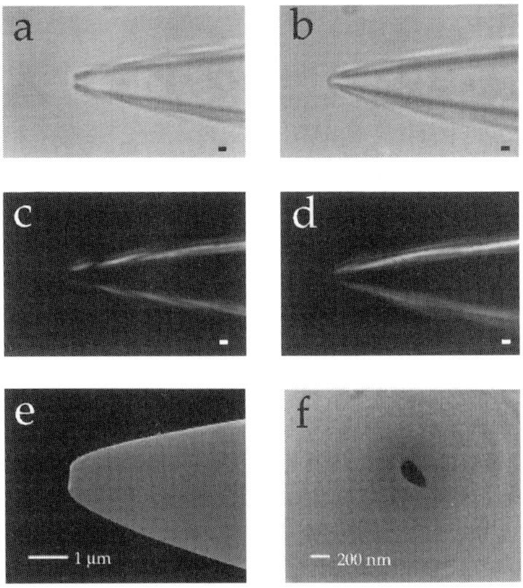

FIG. 2. Recording pipettes before and after fire-polishing. (a) Bright-field image of an unpolished pipette. (b) Bright-field image of a polished pipette. (c) and (d) Polarized light images of the pipettes in parts (a) and (b), respectively. (e) Scanning electron micrograph of a polished pipette (side view) showing the bluntness of the tip. (f) End view of the pipette in part (e) at higher power showing an elliptical hole in the tip. [Reprinted with permission from M. B. Goodman, D. H. Hall, L. Avery, and S. R. Lockery, *Neuron* (in press) (1998).]

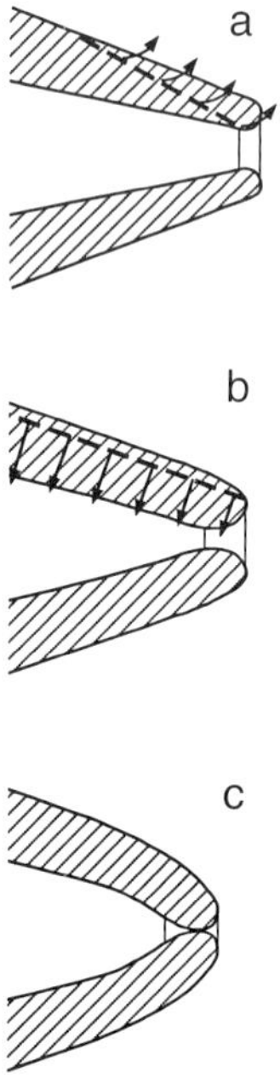

FIG. 3. Visual cues used in fire-polishing. (a) Diagram (side view) of an unpolished pipette tip showing the location of the interference fringe closest to the tip and how this fringe moves during the first polishing step. (b) The location of the fringe parallel to the outer pipette wall and how it moves during the second polishing step. (c) A well-polished pipette showing the characteristic pursed-lips profile.

and polished pipettes. Under polarized light, contrast between the pipette and the background is enhanced, with the pipette walls appearing bright against a dark background. The increase in contrast is similar to that achieved by fiber-optic illumination of the pipette walls.[12] In addition, a series of interference fringes is readily apparent near the tip. Interference fringes are used to provide visual feedback during the polishing process.

The pipette is polished initially to a tip opening of about 400 nm by heating it until the interference fringe closest to the tip (Fig. 3a) disappears. The pipette is then polished to its final size by heating it a second time until the fringe parallel to the outer wall of the pipette tip (Fig. 3b) migrates into the lumen. The walls at the tip of a well-polished pipette look like pursed lips (Fig. 3c). Scanning-electron micrographs of polished pipette tips show they are extremely blunt, having almost flat faces with a tiny elliptical hole in the center (Figs. 2e and 2f). The length of the major axis

[12] J. A. Fox, *J. Neurosci. Meth.* **15,** 239 (1985).

of the ellipse is 262 ± 64 nm (mean ± SD, $n = 4$); the length of the minor axis is 110 ± 39 nm (mean ± SD, $n = 4$).

To minimize accumulation of dust particles in the lumen, we use pipettes within 2 hr of fire-polishing. Pipettes are back-filled with internal saline. Bubbles near the tip are removed by applying pressure (17 kPa) to the back of the pipette. Usable pipettes have resistances between 5 and 15 MΩ.

Solutions

The ionic composition of the fluid surrounding *C. elegans* neurons *in vivo* is unknown. Candidate extracellular salines were assayed previously[10] by observing pharyngeal pumping and its electropharyngeogram in dissected animals. We use a generic extracellular saline that contains (m*M*): NaCl (145), KCl (5), $CaCl_2$ (1), $MgCl_2$ (5), HEPES (10), and D-glucose (20), pH 7.2. The apparent health of exposed neurons depends on the osmolarity of the external saline. By trial and error, we settled on a value of 315–325 mOsm. This is approximately 15 mOsm greater than both M9 medium and *C. elegans* culture medium.[13,14]

The ionic composition of *C. elegans* neuronal cytoplasm is also unknown. We use a generic intracellular saline composed of (m*M*): potassium gluconate (125), KCl (18), $CaCl_2$ (0.7), $MgCl_2$ (2), EGTA (10), and HEPES (10), pH 7.2, 310–315 mOsm. To make K^+-free saline, K^+ is replaced with *N*-methyl-D-glucamine.

Patch-Clamp Setup

Because the cell bodies of *C. elegans* neurons are unusually small, recording from them requires high magnification, exceptional mechanical stability, and specialized electronics. The following steps are taken to improve the optics, mechanical stability, and electronics of the patch-clamp setup.

Optics

The microscope is equipped with differential interference contrast (DIC) optics and a 63× oil immersion objective with a numerical aperture of 1.4. The working distance of this objective is 260 μm. Since a No. 0 coverslip is 100 μm thick, and the worm is 15–80 μm thick, the agarose pad must be quite thin. We have also used a long working distance, water

[13] L. G. Edgar, *in* "*Caenorhabditis elegans:* Modern Biological Analysis of an Organism" (E. H. F. and D. C. Shakes, eds.), p. 303. Academic Press, New York, 1995.

[14] C. A. Shelton and B. Bowerman, *Development* **122,** 2043 (1996).

immersion objective (Zeiss 40×/0.75, 1.9 mm working distance).[10] We prefer the 63×/1.4 objective, however, because of its greater resolving power, which is important for watching the membrane patch while trying to break in.

Mechanics

Recording chambers made of acrylic (e.g., Plexiglas), while adequate for recording from large neurons, cannot be used for recording from *C. elegans* neurons. This is because acrylic chambers flex in response to small thermal fluctuations, causing significant movements along the optical axis of the microscope. To avoid this problem, we use a chamber made from a glass plate (76 × 76 × 1 mm), since the coefficient of thermal expansion of glass is much less than acrylic. A hole (18 mm in diameter) is cut in the center of the plate. The coverslip holding the worms is sealed to the underside of the plate with the worms centered in the hole.

To minimize drift of the recording pipette, we mount the amplifier headstage on a piezoelectric manipulator (MP-300, Sutter Instruments, Navaro, CA). This manipulator has a single-step mode with submicron resolution, which we find useful for positioning the pipette on small cells. The manipulator is mounted on the microscope stage to minimize vibration and relative movement between the pipette and the preparation.

To obtain whole-cell recordings, we find it critical to prevent the pipette tip from moving in response to the suction pulse used to break in. This is done by three modifications of the usual arrangement of the pipette holder, air line, and headstage: (1) The hole in the end cap of the pipette holder (EH-U1, E. W. Wright, Guilford, CT) is drilled to match the outside diameter of the pipette as closely as possible. The final hole size fits the pipette snugly but allows it to slide down the hole without jamming. (2) Suction is delivered by a stainless steel air line connected to the pipette holder and anchored to the manipulator and the microscope. The steel air line is interrupted with a short section of Silastic tubing so the manipulator can move. (3) The pipette is clamped to a stable support attached to the manipulator.[15] The design of the pipette clamp is described in a separate paper.[16] These modifications also extend the duration of electrical recordings by reducing slow mechanical and thermal drift in the pipette holder.

Electronics

Most commercial patch-clamp amplifiers are not designed for recording from small cells, which have a very high input resistance (about 5 GΩ) and

[15] F. Sachs, *Pflug. Arch.* **429,** 434 (1995).
[16] S. R. Lockery, *J. Neurosci. Meth.,* in preparation (1998).

a very low input capacitance (about 1 pA). For recording from *C. elegans* neurons, therefore, we modify a well-known, commercial patch-clamp amplifier (Axopatch 200A, Axon Instruments, Foster City, CA) in four respects, as described next.

1. To record the voltage response to a current pulse accurately, the time course of the pulse should be as step-like as possible and its rise time should be small with respect to the membrane time constant of the cell. Time course and rise time of the pulse are measured by adjusting the fast pipette compensation in voltage clamp, then passing current steps across a model *C. elegans* neuron (input resistance, 5 GΩ; input capacitance, 0.6 pF; series resistance 40 MΩ) and monitoring the time course of injected current. Initially, the time course is multiexponential and rises to one-half its maximum value (T.5) in 10.5 ms. Reducing the value of the feedback capacitor in the current-clamp circuit (Appendix, modification 1) reduces T.5 to 25 μs. Decreasing the rise time of the current command drastically reduces the stability of the current clamp circuit, however. In particular, we cannot completely compensate pipette capacitance without overdriving the current clamp circuit and causing oscillations. Immediately before contacting the cell, therefore, stability in current clamp is tested for each pipette. This is done by first adjusting the pipette capacitance compensation in voltage clamp at high gain and bandwidth (50 kHz). We then switch to current clamp. Instability can sometimes be observed directly as an oscillation in the voltage trace, but more often it appears as amplifier saturation. If necessary, capacitance compensation is fine-tuned in current clamp until the voltage trace is stable. We are careful to keep the bath level constant, since small changes in level affect pipette capacity compensation.

2. Because *C. elegans* neurons have input resistances on the order of gigohms. Membrane potential is especially sensitive to current injection. Accordingly, the current-clamp command sensitivity is reduced from 2 nA/V to 100 pA/V (Appendix, modification 2).

3. A brief, high-voltage pulse (zap) is sometimes used to break in. To reduce the likelihood of destroying the seal as well as the membrane patch, we reduce the minimum duration of the zap pulse to 60 μs and decreased its amplitude to 7 V (Appendix, modification 3). Zapping is most effective in combination with gentle suction.

4. Finally, because *C. elegans* neurons have input capacitances of about 1 pF, we increase the sensitivity of the potentiometer for whole-cell capacitance compensation to 1 pF per turn. The sensitivity of the potentiometer for series resistance compensation is kept at 10 MΩ per turn (Appendix, modification 4).

Recording from Neurons Labeled with Green Fluorescent Protein

Caenorhabditis elegans neurons can be identified by size and position inside the animal but these cues are lost when neurons are exposed in the slit-worm preparation. We are able to identify exposed neurons, however, by using strains of worms in which single neurons [lin-15(n765ts);*gcy-5*::GFP[17]] or small sets of neurons (ceh-23::GFP[18]) express GFP. In these experiments, the microscope is equipped for epifluorescence with a 50-W Hg lamp and standard filter set for visualizing GFP.[5] The pipette is brought up to the labeled cell under simultaneous transmitted (DIC) and epifluorescent illumination. To visualize the pipette and the labeled cell optimally, we adjust the intensity of the illumination by rotating a polarizing filter placed between the transmitted light source and the iris diaphragm.

Tight-Seal Whole-Cell Recording

Sealing and Breaking In

To keep the pipette tip clean, we apply positive pressure (0.4 kPa) continuously to the back end of the pipette. Immediately after adjusting the pipette capacitance compensation and testing for stability in current clamp, the pipette is brought into contact with the cell body of the target neuron. Contact is detected as an increase in pipette resistance of about 10%. At the moment of contact, positive pressure is released. In some cases, a gigohm seal forms spontaneously. In others gentle suction is used, alone or in combination with a negative voltage command (−60 mV) to the pipette. Typical seal resistances are 4–11 GΩ. As noted previously,[10] single-channel activity is apparent in some patches.

The whole-cell configuration is achieved by applying stronger suction to rupture the membrane patch. Zaps are used only as a last resort. GFP-labeled neurons are about twice as easy to break into as unlabeled neurons. In the strain we have used most extensively [lin-15(n765ts);gcy-5::GFP], 56% ($n = 42$) of the seals we obtain lead to successful whole-cell recordings. The success rate for unlabeled cells is 31% ($n = 100$). The patch can also be ruptured by focusing a pulsed dye laser (440-nm VSL-337, Laser Science, Inc., Newton, MA) on the patch through the microscope objective. The laser beam is brought to the microscope by a fiber-optic positioning system (Micropoint, Photonics Instruments, Arlington Heights, IL).

[17] S. Yu, L. Avery, E. Baude, and D. L. Garbers, *Proc. Natl. Acad. Sci. U.S.A.* **94,** 3384 (1997).
[18] W. C. Forrester, E. Perens, J. A. Zallen, and G. Garriga, *Genetics* **148,** 151 (1998).

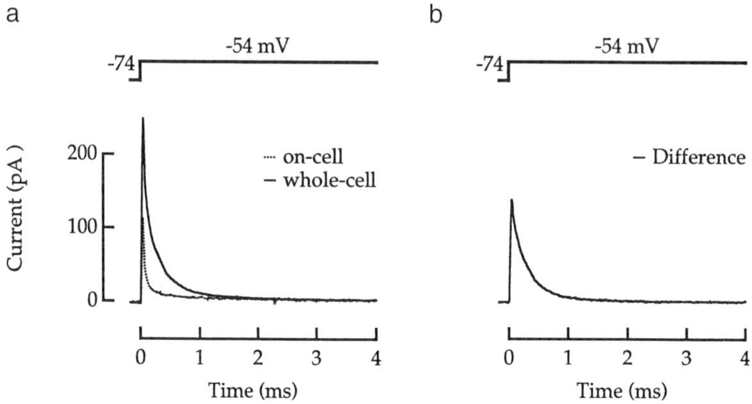

FIG. 4. Voltage-clamp capacity-current transients. (a) Transients recorded in the on-cell (dotted line) and whole-cell (solid line) configurations in response to a 10-ms, 20-mV step from −74 mV. The transient recorded in the on-cell mode represents the pipette capacitance that remained after compensation. The transition from the on-cell to the whole-cell configuration coincided with a detectable change in the capacity transient. For clarity, only the first 4 ms are shown. An equal-amplitude, opposite-polarity transient was apparent on repolarization to −74 mV. (b) The difference current obtained by subtracting the on-cell transient from the whole-cell transient in part (a) to eliminate uncompensated pipette capacitance. This current was used to calculate whole-cell capacitance, access resistance, and clamp speed. [Reprinted with permission from M. B. Goodman, D. H. Hall, L. Avery, and S. R. Lockery, *Neuron* (in press) (1998).]

The perforated patch technique[19] is an obvious alternative to rupturing the patch when trying to record from small cells. Such recordings can be made from *C. elegans* neurons with pipettes containing amphotericin (Lockery, unpublished, 1994). Using this technique, however, it can take tens of minutes to gain access to the cell. This is a concern because recordings must be made in the first hour after gluing.

Whole-Cell Capacitance and Access Resistance

Although *C. elegans* neurons have little capacitance, the transition from the on-cell to the whole-cell recording configuration coincides with a detectable change in the capacity transient elicited by a voltage-clamp pulse (Fig. 4a). The residual capacitance of the pipette and electrode holder, 1.16 ± 0.25 pF (mean ± SD, $n = 10$), is comparable to the whole-cell capacitance of the neuron (0.5–3 pF), however. Thus, whole-cell capacitance (C_{in}) and access resistance (R_a) are estimated from the difference current obtained

[19] J. Rae, K. Cooper, G. Gates, and M. Watsky, *J. Neurosci. Meth.* **37,** 15 (1991).

by subtracting the current elicited by a 20-mV test pulse in the on-cell configuration from the current recorded in the whole-cell configuration (Fig. 4b). The value of C_{in} is calculated by dividing the integral of the difference current by 20 mV. The value of R_a is estimated as τ/C_{in}, where τ is the time constant of the decay of the difference current; R_a is 19 $\pm$ 9 MΩ (mean $\pm$ SD, n = 41).

Membrane Current and Membrane Potential

An example of a whole-cell recording obtained from a neuron in the head of a wild-type L1 animal is shown in Fig. 5a. A decaying outward current is observed at potentials greater than about 6 mV. Hyperpolarizing steps elicited little change in net membrane current. Figure 5b shows the voltage response in the same cell to a family of current pulses. As expected from the absence of hyperpolarizing-activated membrane current (Fig. 5a), the amplitude of the response to hyperpolarizing current pulses is proportional to pulse amplitude, and its time course is approximately exponential. The response to depolarizing current pulses is also graded with amplitude, but the change in membrane potential is smaller than that evoked by a hyperpolarizing pulse of equal amplitude. This is likely to reflect activation of outward current by depolarization.

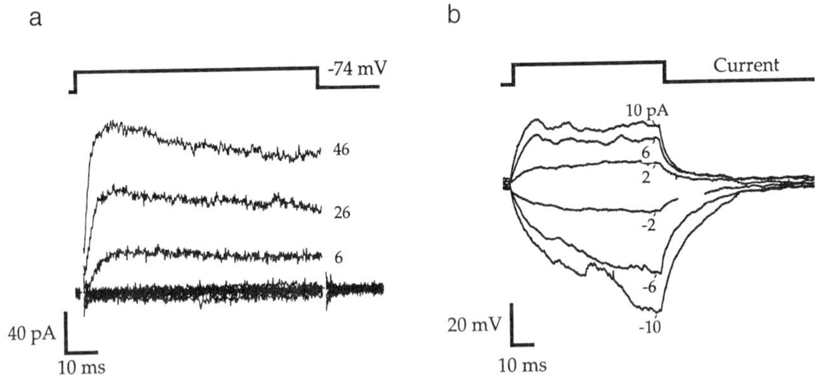

FIG. 5. Whole-cell voltage-clamp and current-clamp recordings from the cell shown in Fig. 4. (a) Membrane current recorded in response to eleven 80-ms voltage steps between −154 and +46 mV in 20-mV increments. The holding potential was −74 mV. An outward current was apparent in response voltage steps to more than +6 mV; no change in membrane current was apparent for voltage steps less than +6 mV. (b) Membrane voltage recorded in response to six 75-ms current steps between −10 and +10 pA in 4-pA increments. Membrane voltage is expressed relative to the zero-current potential, which was −9 mV. In the trace marked "−2," a voltage artifact produced by the perfusion pump was excised, producing a gap in the trace.

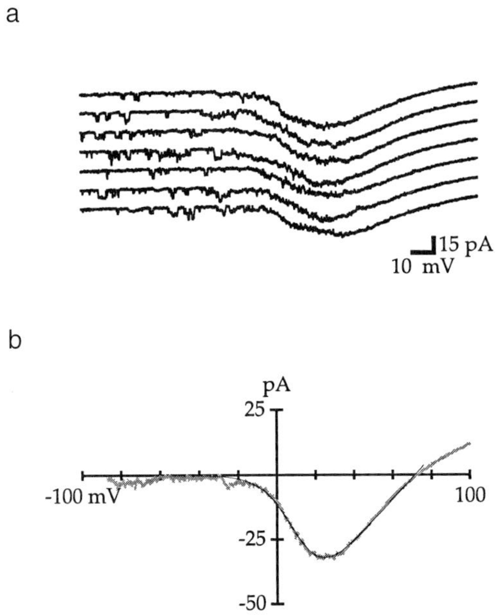

FIG. 6. Whole-cell and putative single-channel currents recorded in the same cell. (a) Current in response to a voltage ramp from −112 to 88 mV (300-ms duration). The holding potential between ramps was −92 mV. The step from the holding potential to −112 mV elicited a capacity transient that was omitted from the traces. The pipette contained K^+-free saline and the cell was superfused by a solution containing (m*M*): $BaCl_2$ (110) and Cs–HEPES (10), pH 7.2. (b) Average of the seven traces in part (a). Average current was well fit by the Boltzmann equation, $I = G_{max}(V - E_r)/(1 + \exp\{(V_0 - V)/V_s\})$, with $G_{max} = 0.82$ nS, $E_r = 71$ mV, $V_0 = 11.6$ mV, and $V_s = 8.2$ mV. [Reprinted with permission from reference M. B. Goodman, D. H. Hall, L. Avery, and S. R. Lockery, *Neuron* (in press) (1998).]

Single-channel currents are occasionally observed during whole-cell recordings. These events are most apparent under conditions that minimized outward currents. One such experiment is shown in Fig. 6a. The pipette contains K^+-free saline and the cell is superfused with an external solution designed to block K^+ currents and enhance $Ca2^+$ currents. At voltages hyperpolarize to −50 mV, single-channel openings can be resolved. At higher voltages the traces become noisy, suggesting that the number of open channels increases. Averaging these traces produces the macroscopic current–voltage curve shown in Fig. 6b. These records resemble recordings from on-cell patches in crustacean muscle that contain multiple Ca^{2+} channels.[20]

[20] C. Erxleben and W. Rathmayer, *J. Gen. Physiol.* **109,** 313 (1997).

Interpreting Patch-Clamp Recordings from Small Cells

There are advantages and disadvantages to recording from small cells.[21] These are discussed next in the context of recording from *C. elegans* neurons.

Voltage Clamp

Voltage Errors Due to Series Resistance. One advantage of recording from small cells is that the membrane currents are also typically small. This reduces the discrepancy between membrane voltage and command voltage when current flows across R_a. Voltage errors in our recordings are at most 10 mV even without series resistance compensation.

Clamp Speed. Without series resistance compensation, the time constant for charging the membrane capacitance in our recordings is 7–118 μs, with a median of 21 μs ($n = 39$). This indicates that in most cases we have rapid control of membrane voltage even in uncompensated recordings.

Filtering of Whole-Cell Voltage-Clamp Currents. In voltage clamp, R_a and C_{in} produce a low-pass filter with a cut-off frequency, F_c, of approximately $1/(2\pi R_a C_{in})$. Because C_{in} in *C. elegans* neurons is small, currents should be filtered less than in larger cells with comparable R_a. In our recordings, F_c for an average *C. elegans* neuron ($R_a = 20$ MΩ, $C_{in} = 0.8$ pF) is 10 kHz. Higher values of F_c can be achieved with whole-cell capacitance compensation. Thus, it should be possible to resolve both fast macroscopic and single-channel currents even in whole-cell recordings.

Current Clamp

In current clamp, R_a, and uncompensated pipette capacitance, C_p, produce a low-pass filter with $F_c \approx 1/(2\pi R_a C_p)$. In large neurons, where $C_{in} \gg C_p$, filtering by the pipette is small compared to filtering by the cell itself. In *C. elegans* neurons, however, $C_{in} \approx C_p$, so filtering by the pipette could be significant. In our recordings, we estimate F_c is about 8 kHz and can be increased by reducing pipette capacitance. Steps that reduce pipette capacitance include using borosilicate or quartz glass, applying heavier Sylgard coatings, and reducing the depth of the recording bath.

Estimating Passive Membrane Properties of Small Neurons

Input resistance, membrane potential, and membrane time constant are difficult to measure in small cells. This is because one cannot make the usual assumptions that input resistance, R_{in}, is much greater than seal

[21] P. H. Barry and J. W. Lynch, *J. Mem. Biol.* **121,** 101 (1991).

resistance, R_s, or that input capacitance, C_{in}, is much greater than uncompensated pipette capacitance, C_p.[21]

Input Resistance. Any estimate of R_{in} in a *C. elegans* neuron is at best a lower bound. This is because R_s is of the same order of magnitude as R_{in}, and R_s and R_{in} act in parallel to determine the response of the cell to a voltage or current pulse. A more accurate estimate may be possible, however, using the time course of single-channel currents in cell-attached patches.[21]

Membrane Potential. A related effect of the similarity between R_{in} and R_s is a reduction in the magnitude of the cell's zero-current (resting) potential, V_m. Because R_s and R_{in} act in parallel, the apparent V_m is the weighted average of the true V_m and the reversal potential associated with R_s, which is approximately 0 mV.

Membrane Time Constant. The quantity τ_m is often estimated from the voltage response to a current impulse or step (charging curve). In the case of an isopotential cell recorded with an ideal electrode ($R_{in} \gg R_a$; $C_{in} \gg C_p$), $\tau_m = R_{in}C_{in}$. For *C. elegans* neurons, however, $C_{in} \approx C_p$, so the apparent τ_m is increased to $R_{in}(C_{in} + C_p)$. At the same time, however, the effective value of R_{in} is decreased by R_s, tending to decrease the apparent τ_m. Similar considerations apply to nonisopotential cells.[22] Thus, estimates of τ_m obtained from charging curves in *C. elegans* neurons are likely to be unreliable.

Spatial Control of Voltage

Recordings from the unbranched, bipolar neuron ASER indicate that this cell could be well space clamped at all larval stages.[11] Most neurons in the head ganglia of *C. elegans* have processes with approximately the same dimensions as those of ASER and are either bipolar or monopolar. Assuming that the electrical properties of ASER are not unique, these two types of neurons should also be well space clamped. Whether a third type of neuron, with a single process running the length of the body, can also be well space clamped remains to be seen.

Prospects

Single Channels in Whole-Cell Recordings

Single-channel currents are rarely resolved in whole-cell recordings. In such recordings from *C. elegans* neurons, however, we found conditions

[22] G. Major, J. D. Evans, and J. J. Jack, *Biophys. J.* **65**, 423 (1993).

that revealed single-channel events. This means it should be possible to study the microscopic and macroscopic properties of a current in a single recording. Also, because *C. elegans* neurons are likely to be well space clamped, it may be possible to record single-channel currents originating at synapses and other sites distant from the cell body.

Identified Neurons

A great advantage of invertebrate systems is the presence of identifiable neurons. Identification can be laborious, however, since neurons often must be identified by several criteria at once, including their intrinsic electrical properties, morphology, and connections to other neurons. In GFP-labeled strains of *C. elegans,* neurons can be identified by direct observation. This saves time and makes it possible to manipulate a neuron's membrane currents in ways that obscure the intrinsic electrical properties by which the neuron would otherwise be identified. It is difficult to assess the effects GFP might have on neuronal physiology in *C. elegans* because we did not record from the same neuron with and without GFP expression. In mammalian expression systems, however, the biophysical properties of exogenously expressed ion channels are unaffected by the presence of GFP.[23,24] In *C. elegans,* the amplitude and time course of net membrane current in GFP-labeled neurons were not unusual when compared with unlabeled neurons, suggesting that the effects of GFP, if any, are not dramatic.

Conclusion

This chapter describes a method for recording from neurons in the head ganglia of *C. elegans* using the slit-worm preparation. By similar means we have also exposed for recording the cell bodies of neurons in the ventral cord and tail ganglia. It should be possible, therefore, to record from almost any neuron in *C. elegans.*

The ability to record from identified neurons in *C. elegans* opens new avenues of research in a system already well known for its complete neuronal wiring diagram,[1] developmental cell lineage,[25] and genomic sequence (due in 1998). For example, it is now possible to combine both classical and molecular genetics with electrical recordings from identified neurons to study the genetic control of neural function and its development. In

[23] P. A. Doevendans, K. D. Becker, R. H. An, and R. S. Kass, *Biochem. Biophys. Res. Comm.* **222,** 352 (1996).

[24] J. Marshall, R. Molloy, G. W. Moss, J. R. Howe, and T. E. Hughes, *Neuron* **14,** 211 (1995).

[25] J. E. Sulston and H. R. Horvitz, *Devel. Biol.* **56,** 110 (1977).

addition, we can now study how identified neurons function in the neural circuits underlying particular behaviors. It should be possible, therefore, to trace the connections between genes and behavior in *C. elegans* more completely than before.

Appendix

This appendix describes circuit modifications to the Axopatch 200A for recording from neurons with high input resistance and low input capacitance. The manufacturer states that the same modifications will work on the Axopatch 200B.

1. The rise time of the current command was increased by reducing capacitor C71 to 330 pF.
2. Current-clamp command sensitivity was reduced by changing R208 to 392 kΩ.
3. Zap duration and amplitude were adjusted by changing C2 to 0.01 μF, R143 to 2.0 kΩ, C3 to a 5.1-kΩ resistor, and R78 to a 0.1-μF capacitor.
4. To increase the sensitivity of the whole-cell capacitance compensation, without altering the sensitivity of the series resistance compensation, R50 and R51 were changed to 20 kΩ; R52 and R53 were changed to 3.92 kΩ; and C63 was changed to 0.01 μF.

Acknowledgments

We thank D. Raizen, L. Avery, and K. Breedlove for first demonstrating that on-cell patch-clamp recordings can be made from *C. elegans* neurons, J. Thomas for the idea of using a laser beam to break into cells, M. Schaffer for scanning electron microscopy, and D. Garbers and C. Bargmann for the gift GFP-labeled strains. Portions of the text appeared in M. B. Goodman, D. H. Hall, L. Avery, and S. C. Lockery, *Neuron* (in press) (1998), and are reprinted here with permission. Supported by NSF, NIMH, ONR, The Sloan Foundation, and The Searle Scholars Program.

[14] Low-Noise Patch-Clamp Techniques

By RICHARD A. LEVIS and JAMES L. RAE

Introduction

It has now been more than 20 years since the patch-clamp technique was introduced.[1] However, it was not until the discovery[2,3] that application of suction to the interior of a clean heat-polished pipette pressed gently against the membrane of certain cells often resulted in formation of a membrane–glass seal with a resistance measurable in gigohms ($10^9\ \Omega$) that the technique gained wide acceptance. The high resistance of such a seal (commonly called a "gigaseal") resulted in dramatic reductions of background noise levels and sparked an intense period of methodological investigation that developed electronic and electrode technology to take full advantage of the possibilities offered by the gigaseal (see, e.g., Refs. 4–6). In a relatively short period of time following the discovery of the gigaseal the patch-clamp technique quite literally revolutionized the investigation of membrane electrophysiology. A variety of "configurations" were discovered almost immediately, that is, on-cell patches, excised patches of both inside-out and outside-out configuration, and the whole-cell configuration.[4] Refinements in electronics (see, e.g., Refs. 4–6) were also very rapid, leading to patch-clamp amplifiers with noise levels that at the time were adequate to take advantage of the existing pipette technology. Since then progress in patch-clamp electronics has been more gradual, but steady improvements in both noise levels and convenience of use have continued.

Initially it was believed that only certain types of glass and certain cells were suitable for the patch-clamp technique. However, it was rapidly realized that just about any glass pulled into appropriate pipettes could form gigaseals with just about any type of cell (provided only that reasonable access to the cell membrane is available). Rae and Levis[6,7] studied a variety

[1] E. Neher and B. Sakmann, *Nature* **260,** 799 (1978).
[2] F. J. Sigworth and E. Neher, *Nature* **287,** 447 (1980).
[3] E. Neher, *in* "Techniques in Cellular Physiology" (P. F. Baker, ed.) p. 1. Elsevier, Amsterdam, 1982.
[4] O. P. Hamill, A. Marty, E. Neher, B. Sakmann, and F. J. Sigworth, *Pflugers Arch.* **391,** 85 (1981).
[5] F. J. Sigworth, *in* "Single Channel Recording" (B. Sakmann and E. Neher, eds.), 2nd Ed., p. 95. Plenum, New York and London, 1995.
[6] J. L. Rae and R. A. Levis, *Mol. Physiol.* **6,** 115 (1984).
[7] J. L. Rae and R. A. Levis, *Methods Enzymol.* **207,** 66 (1992a).

0076-6879/98 $25.00

of different glasses in order to select those that would produce the lowest noise. As expected, it was found that glasses with the least dielectric loss produced the lowest noise. No major differences were ever clearly demonstrated in the ability of different glasses to form seals. It was clear from these early studies that quartz would be ideal for the fabrication of low-noise patch pipettes. However, quartz softens at 1600° and it was not until 1992 that it became possible to pull conveniently quartz patch pipettes. The use of quartz pipettes[8] and other strategies for reducing pipette noise[9,10] have now demonstrated that the noise of even the most modern electronics can dominate total noise in the best of measurement situations.

This article focuses on noise performance in patch and whole-cell voltage clamping. Background noise arises from a variety of different sources, each of which has to be understood if it is to be effectively minimized. Ultimately, the level of background noise determines what size and duration signals can be resolved by these techniques. Low noise is required to allow filter bandwidth to be increased with the resulting increase in time resolution. Most noise sources in the patch-clamp technique are uncorrelated, which means that they add in an rms (root mean square) fashion. This means that the largest individual noise source or sources can often dominate total noise. Nevertheless, low-noise recordings require attention to just about every detail of the technique. Following a brief introduction to some of the basics of the patch-clamp technique, such details are the major part of this article.

Some Patch-Clamp Basics

Pipettes

The appropriate fabrication of patch pipettes is of central importance to many different aspects of the patch-clamp technique. These include seal formation, the expected size of patch, the quality (resistance, stability) of the seal, and, of course, the background noise levels that can be achieved. The general procedures for fabricating patch pipettes have been described elsewhere[7,11] and are not presented here. Instead we briefly review some of the basic features of pipettes that are most important to performance.

[8] R. A. Levis and J. L. Rae, *Biophys. J.* **65,** 1666 (1993).
[9] J. L. Rae and R. A. Levis, *Pflugers Arch.* **42,** 618 (1992b).
[10] K. Benndorf, *in* "Single Channel Recording" (B. Sakmann and E. Neher, eds.), 2nd Ed., p. 129. Plenum, New York and London, 1995.
[11] A. Cavalie, R. Grantyn and H. D. Lux, *in* "Practical Electrophysiological Methods" (H. Kettenmann and R. Grantyn, eds.), p. 171. Wiley-Liss, New York, 1992.

The first issue that the experimenter must resolve is the type of glass to be used. Quartz is the best selection for very low noise measurements, but using it is expensive. A few other low-loss glasses are readily available, and, although their dissipation factors are much higher than that of quartz, these can produce very good results in many situations. For whole-cell measurements the electrical characteristics of the glass are less important, although there is no need these days to ever use very lossy glasses such as the soda-lime glasses that were once very popular. All pipettes require a coating with a suitable elastomer for low-noise applications. This may sometimes seem unnecessary in some whole-cell situations. Nevertheless, we believe that even in such cases applying a light elastomer coating is a good habit to form. Many different geometries of pipettes have been used. The purpose of these variations has sometimes been to reduce noise and at other times to promote the formation of desirable seals, appropriate patch size, and acceptable access resistance (in whole-cell or giant-patch situations). One of the geometrical considerations that must be addressed is the ratio of the outer diameter (d_o) to inner diameter (d_i) of the tubing used prior to pulling. In terms of noise, large d_o/d_i ratios are usually desirable in small patch measurements, whereas smaller d_o/d_i ratios may be better selections for whole-cell recording, since these make it easier to produce low-resistance pipettes. Heat polishing of pipettes can be used to great advantage in many situations; this is particularly true when making whole-cell or large patch measurements. However, it has been shown for quartz pipettes (where heat polishing is impractical due to the high melting temperature of quartz) and for very small tipped pipettes made from quartz or other glasses (where visualization of the tip becomes difficult or impossible) that seals can readily be formed without heat polishing. Issues concerning pipette noise are considered extensively in this article.

Seal Formation

Of course the formation of a high-resistance membrane–glass seal is a prerequisite for low-noise patch-clamp recordings. A high-resistance seal dramatically reduces background noise and enhances recording stability. The mechanisms involved in seal formation are not completely understood (although ideas have been proposed). Thus empirical information based on experience has guided efforts to form efficiently the highest resistance and most stable seals possible. In general, the pipette is pressed against the surface of the selected cell until a noticeable (about a factor of 1.5–2 is typical) change in resistance of the pipette occurs. Pipette resistance is usually monitored by applying a small periodic square-wave potential to the pipette and measuring the resulting current; many software packages for patch clamping have automated this procedure to provide a rapidly updating digital readout of resistance. On occasion seals will form spontaneously when the pipette presses against the cell membrane. More often, however, suction must be applied to the interior of the pipette. A gigaseal will then usually form, although the process may be sudden or gradual (requiring up to about 30 sec). It is generally found that smaller tipped pipettes form higher resistance seals, although there is a great deal of scatter in the value of seal resistances even among apparently identical pipettes. High-resistance seals (e.g., >50 GΩ) are very important to achieving low-noise measurements, and it is to be expected that the higher the seal resistance the lower the noise attributable to the seal will be. Seal noise is difficult to study and has never been characterized in great detail. The noise of the seal is most important at relatively low bandwidths. Thus a 10 GΩ seal may be adequate if you plan to record at bandwidths of 20 kHz or more, while a seal well in excess of 100 GΩ is often desirable when recording at bandwidths of less than 1 kHz.

Configurations

Cell-Attached Patch. The first and most basic configuration is the cell-attached patch. In this configuration a seal is formed with the cell and a membrane patch is thus isolated. Clearly the rest of the cell is then in series with the patch membrane (i.e., patch currents must flow through the rest of the cell membrane). This is not a problem in most situations, but could occasionally produce unwanted effects (e.g., when measuring fairly large currents in a patch on a very small cell). In addition, the on-cell patch does not allow you to know accurately the transpatch potential because of the unknown cell membrane resting potential. Finally, this configuration does not allow changing the ionic composition on both sides of the patch membrane. Thus cell-attached recordings are most commonly used when the channel being studied requires some unknown cytoplasmic factor or factors for normal gating; such factors would be lost if the patch was excised from the cell.

Excised Patches. Excised patches allow easy access to one side of the patch and allow precise control of the transpatch potential. Such patches are also the best choice for very low noise recording since they can readily be withdrawn toward the surface of the bath minimizing immersion depth. There are two basic types of excised patches.

INSIDE-OUT PATCHES. Inside-out patches are easily formed by simply withdrawing the pipette from the cell surface after a seal has been formed. This configuration allows easy access to the intracellular side of the membrane. The most common difficulties associated with inside-out patches are the loss of cytoplasmic factors that may be involved in modulating behavior of some channels and the possibility that a closed vesicle may form.[4]

OUTSIDE-OUT PATCHES. Outside-out patches are formed by disrupting the patch membrane after a seal has been formed and then withdrawing the pipette slowly from the cell surface. In most cases a new patch will be formed with the outer membrane surface facing the bath. This allows easy access to the extracellular face of the patch membrane. As was the case in inside-out patches the loss of cytoplasmic factors affecting some channels may be a problem. In addition, there is no way of knowing before excision what the patch will contain since the original patch with its channels has been destroyed. Finally, in our experience forming stable outside-out patches for high-quality measurements is somewhat more difficult than forming inside-out patches.

Whole-Cell Recording. Whole-cell recording is a powerful technique for measuring the currents from an entire cell. This configuration is normally easily obtained after a seal has been formed by disrupting the patch membrane with additional suction or a brief large voltage pulse. An alternative method is the perforated patch technique.[12,13] A possible shortcoming of the "traditional" whole-cell technique in some situations is the loss of cytoplasmic factors from the cell interior; this can usually be avoided by the perforated patch approach to whole-cell measurements. It is also common to find that the access resistance measured after patch disruption (or perforation) is higher than the original resistance of the pipette. Series resistance compensation is often very important when currents are large or wide bandwidth recordings are desired.

Giant Patches. Gigaseal formation becomes progressively less likely as the size of the pipette tip increases. However, application of a hydrocarbon coating to the pipette tip enhances seal formation and has allowed the formation of patches with diameters in the range of 10–40 μm with patch capacitances of 2–15 pF and seal resistances in the range of 1–10 GΩ.[14,15] Giant patches can be formed in both cell-attached and excised configurations. The noise of such patches is likely to be dominated at high frequencies by the thermal voltage noise of the pipette in series with the large patch capacitance and at lower frequencies by the relatively low seal resistances usually obtained (plus any noise associated with the membrane itself). However, in most cases relatively large signals are expected from such

[12] R. Horn and A. Marty, *J. Gen. Physiol.* **92,** 145 (1988).

[13] J. L. Rae, K. Cooper, P. Gates, and M. Watsky, *J. Neurosci. Methods* **37,** 15 (1991).

[14] D. W. Hilgemann, *Pflugers Arch.* **415,** 247 (1989).

[15] D. W. Hilgemann, *in* "Single Channel Recording" (B. Sakmann and E. Neher, eds.), 2nd Ed., p. 307. Plenum, New York and London, 1995.

patches and thus the signal-to-noise ratio can remain very favorable for many measurements.

Electronics

Patch-clamp electronics have been described in detail in many previous publications by ourselves and others (see e.g., Refs. 5, 6, and 16) and these details are not repeated here. Instead only a few comments that relate to the material of this paper are summarized.

Headstage Amplifier. The "headstage" amplifier is the most important part of the patch clamp in terms of noise. Two basic varieties are available from several manufacturers, namely, resistive feedback headstages and capacitive feedback headstages. Capacitive feedback offers lower noise and can produce wider bandwidth. The noise of a specific capacitive feedback amplifier is described in detail later. Many amplifiers contain two or more different feedback elements for different situations. Capacitive feedback or very high valued resistive feedback (typically 50 GΩ) is intended for small patch measurements and provides the lowest noise. Lower valued resistors are provided for whole-cell situations (typically 500 MΩ for cells of up to 100 pF and 50 MΩ for larger cells) and might also be used with large patches. The noise of the headstage with a 500-MΩ feedback resistor is considerably higher than that with a 50-GΩ resistor or capacitive feedback.

Capacity Compensation. Capacity compensation is provided to cancel transients resulting from the charging of the capacitance of the pipette and its holder (plus other sources of capacitance at the headstage input) when the potential is changed. Most patch clamps provide two time constants to cancel such transients. The second slower time constant is primarily needed to deal with the lossy capacitance of glass pipettes. We have shown that this component is smallest in low-loss glasses.[7] The slow component is *not* well described by a single exponential and therefore cannot be completely canceled by the compensation circuits provided in commercial patch clamps. Minimization of this component is therefore best accomplished by using low-loss glasses (soda-lime glasses in particular should be avoided) and coating pipettes with low-loss elastomers; quartz produces the least amount of such a slow component. For low-noise applications minimization of capacitance is very important. Note also that in addition to the noise arising from the capacitance at the headstage input, the capacity compensation circuitry can also add noise of its own.

Whole Cell Compensations. When performing whole-cell measurements series resistance compensation is often very important. This is discussed in

[16] R. A. Levis and J. L. Rae, *Methods Enzymol.* **207,** 18 (1992).

some detail later. In addition to series resistance compensation, commercial patch-clamp amplifiers also provide compensation for the whole-cell capacitance. This is important in eliminating transients when membrane potential is changed. Of course, the capacitance of the pipette (etc.) also needs to be compensated in this situation.

Low-Noise Recording Techniques

Noise in Single-Channel Measurements

Overview of Patch-Clamp Noise. The noise associated with single-channel patch-clamp measurements has a power spectral density (PSD) that can generally be described by

$$S_{pc}{}^2 = a_0/f + a_1 + a_2 f + a_3 f^2 \qquad \text{amp}^2/\text{Hz} \tag{1}$$

where a_0, a_1, a_2, and a_3 are coefficients that describe the contribution of each noise term to total noise power and f is the frequency in hertz. The rms noise resulting from this PSD can be obtained by integrating from a low-frequency cutoff of B_0 to a high-frequency cutoff of B (Hz) and taking the square root of the result. This yields:

$$I_{pc} = (c_0 a_0 \ln(B/B_0) + c_1 a_1 B + c_2(a_2/2)B^2 + c_3(a_3/3)B^3)^{1/2} \qquad \text{amp rms} \tag{2}$$

where c_0, c_1, c_2, and c_3 are coefficients that depend on the type of filter used, and it has been assumed that $B \gg B_0$ [so that B_0 has been ignored in the last three terms of Eq. (2)]. The coefficient c_0 can usually be taken to be 1.0 without introducing too much error. In considering $1/f$ noise it is more important to determine B_0, the effective low-frequency cutoff of the measurement. Because high-pass filters are not used in most single-channel measurements, B_0 must instead be deduced on the basis of the duration of the measurement (the longer the duration, the smaller B_0). The other coefficients, c_1, c_2, and c_3, need further explanation. For a "brickwall" filter (i.e., a filter that rolls off extremely rapidly at frequencies above its corner frequency) these coefficients are all essentially 1.0. However, for filters with desirable time-domain characteristics the coefficients are larger than 1.0. For an eight-pole Bessel filter (which is the most commonly used type for time domain measurements) these coefficients are approximately $c_1 \approx 1.05$, $c_2 \approx 1.3$, and $c_3 \approx 2.0$.

From Eqs. (1) and (2) it can be seen that patch-clamp noise PSD is generally described by terms that include white noise and terms that vary with frequency as $1/f$ (more precisely $1/f^{\alpha}$ where α is usually near 1.0), f, and f^2. These terms give rise to terms that contribute to total rms noise in proportion to $[\ln(B/B_0)]^{1/2}$, $B^{1/2}$, B, and $B^{3/2}$.

The $1/f$ noise arises from the patch-clamp amplifier and possibly from the seal and the patch itself; in many situations $1/f$ noise can be neglected in that its contribution to overall noise is often very small. White noise arises from the patch-clamp amplifier, from the seal, and from the patch membrane. The f noise arises primarily from lossy dielectrics; these include aspects of the patch-clamp amplifier, and (usually more importantly) the pipette and its holder. The f^2 current noise arises from voltage noise (white) in series with a capacitance. There are several sources of this type of noise in typical patch-clamp recording situations; these include the patch-clamp amplifier, capacitance added to the amplifier input by the holder and pipette, distributed RC noise, and noise arising from the pipette resistance in series with the patch capacitance (R_e-C_p noise).

Clearly $1/f$ noise is most important when bandwidth is very limited (e.g., in the measurement of very small currents) and it generally sets the limit on how much noise may be reduced by restricting bandwidth. White noise is also most important at relatively low bandwidths. Noise types f and particularly f^2 become progressively more important as bandwidth increases.

It is important to note that most noise sources involved in patch-clamp measurements are uncorrelated. This means that they will add together in an rms fashion; that is, considering three uncorrelated noise sources contributing to total noise with rms values in a particular bandwidth denoted by e_1, e_2, and e_3, total rms noise (e_T) is then given by

$$e_T = (e_1^2 + e_2^2 + e_3^2)^{1/2} \qquad (3)$$

An important aspect of this is that the largest individual source of noise can dominate total noise. Thus if $e_1 = 2$ and both e_2 and $e_3 = 1$, then $e_T = 2.45$, which is only 22% more than e_1 alone. This type of information must be remembered when judging the importance of various noise sources in different situations and in determining appropriate compromises between different types of noise when this is required.

The theoretical aspects of individual noise sources are now described beginning with the patch-clamp headstage amplifier.

Patch-Clamp Amplifier. The first noise source to consider is that of the patch-clamp amplifier itself. Considerable progress has been made in recent years to reduce the noise of patch-clamp electronics for single-channel measurements. This has been primarily due to the introduction of capacitive feedback amplifiers; such amplifiers are now available from several manufacturers. Prior to the use of capacitive feedback, patch-clamp amplifiers relied on high-valued resistors as the feedback element; for single-channel measurements the use of a 50-GΩ resistor has been the most common selection. A variety of disadvantages to such resistors have been described elsewhere.[7,16] Briefly, such resistors have a very limited frequency response and thus require "boost" circuits[5,6] to restore the high-frequency components of the measured signal (and, of course, of the noise). Often the frequency response is not characterized by a simple one-pole *RC* rolloff, necessitating complex boost circuits and/or less than perfect corrected responses. Boost circuits need to be retuned periodically because the characteristics of the resistor may vary somewhat over time. In addition, high-valued resistors sometimes show relatively high voltage and temperature coefficients of resistance (i.e., resistance changes slightly with temperature and the voltage across the resistor). This can lead to nonlinearities and small changes of the boosted response with changes in temperature and signal amplitude. Nevertheless, adequate dynamic performance can be achieved with usable bandwidths in excess of 30 kHz. However, more importantly for the present discussion, all high-valued feedback resistors currently available that we are aware of display considerably higher noise than simply the expected thermal current noise.[7,16] The result of all this is that resistive feedback amplifiers typically display open-circuit noise of about 0.25-pA rms in a 10-kHz bandwidth (eight-pole Bessel filter). Capacitive feedback amplifiers can have as little as about half this much noise.

The amplifier that we use for low-noise measurements is the Axopatch 200B (Axon Instruments, Foster City, CA). We deal specifically with one of these instruments and its noise in this discussion so that examples later in the chapter can be associated with specific numerical values. However, the principles described also apply to other amplifiers of this general type made by this and other manufacturers. The input referred open-circuit noise PSD, S_{hs}^2, of the instrument used by one of us is very well fit by the following equation:

$$S_{hs}^2 = 1.9 \times 10^{-32} + 3.5 \times 10^{-35} f + 1.3 \times 10^{-38} f^2 \qquad \text{amp}^2/\text{Hz} \quad (4)$$

where f is the frequency in hertz. This was the lowest noise instrument of the first 10 or so manufactured. The noise PSD is adequately described by a white noise component and components that rise with increasing frequency as f and f^2. This amplifier displays very little $1/f$ current noise, and whatever amount is present is difficult to quantify since it requires very long measurement times and is thus subject to interference arising from mechanical vibrations (and periodic resets). The maximum value of any $1/f$ current noise component of the open-circuit amplifier PSD is $\sim 2 \times 10^{-32}/f$, indicating a $1/f$ corner frequency of about 1 Hz or less.

The white noise term of Eq. (4) is equivalent to the thermal voltage noise of an 850-GΩ resistor; it arises from the shot noise of the input junction field effect transistor (JFET) and from noise associated with the differentiator, which follows the integrating headstage amplifier. In fact, the differentiator is the largest contributor to this noise, producing about 1.1–1.2 $\times$ 10^{-32} amp^2/Hz.[17] The JFET in the Axopatch 200B is cooled to about $-20°$ and its gate leakage current can be calculated to be about 0.02 pA, contributing about 7 $\times$ 10^{-33} amp^2/Hz to the white noise term.

The *f* noise component of Eq. (4) arises primarily from lossy dielectrics associated with the input; these dielectrics include packaging, capacitors, and some contribution from the JFET itself (quite possibly associated with the surface passivation layer). A small contribution to this term arises from $1/f$ voltage noise of the input JFET in series with capacitance associated with the input. The f^2 noise term of Eq. (1) arises primarily from the white noise component of the input voltage noise of the JFET in series with the capacitance associated with the input. This capacitance is dominated by the input capacitance of the JFET itself, but also includes strays, the feedback capacitor and the capacitor used to inject compensation signals, and capacitance associated with the input connector. Smaller contributions to this term arise from noise associated with compensation signals and a term arising from the differentiator.

Equation (4) can be integrated over a bandwidth (i.e., *DC* to *B* Hz) to provide an equation for the variance as a function of frequency. The square root of this result is the input referred rms noise of the open-circuit amplifier (here called i_{hs}) as a function of bandwidth:

$$i_{hs} = (1.9 \times 10^{-32} c_1 B + 1.75 \times 10^{-35} c_2 B^2 + 4.3 \times 10^{-39} c_3 B^3)^{1/2} \text{ amps rms} \tag{5}$$

where c_1, c_2, and c_3 are coefficients that depend on the type of filter used. As mentioned previously, for an eight-pole Bessel filter these are approximately $c_1 \approx 1.05$, $c_2 \approx 1.3$, and $c_3 \approx 2.0$.

Thus this particular amplifier has open-circuit noise of approximately 7-, 41-, and 105-fA rms in bandwidths of 1, 5, and 10 kHz (eight-pole Bessel filter), respectively.

Most noise sources encountered in actual patch-clamp recordings are uncorrelated and therefore simple rules of rms addition apply [see Eq. (3)]. However, this is not the case with noise arising from the input voltage

[17] The noise contribution of the differentiator can be changed by changing its feedback resistor. This will also change the overall gain and bandwidth of the integrator/differentiator combination. For example, increasing the feedback resistor by a factor of 10 will increase the gain by the same factor and decrease the available bandwidth by a factor of 3.16 ($10^{1/2}$). The PSD of the white noise contribution will fall by a factor of 10. Decreasing the feedback resistor will have the opposite effect—decreased gain, increased bandwidth, and increased differentiator noise contribution.

noise, e_n, of the JFET and capacitance added to the input by the addition of the holder and pipette. This noise is perfectly correlated with noise arising from other capacitance at the input (JFET capacitance, strays, feedback and injection capacitors, capacitance of the input connector) so the usual rules of rms addition of uncorrelated noise sources do not apply in this case. This noise has a PSD that rises as f^2 and adds to the f^2 term of Eq. (4). The addition of 2 pF of capacitance (a reasonable number for a small holder and pipette with a moderate depth of immersion) at the input will increase this term to approximately $1.9 \times 10^{-38} f^2$ amp^2/Hz, and this in turn would increase the rms noise of the particular headstage considered here in bandwidths of 5 and 10 kHz (eight-pole Bessel filter) to 47- and 123-fA rms, respectively. Obviously less capacitance will lead to smaller increases in noise and more capacitance would lead to larger noise increments. Remember that the holder and pipette will also contribute other types of uncorrelated noise (see later discussion). However, we will include in the noise of the headstage the noise arising from e_n and *all* capacitance at the input, including the capacitance of the holder and pipette.

Holder Noise. The addition of a traditional holder to the headstage input adds noise due to its capacitance in series with e_n as just described, and because of the lossiness of this capacitance. The lossy capacitance adds *dielectric noise.* The magnitude of this noise is dependent on the size and geometry of the holder and on the material from which the holder is constructed. The most important parameter of the holder material is its dissipation factor. In general a capacitance, C_d, with a dissipation factor D will produce dielectric noise with a PSD, S_d^2, given by

$$S_d^2 = 4kTDC_d(2\pi f) \qquad \text{amp}^2/\text{Hz} \tag{6}$$

where k is the Boltzmann constant and T is absolute temperature. Over a bandwidth of B Hz this will produce an rms noise current given by

$$i_d = (4kTDC_d c_2 \pi B^2)^{1/2} \qquad \text{amp rms} \tag{7}$$

The best materials commonly used to fabricate pipette holders are polycarbonate and Teflon. Teflon has a lower dissipation factor ($\sim 2 \times 10^{-4}$), but displays piezoelectric and space charge effects. The dissipation factor of polycarbonate is higher ($\sim 10^{-3}$), but in actual practice either material has produced acceptable results. Lucite holders should be avoided for low-noise measurements because the dissipation factor of Lucite is rough 30–40 times higher than that of polycarbonate.

As described by Levis and Rae,[8] precise calculations of the dielectric noise introduced by a pipette holder are complicated by the rather complex equivalent circuit presented by most holders. Rough calculations have indicated that a small polycarbonate holder with a capacitance of 0.6 pF is predicted to produce approximately 15-fA rms noise in a bandwidth of 5 kHz; a larger holder (measured capacitance ~1.5 pF) should produce about 25-fA rms noise in this bandwidth. These numbers are in good agreement with actual measurements of holder noise. With a small polycarbonate holder attached to the headstage considered here, noise increases to about 46-fA rms in a 5-kHz bandwidth (eight-pole Bessel filter). A 1.5-pF polycarbonate holder should increase this value to about 52-fA rms.

Benndorf[10] has reported the use of a metal body pipette holder of very small size. Such a holder will not produce significant dielectric noise, but will only slightly add to the capacitance (almost lossless) at the headstage input. Benndorf does not report the amount of capacitance added by this holder, but its size suggests that it is a very small increment. This design doubtlessly represents the lowest noise possible from a holder, but given the small increment in noise associated with more traditional holders and the moderate amount of inconvenience that seems to be associated with the use of the tiny metal holder (and extremely short pipettes fixed with wax to the holder), its utility is not clear except in the most demanding of applications. Benndorf[10] also reports $1/f$ noise associated with polycarbonate holders; we have not found any significant amounts of such noise in the holders that we use.

Note that it is imperative to keep holders clean to achieve the low-noise levels described.

Pipette Noise. The noise associated with pipettes has been discussed extensively in previous publications by ourselves and others.[7,8,10,16] The following review adds only a few new features to the theoretical aspects of pipette noise, although dielectric noise and particularly distributed RC noise are considered in greater detail than in the past, refining and in some cases modifying previous conclusions. This article attempts to bring together a wide range of theoretical and practical information concerning pipette noise in a convenient format.

Several different mechanisms contribute to the noise arising from the pipette. Figure 1 shows a simplified circuit representation of the four most important noise mechanisms resulting from the pipette per se. We first summarize all pipette noise sources and then describe the most important of these in greater detail. Both theoretical and practical issues relating to noise minimization are considered.

1. The pipette adds capacitance to the input of the amplifier. This depends on the length of the electrode, its geometry (especially wall thickness), the use of elastomer coatings, and the depth of immersion. This capacitance is in series with e_n and produces f^2 noise by the mechanism described earlier.

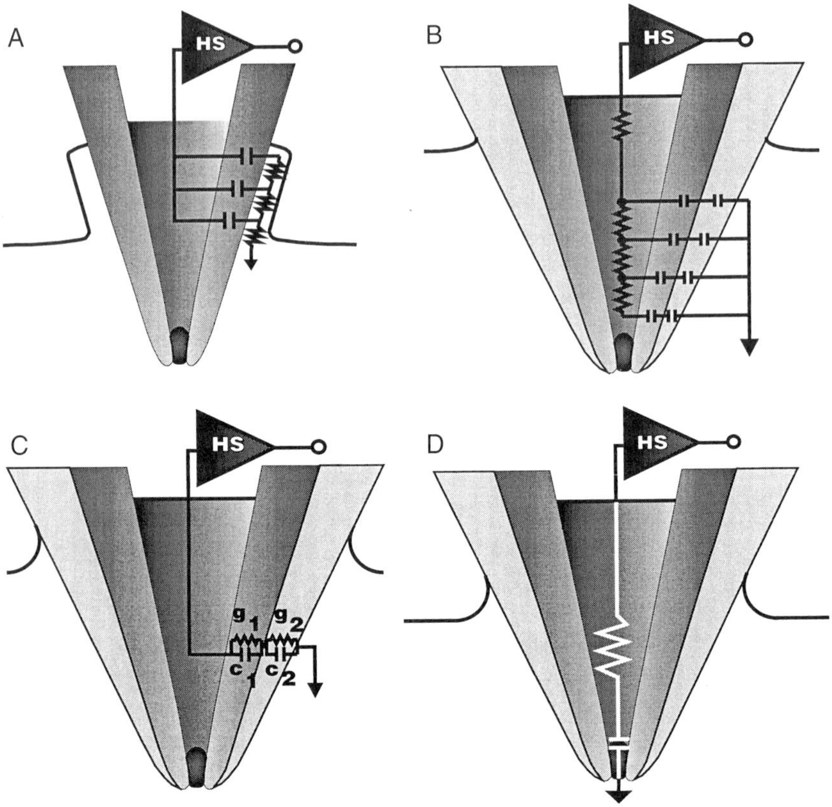

FIG. 1. Simplified circuit representation of the four most important noise mechanisms associated with the patch pipette per se. (A) Thin-film noise arising from the thermal voltage noise of the distributed resistance of a thin film of solution on the pipette exterior in series with the capacitance of the pipette wall. Similar films may also form within the pipette and in the holder. Note that in part (A) the pipette is shown without an elastomer coating; such a coating can essentially eliminate exterior thin film noise. In parts (B), (C), and (D) the pipette is shown with an elastomer coating. (B) Distributed *RC* noise arising from the thermal voltage noise of the distributed resistance of the pipette filling solution in series with the distributed wall capacitance of the immersed portion of the pipette. (C) Dielectric noise of an elastomer coated pipette arising from the series combination of the glass pipette itself (g_1, C_1, where $g_1 = \omega C_1 D_1$, $\omega = 2\pi f$) and the elastomer (g_2, C_2, where $g_2 = \omega C_2 D_2$). (D) R_e-C_p noise arising from the thermal voltage noise of the entire pipette resistance in series with the capacitance of the patch. This figure does not include noise arising from the pipette capacitance in series with the input voltage noise, e_n, of the headstage amplifier, or noise associated with the seal. See text for further details.

2. The capacitance of the immersed portion of the pipette also will produce dielectric noise (Fig. 1C). The amount of this noise depends on the type of glass used, the wall thickness of the glass, the type and thickness of the elastomer coating, and the depth of immersion of the pipette. Dielectric noise produces a PSD that rises linearly with increasing frequency (f noise).

3. Distributed *RC* noise (Fig. 1B) is the name we use to describe the current noise arising from the thermal voltage noise of the distributed resistance of the pipette filling solution in series with the distributed capacitance of the pipette wall. The capacitance is distributed more or less evenly along the immersed portion of the pipette (assuming a roughly constant ratio of inner to outer diameter of the pipette, see later section). The resistance is primarily located near the tip; however, significant resistance remains in regions distal to the tip. The thermal voltage noise of the distributed pipette resistance in series with the distributed capacitance of the immersed portion of the pipette wall produces noise with a PSD that rises with increasing frequency as f^2 over the range of frequencies of interest to patch clamping. Distributed *RC* noise is dependent on the geometry of the pipette and the thickness of its walls, the thickness of elastomer coating, the resistivity of the filling solution and the depth of immersion. It is virtually independent of the type of glass used except insofar as this selection affects wall capacitance (primarily due to the dielectric constant of the glass).

4. R_e-C_p noise (Fig. 1D) is the term we use to describe noise arising from the total resistance of the pipette, R_e, in series with the capacitance of the patch, C_p. This noise clearly depends on the value of the pipette resistance and of the patch capacitance. It is usually minimized by using small-tipped pipettes. R_e-C_p noise produces noise that rises as f^2 over the range of frequencies important to patch voltage clamping.

5. Thin-film noise (Fig. 1A) is produced by films of solution that can form on the outer surface of an uncoated pipette as it emerges from the bath. Such a film can have a very high distributed resistance that is in series with the distributed capacitance of the pipette wall. The noise expected from such a film should rise at low to moderate frequencies and then level out at frequencies in the range of kilohertz to tens of kilohertz. Such a thin film can be a very significant source of noise. Fortunately, however, noise from such films on the external pipette surface can be essentially completely eliminated by coating the pipette with a suitable elastomer. Suitable elastomers present a hydrophobic surface that prevents the formation of external films. It is also possible for such films to form inside the pipette or its holder. Such internal films can be prevented by layering a millimeter or so of silicone or paraffin oil on top of the filling solution. However this is not

normally necessary if excess fluid is carefully cleaned from the back of the pipette by suction and (if necessary) drying the pipette with a jet of air. Maintaining the holder free from solutions is very important.

6. Seal noise is probably the least understood source of noise associated with the pipette. It is clearly minimized by the highest possible seal resistance, but in our experience seal noise is often somewhat unpredictable. Seal noise is considered in greater detail later.

The minimization of pipette noise is often relatively straightforward, and often requires only a small amount of additional effort (although it can become expensive if quartz pipettes are selected). Most necessary precautions are simple and reasonably intuitive. Thus, for example, short pipettes are always advantageous, as is the use of thick-walled glass and a heavy coating of a low-loss elastomer as close as possible to the tip. Shallow depths of immersion will minimize noise whenever this is compatible with the experiment being undertaken. Cleanliness is obviously important. In some cases, however, ultimate minimization of noise is not always compatible with a particular type of measurement. For example, very small-tipped pipettes will reduce R_e-C_p noise and tend to produce the highest seal resistance; they also reduce the likelihood of the patch containing charge translocating processes other than those to be measured. Of course, tiny patches also reduce the likelihood of the patch containing the channel to be studied. Thus this strategy is not always appropriate. In the following discussion we attempt to consider both theoretical and practical methods of achieving the lowest possible noise in various situations.

Dielectric Noise. Dielectric noise results from thermal fluctuations in lossy dielectrics. The magnitude of this noise can be related to the real part of the admittance (the loss conductance) of the dielectric material. Dielectric noise is often the dominant source of noise associated with the pipette. This is particularly true if quartz is not used for pipette fabrication. The equations describing the PSD and rms noise of a single dielectric have already been presented [Eqs. (6) and (7)]. These equations clearly show that dielectric noise depends on the capacitance in question (here the pipette) and on the dissipation factor associated with this capacitance. The dissipation factor and dielectric constant of several glasses are shown in Table I; data for two elastomers are also included. This list is much shorter than lists published previously because many glasses have become scarce or are now completely unavailable. Clearly quartz has the lowest dissipation factor (by a wide margin) and it also has the lowest dielectric constant (by a much smaller margin). Soda-lime glasses (such as 0080) should be avoided for low-noise measurements due to their high dissipation factor and dielectric constant.

TABLE I
DISSIPATION FACTORS AND DIELECTRIC CONSTANTS OF SELECTED GLASSES AND ELASTOMERS

Sample	Dielectric constant	Dissipation factor
Shott 8250	4.9	0.0022
Shott 8330	4.6	0.0037
7740	5.1	0.005
7052	4.9	0.003
Quartz	3.8	10^{-5}–10^{-4}
Soda-lime glass (0080)	7.2	0.009
Sylgard #184	2.9	0.002
R-6101	Not known	0.00025[a]

[a] The dissipation factor listed for R-6101 is an unpublished value provided by the manufacturer; we have not attempted to verify this value in independent tests.

Dielectric noise of the pipette can contribute a large fraction of the total pipette noise, and thus of the noise of the measurement. For example, using Eq. (7), a capacitance, C_d, of 3 pF with a dissipation factor, D, of 0.01 (appropriate for some soda-lime glasses) will produce about 0.22-pA rms noise in a 5-kHz bandwidth. On the other hand, it is certainly possible to minimize dielectric noise. Thus, for example, with C_d of 1 pF and D of 0.0001 (appropriate for quartz), dielectric noise would be only about 13-fA rms in the same bandwidth.

Coating a pipette with a low-loss elastomer such as Sylgard 184 or R-6101 (K. R. Anderson, Santa Clara, CA) will in most cases reduce dielectric noise; quartz is a notable exception to this as described later. Such coating is absolutely necessary for low-noise recordings because it will essentially eliminate thin-film noise and will also reduce distributed *RC* noise (see later discussion). Of course, the noise of a pipette coated with an elastomer can no longer generally be described by the simple equations presented [Eqs. (6) and (7)] because the coated pipette is the series combination of two different dielectrics (the glass and the elastomer). Levis and Rae[8] have presented equations that describe the noise of two dielectrics in series and described expected and measured results for quartz pipettes in considerable detail. The equation for the rms noise of two dielectrics with capacitances C_1 and C_2 and associated dissipation factors D_1 and D_2 is well approximated by the following if $D_1, D_2 \ll 1$ (as is the case here):

$$i_d = [4kTc_2\pi B^2(D_1C_1C_2^2 + D_2C_2C_1^2)/(C_1 + C_2)^2]^{1/2} \quad \text{amp rms} \quad (8)$$

where B is the bandwidth in Hertz and c_2 is a coefficient that depends on the type of filter used ($c_2 \approx 1.3$ for an eight-pole Bessel filter).

Equation (8) can be used to demonstrate a number of conclusions concerning glass type and elastomer coating and the thickness of both the glass and the coating. In the first place it is useful to consider the expected values of the capacitances in Eq. (8). We consider C_1 and D_1 to represent the glass and C_2 and D_2 to represent the elastomer. The capacitance of a tapering cylinder that preserves the ratio of inner diameter, d_i, to outer diameter, d_0, is proportional to $1/\ln(d_o/d_i)$. This can be used to give a rough estimate of capacitances C_1 (glass) and C_2 (elastomer). For a glass with a dielectric constant of 4 (slightly more than quartz and slightly less than most borosilicates) the capacitance is approximately (0.22 pF/mm)/$\ln(d_o/d_i)$. For an elastomer with a dielectric constant of 3, the capacitance is about (0.17 pF/mm)/$\ln(d_o/d_i)$, where d_o and d_i now refer to only the elastomer coating. Glasses commonly used in patch clamping have d_o/d_i ratios ranging from about 1.2 to 4 or more. These values would produce capacitances ranging from about 1.2 pF/mm of immersion (for $d_o/d_i = 1.2$) to 0.16 pF/mm of immersion (for $d_o/d_i = 4$). Coatings of elastomers can be built up to fairly large thickness, although formation of very thick coats near the tip of the pipette is difficult due to the tendency of the elastomer to flow away from the tip prior to curing (see later discussion). An elastomer coating with $d_o/d_i = 1.5$ (here d_o refers to the outer diameter of the elastomer coat and d_i refers to its inner diameter, which, of course, corresponds to the outer diameter of the glass) would have a capacitance of about 0.4 pF/mm of immersion; for $d_o/d_i = 2$, this would decrease to about 0.25 pF/mm. The general advantages of thick-walled glass and of heavy elastomer coating are thus immediately clear, as is the advantage of shallow depths of immersion. Thick-walled glass, heavy coats of elastomer, and shallow depths of immersion reduce the pipette capacitance. Unfortunately, the numbers just presented are highly approximate because the ratios d_o/d_i of the walls of the pipette near the tip and particularly of the elastomer coating are not uniform. In the case of the pipette, there always appears to be some thinning of the glass near the tip. The extent of such thinning depends on the glass type and more importantly on the geometry to which the glass is pulled. Sharp-tipped pipettes with a small cone angle appear to most closely preserve the d_o/d_i ratio as the pipette is pulled (d_o/d_i ratios at the tip of >95% of the initial tubing ratio have been reported[10]), although not too much thinning can be preserved even with somewhat blunter tipped pipettes. Large-tipped pipettes (prior to fire polishing) with pronounced bullet shapes often produce quite considerable thinning near the tip. Thus the estimates of pipette capacitance listed earlier are generally lower than the actual capacitances observed. For example, with $d_o/d_i = 2$ quartz tubing, pulled to pipettes with roughly a 1-μm tip diameter and a resistance of ~5–10 MΩ, Levis and Rae[8] found that the d_o/d_i ratio decreased to about 1.4–1.5

within about 1 mm of the tip. However, even in such cases the *relative* improvements resulting from thicker walled tubing are more or less as predicted. For elastomer coating the nonuniformity in d_o/d_i is generally even more pronounced; thick coats are easy to build up a few hundred microns back from the tip but are much more difficult to produce near the tip. Levis and Rae[8] have described a method that can produce good results all the way to the tip, but even so there is pronounced thinning of the elastomer in the regions nearest the tip. Thus, as was the case with the glass, the estimates of elastomer capacitance given here should be thought of as simple approximations. Details of variations in wall thickness of the glass and of the elastomer coating are presented for some specific pipette geometries in the discussion of distributed *RC* noise.

The dissipation factors of quartz and several other glasses as well as those of two elastomers are listed in Table I. From this table it is clear that quartz has a dissipation factor that is at least 20–30 times lower than that of any other available glass. Pulling quartz into patch pipettes only became possible after the introduction of a laser-based puller (P-2000, Sutter Instruments, Novato, CA) in 1992. Levis and Rae[8] extensively studied the properties of quartz pipettes and concluded that they were the best selection for ultra-low-noise recordings due to low dielectric noise. These conclusions remain valid at the time of this writing, although it should be noted that the geometry of the pipettes investigated in that study was restricted to relatively blunt tapered pipettes (R_e typically about 5–10 MΩ for a ~1-μm tip diameter) and tubing with d_o/d_i ratios was generally in the range of 1.4–2.0 (although a ratio of 3.0 was used in a few experiments at that time). An important conclusion of that study with regard to dielectric noise was that an elastomer coating generally *increased* the dielectric noise of quartz pipettes, but that heavily coated quartz pipettes still showed significantly less dielectric noise than pipettes fabricated from any other type of glass. Theoretical predictions and actual measurements showed that for quartz pipettes fabricated from tubing with a d_o/d_i ratio of 2, coated with Sylgard 184 all the way to the tip, and immersed in the bath to a depth of ~1.8 mm, dielectric noise attributable to the pipette was about 35-fA rms in a 5-kHz bandwidth. Other measurements suggested that dielectric noise can be as low as 15-fA rms in this bandwidth for similar pipettes with a shallower depth of immersion. We have now occasionally used quartz tubing with $d_o/d_i = 4$ and, particularly with pulling techniques that attempt to preserve this ratio near the tip, we believe that these pipettes can display even less dielectric noise.

Examination of Eq. (8) shows why coating with Sylgard 184 with a dissipation factor of 0.002 will improve the dielectric noise of pipettes fabricated from glasses other than quartz (which have dissipation factors higher than that of Sylgard), but will actualy increase the dielectric noise of a pipette fabricated from quartz (which has a dissipation factor at least 20 times less than that of Sylgard). For light coats of Sylgard, C_2 (the capacitance of the immersed portion of the elastomer coating) will generally be more than C_1 (the capacitance of the immersed portion of the glass). With heavy coats of elastomer C_2 can become smaller than C_1, although our experience indicates that for glass with an initial d_o/d_i ratio of 2 (and a pulled ratio of ~1.4–1.5 near the tip) that even with the best techniques it is difficult to produce a value of C_2 less than about $C_1/3$ for an immersion depth of ~2 mm. With shallower depths of immersion and/or thicker walled glass, C_2 and C_1 become more comparable even with very heavy elastomer coatings. In fact, for glass with d_o/d_i more than 2–3 at the tip it is probably not possible to achieve $C_2 < C_1$ in the tip region. Thus there are restrictions to the ratio of C_1 to C_2 that depend on the d_o/d_i ratio of the glass tubing, the geometry of the pipette, and the method elastomer coating. In fact, use of Eq. (8) for the entire immersed portion of the pipette is generally not appropriate if the d_o/d_i ratio of the glass and/or elastomer is not uniform. Instead, more accurate predictions can be made considering "sections" of the immersed portion of the pipette that are sufficiently short that the d_o/d_i ratio of the glass and elastomer can be considered to be constant in each section. Equation (8) can then be applied to each section and the results added together rms. Nevertheless, the equation with reasonable values of C_1 and C_2 for the entire immersed region can still be used to give good estimates of the amount of noise expected.

Equation (8) generally indicates that for any glass when the elastomer coating is thin the dielectric noise will depend primarily on the characteristics of the glass, whereas when the elastomer coating is very thick dielectric noise will become more dependent on the characteristics of the elastomer. However, the restrictions on the relative values of C_1 and C_2 must be borne in mind (and see later discussion of measured results). We begin by considering the situation for glasses other than quartz. The best glasses other than quartz that are presently readily available have dissipation factors in the range of 0.0022–0.005. These values are comparable to (or only somewhat higher than) the dissipation factor of Sylgard. Clearly, if $D_1 \approx D_2$ then Eq. (8) can be approximated by $[4kTc_2\pi B^2DC_1C_2/(C_1 + C_2)]^{1/2}$, where $D = D_1 = D_2$. This is simply the expected expression for two capacitors in series with identical dissipation factors. Thus, for example, if $C_2 = C_1$, then the overall pipette capacitance will be halved and the rms dielectric noise of the pipette will be reduced by 0.707×. In fact, since the dissipation factor of the glasses other than quartz is somewhat worse than that of Sylgard 184, the anticipated improvement will be somewhat greater in terms of dielectric noise.

For quartz the situation is different. If an uncoated quartz pipette has an immersion capacitance, C_1, of 1 pF and a dissipation factor of 0.0001, then from Eq. (7) it is predicted that its dielectric noise would be ~13-fA rms in a 5-kHz bandwidth (eight-pole Bessel filter). Coating with an elastomer with $D_2 = 0.002$ will generally *increase* the dielectric noise of such a pipette. For example, with a light coat of Sylgard 184 with a value of C_2 of 3 pF, dielectric noise calculated from Eq. (8) is predicted to be nearly 27-fA rms in the same bandwidth. With C_2 decreased to 1 pF (i.e., a thicker coat of elastomer), dielectric noise actually increases slightly more to ~29-fA rms in a 5-kHz bandwidth. Finally, with a very heavy coat of Sylgard bringing C_2 down to 0.3 pF it is predicted that dielectric noise will fall somewhat to about 24-fA rms in this bandwidth. Of course, this is still higher than the predicted value for the uncoated quartz pipette, but it must be remembered that without an elastomer coating thin-film noise will be present and that this will almost certainly be far worse than the small penalty in terms of dielectric noise (in addition the elastomer coating reduces distributed *RC* noise; see later discussion). Note that by the time C_2 has been reduced to 0.3 pF (30% of C_1), the predicted dielectric noise depends more on the characteristics of the elastomer than on that of the quartz. Thus, a pipette with an immersion capacitance of 0.3 pF and a dissipation factor of 0.002 (i.e., the characteristics of the elastomer coating) is predicted from Eq. (7) to have dielectric noise of about 31-fA rms, which is not much more than the predicted performance of the coated quartz pipette. Note that if an elastomer with a dissipation factor significantly less than that of Sylgard 184 can be found, then it can improve the performance of quartz pipettes (and, of course, pipettes fabricated from other glasses) even if its dissipation factor is still more than that of quartz (see Levis and Rae[8]). On paper, Dow Corning R-6101 might be such an elastomer since it is claimed by the manufacturer to have a dissipation factor of 0.00025. We believe this elastomer has some advantages, but to date significant improvements in noise has not been one of them. Tests indicate that it performs at least as well as Sylgard 184 in terms of noise reduction, but not significantly better in most situations. Note, however, that R-6101 tends to form thinner coats than does Sylgard 184; this is particularly true in the first millimeter behind the tip. This may be due to the longer time needed to cure R-6101, and may also have reduced the effectiveness of this coating in most of our tests to date.

Before leaving the subject of dielectric noise, it is important to note that the theory just presented has not always been entirely successful in predicting dielectric noise of all pipettes. As already described, we[8] found good agreement between actual measurements and the theoretical predictions of Eq. (8) for Sylgard 184 coated quartz pipettes (~35-fA rms in a

5-kHz bandwidth for a 1.8-mm depth of immersion). In that paper we also reported that similar pipettes fabricated from Corning 7052 and 7760 could display dielectric noise as low as ~70-fA rms (with somewhat higher values being more typical). This is more than would be predicted from Eq. (8). Similarly, Benndorf[10] found that actual measured dielectric noise (fitted f noise component attributable to pipette immersion) considerably exceeded theoretical predictions for pipettes fabricated from Schott 8330 (Duran). These pipettes differed very significantly in geometry from those we have used in most of our measurements, having very thick walls and small rapier-like tips. Nevertheless, the important point here is the discrepancy between theory and measurement. From the data of Benndorf[10] it can be calculated that for fits to actual measurements of the noise of a Sylgard coated pipette made from $d_o/d_i = 4$ Schott 8330 tubing with an immersion depth of 1 mm the f noise component (presumably dielectric noise of the pipette) attributable to immersion produced nearly 90-fA rms of noise in a 5-kHz bandwidth (eight-pole Bessel filter). This is roughly three times more than the theoretical prediction for an *uncoated* pipette with the same d_o/d_i ratio and same depth of immersion as judged from Ref. 10. The conclusion seems clear enough in terms of actual measured performance, a heavily Sylgard coated quartz pipette with an initial d_o/d_i ratio of 2.0 prior to pulling (1.4–1.5 near the tip after pulling) and an immersion depth of 1.8 mm produced measured dielectric noise (f noise) that is only somewhat more than one-third of the dielectric noise produced by a Sylgard coated borosilicate pipette with $d_o/d_i = 4$ prior to pulling (and apparently about 3.85 at the tip after pulling) with an immersion depth of only 1 mm. These results, coupled with our own measurements, seem to indicate clearly that quartz pipettes are significantly better in terms of dielectric noise reduction than pipettes fabricated from other glasses regardless of the elastomer coating. We cannot currently explain the difference between theoretical predictions and measured results of pipettes made from glasses other than quartz. In pipettes made from borosilicates of the geometry that we have used most frequently (and coated with Sylgard) the discrepancy is not quite as large as that just described from Benndorf.[10]

Root mean square dielectric noise in any particular bandwidth is expected to vary approximately with immersion depth, d, as $d^{1/2}$. This relationship would be precise if the d_o/d_i ratios of the glass and elastomer coating were constant over all of the immersed regions considered. This is because rms dielectric noise varies as the square root of capacitance. The nonuniformities already discussed (and see later discussion) will make the actual variation with immersion depth depart from this expectation to some extent. However, estimates based on detailed models of quartz pipettes with moderate to heavy Sylgard coating (with d_o/d_i ratios based on actual measurements under a microscope) indicate that dielectric noise should not depart by more than about ±10% from the $d^{1/2}$ prediction.

Distributed RC Noise. Distributed *RC* noise arises from the distributed capacitance of the immersed portion of the pipette and the distributed resistance of its filling solution. The capacitance of the pipette is distributed reasonably evenly over the immersed portion. This statement must be qualified by the comments already made concerning likely thinning of the pipette wall near its tip and nonuniformities in the thickness of the elastomer coating. On the other hand, the resistance of the pipette is certainly very nonuniform. Most of the pipette resistance resides in regions near the tip. However, there is still considerable resistance in the regions distal to the tip. All of the pipette resistance produces thermal voltage noise, and this noise in series with the capacitance of the pipette produces current noise with a power spectral density that rises as f^2 over the range of frequencies important to patch clamping.

Figure 1B shows a very much oversimplified equivalent circuit representing distributed *RC* noise. This figure approximates the distributed situation with only four resistors (and their associated thermal voltage noise, not shown) in series with four capacitive elements, each made up of the capacitance of a section of the pipette wall and of the overlying elastomer. Obviously a more accurate distributed circuit would have many more resistive and capacitive elements. In any such circuit the first resistor (uppermost resistor in Fig. 1B) would represent all of the resistance of the portion of the pipette not immersed in the bath. The other resistors and capacitors would represent segments of the immersed portion of the pipette. In theoretical estimates of distributed *RC* noise, we have used such an equivalent circuit with many segments. (As many as 20,000 segments have been tested, but results that have converged to within about 10% can often be obtained with as few as ~30 *RC* segments.) However, such estimates may be crude because in order to make them accurate the values of the various elements must be known and this generally requires some sort of model of the pipette geometry that is likely to be oversimplified. We consider more realistic pipette models later, but we begin with some very simplified models of pipette geometry because these can form the basis for a more intuitive understanding of distributed *RC* noise. As an example of such a simple model, consider a pipette in which the tip and much of the shank are modeled as being of conical shape with a cone angle of 5.7 deg. If the electrode is filled with a solution with a resistivity of 50 Ωcm and has a tip diameter of 1 μm, then the expected resistance of the pipette is about 6.4 MΩ. Note that for simplicity it is assumed that this conical shape continues to a distance 5 mm back from the tip and that the electrode presents no further resistance beyond this point (e.g., the Ag|AgCl wire extends at least this far into the pipette and effectively shunts any additional resistance). Such an electrode would have an internal diameter of 100 μm at a distance 1 mm back from the tip, 200 μm at a distance 2 mm from the tip, increasing to 500 μm at a distance of 5 mm. It is simple to calculate the resistance of sections of such a pipette due to its idealized geometry. Thus approximately 3.2 MΩ (half of the total resistance) resides within the first 10 μm from the tip. Another 2.13 MΩ resides in the region from 10 to 50 μm from the tip, and about 480 kΩ occurs in the next 50 μm (i.e., from 50 to 100 μm). In the region from 100 to 200 μm the additional resistance is about 280 kΩ; 180 kΩ occurs in the region from 200 to 500 μm; an additional 62 kΩ occurs from 500 to 1000 μm. Clearly resistance per unit length continues to decrease at distances further and further from the tip, however, another 51 kΩ occurs in the region from 1 to 5 mm beyond the tip. To estimate distributed *RC* noise for this simplified pipette it is also necessary to estimate the capacitance of the immersed portion of the pipette. Here we assume that the capacitance per unit length is constant (even though this is clearly an oversimplification as already described). Table II summarizes predicted results for different immersion depths and two different capacitances per mm of immersion. These are 1 pF/mm of immersion (corresponding roughly to a value of d_o/d_i of 1.25 for a dielectric constant of 4) and 0.25 pF/mm of immersion (corresponding to $d_o/d_i \approx 2.5$). All values are rms distributed RC noise for a bandwidth of 5 kHz (eight-pole Bessel filter). Noise for different bandwidths can be easily calculated by remembering that distributed *RC* noise will vary as $B^{3/2}$ (e.g., for a 10-kHz

TABLE II

PREDICTED DISTRIBUTED RC NOISE IN A 5 kHz BANDWIDTH[a]

Depth of immersion	1.0 pF/mm	0.25 pF/mm	Resistance of pipette not in bath (kΩ)
50 μm	15-fA rms	4-fA rms	1050
100 μm	23-fA rms	6-fA rms	569
200 μm	34-fA rms	8.5-fA rms	292
500 μm	54-fA rms	13.5-fA rms	113
1.0 mm	76-fA rms	19-fA rms	51
2.0 mm	102-fA rms	25.5-fA rms	19
3.0 mm	118-fA rms	29.5-fA rms	8.5

[a] The pipette geometry is assumed to be a simple cone with an angle of 5.7 deg and a constant d_o/d_i ratio producing an immersion capacitance of 1 or 0.25 pF/mm. A bandwidth of 5 kHz (−3 dB, eight-pole Bessel filter) is assumed. See text for further details.

bandwidth multiply all values by a factor of 2.83). For each depth of immersion the table also lists the amount of resistance for the portion of the pipette *not* immersed in the bath.

As expected the values of distributed *RC* noise fall by a factor of 4 for the 4-fold reduction in capacitance (all else being equal, rms values of distributed *RC* noise varies linearly with capacitance and as $R^{1/2}$ with resistance, *R*, provided that the resistance and capacitance change is uniform throughout the pipette—changes in pipette resistance due to tip diameter only have relatively little effect on distributed *RC* noise; see later discussion). Note that the reduction in capacitance can be brought about by either thicker walled glass tubing or heavy elastomer coating (but note that for these predictions to remain valid the geometry of the pipette lumen must be unchanged). Distributed *RC* noise does not depend on the type of glass used, except for the small dependence on dielectric constant. Note also that there is a large variation of distributed *RC* noise with depth of immersion. In fact, for this particular geometry the rms noise varies roughly as the square root of the depth of immersion (the relationship is somewhat steeper for small depths of immersion below about 500 μm and somewhat more shallow for depths of immersion greater than about 1 mm). Finally it should be pointed out that the predicted noise values are quite sensitive to the resistance in the portion of the pipette *not* immersed in the bath when the depth of immersion is relatively deep. Thus if for whatever reason this resistance were increased by 50 kΩ (so that, for example, it became 58.5 kΩ for an immersion depth of 3 mm), the rms noise would increase by a factor of about 1.6 for an immersion depth of 3 mm (to nearly 200-fA rms for the 1 pF/mm pipette), by a factor of about 1.4 for a 2-mm depth of immersion, and by a factor of 1.2 for a 1-mm depth of immersion. However, for a 200-μm depth of immersion the noise would only increase by about 4%.

The results just described provide some insight into distributed RC noise and suggest obvious methods to minimize it. It is important to realize, however, that these results are highly dependent on the particular geometry chosen and therefore on the oversimplified model considered. Nevertheless, it is very clear that distributed *RC* noise can be minimized by using thick-walled pipettes (glass plus elastomer) and shallow depths of immersion. Because coating the tip region of the pipette with thick layers of elastomer is very difficult, thick-walled glass, pulled so as to preserve the d_o/d_i ratio as much as possible, is probably the most convenient and practical method of reducing the capacitance of the immersed portion of the pipette near the tip. Heavy elastomer coating can further reduce pipette capacitance even when thick-walled glass is used. Shallow depths of immersion can also be important to minimizing distributed RC noise, but this may not always be possible, and besides the precise depth of immersion can be affected by a meniscus of solution as the pipette emerges from the bath (see also predictions for more realistic elastomer coating below).

Because of uncertainties involved in theoretical estimates of distributed *RC* noise, Levis and Rae[8] attempted to measure this noise directly. These measurements relied on the fact that changing the ionic strength of the filling solution would affect the pipette resistance but not its capacitance. Thus noise was measured for pipettes coated with Sylgard only roughly above the point where the pipette entered the bath (to maximize distributed *RC* noise, in these studies this was about 2 mm back from the tip) with ionic filling solutions varying from 5 m*M* to 1.5 *M*. Quartz pipettes with an initial (prior to pulling) d_o/d_i ratio of 2.0 were used in these investigations and the pipettes were sealed to Sylgard at an immersion depth of about 1.8 mm. It was concluded that the distributed *RC* noise for the pipette geometry used had a PSD of about $2.5 \times 10^{-38} f^2$ amp^2/Hz for 150 m*M* NaCl filling solution. This would indicate an rms noise of $(8.3 \times 10^{-39} c_3 B^3)^{1/2}$ where c_3 is as usual a coefficient that depends on the type of filter used ($c_3 \approx 2$ for an eight-pole Bessel filter). This indicates an rms noise of about 45-fA rms in a 5-kHz bandwidth. Of course, estimating distributed *RC* noise required that other sources of noise be estimated and subtracted from the measurement; we relied heavily on the fact that in this situation the major source of f^2 noise associated with immersion of the pipette should be distributed *RC* noise. The need to separate noise components introduces some uncertainty into the estimate of each component. These results were in reasonable agreement with (or slightly less than) expectations based on theoretical predictions for the pipette geometry used in these studies. More precise simulations of distributed *RC* noise for pipettes of geometry similar to those used by Levis and Rae[8] are considered later. First, however, a few more simulations of simplified geometries are considered.

It is important to realize that pipette geometry involved in determining pipette resistance can have very significant effects on distributed *RC* noise even if the d_o/d_i ratio of the pipette (glass plus elastomer) is kept constant. This can be appreciated by comparing idealized geometries of the type already described with different cone angles. For this purpose we consider cone angles of 12, 6, 3 and 1.5 deg. In all cases the tip diameter is assumed to be 0.5 μm (note that tip diameter per se has relatively little effect on distributed *RC* noise in these models; see later discussion). The capacitance of the pipette was assumed to be 0.5 pF/mm of immersion (a constant d_o/d_i ratio has been assumed over the immersed region). Table III reports predicted rms distributed *RC* noise in a 5-kHz bandwidth (eight-pole Bessel filter) for each of the pipette geometries. Once again it is assumed

TABLE III
PREDICTED DISTRIBUTED RC NOISE FOR PIPETTES WITH DIFFERENT CONE ANGLES[a]

Depth of immersion	Cone angle			
	12°	6°	3°	1.5°
50 μm	4.0-fA rms	7.8-fA rms	14-fA rms	25-fA rms
100 μm	5.8-fA rms	11.3-fA rms	22-fA rms	40-fA rms
200 μm	8.4-fA rms	16.5-fA rms	32-fA rms	60-fA rms
500 μm	13-fA rms	26-fA rms	52-fA rms	100-fA rms
1.0 mm	18-fA rms	37-fA rms	72-fA rms	145-fA rms
2.0 mm	25-fA rms	49-fA rms	97-fA rms	192-fA rms
3.0 mm	28-fA rms	56-fA rms	112-fA rms	222-fA rms

[a] A bandwidth of 5 kHz (−3 dB, eight-pole Bessel filter) is assumed for all values listed. Cone angles are assumed to be constant over the relevant portion of the pipette, and the capacitance of the pipette is assumed to be 0.5 pF/mm of immersion. Tip diameter is 0.5 μm. See text for further details.

that there is no further resistance after a distance 5 mm back from the tip.

Large cone angles can clearly be useful in minimizing distributed *RC* noise, but may not always be compatible with other requirements. For example, in our experience large cone angles (e.g., 12 deg) are difficult to achieve over extended regions of the pipette with thick-walled glass. Note that for filling solutions with a resistivity of 50 Ωcm the resistance of these electrodes would be about 6, 12, 24, and 49 MΩ for cone angles of 12, 6, 3, and 1.5 deg, respectively. Had the tip diameter been increased to 1 μm these resistances would have been cut in half, however, the distributed RC noise in no case would have decreased by more than about 6%. The largest decreases (i.e., ~6%) would occur for very shallow depths of immersion; for immersion depths of 1 mm or more the reduction would be less than 2%. Increasing the tip diameter to 2 μm (which would drop the total electrode resistance to only about 25% of the values listed above) would have decreased distributed *RC* noise by less than 15% in all cases, and for depths of immersion of 1 mm or more the decrease is considerably less. Decreasing the tip diameter below 0.5 μm similarly only produces relatively small increases in distributed RC noise. Thus it is very important to note that electrode resistance by itself is not a reliable indicator of anticipated distributed RC noise. Instead it is the overall geometry of the electrode that must be considered. The reason that tip diameter has such a small effect on distributed *RC* noise is that despite the fact that most of the resistance of the pipette resides very close to the tip, very little of the pipette immersion capacitance is in this region. The *relative* amount of capacitance near the tip is greater when the immersion depth is small than when the pipette is deeply immersed in the bath. This accounts for the larger effects of tip diameter on distributed RC noise with shallow depths of immersion.

Once again it must be emphasized that the geometries just considered are highly oversimplified, so that the values listed in Table III should only be thought of as guidelines for the amount of variation of distributed *RC* noise that is possible with different geometries. Remember also that these values have assumed a particular capacitance per millimeter of immersion (0.5 pF in this case) and that rms distributed *RC* noise will scale linearly with this capacitance. Moreover, it has been assumed that the capacitance per unit length is constant. This assumption may not be unreasonable relatively near the tip where building up a heavy elastomer coat is difficult, but remember that at distances about 0.5–1 mm from the tip building up heavy coats of elastomer is quite easy and this will reduce the capacitance per unit length in these regions. It is also clear that the assumption of a single cone angle from the tip back to a distance 5 mm behind the tip is not realistic, particularly for the smallest cone angles (e.g., for the 1.5-deg cone angle this means an inner diameter of only 131 μm at a distance of 5 mm from the tip). In addition, the assumption that there is no resistance beyond 5 mm from the tip is clearly an oversimplification. In some pipette geometries the internal diameter has almost reached the original inner diameter of the tubing by a distance of as little as 3 mm, and it would be better to assume no significant resistance sooner; in other pipettes there may still be significant resistance even further back from 5 mm from the tip. Thus the numbers in Table III are not meant to be representative of real pipettes. Instead, they are intended to show the large variations of distributed *RC* noise that are possible with differing pipette geometries.

Benndorf[10] has recently investigated distributed RC noise for pipettes with very shallow cone angles (at least for the first 200 μm from the tip), thick walls (d_o/d_i as much as 8), and very small tip openings (~0.2 μm). These pipettes generally had resistances of 50–90 MΩ when filled with 200% Tyrode solution (specific resistance $\approx$ 26 Ωcm). The pipettes were fabricated from Duran (Schott 8330). This geometry of pipette was used for a variety of reasons, including (1) very high seal resistances (up to 4000 GΩ) were obtained with very small tipped pipettes and (2) the slender pipette geometry near the tip is very effective in preserving the d_o/d_i ratio during pulling. Benndorf measured the resistance of a "prototype pipette" over the first 200 μm behind the tip and used these data (presumably plus a largely unspecified model of the rest of the pipette) to calculate distributed *RC* noise. This pipette (which was typical of those used in that study) had a very shallow cone angle which, over the range from ~80 to 200 μm, can be estimated to be roughly 0.6 deg, but increases somewhat nearer the tip.

No data are presented concerning the electrode geometry further than 200 μm from the tip, except that it is noted that the fitted value of the resistance after breaking the pipette off 200 μm from the tip ("R_{200}") was 2 MΩ.[18] A number of theoretical calculations of distributed RC noise are presented for different d_o/d_i ratios and immersion depths ranging from ~10 μm to 1 mm. Unfortunately the details of these calculations—particularly in terms of assumed pipette geometry at distances greater than 200 μm form the tip—are not presented. Both uncoated and Sylgard coated pipettes were considered and the assumption was made that the d_o/d_i ratio was constant in both cases. It was concluded that distributed *RC* noise increases rapidly with immersion depth up to a depth of about 200 μm, but that further immersion had relatively little effect on predicted distributed *RC* noise. Indeed, for immersion depths greater than 300–400 μm there seems to be essentially no increase in predicted distributed *RC* noise with further immersion of the pipette. This behavior is quite different than the theoretical predictions we described earlier (see Tables II and III). Considering Table II it can be seen from the tabulated data that the predictions of the simple model considered suggest that distributed *RC* noise continues to increase significantly as immersion depth increases. Clearly the slope of the variation of distributed *RC* noise with immersion depth decreases with increasing depth of immersion, but, for example, increasing immersion depth from 200 μm to 1 mm increases the predicted distributed *RC* noise by a factor of somewhat more than 2. However, as already noted the assumptions of a constant cone angle and particularly of a constant capacitance per millimeter of immersion are not realistic (although the latter assumption was also made by Benndorf[10]).

To clarify the expected noise, we studied the geometry of a typical pipette of the type we most often use in greater detail. This pipette was fabricated from d_o/d_i = 2.0 quartz tubing and pulled on the Sutter P-2000. Its tip diameter was approximately 0.5 μm. We carefully measured the inner and outer diameter at various distances from the tip. We found that the interior of the pipette could be very reasonably approximated out to a distance of about 3 mm from the tip by three different cones. The first cone going from the tip to a distance 100 μm behind the tip had an angle of ~6.6 deg (bringing the inner diameter to about 12 μm at 100 μm from the tip). The second cone extended from 100 to 600 μm from the tip and

[18] This value is somewhat surprising, since if the same cone angle continued from 200 to 300 μm from the tip this region alone would add approximately 5 MΩ to the total resistance of the pipette. To achieve a value of R_{200} of only 2 MΩ, the cone angle would have to increase dramatically after 200 μm, although there is no indication of the beginnings of such an increase in the data shown.

had a more shallow angle of only ~2.7 deg (bringing the inner diameter to about 36 μm at a distance of 600 μm from the tip). The third cone extends from 600 μm to about 3 mm and had an angle of 6.1 deg (bringing the inner diameter to somewhat more than 290 μm at a distance of 3 mm from the tip). Beyond 3 mm, the inner diameter quickly increased toward that of the original tubing (750 μm) so that it was within roughly 10% of this diameter at 5 mm from the tip. At all distances from the tip out to 3 mm from the tip this relatively simple model gave inner diameters that agreed with actual measurements of the pipette to better than ±5%. Several other pipettes pulled with the same settings were also examined and found to be very similar to that just described. The outer diameter of the pipette was modeled as having a d_o/d_i ratio of 1.5 in the first cone, 1.55 in the second cone, and 1.6 in the third cone. This produced reasonable agreement with measured ratios, but not as precise as the agreement between model and measured inner diameters. In particular, it was found in some pipettes that the thinning of the pipette wall during pulling could be greater on one side of the pipette than on the other. This is presumably due to the characteristics of the laser puller, but was not taken into account in this model.

The Sylgard coat was applied by the dip method described by Levis and Rae.[8] Only a single dip was used for the data to be considered at this point, although prior to this dip a ring of Sylgard was painted and cured about 2 mm back from the tip. The thickness of this coating was then carefully measured under the microscope at various distances from the tip. It was found that at the tip in this situation the d_o/d_i ratio of the Sylgard coat (i.e., the outer diameter of the coated pipette divided by the outer diameter of the glass) was only about 1.1, while at 2–3 mm from the tip this ratio can be 4 or more. The approximate d_o/d_i ratios of the Sylgard coat at various distances from the tip were as follows:

Distance from tip (μm)	d_o/d_i Ratio of Sylgard
50	1.12
100	1.16
200	1.2
300	1.35
500	1.6
1000	2.2
2000	3.4
3000	4.6

The form of the model pipette with its Sylgard coating is shown in Fig. 2. Note that Fig. 2 shows the diameter of the inner and outer walls of the quartz and the outer edge of the Sylgard coating as a function of distance from the tip, but the radial and longitudinal scales are different. Figure 2A shows the model for the first 3 mm beyond the tip and Fig. 2B shows an expanded view of the first 1 mm. While the wall of the quartz pipette model has a reasonably constant d_o/d_i ratio (somewhat more so than the actual pipette), the Sylgard coating can easily be seen to be extremely nonuniform.

This model was then used to compute distributed *RC* noise in an uncoated pipette and a pipette with a Sylgard 184 coating as shown in Fig. 2. The resistance was computed on the basis of the inner diameter of the model assuming that the specific resistance of the filling solution was 50 Ωcm. Computations were made using segment lengths of only 0.2 μm for up to 3 mm back from the tip (the largest depth of immersion considered). The resistance from 3 to 5 mm from the tip was estimated to be 4 kΩ, and it was assumed that there was no additional resistance beyond 5 mm from the tip. The total pipette resistance is estimated to be ~11.5 MΩ. The capacitance per unit length in the case of an uncoated pipette was somewhat different in each cone; namely, 0.52 pF/mm in the first cone, 0.48 pF/mm in the second cone, and 0.45 pF/mm in the third cone. In the case of the Sylgard coated pipette, the capacitance per unit length was set in 13 different regions giving a good approximation to the measured data. These regions were of shorter length near the tip and increased in length at greater distances from the tip. The average d_o/d_i ratio of the Sylgard coating in each region was approximated and this information combined with the d_o/d_i ratio of the quartz pipette of the model was used to determine the capacitance per unit length in each region. This varied from about 0.39 pF/mm at the tip to as little as 0.09 pF/mm at 2.6 to 3 mm from the tip.

The results of these simulations are summarized in Fig. 3 for immersion depths up to 3 mm; the bandwidth considered is 5 kHz (eight-pole Bessel filter), but recall that any other bandwidth can be determined by remembering that distributed *RC* noise varies as $B^{3/2}$. Upper curve (a) in Fig. 3 shows the prediction for the uncoated pipette and lower curve (b) shows the prediction for the Sylgard-coated pipette. It is worthwhile to compare these results to the measured data from Levis and Rae[8] at an immersion depth of 1.8 mm (the depth used in that study). At this depth the uncoated pipette is predicted to have distributed *RC* noise of about 55-fA rms and a current noise PSD of $3.6 \times 10^{-38} f^2$ amp^2/Hz. This is about 40% higher than the estimated PSD of $2.5 \times 10^{-38} f^2$ amp^2/Hz reported by Levis and Rae,[8] and the rms noise predicted by the simulation is about 20% higher than the 45-fA rms estimated in that study for the same bandwidth and depth of immersion. However, the agreement seems reasonable considering likely differences in the geometry of the pipettes in the two cases (the pipettes used here were similar to—but certainly not identical to—those used in

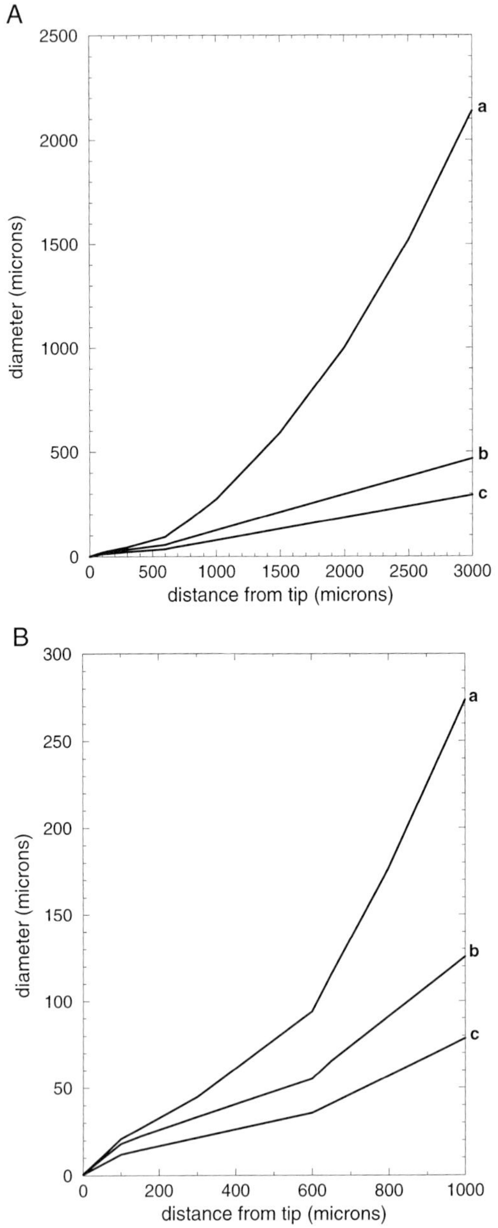

FIG. 2. Geometry of the pipette model used to compute distributed *RC* noise shown in Fig. 3. Note the plots of the diameter (not radius) of the pipette (inner and outer diameter) and of the elastomer coating (outer diameter). The geometry of the pipette and its elastomer coating is based on detailed microscopic measurements of an actual pipette pulled from d_o/

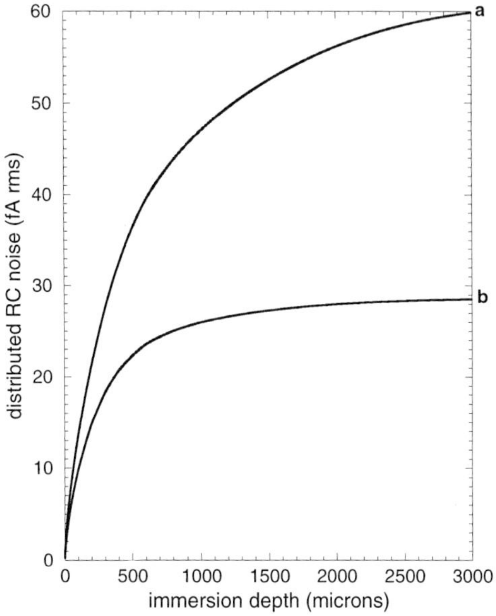

FIG. 3. Predictions of rms distributed *RC* noise as a function of immersion depth for a bandwidth of 5 kHz (−3 dB, eight-pole Bessel filter) for a pipette with the geometry illustrated in Fig. 2. The pipette is assumed to be fabricated from quartz (dielectric constant of 3.8) and the elastomer coating is Sylgard 184 (dielectric constant of 2.9). Since the dielectric constant of borosilicate glasses is only about 20–30% higher than that of quartz, predictions for such glasses pulled to the same geometry would only be somewhat higher. Curve (a) is predicted rms distributed *RC* noise for an *uncoated* pipette of the geometry shown in Fig. 2 and curve (b) is predicted rms distributed *RC* noise for the same pipette with a coating of Sylgard 184 as illustrated in Fig. 2. See text for further details.

the previous study). However, in the previous study we predicted that the PSD of distributed *RC* noise of the pipette would vary as C_e^2, where C_e is the capacitance of the immersed portion of the pipette, including both the glass and the elastomer; specifically we suggested that for the pipettes used at the time the PSD of distributed *RC* noise might be estimated by $\sim 10^{-14} C_e^2 f^2$ amp^2/Hz. For a 2-mm depth of immersion the model pipette with its Sylgard coating has a capacitance, C_e, of about 0.35 pF. Based on

---

d_i = 2 quartz tubing. (A) First 3 mm of the pipette and its elastomer coating. (B) Expanded view of the first 1 mm of the model pipette. In both parts (A) and (B) curve (a) is the outer diameter of the Sylgard coat, (b) is the outer diameter of the quartz pipette, and (c) is the inner diameter of the pipette. See text for further details. [Reprinted from R. A. Levis and J. L. Rae, The use of quartz pipettes for low noise single channel recording. *Biophys. J.* **65,** 1666–1667 (1993).]

our previous suggestion, this would produce an estimated PSD of about $1.2 \times 10^{-39}f^2$ amp^2/Hz at this depth of immersion and thus predicts about 10-fA rms of noise in a 5-kHz bandwidth. However, the present simulation indicates a PSD of about $9 \times 10^{-39}f^2$ amp^2/Hz for this depth of immersion and predicts about 28-fA rms in a 5-kHz bandwidth. The reason for this discrepancy is simple: in forming our earlier prediction we[8] made the assumption of a uniform d_o/d_i ratio, and thereby failed to take into account the extreme nonuniformity of the coating of Sylgard. The Sylgard coating is quite thin near the tip of the pipette and only becomes comparable to or greater than the quartz pipette wall in terms of d_o/d_i ratio at distances of 500 μm from the tip. Because of this most of the distributed *RC* noise in the elastomer-coated pipette arises relatively near the tip and it significantly exceeds the previous predictions for a Sylgard-coated pipette of roughly this geometry.

Examination of the two curves in Fig. 3 shows that for the uncoated pipette there is a significant variation of distributed *RC* noise with immersion depth all the way to 3 mm. In fact, the form of the curve is quite similar to that produced by the simpler model used for computing Table II. On the other hand, the Sylgard-coated pipette shows relatively little increase in distributed *RC* noise at immersion depths beyond about 1 mm. The reason for this (as just explained) is that the Sylgard coating is very thin near the tip but becomes quite thick at distances beyond 0.5–1 mm from the tip. The reduced variation of distributed *RC* noise with immersion depths greater than ~1 mm in the model pipette with Sylgard coating is somewhat reminiscent of the theoretical calculations presented by Benndorf.[10] However, note that Benndorf's predictions show even less variation at distances beyond 200 μm from the tip, and that this low variation cannot be accounted for by the nonuniformity of an elastomer coat. Benndorf[10] assumed that the d_o/d_i ratios in his pipette models were completely constant with and without elastomer coatings.

We have also investigated methods of building up heavier coats of elastomer and pipettes pulled from $d_o/d_i = 4$ quartz tubing. Our investigations of very thick-walled quartz pipettes have to date been rather limited, but with settings that are essentially the same as those used to pull the pipette approximated in Fig. 2, we achieve a pipette with a d_o/d_i ratio that varies from about 2.5 to 3.5 over the first 3 mm from the tip. The area within 100 μm of the tip actually has a slightly higher d_o/d_i ratio than somewhat more distal regions, but we feel that approximating the d_o/d_i ratio as 2.7 is reasonable for the entire first 3 mm. Based on measurements of the actual pipette we concluded that the interior of the pipette could be modeled quite reasonably over the first 3 mm as three conical sections. The first cone had an angle of 2 deg and extended from the tip to 350 μm; the second cone extended from 350 μm to 1 mm and had an angle of 4.2 deg; the third cone extended from 1 to 3 mm and had an angle of only 1.4 deg. The geometry of this pipette model is shown in Fig. 4. (Figure 4 also shows a very heavy Sylgard layer described below.) Just prior to 3 mm from the tip the inner diameter of the actual pipette began to increase rapidly, reaching about 90% of the initial inner diameter of the tubing (375 μm) by 5 mm back from the tip. The tip diameter was taken to be 0.5 μm and the resistance of the pipette filling solution (specific resistance = 50 Ωcm) from 3 to 5 mm was taken to be 12 kΩ, and, as before, it was assumed that there was no further resistance beyond 5 mm from the tip. The total predicted resistance of the pipette is about 35 MΩ. Predictions of distributed *RC* noise (rms noise in a 5-kHz bandwidth established by an eight-pole Bessel filter) for an *uncoated* pipette with this geometry are shown as the uppermost curve (a) of Fig. 5. It can be seen by comparison with Fig. 3 that the predicted distributed *RC* noise in this case is only somewhat less than for the uncoated pipette pulled from $d_o/d_i = 2$ tubing. At an immersion depth of 1.8 mm the predicted noise of this pipette is about 44-fA rms. It can also be seen in this case that the dependence of distributed *RC* noise on immersion depth is greater for the thick-walled pipette. The reason that noise only decreased slightly despite a significant increase in the d_o/d_i ratio is the much smaller bore of the pipette, which has only reached an inner diameter of 109 μm at a distance of 3 mm back from the tip; this also explains the steeper dependence of the noise on immersion depth. Thus, an increased d_o/d_i ratio by itself may not be very useful in reducing distributed *RC* noise—the geometry of the pipette lumen must also be considered.

On the other hand, heavier elastomer coatings than those considered above can continue to reduce distributed *RC* noise significantly. We have used the following technique to build up heavier coats of Sylgard 184. Paint a relatively heavy blob of elastomer entirely around the shank of the pipette just below where the pipette begins to taper. Then place the tip of the painted pipette into the prewarmed blowing air from a heat gun with the tip *pointing at the ground* and twirl the pipette around its long axis until the elastomer is cured. This results in an elastomer with a large blunt front edge about halfway up the tapered region of the pipette. Repeat this entire procedure one or two more times, each time placing the elastomer blob just in front of (i.e., toward the tip) of the blunt end of the cured elastomer. Then, if desired, paint elastomer over the outside of all the existing elastomer simply to obtain a uniform thickness over the entire coated area. Finally, any uncoated area at the tip can be coated using our previously described[8] "tip dip" method. With practice, the entire process can be completed in about 3–4 min. This can produce quite striking results as illustrated in Fig. 4. This figure shows the inner and outer diameter of the model of

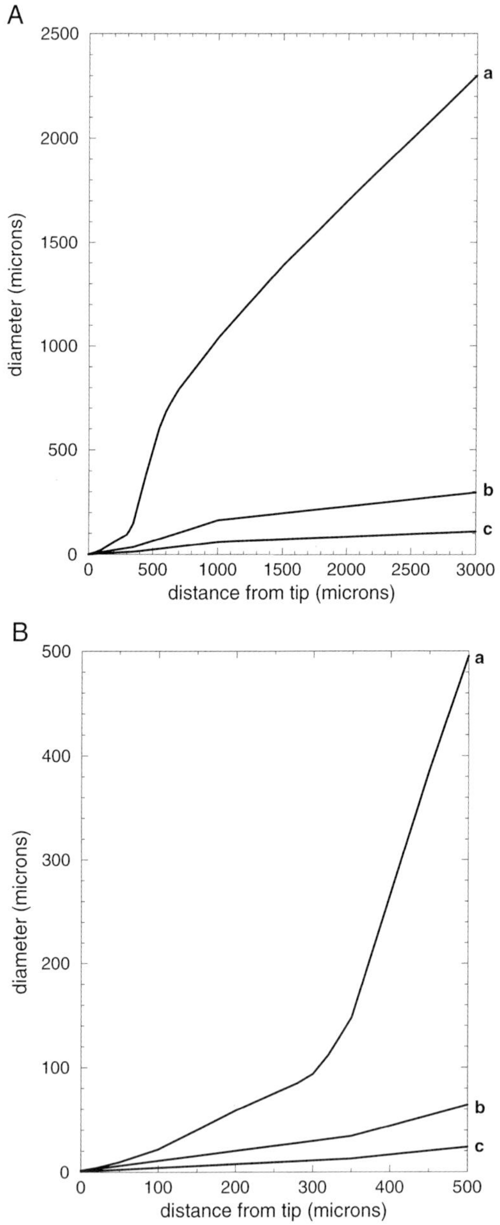

FIG. 4. Geometry of the pipette model used to compute distributed *RC* noise shown in Fig. 5. As in Fig. 2, this figure plots the diameter (not radius) of the pipette (inner and outer diameter) and of the elastomer coating (outer diameter). The geometry of the pipette and

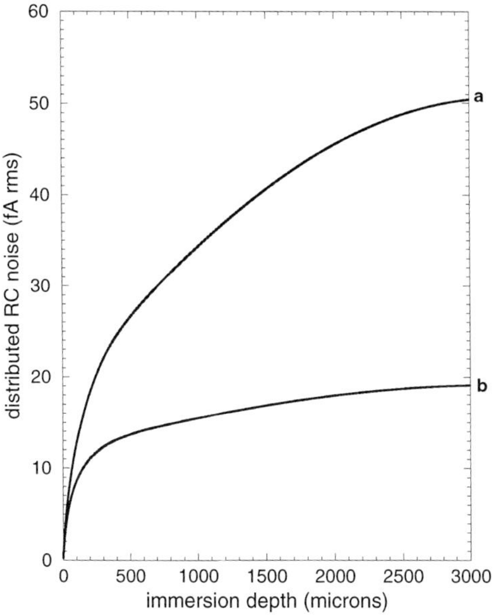

FIG. 5. Predictions of rms distributed *RC* noise as a function of immersion depth for a bandwidth of 5 kHz (−3 dB, eight-pole Bessel filter) for a pipette with the geometry illustrated in Fig. 4. As in Fig. 3, the pipette is assumed to be fabricated from quartz (dielectric constant of 3.8) and the elastomer coating is Sylgard 184 (dielectric constant of 2.9). Since the dielectric constant of borosilicate glasses is only about 20–30% higher than that of quartz, predictions for such glasses pulled to the same geometry would only be somewhat higher. Curve (a) is predicted rms distributed *RC* noise for an *uncoated* pipette of the geometry shown in Fig. 4 and curve (b) is predicted rms distributed *RC* noise for the same pipette with a coating of Sylgard 184 as illustrated in Fig. 4. Note that a Sylgard coat with the same d_o/d_i ratio as that considered here applied to a pipette with the geometry shown in Fig. 2 is predicted to produce somewhat *less* distributed RC noise than is shown in curve (b). However, the reduction is only 15%. See text for further details.

the pipette pulled from $d_o/d_i = 4$ quartz tubing described above and the outer diameter of a heavy layer of Sylgard as measured following application to such a pipette by the procedure just described. Figure 4A shows the first 3 mm of the pipette and Fig. 4B shows an expanded view of the first 500

---

its elastomer coating is based on detailed microscopic measurements of an actual pipette pulled from $d_o/d_i = 4$ quartz tubing. The heavy elastomer coat was produced by the method described in the text. (A) First 3 mm of the pipette and its elastomer coating. (B) Expanded view of the first 500 μm of the model pipette. In both parts (A) and (B) curve (a) is the outer diameter of the Sylgard coat, (b) is the outer diameter of the quartz pipette, and (c) is the inner diameter of the pipette. See text for further details.

μm. Although we have been unable to get the Sylgard coat to have a d_o/d_i ratio greater than about 1.1 at the tip itself, the d_o/d_i ratio has increased to more than 1.5 at a distance of 50 μm back from the tip and to 2.0 at 100 μm from the tip. The Sylgard layer increases dramatically after about 300 μm with d_o/d_i ratios over 5.0 occurring at distances beyond 500 μm from the tip. Thus, except very close to the tip, it is possible to produce very heavy elastomer coatings. Note that such heavily coated pipettes could be hard to work with, since the overall pipette diameter becomes quite large quite quickly. However, with a little practice the shape of the elastomer coating can be adjusted fairly easily. It is important to note that although the d_o/d_i ratio of the elastomer has been increased at all distances more than a few microns from the tip, this ratio remains extremely nonuniform.

Results of simulations of distributed *RC* noise for the pipette pulled from $d_o/d_i = 4$ quartz tubing and the Sylgard layer illustrated in Fig. 4 are also presented in Fig. 5, curve (b). Immersion depths up 3 mm are shown and, as in the uncoated curve, a 5-kHz bandwidth is assumed. It can easily be seen that distributed *RC* noise is greatly reduced by the heavy elastomer coating at all depths of immersion, with a predicted value of less than 18-fA rms at an immersion depth of 1.8 mm. However, to reach values less than 10-fA rms the depth of immersion would have to be less than 200 μm, and this is often impractical (probably particularly so with a pipette with so much elastomer coating). Different pipette geometries pulled from $d_o/d_i = 4$ tubing can yield somewhat better results, but the noise levels already described are probably small enough for even very demanding applications.

Finally, note that, as expected, a very heavy coat of Sylgard 184 similar in appearance (approximately the same d_o/d_i ratio for the elastomer coating) to that shown in Fig. 4 can further reduce the distributed *RC* noise of pipettes of geometries other than that shown in Fig. 4 for $d_o/d_i = 4$ quartz. For example, a simulation with the pipette geometry illustrated in Fig. 2 (i.e., a pipette pulled from $d_o/d_i = 2$ quartz tubing) with a heavy Sylgard coating with essentially the same d_o/d_i ratio of the elastomer coating shown in Fig. 4 actually produced slightly less distributed *RC* noise than that predicted for the pipette pulled from $d_o/d_i = 4$ quartz tubing. The predicted improvement ($\sim$15-fA rms for a 5-kHz bandwidth at an immersion depth of 1.8 mm) was not very large, although the variation of noise with immersion depth was also reduced. This indicates that for distributed *RC* noise both overall (glass plus elastomer) wall thickness and the geometry of the pipette lumen need to be considered. The $d_o/d_i = 2$ quartz tubing pulled to pipettes similar to that shown in Fig. 2 and heavily coated with elastomer is adequate in terms of distributed *RC*, although somewhat less dielectric noise is predicted for pipettes pulled from $d_o/d_i = 4$ quartz.

Our theoretical and experimental results suggest that for pipettes of the geometry we most commonly use pulled from tubing with $d_o/d_i = 0.2$ and with a moderate to heavy elastomer coating distributed *RC* noise can be kept to below 30-fA rms in a 5-kHz bandwidth for a depth of immersion of at least 2 mm. This figure can be roughly cut in half with extremely heavy Sylgard coatings such as those illustrated in Fig. 4. The expected noise depends only slightly (due to different dielectric constants) on the type of glass. With $d_o/d_i = 4.0$ tubing distributed *RC* noise may be reduced somewhat, although this depends on the geometry of the pipette and the degree of elastomer coating. Smaller depths of immersion can reduce distributed *RC* noise further.

It is worthwhile to compare the expected magnitudes of distributed *RC* noise and dielectric noise. While distributed *RC* noise does not depend significantly on the type of glass used, glass type seems to be even more important to dielectric noise than is theoretically predicted. With quartz pipettes (elastomer coated, initial d_o/d_i ratio = 2) at an immersion depth of about 2 mm measured dielectric noise is about 30- to 35-fA rms in a 5-kHz bandwidth. More precise calculations of dielectric noise for the model quartz pipettes shown in Figs. 2 and 4 using distributed models that take into account the nonuniformities in d_o/d_i ratio of the quartz and the Sylgard predict dielectric noise of ~27-fA rms and ~17-fA rms for the pipettes of Figs. 2 and 4, respectively, at a 2-mm depth of immersion. Dielectric noise in both cases varies roughly (but not precisely) as the square root of immersion depth. On the other hand, with borosilicate glasses, even with heavy elastomer coatings, measured dielectric noise is generally ≥70-fA rms in a 5-kHz bandwidth, which is significantly more than would be theoretically predicted. Thus, at this bandwidth and a 2-mm depth of immersion, dielectric noise is comparable to distributed *RC* noise for quartz and significantly higher than distributed *RC* noise for borosilicate pipettes of the geometries considered here. Fortunately, minimizing distributed *RC* noise and dielectric noise generally requires the same basic strategy, that is, use of thick-walled glass (tubing $d_o/d_i \geq 2$), heavy elastomer coating, and shallow depth of immersion. However, the degree to which each of these measures reduces distributed *RC* noise and dielectric noise is sometimes different. For example, geometry of the pipette lumen must be considered in the case of distributed *RC* noise, but is essentially unimportant to dielectric noise. Large cone angles can be useful in reducing distributed *RC* noise, but will not directly affect dielectric noise; moreover, large cone angles may not be practical or possible with thick-walled tubing, which is otherwise beneficial to the reduction of both distributed *RC* noise and dielectric noise. Nevertheless, the same basic strategy generally applies to minimizing both types of noise. However, there are two major distinctions between factors determining distributed *RC* noise and dielectric noise that must be remembered: (1) effects of glass type and (2) variation of noise with bandwidth. Glass type has already been considered in detail, but it is worth recalling that the use of quartz has been shown to be important to minimizing dielectric noise, but is largely unimportant to the reduction of distributed *RC* noise. The bandwidth of the measurement will affect the *relative* magnitudes of these noise types since rms dielectric noise increases with increasing bandwidth (B) linearly, while distributed *RC* noise increases as $B^{3/2}$. For each doubling of the bandwidth, dielectric noise increases by a factor of 2 while distributed *RC* noise increases by a factor of 2.83. Thus, as bandwidth increases the importance of distributed *RC* noise relative to dielectric noise increases.

R_e-C_p NOISE. R_e-C_p noise arises from the entire (lumped) resistance of the pipette, R_e, in series with the capacitance of the patch, C_p. Of course, in most single-channel measurement situations the capacitance of the patch is very small. Nevertheless, this small capacitance is in series with the thermal voltage noise of a large resistance and can therefore sometimes produce significant amounts of noise. Over the frequency range of interest to patch clamping the PSD of R_e-C_p noise, $S_{ep}{}^2$, rises with increasing frequency as f^2. This PSD is given by:

$$S_{ep}{}^2 = 4\pi^2 e_e{}^2 C_p{}^2 f^2 \qquad \text{amp}^2/\text{Hz} \tag{9}$$

where $e_e{}^2 = 4kTR_e$ is the thermal voltage noise PSD of the pipette resistance. This equation can be integrated over a bandwidth B (DC to B Hz) to give an expression for the rms noise, i_{ep}, attributable to this mechanism:

$$i_{ep} = \{(4/3)\pi^2 c_3 e_e{}^2 C_p{}^2 B^3\}^{1/2} \qquad \text{amp rms} \tag{10}$$

where c_3 is a coefficient that depends on the type of filter used as described previously, and R_e can range from as little as ~1 MΩ to many tens of megohms for typical patch-clamp measurements. For traditional patches C_p is expected to fall in the range of a few to perhaps 300 fF. Although the relationship is not perfect, it is expected that higher resistance pipettes with smaller tips will have smaller patches with less capacitance. It can be seen from Eq. (10) that the rms value of R_e-C_p noise depends linearly on C_p but on the square root of R_e. Thus this noise is minimized by small patches even if these are associated with high-resistance pipettes. An unfavorable example of R_e-C_p noise would be a 2-MΩ pipette with a 300-fF patch. From Eq. (10) it can easily be calculated that this would result in nearly 100-fA rms noise in a bandwidth of 5-kHz (eight-pole Bessel filter). This could then be a significant source of noise to the overall measurement. On the other hand, a 10-MΩ pipette with a 10-fF patch would produce only about 6-fA rms of noise in this bandwidth. The most extreme situation that we are aware of is that reported by Benndorf[10] in which pipettes with resistances on the order of 50–100 MΩ (tip diameter ~0.2 μm) were used to form patches with a capacitance that should typically be less than 1 fF. Such a patch and pipette would produce less than 2-fA rms noise in a 5-kHz bandwidth. Either of the last two examples clearly produces noise that is negligible in comparison to other sources of noise associated with the pipette. Thus in most cases R_e-C_p noise is easy to make sufficiently small to be ignored. Of course, with giant patches that can have capacitances significantly greater than 1 pF, R_e-C_p noise can become a very important, often dominant, source of noise. Even though the resistance of the large-tipped pipettes used in giant-patch recordings is much less than that of more "typical" patch pipettes, the increased patch capacitance more than makes up for this in terms of noise. Thus, for a large-tipped pipette with R_e = 100 kΩ and C_p = 10 pF, R_e-C_p noise is expected to be more than 0.7-pA rms in a 5-kHz bandwidth. In such cases, however, the signals being measured are usually relatively large and thus higher noise levels can be tolerated.

SEAL NOISE. The *minimum* noise of a gigaseal is readily determined as its expected thermal current noise. This would produce a PSD of $4kT/R_{sh}$, where R_{sh} is the *DC* seal resistance, and an rms noise given by $(4kTc_1B/R_{sh})^{1/2}$. Seals can range from a few gigohms up to as much as 4 TΩ (4×10^{12} Ω). Thus, a seal with a resistance of only 2 GΩ would produce a minimum of about 0.2-pA rms noise for a 5-kHz bandwidth while a 4-TΩ seal would produce a minimum of only about 5-fA rms in this bandwidth. It is clear that high-resistance seals are a prerequisite for very low noise patch-clamp recordings. Unfortunately, however, it is not clear that the noise associated with a seal is well described by its minimum thermal current noise. We have presented evidence that the noise of at least some seals (resistance range 40–100 GΩ) is indistinguishable from the thermal current noise expected on the basis of the seal resistance.[9] However, it certainly seems that seal noise can often exceed this minimum amount.

More generally, the expected PSD of the membrane–glass seal for zero applied voltage should be given by $4kT\,\mathrm{Re}\{Y_{sh}\}$, where $\mathrm{Re}\{Y_{sh}\}$ is the real part of the seal admittance. The minimum value of $\mathrm{Re}\{Y_{sh}\}$ is $1/R_{sh}$. Of course, the measured "seal resistance" is actually the parallel combination of the seal and the patch membrane (with all known channels closed). In most situations we do not expect that the membrane itself will contribute much to the measured resistance (e.g., a 10-fF patch with a specific resistance of 20 kΩcm^2 would have a resistance of 2 TΩ), but the membrane contribution to apparent seal noise should be remembered, and it should be borne in mind that the membrane may also contain other charge translocating processes that may be thought of as contributing to what would commonly be called seal noise. In any case, the noise of the seal/patch may very well exceed the minimum estimate described earlier. Because noise often varies in unexplained ways from one patch to the next (even when all measured parameters seem similar) it is tempting to blame such variations on the seal. This seems particularly reasonable since noise measured with pipettes sealed to Sylgard is generally quite consistent and in good agreement with theoretical predictions. On the other hand, noise from actual patches shows much more variation (although general trends are as expected, see, e.g., Levis and Rae[8]). The only difference between a pipette sealed to Sylgard and an actual patch should be the seal/patch; other sources of pipette noise should be the same in both cases. Thus, it frequently seems reasonable (if anecdotal) to attribute noise variations in membrane patches to unexplained differences in the seal/patch.

Benndorf[10] has suggested that seals may produce shot noise. Shot noise is associated with current flow across a potential barrier. We know of no obvious reason to expect shot noise to be generated by the seal, although since the precise nature of the membrane–glass seal is not known, shot noise cannot be ruled out.

Despite the uncertainties in regard to seal noise it is clear that it is minimized by high-resistance seals. Benndorf[10] has reported that very small tipped pipettes (opening diameter ~0.2 μm) resulting in tiny patches can produce seals in the range of 1–4 TΩ. He also suggests that seal resistance should be linearly inversely related to pipette pore diameter. We have certainly noticed a correlation between small-tipped pipettes and high-resistance seals, but it is not altogether precise. We have frequently obtained seals with resistances in the range of 100–200 GΩ with pipettes with resistances of ~5 MΩ. Extremely high resistance seals should be of considerable importance to noise minimization at low to moderate bandwidths (provided, of course, that the electronics used can take advantage of such low noise at these frequencies). Thus, for example, with the amplifier considered here, and a 1-TΩ seal, total noise at bandwidths of 100 Hz and 1 kHz could be as little as 2- and 9-fA rms, respectively (of course, somewhat higher values are more likely in most cases). As bandwidth increases the contribution of seal noise to total noise will decrease because the rms value of seal noise should vary with bandwidth as $B^{1/2}$, whereas rms dielectric noise will vary as B, and distributed RC noise and R_e-C_p noise (as well as much of the noise of the amplifier) will produce rms noise that varies as $B^{3/2}$.

The simplest strategy for obtaining high-resistance seals and thereby minimizing seal noise is to use relatively small-tipped pipettes. The size of the tip, of course, will also be dependent on the type of measurement being undertaken (e.g., when studying channels with a low density in the cell membrane, very small tipped pipettes can become quite frustrating). Seal noise may sometimes be as low as the predicted thermal current noise of the DC seal resistance, but it certainly seems that it can frequently exceed this lower limit.

SUMMARY AND STRATEGIES. At this point it is useful to summarize the noise sources described and to provide an indication of the magnitudes of these noises that can be expected in "typical" low-noise recording situations and in "best case" situations. The total rms noise, i_T, of a single-channel patch-clamp recording in a particular bandwidth can be summarized as

$$i_T = (i_{hs}^2 + i_h^2 + i_d^2 + i_{RC}^2 + i_{ep}^2 + i_{sh}^2)^{1/2} \quad \text{amp rms} \tag{11}$$

Where i_{hs} is the rms noise of the headstage amplifier, *including* any correlated noise arising from e_n in series with capacitance of the pipette and its holder; i_h is the (uncorrelated) noise of the holder; i_d is the dielectric noise of the pipette; i_{RC} is the distributed *RC* noise of the pipette; i_{ep} is R_e-C_p noise; and i_{sh} is the noise of the seal, including any noise arising from the patch membrane itself.

Headstage noise obviously depends on the amplifier being used; the noise of the amplifier one of us uses has already been described. The open-circuit noise of this amplifier in a 5-kHz bandwidth (eight-pole Bessel filter) is 41-fA rms. This noise increases to about 44, 47, and 50-fA rms in this bandwidth with the addition of 1, 2, and 3 pF, respectively, of capacitance to the input due to this capacitance in series with e_n. This represents the capacitance of the holder and pipette, but ignores other sources of noise associated with these capacitances.

Holder noise is minimized by using small holders made from low-loss materials (or a small metal holder such as that described by Benndorf[10]). These can be purchased commercially or custom made; in either case the noise performance of the holder should be measured (with and without a pipette with its tip just above the bath) to ensure acceptable results, and its noise should then be monitored periodically. In our experience, custom-made holders can often outperform those that are commercially available. However, holder noise is usually sufficiently small that the added inconvenience of a custom-made holder is normally not necessary. If holder noise increases, the holder must be cleaned. The addition of a small polycarbonate holder with about 0.6 pF of capacitance should add about 15-fA rms of noise in a 5-kHz bandwidth; a larger (~1.5-pF) polycarbonate holder adds about 25-fA rms. Taking into account the capacitance of the holder added to the input, total noise with such holders will increase to about 45- to 46-fA rms for the small holder and about 52-fA rms for the larger holder with the low-noise headstage amplifier considered here.

Pipette noise in general is minimized by using short pipettes (although we have never used extremely short—as little as 8 mm—pipettes such as those favored by Benndorf[10]) fabricated from thick-walled glass tubing ($d_o/d_i \geq 2$) and heavily coated with a low-loss elastomer as close to the tip as possible. Shallow depths of immersion also minimize pipette noise. Small tip openings are also generally desirable for the lowest noise recordings because this will tend to minimize seal noise and R_e-C_p noise.

In the case of dielectric noise, the best approach to minimization is the use of quartz. However, reasonable results can also be achieved with other low-loss glasses. Regardless of the type of glass used, thick-walled pipettes are desirable as are techniques of pulling that attempt to preserve the d_o/d_i ratio as closely as possible near the tip. Heavily coating the pipette with a low-loss elastomer will reduce the noise of pipettes made from glasses other than quartz, but may actually increase the dielectric noise of a quartz pipette. But note that a very heavy elastomer coating with a quartz pipette is predicted to produce less dielectric noise than a moderate coating—even though both should produce more dielectric noise than is predicted for an uncoated quartz pipette. Even so, quartz pipettes heavily coated with Sylgard 184 or R-6101 have been measured to produce only about half the dielectric noise of similar pipettes made from other glasses. For an ~2-mm depth of immersion dielectric noise of quartz pipettes ($d_o/d_i = 2$ prior to pulling) has been measured to be about 35-fA rms in a 5-kHz bandwidth. For pipettes of essentially the same geometry, elastomer coating, and depth of immersion made from other low-loss glasses, we have typically found that dielectric noise is 70-fA rms or more in this bandwidth. Recalling that dielectric noise varies linearly with bandwidth, these numbers will double for a 10-kHz bandwidth. The rms dielectric noise should vary roughly as the square root of immersion depth and also roughly as the square root of pipette capacitance. With thicker walled glass and shallower depths of immersion dielectric noise can be reduced further. For example, with $d_o/d_i = 4$ quartz and an immersion depth of 500 μm it should be possible to keep dielectric noise to as little as 10-fA rms in a 5-kHz bandwidth. The fact that dielectric noise from elastomer-coated pipettes fabricated from glasses other than quartz exceeds theoretical predictions deserves further investigation.

Distributed *RC* noise is highly dependent on the geometry of the pipette, but almost independent of the type of glass used. Several examples of distributed *RC* noise have already been presented. From these it can be seen that for pipettes of the general geometry that we use most frequently (see Fig. 3), it should be possible in a 5-kHz bandwidth to keep distributed RC noise less than 30-fA rms for an immersion depth of ~2 mm with a moderate to heavy coating of Sylgard 184 (or other suitable low-loss elastomer). With an extremely heavy Sylgard coating it should be possible to reduce distributed RC noise to as little as 15-fA rms under the same conditions. Thus distributed RC noise can be made smaller than dielectric noise even in quartz pipettes at a bandwidth of 5 kHz; for pipettes made from glasses other than quartz (which will have little effect on distributed RC noise, but significant effects on dielectric noise) at this bandwidth dielectric noise is likely to exceed distributed RC noise of an optimal pipette by a factor of as much as 4–5. Remember, however, that rms distributed RC noise will vary with bandwidth (B) as $B^{3/2}$, while dielectric noise varies as B. This means that at lower bandwidths distributed RC noise becomes less and less important, while at wider bandwidths distributed RC noise becomes progressively more important. If rms distributed RC noise is two times smaller than dielectric noise at a 5-kHz bandwidth, it will become equal to rms dielectric noise at a bandwidth of 20 kHz. Thus, the bandwidth of the measurement to be undertaken must also be considered. Fortunately, the steps taken to minimize distributed RC noise and dielectric noise are generally the same. The only significant exception to this is the decision whether or not to use quartz. The major advantage of quartz relates to dielectric noise, consequently at very high bandwidths (as well as very low bandwidths where neither dielectric noise or distributed RC noise are likely to dominate overall noise) the advantages of quartz over low-loss borosilicates may become unimportant. However, from our experience the bandwidths in question are in excess of 50 kHz.

Distributed RC noise is generally minimized by using thick-walled glass (tubing $d_o/d_i \geq 2$), heavy coats of elastomer extending as close to the tip as practical, and shallow depths of immersion. The benefits of very thick walled glass (e.g., $d_o/d_i = 4$) are dependent on the geometry of the pipette and may not be as large as expected in many actual situations. Effects of immersion depth are minimized by very heavy coatings of elastomer.

R_e-C_p noise is usually only expected to become significant relative to other noise sources when C_p is in the range of roughly 0.1 pF or higher. Such patches are sometimes necessary (e.g., when studying channels with a low density in the cell membrane), but can bring with them a noise penalty. In situations where relatively large patches are necessary it is clearly advantageous in terms of R_e-C_p noise to use the lowest resistance pipettes possible. However, for the lowest noise recordings smaller patches are best able to avoid R_e-C_p noise. In a 5-kHz bandwidth, R_e-C_p noise can range from a few fA rms for patches with $\leq$10 fF of capacitance up to perhaps 100-fA rms for patches in the range of 200–300 fF. For giant patches ($C_p > 1$ pF) R_e-C_p noise is likely to dominate overall noise at bandwidths above a few kilohertz (provided of course that other noise sources have been minimized).

The noise of a high-resistance seal is normally most important at low bandwidths (say, a few kilohertz or less). However, as described earlier, there is reason to suspect that seal noise may be somewhat unpredictable. It has been shown that under some circumstances seal noise cannot be distinguished from the expected thermal current noise of the *DC* seal resistance, R_{sh}. This is the minimum amount of seal noise possible, and amounts to 40-, 20-, and 9-fA rms in a 5-kHz bandwidth for seal resistances of 50, 200, and 1000 GΩ, respectively. Higher noise may well occur for such seal resistances, but it is clear that higher resistance seals produce less noise. Small tip openings tend to produce the highest resistance seals, but we have often obtained seal resistances in the 100- to 200-GΩ range with tip openings of about 1 μm.

It is of some interest to predict what sort of noise can be expected for the best measurements presently possible, as well as what is reasonable to expect in more "typical" low-noise situations. For an eight-pole Bessel filter it is reasonable to approximate best case pipette noise (excluding noise associated with pipette capacitance in series with e_n and also excluding noise of the seal/patch) for a relatively small-tipped (roughly 0.5 μm) quartz pipette with a heavy elastomer coating and a depth of immersion of ~1 mm:

$$\{1.5 \times 10^{-35}B^2 + 2 \times 10^{-39}B^3\}^{1/2} \qquad \text{amp rms}$$

where the B term is dielectric noise, and the $B^{3/2}$ term is distributed *RC* noise and a smaller contribution from R_e-C_p noise. To this noise it is necessary to add the noise of the amplifier (including correlated noise arising from holder and pipette capacitance in series with e_n), the uncorrelated noise of the holder, and the noise of the seal/patch. Since the seal/patch is harder to predict, we begin by adding the noise of the amplifier described earlier and a small low-noise holder. These will contribute roughly the following rms noise (eight-pole Bessel filter is assumed):

$$\{2 \times 10^{-32}B + 3.2 \times 10^{-35}B^2 + 1.1 \times 10^{-38}B^3\}^{1/2} \qquad \text{amp rms}$$

Total best case noise is then

$$\{2 \times 10^{-32}B + 4.7 \times 10^{-35}B^2 + 1.3 \times 10^{-38}B^3 + \text{seal/patch noise}\}^{1/2} \qquad \text{amp rms}$$

This then predicts a best case noise of the rms addition of 54-fA rms + seal/patch noise for a 5-kHz bandwidth and 134-fA rms + seal/patch noise for a 10-kHz bandwidth. If the seal/patch noise was as little as 20- and 28-fA rms in bandwidths of 5 and 10 kHz, respectively, total noise could be as little as 58- and 137-fA rms in these bandwidths. These values are only very slightly less than the best noise we have ever achieved with real patches.

For a Schott 8330 or 8250 borosilicate pipette of the same general geometry as that shown in Fig. 2 and assuming dielectric noise of 70-fA rms in a 5-kHz bandwidth, best case noise would increase to about 86- and 188-fA rms in 5- and 10-kHz bandwidths respectively.

"Typical" noise of a quartz or borosilicate patch pipette in a situation where all of the low-noise practices described have been followed and a very high resistance seal has been achieved can be estimated to be perhaps 30–50% higher than the figures just quoted. For seals of roughly 100 GΩ and Sylgard-coated quartz pipettes pulled from $d_o/d_i = 2$ quartz and tip openings of 0.5–1.0 μm, we have found that average noise is about 85- to 90-fA rms in a 5-kHz bandwidth. However, as already noted, it is certainly not an infrequent occurrence to have done everything right, achieved a high-resistance seal, and ended up with noise significantly higher than this. Much of this variability may be due to the seal, but it also seems likely that the other noise factors considered here show significant variation from one patch to the next.

Noise in Whole-Cell Patch-Clamp Measurements

Whole-cell measurements are subject to all of the same noise sources described in regard to single-channel recording situations with the exception of R_e-C_p noise. Of course, R_e-C_p noise is replaced in the whole-cell situation by noise arising from the pipette/access resistance in series with the whole-cell capacitance, C_m. This is a much larger source of noise since the cell capacitance is many times larger than the capacitance of the patch. In the case of R_e-C_p noise the time constant formed by R_e and C_p is sufficiently small to be ignored in almost all patch-clamp situations (R_e-C_p should be on the order of about 1 μs or less in most cases). However, in the whole-cell situation the time constant R_s-C_m (where R_s is the series resistance, normally dominated by the pipette) can be as high as a few milliseconds and is typically on the order of 100–300 μs (e.g., 210 μs for a more or less "typical" 30-pF cell with a total series resistance of 7 MΩ). Note that the series resistance is called R_s even though it is usually dominated by the resistance of the pipette (called R_e throughout this article). This is to distinguish the total access or series resistance from the pipette resistance measured prior to sealing and achieving the whole-cell configuration. Measured series resistance is normally higher than the initial resistance of the pipette. This can result from partial clogging of the pipette tip as well as from other sources of resistance in series with the membrane.

Noise arising from R_s and C_m has a PSD, $S_{sm}{}^2$, given by

$$S_{sm}{}^2 = (4\pi^2 f^2 e_s{}^2 C_m{}^2)/(1 + 4\pi^2 f^2 \tau_{sr}{}^2) \qquad \text{amp}^2/\text{Hz} \tag{12}$$

where $\tau_{sr} = R_{sr}$-C_m and R_{sr} is the uncompensated portion of the series resistance and $e_s{}^2 = 4kTR_s$ is the PSD of the thermal voltage noise of R_s.

The fraction of series resistance that is compensated will be denoted by α ($0 < \alpha < 1$) and $\beta = 1 - \alpha$; thus $R_{sr} = \beta R_s$.

Series resistance compensation is very important to the dynamic characteristics of whole-cell voltage clamping as has been described previously (see, e.g., Refs. 16 and 19). The effects of IR drops producing voltage errors are well known and are not discussed here. Instead, the most important point we wish to emphasize (and have described previously on several occasions) is that in the absence of series resistance compensation the actual maximum usable bandwidth of a whole-cell recording is limited to $1/(2\pi\tau_s)$, where $\tau_s = R_s\text{-}C_m$. This is because the series resistance and the cell capacitance effectively form a one-pole low-pass *RC* filter at this frequency. This restriction can be quite severe; for example, with $R_s = 10\ \text{M}\Omega$ and $C_m = 100\ \text{pF}$, $1/(2\pi\tau_s) \approx 160$ Hz. Series resistance compensation increases this maximum usable bandwidth to $1/(2\pi\tau_{sr})$, so that with the same parameters just considered but with 80% series resistance compensation $1/(2\pi\tau_{sr}) \approx 800$ Hz; 90% compensation will extend this to 1600 Hz. Of course, most situations are better than the example just considered, but even with $R_s = 5\ \text{M}\Omega$ and $C_m = 30$ pF, $1/(2\pi\tau_s) \approx 1.06$ kHz, and, for example, 65% series resistance compensation is needed to extend the maximum usable bandwidth to 3 kHz. Setting an external filter bandwidth to anything higher than $1/(2\pi\tau_{sr})$ will essentially add no new information, but it will add additional noise. (Of course, this can be corrected after the fact by the use of a digital filter.)

Examination of Eq. (12) shows that the PSD of the noise arising from R_s and C_m will initially rise as f^2 at frequencies below $1/(2\pi\tau_{sr})$ and will then eventually plateau at frequencies above $1/(2\pi\tau_{sr})$. Without series resistance compensation ($R_{sr} = R_s$), the value of the PSD once this plateau is reached will be $4kT/R_s$, which is the thermal current noise of R_s. With R_s compensation the plateau reaches a level given by $4kT/\beta^2 R_s$. This is a quite significant amount of noise, so that it can be seen that there is a sizable noise penalty (with little if any new information about the signal) for setting the external filter to a bandwidth greater than $1/(2\pi\tau_{sr})$.

Below frequencies of about $1/2\pi\tau_{sr}$, Eq. (12) can be approximated by $4\pi^2 f^2 e_s^2 C_m^2$ and the rms noise current arising from R_s and C_m is then given by $\{(4/3)\pi^2 c_3 e_s^2 C_m^2 B^3\}^{1/2}$, which (noting that $e_s^2 = 4kTR_s$) can also be written as $\{1.33\pi^2 c_3 (4kTR_s) C_m^2 B^3\}^{1/2}$. To consider a more or less typical example, we assume that $R_s = 7\ \text{M}\Omega$ and $C_m = 30$ pF and that enough series resistance compensation is used to justify any bandwidths mentioned. This will produce rms noise of approximately 0.15, 1.6, 8.5, and 18.3 pA in bandwidths

[19] A. Marty and E. Neher, *in* "Single Channel Recording" (B. Sakmann and E. Neher, eds.), 2nd Ed., p. 31. Plenum, New York and London, 1995.

(eight-pole Bessel filter) of 200 Hz, 1 kHz, 3 kHz, and 5 kHz respectively. All of these values are much higher than the noise described earlier for the amplifier plus pipette, even when the fact that the amplifier for whole-cell measurements normally uses a 500-MΩ feedback resistor is taken into account. Whole-cell amplifiers typically have noise of less than 0.2-pA rms in a 1-kHz bandwidth and less than 0.5-pA rms in a 5-kHz bandwidth. Because of the increased noise of the patch clamp itself, and more importantly the noise resulting from R_s and C_m, the characteristics of the pipette are clearly much less critical to the noise performance in whole-cell situations. Even with a very favorable situation for low-noise whole-cell recording it can be appreciated that the noise of R_s in series with C_m should still dominate total noise at bandwidths above about 1 kHz. As an example of such a favorable situation consider $R_s = 5$ MΩ and $C_m = 10$ pF. In this case the noise from R_s and C_m will be about 0.46- and 5.2-pA rms in bandwidths of 1 and 5 kHz, and exceeds the rms current noise of a 500-MΩ resistor at all bandwidths above about 390 Hz.

Despite these conclusions it seems wise to continue to follow good low-noise practices even with whole-cell pipettes. This would include using low melting temperature glasses with reasonably low dissipation factors (Schott 8250 is a good selection) and at least a light to moderate coat of a suitable elastomer. Among other things this will minimize the size of the pipette capacity transient and especially reduce the amplitude of the slow component of this transient (see Rae and Levis[7]). However, thick-walled glass is usually unnecessary (and may be detrimental, see later discussion), and we doubt that quartz will find significant use in fabricating whole-cell pipettes. The most important characteristic of the pipette in terms of noise in whole-cell situations is its resistance, which normally dominates R_s. This is because of the role it plays in producing noise in conjunction with C_m. In the range of bandwidths below $1/(2\pi\tau_{sr})$ it can be seen that rms noise arising from R_s and C_m depends linearly on C_m and on $R_s^{1/2}$. Reducing either R_s or C_m will reduce noise, but in general C_m is not under the experimenter's control. Besides, if you are studying a particular channel type with a density per unit membrane area that is constant among cells of different sizes, then it is clear that the signal as well as the rms R_s-C_m noise will both scale linearly with C_m, so that in this situation signal-to-noise ratio does not depend on C_m. Thus, in most cases the only practical way of reducing noise (and increasing signal-to-noise ratio) is to minimize R_s, and this means minimizing R_e.

In most situations it is advantageous to use pipettes with tip openings as large as can conveniently form seals with the cells you are using. In addition, it is beneficial to pull pipettes with relatively large cone angles. This is often most easily accomplished by pulling pipettes with quite large tips prior to heat polishing and then using heat polishing to achieve the desired final opening diameter.[7] Thin-walled tubing often makes this process easier, and as already noted the penalties in terms of dielectric and distributed *RC* noise of the pipette are generally unimportant in whole-cell situations.

Thus, in terms of achieving low noise, the best strategy for whole-cell recording is to use relatively large-tipped pipettes with large cone angles. It is also important to remember that the bandwidth of the measurement should not normally exceed $1/(2\pi\tau_{sr})$ since such unnecessary bandwidths will simply add noise without adding significant new information.

Conclusion

This article has attempted to describe both theoretical and practical aspects of noise reduction for single-channel and whole-cell patch-clamp recordings. In the case of single-channel (small patch) measurements it has been shown that the most important sources of noise arising from the pipette can be sufficiently minimized so that even the quietest electronics presently available can still dominate total noise.

For whole-cell measurements, minimizing the resistance of the pipette (which normally dominates total series resistance) is the best way to reduce noise and to ensure a good signal-to-noise ratio.

Low-noise measurements can require some patience and some perseverance, but there is nothing magic about achieving the levels of performance we have described in this article. The best of results are relatively infrequent (for reasons that are not altogether clear), but very good results can become more or less routine. The techniques described and the understanding of the noise sources that we have attempted to provide are the first steps toward achieving such results. The habits formed in ensuring the best possibility of low-noise measurements are good ones and should improve the quality of patch-clamp measurements in general.

[15] Giant Membrane Patches: Improvements and Applications

By DONALD W. HILGEMANN and CHIN-CHIH LU

Introduction

Why make large-diameter, tight-seal patches? The "giant" membrane patches described in this article (20–40 μm in diameter; 4–15 pF) were developed to study the function and regulation of electrogenic membrane transport mechanisms with free access to the cytoplasmic side. Rigorous control of solution composition is possible on both membrane sides. And this is an absolute prerequisite to study transporter function with the goal of developing meaningful transport models. For regulation studies, free access to the cytoplasmic side facilitates tremendously the study of regulatory factors and enzymes. Giant patches are also advantageous for studies of ion channel currents in many circumstances; for example, when channels have a low density or low expression, for resolving charge-moving reactions, and whenever fast voltage clamp of large currents is desired. An optimized voltage clamp allows resolution of charge-moving reactions on a 1-μs time scale.[1]

What are the limitations of the giant patch methods? First, seal formation requires the presence of divalent cations and chloride, either in the pipette solution or the bath solution. Chloride can be replaced by other anions after seal formation, but the presence of divalent cations (≥0.5 m*M* magnesium or calcium) on one membrane side is essential to maintain seals. Second, giant excised patches may not tolerate large holding potentials, which smaller patches tolerate well. In this regard, we report on improvements of seal stability in this article. Third, formation of giant outside-out patches (i.e., >10 μm in diameter) is in general not successful. Nevertheless, it is possible routinely with *Xenopus* oocytes under some conditions.[2] Fourth, giant patches can develop significant "rim current" (i.e., membrane components that are not well voltage clamped) which will distort the voltage and time dependence of currents studied. In general, this may be the case when sealing conditions favor a patch configuration with a membrane rising high up into the pipette.

[1] C. C. Lu, A. Kabakov, V. S. Markin, S. Mager, G. A. Frazier, and D. W. Hilgemann, *Proc. Natl. Acad. Sci. U.S.A.* **92,** 11220 (1995).

[2] J. Rettinger, L. A. Vasilets, S. Elsner, and W. Schwarz, *in* "The Sodium Pump" (E. Bamberg and W. Schoner, eds.), p. 553. Steikopff Verlag, Darmstadt, 1994.

0076-6879/98 $25.00

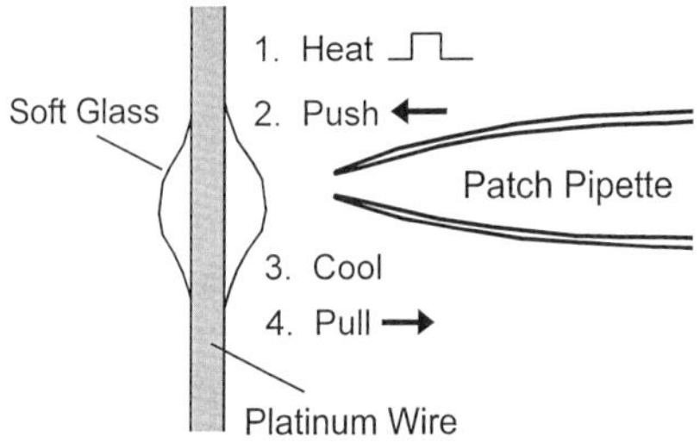

FIG. 1. Fabrication of giant patch electrodes. A large (0.4- to 0.7-mm) diameter platinum wire is used for the microforge. Soft glass (lead or soda glass) is melted onto the wire. (1) Current is applied via a foot switch until the pipette tip begins to recede, and then current is switched off. (2) The pipette tip is pushed into the molten glass and (3) allowed to cool. (4) The pipette is cut by pulling back, followed by melting and repetition of (1–4) until the desired shape is achieved.

We refer readers who are implementing giant patch techniques for the first time to another article as a primary information source.[3] Here, we provide first an overview of the methods and, second, a description of new developments and applications that may be of general utility.

Overview of Giant Patch Methods

In our experience, large-diameter patches can be formed and excised from all cells that allow gigohm seal formation with conventional techniques. Cell preparation, pipette fabrication, pipette coating, and seal techniques are all important for success.

Briefly, conventional borosilicate patch pipettes are prepared with a standard patch pipette puller. The tips are cut and melted to the desired shapes using a microforge, as sketched in Fig. 1. A bead of soft glass, either lead glass or soda glass, is melted on a large-diameter (0.4- to 0.6-mm) platinum wire. Then, using a foot-switch control, current to the wire (10–13 A) is switched off, the pipette tip is pushed into the molten glass, and it is cut spontaneously as the cooled glass bead retracts or by a manual pull. The pipette is then heavily polished. Usually, these steps are repeated to obtain pipette tips with relatively steep descents to the tips and thick walls to minimize pipette capacitance. Before back-filling the pipettes with the desired extracellular solution, the tip is dipped into a hydrocarbon mixture with syrup-like consistency to facilitate seal formation and stabilize patches. The mixture is prepared by heating and mixing for several minutes

[3] D. W. Hilgemann, *in* "Single-Channel Recording" (B. Sakmann and E. Neher, eds.), 2nd Ed., p. 307. Plenum Press, New York, 1995.

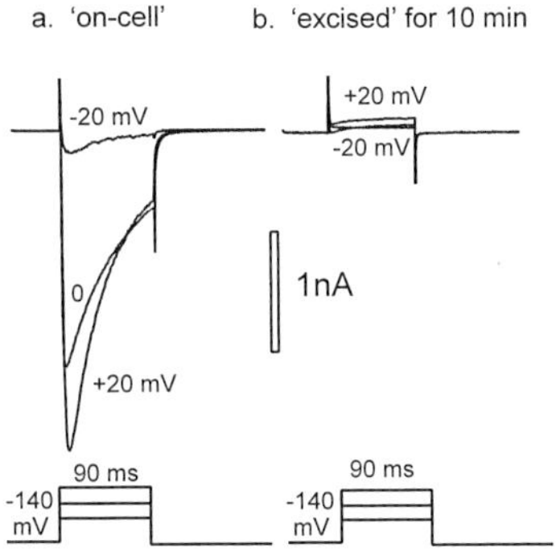

FIG. 2. N-type calcium current in a giant *Xenopus* oocyte membrane patch. Holding potential is −140 mV. Voltage pulses to −20, 0, and 20 mV. (a) "On-cell" records in isotonic KCl solution. (b) Records from the same patch 10 min after excision. See text for details.

equal parts of light and heavy mineral oil (Sigma, St. Louis, MO) with about 10% shredded Parafilm. The general techniques and precautions of conventional patch seal formation will allow formation of giant patches, although greater patience and attention to details may be required for giant patches.

Improvement of Seal Stability

Over several years we have tested many other hydrocarbon mixtures in comparison to that just described. Only one change has clearly improved seal quality and stability. While heating the mineral oil/Parafilm mixture just described, approximately 5–10% of a "superbonding" sticky wax adhesive (Kindt-Collins Company, Cleveland, OH) is added and stirred until consistent ("complete mixture"). This addition causes no evident changes of the properties of ion currents routinely studied in our laboratory. As illustrated in Fig. 2, however, excised patches can routinely be maintained at more negative membrane potentials and for longer times than is possible with the original sealing mixture.

Figure 2 shows records of the N-type calcium current using the "complete mixture" for sealing. The three N-channel subunits, α1B, β3, and α2-δ, were expressed in *Xenopus* oocytes.[4] The bath solution was isotonic KCl with 5 m*M* EDTA and no divalent cations, and the pipette solution contained 20 m*M* $BaCl_2$. In "on-cell" configurations (Fig. 2a), the currents can be studied for >30 min with holding potentials of about −140 mV (or more). After patch excision (Fig. 2b), the current runs down in about 1 min,

[4] P. T. Ellinor, J. F. Zhang, W. A. Horne, and R. W. Tsien, *Nature* **372,** 272 (1994).

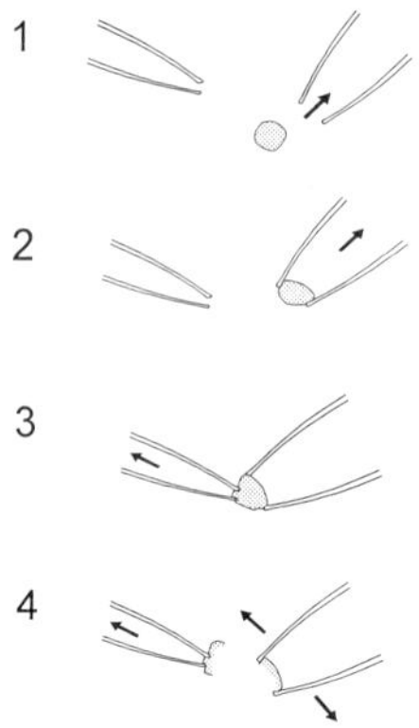

FIG. 3. Formation of half-cell giant excised patch. Cells employed should be easily removed from the dish surface by suction. A second suction pipette with a 3- to 5-μm tip is mounted in the dish. (1) The chosen cell is sucked briskly onto the pipette tip, which is a few microns smaller in diameter than the cell. (2) Gentle suction and rapid movements of the pipette in each direction favor gigohm seal formation. (3) The patch pipette with sealed cell is moved to the dissection pipette tip, and suction is used to attach the second pipette. (4) Lateral motion of the patch pipette dissects away the patch without disrupting the seal.

but the patch itself is stable for periods >20 min at holding potentials of -140 mV. Patches with similar stablity can be obtained using 2 m*M* calcium in the pipette, instead of barium.

Half-Cell Giant Patch

Cultured cell lines, instead of *Xenopus* oocytes, are being used increasingly for expression of ion transporters and channels. Figure 3 describes our implementation of giant membrane patch techniques for small cells, such as HEK and BHK cells. The techniques can be used for both stable and transiently transfected cells. In the latter case, coexpression of the jellyfish green fluorescent protein (GFP) allows reliable identification of expressing cells via fluorescence microscopy.[5]

Cells can be removed from dishes by either enzymatic (trypsin) or mechanical means. Seal formation is easier after enzymatic removal, but mechanically removed cells are also usable. To avoid cell adhesion to dishes, before sealing, the dishes are prewashed with a solution containing 1 mg/ml fatty acid-free albumin. Relatively large cells are advantageous. However, this must be weighed against the experience that smaller cells often have higher expression densities, especially with transient transfection.

[5] J. Marshall, R. Molloy, G. W. Moss, J. R. Howe, and T. E. Hughes, *Neuron* **14,** 211 (1995).

Techniques to form and excise patches are described schematically in Fig. 3. Usually, a calcium-free bath solution with 1–10 m*M* EGTA and 2–4 m*M* magnesium is employed during seal formation. Thus, the cytoplasmic membrane face is not exposed to calcium when the patch is excised. Pipettes are prepared with 10- to 15-μm inner diameters, just a few micrometers smaller than cell diameters. We use mouth-control of pressure to the pipette and have not been successful using automated pressure control methods. The chosen cell is approached with just enough positive pressure, applied to the patch pipette, so that the cell is visibly deformed by solution streaming. When the pipette is aimed directly at the cell within 10–25 μm, a negative pressure is applied [(1), Fig. 3] to briskly suck the cell onto the pipette tip. With gentle pulses of negative pressure, gigohm seals are formed [(2), Fig. 3] with >50% success rate. About 30% of the cell membrane (3 pF) is voltage clamped, based on a comparison of giant patch and whole-cell capacitance. This on-cell configuration may be advantageous for a number of purposes, particularly if it is possible to "permeablize" one-half of the cell with ionophores or other methods without loss of cytoplasmic protein.

To excise the patch, a second smaller suction pipette is used to dissect the exposed part of the cell [(3), Fig. 3]. In our procedure, the second pipette is mounted on the side of the recording chamber employed.[6] The patch pipette tip, with sealed cell, is moved next to the second pipette, and negative pressure is applied to the suction pipette to hold the cell. Then, a side-to-side movement of the patch pipette [(4), Fig. 3] is used to disrupt and remove progressively the exposed half-cell. Finally, the excised patch is moved to the recording chamber for experimentation. Patch stability and success rates are both comparable to those achieved with giant oocyte membrane patches.

Bath Solution Switching

Reliable solution switching on both membrane sides is essential for many purposes, and several different methods can be used. Our routine method for switching "bath" solutions (i.e., on the cytoplasmic side of inside-out patches) in a small recording chamber provides temperature control and allows either pulsatile or constant solution flow across the patch.[6] However, it does not allow computer control of the solution switch, and the switch speeds are not very fast (approximately 0.4 sec).

For faster, temperature-controlled bath solution switches, we use the

[6] A. Collins, A. V. Somlyo, and D. W. Hilgemann, *J. Physiol.* (*Lond.*) **454,** 27 (1992).

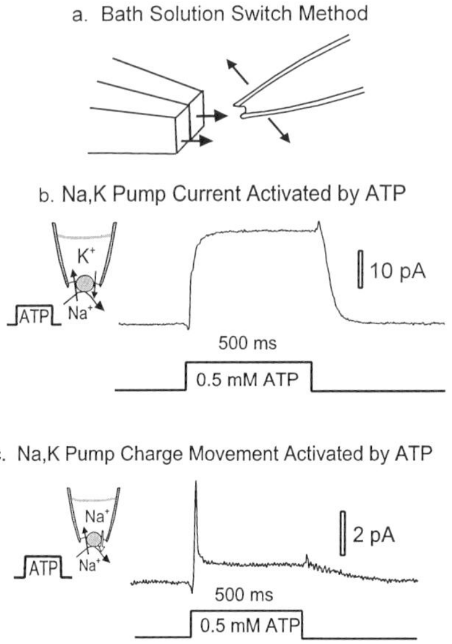

FIG. 4. a. Fast cytoplasmic (bath) solution switch method. a. Parallel, temperated solution streams are formed by gravity flow through water-jacketed solution lines which terminate in fused square (1-mm) pipettes. The patch pipette with headstage is more easily moved, and with less electrical artifacts, than the water jacket with parallel outlets because the latter is bulky. b. Activation of Na/K pump current by ATP. c. Activation of Na/K charge movement by ATP.

method described in Fig. 4a with parallel solution streams. The principle is to move the patch pipette between two gravity-driven solution streams, similar to techniques used by others with excised patches[7,8] and whole cells.[9] Because our temperature-controlled recording chamber is bulky, we move the patch pipette and head stage instead. We use fused 1-mm square glass tubes (Frederich & Dimmok, Millville, NJ) for this purpose. They are advantageous, compared to theta-shaped pipettes, because placement of the patch pipette tip is less critical. Also, multiple parallel solution streams can be arranged to allow application of multiple solutions in a short time.

For two parallel pipettes, we pull the fused tubes to a tip on a standard patch pipette puller, and then cut the outlet to about 250 μm using the technique described earlier. The opposite ends of the tubes are cut to give a total tube length of about 1 cm, and polyethylene tubes are glued into

[7] R. S. Brett, J. P. Dilger, P. R. Adams, and B. Lancaster, *Biophys. J.* **50,** 987 (1986).

[8] P. Jonas, *in* "Single-Channel Recording" (B. Sakmann and E. Neher, eds.), 2nd Ed., p. 231. Plenum Press, New York, 1995.

[9] K. W. Spitzer and J. H. Bridge, *Am. J. Physiol.* **256,** C441 (1989).

them for solution delivery. For temperature control, the polyethylene tubes are water jacketed up to their entrance into the short end piece of square tubes. We then use a piezoelectric-type manipulator (Burleigh, Fishers, NY) with a custom-built motion amplifier to move the patch pipette laterally, together with the patch-clamp headstage, by up to 0.8 mm in 3 ms. Although relatively long-travel stages (200 μm) are now available, the still longer movements achieved with additional levering are advantageous in many applications. From the mechanical standpoint, solution switches are easily carried out in less than 1 ms, but the speed of solution switches is limited by diffusion rates up to the membrane. Depending on the size and configuration of the giant patch, half-times determined from measurements with ion channels range from 10 to 200 ms.

Figures 4b and c illustrate the use of solution switching to study sodium–potassium pump currents and charge movements in excised cardiac patches at 37°. In Fig. 4b, the sodium–potassium pump current is turned on and off by application and removal of 0.5 m*M* Mg-ATP in the presence of 5 m*M* extracellular potassium and 20 m*M* cytoplasmic sodium. Note the approximately 40-ms half-time of the current activation and 70-ms half-time for current decay. This is typical for an inside-out patch with a diameter of about 25 μm. Since the half-maximal ATP concentration is about 100 μ*M*, the activation of current is faster than the expected time course of ATP equilibration at the membrane. With high-affinity mechanisms, such as the sodium–potassium pump, therefore, it is possible to accomplish a much faster activation than can be expected from diffusion times up to the membrane (e.g., 10 times faster with 5 m*M* ATP than shown here).

The isolation of charge movements using cytoplasmic solution switches is described for the sodium–potassium pump in Fig. 4c. In the presence of low (20 m*M*) extracellular sodium and the absence of extracellular potassium, the major charge movement of the sodium pump cycle can be isolated by a cytoplasmic ATP jump in the presence of cytoplasmic sodium. The charge movement or current transient is absent when ouabain is included in the pipette (not shown); it corresponds with good certainty to the outward transport and release of sodium ions by the pump, subsequent to phosphorylation.[10] There is only a small steady-state pump current in this circumstance; it depends on the presence of cytoplasmic sodium and probably corresponds to a slow 3Na–2Na transport cycle.[10] The lack of a returning charge movement probably reflects an electroneutrality of the returning steps which move two sodium ions together with coordinating negatively charged groups through membrane electrical field. In general, the time courses of the charge movements resolved in this way reflect diffusion time courses, not

[10] P. Lauger, *in* "Electrogenic Ion Pumps," Sinauer Associates, Sunderland, Massachusetts, 1991.

transporter kinetics. Nevertheless, this methodological approach allows a valid isolation, identification, and quantification of partial reactions of transporters with relatively few artifacts and excellent quantitative resolution. One great advantage with respect to "caged" compounds is that any substance or ion can be "flashed" by solution switching.

Pipette Solution Switching

Solution switching in the pipette is more challenging, and Fig. 5 shows the positive-pressure method that we currently favor. In our experience, this method is improved considerably in its reliability and ease of use from previously published methods.[3,11,12] Flexible, quartz (fused-silica) tubing (Polymicro Technologies, Phoeniz, AZ) is used to change solution in the pipette tip by positive pressure. Perfusion solution reservoirs are made from 100-μl "yellow" pipette tips, as described in detail in the legend to Fig. 5a. Using our standard (bullet-shaped) patch pipettes, the 100- and 150-μm-diameter tubes (available with 40- to 70-μm inner diameters) can be placed within 200 μm of the patch pipette orifice without pulling the tubes to a tip on a flame. Rather, the tubing is cut at a right angle by simply fusing it into a bead of molten glass on the microforge, cooling it, and breaking it by a quick pull.

As illustrated in Fig. 5b, the quartz tube (QT) is brought into the patch pipette tip (PP) via a polyethylene tube (PET) sleeve, fitted snugly through a hole drilled in the side of the pipette holder. The quartz tube is positioned to within 200 μm of the pipette orifice just before making a seal. During preparations, the other end of the quartz tube is fed through a fine (28.5-gauge) "insulin" syringe needle (N) and held in place at the back of the needle with superbonding sticky wax (Kindt-Collins Company, Cleveland, OH). The tube is first threaded through the needle and fixed with wax so that the tube protrudes slightly. Then the tube is retracted so that the quartz end resides just within the needle. The needle is mounted just above the patch-clamp headstage on a mechanical mount that does not impact the patch pipette tip. For solution delivery, the needle is inserted through a silicone membrane (SM) with a fine opening, entrance to which is facilitated by a pipette tip guide (see Fig. 5a). The reservoirs are fitted with caps that connect via polyethylene tubing (PET) to a syringe (S) through which air pressure can be applied to the pipette tip (PT). The seals formed at the needle/silicone interface withstand high pressures, so that small-diameter perfusion tubes can be employed with this method.

[11] M. Soejima and A. Noma, *Pflugers Arch.* **400,** 424 (1984).

[12] J. M. Tang, J. Wang, and R. S. Eisenberg, *Methods Enzymol.* **207,** 176 (1992).

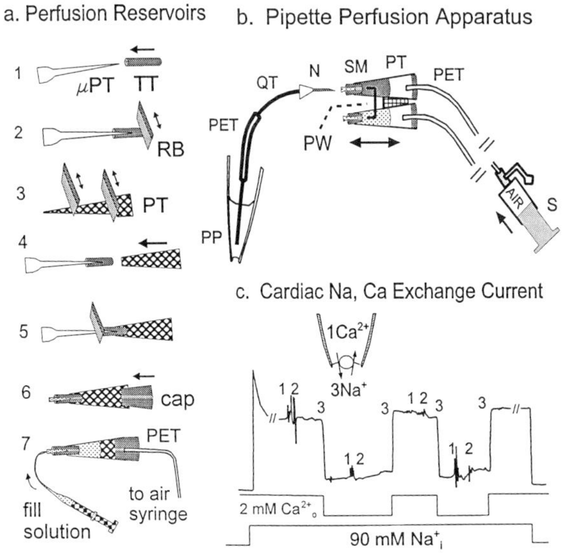

FIG. 5. Positive-pressure pipette perfusion method. (a) Fabrication of solution reservoirs. Standard 100-μm pipette tips are used as reservoirs. Silicone or Tygon tubing can be used to make a membrane through which the needle with quartz tube can be inserted; we use a Tygon tube with 2-mm outer diameter and 0.2- to 0.3-mm inner diameter (Cole-Palmer Instrument Company, Vernon Hills, IL). To construct a guide for the needle, a 10-μl small pipette tip is inserted into one end of the Tygon tube (1), and an approximately 1.5-mm length of tube is cut off with a razor blade (RB, 2). A 100-μl (yellow) pipette tip (PT) is trimmed (3) so that the pipette tip with Tygon tube can be pressed very tightly into it (4). The smaller pipette is then cut off with a razor blade, leaving a needle guide of about 1 mm (5). Several such reservoirs are prepared, and they are fitted with plastic caps (CAP) through which positive pressure can be applied via polyethylene tubes (6). The reservoirs are back-filled with 15–30 μl of the desired pipette solution from polyethylene syringes pulled to a fine tip (6). (b) Pipette perfusion method. A small-diameter quartz tube (100–150 μm; QT) is placed in the tip of the patch pipette (PP) through a polyethylene tube (PET) fitted snugly through a hole in the patch pipette holder. The opposite end of the quartz tube is fixed in a small-diameter (28.5-gauge) syringe needle (N). To change solutions, the solution reservoir is removed and a new one attached to the needle, followed by application of positive pressure through the connected syringe (S). To avoid the application of solutions at an electrical potential sufficient to disrupt giant patches, the pipette tips (PT) used as solution reservoirs can be connected by platinum wire. In this case, small holes are made in the tips, and the wire is inserted and glued into place. Alternatively, the individual reservoirs can be discharged to the potential of the bath solution just before connection to the needle. This is done by connecting the reservoirs to another needle that is connected electrically to the recording chamber through a polyethylene tube. (c) Cardiac Na–Ca exchange current. The current record illustrates the activation of cardiac sodium–calcium exchange by cytoplasmic sodium and extracellular calcium. As indicated below the record, either the cytoplasmic sodium was exchanged for cesium, or the extracellular (pipette) calcium was exchanged for magnesium via pipette perfusion. The indicated disturbances in the record correspond to the removal of a solution reservoir (1), the attachment of a new one (2), or the application of positive pressure to deliver the new solution.

To avoid the transient application of large electrical potentials during the connection of new solutions to the pipette, the solution reservoirs employed in an experiment are connected by platinum wires. Alternatively, reservoirs can be discharged adequately by connecting them briefly to a long solution bridge in communication with the recording solution (see Fig. 5b). Without one of these precautions, patches are often destroyed when changing solution reservoirs during experiments.

The use of this pipette perfusion technique for recording cardiac Na–Ca exchange current is described in Fig. 5c. As indicated below the record, outward exchange current was activated under our standard conditions[6] at 37° by application of cytoplasmic sodium in the presence of extracellular (pipette) calcium. The current was then turned off and on twice by pipette perfusion before being turned off at the close of the record by removal of cytoplasmic sodium. The pipette perfusion speeds achieved are on the order of a few seconds, which is slower than achieved with another device described previously.[2]

Capacitance Measurements

Capacitance measurements can be used for many purposes, the most common being to quantify membrane area and changes of membrane area.[13] Giant membrane patches are advantageous for such measurements, owing to the speed of voltage clamp. The reason is that the actual membrane potential will remain nearly synchronized with the applied membrane potential up to rather high frequencies of sinusoidal perturbation used to quantify membrane capacitance. Here, we wish to illustrate the use of capacitance measurements as a means to probe conformational changes of membrane proteins. The basis for this application is that increasing numbers of membrane proteins can be expressed at high density in a variety of expression systems. Any reaction in a membrane protein that moves charged groups within the membrane electrical field is a source of membrane capacitance. The high sensitivity of capacitance measurements means that a wide range of membrane protein conformational changes might be monitored with this approach, including membrane proteins besides ion transporters.

Figure 6 illustrates the use of capacitance measurements to monitor cytoplasmic chloride binding by the GAT1 (GABA, sodium, chloride) cotransporter, expressed in *Xenopus* oocytes.[1] When chloride is applied to the cytoplasmic surface there is an approximate 1% fall of membrane

[13] K. D. Gillis, *in* "Single-Channel Recording" (B. Sakmann and E. Neher, eds.), 2nd Ed., p. 155. Plenum Press, New York, 1995.

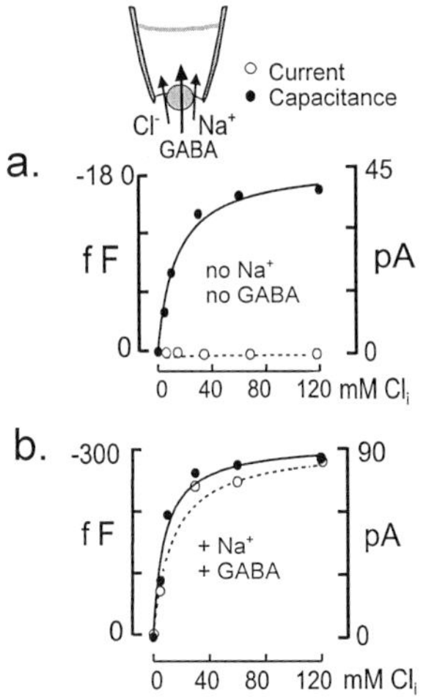

FIG. 6. Capacitance measurements can be used to monitor cytoplasmic chloride binding to the oocyte-expressed GAT1 [GABA (γ-aminobutyric acid)/sodium/chloride] cotransporter. (a) In the absence of cytoplasmic GABA and sodium, application of chloride to the cytoplasmic membrane surface induces a capacitance decrease (~1% fall of membrane capacitance with 120 m*M* chloride) in a concentration-dependent manner. No transport current is activated under this condition. (b) In the presence of cytoplasmic GABA and sodium, a similar capacitance decrease can be induced by cytoplasmic chloride with simultaneous activation of the transport current.

capacitance, monitored with a 1- to 20-kHz perturbation of 1-mV amplitude. This signal is absent in control membrane. It must correspond to the suppression of a charge-moving reaction in the protein by chloride binding. As shown in Fig. 6a, the capacitance drop ($K_d \approx 12$ m*M*) can be monitored in the absence of cotransported substrates. The importance of the measurement is that ion binding can be monitored in states where transport is not possible. As shown in Fig. 6b, the capacitance change can also be monitored in the presence of cosubstrates, and the capacitance signal has a very similar chloride dependence to that of the transport current.

Phospholipid Exchange in Giant Patches

It is possible to change the phospholipid composition of the cytoplasmic membrane leaflet in giant patches via application of phospholipids to the pipette tip.[14] Also, the simple application of sonicated phospholipid vesicles to the cytoplasmic face is effective, and in this way the modulation and

[14] A. Collins and D. W. Hilgemann, *Pflugers Arch.* **423,** 347 (1993).

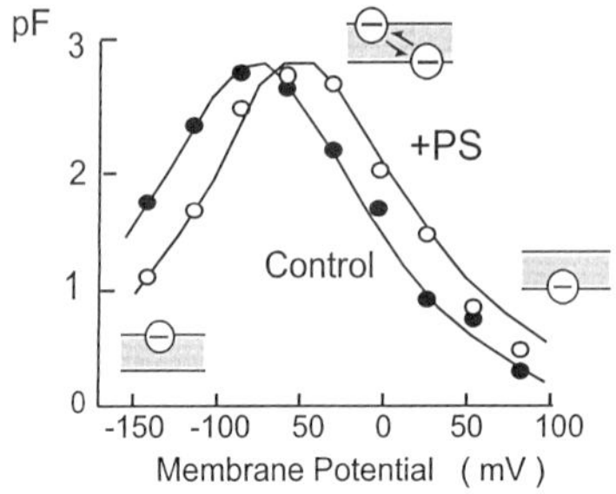

FIG. 7. Membrane capacitance induced by application of 3 μM dipicrylamine to the extracellular side of a giant *Xenopus* oocyte membrane patch. The sinusoidal perturbation frequency is 1 kHz, and the perturbation magnitude is 1 mV. As illustrated, hydrophobic anions will locate primarily to the extracellular side at negative potentials and to the cytoplasmic side at positive potentials. Application of phosphatidylserine (PS) vesicles to the cytoplasmic side at an apparent concentration of 0.3 mM induces a 35-mV shift of the capacitance–voltage relation to more positive potentials, as expected if substantial amounts of PS become incorporated into the membrane.

regulation of ion transport by membrane lipids can be studied. Unfortunately, it is difficult to know to what extent the membrane composition is actually changed.

Figure 7 illustrates the use of capacitance measurements to measure the extent of membrane composition change when sonicated phospholipids, which bear an ionic charge, are applied. In the case of negatively charged phospholipids, significant incorporation of lipids should produce a negative electrical potential at the membrane surface, which will be experienced by a charge in the membrane field of the bilayer. Although the magnitudes of surface potentials in natural membranes are not well established, it is possible to monitor changes of surface potentials by several means. The use of the hydrophobic anion, dipicrylamine, is described in Fig. 7, whereby membrane capacitance is monitored to determine the relative probability of charges being positioned on one side or the other of the membrane field.[1]

Dipicrylamine[15] (3 μM) is included in the pipette solution, whereby it accumulates on both sides of the membrane because it crosses the membrane spontaneously. Because it translocates across the membrane faster than it dissociates from either membrane side, it accumulates on the outside of the membrane at negative potentials and on the cytoplasmic side at positive potentials (see illustrations in Fig. 7). The rates of dipicrylamine translocation are 3000–25,000 per second over the potential range of Fig. 7, so that the charge movement of anion translocation contributes to mem-

[15] R. Benz, P. Lauger, and K. Janko, *Biochim. Biophys. Acta* **455,** 701 (1976).

brane capacitance when capacitance is measured with perturbation frequencies of 1 kHz, as in Fig. 7. The voltage dependence of the anion-associated capacitance is bell shaped because, at extremes of potential, the anions are located predominantly on one side of the membrane or the other and cannot be moved by small perturbations of the membrane potential. The peak dipicrylamine capacitance in the example in Fig. 7 corresponds to about 20% of the total membrane capacitance, and this can be increased many fold by increasing the dipicrylamine concentration.

When negatively charged phosphatidylserine (PS) vesicles are applied from the cytoplasmic side, the voltage dependence of the dipicrylamine capacitance is shifted to more positive potentials. The magnitude of the shift is enhanced in Fig. 7 by the use of a cytoplasmic solution with ionic strength reduced by 50%. This is an expected result if incorporation of negatively charged lipids is significant, and from surface potential theory,[16,17] the 35-mV shift observed requires that PS incorporation amounted to about 10% of total cytoplasmic phospholipids.

The same types of experiments gave no evidence for major changes of membrane composition when vesicles were applied by pipette perfusion to the extracellular side. The reason is evidently that hydrophobic agents, including phospholipids, dipicrylamine, other hydrophobic ions, and several fluorescent probes, incorporate much better from the cytoplasmic side than from the extracellular side. For example, >10 times smaller dipicrylamine concentrations give the same magnitude of signals when applied from the cytoplasmic side. This situation is in fact reflected in the dipicrylamine capacitances, shown in Fig. 7, in that the peak of the dipicrylamine capacitance occurs at quite negative potentials (−30 to −80 mV); dipicrylamine is relatively repelled from the extracellular leaflet and attracted to the cytoplasmic leaflet. One possibly important factor for this asymmetric behavior is the high density of glycosylated lipids in the outer leaflet.

A second conclusion from measurements of this type is that the incorporation of phospholipids via application of phospholipid vesicles takes place by an exchange process. First, the membrane capacitance does not change, or changes only very little, when the membrane composition changes substantially. Second, the effects of PS could be reversed by applying neutral phospholipids, such as phosphatidylcholine, over 5 to 10 min (not shown). An involvement of membrane-associated phospholipid exchange proteins would be an attractive interpretation. However, similar experimental results have been obtained after treatment of patches with trypsin, which presum-

[16] S. McLaughlin, G. Szabo, and G. Eisenman, *J. Gen. Physiol.* **58,** 667 (1971).

[17] S. McLaughlin, *Curr. Top. Membr. Transp.* **9,** 71 (1977).

ably would destroy exchange proteins. Based on observations with fluorescent phospholipids, we favor the idea that phospholipids bind first to the wall of the pipette and then diffuse into the membrane across the glass–membrane boundary. Physical factors might favor the extrusion from the patch of a roughly equal number of lipids to the pipette wall.

Perspectives

This volume provides readers with an in-depth impression of some electrophysiologic methods available at this time. The giant excised membrane patch will be an attractive method whenever a large membrane area is required with fast voltage clamp and control of solutions on both membrane sides. Disadvantages still include a relative instability with large holding potentials, loss of cytoplasmic components, disruption to a greater or lesser extent of cytoskeleton, and the relative difficulty of some of the methods. We foresee many new applications from the ability to perform half-cell voltage clamp and from the increasing prospects to maintain and/or reconstitute submembrane processes such as membrane insertion and retrieval, as well as calcium release from submembrane stores.

Acknowledgments

We thank Siyi Feng and Quisheng Tong for discussions and technical help. We thank Ilya Bezprozvanny and Anton Maximov for collaboration on the experiments with N-type calcium channels.

[16] Electrophysiologic Recordings from *Xenopus* Oocytes

By WALTER STÜHMER

Introduction

The *Xenopus* oocyte as a heterologous expression system has gained tremendous popularity in recent years. It is a reliable expression system for membrane-bound proteins, in particular ion channels, membrane receptors, and transport systems. This article describes electrophysiologic methods applied to *Xenopus* oocytes, as well as a discussion of advantages and limitations of the method. Other expression systems are discussed elsewhere in this volume.[3,31] The large size of *Xenopus* oocytes, which allows easy injections of RNA, DNA, chemical compounds, and even the insertion of

0076-6879/98 $25.00

pipettes containing a patch (patch cramming), is a clear advantage of using oocytes as an expression system. In addition, low-noise recordings from macropatches containing a large number of membrane proteins allows experiments that are currently difficult to obtain from any system other than the *Xenopus* oocytes.

Oocytes from *Xenopus* have been used successfully to translate messenger RNAs (mRNA) into the respective membrane proteins, including post-translational modifications.[1-4] Genetic engineering techniques allow for mutations at specific loci of the message coding for the membrane-bound proteins and thus to study structure–function relationships in the respective proteins.[5-11]

Methods used to maintain *Xenopus laevis* and their oocytes, to microinject RNA, to remove the follicular cell layer, and to express the various membrane-bound proteins in *Xenopus* oocytes are described elsewhere in this volume.[12,13]

Preparation and Handling of RNA

The injection of sufficient amounts of RNA coding for membrane-bound proteins to be studied is a prerequisite for their functional expression. This RNA may be of diverse origin: total mRNA or poly(A) mRNA extracted from native tissue samples or cDNA-derived mRNA (cRNA). The latter allows for the manipulation of the genetic sequence but has the disadvantage that only homomultimeric protein complexes will be formed on injection of a single cRNA sequence. Most membrane-bound proteins form complexes not only with other membrane-bound and cytoplasmic

[1] J. B. Gourdon, C. D. Lane, H. R. Woodland, and G. Marbaix, *Nature (Lond.)* **233,** 277 (1971).
[2] C. D. Lane, *Curr. Top. Dev. Biol.* **18,** 89 (1983).
[3] F. L. Theodoulou and A. J. Miller, *Mol. Biotechnol.* **3,** 101 (1995).
[4] F. L. Theodoulou and A. J. Miller, *Methods Mol. Biol.* **49,** 317 (1995).
[5] K. Sumikawa, M. Goughton, J. S. Emtage, B. M. Richards, and E. A. Barnard, *Nature (Lond.)* **292,** 862 (1981).
[6] R. Miledi, I. Parker, and K. Sumikawa, *Proc. R. Soc. Lond. B* **216,** 509 (1982).
[7] E. A. Barnard, R. Miledi, and K. Sumikawa, *Proc. R. Soc. Lond. B* **220,** 131 (1983).
[8] C. B. Gundersen, R. Miledi, and I. Parker, *Proc. R. Soc. Lond. B* **220,** 131 (1983).
[9] L. C. Timpe, T. L. Schwarz, B. L. Tempel, D. M. Papazian, Y. N. Jan, and L. Y. Jan, *Nature (Lond.)* **331,** 143 (1988).
[10] A. Mikami, K. Imoto, T. Tanabe, T. Niidome, Y. Mori, H. Takeshima, S. Narumiya, and S. Numa, *Nature (Lond.)* **340,** 230 (1989).
[11] U. B. Kaupp, T. Niidome, T. Tanabe, S. Terada, W. Bönink, W. Stühmer, N. J. Cook, K. Kangawa, H. Matsuo, T. Hirose, T. Miyata, and S. Numa, *Nature (Lond.)* **342,** 762 (1989).
[12] A. L. Goldin, *Methods Enzymol.* **207,** 266 (1992).
[13] C. Methfessel, V. Witzemann, T. Takahashi, M. Mishima, and S. Numa, *Pfluegers Arch.* **407,** 577 (1986).

proteins, but also with cytoskeletal elements,[14] so that care should be taken in the interpretation of data obtained from homomeric and/or incomplete assemblies.

Total and poly(A) mRNA injection has the advantage of containing all the native mRNAs coding for the protein complexes, but also contains mRNA coding for proteins other than those of interest. This might cause interference (e.g., in the case of voltage-gated channels where the resulting currents on depolarization might contain Na, K, and Ca components), and the mRNA will be at a lower concentration than desired. The relative abundance of the desired mRNA can be increased by size fractionation.[15,16] However, this procedure will also deplete mRNA that might code for associated proteins required for functional expression. A tissue sample of about 1 g yields 10–100 μg of poly(A) mRNA. Extraction procedures have been described for total RNA[17,18] and poly(A) mRNA.[19]

The mRNA from tissue extracts should be injected at concentrations of 1–10 μg/μl, and the cRNA can be injected at concentrations of 0.1–1.0 μg/μl, both in aqueous solution. Extreme care should be exercised at all stages to avoid and/or inhibit RNases, such as using baked glassware (e.g., 250° for 6 hr), wearing sterile gloves, and using diethyl pyrocarbonate-treated, autoclaved doubly distilled water (DEPC–H_2O) for all solutions. Also, care should be taken to keep the solution free of particles that could clog the injection pipettes. Thus, the mRNA should be centrifuged to precipitate particles suspended in solution. Silanized Eppendorf tubes of 1 or 0.5 ml are best suited as containers for mRNA. The mRNA should be stored at −80°. For brief periods of time, storage at −20° is allowable. In fact, even room temperature is admissible for storage if the preparation is truly free of RNases. In general, it is advisable to aliquot the mRNA and/or cRNA in portions sufficient for the injections on a single day.

mRNA is best stored and shipped precipitated in ethanol. This precipitation is best carried out as follows: 0.1 volume of 20% (v/v) potassium acetate and 2.5 volumes of cold 100% ethanol are added to 1 volume of aqueous mRNA solution. After mixing by inversion of the tube, the mRNA is allowed to precipitate for at least 30 min at −70°. The mRNA can be stored

[14] M. Sheng and E. Kim, *Curr. Opin. Neurobiol.* **6,** 603 (1996) or chapter [56] in this volume.

[15] P. Fourcroy, *Electrophoresis* **5,** 73 (1984).

[16] H. Lübbert, B. J. Hoffman, T. P. Snutch, T. van Dyke, A. J. Levine, P. R. Hartig, H. A. Lester, and N. Davidson, *Proc. Natl. Acad. Sci. U.S.A.* **84,** 4332 (1987).

[17] J. M. Chirgwin, A. E. Pryzybyla, R. J. MachDonnald, and W. J. Rutter, *Biochemistry* **18,** 5924 (1979).

[18] P. Dierks, A. van Ooyen, N. Mantel, and C. Weissmann, *Proc. Natl. Acad. Sci. U.S.A.* **78,** 1411 (1981).

[19] H. Aviv and P. Leder, *Proc. Natl. Acad. Sci. U.S.A.* **69,** 1408 (1972).

as such; however, it is more convenient to keep it in pure ethanol after decreasing the salt concentration. To do this, the tube containing the mRNA is incubated at room temperature for approximately 5 min, then centrifuged at 12,000*g* for 15 min at 4° in a cooled centrifuge (e.g., Hettich centrifuge, Eppendorf centrifuge, or Heraeus Biofuge A). The supernatant is then carefully removed with a sterile suction pipette, and the pellet washed with 2 volumes of 70% (v/v) ethanol. A centrifugation of 10 min will pellet the mRNA again. After removing the salt-containing supernatant the mRNA is stored in 100% ethanol. Care should be taken to avoid touching the mRNA pellet with the suction pipette and not to leave the tube open for extended periods to minimize contact with airborne RNases. To recover the precipitated mRNA it is thawed and centrifuged for 10 min at 12,000*g*, the supernatant removed, and the opening of the tube covered with Parafilm into which three or four holes are poked with a sterile needle. The mRNA is then dried under reduced pressure in a desiccator for 4–10 min. Then the mRNA can be carefully resuspended in DEPC–H_2O at the desired concentration for use. The best method for drying the RNA prior to resuspending it in water is to use a Speed-Vac or any other centrifugation under vacuum, so that the ethanol evaporates at the same time that the pellet is maintained at the bottom of the Eppendorf tube.

Injection of Messenger RNA

For transferring the mRNA from the Eppendorf tube to the injection pipette, transpipettor tubes (disposable micropipettes; Brand) or capillaries are used. The transpipettor tubes should be made hydrophobic by flushing them in a solution consisting of 50% (v/v) ether and 50% (v/v) silane (dimethyldichlorosilane, Fluka, Ronkonkoma, NY; *caution*: health hazard). After allowing the ether to evaporate under a hood, the tubes are baked at 180° for 2 hr to inactivate any remaining RNases. Tubes treated in this way can be stored under sterile conditions for several months. A micrometer-controlled syringe, filled with mineral oil to reduce the air volume and evaporation of water from the mRNA solution, is used to load the transfer pipette with a few microliters of mRNA solution.

Injection pipettes can be pulled using a standard pipette puller. They should have a long shank for estimating the volume of mRNA solution loaded and ensuring that the oocytes are successfully injected. If pulled shortly before use, the injection pipettes will be RNase free owing to the high temperature during the pulling process. The tips of the injection pipettes are broken under a microscope to an opening diameter of about 10 μm. Usually this will leave a rough and uneven rim, which might lesion the oocytes when penetrating them. This can be avoided by fire polishing the rim and pulling a sharp tip. This is done by approaching the tip of the injection pipette with a microfilament that has a small drop of molten glass, preferably of the same type as the injection pipette, on the tip. A sharp, needle-like protrusion can be obtained by rapidly removing the glass-covered filament after touching the injection pipette. Radiation heat during this process is sufficient to smooth the ragged rim. The injection pipettes are filled under stereomicroscopic control by suction applied through a conventional syringe. For this purpose, mRNA solution is extruded to form a droplet just outside of the transfer pipette by means of the micromanipulator-driven syringe. At this stage, the injection pipette is filled by application of suction with approximately 500 nl of mRNA solution, sufficient for injecting about 10 oocytes. After this, it is convenient to draw the remaining mRNA solution a few millimeters back into the transfer pipette by slight suction. This procedure decreases evaporation of water from the mRNA solution, which will be used for the next batch of oocytes to be injected. A hand-driven coarse manipulator is sufficient to maneuver the injection pipette. Grooves carved into a thin Perspex plate fixed to the bottom of the injection chamber support the oocytes during injection. Alternatively, a scratched petri dish or one covered with either silicone curing agent (RTV615, General Electric, Waterford, NY) or agar can be used to fix the oocytes. The fixation of the oocytes is achieved by reducing the amount of solution in the injection chamber, so that the surface tension holds the oocytes in the grooves. Manipulation of oocytes is performed with the aid of Pasteur pipettes whose tips have been broken to enlarge the opening and then fire polished. A slight kink in the pipette tips assists in handling the oocytes during transport.

The mRNA is injected in aliquots of about 50 nl per oocyte. There are many procedures to achieve this, and very simple manual syringe systems or injection machines can be used. More sophisticated injection machines such as the Eppendorf microinjector or the Drummond oocyte injector are helpful when injecting large quantities of oocytes at a time.

Incubation

In most cases the large stage V and VI oocytes[20] can be used. The oocytes are incubated in small petri dishes in Barth's medium [in m*M*: 84 NaCl, 1 KCl, 2.4 $NaHCO_3$, 0.82 $MgSO_4$, 0.33 $Ca(NO_3)_2$, 0.41 $CaCl_2$, 7.5 Tris-HCl, pH 7.4] with additions of penicillin and streptomycin (100 U/ml and 100 μg/ml, respectively). Some protocols use gentamicin (50–100 μg/ml). The optimal incubation temperature is 19°. Barth's medium should

[20] J. N. Dumont, *J. Morphol.* **136,** 153 (1972).

be changed every day using sterile pipettes. Oocytes having a blurred delimitation between the animal and vegetal pole should be removed.

Removal of Follicular Cell Layer

The follicular cell layer can be removed either before or after injection of the mRNA. It can be removed by mere mechanical means, but partial or total digestion of the follicular cell layer aids in preventing mechanical damage of the oocytes. Mild digestion is achieved in Ca^{2+}-free Barth's medium containing 1 mg/ml collagenase I (e.g., Sigma, St. Louis, MO, type I) for 1 hr at room temperature. The follicular cell layer is then removed mechanically using two pairs of forceps (No. 5, Dumont & Fils, Montignez, Switzerland). Alternatively, the ovaries can be incubated for 2–3 hr in the collagenase-containing solution under shaking until the oocytes are dispersed. It is important to wash (three to four times) the oocytes extensively in Barth's medium (approximately 5 ml volume) after collagenase treatment to arrest the enzymatic reaction.

There appears to be no *ad hoc* rule to anticipate when the various proteins will be inserted in the oocyte membrane following injection. Therefore, it is necessary to assay for protein expression on a daily basis. In general, this entails two-electrode voltage clamping for ion channels and electrogenic pumps. If no patch-clamp experiments are intended, the vitelline envelope need not be removed. However, the removal of this last barrier to the plasma membrane makes insertion of electrodes easier, improves access and washout of solutes, and fixes the oocytes to the bottom of the experimental chamber (we use new 35-mm petri dishes as experimental chambers). Methods for removing the vitelline layer are described later in the section on patch-clamp recording and in Ref. 13. On the other hand, the vitelline layer stabilizes the oocyte membrane. This is particularly important when using high flow rates.

Electrophysiologic Measurements

For most electrophysiologic recordings from oocytes, the membrane potential must be under control. This is achieved by clamping the oocyte to predefined potentials using a conventional two-electrode voltage clamp.[21,22] Specifically, one intracellular electrode is used to record the actual intracel-

[21] T. G. Smith, J. Lecar, S. J. Redmann, and P. W. Gage, eds., *in* "Voltage and Patch Clamping with Microelectrodes," American Physiological Society, Bethesda, Maryland, 1985.

[22] N. B. Standen, P. T. A. Green, and M. J. Whitaker, eds., "Microelectrode Techniques: The Plymouth Workshop Handbook." The Company of Biologists, Cambridge, Massachusetts, 1987.

lular potential, and the second electrode is used to pass current so as to maintain the desired potential. This is achieved using a feedback circuit, which is the main component of the voltage clamp. The current needed to maintain a given potential is the measured parameter. There are two means of measuring this current: as current flowing to ground through the grounding electrode (using a virtual ground amplifier) or as the current flowing through the current electrode. The intracellular potential recording electrode has such a high impedance that no current flows through it. Most voltage-clamp setups also use a potential recording electrode (reference electrode) for the bath solution. This avoids polarization errors that arise from the current passing through the ground electrode. In this configuration, the transmembrane potential is taken as the difference between the intracellular potential electrode and the reference electrode. Therefore, the small error introduced due to the possibility that the bath does not remain at ground potential is corrected for. A block diagram of the main components of a voltage clamp is illustrated in Fig. 1.

In addition to the voltage clamp itself, a pulse generator is required for adjusting the voltage-clamp amplifier and to perform experiments that require measurements in response to changes in membrane potential. This pulse generator may be anything from a simple waveform generator to a computer-controlled pulse generator. If the voltage clamp does not provide a built-in filter, a Bessel low-pass filter is required to reduce the high-frequency noise and as an antialiasing filter if the data are to be recorded digitally. The data recording equipment may be a simple chart recorder (for slow events), an instrumentation tape recorder, or a computer-based data recording system. In principle, all the instrumentation will be quite

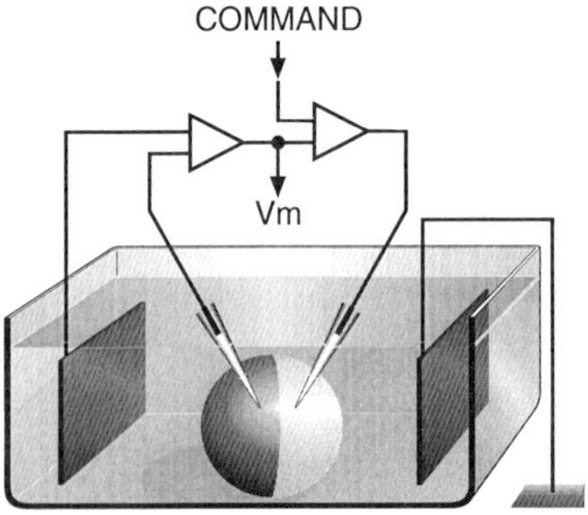

FIG. 1. Schematic diagram of the main components of a two-electrode voltage clamp. The output of the left amplifier–comparator (V_m) is the difference in potential between the bath and the oocyte electrodes. V_m is compared with the command potential, and any difference is injected into the oocyte through the current electrode. This current flows to ground through the bath ground electrode. If V_m equals the command potential, no current flows to ground.

similar to the one used for patch-clamp recording, and details are provided in [14] in this volume.

The two-electrode voltage clamp in oocytes is performed in accordance with techniques used for smaller cells.[21,22] The main differences reside in the larger size of the oocytes, which implies a higher membrane capacitance. This introduces no complication provided that only slow or no changes in potential are required for the experiments. However, when measuring voltage-gated channels, the speed of the clamp and the probability of compensating for capacitive transients are of prime importance. The time constant for charging the membrane and hence the speed of the voltage clamp depend critically on the electrode resistance through the relation $\tau = RC/A$, where R is the current electrode resistance, C the cell capacitance, and A the gain of the feedback loop of the voltage clamp. For time-critical applications, the cell capacitance can be reduced by using smaller (stage III) oocytes.[23]

Intracellular Electrodes

Standard intracellular electrodes, whose tips have been broken, have resistances below 1 MΩ and are quite appropriate for intracellular electrodes in oocytes. Electrode pipettes are made from capillary glass containing a thin filament, which ensures that the electrode filling solution reaches the tip. We use glass from Clark (Reading, England), types GC120TF-10 (thin), GC150F-10 (medium), and GC200F-15 (thick). Pulling is best done using a standard intracellular electrode puller (David Kopf Instruments, Tujunga, CA) or a programmable electrode puller. To prevent the pipette filling solution from climbing up the back end of the electrode, a small amount of dental wax or sticky wax is applied to the back end of the pipette after moderate heating just sufficient to melt the wax. The melted wax seeps a few millimeters into the pipette along the filament during application. After this, the pipettes can be back-filled either with a solution containing 2 *M* KCl or with 0.5 *M* potassium sulfate (or aspartate), containing at least 30 m*M* KCl. The 2 *M* KCl will be useful for most applications. The sulfate or aspartate solutions are to reduce the Cl^- load to the oocytes, and the KCl is needed for the silver/silver chloride electrodes.

The electrode pipettes can be stored for several days in a covered receptable containing water on the bottom to ensure high humidity. The electrode tips are normally in the submicron range and need to be broken at the very tip to decrease the resistance to the megohm range. This is

[23] D. S. Krafte and H. A. Lester, *Methods in Enzymology* **207,** 339 (1992).

done either under a separate microscope or by simply jamming the electrode against the bottom of the recording chamber before an experiment. Typical tip diameters will be in the range of 1–5 μm, giving resistances in the range of 2–0.4 MΩ. The electrical contact to the electrode filling solution is made through a silver chloride electrode. The silver chloride electrodes can be made from chlorinated silver wire, from silver wire immersed in melted silver chloride, or from a silver/silver chloride pellet. The lifetimes of the electrodes mentioned as well as their diameters are given in ascending order.

Most voltage-clamp amplifiers incorporate an electrode resistance measurement. That of Polder (NPI Electronics, Tamm, Germany) allows electrode resistance measurements even with the electrodes positioned inside the oocyte, with a direct readout in megohms. Alternatively, the electrode resistance can be measured by applying current pulses (ΔI) through the Ag|AgCl electrode and measuring the potential jump (ΔV) induced. The electrode resistance will be $\Delta V/\Delta I$.

Provided the electrodes retain their low resistance, they can be reused for several oocytes. This is particularly true for electrodes made out of quartz glass using an infrared laser as a heating stage. Clogged electrodes can be cleared by applying pneumatic backpressure with a syringe. This process can be monitored by the extrusion of the electrode filling solution (which has a different refraction index) under the microscope. A patch-clamp pipette holder can be used as an electrode pipette holder. For long-term experiments, the electrode filling solution should be of such a volume that the hydrostatic pressure approximately compensates the pressure from the surface tension within the pipette. If the hydrostatic pressure in the electrode allows a large net outflow of solution, the oocytes will bulge around the site where the electrode is inserted as the oocytes accumulate KCl. If, on the contrary, there is a net inflow, the electrode resistance will increase as intracellular fluid enters the pipette. A very slight net outflow is the best choice, since this will stabilize diffusion potentials at the electrode tip.

Owing to the large size of the oocytes, the electrodes can be positioned with simple, coarse manipulators. It is important, however, that the setup be free of vibrations because otherwise oscillations will cause damage to the membrane where the electrodes penetrate the oocyte. A vibration-isolated table is usually not necessary in buildings where the floor is sufficiently quiet. This can be tested by observing the electrode tip under the setup microscope while someone jumps on the floor near the setup. A Faraday cage is usually not necessary for normal two-voltage electrode clamping, provided that standard grounding techniques are used (see [14] in this volume). The microscope can be a very simple one, either inverted or noninverted. Low magnification (5–10× objective) provides a comfortable working distance. A simple setup can also be built using a simple large magnifying lens.

Adjusting Voltage Clamp

Once the electrodes are in solution, the offset of the potential electrode should be cancelled to within ±1 mV. It is advisable to test this adjustment after every experiment to ensure that there was no drift and that the potentials to which the oocyte had been clamped were correct. The electrodes are then positioned with the manipulators as depicted in Fig. 2. The use of the following (or any equivalent) procedure is advisable owing to the fact that the oocytes are opaque, so that the actual site of oocyte penetration is not visible, particularly when using an inverted microscope.

The positioning is done as follows. First, the plane even with the bottom of the oocyte is brought into focus (make use of particles or scratches on the bottom of the chamber to focus properly). Then the rim of the oocyte is brought into focus, and the distance (*a* in Fig. 2) can be estimated by the amount of refocusing necessary. Then the focus is brought up by about the same distance (*b* in Fig. 2) so that the focal plane is even with the top rim of the oocyte. The electrodes are brought into the positions shown in the bottom part of Fig. 2, where the electrode tips should be in focus. Axial movement of the electrodes will now ensure that the electrodes will penetrate the oocyte correctly. Obviously, for this procedure to work, it is important to have the ability to move the electrodes axially under a tilt of about 45 deg. To prevent oocytes from moving away during penetration, the oocyte can be supported by initially advancing the current electrode

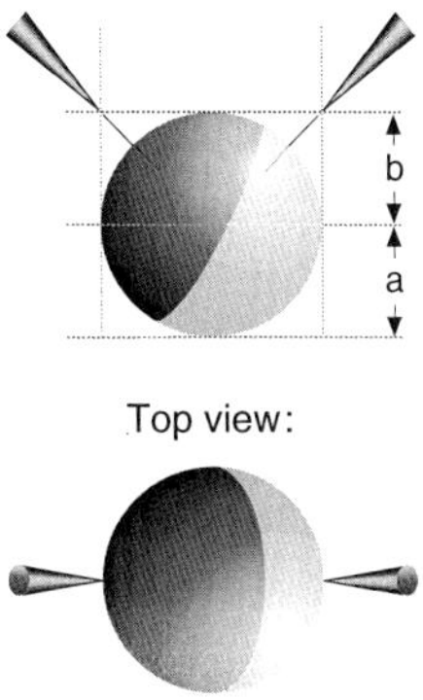

FIG. 2. Side view and top view of the oocyte with the intracellular electrodes positioned for penetration.

until the oocyte starts to move away. Then the potential electrode is inserted. A rough bottom (polypropylene mesh, Fisher, Pittsburgh, PA) or just simple scratches on the bottom of the experimental chamber are helpful in holding oocytes in place during electrode penetration. (However, the visibility using inverted microscopes will suffer great deterioration.) The whole problem of perforating the vitelline envelope with microelectrodes and fixation of oocytes to the experimental chamber can be overcome if the vitelline envelope is removed as described later for patch clamping. This also makes the membrane more accessible to changes in the extracellular solution.

Penetration of the potential electrode can be monitored through the resting potential recorded. The resting potential of oocytes can be anywhere in the range of −20 to −80 mV. Monitoring the penetration of the current electrode is more difficult. The oocyte voltage clamp from Polder allows the measurement of the potential through the current electrode in the current clamp mode. Therefore, insertion of the current electrode is monitored in a manner analogous to that of the potential electrode. The Polder amplifier also features an audio monitor for the potential measured by either electrode, greatly simplifying electrode penetration while looking through the microscope. With most other amplifiers, penetration of the current electrode is monitored using the current clamp mode. For this, a current pulse is passed through the current electrode, which causes a change in potential within the oocyte as soon as the electrode penetrates the cell. This pulsed change in cell potential is monitored through the potential electrode, which must be inserted first. The amplitude of the potential response will depend on the input resistance of the oocyte, which can be obtained by evaluating $\Delta V/\Delta I$. It will be low for leaky oocytes or for oocytes expressing large quantities of exogenous channels having high probabilities of being open under the given experimental conditions.

Once the two electrodes are inserted, the actual feedback circuit for clamping the transmembrane potential may be closed. To monitor the adjustments necessary for this step, voltage pulses of about 10 mV are given as command potentials to the voltage clamp, and the current response is observed on an oscilloscope. Before closing the clamp, the gain of the feedback loop is set to a minimal value, and the holding potential is set to the desired value. Closure of the clamp will cause the membrane potential to approach the setting of the holding potential. Holding currents needed to maintain a potential of −100 mV are in the range of 100 to 200 nA, with larger currents being an indication of a leaky oocyte. The voltage-clamp feedback gain can now be increased carefully. Higher gain will speed up the time response of the voltage clamp, as seen from the decrease in the time constant of the current response. Too much gain, however, will cause oscillation and normally irreversible damage to the oocyte and current electrode. This can be avoided by either limiting the current output or by using an automatic oscillation shutoff feature. Both possibilities are provided in the Polder clamp, and many other commercially available clamps have provisions for current limiting.

Improving Frequency Response

The price for improved frequency response is an increase in noise. Therefore, no more gain than necessary should be used. The frequency response can be improved by higher gain, using series resistance compensation (not routinely necessary) and compensating for the potential electrode capacitance (capacitance neutralization). Electrode capacitance can be minimized by using a minimal solution level, just sufficient to cover the oocyte. Capacitive coupling between the two recording electrodes should be avoided, but the laborious procedures used for small cells like silver paint coating and subsequent isolation of the current electrode (effectively providing a grounded shield down to the very tip, see Refs. 21 and 22) are normally not required. For critical wide bandwidth applications, a simple grounded metallic shield around the current electrode reaching as close to the solution surface as possible is usually sufficient to reduce capacitive coupling between the electrodes.

Capacitance transients can be largely compensated by adding the differentiated command voltage step with appropriate amplitude and time constant to the current trace. Two components are normally sufficient to compensate more than 90% of the capacitance transients. This compensation is only needed when recording from fast, voltage-activated channels and will in general decrease the signal-to-noise (*S/N*) ratio.

Another source of slow and unfaithful response in two-electrode voltage-clamped oocytes is the resistance of the cytoplasm, which contributes to the series resistance. When passing current, there will be a potential gradient between the tip of the intracellular electrode and the oocyte membrane. The intracellular potential electrode will sample a voltage anywhere between these two potentials, and will therefore not be correct. Assuming an even distribution of ion channels and of the cytoplasm resistance, this error would be abolished by placing the current electrode in the center of the oocyte and the potential electrode close to the membrane. However, this theoretical situation is practically not possible due to other constraints such as the nucleus, electrode damage to the membrane on deep penetrations, etc. One approach to the problem is the use of a switched voltage clamp, where the potential is measured only after switching off the current injection. This technique requires careful adjustment of all the time constants involved. Once setup, it allows the measurement of very fast events such as sodium gating currents.

Improving Signal-to-Noise Ratio

For slower applications, several procedures may be helpful in improving the *S/N* ratio. For example, the bandwidth of the feedback loop can be limited. Some amplifiers provide a feedback limited in bandwidth to 10 Hz; other units provide a setting having a high dc feedback gain. Otherwise, the lowest gain setting can be used. This, however, has the disadvantage that the clamp error will increase. The optimal solution in this case is to use an integrating feedback, which will eliminate the feedback error. This feature can even be used for fast potential-gated channels if the rise time of the pulse is adjusted accordingly. The Polder amplifier provides all the features described here.

Responses should be filtered using the lowest cutoff frequently possible. The limits of current resolution will depend on this filter setting. The range of measurable current amplitudes spans from several tens of nanoamperes for slow processes to several hundred nanoamperes for the fast voltage-activated channels. Another limitation of current resolution is the presence of endogenous channels and carriers in the oocyte membrane. For instance, depolarizations much above +30 mV lead to the development of a slow outward current. Part of this current arises from Ca^{2+} entry, which gates Ca^{2+}-activated Cl^- channels. Since there are two reversal potentials involved, the IV characteristics of this current is complex, but in general the outward current will be larger than the inward current, because a single Ca^{2+} ion will gate the passage of many Cl^- ions, particularly at the high electromotive during forces of large depolarizations. Several other channel types are also present to various degrees, and great care should be taken to prevent these currents from contaminating the desired signals.

Changes in Solution

In general, the two-electrode voltage clamp is stable for several hours, and extracellular solution changes are well tolerated. The perfusion can be simply gravity driven, with a suction pipette used for level control. While changing solutions manually with a Pasteur pipette or a syringe, the clamp feedback gain should be reduced, since otherwise some amplifiers tend to oscillate, particularly when close to a critical setting. Changes in fluid level will cause changes in electrode shunt capacitance and require readjustment of the capacitance compensation and neutralization. In some cases a dependence of current magnitude on solution flow and fluid level has been observed. The reason for this is not quite clear, but it could be due to a

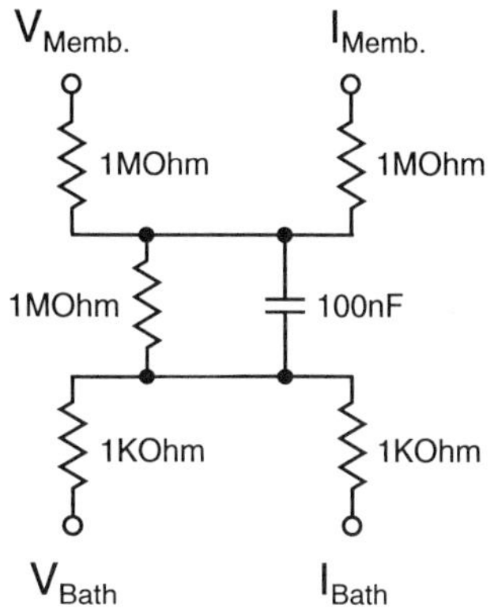

FIG. 3. Simple equivalent circuit for the electrical impedance of an oocyte. The upper part corresponds to the intracellular side.

hydrostatic access resistance dependence on the microvilli present to various degrees (depending on developmental stage) on the oocyte membrane. For critical applications, where exact control of fluid level is mandatory, a fluid level controller (e.g., MPCU, Adams and List, Westbury, NY) is very useful. The intracellular medium can be modified by injection of concentrated solutions. Up to 50 nl of solutions can be injected while under voltage clamp with injection pipettes similar to the ones described for mRNA injection, provided that the oocyte is free of the vitelline envelope. For rapid solution changes, including control of intracellular solutions, refer to the paper by Costa *et al.*[24]

Problems

In most cases, problems in recording currents from oocytes clamped with two electrodes arise from the electrodes themselves. Therefore, the possibility of measuring the electrode resistance during the experiment is very helpful. Any increase in electrode resistance deteriorates the clamp performance. This can often lead to oscillations. Further, increases (and decreases, when using capacitance neutralization) in electrode shunt capacitance can cause oscillations. This is particularly relevant when changing the solution levels. Most of the problems (and some cures) related to the electrodes and oocyte preparation were dealt with earlier.

But how can the performance of the voltage clamp itself be tested? Figure 3 is a schematic diagram of the most simple equivalent circuit of an oocyte, including its electrodes. It assumes intracellular electrode resistance

[24] A. C. Costa, J. W. Patrick, and J. A. Dani, *Biophys. J.* **67,** 395 (1994).

of 1 MΩ, and the membrane resistance (1 MΩ) and capacitance (100 nF) correspond to average values encountered in oocytes. The 1-kΩ resistances largely represent the bath electrode resistances. It is convenient to incorporate the circuit of Fig. 3 into a small (grounded) box with connections that can easily substitute the ones leading to the electrodes used in a real experiment. The electrical equivalent of an oocyte is also very helpful in getting acquainted with the voltage-clamp amplifier being used. The equivalent circuit does not emulate the diffusion potential of 50–60 mV from real electrodes filled with a 2 *M* KCl solution.

Patch-Clamp Recording from Oocytes

Prior to patch clamping on oocytes, the vitelline envelope must be removed. This is done mechanically following osmotic shrinkage of the oocyte in a solution containing (in m*M*): 200 potassium aspartate, 20 KCl, 1 $MgCl_2$, 5 EGTA–KOH, 10 HEPES–KOH, pH 7.4. After incubating for 3–10 min in this solution, the transparent vitelline envelope becomes visible and can be removed mechanically with two pairs of blunt-tipped forceps. For best visibility of the elevated vitelline envelope, the combined use of illumination from the side and below is helpful. Care should be taken to use a cold light source to avoid heating of the oocytes. In any case, a cooled stage is convenient. Extreme care must be taken to avoid any damage to the cellular membrane since the bare oocytes are very fragile; they are particularly sensitive to air exposure. After a brief wash in normal frog Ringer's solution [NFR (in m*M*): 115 NaCl, 2.5 KCl, 1.8 $NaCl_2$, 10 HEPES, pH 7.2], the oocytes are transferred to the final experimental chamber. The bare oocytes will attach to any clean surface within minutes. Thereafter, any brusque movement of the oocyte will cause membrane damage. More details of the procedure are described in Methfessel *et al.*[13]

Patch Pipettes and Macropatches

Pipettes for single-channel recording are similar to the ones used for small cells. However, a thick coating of a silicone curing agent (RTV615, General Electric, Waterford, NY), high enough up the pipette, is necessary because the large size of the oocytes requires deeper immersion of the pipette in the solution.

One main advantage of using *Xenopus* oocytes as a source of biological material rich in transmembrane proteins for electrophysiologic studies is that macropatches can be easily obtained.[25] The pipettes for macropatches

[25] W. Stühmer, C. Methfessel, B. Sakmann, M. Noda, and S. Numa, *Eur. Biophys. J.* **14,** 131 (1987).

will have tip opening diameters of 2–8 μm, giving resistances of 0.6–2 MΩ, depending also on taper shape and pipette solution. The taper should be as short as possible to avoid unnecessary access resistance. A hard aluminum silicate glass such as the one supplied by Hilgenberg (Malsfeld, Germany), or the type 7052 (e.g., A-M Systems, Everett, WA), gives smooth rims for macropipettes and low noise.

Seal Formation

The formation of a seal on oocytes is similar to seal formation in other cells for the small pipettes usually used for single-channel recording.[13] For macropatches, however, seal formation can be very slow (3–5 min). Also, the suction used to obtain seals in much smaller than when using small pipettes. It is convenient to be able to measure the pressure in the pipette. Positive pressures when entering the solution range between 10 and 40 mm H_2O; the suction during seal formation is in the range of 100–200 mm H_2O. The suction should be stopped when a seal resistance larger than 1 GΩ is achieved. The MPCU (Lorenz Messgerätebau, D-37191 Lindau, Germany) fluid level controller provides an adjustable pressure outlet that can be used to measure and control pipette pressure. A slight depolarization (10–20 mV) of the patch is sometimes helpful. For macropatches seal formation becomes particularly slow when the seal resistance crosses the 10- to 20-MΩ range. Some oocytes will form gigohm seals as quickly as small cells do. Seal resistance up to 100 GΩ can be obtained with small and large pipettes, having less than 200-fA root mean square (rms) of current noise at 3 kHz. To avoid suspended particles, the solutions used in the bath and in the pipette should be filtered.

The most disturbing conductance in some oocytes is an endogenous stretch-activated channel with a conductance of about 40 pS. It can be easily identified by applying gentle suction to the patch pipette, which will cause the channel activity to increase. The open probability is also potential dependent, increasing with increasing transmembrane potential. Concentrations of 10–100 μM gadolinium will block these channels to a large extent.[26] These channels appear to be totally absent in more than 50% of the oocytes originating from the same animal.

For long depolarizations to very positive potentials, a slow outward current develops. This current is partially carried by Cl^- through Ca^{2+}-activated Cl^- channels and by a slowly activating potassium conductance. The density and number of these channels are highly variable from oocyte to oocyte and are highly variable from one pole to the other.[27] Chloride ions

[26] X. C. Yang and F. Sachs, *Biophys. J.* **53,** 412a (1988).

[27] M. Gómez, W. Stühmer, and A. B. Parekh, *J. Physiol.* **502,** 569 (1997).

can be eliminated from the solution by substitution with methanesulfonic acid.[10,23] Alternatively, the current can be blocked by 0.3 m*M* niflumic acid or by 0.5 m*M* flufenamic acid.[28] Injection of EGTA or BAPTA (50 nl/oocyte at 0.1 *M*) also reduces the early response to inflowing Ca^{2+}. The most stringent requirements regarding the absence of any ion conductance apply when measuring gating currents. A detailed description of methods used for these particular measurements is found in [19] in this volume.

Other Configurations

Inside-out patches are obtained by abrupt withdrawal of the patch pipette from the oocyte after seal formation, in a solution normally containing high K^+ and no Ca^{2+} [e.g., (in m*M*), 100 KCl, 10 EGTA, 10 HEPES, pH 7.4]. Oocytes will survive for several hours in this solution as external medium, and several patches can be obtained from the same oocyte. Some batches of oocytes, usually from the same animal, do not survive the excision procedure. On rare occasions vesicles are formed during patch excision. This is in most cases a consequence of a slow excision. These vesicles are difficult to detect but are indicated by the appearance of a slow capacitive transient afer excision. Therefore, before patch excision, the capacitance transient should be compensated and observed carefully during excision. With small patches, it is possible to disrupt the vesicle by brief exposure to air. With macropatches, this procedure leads to destruction of the patch in most cases. A preferred method of disrupting the vesicle is to carefully advance the pipette toward a small, freshly made silicone (RTV615) sphere until a change in the capacitive transient is observed. This leads to rupture of the outer membrane of the vesicle and leaves the inside-out membrane intact in about 95% of all cases. Outside-out patches are obtained in a manner similar to the one used for small cells, for example, rupture of the patch after seal formation and slow withdrawal of the pipette. On rupture of the patch, the patch resistance will be very small due to the low resistance of the whole oocyte. Therefore, it is impossible to discern the rupture of the seal from the whole-oocyte configuration. The survival of the seal will only be assessed after the successful formation of the outside-out configuration.

Most patches will tolerate extensive solution changes. The stability of even large patches may be due to the fact that the membrane forms an omega shape within the pipette tip. The tendency to form this omega-shaped patch is increased in oocytes after long incubation times (7 or more days after injection of the mRNA). Formation of an extensive omega

[28] J. P. Leonard and S. R. Kelso, *Neuron* **4,** 53 (1990).

patch can be detected during seal formation by the development of a slow capacitance transient, which cannot be compensated well, leaving a biphasic capacitive transient at best. These patches are not suitable for the study of kinetic effects since solution changes, as well as potential changes, will not have unrestricted access to the membrane regions in close contact with the pipette wall. Also, ion accumulation effects should be considered in these cases.

Patch Cramming. The large size of *Xenopus* oocytes permits introduction of a patch-containing pipette into an oocyte. This technique is called *patch cramming.*[29] The patch can be introduced into the same oocyte from which the patch was excised, or into another oocyte which has been exposed to different conditions (*cross-cramming,* see Ref. 30). The technique is very useful to study regulation of ion channels and transporters. Simple, coarse manipulators will suffice, and the patch-containing pipette should be jammed sufficiently into the oocyte to warrant oocyte penetration. Obviously, the oocytes to be crammed have to be without a vitelline envelope. Surprisingly enough, the patch survives the cramming procedure in 80–90% of the trials, and repeated cycles of exposure of the patch to intracellular and extracellular conditions can be obtained. The latter is particularly important for controls and because the channel densities vary from one patch to another.

Cut-Open Oocyte. The disadvantages of low current amplitudes and/or inability to change the intercellular solution can be circumvented by using the cut-open oocyte technique.[24,31] In this technique, defolliculated oocytes are placed in a perspex chamber with three compartments: the bottom for recording, the middle as a guard partition, and the top for "intracellular" potential recording, current injection, and solution exchange. The top compartment is cut open and the various compartments are isolated from each other on the extracellular side of the oocyte by a vaseline seal. The method is ideally suited to record small current densities since a significant amount of membrane is voltage clamped. One disadvantage of this system is the poor voltage control of the rather large fraction of membrane under the vaseline seal. This leads to contributions to the membrane current that are poorly voltage clamped.

Gigapatches. Gigapatches or giant membrane patches are patches having diameters of 10–40 μm.[32] This technique has proven to be particularly

[29] R. H. Kramer, *Neuron* **2,** 335 (1990).

[30] A. B. Parekh, H. Terlau, and W. Stühmer, *Nature* **364,** 814 (1993).

[31] E. Perozo, D. M. Papazian, E. Stefani, and F. Bezanilla, *Biophys. J.* **62,** 160 (1992).

[32] D. W. Hilgemann, *in* "Single-Channel Recording," 2nd Ed. (B. Sakmann and E. Neher, eds.), p. 307–327. Plenum Press, New York and London (1995) or chapter [15] in this volume.

useful for the recording of currents present only at very low densities, such as currents carried by transporters.

Distribution and Amounts of Ion Channels

The distribution of channels is not uniform over the membrane of the oocyte. The magnitude of variation of the channel densities depends on the channel type. In our experience, sodium channels seem to be more uniformly distributed than potassium channels. Therefore, prediction of current size in a patch from the measured current size using a two-electrode voltage clamp is more accurate for the former channel type. Whole oocyte currents of about 10 μA should yield patch currents of the order of 50–100 pA. When recording potassium currents from oocytes (expressing more than 10 μA of whole-cell K^+ current), some patches have practically no current, whereas patches from other regions exhibit large currents. Therefore, several regions must be explored with various patches in order to localize an area of high channel density. The regions of high channel density are relatively large, so that four measurements around the accessible portion of an oocyte are usually sufficient for localizing such a region. It should also be considered that the native calcium-dependent chloride currents are distributed quite unevenly.[27]

The injection of mRNA affects the level of expression of other proteins in oocytes. For one, endogenous currents such as the calcium-activated chloride current and a hyperpolarization-activated cation-selective current are augmented on expression of heterologous membrane proteins.[33] Also, the expression of some β subunits causes increases in the level of expression of the corresponding main (α) subunits.[34] This indicates that β subunits act as chaperones for their respective main subunits.

Seasonal Variations

There are clear variations in the condition of the oocytes and the levels of protein expression. In the northern hemisphere, ion channels tend to give improved results during the winter and ionic pumps appear to function better during the summer. In general, oocytes are more amenable for patch formation during the winter. The best results obtained for voltage-dependent channels are during the period from November to April. This period surprisingly cannot be shifted by using *Xenopus* acutely imported from South Africa, controlling the temperature, light conditions, etc.

[33] T. Tzounopoulos, J. Maylie, and J. P. Adelman, *Biophys. J.* **69,** 904 (1995).

[34] K. McCormack, T. McCormack, M. Tanouye, B. Rudy, and W. Stühmer, *FEBS Lett.* **370,** 32 (1995).

Using Two-Electrode and Patch-Clamp Amplifiers Simultaneously

When, for example, particular Na^+ channels without or with only very slow inactivation are expressed at high density, the oocytes will spontaneously depolarize or even tend toward the sodium reversal potential. This persistent depolarization can be avoided during patch-clamp experiments by holding the cell at hyperpolarizing potentials with a two-electrode voltage clamp. A high holding potential will also reduce the current required to hyperpolarize the patch, particularly with low-resistance seals. However, the two-electrode voltage clamp introduces noise into the system. This noise can be drastically reduced by limiting the gain and/or the bandwidth of the feedback amplifier of the two-electrode amplifier. Some two-electrode amplifies have a mode that reduces the bandwidth of the feedback loop to 10 Hz (e.g., Polder), a speed sufficient to compensate for slow drifts in membrane potential.

Comparison of Recordings from Two-Electrode Voltage Clamps and from Macropatches in Oocytes

The two-electrode clamp of oocytes offers a series of advantages over patch-clamp recording from macropatches. It is simpler, more stable, allows recording at lower channel densities, and the extracellular solution is easily changed. In addition, it is not as sensitive to varying oocyte conditions since the formation of gigaseals is not required. However, the kinetics of fast processes cannot be resolved owing to the high cell capacitance in combination with an upper limit of resistance of the current electrode. The temporal resolution is of the order of 200–1000 μs. In addition, kinetics will depend on several parameters of the voltage-clamp amplifier (i.e., rise time limit or dc gain) as well as on the morphology of the oocyte membrane, which is extensively invaginated (microvilli). For example, extreme discrepancies can be obtained when recording from inactivating *Shaker* potassium channels in the two-electrode clamp mode. In some batches of oocytes these currents will show practically no inactivation. Currents from macropatches or ensemble averages from single-channel currents from the same oocytes will have reproducible, fast inactivating kinetics. Therefore, extreme caution must be taken when attempting to analyze channel kinetics from two-electrode data.

Macropatches achieve a high temporal resolution by electrically isolating a small area of membrane through pipettes with resistances of the order of 1 MΩ. Therefore, the temporal resolution is of the order of 50–200 μS. The kinetics of fast processes are much more reproducible, probably because the microvillar structure is made more accessible in the membrane patch under the pipette. The solutions on both sides of the membrane are well defined, and exchange of the "intracellular" solution is possible in inside-out patches. Exchange of the extracellular solution is more difficult. This requires either the perfusion of the patch pipette or recording in the outside-out configuration. Recording from single channels naturally provides more information on the actual conformational changes between the open and closed states. Macropatches allow the characterization of ionic currents equivalent to the whole-cell configuration. Therefore, each of the various recording modes has its own advantages (and disadvantages), and a combination of several modes will, in most cases, allow a detailed characterization of the parameters under study.

[17] Cut-Open Oocyte Voltage-Clamp Technique

By ENRICO STEFANI and FRANCISCO BEZANILLA

Introduction

The *Xenopus* oocyte is widely used as an expression system for ion channels.[1] We have developed the cut-open oocyte voltage-clamp technique (COVG) that has relatively low current noise (ca. 1-nA rms at 5 kHz), can charge the membrane capacity in 20–40 μs, and makes it possible to control the intracellular milieu by internal perfusion of the oocyte.[2,3] Another feature is that stable recordings lasting for several hours can be easily obtained. These properties allow the adequate resolution of the time course of fast ionic and gating charge currents. In this chapter, we describe this technique in detail.

Molecular Biology, Oocytes Preparation, and Injection of cRNAs

For K^+ channels the following clones are used: *Shaker* H4 (*Sh*H4),[4,5] the noninactivating Δ6-46 deletion mutant *Shaker* H4 IR (*Sh*H4 IR),[6] and the nonconducting versions with the W434F pore mutations (*Sh*H4 IR W434F).[7] For these clones, cRNA is synthesized from each cDNA construct by linearization of the pBluescript plasmids (Stratagene, Madison, WI) with

[1] N. Dascal, *CRC Crit. Rev. Biochem.* **22,** 317 (1987).
[2] M. Taglialatela, L. Toro, and E. Stefani, *Biophys. J.* **61,** 78 (1992).
[3] E. Stefani, L. Toro, E. Perozo, and F. Bezanilla, *Biophys. J.* **66,** 996 (1994).
[4] A. Kamb, J. Tweng-Drank, and M. A. Tanouye, *Neuron* **1,** 421 (1988).
[5] B. L. Tempel, D. M. Papazian, T. L. Schwarz, Y. L. Jan, and L. Y. Jan, *Science* **237,** 770 (1987).
[6] T. Hoshi, W. N. Zagotta, and R. W. Aldrich, *Science* **250,** 533 (1990).
[7] E. Perozo, R. MacKinnon, F. Bezanilla, and E. Stefani, *Neuron* **11,** 353 (1993).

0076-6879/98 $25.00

*Eco*RI and transcription with T7 RNA polymerase (Promega, La Jolla, CA) using mMessage mMachine (Ambion Inc., Austin TX).

For Ca^{2+} channels, the cardiac α_{1C} and the β_{2a} clones are used.[8,9] In α_{1C} the amino-terminal (ΔN60) deletion mutant is used because it can express large Ca^{2+} currents with properties similar to the full-length α_{1C} clone.[10,11] We refer to this amino-terminal deletion as α_{1C}. The plasmids containing cDNA fragments encoding the α_1 and β subunits are digested with *Hin*dIII. The linearized templates are treated with 2 μg proteinase K and 0.5% (w/v) sodium dodecyl sulfate (SDS) at 37° for 30 min, then extracted twice with phenol/chloroform, precipitated with ethanol, and resuspended in distilled water to a final concentration of 0.5 μg/μl. The cRNAs are transcribed from 0.5 μg of linearized DNA template at 37° with 10 units of T7 RNA polymerase (Boehringer Mannheim, Indianapolis, IN) in a volume of 25 μl containing 40 m*M* Tris-HCl (pH 7.2), 6 m*M* $MgCl_2$, 10 m*M* dithiothreitol (DTT), 0.4 m*M* each of ATP, GTP, CTP, and UTP, and 0.8 m*M* 7-methyl-GTP. The transcription products are extracted with phenol and chloroform, twice precipitated with ethanol, and resuspended in doubly distilled water to a final concentration of 0.2 μg/μl.

Xenopus laevis (NASCO, Modesto, CA) oocytes (stages V–VI) are used. One day before injection of the cRNA, the oocytes are collected and treated with 200 units/ml collagenase (GIBCO-BRL, Grand Island, NY) in a Ca^{2+}-free solution to remove the follicular layer. They are injected with 50 nl of cRNA at 0.05–1 μg/μl depending on the clone and batch of RNA. They are maintained at 19° in an amphibian saline solution supplemented with 50 mg/ml gentamicin (GIBCO) for 2–7 days before experiments.

Solutions

Solutions are made by mixing stock isotonic (240 mOsm) solutions of the main cation buffered at pH 7.0 with 10 m*M* 4-(2-hydroxyethyl)-1-piperazineethanesulfonic acid (HEPES). We use Glu for glutamate, MES for methanesulfonic acid, TEA for tetraethylammonium, NMDG for *N*-methyl-D-glucamine. A typical external solution is NaMES 105 m*M*, $Ca(MES)_2$ 2 m*M*, KMES 2 m*M*, and HEPES 10 m*M*. It is abbreviated as

[8] P. Ruth, A. Rohrkasten, M. Biel, E. Bosse, S. Regulla, H. E. Meyer, V. Flockerzi, and F. Hofmann, *Science* **245,** 1115 (1989).

[9] E. Perez Reyes, A. Castellano, H. S. Kim, P. Bertrand, E. Baggstrom, A. E. Lacerda, X. Y. Wei, and L. Birnbaumer, *J. Biol. Chem.* **267,** 1792 (1992).

[10] X. Y. Wei, A. Neely, R. Olcese, E. Stefani, and L. Birnbaumer, *Biophys. J.* **66,** A128 (1994).

[11] X. Wei, A. Neely, R. Olcese, W. Lang, E. Stefani, and L. Birnbaumer, *Recept. Channels* **4,** 205 (1996).

NaMES Ca2 K2. For recording gating currents in the absence of K^+, external solutions are NaMES Ca2 or NMDG-MES Ca2, whereas the solution in the bottom chamber is NaMES 110, $MgCl_2$ 2, EGTA 0.1 or NMDGMES 110, $MgCl_2$ 2, EGTA 0.1. The microelectrode solution is NaMES 2.7 *M*, Na_2-EGTA 10 m*M*, and NaCl, 10 m*M*. The microelectrode solution is K^+-free to prevent K^+ leakage from the microelectrode into the oocytes while washing the internal K^+. For Ca^{2+} channel recordings, external solutions are (in m*M*): $BaCl_2$ 10, NaCl 96, HEPES 10 (NaCl Ba10), $CaCl_2$ 5, NaCl 110, HEPES 10 (NaCl Ca5) and $Ca(MES)_2$ 5 NaNES, HEPES 10 (NaMES Ca5). To prevent contamination by Ca^{2+}- or Ba^{2+}-activated Cl^- currents prior to the recordings, 100 nl of 50 m*M* Na_4-BAPTA [1,2-bis(*o*-aminophenoxy)-ethane-*N,N,N',N'*-tetraacetic acid] is injected into the oocytes.[12] Experiments are performed at room temperature.

Electrical access to the cytoplasm of the oocyte is achieved by applying a K-Glu or NMDG-Glu solution supplemented with 0.1% of saponin in the bottom chamber. In some cases, to remove the internal K^+, the oocyte is internally perfused (1 ml/hr) with NMDG-MES or NMDG-Glu with $MgCl_2$ 2 and EGTA 0.1 added via a glass pipette (20- to 50-μM diameter at the tip) connected to a syringe pump and inserted into the bottom of the oocyte. All solutions are buffered to pH 7.0 with 10 m*M* HEPES.

Schematic Diagram of the COVG Chambers: Mounting Procedure

Figure 1A shows a schematic diagram of the COVG chamber (Fig. 1A, upper) and amplifier (Fig. 1A, lower). The oocyte is placed in a triple compartment Perspex, with a diameter of 700 μm for the top and bottom rings. Three chambers are made: (1) the upper or recording chamber with the oocyte upper domus, which has the oocyte membrane under clamp; (2) the middle chamber is the guard; and (3) the bottom chamber injects current intracellularly. The oocyte bottom can be cut or permeabilized by 0.1% saponin. The perfusing pipette is introduced into the oocyte through this bottom chamber.

Figure 1B shows an actual chamber with the microelectrode (ME) and the agar bridges (AB) in position. The main chamber (or bottom chamber, BC) is molded in heat-conductive epoxy and mounted on top of two Peltier devices (Cambion Thermoelectric Devices, Cambridge, MA) that dissipate heat on a brass block (CB) with circulating cold water. The temperature of the chamber is measured by a thermistor and compared to the desired value and the difference is amplified and fed through power transistors to the Peltier cooler (PC) in a negative feedback arrangement to maintain

[12] A. Neely, R. Olcese, X. Wei, L. Birnbaumer, and E. Stefani, *Biophys. J.* **66,** 1895 (1994).

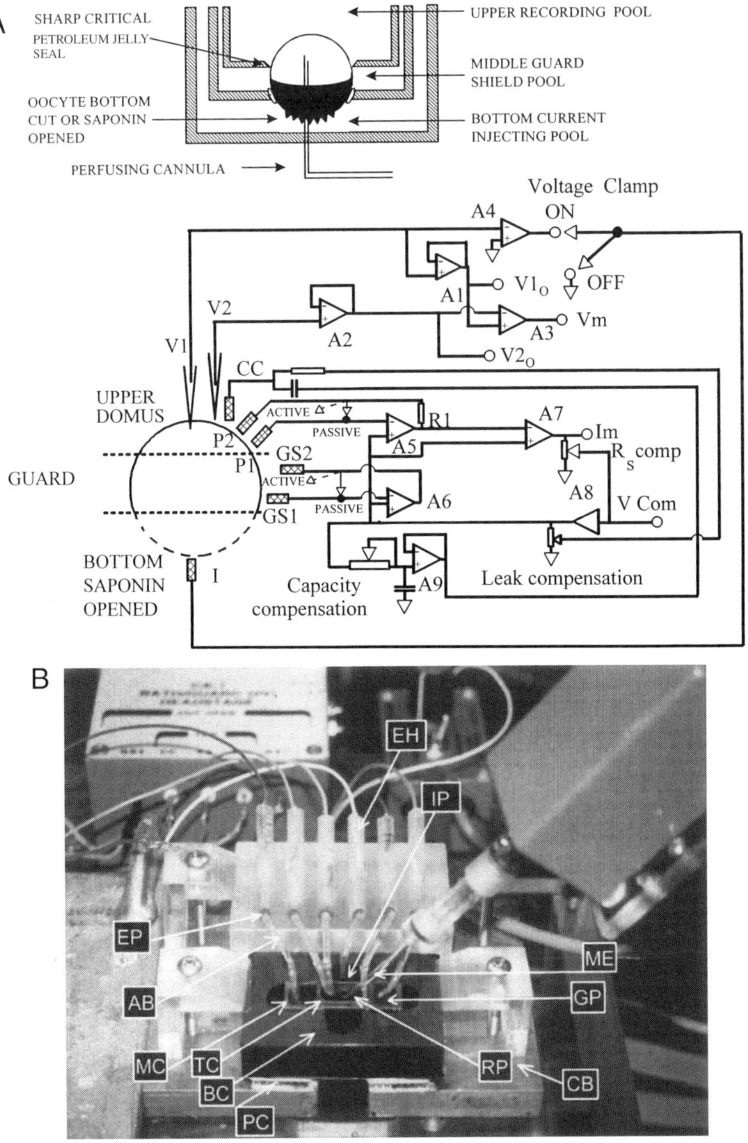

FIG. 1. (A) Schematic diagram of the circuitry and layout of the COVG technique. Control amplifiers (A1, A4, A5, and A6) were OPA 102 (Burr-Brown, Tucson, AZ). (B) Experimental chamber with electrodes in position. The dome of the oocyte is seen in the center of the recording pool (RP). AB, agar bridge; BC, bottom chamber; CB, cooling brass block; EH, electrode holder containing the silver–silver chloride coil; EP, pool with 1 *M* NaCl in content with the Ag|AgCl electrode and where one end of the agar bridge rests; GP, guard pool; IP, current injection pool; MC, middle chamber; ME, microelectrode; PC, Peltier device; RP, recording pool; TC, top chamber.

the temperature constant. The middle (MC) and the top (TC) chambers fit into the main slot of the bottom chamber.

At the beginning of the experiment, the upper division (TC) of a Perspex chamber is lifted and the chamber filled with extracellular Barth's solution. A defolliculated oocyte is placed in the hole of the middle chamber (MC) and a petroleum jelly rim (ca. 300 μm thick) is placed between the oocyte membrane and the edge of the hole. The size of this petroleum jelly rim is not critical for clamp performance. To improve the seal resistance, hydrostatic pressure is applied to the oocyte by adding solution to the upper chamber. Once the oocyte is secured, the top chamber (TC) is positioned on top of the oocyte in such a way that the hole coincides with the upper pole of the oocyte. This procedure establishes three electrically independent pools: (1) a lower pool to inject current (IP) and to have access to the cell interior, (2) a middle or guard pool (GP) to connect an electronic guard shield, and (3) an upper or recording pool (RP), which contains the oocyte membrane to clamp. Connections between pools and the electronic components are made via agar bridges (AB) and nonpolarizable Ag|AgCl electrodes immersed in 1 *M* NaCl (EP).

Schematic Diagram of COVG Amplifier

The circuit (Fig. 1A, lower) consists of three active clamps: amplifier A4 maintains the oocyte interior at ground as measured by the intracellular microelectrode tip V1, and amplifiers A5 and A6 clamp to the command pulse at the top and the middle (guard) chambers, respectively. Amplifiers A5 and A6 can be set in *passive mode* with the negative input and output short-circuited or in *active mode* with this connection open. In active mode, they function at the maximum gain. The voltage clamp is turned OFF or ON by the switch at the output of amplifier A4. In voltage-clamp OFF mode, the bottom chamber is directly connected to ground. The oocyte domus transmembrane potential is differentially recorded between V1 and V2. The microelectrodes used have a resistance of 1 MΩ or less, which sets a time constant near 1 μs in the voltage recording system with 1 pF as the input capacity of A1.

The system has five bath electrodes, which are connected with agar bridges (typically 120 m*M* NaMES in 3% agar) to individual small pools with coiled chlorided silver electrodes. The small pools with the chlorided silver electrodes are filled with 1 *M* NaCl. The silver electrodes are made of ca. 5 cm of thin (1-mm) coiled silver and are chlorided by leaving them overnight at 5% commercial bleach. The opposite end from the pools of the silver wires are directly connected to the amplifier headstage inputs. Electrodes P1 (top recording chamber) and GS1 (guard shield chamber) are voltage recording, and P2 and GS2 deliver the current to maintain P1 and GS2 at the command potential. The recording electrodes P1 and GS1 are placed as close as possible to the oocyte in order to control membrane potential reducing IR voltage drops across the solution. The recording top pool is maintained at the command potential through a current–voltage converted amplifier A5 (via P1 and P2 with R1 ranging between 0.1 and 10 MΩ), while the potential in the oocyte interior is actively clamped to ground by the feedback amplifier A4 in voltage-clamp ON position. The feedback loop in amplifier A4 is established during voltage-clamp ON. This loop is established between the intracellular microelectrode V1 and the bottom or injection pool IP via the current injection electrode I, which is in electrical continuity with the tip of the microelectrode V1 inside the cell. The electrical continuity is established by saponin permeabilization or by direct rupture of the oocyte bottom membrane. The middle (guard) pool, actively clamped to the command potential by amplifier A6 via electrodes GS1 and GS2, establishes an electrical guard between the top and the lower pools. This eliminates any voltage difference between the middle and the top compartment nulling leakage currents through the seals. Linear capacitance transients are compensated for by injecting current transients through a third bath electrode (CC) placed in the top (recording) pool (A9). The compensation has three independent time constants and amplitudes, and a phase control. The capacity compensation increases the dynamic range and prevents amplifier saturation of the control amplifiers (A4, A5, and A6). In addition, the system provides a compensation circuitry for the series resistance (R_s) in which a voltage proportional to the membrane current is added to the command pulse summing junction for A5 and A6 control amplifiers.[13] R_s is compensated by adjusting the speed of the capacity transient, which was set prior to oscillation. Typical values for R_s compensated are ca. 100 Ω as calculated from the time constant reduction of the capacity transient. With a sharp microelectrode penetration R_s compensation is generally not needed (see Fig. 4).

Agar Bridges

Figure 2 shows the electrical properties of the agar bridges used to connect the oocyte chambers with the NaCl pools. The agar bridges are made of glass capillary tubes with a 1.5-mm inner diameter. They are bent at both ends to place them into the oocyte chambers and the NaCl pools. To perform this measurement the voltage clamp was in the OFF position and passive mode. We use two bridges in series, one connected to P1 and

[13] A. L. Hodgkin, A. F. Huxley, and B. Katz, *J. Physiol.* **116,** 424 (1952).

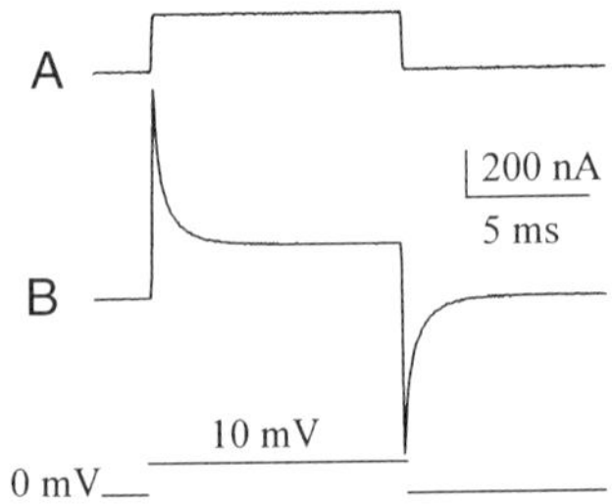

FIG. 2. Electrical properties of agar bridges. Agar bridges were filled with NaMES 120 HEPES 10 m*M* dissolved in 1% agar. The recordings were performed with the voltage clamp OFF under *passive mode* conditions with the negative input and output of A5 and A6 amplifiers short circuited. A command pulse of 10 mV was applied. (A) This trace corresponds to the current through a regular agar bridge for a 10-mV pulse. (B) The agar bridge has an internal platinum wire. Note that the introduction of the platinum wire increases the capability of passing high-frequency currents. The steady-state resistance of both bridges in series 49 kΩ, which gives about 25 kΩ per bridge. Pulse from 0 to 10 mV; bath solution NaMES Ca2 sampling time 40 μs; filter 5 kHz; R_s uncompensated; temperature 22°.

the other to I. The upper trace (Fig. 2A) corresponds to the current through the two agar bridges in series for a 10-mV pulse. In Fig. 2B both agar bridges have an internal platinized platinum wire. The introduction of the platinum wire increases the capability of passing high-frequency currents without changing the steady-state resistance, which in this case was 40 kΩ for both bridges in series. To avoid metal junction potential, the platinum wire should not be in direct contact with the pool NaCl solution or the solutions in the oocyte chamber. Because the recording bath electrodes P1 and GS1 do not pass currents, bridges with platinum wire are less critical. The bridges with platinum wire should be used for the current-injecting electrodes to improve the dynamic response of the system (P2, GS2, CC, and I).

To adjust the junction potentials, the voltage clamp should be OFF, in passive mode, with a holding potential (HP) of 0 mV. This procedure is performed prior to the mounting of the oocyte with the chambers in place. First, we place the I bridge in the bottom chamber and the P1 and P2 bridges in the top chamber, then we zero the current with the offset adjustment of P1. Second, we add the GS1 and GS2 bridges and zero the current with a GS1 offset adjustment. Both adjustments are updated until the current is zero.

Sequence of Saponin Permeabilization

The sequence of saponin permeabilization in a control oocyte is illustrated in Fig. 3, which shows voltage (v) and current (i) recordings to 40-

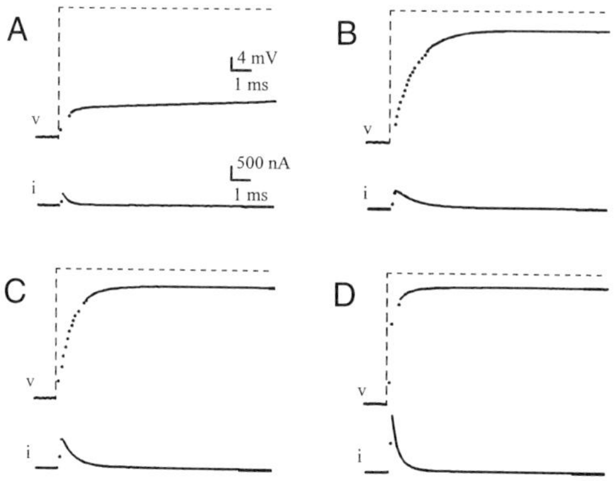

FIG. 3. Sequence of saponin permeabilization. A to C are voltage (v) and current (i) traces with voltage clamp in the active mode and turned OFF. Dashed lines are the 40-mV command pulse traces. (A) Before and (B) after saponin permeabilization. In (C) and (D), after saponin permeabilization, a black-platinized 100-μm-diameter silver wire was introduced from the bottom of the oocyte. In (C), ca. 700 μm in depth and in (D), just before rupture of the upper domus. HP = −90 mV, 40-mV command pulses. The voltage across the membrane domus was recorded between V1 and V2. Capacity transients were fitted to a single exponential function with time constants of 1.25 ms in (B), 0.76 ms in (C), and 0.40 ms in (D). Diameter of the hole in the upper chamber is 700 μm. External and guard solution NaMES Ca2; bottom solution KGLU + EGTA 1 m*M*; sampling time 50 μs per point; filter 3 kHz; 1 sweep per trace; R_s uncompensated; temperature 22°.

mV command pulses from a HP of −90 mV. To evaluate the changes in the oocyte electrical properties during the opening of the oocyte bottom, the amplifier is set in voltage-clamp OFF position and active mode. The top and guard compartments are actively clamped by A5 and A6, while the bottom compartment is connected to ground. In this configuration, prior to saponin permeabilization, the upper domus membrane is charged via the oocyte cytoplasmic resistance, which is in series with the capacity and the resistance of the oocyte membrane facing the bottom chamber. In this case, the voltage responses (Fig. 3A, v) to the applied voltage command pulses (dashed traces) are greatly attenuated by this combined access resistance. The clamp error is the difference between the command pulse (dash traces) and the voltage responses (v). In Fig. 3B, after perfusing the bottom chamber with KMES + saponin (0.1%), the electrical opening of the oocyte bottom becomes evident due to a dramatic increase of the voltage responses due to the elimination of the bottom membrane resistance. The saponin permeabilization takes about 30–60 sec. In this condition, the upper domus membrane capacity is charged via the oocyte cytoplasmic resistance in series with the I bridge resistance. The resulting resistance is called *access resistance.* In this experiment, with a capacity of 39 nF and a capacity transient time constant of 1.25 ms, the calculated access resistance is 32 kΩ, of which about 5% is from the I bridge. The I bridge in this experiment is filled with 1 *M* KCl in 3% agar and has a resistance of 2.5 kΩ. The oocyte cytoplasmic resistance is short circuited by the introduction of a vertical platinized silver wire of 100-μm diameter through the oocyte bottom. To prevent metal junction steady current, the silver wire is grounded via a 1-μF capacitor. With the penetration of the silver wire ca. 700 μm in depth (Fig. 3C) and just before rupture of the upper membrane (Fig. 3D), the capacity charging time constant measured from the capacity transient is reduced from 1.25 ms in Fig. 3B to 0.76 ms in Fig. 3C and 0.4 ms in Fig. 3D. The calculated cytoplasmic access resistance is reduced from 32 kΩ to 19.5 and 10.3 kΩ, respectively. As discussed in Figs. 4 and 5, to achieve speed it is necessary to establish an active voltage clamp in V1 with the voltage clamp ON. In this case, we clamp the potential at the tip of the recording microelectrode V1 and the major components of the access resistance will reside in the cytoplasm between the V1 microelectrode tip and the inner domus membrane surface.

Clamp Speed as Judged from Capacity Transients

The saponin permeabilization sequence shown in Fig. 3 was performed with the microelectrode inserted across the domus oocyte membrane. The routine procedure is to perform the oocyte permeabilization with V1 and

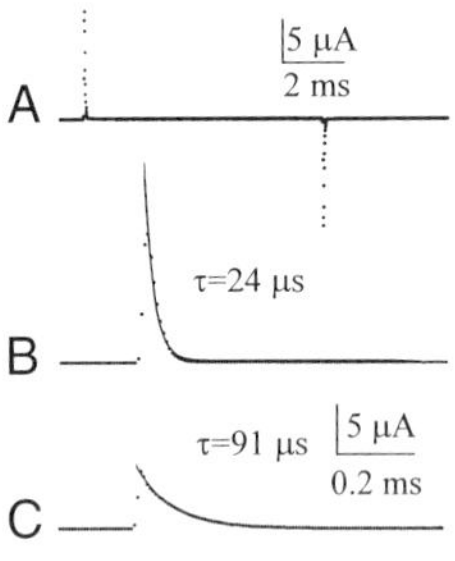

FIG. 4. Speed of the capacity transients as a function of the microelectrode penetration. Recordings in voltage clamp ON under active conditions. (A) and (B) are membrane currents for a voltage pulse from −50 to −30 mV, after a sharp penetration of V1. The charging time constant was 24 μs. In (C), V1 was introduced ca. 50 μm in depth. Note the slower time constant with the deeper microelectrode insertion. The time integrals in (A) and (B) are 578 and 567 pC, respectively, which correspond to membrane capacities of 28.9 and 28.4 nF, respectively. Diameter of the hole in the upper chamber is 700 μm, 20-mV pulse from −50 to −30 mV; external and guard solution NaMES Ca2; bottom solution KGLU + EGTA 1 m*M*; sampling time 8 μs per point; filter 25 kHz; R_s uncompensated; saponin opened oocyte; temperature 22°.

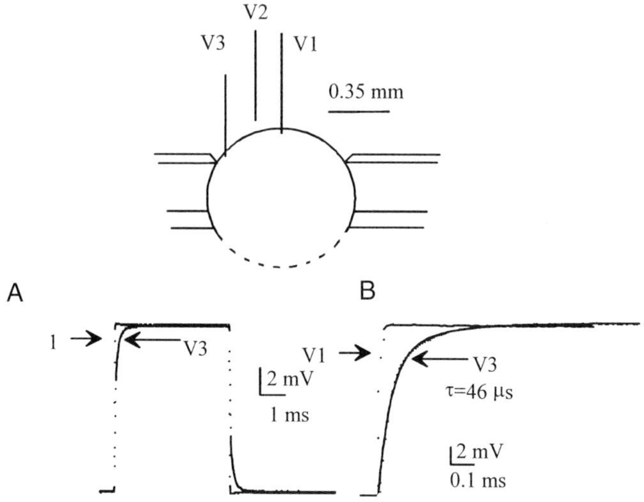

FIG. 5. Voltage inhomogeneity across the oocyte domus. The diagram shows the geometrical arrangement of the recording conditions. The voltage is recorded as V1–V2 and V3–V2. (A) and (B) (expanded scale) are the recorded transmembrane voltage responses (20-mV command pulses) of microelectrodes V1 and V3 in reference to a third external microelectrode (V2) placed in the vicinity. Diameter of the hole in the upper chamber is 700 μm. External and guard solutions NaMES Ca2; bottom solution KGlu + EGTA 1 mM; saponin opened oocyte; sampling time 10 μs per point; filter 10 kHz; four sweep averages; R_s uncompensated; saponin opened oocyte; temperature 22°.

V2 in the bath, but not V1 inside the cell. The oocyte bottom opening can be followed by the progressive increase in size of the capacity transients recorded in the active mode and with the voltage clamp OFF. Once the oocyte bottom is opened, a HP of −90 mV is typically applied and the V1 (A1 amplifier offset) and V2 (A2 amplifier offset) voltage offsets are adjusted to 90 mV in V1 and V2 outputs ($V1_o$ and $V2_o$), and V1–V2 = V_m should read 0 mV. The penetration of V1 across the domus membrane reads the membrane potential in V_m and a voltage response as in Fig. 3B, v is recorded. Typically, a transmembrane potential around −50 mV is recorded in healthy oocytes. At this moment the voltage clamp can be turned ON and the potential in V_m has to be adjusted to the HP (−90 mV in this case) by the offset of amplifier A4. In this configuration the fast capacity transients shown in Fig. 4 can be recorded.

The clamp speed is critically dependent on the sharpness of the microelectrode penetration. As previously mentioned, with the voltage clamp ON, the voltage at the tip of the V1 microelectrode is controlled, and the series resistance arises from the distributed resistance between the microelectrode tip and the oocyte membrane of the upper domus. The clamp speed can be evaluated from the time constant of decay of the capacity transient, which in optimal conditions with a 700-μm diameter domus is about 25 μs (Figs. 4A and 4B). In Fig. 4C, the intracellular microelectrode V1 is introduced ca. 50 μm in depth into the cytoplasm. The time constant of decay of the capacity transient increases from 24 to 91 μs. The time integrals of the capacity transient in Figs. 4A and 4B are maintained nearly constant at 578 and 567 pC, which corresponds to membrane capacities of 28.9 and 28.4 nF, respectively (20-mV command pulse). The effective resistance can be calculated by dividing the time constant by the capacity. The increase in the time constant from 24 to 91 μs corresponds to an increase in the effective series resistance from 0.83 to 3.1 kΩ. In summary, the value of R_s is critically dependent on the depth of penetration of the clamp control microelectrode V1 in the oocyte cytoplasm. In fact, the main cause of the variability of clamp speed depends on the sharpness and the depth of the V1 microelectrode penetration and the resulting R_s value. Thus, to achieve maximum speed and to minimize R_s, the tip of the control V1 microelectrode has to be as close as possible to inner oocyte membrane. The geometrical conditions, as in Fig. 5 (see next section), set a limit of about 25 μs for the capacity transient time constant and about 50 μs to achieve spatial uniformity in the absence of large ionic currents. These values for the clamp speed should be adequate to obtain reliable measurements of oocyte ionic and gating currents.

Voltage In-Homogeneity in the Oocyte Upper Domus

Figure 5 illustrates typical recordings of passive properties obtained with the COVG technique and a scheme of the microelectrode positioning. Records are obtained in the absence of R_s compensation. A shows voltage recordings between the intracellular microelectrodes V1 and V3 in reference to V2 located just outside the oocyte membrane (B, expanded scale). The time constant of the capacity transient is 24 μs (Fig. 4). The voltage response in V1 (voltage-clamp control microelectrode) follows the command pulse and reaches steady state in three 8-μs sample points. In contrast, the voltage response in V3 placed close to the edge has a slower time constant (46 μs). This observation indicates that the recorded membrane currents are generated from membrane regions that reached isopotentiality between a few microseconds and 46 μs after the beginning of the pulse. As a consequence, the recorded current is a weighted average of currents collected at different distances from the V1 controlling microelectrode. The capacity currents that cross the membrane near V1 have a faster time constant than the capacity currents that cross the membrane further away near the edge of the domus. This radial decay of voltage during high frequencies from the V1 electrode explains why the decay time constant of the capacity current is faster (24 μs) than the time constant of the voltage jump at V3 (46 μs). Spatial uniformity can be improved by reducing the diameter of the hole in the upper recording chamber; however, it makes the edge effects more important. Typical values for a 700-μm upper whole diameter are 14- to 32-μs time constant for the capacity transient.

Nonlinear Capacity Currents in Control Oocytes

Records in Fig. 6A illustrate the linearity of the oocyte membrane and the virtual absence of gating currents and/or nonlinear capacity components (<5 nA) in noninjected oocytes in the explored voltage range (−140 to +50 mV). For comparison, peak gating currents from oocytes expressing *Shaker* K^+ channels can measure at 0 mV, up to 10 μA (see Fig. 8). In some batches of oocytes (10 out of more than 100) we recorded small (5- to 25-nA) nonlinear currents that were blocked by 100 μM ouabain,

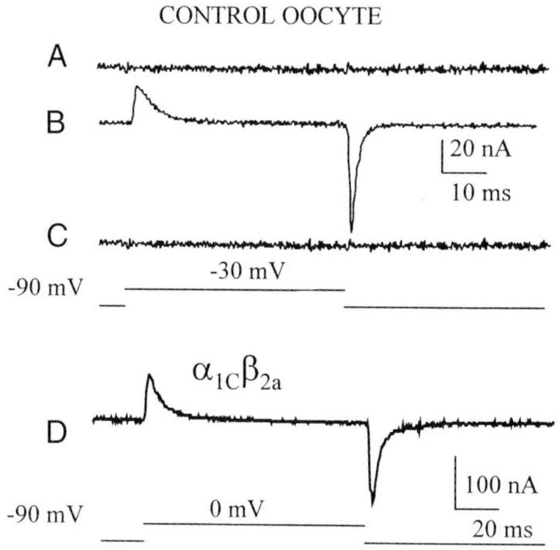

FIG. 6. Endogenous nonlinear capacity in control oocytes and gating currents in Ca^{2+} channel expressing oocytes. (A) Control oocyte. Subtracted records (HP = −90 mV, P/−4, and SHP = −120 mV) in external NaMES Ca2. (B) Control oocyte from another batch. Nonlinear transient currents to −30-mV pulse potential in subtracted records (HP = −90 mV, *P*/−4, and SHP = −120 mV). The external solution was NaMES Ca2 and the bottom solution KGLU EGTA. (C) Same pulse and conditions as in part (B), but after the addition of 100 μM ouabain to the external solution. (D) Gating currents in $\alpha_{1C}\beta_{2a}$ injected oocyte in NaMES Co 2; Pulse from −90 to 0 mV, P/−4, and SHP = −90 mV). Diameter of the hole in the upper chamber is 700 μm. In all records, bottom solution KGLU + EGTA 1 mM; sampling time 100 μs per point; filter 2 kHz; R_s uncompensated; saponin opened oocyte; temperature 22°.

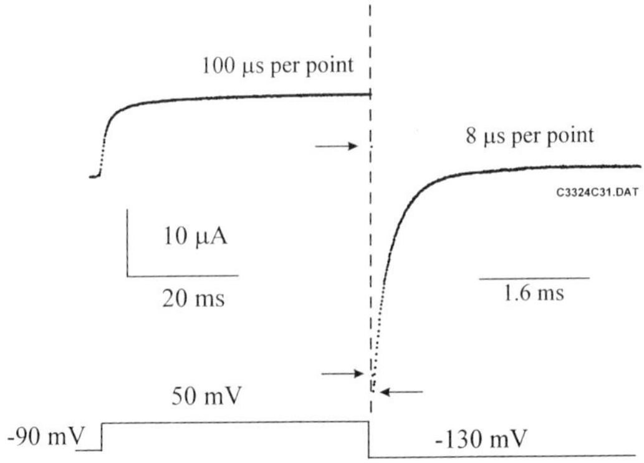

FIG. 7. Clamp speed measured from the jump of tail currents in isotonic external K^+. The trace is from *Sh*H4-IR in external KMES Ca2. The pulse is from −90 to 50 mV and returned to −130 mV; P/−4 and SHP = −130 mV. The record was acquired at 100 μs per point (before the vertical dashed line) and at 8 μs per point (after the vertical dashed line), and filtered at 20 kHz. The peak tail current is established in three sample points (horizontal arrows). Diameter of the hole in the upper chamber is 700 μm. External and guard solution KMES Ca2; bottom solution KGLU + EGTA 1 m*M*; saponin opened oocyte; R_s uncompensated; temperature 22°.

and suppressed by replacing external Na^+ with K^+ or TEA (Figs. 6B and C); these currents have been attributed to the Na/K pump.[14] Due to their small size these nonlinear currents, when present, cannot significantly contaminate large K^+ channel gating charge records. However, they can contaminate in a significant way small gating currents like the ones recorded from Ca^{2+} channels as shown in Fig. 6D. Thus, it is advisable when recording gating currents, to check for these endogenous gating currents in control oocytes, and the use of Na-free salines or ouabain.

The records also illustrate the low current noise, typically 1.2-nA rms at 5 kHz for a domus membrane capacity of 40 nF with a capacity noise ratio of ca. 40 F/A. An equivalent level can be measured with giant patches, typically at 5 kHz 0.5-pA rms for a 10-pF capacity that gives a capacity noise ratio of ca. 20 F/A.

K^+ Tail Currents in High External K^+

Figure 7 shows KMES Ca2 ionic currents elicited by pulses to 50 mV followed by postpulses to −130 mV. The measurements of tail currents can give a critical evaluation of the clamp settling time and spatial unifor-

[14] R. F. Rakowski, *J. Gen. Physiol.* **101,** 117 (1993).

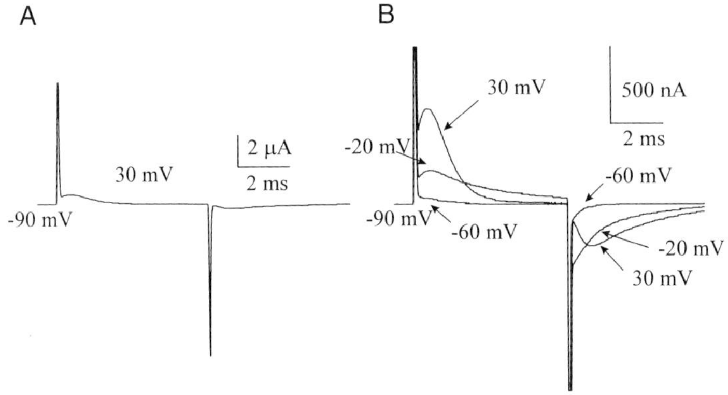

FIG. 8. Unsubtracted gating currents in *Shaker* K^+ channels. Traces are unsubtracted gating current records from *Sh*H4-IR-W434F clone; HP = −90 mV. (A) Pulse to 30 mV at a lower gain to illustrate to full capacity transient. (B) Pulse to −60, −20, and 30 mV at higher gain with truncated capacity transients. Diameter of the hole in the upper chamber is 700 μm. External and guard solution NaMESCa2; bottom solution KGLU + EGTA 1 m*M*; saponin opened oocyte; temperature 22°. Sampling time 12 μs per point; filter 10 kHz; 12 sweep averages per trace. R_s was minimally compensated.

mity. The record is acquired at 100 μs per point (before the vertical dashed line) and at 8 μs per point (after the vertical dashed line), and filtered at 20 kHz. Note that the current jump at the end of the pulse to −130 mV is quasi-instantaneous with few sampling points. This current jump is a good indication of the system speed. Tail currents are instantaneous at the voltage jump and the traces do not show humps, which indicate that the membrane potential during the postpulse is uniform.

Unsubtracted Gating Charge Records

To record charge movement we use the mutation W434F, which we have shown eliminates ionic conduction in *Sh*H4-IR clones without major effects on gating current voltage dependence and kinetics.[15] Large gating currents (1–10 μA at 0 mV) could be recorded 2–4 days after injection of *Sh*H4-IR-W434F cRNA. The large expression of these channels in conjunction with the high time resolution of the clamp allowed the recording of unsubstracted gating currents after a large fraction of the linear membrane capacity has been charged. Figure 8 illustrates the main properties of *Sh*H4-IR-W434F K^+ gating currents in unsubtracted records at a fast time resolution, from HP = −90 mV to different potentials. Figure 8A shows the

[15] E. Perozo, R. MacKinnon, F. Bezanilla, and E. Stefani, *Neuron* **11**, 353 (1993).

capacity transient followed by the gating currents at a lower gain to illustrate the full capacity transient. Figure 8B shows records to −60, −20, and 30 mV at higher gain with truncated capacity transients. Gating current records show in the ON transients a monotonic decay for the pulse to −60 mV, and the appearance of a defined rising phase at −20- and 30-mV pulse potentials. The decay of the OFF transients becomes progressively slower with depolarizing pulses and shows an OFF rising phase for the pulse to 30 mV.

Gating Currents After Analog Compensation of Linear Resistive and Capacity Components

Linear components can be analogically compensated in a linear voltage region. The linear voltage region depends on the expressed clone. For *Shaker* K^+ channels the compensation of the linear components is performed by holding the potential at −140 mV or close to 0 mV. After the compensation, the cell is repolarized to a HP of −90 mV. The analog compensation can be switched ON and OFF. The record in Fig. 8B is without compensation. The analog compensation is also used with control subtracting pulses since it prevents amplifier saturation.

Recordings of K^+ Channel Gating Currents After Removing Internal K^+ by Intracellular Perfusion

Traces in Fig. 9 are records from a *Sh*H4 clone at different times (left numbers) after starting the internal perfusion with NMDGMES $MgCl_2$ 2 and 0.1 EGTA at 1 ml/hr via a glass pipette with a 50-μm diameter tip to a syringe pump, inserted into the bottom of the oocyte. The solutions used are K^+ free. External and guard solution is NaMES Ca2 and the bottom solution is NMDGMES $MgCl_2$ 2 and EGTA 0.1. The solutions in the three chambers are frequently replaced to eliminate the K^+ contamination from the oocyte. The intracellular electrode is filled with NaMES 2.7 *M*, NaCl 10 m*M*, EGTA 10 m*M*, and HEPES 10 m*M*. The oocyte was saponin opened prior to the introduction the perfusing pipette through the oocyte bottom. The perfusing pipette with a constant flow rate of 1 ml/hr was mounted in a micromanipulator and gradually inserted through the oocyte bottom. The traces show the progressive reduction of the outward ionic K^+ currents and the unmasking of the ON and OFF gating currents.

Gating Currents from Cut-Open Oocyte and Cell-Attached Macropatches

Figure 10 displays subtracted ($P/-4$, SHP = −120 mV) with the COVG technique in external NaMES Ca2 (Fig. 10A) and subtracted ($P/4$, SHP =

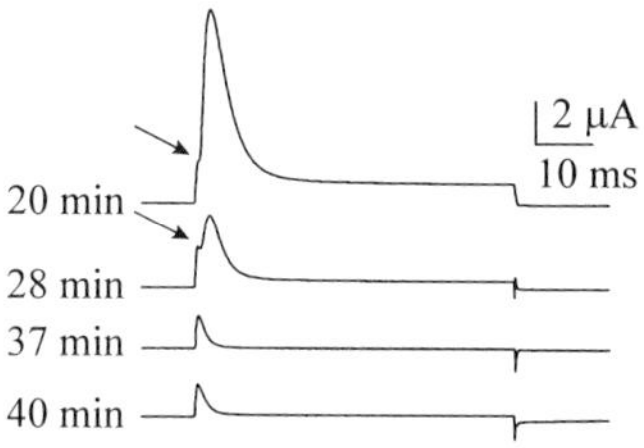

FIG. 9. Recordings of K^+ channel gating currents after removing internal K^+ by intracellular perfusion. Traces are records from the *Sh*H4 clone at different times (left numbers) after starting the internal perfusion with NMDGMES $MgCl_2$ 2 and 0.1 EGTA at (1 ml/hr) via a glass pipette with a 50-μm diameter tip to a syringe pump, inserted into the bottom of the oocyte. The intracellular electrode was filled with NaMES 2.7 *M*, NaCl 10 m*M*, EGTA 10 m*M*, and HEPES 10 m*M*. Pulses from −90 to 0 mV; unsubtracted records with linear components analogically compensated. The arrows mark the separation between the gating and ionic currents. Diameter of the hole in the upper chamber is 700 μm. External and guard solution NaMESCa2; bottom solution NMDGMES $MgCl_2$ 2 and EGTA 0.1. The oocyte was saponin opened prior to the introduction of the perfusing pipette through the oocyte bottom. Sampling time 40 μs per point; filter 5 kHz; R_s uncompensated; temperature 22°.

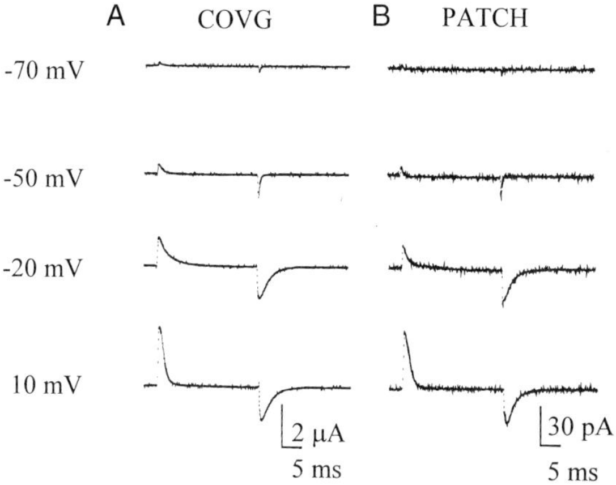

FIG. 10. Comparison of gating currents in K^+ channels recorded with the COVG and patch-clamp techniques. Traces are gating current records from the *Sh*H4-IR-W434F clone; HP = −90 mV to different test potentials. When using the COVG technique (A) the external and guard solution was NaMES Ca2 and the bottom solution KGLU + EGTA 1 m*M*; saponin opened oocyte. Diameter of the hole in the upper chamber is 700 μm. In cell-attached macropatch the bath solution was KMES + EGTA 1 m*M* and the pipette solution NaMESCa2. Sampling time 50 μs per point; filter 5 kHz. In (A) $P/-4$, SHP = −120 mV; in (B) $P/4$, SHP = 20 mV. R_s was minimally compensated in (A). Temperature 22°.

20 mV) in a cell-attached macropatch (6-μm internal diameter). The bath solution is isotonic K^+ (KMES + EGTA 1 m*M*) that sets the membrane potential close to 0 mV. The solution in the patch pipette is NaMES Ca2. Gating current records obtained in cell-attached macropatches are very similar in their time course and voltage dependency to those obtained with the COVG. For example, for a pulse to 10 mV and at 22°, the decay time constant is 1.4 ms with COVG and 1.7 ms with macropatch. Gating current records with the COVG technique show, however, some differences from those obtained with macropatches after excision; namely, the OFF gating current becomes progressively slower to the extent that in some cases it is difficult to separate it from the baseline noise.

Ca^{2+} Channel Currents After Removing Contaminating Ca^{2+}-Activated Cl^- Currents

Because prominent Ca^{2+}-activated Cl^- currents in *Xenopus* oocytes contaminate Ca^{2+} currents when Ca^{2+} is used as the charge carrier, we

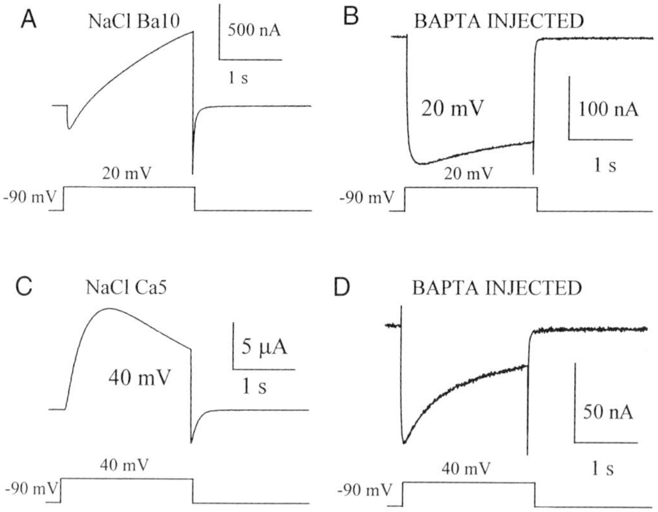

FIG. 11. Isolation of Ca^{2+} currents from contaminating Ca^{2+}-activated Cl^- currents in Ca^{2+} channel expressing oocytes. Traces are from oocytes expressing the $\alpha_{1C}\beta_{2a}$ clone. *Left*: (A and C) are from non-BAPTA-injected oocytes in external 10 m*M* Ba^{2+} (A) and 5 m*M* Ca^{2+} (C). *Right*: After the injection of 100 nl of 50 m*M* BAPTA(Na)$_4$ in 10 m*M* Ba^{2+} (B) and 5 m*M* Ca^{2+} (D). Linear components were subtracted by the P/−4 method from −90 mV. Diameter of the hole in the upper chamber is 700 μm. External and guard solutions NaCl Ba10 in parts (A) and (B) and NaCl Ca5 in parts (C) and (D); bottom solution KGLU + EGTA 1 m*M*; sampling time 2 μs per point; filter 1 kHz; R_s uncompensated; saponin opened oocyte; temperature 22°.

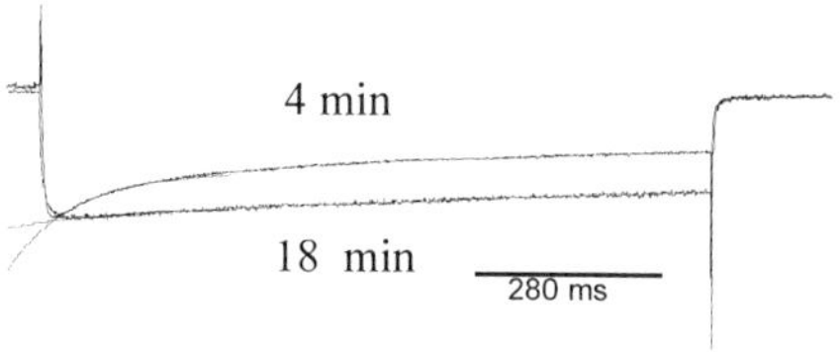

FIG. 12. Removal of Ca^{2+}-dependent inactivation by the intracellular perfusion of a fast Ca^{2+} chelator. Superimposed currents at 4 and 18 min after starting the perfusion with BAPTA 500 m*M* at the speed of 1 ml/hr. The currents were elicited by pulsing to 10 mV from a HP of −90 mV in an oocyte injected with $\alpha_{1C}\beta_{2a}$ (P/−4, SHP = −90 mV). Diameter of the hole in the upper chamber is 700 μm. External and guard solution NaMES Ca10; bottom solution KGLU + EGTA 1 m*M*; sampling time 2 ms per point; filter 1 kHz; saponin opened oocyte; temperature 22°.

eliminate Ca^{2+}-activated Cl^- currents by injecting a fast Ca^{2+} channel chelator BAPTA. Using Cl^- in the external solution (NaCl Ba10) results in the activation of outward currents preceded by an inward current (Fig. 11A), and, as expected for Ca^{2+}-dependent Cl^- currents, replacing 10 m*M* Ba^{2+} with 5 m*M* Ca^{2+} (NaCl Ca5, Fig. 11C) greatly potentiates these currents. Even in the absence of external Cl^- and with Ba^{2+} as the charge carrier, inward currents can be contaminated by the outward movement of Cl^-. This contamination is more prominent during tail current measurements at negative potentials due to the large outward driving force for intracellular Cl^-.[16]

The experiments in Fig. 11 show that the injection of the Ca^{2+} buffer BAPTA prevents the activation of Cl^- currents in oocytes expressing cardiac Ca^{2+} channels. In Figs. 11B and 11D, Ba^{2+} and Ca^{2+} currents from an oocyte injected with 37 nl of 50 m*M* BAPTA (approximately 1850 pmol) are recorded in the presence of external Cl^-. With intracellular BAPTA, even in the presence of extracellular Cl^-, there are no outward currents during the pulse, and deactivation tail currents are fast, indicating the currents are no longer contaminated by Cl^- influx. In conclusion, these experiments indicate that unless internal free Ca^{2+} is reduced by a Ca^{2+} buffer, Ba^{2+} as well as Ca^{2+} currents can be contaminated by the activation of endogenous Cl^- currents, which can take place even in the absence of external Cl^-. In these conditions inward Ca^{2+} currents are not maintained during the pulse, unmasking a Ca^{2+}-dependent inactivation process.

Removal of Ca^{2+}-Dependent Inactivation in $\alpha_{1C}\beta_{2a}$ Currents by Intracellular BAPTA Perfusion

The perfusion system can be used to test the accessibility of the Ca^{2+} inactivating site to Ca^{2+} buffers. Figure 12 shows the effect of perfusing

[16] A. Neely, R. Olcese, X. Wei, L. Birnbaumer and E. Stefani, *Biophys. J.* **66,** 1895 (1994).

high BAPTA concentration on the Ca^{2+}-dependent inactivation rates. The internal perfusion is started after saponin permeabilization (BAPTA 500 mM at the speed of 1 ml/hr). As the perfusion progresses, Ca^{2+} current decay becomes slower. The records in Fig. 12 are superimposed scaled traces at different times after the perfusion.

Summary and Conclusions

In this article, we described the use and limitations of the COVG technique. The advantages of the present method as follows:

1. *High-frequency response and low noise recording* (24-μs time constant and 1.2-nA rms at 5 kHz). This allows the accurate description of small gating currents and fast activation or deactivation of ionic currents to be adequately resolved. Currents up to 20–30 μA can be adequately clamped.
2. *Stable recording conditions lasting for several hours.* Even though this technique has internal access, rundown is minimized compared to excised patches.
3. *Access to the cell interior.* The intracellular medium can be exchanged with various solutions. This was of invaluable help in recording K^+ gating currents, an experimental condition that required the complete blockade of all ionic currents. This advantage will make the present method suitable for the study of channel modulation by second messengers and drugs. In addition, channel selectivity properties can be defined by ion substitution experiments.
4. *Channel localization.* A study of the membrane compartmentalization of channels can be easily obtained with this method, since it allows the voltage clamping of different regions of the oocyte surface.

Acknowledgments

This work was supported by NIH grants GM52203 and AR38970 to E.S. GM30376 to F.B.

[18] Cut-Open Recording Techniques

By SHUJI KANEKO, AKINORI AKAIKE, and MASAMICHI SATOH

Introduction

Xenopus oocytes are one of the most widely utilized expression systems for voltage-dependent ion channels. The main electrophysiologic technique currently used is the two-microelectrode voltage-clamp method for the conventional recording of whole oocyte currents. This method allows recording of large current responses (>10 μA) by means of high-gain amplifiers and low-resistance electrodes for current injection. However, the kinetics of fast processes cannot be resolved owing to the high cell capacitance in combination with an upper limit of resistance of the current electrode. The kinetics are also affected by several parameters of the voltage-clamp amplifier and by spontaneous increases in the microelectrode resistance during the long times required for recording.

Alternatively, patch-clamp current recording[1] can be applied to oocytes for the detection of single- or multichannel activities. However, it is technically difficult and time-consuming to remove the vitelline envelope by shrinking oocytes in a high-osmotic solution to form gigaseal patches directly on the plasma membrane. Moreover, the gigaseal patches are difficult to maintain for long periods of time.

This chapter describes a new technique for voltage-clamp recording of *Xenopus* oocytes that enables fast and stable monitoring of transmembrane current as well as intracellular perfusion by cutting a part of the plasma membrane to open the cytoplasm to the recording chamber, which is filled with an artificial intracellular solution. The original method was developed and described by E. Stefani and colleagues,[2] and we have made a modification by exchanging intracellular solution using a push–pull cannula. These are the main features of this technique:

1. *High-frequency response and relatively low current noise.* Fast activation and deactivation of ionic currents can be precisely evaluated.
2. *Stable recording conditions lasting for several hours.* This is a marked advantage over the conventional and patch-clamp recording.

[1] C. Methfessel, V. Witzemann, T. Takahashi, M. Mishina, S. Numa, and B. Sakmann, *Pflügers Arch.* **407,** 577 (1986).
[2] M. Taglialatela, L. Toro, and E. Stefani, *Biophys. J.* **61,** 78 (1992).

0076-6879/98 $25.00

3. *Control of the ionic composition of both the internal and external media.* Although it is possible to study the action of compounds on the internal surface of oocytes using inside-out patches or by injecting drugs under a whole-cell clamp, the cut-open configuration enables the desired manipulation of the internal milieu of oocytes easily.

These features allow reliable measurements of gating currents and fast ionic currents of *Shaker* K^+ channels in single and unsubtracted traces.[3,4] Moreover, the easy access to the intracellular milieu is suitable for the study of channel modulation by second messengers and drugs. This method has been applied for the studies of transient K^+ channels,[5] rapid delayed rectifier K^+ channels,[6] voltage-dependent Ca^{2+} channels,[7] Na^+-glucose cotransporters,[8] and Na^+-independent amino acid transporters.[9]

Cut-Open Procedures

The cut-open technique is based on partitioning an oocyte into three compartments (Fig. 1). The bottom compartment is used to access the inside of the oocyte, which is held at zero potential (virtual ground). The middle and top compartments are voltage clamped to the same command voltage, but each by a separate voltage-clamp circuit with separate sets of electrodes. Membrane currents are recorded only from the top compartment. The middle compartment serves as a guard to restrict possible voltage error to the border between the middle and bottom compartments. Because the inside of the oocyte is held at virtual ground and the outside is voltage clamped to get the correct membrane potential, the command voltage applied to the middle and top compartments is inverted from the actual external command input.

Intracellular dialysis can be done by permeabilizing membranes facing the bottom compartment with saponin (0.1–0.5%), or more actively by using a perfusing glass pipette, which is attached to a syringe pump and inserted into the oocyte from the chamber bottom. However, the use of a push–pull cannula is highly recommended to avoid the "explosion" of the

[3] E. Perozo, D. M. Papazian, E. Stefani, and F. Bezanilla, *Biophys. J.* **62,** 160 (1992).

[4] F. Bezanilla, E. Perozo, D. M. Papazian, and E. Stefani, *Science* **254,** 679 (1991).

[5] R. L. Rasmusson, Y. Zhang, D. L. Campbell, M. B. Comer, R. C. Castellino, S. Liu, and H. C. Strauss, *J. Physiol. Lond.* **485.1,** 59 (1995).

[6] S. Wang, M. J. Morales, S. Liu, H. C. Strauss, and R. L. Ramusson, *FEBS Lett.* **389,** 167 (1996).

[7] T. Schneider, X. Wei, R. Olcese, J. L. Constantin, A. Neely, P. Palade, E. Perez-Reyes, N. Qin, J. Zhou, and G. D. Crawford, *Recept. Channels* **2,** 255 (1994).

[8] X.-Z. Chen, M. J. Coady, and J.-Y. Lapointe, *Biophys. J.* **71,** 2544 (1996).

[9] M. J. Coady, F. Jalal, X.-Z. Chen, G. Lemay, A. Berteloot, and J.-Y. Lapointe, *FEBS Lett.* **356,** 174 (1994).

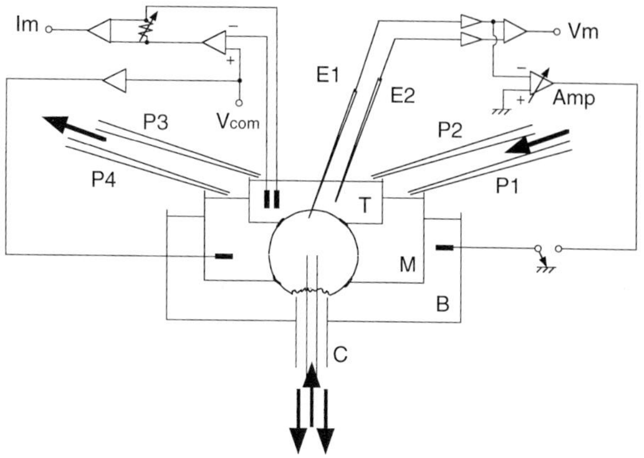

FIG. 1. Schematic drawing of the perfusion chamber and electronic circuits. A transverse section of an oocyte mounted between the top (T) and middle (M) compartments is shown. A push–pull cannula(C) is inserted into the oocyte to cut open the cytoplasm to the bottom (B) chamber where internal solution is filled. The two rims of chambers touching oocyte membranes are sealed with thin layer of petroleum jelly, and the three compartments (T, M, and B) are electrically separated. The surface of oocyte exposed to the top pool is voltage clamped to ground by means of two glass electrodes (E1 and E2; for monitoring the transmembrane potential, V_m) and a high-gain differential amplifier (Amp). Membrane current (I_m) is measured by another differential amplifier as a voltage drop across the feedback resistance. The middle pool is also voltage clamped to the command potential (V_{com}) to improve electrical isolation. The top and middle pools are continuously perfused with external solution by two inlet (P1 and P2) and two outlet (P3 and P4) pipettes.

oocyte during the perfusion by resealing of a ruptured membrane with the pipette and a consequent increase in the intracellular pressure.

Instruments

The Dagan CA-1a amplifier is the commercial system designed for cut-open recording. A Perspex perfusion chamber, Ag|AgCl electrodes and manifold wells are also available from Dagan. Two-electrode voltage clamping can also be performed by this system.

For basic current recording:

- Amplifier: Dagan (Minneapolis, MN) oocyte clamp CA-1a
- Chamber: CCP-2D (with needle holder) or CC-1D (without needle holder) Ag|AgCl electrodes and manifold wells: PL-6
- Ordinary electrophysiologic equipment (including oscilloscope, pulse generator, glass electrode puller, mechanical micromanipulator, and data acquisition system)
- Set of agar bridges (e.g., 2–3% agar in 1 *M* NaCl) connecting the chamber pools and manifold wells

The options for intracellular perfusion:

Push–pull cannula: Plastic Products Company (Roanoke, VA) C315ICP

Microinjection pump: BAS Inc. (Tokyo, Japan) CMA100 with push–pull adapter

Oocyte Preparation

Procedures for isolation of oocytes from *Xenopus laevis,* RNA injection, and culture of oocytes are well-documented elsewhere.[10] Fully grown, stage V–VI oocytes are suitable for placing onto the holes of the CCP-2D chamber. Follicular cells surrounding oocytes should be carefully removed before recording by collagenase treatment. If the follicular layer surrounding oocyte is still present, remove it with fine forceps (No. 5, Dumont & Fils, Montignex, Switzerland).

Initial Settings Prior to Mounting

1. Assemble the perfusion chambers (Fig. 2). If you use the push–pull cannula for internal perfusion, the tip of the cannula should be placed just below the hole of the middle chamber so as not to break the oocyte during mounting. Fill all of the compartments of the chamber with external solution. The flow rate of the external solution should be set to a minimum.

2. Set AgCl electrodes in the manifold wells and fill them with salt solution (e.g., 1 *M* NaCl). Connect AgCl electrode cables to the headstages of the amplifier. Connect the wells in the electrode manifold to the chamber assembly with agar bridges as indicated in the manual for the CA-1a amplifier.

3. Prepare two glass microelectrodes filled with 3 *M* KCl. The resistance should be 100–200 kΩ for best clamp performance. Such an electrode can be made by scratching the electrode tip with a piece of Kimwipe gently. The extracellular electrode (E2 in Fig. 1) can be omitted.

4. Set aside the agar bridges dipped in the top pool. Remove the top pool from the chamber assembly until an oocyte is mounted.

5. With an 1-ml syringe and 26-gauge, polished-tipped needle filled with petroleum grease, apply a thin annulus of grease around the hole in the floor of the middle chamber.

6. The bottom rim of the top chamber hole should be covered with a very thin layer of petroleum jelly. The thickness (approximately 50 μm) of this second petroleum jelly rim is critical to achieve an optimal frequency

[10] H. Soreq and S. Seidman, *Methods Enzymol.* **207,** 225 (1992).

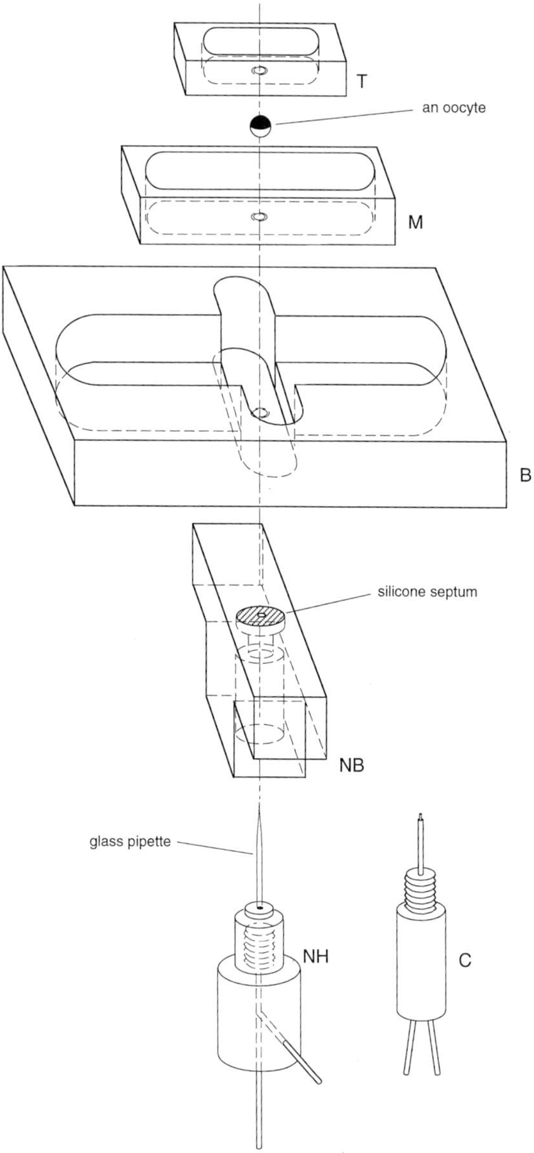

FIG. 2. Cut-open chamber and push–pull cannula. The Dagan CCP-2D Perspex chamber is composed of a basement having a bottom chamber (B), two boats for top (T) and middle (M) pools having small holes for oocyte mounting, a needle blanket (NB) fixed to the basement by studs, and a needle holder (NH) for intracellular perfusion. A push–pull cannula for brain microdialysis (C) can be used as an alternative to the genuine needle holder to improve intracellular perfusion (see text).

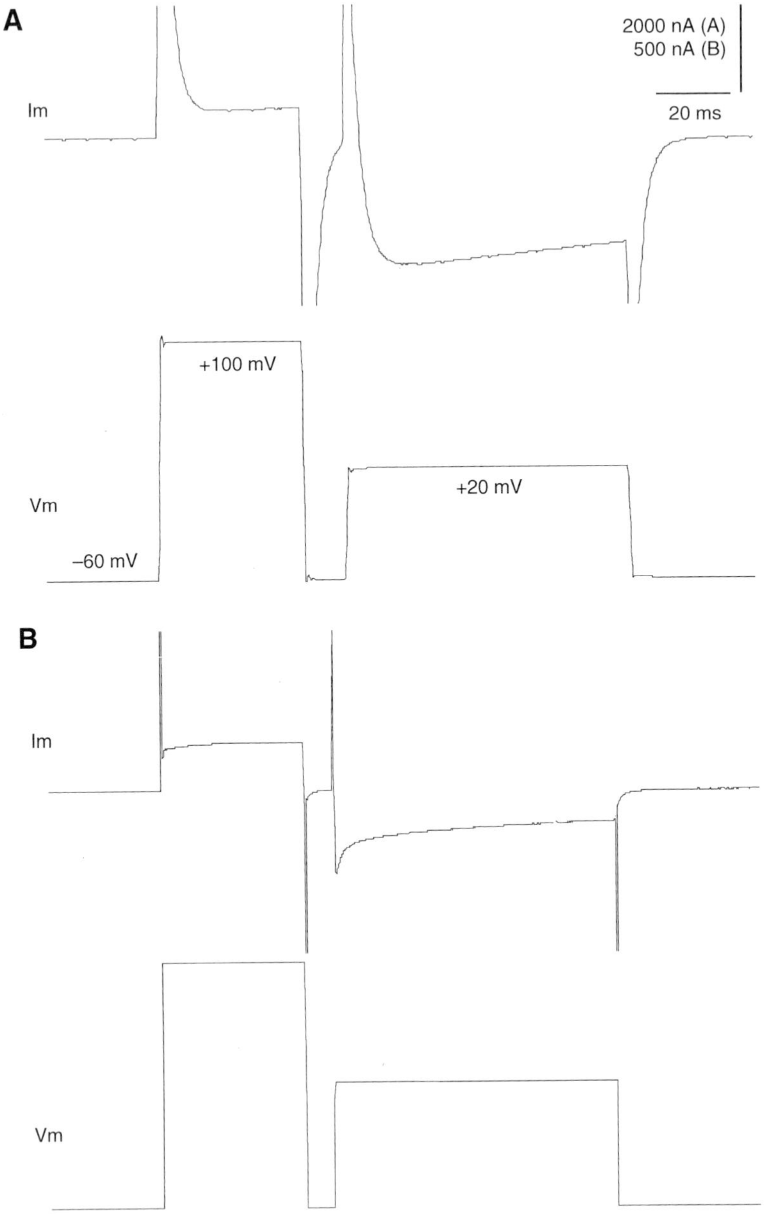
A
2000 nA (A)
500 nA (B)
20 ms
Im
+100 mV
Vm
+20 mV
–60 mV
B
Im
Vm

response of the clamp and to prevent the artifactual recording of currents arising in the oocyte membrane under this seal.[2]

Mounting the Oocyte

1. Select an oocyte and move it to the middle pool of the chamber assembly. Place the oocyte onto the petroleum jelly donut, pigmented side (animal hemisphere) up, so you can see the tips of the glass electrode well in the dark background.

2. Observe the I_m signal output on the oscilloscope. This is done by switching TEST to ON on the front panel of the CA-1a amplifier.

3. Gently press the oocyte to seat it onto the hole. The I_m signal will become smaller as the seal resistance increases.

4. Place the top pool into the chamber and position the hole directly above the oocyte. Fill the top pool with external solution. Push the top pool with your finger until it comes into contact with the oocyte.

5. Reposition the agar bridges as they were. Start the perfusion of external solution (2 ml/min in our experiment). The I_m signal should be at a minimum now (indicating maximal resistance between pools).

Cut-Open and Permeabilizing the Oocyte

1. Exchange the lower pool solution with internal experimental solution. The solution change is easily done with a Pasteur pipette.

2. Carefully screw up the push–pull cannula until it breaks the bottom of the oocyte.

3. Turn on the microinjection pump. The optimal flow rate for rapid intracellular dialysis is 2 μl/min. Check that the volume of oocyte does not change.

Impaling and Voltage Clamping the Oocyte

1. Mount two microelectrodes to headstages and position the tip in the top pool near the oocytes. Check the resistance of the microelectrodes (<1 MΩ).

FIG. 3. Nonsubtracted and noncompensated traces of membrane potentials and transmembrane currents flowing through voltage-dependent Ca^{2+} channels expressed in *Xenopus* oocytes recorded by the (A) whole-cell and (B) cut-open methods. Recordings were made by the whole-cell clamp method using Nihon Kohden CEZ-1200 amplifier (compliance voltage $\pm$100 V) or by the cut-open method using a Dagan CA-1a amplifier. Oocytes were injected with RNAs for Ca^{2+} channel α_{1A} and β subunits, and clamped in 50Ba solution. The same test command pulses from -60 to $+20$ mV and a preceding prepulse to $+100$ mV were applied. Note the difference in the wave form of I_m and the speed and accuracy of the voltage clamp.

2. Adjust the voltage offset of the two electrodes to the negative of holding potential. Note that the meter should read +60 mV if the desired holding potential is −60 mV since the top pool is always voltage clamped to 0 mV.

3. Impale the oocyte with one electrode. The difference between the voltage of two electrodes is the oocyte membrane potential.

4. Activate the voltage clamp and start recording.

Recording of Voltage-Dependent Ca^{2+} Channel Current

We have used the cut-open recording to investigate the intracellular modulation of N-type Ca^{2+} channels by G protein $\beta\gamma$ subunits.[11] The intracellular perfusion technique reveals that tonic inhibition of N-type channels by G protein is washed away by replacing cytosol with Ca^{2+}-free solution, and that the disinhibition is partly inhibited by GTP.

Experimental Procedures

Recording was made using *Xenopus* oocytes coexpressing rabbit BIII (voltage-dependent Ca^{2+} channel α_{1B} subunit) and human Ca^{2+} channel β subunit of β_{1B} subtype. The solutions used in the present study were of the following composition (m*M*):

Frog Ringer's solution for the recording of endogenous Cl^- channel current: NaCl 115, KCl 2.5, $CaCl_2$ 1.8, HEPES 10, pH 7.4, with NaOH

50Ba solution for the recording of Ca^{2+} channel current carried by Ba^{2+}: $Ba(OH)_2$ 50, Tris base 50, CsOH 5, pH 7.6, with methanesulfonic acid

Intracellular solution for the recording of Ca^{2+} channel current: CsOH 110, EGTA 10, $CaCl_2$ 1, $MgCl_2$ 1.6 HEPES 10, pH 7.4, with methanesulfonic acid

Comparison Between Conventional and Cut-Open Recording

Figure 3 shows a comparison of the speed and accuracy of the voltage clamp between conventional whole-cell clamping and cut-open recording. In the whole oocyte recording (Fig. 3A), an inadequate feedback phase setting induced a small overshoot in the V_m trace and a large capacitive discharge in the I_m trace. Because it took about 5 ms to reach the accurate command potentials, the activation kinetics of inward Ba^{2+} current was apparently slowed. In the cut-open recording (Fig. 3B), the voltage response followed the command pulse much more closely, so that the fast-inactivating

[11] S. Kaneko, Y. Mitani, T. Hata, M. Kikuwaka, A. Akaike, and M. Satoh, *Soc. Neurosci. Abst.* **22,** 344 (1996).

current component could be observed. According to the original report,[2] the time constant of the decay of the capacity transient should be 20–100 μs. The relatively slow charging process may result from currents arising from a poorly clamped membrane area located under the seal since the

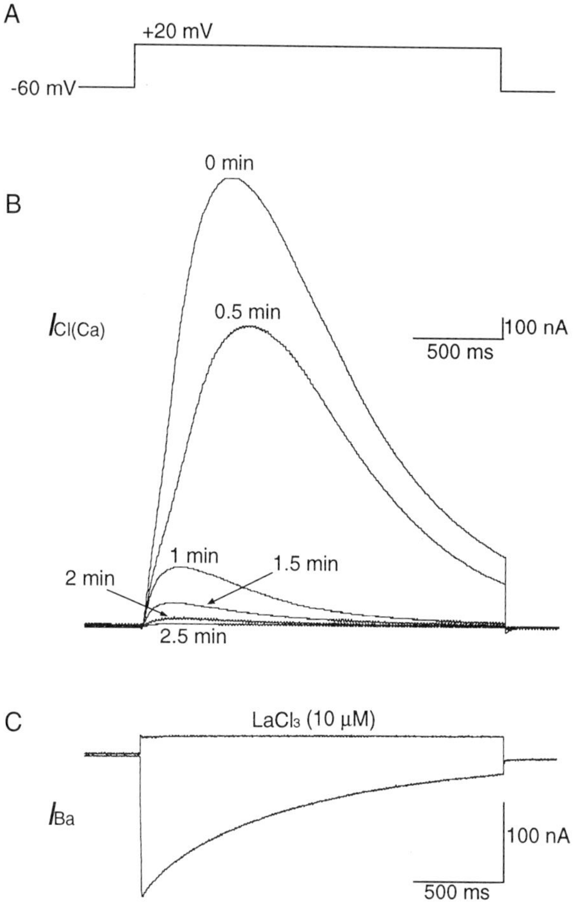

FIG. 4. Reduction rate of Ca^{2+}-mediated Cl^- current ($I_{Cl(Ca)}$) by intracellular perfusion with Cs-EGTA solution and isolation of Ba^{2+} current (I_{Ba}) in the cut-open recording from an oocyte expressing cloned N-type Ca^{2+} channels. Application of depolarizing pulse (A) to oocytes expressing Ca^{2+} channels triggers Ca^{2+} entry, which induces an outward current, $I_{Cl(Ca)}$ (B). The $I_{Cl(Ca)}$ was disappeared during the intracellular perfusion with EGTA buffer. The time-course of EGTA effect will be slower than the exchange rate of internal solution because of the slow rate of EGTA-Ca^{2+} association[14]. The La^{3+}-sensitive, Ca^{2+} channel current carried by Ba^{2+} (C) can be isolated after changing the external perfusate to 50Ba solution. Nonsubtracted and capacitive charge-compensated recordings.

petroleum jelly rim around the oocyte in the upper compartment was too thick. The series resistance and capacitive charging current, which are not processed in Fig. 3, can be subtracted or compensated using the signal conditioning circuits of the CA-1a amplifier.

Speed of Intracellular Perfusion

A large outward Cl^- current response is observed by an application of depolarizing step to the oocyte expressing voltage-dependent Ca^{2+} channels in a Ca^{2+}-containing solution. This response is designed $I_{Cl(Ca)}$ since it is mediated by an influx of Ca^{2+}, which activates endogenous Ca^{2+}/calmodu-

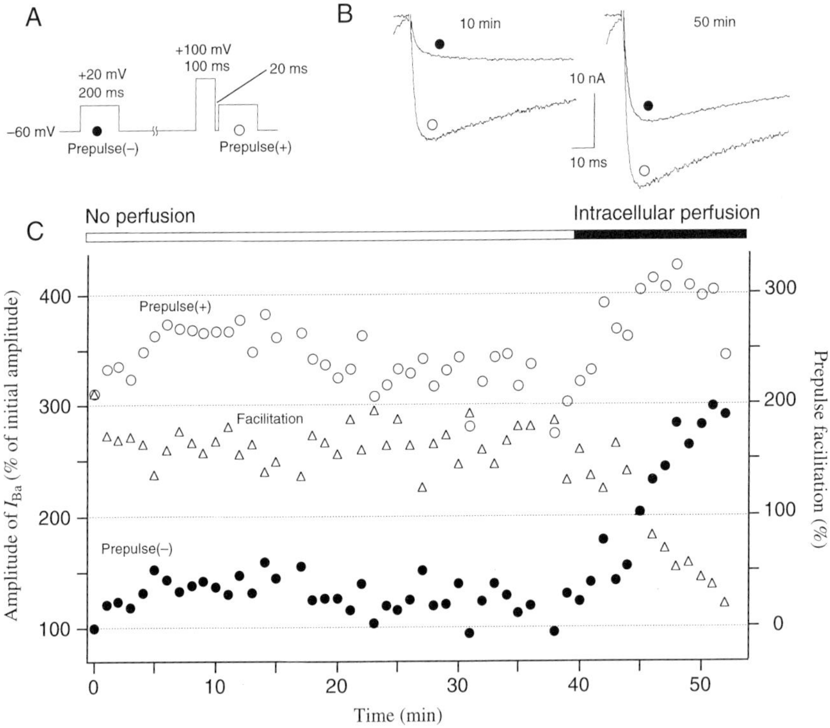

FIG. 5. Stability of prepulse facilitation of an N-type channel current before intracellular perfusion with the Cs-EGTA solution of the cut-open recording. After the bottom of oocyte was punctured by the push–pull cannula, cut-open recording of I_{Ba} was started without intracellular perfusion. (A) A large depolarizing prepulse was applied before and after recording of I_{Ba}, (B) to evaluate prepulse-induced facilitation of an N-type channel current. (C) The amplitudes of I_{Ba} and the ratio of prepulse facilitation are plotted. After a 40-min recording, intracellular perfusion with Cs-EGTA solution was started.

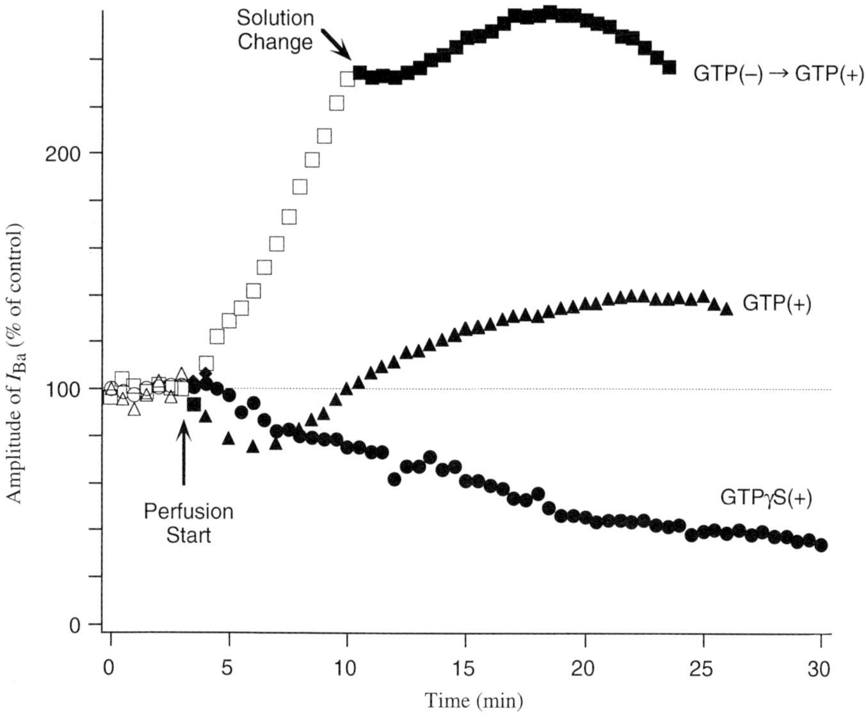

FIG. 6. Changes in the amplitude of I_{Ba} flowing through N-type channels after intracellular perfusion with or without GTP analogs (100 μM). The amplitude of an *N*-type channel current showed an immediate "run-up" by internal perfusion. Supplementation of the internal solution with GTP (100 μM) caused a slight decrease followed by a gradual increase in the amplitude. Perfusion with GTPγS (100 μM) caused irreversible activation of G proteins and a decrease in the amplitude of I_{Ba}.

lin-dependent Cl^- channels.[12,13] When $I_{Cl(Ca)}$ was recorded in the cut-open configuration and the cytoplasm of the oocyte was perfused with EGTA-containing solution by the push–pull cannula at a rate of 2 μl/min, the amplitude of $I_{Cl(Ca)}$ was rapidly decreased within 2–3 min (Fig. 4). Considering that the volume of oocyte is estimated to be about 0.5 μl and that the association kinetics of EGTA to Ca^{2+} is relatively slow,[14] the cytoplasmic solution may be completely changed within a minute.

[12] N. Dascal, B. Gillo, and Y. Lass, *J. Physiol. Lond.* **366,** 299 (1985).

[13] I. Ito, C. Hirono, S. Yamagishi, Y. Nomura, S. Kaneko, and H. Sugiyama, *J. Cell. Physiol.* **134,** 155 (1988).

[14] R. Y. Tsien, *Biochemistry* **19,** 2396 (1980).

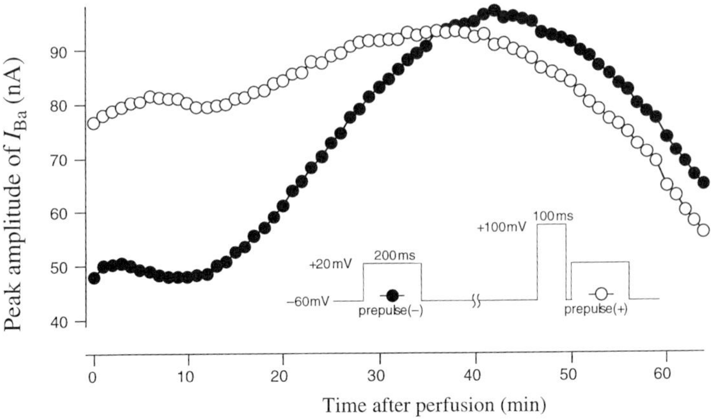

FIG. 7. Typical run-up and run-down of N-type channel current observed when oocytes were internally perfused with GTP (100 μM)-containing solution. After the run-up of the amplitude without prepulse, it became almost equivalent to that with prepulse, indicating that run-up is caused by "washout" of tonic G-protein-mediated inhibition. In contrast, run-down occurred along with the decrease in both amplitudes.

Effect of Intracellular Perfusion with GTP on Tonic Inhibition by G Proteins

When a cut-open recording was made without intracellular perfusion in an oocyte expressing N-type channels, application of a strong prepulse increased the amplitude of the Ca^{2+} channel current (I_{Ba}) and increased the speed of activation of the channel opening (Fig. 5). Because depolarization-sensitive inhibition and slowing is the specific feature of inhibition of neuronal Ca^{2+} channels by G protein $\beta\gamma$ subunits,[15] this observation indicates that exogenously expressed N-type channels are tonically inhibited and slowed by oocyte-intrinsic $G\beta\gamma$. The prepulse-induced facilitation in the amplitude was stable until cytoplasmic perfusion began. During perfusion with an internal solution, the amplitudes of I_{Ba} before prepulse were rapidly increased; however, those after prepulse were relatively unchanged, resulting in a decrease in the magnitude of prepulse facilitation. After 10 min of internal perfusion (at a time of 50 min in Fig. 5B), application of the prepulse still increased the amplitude of I_{Ba}, but did not apparently increase the speed of the activation kinetics of I_{Ba}, indicating that the inhibition of amplitude and the slowing of the activation kinetics can be separable under this condition.

[15] A. C. Dolphin, *TINS* **19,** 35 (1996).

Although the "washout" of cytoplasmic solution led to a rapid increase in the amplitude of channel current, an addition of GTP (100 μM) to the internal solution slowed the rate of increase (Fig. 6). The perfusion with GTPγS caused a continuous decrease in the amplitude of I_{Ba}. Such a decrease was also seen in the initial period of the perfusion with GTP, suggesting that an additional population of heterotrimeric G proteins was activated and dissociated by these guanine nucleotides. The relative slowness in the increase of I_{Ba} amplitude in the presence of GTP may reflect the balance between the removal of G$\beta\gamma$ from the cytoplasm by internal perfusion and GTP-induced liberation of G$\beta\gamma$ from the plasma membranes.

In the recording of voltage-dependent Ca^{2+} channels, the duration of stable recording is restricted by "run-down" of the channel current. No run-down was observed without internal perfusion of the oocyte up to 2 hr, however, when oocytes were internally perfused with the GTP-containing solution, a rapid run-down was observed 30–40 min after the beginning of perfusion (Fig. 7). The decay of I_{Ba} was not coincident with the "run-up" (washout of tonic inhibition) of I_{Ba}. Therefore, the mechanisms of Ca^{2+} channel modulation by G proteins and other unknown substances from the intracellular side can be investigated using the cut-open method with intracellular perfusion.

Acknowledgments

The authors thank Dr. Yasuo Mori for providing plasmid carrying Ca^{2+} channel α_1 subunit, and we thank former and present colleagues Takahiro Hata, Yasuyuki Mitani, Masanobu Kikuwaka, and Mariko Kinoshita. This work was supported by Grant-in-Aid from the Ministry of Education, Science, Sports and Culture, Japan.

[19] Gating Currents

By FRANCISCO BEZANILLA and ENRICO STEFANI

Introduction

Several membrane transport mechanisms are voltage dependent. These include voltage-dependent ion channels and some transporters and pumps. Regardless of the structure of the molecule, the membrane voltage exerts its effect by acting on a "voltage sensor." Hodgkin and Huxley[1] predicted the existence of a voltage sensor when they stated: "It seems difficult to

[1] A. L. Hodgkin and A. F. Huxley, *J. Physiol. (Lond.)* **117,** 500 (1952).

0076-6879/98 $25.00

escape the conclusion that the changes in ionic permeability depend on the movement of some component of the membrane which behaves as though it had a large charge or dipole moment." Thus, the voltage sensor is the machinery that mediates the transduction between the membrane potential and whatever change may occur in the membrane to initiate or terminate ion conduction.

A simple view of the voltage sensor consists of an embedded charge in the channel protein that can move in response to changes in the membrane electric field, and that movement leads to the opening or closing of the conduction pathway. An immediate consequence of the charge movement is that it generates a current and, because this current is connected to the channel gating process, it is called the *gating current.* The most general property of the gating current is that it is transient because the charge that moves is confined in the membrane field, and for long periods will remain in a stable position. In this regard, gating currents share the properties of capacitive currents and may be considered to be displacement currents produced by dielectric relaxation.

At the single-molecule level, gating currents represent the rate of displacement of individual charges or dipole reorientation. In the original Hodgkin and Huxley[1] formulation, the opening of the potassium channel occurs when four independent gating particles are located simultaneously in the active position. Furthermore, each particle can only be in one of two stable positions, the rate of translocation from the resting to the active position is increased by depolarization, and the reverse rate is increased by hyperpolarization. Figure 1 illustrates a simulation based on these assumptions. Because each charged particle may only dwell in one of two positions, the transition between them is instantaneous, producing an infinitely large current shot (δ function, see Fig. 1, single channel gating shots). The integral of this current is the total charge moved times the fraction of the field it traverses. In this simulation, the result of a single step of membrane potential started from a very negative value (-120 mV) to a positive value ($+50$ mV) is shown for the single gating shots and the resultant single-channel current. The ensemble current of shots for 1000 pulses is shown as the gating current and the resultant macroscopic current as the macroscopic ionic current. This type of simulation sets the stage for the conditions to record these different types of current from real cells because, depending on the kinetics and the amount of gating charge per channel, the relative magnitudes of the different currents vary widely.

Studying Gating Currents

When the voltage-dependent channel opens, ionic current flows can be detected during voltage clamp. If the channel had only two physical

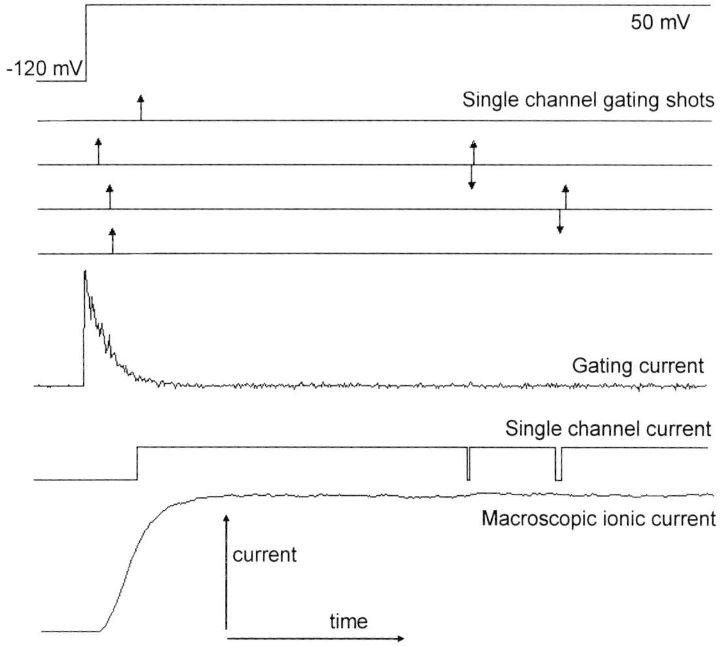

FIG. 1. The predicted ionic and gating events for a channel with four independent gating particles, each obeying first-order kinetics with voltage-dependent rate constants. The pulse goes from −120 to 50 mV, and the single-channel gating shots show the sequence of events for one trial. The resultant single-channel current is shown for the same trial. The average of the gating shots for 1000 trials is shown in the trace labeled "Gating current," while the average of the single-channel openings for the same number of trials is shown in the "Macroscopic ionic current" trace.

states—closed or open—the information obtained from recording the macroscopic ionic current would be enough to infer the voltage dependence of the transition. However, channels exhibit many physical states, most of which are several steps away from the open state. In that case, the information obtained from the macroscopic currents or even from single-channel recordings will be incomplete and inferring the properties of transitions between closed states will be more difficult the further that transition occurs from the open state. Gating currents, on the other hand, include the direct contributions of those far removed charge carrying transitions and their characterization helps in elucidating the physical states and transitions the channel undergoes from the closed to the open states. In fact, most of the charge carrying transitions occur before the actual opening of the channel; therefore, the study of gating currents is necessary if we are to obtain a full characterization of the voltage-dependent process.

Charge Per Channel

The time integral of the gating current measures the product $z\,\delta$ where z is the charge displaced and δ is the fraction of the field it traverses. The voltage dependence of a channel will depend on the total number of charges displaced, and this number has been found recently for several types of channels using two different methods. The first method makes the measurement of the total charge (Q_t) by measuring the asymptotic values of the charge versus potential (Q–V) curve from the time integral of the gating currents. The charge per channel is computed by dividing Q_t by N, the number of channels present, normally estimated by noise analysis of the ionic currents[2] or counting the channels with a specific toxin.[3] The second method estimates the charge per channel from the derivative with respect to voltage of the logarithm of the open probability versus voltage relation (P_0–V curve). This estimation is only accurate when the derivative is computed at very low values of P_0, which can be done by single-channel recording[4] or by extending the P_0–V curve using the Q–V curve.[5,6] The values measured for Na^+ and K^+ channels vary between 11 and 13 e_0.

Detection Problem

In the idealized case, the magnitude of the single-channel shot (see simulation of Fig. 1) is predicted to be a delta (δ) function. In practice, the transition will not be instantaneous, but it will appear so to the observer due to bandwidth limitations. In fact, the shots should be recorded as the impulse response of the filter used, and the predicted magnitude is about 1 fA for a 5-kHz bandwidth, which is below the resolution of present technology, if normal low-pass recording techniques are used. The ensemble of many channels will give a signal whose noise can be analyzed to infer the size of the elementary event (see later discussion).

When the recording is done from a membrane that contains a large number of channels, the contribution of the shots of the individual channels sums to a larger signal called the gating current (see Fig. 1), which can be experimentally recorded. The magnitude of the gating current depends on the number of charges moving per channel, the total number of channels, and the kinetics of the charge transition. These parameters have important consequences for the experimental detection of gating currents because

[2] N. E. Schoppa, K. McCormack, M. A. Tanouye, and F. J. Sigworth, *Science* **255,** 1712 (1992).
[3] S. K. Aggarwal and R. MacKinnon, *Neuron* **16,** 1169 (1996).
[4] B. Hirshberg, A. Rovner, M. Lieberman, and J. Patlak, *J. Gen. Physiol.* **106,** 1053 (1996).
[5] D. Sigg and F. Bezanilla, *J. Gen. Physiol.* **109,** 27 (1997).
[6] S.-A. Seoh, D. Sigg, D. M. Papazian, and F. Bezanilla, *Neuron* **16,** 1159 (1996).

present techniques can only resolve signals that are not extremely fast, due to bandwidth limitations, and not too slow because of uncertainties and noise of the baseline. For a given total charge, fast kinetics will produce a larger gating current than a slower channel. It is not surprising then that the Na^+ gating currents were recorded before the K^+ channel gating currents[7] because the K^+ channel density in squid axon is lower, and its kinetics are about 10 times slower than the Na^+ channel. After increasing the temperature by about 15°, which speeds up kinetics roughly by a factor of 10, K^+gating currents were successfully recorded.[8] The channel density is a very important factor in detecting gating currents with a good signal-to-noise ratio (*S/N*) because the noise increases with recording area. The best recordings of gating currents have been obtained from preparations with high channel density such as the squid,[7] crayfish,[9] and Myxicola[10] giant axons as well as the eel electroplax.[11]

With the advent of molecular cloning techniques, it is now possible to express cloned channels in oocytes or mammalian cells at even higher densities, and the *S/N* of currents recorded from these expression systems are superior to the native cells. In addition, the cloned channel can be modified by replacing, inserting, or deleting residues that are suspected to be important for the function of the channel. These mutated channels can be expressed and gating currents analyzed to infer how a particular residue that was replaced affected the normal operation of the channel. With this procedure, it has been possible to locate the important residues that make up the voltage sensor in *Shaker* K^+ channels.[3,6]

When the expected size of the gating currents is smaller than the noise of the recording, it is always possible to use signal averaging to improve the *S/N*. The recording of gating currents lends itself to this technique because the events are time locked with the stimulating voltage pulse, allowing the averaging of the time-locked signal in the computer memory.

Separation Problem

Once the charge per channel and the density and kinetics are favorable for detecting gating currents, they must be separated from the membrane capacitive current and the ionic currents. Frequently, these currents are much larger than the gating currents, masking them partially or completely.

[7] C. M. Armstrong and F. Bezanilla, *Nature* **242,** 459 (1973).

[8] F. Bezanilla, M. M. White, and R. E. Taylor, *Nature* **296,** 657 (1982).

[9] J. G. Starkus, B. D. Fellmeth, and M. D. Rayner, *Biophys. J.* **35,** 521 (1981).

[10] B. Rudy, *Proc. Roy. Soc. Lond. B.* **193,** 469 (1976).

[11] S. Shenkel and F. Bezanilla, *J. Gen. Physiol.* **98,** 465 (1991).

The ratio of the peak gating current to the ionic current is dependent on the kinetics of gating because, as explained earlier, gating currents will be larger for faster channels. Thus, for Na^+ channels the ratio is about 1/50 to 1/100, while for K^+ channels it is smaller than 1/200. The ionic currents can be abolished by replacing permeant ions on both sides of the membrane by species that do not pass through the channel to be measured and other channels that might be present in the membrane. Typical cation replacements are tetramethylammonium ion, cesium ion, or *N*-methylglucamine (NMG) ions. Some ions must be tested for effects on gating, such as those seen with tetraethylammonium (TEA) (see later discussion). Details of solutions used can be found in the chapter on the cut-open oocyte voltage clamp.[12] In addition to impermeant ions, it is possible to use blocking agents, but first it should be ascertained that they do not modify gating.

After the ionic currents have been eliminated, the recorded current will include some residual leakage, capacitive currents, and gating currents. The remaining leakage should be as small as possible and should be examined for time-dependent changes. Frequently, time-dependent leaks occur in very leaky membranes and make the separation of the gating current a very difficult or impossible task. The capacitive current is normally separated by using the property that the gating currents, although capacitive themselves, are nonlinearly related to voltage. This is an expected property of buried charges or dipoles that are confined to the membrane field. In contrast, the charging of the membrane capacitance is not expected to show saturation because the charging and discharging are provided by the separation of charge by the virtually infinite reservoir of ions in the solutions surrounding both sides of the membrane. The logic is simple: a positive going pulse and a negative going pulse of the same magnitude produce capacitive currents of the same magnitude and time course but of opposite polarity, regardless of at which base or holding potential (HP) the pulses are given (see Fig. 2A). Therefore, the addition of those currents, synchronized exactly with the beginning of the pulse, will give a null current. However, if a gating charge moves during the positive pulse, but it does not move during the negative pulse (because at that potential HP, the charge movement is saturated), the addition of the currents produced by P and $-$P pulses will give a net current that will reflect the movement of the gating charge during the P pulse, canceling completely the capacitive currents of the membrane. This pulse protocol (Fig. 2A) has been called the $\pm$P procedure.[7] To obtain a reliable measurement of the gating current, it is imperative that there be no gating charge movement in the negative direction from the potential HP. This condition is rarely met becauise it has been found that there is

[12] E. Stefani and F. Bezanilla, *Methods Enzymol.* **293,** [17], (1998) (this volume).

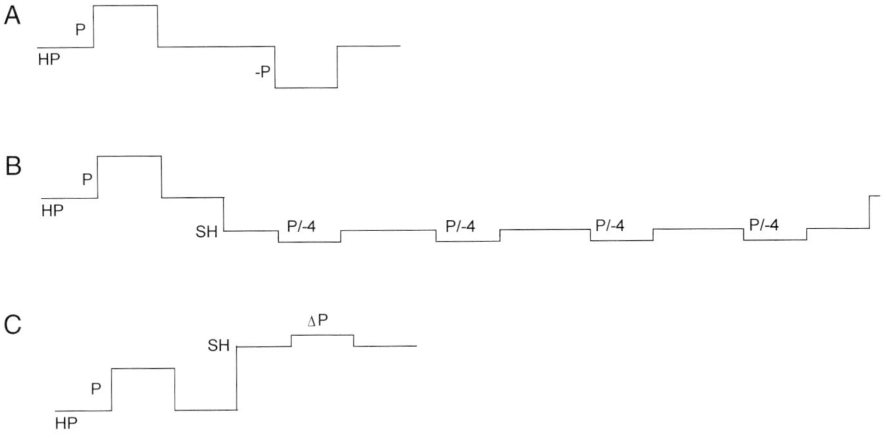

FIG. 2. Subtraction protocols for elimination of the linear capacitive current. HP represents the value of the holding potential. (A) The ±P procedure. (B) The P/−4 procedure with a negative subtracting holding potential (SH). (C) The variable scaling procedure with a positive SH. For details, see text.

a detectable amount of gating charge at potentials more negative than the resting potential, the usual value where the HP is set. To correct this problem, one could move HP to a more negative potential, but when giving a large P pulse, the corresponding −P pulse will reach prohibitive values of membrane potentials, with subsequent breakdown of the membrane. The solution is to step to values of membrane potential where there is no gating charge movement (SH) to give the subtraction pulses for the time needed, and give smaller pulses to prevent membrane breakdown. This is the basis of the P/4 procedure,[13] which has several variants. Figure 2B shows an example with pulses going in the opposite direction to the test pulse P (the P/−4 procedure). The potential at which subtracting pulses are given can be chosen to be more negative than the HP, or more positive than the HP. One should design the direction of the subtracting pulses so that they occur in the region where no gating charge moves. The reason why the P/4 procedure gives four pulses is because the current produced by a single pulse of one-fourth of the amplitude would have to be multiplied by four before it could be subtracted (or added) to the current produced by the P pulse, increasing the noise of the result. In fact, P/4 does not restore the *S/N* but it was selected as a good compromise to limit the number of pulses and the time the membrane is taken to the SH potential.

[13] F. Bezanilla and C. M. Armstrong, *J. Gen. Physiol.* **70,** 549 (1977).

The P/4 procedure can be extended to be P/n, and the SH potential may be negative or positive depending on the saturation voltages of the Q–V curve.

Another general subtraction procedure is shown in Fig. 2C, where a pulse of amplitude ΔP is given at a SH (negative or positive, depending on the position of the Q–V) and the currents produced by both pulses are stored. Later, the current produced by the ΔP pulse is scaled appropriately to do the subtraction. Frequently, this method is done by recording a single ΔP current and used off-line subtraction of any P-pulse-produced currents. This method has the disadvantage of producing extra noise because of the scaling factor when ΔP is smaller than P. It is recommended that several sizes of ΔP be recorded and signal averaging used to minimize their noise contribution. The recording of ΔP separately from each of the P recordings has the disadvantage that any drift in the time course of the capacitive transient during the experiment will affect the recording of the time course of the gating current. This last problem is not present with the P/n technique because the subtraction is done every time a P pulse is given.

When the gating current is small compared to the capacitive current, the subtraction procedure relies on extracting a small quantity from the subtraction of two large quantities. This could produce serious error because small drifts in the preparation would make the subtraction completely worthless. A typical example where this problem can be pervasive happens when recording in the whole-cell configuration. The time course of the capacity transient depends on the membrane capacity and the series resistance, and the latter depends heavily on the access resistance in the whole-cell configuration. A small change in the access resistance, which frequently occurs in experiments done in that configuration, will make the subtraction invalid.

Voltage Clamp

For the study of gating currents of expressed channels, two major techniques have been used to clamp the membrane: the patch clamp and the cut-open oocyte voltage clamp. The two-microelectrode clamp has also been used, but its poor time resolution makes it less appropriate for studying the kinetics of gating currents.

Patch Clamp

In the excised or cell-attached configuration, the patch clamp is an ideal technique because it exhibits low noise and can record currents with high bandwidth. The techniques are described in detail in other articles in this book,[14] but for the purposes of studying gating currents there are a few considerations that must be added. First, as the number of channels increases with the area, it is necessary to record from large areas, and the macropatch or the giant patch techniques are the configurations of choice. The coating of the pipettes with Sylgard for macropatches or with the oil–Parafilm mixture for the giant patches is very important to decrease the capacitance and to maintain a stable capacitance during the recording. In addition, the pipettes should be made with short shanks to minimize the access resistance. Best results are obtained with pipettes with resistance lower than 0.6 MΩ. The dynamic range of the input stage of the patch clamp is governed by the feedback resistance of the current-to-voltage converter when it is of the resistive type. Capacitive transients, even after compensation, may be much larger than the currents to be recorded. To decrease even further the magnitude of the capacitive transient, the command pulse may be slowed down but if after this is done all of these manipulations still produce saturation, the use of a lower gain for the patch-clamp headstage may be required. This is not recommended because noise will increase, leading to deterioration of the quality of the recordings. A good solution to the dynamic range problem is to use an integrating headstage and to synchronize the resetting of the integrator with a pulse preceding the acquisition period.

The attached patch is generally more stable than the excised patch configuration and, in addition, preserves the time course of the gating current by minimizing excision-induced rundown. The problem of replacing the internal solution with impermeant ions in the attached patch configuration is resolved by immersing the oocyte in the solution with impermeant ions and perforating the surface of the oocyte in several places with a pipette, letting it equilibrate to replace the internal contents.

Although it is quite simple to record gating currents with a bandwidth of 10 kHz, we have recently extended the frequency response of the patch amplifier to about 200 kHz and, using giant patches with access resistance below 0.2 MΩ, we have been able to record gating currents revealing a faster event that precedes the rising phase of the normal gating current. This faster transient has a shallow voltage dependence and carries less than 10% of the total charge (see Ref. 15).

The patch-clamp method is the only available techique to record gating current fluctuations because the number of channels can be controlled to be within a range that does not exceed the resolution of the elementary

[14] R. Levis, *Methods Enzymol.* **293,** 218 (1998) (this volume); D. Hilgemann, *Methods Enzymol.* **293,** 267 (1998) (this volume).

[15] E. Stefani and F. Bezanilla, *Biophys. J.* **70,** A143 (1996); ibid *Biophys. J.* **72,** A131 (1997).

event. For the same reason, to count the number of channels to determine charge per channel, the patch clamp is the only technique available.

Cut-Open Oocyte Voltage Clamp

This technique is described in detail in another article in this volume.[12] The large recording area of this technique allows the detection of very small gating currents that would escape detection using the patch-clamp technique. This is particularly advantageous when a mutation is introduced in the molecule under study that cuts down the expression of the protein in the membrane. In addition, it has been found that the effect of patch excision, which frequently produces rundown of the currents, is minimized in the cut-open oocyte technique.

Recording Step

Figure 3 shows a generic setup that we will use to analyze the components that can influence the quality of the recorded gating currents.

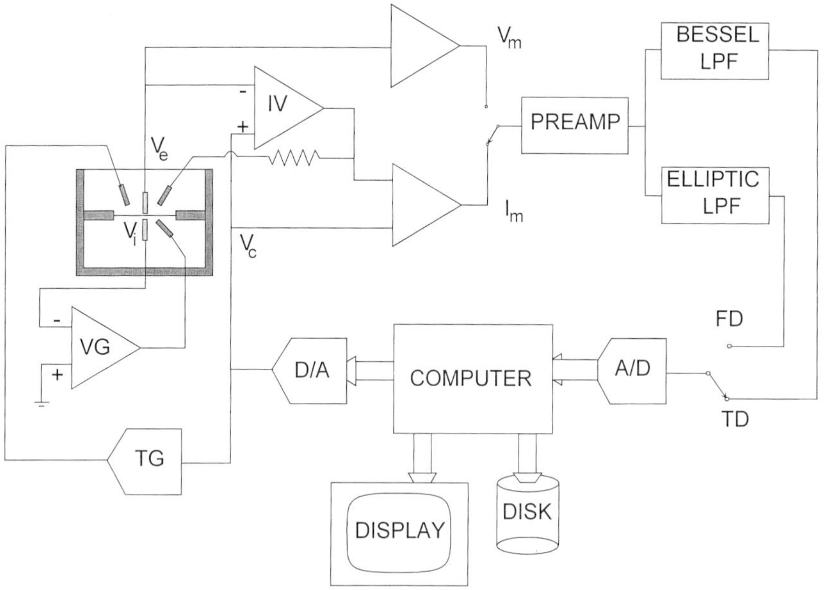

FIG. 3. Schematic diagram of a recording setup to measure gating currents. D/A, digital-to-analog converter; FD, frequency domain mode; I–V, current-to-voltage converter; TD, time domain mode; TG, transient generator; V_c, command voltage; V_e, external voltage sensing electrode; VG, virtual ground control amplifier; V_i, internal voltage sensing electrode.

Membrane and Voltage Clamp

The chamber presented on the left part of Fig. 3 is a general arrangement for a voltage clamp of a piece of membrane. There are many different configurations, depending on the type of cell and voltage clamp used; this particular arrangement is similar to the cut-open oocyte voltage clamp. VG is an amplifier that maintains the inside (bottom chamber) at virtual ground, and *I–V* is a current-to-voltage converter that imposes the command voltage on the external side of the membrane. Notice that in this configuration, the electrode that feeds back from *I–V* into the chamber is different from the voltage-sensing electrode V_e, the same is true for the voltage sensing electrode V_i and the current passing electrode at the output of VG. In a patch configuration, VG is not present and electrode V_e becomes the same as the current passing electrode: this could pose a problem for large currents. In the squid axon voltage clamp, the control is done from the inside, and the outside is maintained at virtual ground by the current-to-voltage converter.

First Amplifier

It is crucial that the current-to-voltage converter (*I–V*) amplifier have enough dynamic range so that it will not saturate during the capacity transient. In addition, it should have wide bandwidth and a fast slew rate. Saturation produces different responses for large pulses as compared to small pulses, invalidating the subtraction procedure. Saturation can be prevented by reducing the gain of the amplifier, but it has the disadvantage that the noise increases. The other possibility is to cancel some of the capacitive transient by injecting a similar current of opposite magnitude into the same side of the chamber using a transient generator (TG). Ideally, the TG is capable of generating transients of 2 or 3 exponential components that the user can manually change until the capacity transient is abolished. The transient generator is fed the same pulse that is used as the command to clamp the membrane and, as the clamp is normally slower than the generator itself, it is convenient to add a phase control to match the rising phase of the transient. The output of the transient generator is fed to the chamber through a different electrode to eliminate interactions with the recording electrode. Again, in a patch-clamp situation, this extra electrode is not implemented. Instead, the transient generator (built-in in commercial units) will feed the current into the summing junction of the current-to-voltage converter.

Preamplifiers

The output signal of the *I–V* converter may be too small to fill the dynamic range of the analog-to-digital converter (A/D) so an amplifier can be added. This amplifier must have a wide bandwidth, low noise, and excellent linearity characteristics. Saturation should be prevented and carefully checked. For frequency-domain applications it is critical that the frequency response (magnitude and phase) be preserved across the whole range of amplification of the preamplifier.

Filters

The filter should be placed after the amplifier to prevent further amplification of the intrinsic noise of the filter. However, in some cases two stages of amplification can be used with the filter between. This arrangement allows preamplification of large transients that still fit within the dynamic range of the filter and use postamplification after the filter has attenuated the large capacitive transient. In time-domain applications, the filter should be of the linear phase type (Bessel or Gaussian) to prevent ringing of transient responses, but in the frequency domain it is more important to have a sharper cutoff and an elliptic filter can be used. In any case, to prevent aliasing errors, the filter must be set to a 3-dB cutoff frequency no more than one-half the sampling frequency, but because the cutoff is gradual, it is better to use a cutoff at about one-fifth of the sampling frequency. In the frequency domain a value close to one-half of the sampling frequency can be used because the frequency response of elliptic filters is sharper. An exact value of the cutoff to eliminate alisiang can be calculated knowing the frequency response of the filter and matching the frequency at which the attenuation of the signal is just beyond the resolution of the A/D converter.

Data Acquisition

Acquisition is carried out with an A/D converter which must be preceded by a sample-and-hold amplifier for typical successive approximation or double ranging converters. This is a critical part of the circuitry and must be selected to be linear, monotonic, and of high resolution. Best results are obtained with 16-bit converters because they allow a larger dynamic range than the typically used 12-bit converters. High resolution is particularly important for subtraction of large transients to extract small gating currents, even after the dynamic range has been enlarged by using a transient generator, and it is imperative for extracting information from noise analysis of gating currents (see later discussion). The signal should be preamplified to maximize the converter dynamic range.

Waveform Generation

The command signal applied to the clamp is generated by a D/A converter. Important characteristics for this converter follow:

1. *Low noise.* Any noise produced by the D/A converter will be differentiated by the membrane capacitance, producing extra noise in the recordings.
2. *Linearity.* Because this is the source for the P and P/*n* pulses, the linearity is critical to achieve good subtraction.
3. *High-frequency response and slew rate.* Again, this is important to match the rising phase of the pulses, regardless of their size, to produce reliable subtraction.
4. *Low glitch power.* Glitches are produced by the asynchronous switching of the bits that form the digital word and they occur with different magnitudes at different values of the input word. Glitches can produce enormous spikes, which may be present in the P pulse and not in the P/*n* pulse, or vice versa, making the subtraction totally unreliable and frequently producing spurious signals that may be confused with gating currents.

Computer Control

Typically, the interface between the analog setup and the digital domain is done under program control using a personal computer. The timing of the A/D and D/A converters must be under the control of a master clock that runs independently of the software. This is critical because timing loops in software, especially in multitasking operating systems, will make the timing completely unreliable. The clock running the A/D conversion can be different from the clock running the D/A converter, but both must be derived from a common source. It is easy to see that if the pulse generator jitters with respect to the acquisition, even by one clock pulse, it can have disastrous consequences in the subtraction, making it impossible to record gating currents.

System Performance

The simplest and most reliable way of testing the setup is to replace the membrane with a model circuit that simulates the membrane resistance, capacitance, series resistance, and electrode resistances. This model circuit is then subjected to the same pulse protocols that are used in the live preparation. Thus, using a P/*n* procedure it should produce a null trace. Lost sample points at the beginning and end of the pulse should be investigated carefully to trace their origin to saturation or nonlinearities of one or more of the components in the system. By measuring the output of each one of the stages, it is possible to trace where there might be a nonlinearity or saturation. One common pitfall is to look at the output of the filter and decide that is still within its dynamic range, but it is quite common for the input of the filter to be saturated by the incoming signal but look smooth on its output because of its filtering action. Another, more subtle, problem is pickup from the fast rising and fall of the command pulse into one of the inputs of the system. This can happen by capactive coupling or by imperfect ground return (ground loops) in the system setup. An especially bad situation occurs when the analog return is drained through a ground that drains digital signals because the contribution of the digital noise will not be proportional to the size of the pulse and the P/*n* procedure will not cancel properly.

Recording of Gating Currents

In native cells, the recording of gating currents is complicated by the presence of other channels, making it difficult to separate the current of the channel of interest. In this regard, the squid axon is a simple preparation because the two channels present at high density have very different kinetics and temperature changes can be used to separate them.[16] In the squid axon the sodium pump is also expressed at very high densities, but its displacement current can be easily dissected from the Na^+ and K^+ gating currents by excluding ATP from the internal medium. The pump transient currents can be studied separately by recording the currents before and after the addition of ouabain, which subtracts the Na^+ and K^+ gating currents.[17]

The study of gating currents is easier in expression systems where the channel being studied is made predominant in the membrane. Molecular biology techniques have made possible the ability to put the coding region of channels in expression vectors so that they can be used by cells as DNA or RNA, for synthesis of the protein and insertion into the membrane. By selecting cells that have a very low background of intrinsic channels, it is possible to study the expressed channel in practical isolation, especially when the expression is optimized. Many channels have been expressed in this way, but the gating currents of only a few channels have been studied. Brain sodium channel gating currents expressed in *Xenopus* oocytes were reported by Conti and Stuhmer.[18] *Shaker* K^+ channel gating currents, also expressed in *Xenopus* oocytes, were reported by Bezanilla *et al.*[19] and calcium channel gating currents were reported by Neely *et al.*[20]

[16] F. Bezanilla, *J. Membr. Biol.* **88,** 97 (1985).
[17] J. Wagg, M. Holmgren, D. C. Gadsby, F. Bezanilla, R. F. Rakowski, and P. De Weer, *Biophys. J.* **70,** A19 (1996).
[18] F. Conti and W. Stuhmer, *Eur. Biophys. J.* **17,** 53 (1989).
[19] F. Bezanilla, E. Perozo, D. M. Papazian, and E. Stefani, *Science* **254,** 679 (1991).
[20] A. Neely, X. Wei, R. Olcese, L. Birnbaumer, and E. Stefani, *Science* **262,** 575 (1994).

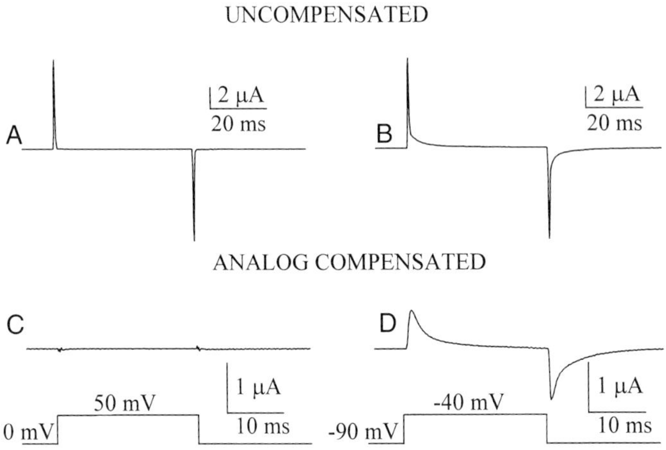

FIG. 4. Gating current recordings with analog compensation of linear components. Traces are records from *Sh*H4-IR-W434F clone. (A) and (B) are membrane currents for pulses from 0 to 50 mV and −90 to −40 mV, respectively. In (C) for the pulse from 0 to 50 mV the leak and capacity transients were analogically compensated with the capacity and leak compensation control of the cut-open oocyte transient generator. In (D), the same compensation as in part (C), pulse from −90 to −40 mV. Sampling time was 100 μs per point, filter 2 kHz. Temperature 22°.

The *Shaker* K^+ channel expresses at very high densities in *Xenopus* oocytes[19] and in HEK cells (Starace, Stefani, and Bezanilla, 1997). In addition, a mutation in the putative pore region (W434F) renders the channel nonconductive, although the gating currents are still preserved.[21] Thus the *Shaker* W434F channel lends itself to detailed studies of the kinetics and steady-state properties of gating charge.[22,23]

The procedures for recording and analyzing gating currents are illustrated using the *Shaker* H4-IR (inactivation removed) W434F clone. The expression of this clone is high enough that gating currents can be observed without subtraction. Figure 4 shows a sample experiment using the cut-open oocyte voltage-clamp technique.[12] The *Q–V* curve of this clone saturated at 0 mV; therefore, the subtraction can be applied at this potential, and this is the correct potential for compensating the capacitive transient with the analog transient generator. In Fig. 4A the current recorded for a pulse from 0 to 50 mV is shown, while in Fig. 4C the current shown at higher gain and time resolution has been recorded after adjusting the transient

[21] E. Perozo, R. MacKinnon, F. Bezanilla, and E. Stefani, *Neuron* **11,** 353 (1993).
[22] F. Bezanilla, E. Perozo, and E. Stefani, *Biophys. J.* **66,** 1011 (1994).
[23] D. Sigg, E. Stefani, and F. Bezanilla, *Science* **264,** 578 (1994).

generator for compensation of most of the capacity current. Figure 4B shows the current recorded for a pulse to −40 mV from a HP of −90 mV. It is clear that the capacitive transient now has a new, slower component in both the ON and OFF positions of the pulse. This new component is the gating current that contributes as much area to the integral as the capacitive transient alone. By using the same analog settings of the transient generator shown in Fig. 4C, the current recorded in Fig. 4D shows the gating current with the linear component removed. Thus, in this preparation it is possible to record gating currents without digital subtraction, such as the P/*n* procedure.

An example of a family of gating current records for the *Shaker* H4-IR W434F mutant recorded at 14° is shown in Fig. 5. In this case, the capacity transient was balanced holding the membrane at 0 mV and applying a small 20-mV test pulse. Subsequently, the holding potential was changed to −90 mV and a series of pulses was applied from a prepulse potential of −130 mV and returning to a postpulse potential of −130 mV. The pulses went from −125 to 30 mV in 5-mV increments. The gating currents during the test pulse (ON gating current) are small for small depolarizations and progressively increase in amplitude for larger depolarizations. Their decay rate is fast at small depolarizations, but at larger depolarizations it slows down and at even larger depolarizations speeds up again. (Notice that the current trace for the 30 mV crosses the current trace for a depolarization to −10 mV.) The ON gating currents exhibit a rising phase, most obvious at large depolarizations, and a decaying phase that has a single exponential component for small and very large depolarizations but has at least two components for intermediate depolarizations.[22] The components are also obvious in the steady-state properties of the Q–V curve. The time integrals of the ON gating current traces shown in Fig. 5 are presented as the filled circles in the lower panel of Fig. 5. A fit to the sum of 2 two-state Boltzmann distributions is shown as the continuous curve through the points, and the two components are labeled as Q_1 and Q_2. These two steady-state components follow the same voltage dependence of the kinetic components fitted from gating current traces. The gating currents after the test pulses (OFF gating current) have very different kinetics depending on the depolarization during the preceding pulse. For small depolarizations, the charge returns very quickly, as a single exponential function. As the depolarization during the pulse becomes larger, but smaller than −40 mV, the single exponential decay is maintained, and the peak current increases (largest transient off current shown in Fig. 5). However, for larger depolarizations, OFF gating currents become slower, and at even larger depolarizations (beyond −20 mV) all superimpose (see Fig. 5). The appearance of the slow component, which eliminates the fast component, has been traced to the

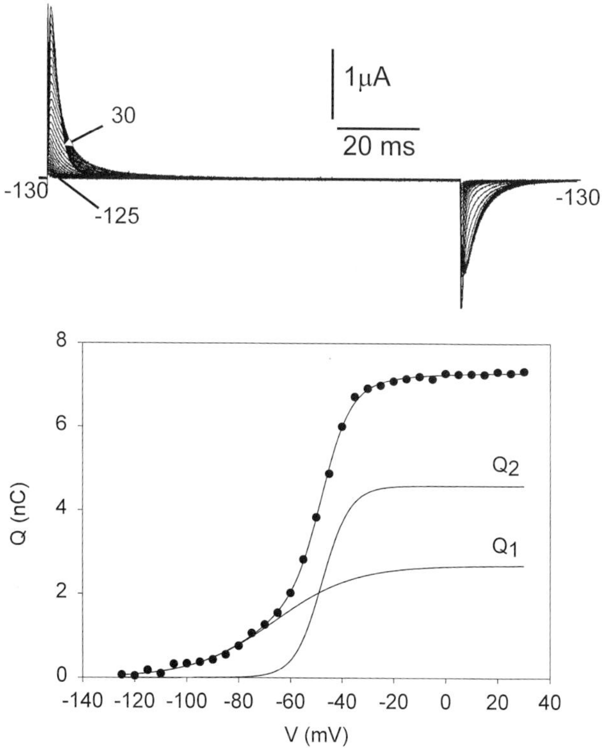

FIG. 5. *Top:* Family of gating current records from *Shaker* H4-IR W434F mutant. Records have been compensated with analog subtraction only (no digital subtraction), which was adjusted at 0 mV. The holding potential was −90 mV and the pulses from −125 to 30 mV were preceded and followed by 50-ms pulses to −130 mV. Sampling time was 50 μs per point and the filter was set to 5 kHz. Temperature was 14°. *Bottom: Q–V* curve, as measured from the records of the top panel. The continuous line is the fit to the sum of two Boltzmann distributions (Q_1 and Q_2) with the following parameters: Q_{1max} = 2.69 nC, Q_{2max} = 4.59 nC, $z_1 = 1.61\ e_0$, $z_2 = 4.65\ e_0$, half points of the distributions: $V_1 = -66.48$ mV and $V_2 = -47.68$ mV.

opening of the channel, and it can be explained by a stabilization of the open state that slows the return of the charge.[22,24]

The return of the charge is even slower in the case of the inactivating (wild-type) *Shaker* channel[19] as expected from the extra stabilization induced by the inactivating ball[25] while it is docked in the channel. In the sodium channel, inactivation also produces charge immobilization,[18,26] which produces an apparent decrease of the OFF charge in relation to the

[24] W. N. Zagotta, T. Hoshi, and R. W. Aldrich, *J. Gen. Physiol.* **103,** 321 (1994).
[25] T. Hoshi, W. N. Zagotta and R. W. Aldrich, *Science* **250,** 533 (1990).
[26] C. M. Armstrong and F. Bezanilla, *J. Gen. Physiol.* **70,** 567 (1977).

ON charge. In fact, the charge has not disappeared, but it returns much more slowly and becomes undetectable due to baseline noise. By returning to more negative potentials, it is possible to observe the slow mobilization of the missing charge with a time course that resembles the time course of recovery from inactivation.[19,26] Note that charge immobilization can also occur as a consequence of a blocking effect of ions on the channel. For example, TEA ions in the internal solution can induce complete immobilization of the charge in the inactivation-removed *Shaker* gating current,[19] but *N*-methylglucamine ions do not induce any apparent immobilization. This result points to a careful consideration of the ion replacement to be used in recording gating currents. Thus, charge immobilization should always be looked on first as a possible ion effect that has to be discarded before deducing that there is an intrinsic immobilization of the charge by the channel molecule.

Elementary Gating Event

As discussed earlier, the elementary event carries a small amount of charge that produces a current too small to detect with conventional techniques. However, the elementary event produced fluctuations in the recorded gating currents. Although these fluctuations are small, the low noise achieved with the patch-clamp technique makes it possible to detect it when recording more than about 10,000 channels in the patch. Conti and Stuhmer[18] succeeded in recording the fluctuations of brain Na^+ gating currents and computed an elementary charge of 2.3 e_0. Crouzy and Sigworth[27] have published a detailed theory of gating current fluctuations where they point out the effects of limited bandwidth and the interpretation of the calculated elementary charge in terms of a general kinetic model.

The conditions for recording fluctuations of gating currents are a low-noise recording setup and a number of channels that is not too large compared to the expected size of the fluctuations, so that the fluctuations can be resolved by the digitization of the A/D converter. The number of channels recorded with the cut-open oocyte technique is of the order of 10^9, too large in comparison to the expected fluctuations. If the expression is made less efficient, the number of channels may be decreased, but the limiting factor will be the background noise. This implies that the only technique available for this determination is the patch-clamp technique in the cell-attached or excised configuration. For the *Shaker* potassium channel, Sigg *et al.*[23] found a value of 2.2 e_0 for the elementary charge. The experiments were done with macropatches using the nonconducting W434F version of the *Shaker* H4-IR clone. Figure 6 shows an experiment done

[27] S. C. Crouzy and F. J. Sigworth, *Biophys. J.* **64,** 68 (1993).

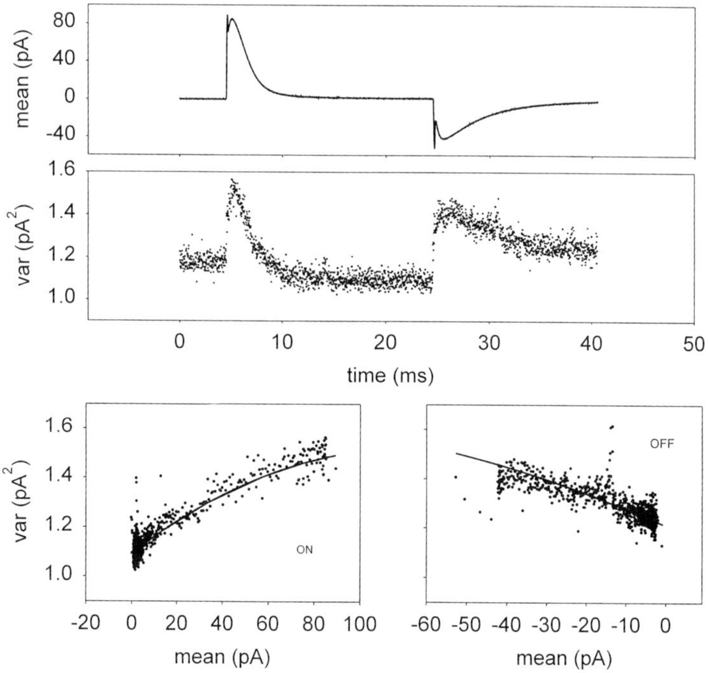

FIG. 6. *Top:* Mean and variance recorded with an attached giant patch from an oocyte expressing *Shaker* H4-IR W434F. Pulse from −120 to 0 mV. The oocyte was perforated in several places and equilibrated with a solution that contained NMG as the only cation. The pipette contained NMG and calcium. A total of 8000 trials were recorded. Bandwidth was 8 kHz. *Bottom:* Mean–variance plots for the ON and OFF positions of the pulse. The computed elementary charge was the same, 2.12 e_0, for both ON and OFF.

with an attached giant patch with an oocyte that was perforated to equilibrate with a solution of isotonic NMG while the patch pipette also contained NMG. The top trace shows the mean, and the second trace shows the variance recorded with a bandwidth of 8 kHz. To eliminate drift effects, the variance was computed by pairs of records. Notice that the variance during the pulse is lower than during the negative potential before the pulse, an expected feature of the shallow Q–V relation at negative potentials of the *Shaker* K^+ channel (see Fig. 5). From the mean–variance plots (Fig. 6, bottom), the estimated elementary charges for the ON and OFF responses were 2.12 e_0, similar to the values reported before. In this case the possible contribution of fast inactivating openings of the W434F mutant, as reported by Yang *et al.*,[28] is not present because there are no permeant ions on either side of the membrane.

[28] Y. Yang, Y. Yan, and F. J. Sigworth, *J. Gen. Physiol.* **109,** 779 (1997).

The full characterization of the frequency dependence of the capacitance at different voltages gives a more detailed description of the voltage dependence of the voltage sensor.

The study of the charge movement in voltage-dependent channels is normally done by eliminating all of the linear components of the capacitance. The presence of proteins in the membrane should add quasi-voltage-independent but frequency-dependent capacitance due to the lossy dielectric properties of buried charges and dipoles. Frequency-domain techniques, which have far better resolution than time-domain techniques, may be used to study this voltage-independent but frequency-dependent capacitance added by the expression of the channels to the oocyte membrane.

Acknowledgments

We thank Abert Cha for reading and commenting on the manuscript. Supported by NIH grants GM30376 to F.B. and GM52203 and AR38970 to E.S.

[20] Calcium Influx during an Action Potential

By J. GERARD G. BORST and FRITJOF HELMCHEN

Introduction

Action potentials open voltage-sensitive calcium channels in excitable cells, leading to an influx of calcium ions.[1] Calcium ions may control, among others, cell excitability, neurotransmitter release, or gene transcription. This chapter deals with two different methods that can be used to quantify how much Ca^{2+} flows into a cell or cell compartment during an action potential. Quantification of Ca^{2+} influx may be a first step toward the construction of a model for the calcium dynamics of cells. It may also serve as a reference to study pathological processes such as cell death during ischemia or amyotrophic lateral sclerosis, where increases in calcium influx have generally been implicated.[2,3]

[1] S. Hagiwara and K. I. Naka, *J. Gen. Physiol.* **48,** 141 (1964).
[2] D. W. Choi, *Trends Neurosci.* **18,** 58 (1995).
[3] E. Louvel, J. Hugon, and A. Doble, *Trends Pharmacol. Sci.* **18,** 196 (1997).

0076-6879/98 $25.00

The two methods that we compare in this chapter are a voltage-clamp method, in which cells are voltage clamped with an action potential waveform command, and a fluorescence method, in which cells are loaded with a high concentration of a calcium-sensitive fluorescent dye. We apply both methods to a "giant" axosomatic terminal, called the calyx of Held, in a slice preparation of the rat brain stem. Each principal cell in the medial nucleus of the trapezoid body (MNTB) is contacted by a single calyx-type terminal. This synapse is part of a fast auditory pathway that is involved in the localization of sound.[4] We were interested in the relationship between calcium influx and neurotransmitter release at this glutamatergic synapse. We found that both methods gave similar values for the number of calcium ions that enter the presynaptic terminal during an action potential. We first discuss the action potential waveform voltage-clamp method.

Action Potential Waveform Voltage Clamp

Introduction

This method was first used to quantify calcium influx into the squid giant terminal during an action potential.[5] An action potential was first recorded in current clamp and subsequently used as a command template in a voltage-clamp experiment.[6] With this paradigm, the voltage-sensitive channels experience the same voltage change as that experienced during a real action potential. If all voltage-sensitive currents, except for the calcium current, are blocked, it becomes possible to record the amplitude and time course of the calcium current in isolation.[5] Integration of this current then gives an estimate for the total calcium influx during an action potential. This method has been used to dissect the contribution of different calcium channel subtypes to the calcium influx during an action potential,[7–10] to study how the time course of the Ca^{2+} influx may shape release,[11] and to study modulation of the calcium currents by neurotransmitters.[7,12] In a

[4] R. H. Helfert and A. Aschoff, *in* "The Central Auditory System" (G. Ehret and R. Romand, eds.), pp. 193–258. Oxford University Press, New York, 1997.

[5] R. Llinás, M. Sugimori, and S. M. Simon, *Proc. Natl. Acad. Sci. U.S.A.* **79,** 2415 (1982).

[6] M. E. Starzak and R. J. Starzak, *IEEE Trans. Biomed. Eng.* **25,** 201 (1978)

[7] J. Arreola, R. T. Dirksen, R. C. Shieh, D. J. Williford, and S. S. Sheu, *Am. J. Physiol.* **261,** C393 (1991).

[8] R. S. Scroggs and A. P. Fox, *J. Neurosci.* **12,** 1789 (1992).

[9] D. P. McCobb and K. G. Beam, *Neuron* **7,** 119 (1991).

[10] D. B. Wheeler, A. Randall, and R. W. Tsien, *J. Neurosci.* **16,** 2226 (1996).

[11] A. N. Spencer, J. Przysiezniak, J. Acosta-Urquidi, and T. A. Basarsky, *Nature* **340,** 636 (1989).

[12] N. J. Penington, J. S. Kelly, and A. P. Fox, *Proc. R. Soc. Lond. B* **248,** 171 (1992).

variation on this method, only the calcium current is pharmacologically blocked after recording an action potential in current clamp.[13] In that case, the current that is needed to maintain the same action potential waveform is the calcium current plus possible calcium-activated currents.

For this method to work, a representative template, good voltage control, subtraction of leak and capacitative currents, and the pharmacologic isolation of the calcium currents are needed. The last three points are not specific for this method, but are important whenever calcium currents are recorded. General techniques for recording whole-cell calcium currents have been reviewed in a previous volume.[14] We briefly discuss some aspects that are relevant to recording calcium currents in the calyx of Held.

Isolation of Calcium Currents

Calcium currents are pharmacologically isolated by blocking sodium channels with tetrodotoxin and potassium currents with external tetraethylammonium and 3,4-diaminopyridine and internal Cs^+. We use gluconate or glutamate as the major internal anion in our solution for these experiments. We have not observed a blocking effect of gluconate[15] on the calcium current. Peak amplitudes are similar when chloride or methyl sulfate are used as the main anion. Furthermore, transmitter release is not reduced after dialysis of the terminal with solutions containing mixtures of gluconate and glutamate.[16,17] A disadvantage of using cesium gluconate, however, is that purified cesium gluconate is not readily available. Commercially available gluconic acid solutions are bulk grade (<95%), making further purification necessary, for example, by repeated precipitation of cesium gluconate in methanol. With these internal and external solutions, an (apparent) reversal potential for the calcium currents more positive than +40 mV is obtained. Additional evidence that there is little contamination by other currents in the physiologically relevant range is the good correspondence of the charges carried by Ca^{2+}, as obtained with the fluorescence method (see later section), and the calcium current integrals.

Voltage Clamp

Good spatial and temporal voltage clamp in the calyx of Held is not easy to obtain. The calcium channels in the calyx of Held gate very fast.[18]

[13] T. Doerr, R. Denger, and W. Trautwein, *Pflügers Archiv.* **413,** 599 (1989).
[14] B. P. Bean, *Methods Enzymol.* **207,** 181 (1992).
[15] A. A. Velumian, L. Zhang, P. Pennefather, and P. L. Carlen, *Pflügers Archiv.* **433,** 343 (1997).
[16] J. G. G. Borst, F. Helmchen, and B. Sakmann, *J. Physiol.* **489,** 825 (1995).
[17] J. G. G. Borst and B. Sakmann, *Nature* **383,** 431 (1996).
[18] J. G. G. Borst and B. Sakmann, *J. Physiol.* **506,** 143 (1998).

If the spatial or temporal voltage clamp is not of sufficient quality, the action potential waveform will be distorted, leading to uninterpretable results.[19] The most important determinant for the quality of the spatial clamp in the calyx of Held is the length of the axon. A long axon often makes voltage clamping impossible. Nevertheless, for short pulses, like an action potential waveform, the problem of poorly clamped axonal calcium channels is much less severe than for long depolarizing voltage steps, as the calcium channels in remote areas will not be activated during short pulses.[18] Because the calyx is a rather thin structure, it is difficult to obtain stable whole-cell recordings with low access resistance. We have circumvented this problem by using two-electrode voltage clamping, in which one patch electrode injects current and the other measures voltage. This configuration makes it possible to compensate a high access resistance (e.g., 40 MΩ) with an increase in the voltage-clamp gain. In addition, fortunately, the membrane voltage does not have to change as fast during the action potential waveform as when tail currents are recorded.

The current that is needed to force the calcium channels to undergo the same voltage change as during a real action potential consists of an active and a passive component. The active, smaller component is due to the opening of calcium channels. The passive component consists of the current that is needed to charge and discharge the cell capacitance plus the current that flows through the leak resistance of the cell:

$$I = C_{\mathrm{m}} \frac{dV_{\mathrm{m}}}{dt} + \frac{V_{\mathrm{m}} - E_{\mathrm{leak}}}{R_{\mathrm{m}}} \tag{1}$$

where C_{m} is membrane capacitance, V_{m} is voltage, E_{leak} is the reversal potential of the leak current, and R_{m} is membrane resistance.

The same two strategies that have been used to subtract these linear components of the membrane current signals during voltage steps can also be applied in the case of action potential waveforms.[5] An estimate for the passive currents can be obtained in the presence of a calcium channel blocker. Very stable recordings are needed in that case because the calcium currents are generally much smaller than the passive currents. The other method uses the scaled response to an action potential that is much smaller than the full action potential for subtraction (Fig. 1). The increase in noise can be minimized by averaging many responses with the small action potential before subtraction. The first derivative of the measured voltage matched the passive current well for the action potential waveform (Fig. 1B). This allows an estimate of the capacitance "seen" by the electrode during the waveform. Ideally, this value is almost as large as the total capacitance,

[19] C. M. Armstrong and W. F. Gilly, *Methods Enzymol.* **207,** 100 (1992).

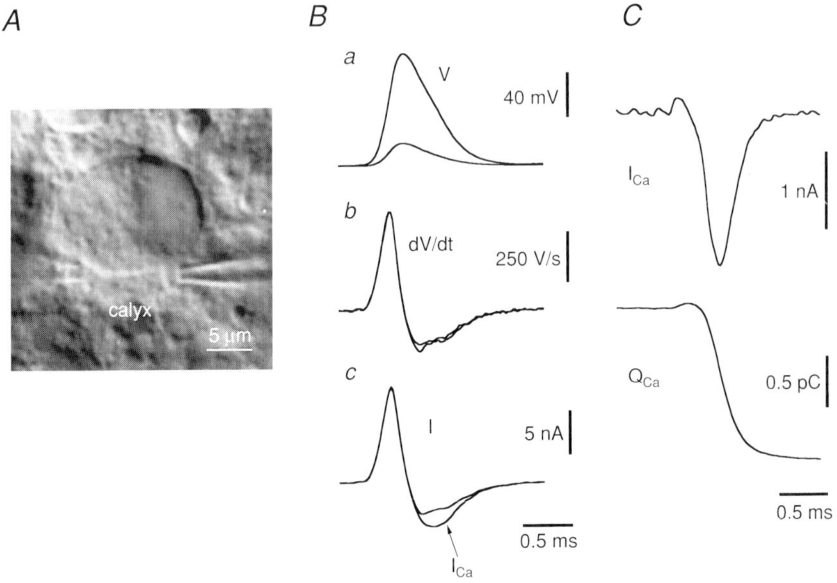

FIG. 1. Voltage clamping a giant presynaptic terminal with an action potential waveform. (A) Infrared differential interference contrast video image of a calyx of Held. For two-electrode voltage clamp, two whole-cell recordings were made. One patch electrode measured voltage and provided a feedback signal for the other electrode, which was used for current injection. [Adapted from J. G. G. Borst and B. Sakmann, *Nature* **383,** 431–434 (1996).] (B) *a:* Command voltage as measured with the voltage electrode. Also shown is the action potential waveform that has been scaled down fivefold and that is used for subtraction of the passive response. Ten current responses to the smaller action potentials were averaged and scaled up fivefold before subtraction. *b:* First derivative of the membrane voltage of the full-sized action potential and of the small action potential after scaling. *c:* Total current. Note the resemblance with the first derivative of the voltage, as expected for the passive component. The current associated with the full action potential is larger during the repolarization phase. Subtraction of the passive component gives the calcium current, shown in (C). [Adapted from F. Helmchen, J. G. G. Borst, and B. Sakmann, *Biophys. J.* **72,** 1458–1471 (1997).] (C) Calcium current during an action potential and integral of the calcium current Q_{Ca}. All data are from the same experiment.

obtained from integration of step responses. In calyces without an axon, this is indeed the case.[18] After subtraction of the passive current, the calcium current remained (Fig. 1C). A small outward current preceded the inward current. This current may be due to imperfect voltage clamp, although gating currents may also contribute. Integration of the subtracted current yields the total calcium influx (Fig. 1C).

Choice of Voltage Template

A difficult problem with this method is the choice of a representative voltage template. Ideally, an action potential is first recorded in current clamp and played back to the same cell in voltage clamp. However, pharmacologic isolation of the calcium currents is difficult to achieve without internal potassium channel blockers, making it necessary to repatch with a different solution or to use pipette perfusion in whole-cell patch-clamp experiments. Also, many voltage-clamp amplifiers are not suitable for current clamp recordings.[20] The alternative is to select a "typical" action potential and to use this action potential as a voltage template in all subsequent experiments, as was done in Fig. 1. This may also be useful with cells in primary isolation, whose action potential shape may have dramatically changed. An action potential that was recorded in a current clamp experiment *in situ* may then serve as the template.[11,21] Alternatively, voltage ramps may be used,[9,10] which have the advantage that they are more easily reproduced by other authors. In either case, the question of whether the voltage template is representative is an important one. In the calyx of Held, action potentials are relatively invariant between cells, making it easier to select a representative template.

Once a suitable template has been recorded, the action potential waveform must be provided to the amplifier as a command potential without much distortion. It is important that the interval between points of the action potential waveform command is clearly less than the sample interval at which the template was recorded. Using twice the corner frequency at which the action potential was recorded as the output frequency may not be sufficient because of the gradual rolloff of RC or Bessel filters at higher frequencies, and because of the absence of any interpolation by the digital/analog (D/A) converter. We use five times the corner frequency to be on the safe side. The software that allows us to use action potential templates is Pulse Control 4.6 (Ref. 22; http://chroma.med.miami.edu/cap/), in combination with IGOR macros (Wavemetrics, Lake Oswego, OR). The flexibility of the macro-based software makes it possible to change the waveform systematically,[18] as well as to perform analyses on-line.

An advantage of the action potential waveform voltage-clamp method is that information about the time course of the calcium current during an action potential is obtained. Alternatively, it may be possible to obtain this information by using cell-attached patch-clamp recordings[23] or by using the time derivative of the fluorescence signal of a calcium indicator (see later discussion). With the action potential waveform method, time delays in the recorded currents introduced by external filters, or the filter formed by the

[20] J. Magistretti, M. Mantegazza, E. Guatteo, and E. Wanke, *Trends Neurosci.* **19,** 530 (1996).
[21] W. J. Song and D. J. Surmeier, *J. Neurophysiol.* **76,** 2290 (1996).
[22] J. Herrington and R. J. Bookman, *in* "Pulse Control V4.0: IGOR XOPs for Patch Clamp Data Acquisition and Capacitance Measurements." University of Miami, Florida, 1994.
[23] M. Mazzanti and L. J. DeFelice, *Biophys. J.* **58,** 1059 (1990).

combination of cell capacitance and access resistance, must be corrected. We have addressed this by making simultaneous current and voltage recordings from the terminal and by using the observation that the action potential in the calyx of Held can be capacitively recorded in the postsynaptic cell.[16,24] Other methods to estimate the delay between the voltage command and the change in the membrane voltage have been discussed in Ref. 25. After correction for delays, the number of open calcium channels can be estimated at any time during the action potential, if the single-channel conductance of the calcium channel is known.

Fluorometric Measurement of Calcium Influx

Introduction

An alternative method to quantify the calcium influx during an action potential is by using a fluorescent calcium indicator. Calcium indicators are generally used to report changes in the intracellular free calcium concentration, $[Ca]_i$. The indicator dyes, which act as exogenous calcium buffers, will distort the time course of the calcium transients if their concentration is not sufficiently small. At very high concentrations the indicator will become the dominant intracellular calcium buffer. In this case, which is referred to as "dye overload," virtually all calcium ions that enter the cytosol are captured by the indicator. Because, in general, the fluorescence change (ΔF) of an indicator is proportional to the number of dye molecules that have captured a calcium ion, under overload conditions ΔF will be proportional to the total calcium current Q_{Ca}:

$$\Delta F = f_{max} \int I_{Ca}\, dt = f_{max}\, Q_{Ca} \tag{2}$$

Therefore, with this approach calcium indicators can also be used to measure calcium fluxes.[26,27] If the proportionality constant f_{max} is known, the total calcium influx can be directly calculated from ΔF. This method is much simpler and requires fewer assumptions than estimating calcium fluxes from changes in the intracellular calcium concentration.[28,29]

One of the advantages of the dye overload technique is that it also is applicable when ions other than Ca^{2+} are contributing to the currents. Indeed, this method was first used to determine the fractional contribution

[24] I. D. Forsythe, *J. Physiol.* **479,** 381 (1994).
[25] F. J. Sigworth and J. Zhou, *Methods Enzymol.* **207,** 746 (1992).
[26] E. Neher and G. J. Augustine, *J. Physiol.* **450,** 273 (1992).
[27] E. Neher, *Neuropharmacology* **34,** 1423 (1995).
[28] S. M. Baylor, W. K. Chandler, and M. W. Marshall, *J. Physiol.* **344,** 625 (1983).
[29] D. W. Tank, W. G. Regehr, and K. R. Delaney, *J. Neurosci.* **15,** 7940 (1995).

of Ca^{2+} to the current through ligand-gated ion channels.[27,30–31a] We applied dye overload to measure the total calcium influx during an action potential in the calyx of Held.[32] In Fig. 2A the fluorescence image of a calyx loaded with 1 mM of the high-affinity indicator Fura-2 (Molecular Probes, Eugene, OR) is shown. At this concentration, Fura-2 overrides the endogenous calcium buffers of the terminal (see later discussion).[16,32] With excitation at 380 nm, the fluorescence emission of Fura-2 decreases after calcium binding. A single presynaptic action potential that was evoked by stimulation of the afferent nerve fiber caused a clear decrease in the total Fura-2 fluorescence of the whole terminal (ΔF_{380}; Fig. 2B). Before ΔF_{380} can be converted to Q_{Ca}, several conditions have to be met. Normalization and calibration of the fluorescence signal are also required. These aspects are discussed in the following sections.

Dye Overload

Dye overload is reached when the indicator effectively outcompetes the endogenous calcium buffers. This suggests the use of a high-affinity calcium indicator, with a dissociation constant in the same range as the resting intracellular Ca^{2+} concentration. So far only Fura-2 has been used. However, in principle, the dye overload technique is not restricted to ratiometric dyes since it requires fluorescence excitation (or collection of fluorescence emission) at a single, calcium-dependent wavelength. Therefore, other high-affinity indicators such as Calcium Green-1, Indo-1, or even EGTA in combination with a pH indicator[33] may also be used.

How much dye is needed to outcompete the endogenous buffers? An absolute value for the concentration that is needed cannot be given because it depends on the relative calcium-buffering capacities of the indicator and of the endogenous calcium buffers. We illustrate this using the single compartment model introduced by Neher and Augustine.[26] This model describes the competition of an indicator (B) with a pool of rapid endogenous calcium buffers (S). The calcium-buffering capacity of a buffer is given by its calcium-binding ratio[27]:

$$\kappa_{\mathrm{B}} = \frac{\partial[\mathrm{CaB}]}{\partial[\mathrm{Ca}]_{\mathrm{i}}} = \frac{K_d}{(K_d + [\mathrm{Ca}]_{\mathrm{i}})^2}[\mathrm{B}]_{\mathrm{T}} \tag{3}$$

[30] R. Schneggenburger, Z. Zhou, A. Konnerth, and E. Neher, *Neuron* **11,** 133 (1993).
[31] Z. Zhou and E. Neher, *Pflügers Archiv.* **425,** 511 (1993).
[31a] G. Veliçelebi, K. A. Staderman, M. A. Varney, M. Akong, S. D. Hess, and E. C. Johnson, *Methods Enzymol.* **293,** [2], 1998 (this volume).
[32] F. Helmchen, J. G. G. Borst, and B. Sakmann, *Biophys. J.* **72,** 1458 (1997).
[33] P. C. Pape, D. S. Jong, and W. K. Chandler, *J. Gen. Physiol.* **106,** 259 (1995).

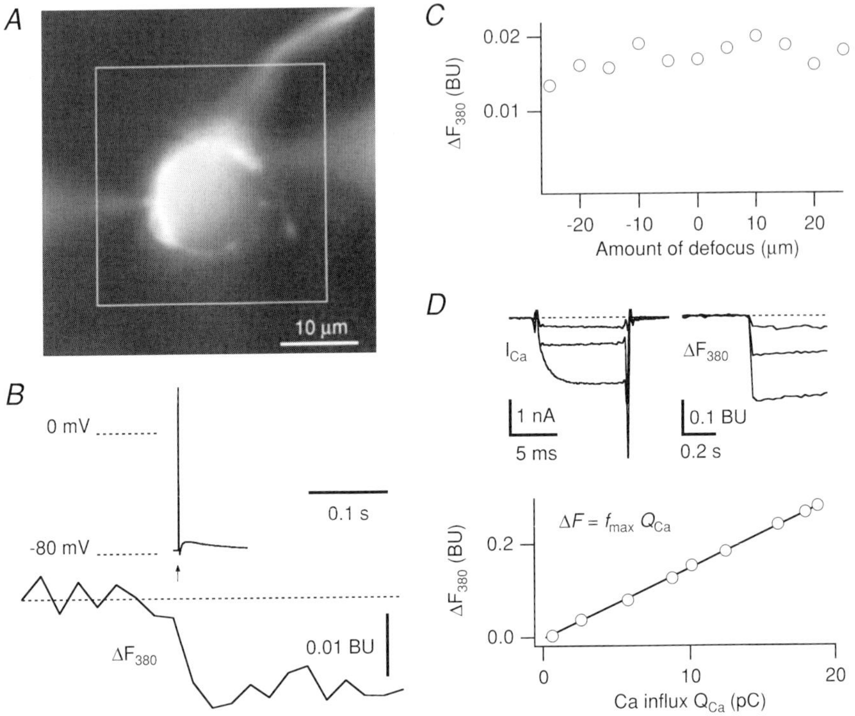

FIG. 2. Fluorometric measurement of the total presynaptic calcium influx during an action potential. (A) Wide-field fluorescence image of a calyx of Held filled with 1 m*M* Fura-2. The rectangle indicates the subregion from which the total fluorescence signal was collected with the CCD camera. (B) Fura-2 fluorescence decrement at 380-nm excitation (ΔF_{380}) evoked by a single presynaptic action potential, which was elicited orthodromically (arrow). Fluorescence is expressed in bead units (BU). (C) Decrements evoked by single action potentials with the microscope defocused by different amounts. (D) Determination of $f_{\max}$. Isolated presynaptic calcium currents (I_{Ca}) were evoked by depolarizing voltage steps and Fura-2 fluorescence decrements were measured simultaneously. The slope of the plot of ΔF_{380} versus the calcium current integral yields $f_{\max}$, which was 15 BU nC^{-1}. [Adapted from J. G. G. Borst and B. Sakmann, *Nature* **383,** 431–434 (1996).]

where [CaB] is the concentration of the bound form, K_d is the dissociation constant, and $[\mathrm{B}]_{\mathrm{T}}$ is the total concentration of the buffer. According to the model, the fluorescence change divided by the total calcium influx (the so-called *F/Q* ratio, *f*) is given by

$$f = \frac{\Delta F}{Q_{\mathrm{Ca}}} = f_{\max} \frac{\kappa_{\mathrm{B}}}{(1 + \kappa_{\mathrm{S}} + \kappa_{\mathrm{B}})} \tag{4}$$

where f depends on the relative sizes of the endogenous calcium-binding ratio κ_S and the exogenous calcium-binding ratio κ_B. It reaches a saturating value f_{max} if κ_B is much larger than κ_S and, in this case, Eq. (4) simplifies to Eq. (2). Thus, dye overload requires $\kappa_B \gg \kappa_S$. To choose a dye concentration that fulfills the overload condition, it is useful to have an estimate of κ_S. However, this is not strictly necessary. As a simple test one can measure the change in fluorescence that is evoked by a more or less constant calcium influx (e.g., during an action potential) in the presence of different dye concentrations. If the dye concentration is sufficiently high, ΔF should not change significantly when the dye concentration is doubled. On the other hand, determination of κ_S is advantageous because the fraction of the incoming calcium ions that is captured by the indicator can be estimated, and the signal-to-noise ratio can be more readily optimized (see later discussion). For many neurons κ_S is between 50 and 150.[27] For the calyx of Held we obtained a value of about 40 by monitoring the effect of Fura-2 on calcium transients evoked by action potentials during Fura-2 loading.[32] Therefore, a concentration of 1 mM Fura-2, which corresponds to a κ_B of around 2500, is sufficient for overload. According to Eq. (4), more than 98% of the calcium ions will be captured by Fura-2 if κ_S is 40. In neurons with a higher endogenous calcium-binding ratio, such as cerebellar Purkinje cells,[34] much higher concentrations of the dye would be needed.

One might think that the higher the dye concentration the better. This is not true for two reasons. First, the indicator cannot catch more than 100% of all incoming calcium ions. When the dye concentration is too high, the signal-to-noise ratio will decrease, as discussed further below. Second, the fluorescence intensity in this case may no longer depend linearly on the dye concentration ([B]). Beer's law states that the fluorescence intensity F of a fluorescent layer of thickness d is proportional to $1-10^{-\varepsilon d[B]}$, where ε is the extinction coefficient of the dye. The linear approximation $F \propto \varepsilon\, d\, [B]$ is only valid if $[B] \ll [\ln(10)\, \varepsilon\, d]^{-1}$. As an example, we consider a small neuron with a diameter of 10 μm. The extinction coefficient of Fura-2 is about 30,000 M^{-1} cm^{-1}; therefore, in this case, [B] should be well below 15 mM.

Normalization and Calibration of Fluorescence Changes

The fluorescence intensity that is measured depends on the illumination intensity as well as the photon-detection efficiency. Because these parameters may change with time, intensities should be normalized to a fluorescent standard. Commonly, the fluorescence intensity of fluoresbrite beads (diam-

[34] L. Fierro and I. Llano, *J. Physiol.* **496,** 617 (1996).

eter 4.5 μm, Polysciences, Warrington, PA) is measured on each experimental day and all intensities are normalized to this value.[26,27,30] Fluorescence therefore is expressed in "bead units" (BU; Ref. 30). The intensity of the beads depends on the bandwidths of the excitation and the emission filters. Because maximal emission is at around 470 nm the dichroic mirror should separate below that wavelength. The bead intensity also depends on the solution, which therefore always should be the same.[27] We dilute the beads 1:200 in distilled water, immobilize the beads by drying a drop of this solution on a coverslip, place the coverslip in distilled water under a water-immersion objective, and subsequently measure the intensity of five beads, each placed in the same region as the terminals. With the CCD camera some pixels could be saturated by the bright beads. This was avoided by slightly defocusing the bead for the measurements.

Conversion of the fluorescence changes to total calcium current Q_{Ca} requires the determination of the proportionality factor f_{max} (in units of BU nC^{-1}). This is achieved by measuring fluorescence changes evoked by isolated calcium currents, because calcium channels have a very high selectivity for Ca^{2+} over other ions under physiologic conditions.[35] We blocked sodium and potassium channels as described earlier, and evoked presynaptic calcium currents with depolarizing voltage steps in voltage clamp. By correlating the decrements in Fura-2 fluorescence with the calcium current integrals we obtained an estimate for f_{max} (Fig. 2D). Usually, the correlation coefficient was very high for depolarizing steps to voltages in the range of -30 to $+30$ mV ($r > 0.99$), indicating good pharmacologic isolation of the calcium currents. It is noteworthy that f_{max} depends on the spectral properties of the experimental setup since the beads and Fura-2 differ with respect to their emission spectrum.[31] The total calcium influx Q_{Ca} during an action potential is calculated by dividing the fluorescence change through f_{max} [Eq. (2)].

Fluorescence Detection

Fluorescence can be detected with a camera, a photomultiplier, or a photodiode. It should be collected with the same efficiency from all cellular compartments that are studied. A region that encloses the relevant structures can be selected as a subregion of the camera (Fig. 2A), with a diaphragm in front of the photomultiplier, or by placement of a small photodiode in the image plane of the microscope (see later discussion). For the determination of f_{max} it is also essential that fluorescence be measured from all structures that contribute to the calcium current measured under voltage

[35] B. Hille, *in* "Ionic Channels of Excitable Membranes," pp. 1–607. Sinauer, Sunderland, Massachusetts, 1992.

clamp. Therefore, the calcium influx into cellular structures from which fluorescence is not collected must be considered. For example, the preterminal axon in Fig. 2A is not fully included within the region of measurement. In this case previous measurements, however, have shown that the calcium signal is very small in the axon compared to the terminal.[16] It is also important that the light emitted from structures that are not in focus be collected as well. To test this, fluorescence decrements can be measured at different focal planes. The decrements should not significantly change if the cell is defocused (Fig. 2C). Fluorescent beads may also be helpful for testing this assumption. Actually, the poor spatial resolution in the Z-axis of the wide-field microscope compared to a confocal or a two-photon microscope is an advantage for overload measurements.

It may even be possible to measure the total calcium influx in neurons with extensive dendrites, using a low magnification objective with a large field of view. In that case it is important to make sure that the indicator is not locally saturated[27] and that it outcompetes the endogenous buffers throughout the whole cell, because equilibrating remote processes of a cell with the dye may take a long time.[36] For such measurements, f_{max} would have to be determined in a different cell type under conditions that differ as little as possible from the measurements, because voltage clamping of cells with extensive dendrites is not possible. The same strategy would have to be followed for other possible applications of this method such as measurement of calcium influx through low-voltage-activated calcium channels during excitatory postsynaptic potentials, estimation of calcium current densities in dendritic segments, and measurement of relative calcium influx into subcellular compartments compared to the whole cell. Another approach that can be used to simplify the interpretation of the fluorescence measurements is to allow calcium influx only locally, either by local application of agonists[37] or by exposing only the structure that is under study to physiologic $[Ca^{2+}]_o$.[38]

Comparison of the Voltage-Clamp and the Fluorescence Method

The two methods described gave a similar value of about 0.9 pC (corresponding to around 35 $fCpF^{-1}$) for the total calcium influx during an action potential in the calyx of Held (23°, 2 m*M* $[Ca^{2+}]_o$). This is remarkable since both methods rely on different assumptions. First, the voltage-clamp

[36] J. Eilers, R. Schneggenburger, and A. Konnerth, *in* "Single-Channel Recording", 2nd Ed. (B. Sakmann and E. Neher, eds.), pp. 213–229. Plenum, New York, 1995.

[37] O. Garaschuk, R. Schneggenburger, C. Schirra, F. Tempia, and A. Konnerth, *J. Physiol.* **491,** 757 (1996).

[38] G. J. Augustine, M. P. Charlton, and S. J. Smith, *J. Physiol.* **367,** 143 (1985).

method requires blocking of sodium and potassium channels, whereas the fluorometric measurements can be done in current clamp mode using physiologic external solution. Second, the same action potential waveform was used for voltage clamping the nerve terminals in all experiments. In contrast, action potentials were elicited by orthodromic stimulation in the dye overload experiments and displayed their normal variability with respect to amplitude and time course. Both methods required patch-clamp recordings with relatively low access resistance (R_{acc}). For reliable dye loading of the nerve terminal, a low access resistance ($\approx 15\ M\Omega$) during the first 10 min of recording is needed.[16] As soon as the terminal is loaded, increases in R_{acc} are no longer critical. In the voltage-clamp measurements the problem of increases in R_{acc} was circumvented by using two-electrode voltage clamp. We will show later that a single electrode may also be sufficient.

The main disadvantage of the overload technique compared to the action potential waveform voltage-clamp technique is that it does not provide information about the time course of the calcium influx. There have been attempts to back calculate the time course of the calcium current by taking the derivative of the fluorescence signal of a calcium indicator [compare Eq. (2)].[27,28,39,40] However, for this method to work, fluorescence has to be detected with submillisecond time resolution, and the kinetics of the dye have to be fast enough to report the time course of the influx. Therefore, only low-affinity dyes such as MagFura-2 or Magnesium Green may be suitable. According to Eq. (3), it is very difficult for these dyes to outcompete the endogenous buffers, so that they capture only a fraction of the incoming calcium ions. In addition, dye depletion close to the channels may occur during calcium influx. Currently, it does not seem possible to simultaneously obtain quantitative measurements of both amplitude and time course of calcium influx using a fluorometric method.

Measuring Modulation of Calcium Influx

With both methods, it becomes possible to study modulation of the calcium influx. To illustrate the possibilities, we show here two simple examples, the dependence of the calcium influx during an action potential in the calyx of Held on extracellular calcium concentration and on temperature (Figs. 3 and 4). A comparison of both methods shows that they give similar values for both manipulations. The signal-to-noise (S/N) ratio is better for the voltage-clamp method than for the fluorescence method, though. The coefficient of variation (standard deviation/mean) was 0.02 for the current integrals during the baseline period shown in Fig. 3B and 0.14 for the fluorescence signals shown in Fig. 3D. In principle, both methods could be

[39] B. L. Sabatini and W. G. Regehr, *Nature* **384,** 170 (1996).
[40] S. R. Sinha, L. G. Wu, and P. Saggau, *Biophys. J.* **72,** 637 (1997).

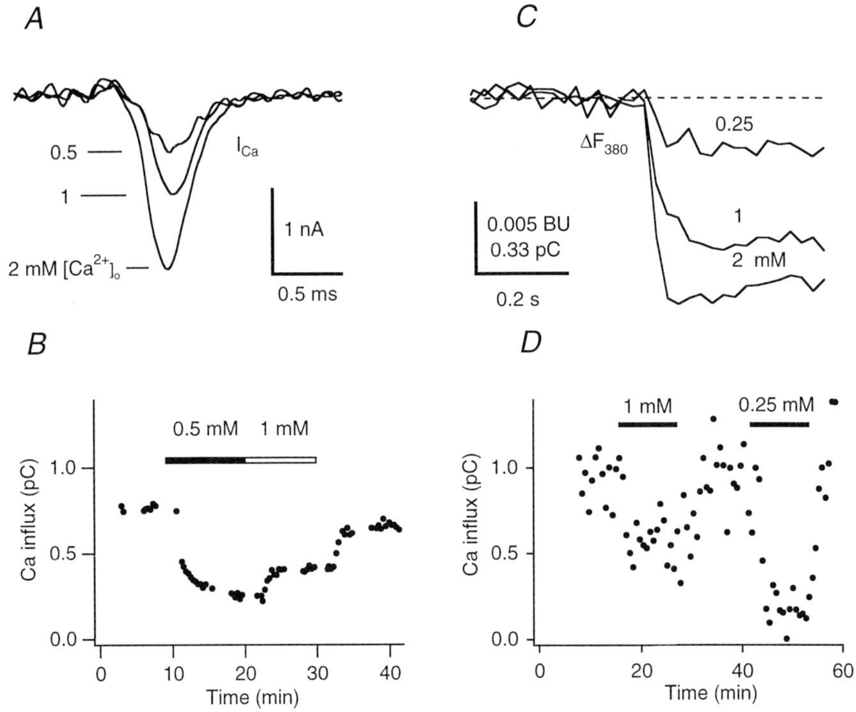

FIG. 3. Dependence of the presynaptic calcium influx during an action potential on the external calcium concentration. (A) Calcium currents during action potentials at different external calcium concentrations ($[Ca^{2+}]_o$), as indicated. (B) Time course after break-in of the calcium current integrals during periods with different $[Ca^{2+}]_o$. Same experiment as shown in Fig. 1. [Adapted from J. G. G. Borst and B. Sakmann, *Nature* **383,** 431 (1996).] (C) Fluorescence decrements evoked by action potentials with different $[Ca^{2+}]_o$ as indicated (each trace is an average of nine sweeps). (D) Time course after break-in of the total calcium influx during periods with different $[Ca^{2+}]_o$ as measured from the fluorescence decrements. [Adapted from F. Helmchen, J. G. G. Borst, and B. Sakmann, *Biophys. J.* **72,** 1458 (1997).]

used to correlate the presynaptic calcium influx during action potentials with the amplitude of the postsynaptic currents in simultaneous pre- and postsynaptic recordings. However, although the fluorescence method is technically easier to perform, the introduction of 1 m*M* Fura-2 into the terminal would reduce neurotransmitter release significantly.[16]

Signal-to-Noise Considerations

Voltage Clamp. Baseline noise in whole-cell experiments depends mainly on access resistance and membrane capacitance.[41] Noise due to

[41] R. A. Levis and J. L. Rae, *Methods Enzymol.* **207,** 14 (1992).

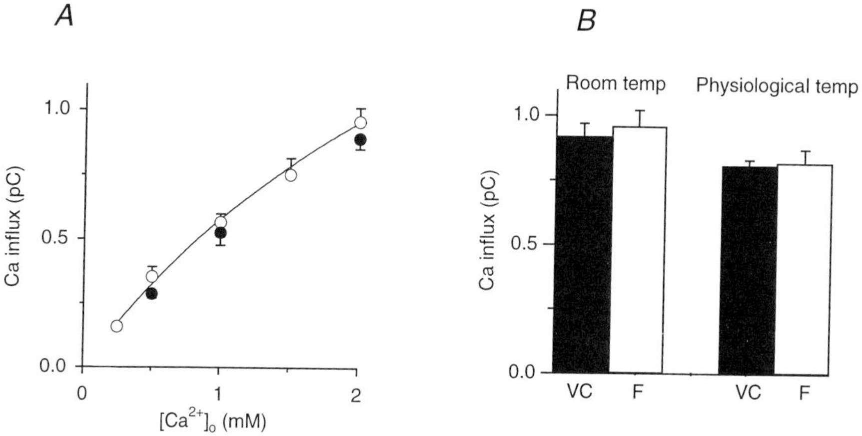

FIG. 4. Comparison of voltage-clamp and fluorescence measurements of presynaptic calcium influx in the calyx of Held. (A) Dependence of total calcium influx during an action potential on the external calcium concentration $[Ca^{2+}]_o$. The voltage-clamp method (filled circles) and the fluorescence method (open circles) gave similar results. [Adapted from F. Helmchen, J. G. G. Borst, and B. Sakmann, *Biophys. J.* **72,** 1458 (1997), and from J. G. G. Borst and B. Sakmann, *Nature* **383,** 431 (1996).] (B) Temperature dependence of the total calcium influx with 2 m*M* $[Ca^{2+}]_o$. Again, the voltage-clamp method (VC) and the fluorescence method (F) gave similar results.

calcium currents will depend, in addition to recording bandwidth, also on open probability, total number of functional calcium channels, and single-channel current.[42] The large calcium currents in the terminal make it possible to obtain a good *S/N* ratio (defined as peak amplitude divided by standard deviation of the baseline noise). Even though the noise with the two-electrode voltage clamp is clearly higher than for whole-cell patch-clamp recordings, *S/N* ratios of >20 (at 5 kHz) are easily obtained. A further reduction in baseline noise would permit to do nonstationary noise analysis of the calcium currents during action potentials or to study the contribution of the variance of the calcium currents to the variance of the synaptic currents.

Fluorescence. Fluorescence changes evoked by action potentials are relatively small, for example, the signal shown in Fig. 2B corresponds to a relative fluorescence change of only 0.7%. Therefore, the fluorescence noise must be reduced. Noise sources include fluctuations in the output of the light source, vibrational noise, dark noise of the photodetector, and photon

[42] S. H. Heinemann and F. Conti, *Methods Enzymol.* **207,** 131 (1992).

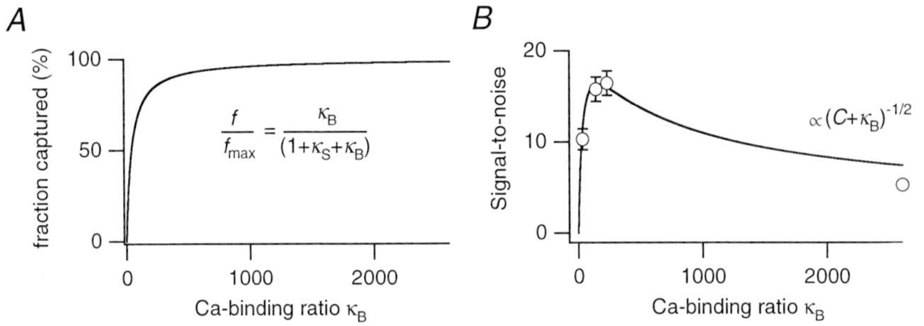

FIG. 5. Signal-to-noise ratio of fluorescence changes evoked by action potentials. (A) The fraction of calcium ions that bind to Fura-2 is plotted as a function of the Fura-2 calcium-binding ratio κ_B. Around half of the calcium ions bind to Fura-2 when κ_B equals the endogenous calcium-binding ratio κ_S, which was assumed to be 40. One m*M* Fura-2 corresponds to a calcium-binding ratio of about 2500. (B) Signal-to-noise ratio of action potential evoked Fura-2 fluorescence decrements versus κ_B. The *S/N* ratio is defined as ΔF divided by the standard deviation of the prestimulus fluorescence trace. Data points were obtained at Fura-2 concentrations of ≈20 μM, 50 μM, 100 μM, and 1 m*M*, respectively. A curve according to Eq. (6) was scaled to fit the three first points (C = 220, κ_S = 40). [Adapted from F. Helmchen, J. G. G. Borst, and B. Sakmann, *Biophys. J.* **72,** 1458–1470 (1997).]

shot noise.[43] Because the signals are so small, it may be important to find the dye concentration that yields the highest *S/N* ratio. We define the *S/N* ratio as the fluorescence change ΔF divided by the standard deviation of the baseline fluorescence σ_F. The percentage of calcium ions captured by the indicator is given by f/f_{max} [Eq. (4)]. It is shown in Fig. 5A as a function of the exogenous calcium-binding ratio κ_B. The endogenous calcium-binding ratio κ_S was assumed to be 40.[32] In the best case of a shot-noise-limited optical recording the variance of the fluorescence signal is proportional to the number of measured photons. However, in brain slices the autofluorescence of endogenous fluorophores[36] results in a relatively high background fluorescence. For example, at our setup the background fluorescence is equivalent to the fluorescence of a terminal loaded with 100 μM Fura-2, corresponding to $\kappa_B \approx 220$. Assuming that background and dye fluorescence are independent, the variance of the total fluorescence is proportional to the sum of the background fluorescence and the indicator fluorescence F, where F is proportional to $[B]_T$, which for small perturbations of $[Ca]_i$ from the resting level itself is proportional to κ_B [Eq. (3)]. Therefore the variance is proportional to:

$$\sigma_F^2 \propto (C + \kappa_B) \qquad (5)$$

[43] J.-Y. Wu and L. B. Cohen, *in* "Fluorescent and Luminescent Probes for Biological Activity" (W. T. Mason, ed.), pp. 389–404. Academic Press, London, 1993.

where C is a constant representing the binding ratio that the indicator would have if its fluorescence were equal to the background fluorescence.

Neglecting other noise sources, it follows that:

$$S/N \text{ ratio} = \frac{\Delta F}{\sigma_F} \propto \frac{\kappa_B}{(1 + \kappa_S + \kappa_B)\sqrt{C + \kappa_B}} \qquad (6)$$

As seen in Fig. 5B, the actual S/N ratios that we obtained with the CCD camera using different concentrations of Fura-2 are reasonably well described by this relationship. For high dye concentrations, $\kappa_B \gg \kappa_S$, the S/N ratio decreases proportional to $(C + \kappa_B)^{-1/2}$ [Eq. (6)]. This decrease occurs because ΔF saturates (the indicator cannot bind more calcium ions than enter the cell) but the noise still increases with increasing dye concentrations. To obtain the exogenous calcium-binding ratio κ_0 for which the S/N ratio is maximal, Eq. (6) can be differentiated with respect to κ_B, yielding:

$$\kappa_0 = \frac{(1 + \kappa_S)}{2}\left(1 + \sqrt{1 + 8\frac{C}{1 + \kappa_S}}\right) \qquad (7)$$

Thus, if background fluorescence can be neglected ($C = 0$), the highest S/N ratio is obtained if cells are loaded with a dye concentration that is equivalent to $\kappa_B = \kappa_0 = 1 + \kappa_S$. In this situation, which has been referred to as "balanced loading,"[44] only 50% of the calcium ions are captured by the indicator. With higher background fluorescence, more dye is needed for the best S/N ratio; for example, with a C equivalent to κ_S, the optimal dye concentration is already twice as large. By definition, $\kappa_B \gg \kappa_S$ during overload, so unless C is much larger than κ_S, the S/N ratio will be clearly suboptimal. To increase the S/N ratio one may choose to work with a lower dye concentration so that only a fraction of the calcium ions is captured and the total calcium influx is calculated from Eq. (4) using an estimate of κ_S. However, κ_S may vary between cells, leading to uncertainties in this calculation.

Simple Photodetector

Another possibility for increasing the S/N ratio of the fluorescence measurements is to use a photodetector with a high quantum efficiency. Because spatial resolution is not needed, an alternative to a CCD camera is a photodiode. Advantages are their low cost, ease of operation and high bandwidth. Compared to photomultipliers, photodiodes have a larger dark current but this is not critical in our case because of the high dye concentrations used. Photodiodes have been widely used to measure calcium signals

[44] Z. Zhou and E. Neher, *J. Physiol.* **469,** 245 (1993).

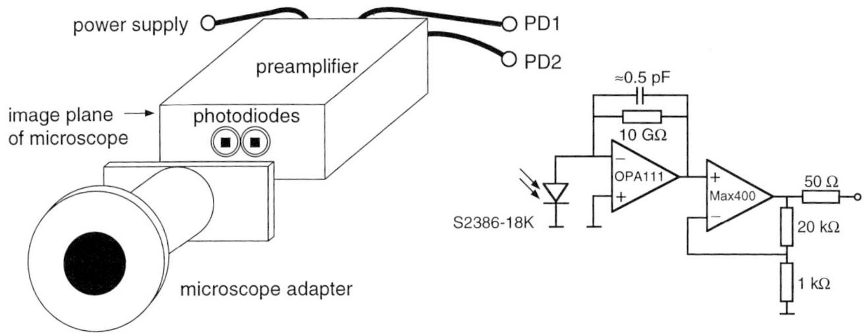

FIG. 6. A simple photodetector to measure calcium influx during an action potential. (Left) Schematic diagram of the photodiode system. The metal box contains the preamplifier. Two silicon photodiodes (S2386-18K, Hamamatsu Photonics, Japan) are located in the image plane of the microscope's camera output port. Active area of the photodiodes is 1.1 × 1.1 mm, corresponding to 27.5 × 27.5 μm in the focal plane of a Zeiss 40× water immersion objective. The quantum efficiency of the photodiode used is 84% at 560 nm. (Right) Diagram of the photodiode preamplifier, consisting of an OPA111 (Burr-Brown, Tucson, AZ) current-to-voltage converter and a Max400 (Maxim Integrated Products, Sunnyvale, CA) differential amplifier that provides a gain of 21. Feedback resistor is 10 GΩ. Stray capacitance is around 0.5 pF, giving a corner frequency of around 300 Hz, as measured from the rise time of the photodiode signal to an LED light source.

from pools of labeled terminals.[45,46] We have constructed a simple photodiode to measure calcium signals from small cell compartments such as the calyx of Held (Fig. 6). Light is collected by the photodiode from a region of similar size as the subregion of the camera shown in Fig. 2A. Simultaneous measurement of a background region is possible using a second neighboring photodiode, although this is not necessary when total influx is measured. With this simple device fluorescence decrements evoked by single action potentials can be resolved using 1 m*M* Fura-2 for overload (Fig. 7).

Simultaneous Voltage-Clamp and Fluorescence Measurements

To illustrate the use of this photodiode we have simultaneously measured calcium influx during an action potential with both methods (Fig. 7). A terminal was voltage clamped using an action potential waveform, as shown in Fig. 1, except continuous whole-cell voltage clamp with a single electrode was used and fluorescence was monitored at the same time with the photodiode. To be able to voltage clamp the terminals with a single electrode, it was important to select terminals with short axons and to keep

[45] L.-G. Wu and P. Saggau, *J. Neurosci.* **14,** 645 (1994).
[46] W. G. Regehr and D. W. Tank, *Neuron* **7,** 451 (1991).

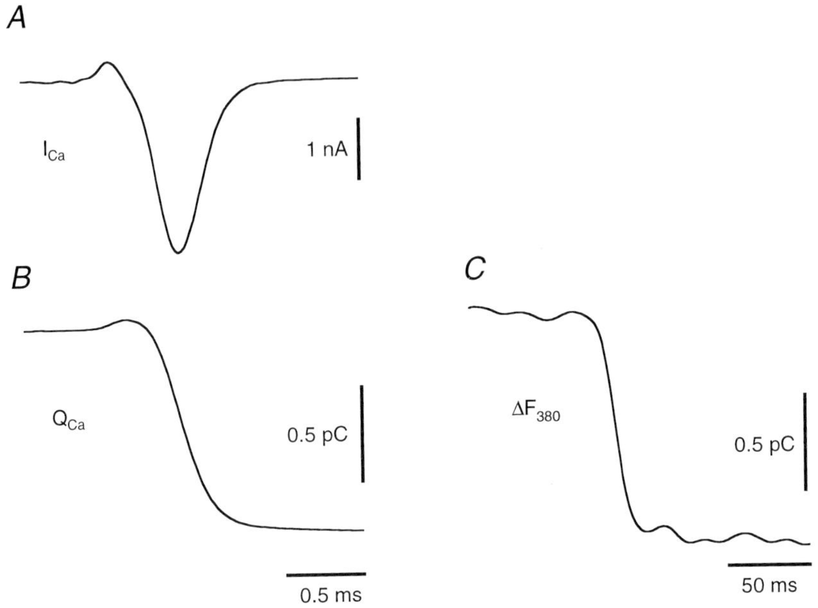

FIG. 7. Simultaneous recording of calcium current during an action potential using the voltage-clamp and fluorescence methods. A terminal was loaded with Fura-2 and voltage clamped with an action potential waveform to compare the calcium influx obtained with both methods. Determination of f_{max} could be done in this case within the same experiment by construction of a current–voltage relation, as shown in Fig. 2D. Continuous single electrode voltage clamp was used. The outputs of the patch-clamp amplifier and the photodiode amplifier were simultaneously digitized with a 16-bit A/D converter (ITC-16 Instrutech, Great Neck, NY) at 50 kHz; the photodiode voltages were decimated to 1 kHz to reduce memory requirements. Access resistance was 8 MΩ, compensated to 90%. Total capacitance was 33 pF. Axon length was around 100 μm. (A) Calcium currents. Filtered at 5 kHz. (B) Integral of response shown in part (A). (C) Fluorescence measurements. Photodiode signal was digitally filtered to 20 Hz. Calibration bar is shown in pC, 0.5 pC corresponded to 0.0056 BU and 73 fA photodiode current. Responses are the average of 20.

the access resistance after electronic compensation below 2 MΩ. Both methods gave similar values for the calcium influx. The *S/N* ratio was again better for the voltage-clamp method than for the fluorescence method.

Conclusions

We have compared two methods to detect the influx of calcium ions during single action potentials and found that they gave similar results in a giant synaptic terminal of the rat brain stem. Both methods have advantageous features. The main advantages of the voltage-clamp method are that it provides information about the time course of the calcium influx, that it can be used in the presence of physiologic intracellular calcium buffer concentrations, that a systematic manipulation of the action potential waveform is possible, and that it gives a better *S/N* ratio than the dye overload technique. The main advantages of the fluorometric method are that it is relatively simple, that measurements are done in current clamp in physiological external solutions, and that it can in principle be used to quantify fluxes in small subcellular compartments.

Acknowledgments

We thank Dr. B. Sakmann for continuous support, R. Rödel and K. Schmidt for construction of the photodiode amplifier and Dr. T. D. Parsons for comments on an earlier version of this manuscript. J. G. G. B. was supported by the E. U. (TMR program).

[21] Combined Whole-Cell and Single-Channel Current Measurement with Quantitative Ca^{2+} Injection or Fura-2 Measurement of Ca^{2+}

By L. DONALD PARTRIDGE, HANNS ULRICH ZEILHOFER, and DIETER SWANDULLA

Introduction

Intracellular free Ca^{2+} is an important second messenger in most cell types and cytoplasmic Ca^{2+} concentration, $[Ca^{2+}]_i$, is tightly controlled by cytoplasmic buffering and regulation of influx and efflux pathways. In excitable cells, cytoplasmic Ca^{2+} couples electrical events with a broad range of effector functions. Voltage-dependent calcium channels provide a direct means for this coupling, while postsynaptic ligand-gated channels provide additional regulated pathways for Ca^{2+} entry. In addition to its well-described roles in excitation–contraction coupling and excitation–secretion coupling, cytoplasmic Ca^{2+} is responsible for excitation–gating coupling for several important classes of ion channels. The broad class of Ca^{2+}-regulated channels includes Ca^{2+}-activated potassium, nonselective (CAN) and chloride channels, and Ca^{2+}-inactivated calcium channels. In addition, these and other channels can be modulated by metabolic processes that depend on Ca^{2+}.

Established techniques are available to measure membrane currents in cells voltage clamped with either intracellular microelectrodes or with patch

0076-6879/98 $25.00

electrodes. These methods have been invaluable in establishing our current understanding of the function of voltage- and ligand-gated channels in the plasma membrane. Techniques for introducing Ca^{2+} into the cytoplasm are also well established and several of these methods have been refined so that reasonably quantitative introductions of Ca^{2+} can be made. Recent years have seen rapid advances in techniques for measuring $[Ca^{2+}]_i$ using fluorescent dyes, but older methods, utilizing aequorin, arsenazo, and ion-sensitive microelectrodes, take these measurements back several decades.

Figure 1 is an example in which these three techniques were combined in an experiment in a single cell. The $[Ca^{2+}]_i$ was changed in a quantitative manner as the result of controlled influx through voltage-gated calcium

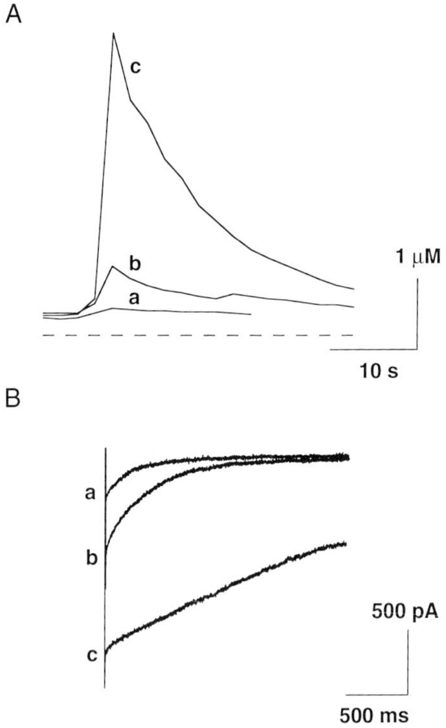

FIG. 1. Ca^{2+} introduction through voltage-gated calcium channels with simultaneous measurement of Ca^{2+}-activated currents and $[Ca^{2+}]_i$. Calcium currents were activated by depolarizing steps from −80 to 0 mV lasting 100 (a), 200 (b), and 500 ms (c). (A) $[Ca^{2+}]_i$ was measured with Fura-2 introduced through the patch pipette, which contained 100 μM Fura-2. Dashed line represents zero $[Ca^{2+}]_i$ level. (B) Ca^{2+}-activated currents were measured as tail currents following the three depolarizing steps. [Adapted from L. D. Partridge and D. Swandulla (1988) Calcium-activated nonspecific cation channels. *Trends in Neurosciences* 11(2):69–72.]

channels. The change in $[Ca^{2+}]_i$ was measured using fluorescent imaging with the Ca^{2+}-sensitive dye Fura-2 (Molecular Probes, Eugene, OR), and membrane (tail) currents were measured. The strength of combining these techniques is apparent in this experiment where it was possible to control $[Ca^{2+}]_i$ quantitatively and observe its effects on the activation of Ca^{2+}-activated currents. Experiments combining measurements of membrane currents and $[Ca^{2+}]_i$ with the controlled introduction of Ca^{2+} into the cytoplasm have provided important insight into cellular Ca^{2+} regulation and signaling. For example, the combination of these techniques has been used to determine cytoplasmic Ca^{2+} buffering,[1,2] mitochondria-dependent Ca^{2+} clearance,[3] resting $[Ca^{2+}]_i$,[4] and Ca^{2+}-induced Ca^{2+} release mechanisms.[5]

We discuss here various combinations of these techniques that allow measurement or control of $[Ca^{2+}]_i$ and simultaneously allow the measurement of membrane currents. These combinations, while adding to the complexity of the experiment and thereby sometimes limiting the available preparations, are necessary in order to study control of $[Ca^{2+}]_i$ and its role in such important neuronal processes as excitation–gating coupling.

Combined Measurement of Membrane Currents with Quantitative Introduction of Ca^{2+}

While inside-out patches allow the most rapid and accurate control of $[Ca^{2+}]$ at the cytoplasmic surface of the membrane, the various methods of introducing Ca^{2+} into the cytoplasm have important advantages in studying Ca^{2+}-activated currents. First, isolated patches remove the channel from its native environment. For instance, the K_D for Ca^{2+} binding to CAN channels is increased when inside-out patches are removed from the cell,[6] Ca^{2+}-activated potassium channels are activated specifically by local Ca^{2+} influx,[7] and CAN channel modulation requires the presence of membrane receptors and cytoplasmic second messenger systems.[8] Second, whole-cell currents cannot be measured in inside-out patches. This eliminates the possibility of observing the simultaneous activation of several Ca^{2+}-activated currents, or the comparison of Ca^{2+}-activated currents through a single channel in the cell-attached configuration with whole-cell currents

[1] E. Neher and G. J. Augustine, *J. Physiol.* (*Lond.*) **450,** 272 (1992).
[2] T. H. Müller, L. D. Partridge, and D. Swandulla, *Pflügers Arch.* **425,** 499 (1993).
[3] J. Herrington, Y. B. Park, D. F. Babcock, and B. Hille, *Neuron* **16,** 219 (1996).
[4] H. J. Kennedy and R. C. Thomas, *Biophys. J.* **70,** 2120 (1996).
[5] I. Llano, R. DiPolo, and A. Marty, *Neuron* **12,** 663 (1994).
[6] Y. Marayama and O. H. Peterson, *J. Membr. Biol.* **81,** 83 (1984).
[7] M. Gola and M. Crest, *Neuron* **10,** 689 (1993).
[8] L. D. Partridge, D. Swandulla, and T. H. Müller, *J. Physiol.* (*Lond.*) **429,** 131 (1990).

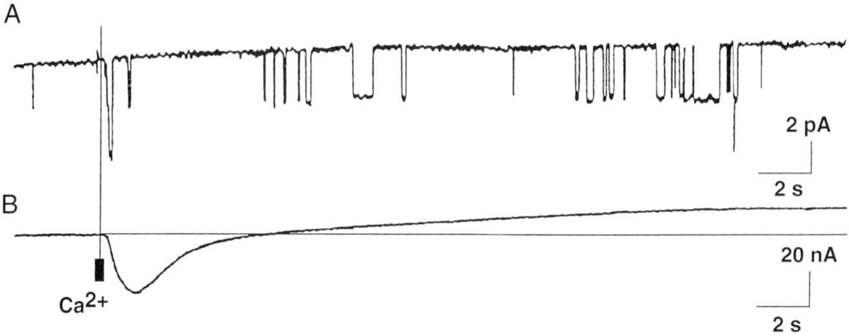

FIG. 2. Quantitative pressure injection of Ca^{2+} in a *Helix pomatia* burster neuron. Simultaneous records were made with (A) a cell-attached patch and (B) a two-electrode voltage clamp. The cell-attached patch contains at least two CAN channels. The patch electrode contained 100 m*M* NaCl and 100 n*M* $CaCl_2$, cell clamped to −50 mV and the patch electrode to +50 mV; patch current was filtered at 150 Hz. Ca^{2+} injection of about 1 pmol. CAN channels continue to open well into the time when the membrane current became outward. [Modified from Fig. 1 in L. D. Partridge and D. Swandulla, *Trends Neurosci.* **11,** 69 (1988) with permission.]

(e.g., Fig. 2). Various methods have been devised for quantitative introduction of Ca^{2+} into the cytoplasm of intact cells. The advantages and disadvantages of some of these methods are considered briefly below.

Iontophoresis

Many ions, including Ca^{2+}, can be introduced into the cytoplasm of cells by iontophoresis from intracellular microelectrodes. A microelectrode containing Ca^{2+} ions is inserted into the cell and Ca^{2+} is introduced by passing a current from the electrode to the cell interior. The iontophoretic current will depolarize the membrane potential unless the cell is voltage clamped or the iontophoretic electrode is part of a "floating current clamp" (see, e.g., Refs. 9 and 10). Because of the requirement for multiple intracellular microelectrodes, this technique is restricted to studies of medium to large neurons. The use of large cells has the advantage of increasing the success of making repeated injections of Ca^{2+} into the same cell. To calculate the amount of Ca^{2+} introduced by iontophoresis, it is necessary to know the transport numbers of the ions in the injection electrode, and the practical determination of these numbers is often difficult. Finally, large Ca^{2+} injections require a large time integral of the injection current. This requires

[9] R. C. Thomas, *J. Physiol.* (*Lond.*) **245,** 20P (1975).
[10] A. Fein and J. S. Charlton, *J. Gen. Physiol.* **70,** 591 (1977).

either a long injection duration or a large current amplitude and Ca^{2+}-containing electrodes have a tendency to clog under these conditions.[11]

Quantitative Pressure Injection

The quantitative pressure injection technique[12,13] has been used effectively for the study of Ca^{2+}-activated currents. In this technique, a Ca^{2+}-containing solution in the tip of an intracellular injection electrode forms a visible interface with an ion-exchange solution in the shank of the electrode. Direct measurement of the movement of this interface is used to determine the amount of Ca^{2+} injected. As with iontophoresis, this technique requires the use of medium to large neurons because of the need for multiple intracellular microelectrodes, although when a single-electrode voltage clamp is used, only two intracellular electrodes are required.[14] Quantitative pressure injection permits the rapid introduction of quantities equal to about 1% of the cell volume into the cytoplasm without change in the membrane potential, even in unclamped neurons, and pressure injections can be successfully repeated many times in the same neuron. This technique has been shown to produce a uniform distribution of Ca^{2+} in the cytoplasm,[12] presumably by transiently overwhelming cytoplasmic Ca^{2+} buffers.

Figure 2 shows an experiment where Ca^{2+} was quantitatively injected into a neuron that was voltage clamped with intracellular microelectrodes and patch clamped in the cell-attached configuration. The whole-cell current shows clearly the activation of multiple Ca^{2+}-activated currents, whereas the single-channel record shows the concurrent activity of two of these Ca^{2+}-activated channels.

Tail Currents

Calcium-activated currents can be studied as tail currents following depolarizing pulses.[15,16] The cell is clamped at a sufficiently depolarized potential to activate voltage-gated calcium channels, and the decaying Ca^{2+}-activated currents are studied during a second test potential. Because tail currents through voltage-activated calcium channels deactivate within a few milliseconds after repolarization, Ca^{2+}-activated tail currents can be recorded in isolation of voltage-activated calcium currents. Because this technique is easily accomplished in the whole-cell configuration of the patch

[11] K. Krnjevic and A. Lisiewicz, *J. Physiol. (Lond.)* **225,** 363 (1972).
[12] G. Hofmeier and H. D. Lux, *Pflügers Arch.* **391,** 242 (1981).
[13] D. Swandulla and H. D. Lux, *J. Neurophysiol.* **76,** 3934 (1985).
[14] L. D. Partridge, *Brain Res.* **647,** 76 (1994).
[15] R. W. Meech and N. B. Standen, *J. Physiol. (Lond.)* **249,** 211 (1975).
[16] H. Tatsumi and Y. Katayasma, *J. Neurosci. Methods* **53,** 209 (1994).

clamp, it can be used on any sized neuron. Figure 1 gives an example of activation of CAN channels as a whole-cell tail current.

This method introduces Ca^{2+} through the plasma membrane into the submembrane compartment adjacent to Ca^{2+}-activated channels and is thus presumably a very physiologic means of activating these channels. Rather substantial depolarization of the cell is necessary to activate sufficiently the voltage-activated calcium channels and these repeated depolarizations may have adverse effects on the neuron under study. To determine accurately the Ca^{2+} influx, it is necessary to block all other voltage-dependent channels so that the measured inward current is carried solely by Ca^{2+} ions. In most central nervous system neurons, potassium currents can be blocked by perfusing the neurons with 120 m*M* CsCl and 10 m*M* tetraethylammonium chloride (TEA-Cl) from the patch pipette. In addition, application of 10 m*M* TEA-Cl from the outside will usually produce a complete block of potassium currents. Most sodium currents can be blocked with 0.5 μM tetrodotoxin (TTX). However, adult rat dorsal root ganglion (DRG) neurons possess TTX-resistant sodium channels and other strategies have to be used to record voltage-gated calcium currents in isolation from sodium currents in these neurons. The presence of channel blocking drugs may also complicate the study of Ca^{2+}-activated channels.

Photolysis of Caged Ca^{2+}

Caged calcium (photolabile calcium chelators) can be used to control quantitatively the $[Ca^{2+}]_i$ in cells (for reviews, see Refs. 17 and 18). Caged compounds currently in use such as nitr compounds (DM-nitrophen), and nitrophenyl-EGTA (NPE) are photolyzed to release calcium by near-UV light. A xenon arc flash lamp or laser can serve as the light source.[18,19] This technique makes it possible to change $[Ca^{2+}]_i$ rapidly and rather uniformly throughout the cell or in a selected subcellular region. This method has advantages over methods such as increasing cell membrane permeability to Ca^{2+} using ionophores, which is slow, or activation of voltage-dependent calcium channels, which causes steep spatial gradients before the intracellular $[Ca^{2+}]_i$ equilibrates. Thorough knowledge of the possible complex chemical properties and photochemistry of caged compounds is required to attribute quantitatively physiologic responses to the changes in $[Ca^{2+}]_i$ caused by photolysis.[18,20] For example, a serious drawback of DM-nitrophen is its binding of other cations such as H^+ and Mg^{2+}. To improve quantifica-

[17] J. H. Kaplan, *Annu. Rev. Physiol.* **52,** 897 (1990).
[18] R. S. Zucker, *Methods Cell Biol.* **40,** 31 (1994).
[19] G. C. Ellis-Davies, J. H. Kaplan, and R. J. Barsotti, *Biophys. J.* **70,** 1006 (1996).
[20] E. Neher and R. S. Zucker, *Neuron* **10,** 21 (1993).

tion, cells may be simultaneously loaded with calcium indicator dyes, which allow direct monitoring of $[Ca^{2+}]_i$. In combination with confocal laser scanning microscopy, photolysis can be restricted to very small volumes. For example, within a millisecond, $[Ca^{2+}]_i$ can be elevated in femtoliter volumes of single dendrites.[21] Combining focal photolysis with calcium imaging and patch clamping provides a practical method to obtain information about Ca^{2+}-induced signaling within subcellular organelles or compartments.[21,22] Multiphoton optical absorption has been used to mediate excitation in laser scanning microscopy. Multiphoton photochemistry confines the uncaging volume to even smaller values. The combination of localized two-photon calcium uncaging with two-photon fluorescence microscopy might open up new ways for the analysis of local calcium signaling (for a review, see Ref. 23).

Combined Measurement of Membrane Currents and $[Ca^{2+}]_i$

Experiments that combine $[Ca^{2+}]_i$ measurements with membrane current recordings have been used to address important questions in cellular electrophysiology (e.g., Refs. 24–27). Such studies have contributed to our understanding of second messenger systems, synaptic mechanisms, and the activation of Ca^{2+}-dependent currents. We concentrate here on one currently very fruitful example where it is necessary to measure $[Ca^{2+}]_i$ while simultaneously measuring membrane current.

Quantitative Measurement of Ca^{2+} Fluxes into the Cell: Fractional Ca^{2+} Currents

Combined whole-cell current and $[Ca^{2+}]_i$ measurements have been used to calculate the permeability of several cation channels to Ca^{2+} (P_{Ca}). The determination of P_{Ca} from increases in $[Ca^{2+}]_i$, however, depends on the knowledge of at least two parameters that are usually not known exactly, namely, the cell volume accessible to Ca^{2+} and the Ca^{2+} buffering capacity of the cytoplasm.

[21] S. S.-H. Wang and G. J. Augustine, *Neuron* **15,** 755 (1995).

[22] P. Lipp, C. Luscher, and E. Niggli, *Cell Calcium* **19,** 255 (1996).

[23] W. Denk and K. Svoboda, *Neuron* **18,** 351 (1997).

[24] G. J. Augustine, *J. Neurosci. Methods* **54,** 163 (1994).

[25] J. A. Verheugen, H. P. Vijverberg, M. Oortgiesen, and M. D. Cahalan, *J. Gen. Physiol.* **105,** 765 (1995).

[26] G. J. Augustine, M. P. Charlton, and S. J. Smith, *J. Physiol.* (*Lond.*) **367,** 143 (1985).

[27] A. Hernandez-Cruz, F. Sala, and P. R. Adams, *Science* **247,** 858 (1990).

In 1993, Neher and co-workers[28,29] introduced a technique that combined whole-cell current recordings with Fura-2 fluorescence recordings for the calculation of fractional Ca^{2+} currents through cation channels. The whole-cell current recording provides a measure of the total current flowing through the cation channels and the fluorescence recording gives a measurement of the influx of Ca^{2+}. This method makes it possible to determine the Ca^{2+} conductance of cation channels at physiologic concentrations of Ca^{2+} and at nonzero electrochemical driving forces. These two features are a significant advantage of this new technique over the more classical methods that require reversal potential measurements under biionic conditions, usually with 100 m*M* $CaCl_2$ outside. Using this method, fractional Ca^{2+} currents have now been measured for a variety of cation channels including native and artificially expressed glutamate and nicotinic acetylcholine receptor channels and also c-GMP-, capsaicin-, and proton-activated ion channels.

Detailed Methods

Most of the techniques discussed here are part of the standard repertoire of cellular electrophysiology (e.g., Refs. 30 and 31). The combination of whole-cell current recordings with microfluorometric measurements of $[Ca^{2+}]_i$ has become a widely used technique, and techniques for introducing Ca^{2+} into the cytoplasm have seen important recent developments. In the following sections, we concentrate on two specific combinations of methods that have been especially productive in the study of Ca^{2+}-activated currents and transmembrane Ca^{2+} flux pathways.

Quantitative Pressure Injection

The fast quantitative injection technique is described in detail in Hofmeier and Lux[12] and is summarized here. Injection electrodes are pulled in two steps. First, the capillary tube is thinned over a length of about 1.5 mm to produce a cylindrical chamber with an inner diameter of about 50 μm. Second, a sharp tip of about 1 μm is produced. The pipette is backfilled with potassium ion exchanger (Corning 477317, Corning Medical, Medfield, MA) and then the injection solution is sucked through the tip into the cylindrical chamber. A visible phase boundary is formed between the ion exchanger and the injection solution and this phase boundary is

[28] R. Schneggenbruger, Z. Zhou, A. Konnerth, and E. Neher, *Neuron* **11,** 133 (1993).

[29] Z. Zhou and E. Neher, *Pflügers Arch.* **425,** 511 (1993).

[30] C. Grynkiewicz, M. Poenie, and R. W. Tsein, *J. Biol. Chem.* **260,** 3440 (1985).

[31] O. P. Hamill, A. Marty, E. Neher, B. Sakmann, and F. J. Sigworth, *Pflügers Arch.* **391,** 95 (1981).

observed under the microscope through an ocular micrometer. Controlled pressure pulses are applied and the resultant changes in the phase boundary represent quantitative injections with a resolution of 10 pl. A small negative current is passed through the injection electrode to prevent Ca^{2+} leak and to indicate successful penetration into a voltage-clamped cell. Typical Ca^{2+} injection solutions contain 100 m*M* $CaCl_2$ and 100 m*M* KCl, pH 7.4 (HEPES buffered).

Quantitative pressure injection has proven to be a powerful tool to activate Ca^{2+}-activated membrane currents in intact neurons where multiple current processes can be compared and where cell-attached single channel currents can be correlated with resultant whole-cell currents.[32,33] These currents have uniformly proved more difficult to activate with procedures that lead to slower and smaller increases in $[Ca^{2+}]_i$. The capability afforded by this technique to inject repeatedly the same known amount of Ca^{2+} has made possible the study of CAN current modulation by a cAMP-dependent second messenger system.[8]

Measurement of Fractional Ca^{2+} Currents

Whole-cell currents through ligand-gated ion channels are recorded using the patch-clamp technique. External and internal solutions are usually designed to maximize the cation current under study and to record voltage-gated calcium currents in isolation from sodium and potassium currents. The following compositions have been successfully used: External solution (in m*M*), 145 NaCl, 10 TEA-Cl, 2.5 KCl, 1.6 $CaCl_2$, 1 $MgCl_2$, 10 HEPES (pH 7.30, adjusted with NaOH); internal solution (in m*M*), 130 CsCl, 20 TEA-Cl, 2 $MgCl_2$, 10 HEPES, 2 Na_2-ATP and 1 K_5-Fura-2 (pH 7.30, adjusted with CsOH).

Under conditions where all incoming Ca^{2+} binds to Fura-2, the fractional Ca^{2+} current, P_f, is given as:

$$P_f = \frac{(\Delta F_{380}/Q)_{\text{cation current}}}{(\Delta F_{380}/Q)_{\text{pure calcium current}}}$$

The term $(\Delta F_{380}/Q)_{\text{pure calcium current}}$ is usually called f_{max}. It is a calibration constant that is identical to the proportionalilty coefficient of the charge carried by Ca^{2+} ions into the cell and the change in emitted fluorescence light intensity due to binding of these Ca^{2+} ions to Fura-2. It must be determined individually for each experimental setup, but is independent of the cell type used[34] and f_{max} can be determined from any pure calcium

[32] T. H. Müller, D. Swandulla, and H. D. Lux, *J. Gen. Physiol.* **94,** 997 (1989).
[33] L. D. Partridge and D. Swandulla, *Pflügers Arch.* **410,** 627 (1987).
[34] E. Neher, *Neuropharmacology* **34,** 1423 (1995).

current. In neurons, this is typically achieved by recording voltage-activated calcium currents, which are assumed to be carried by only Ca^{2+}. Alternatively, inward currents through the cation channel under investigation can be elicited under conditions in which Ca^{2+} is the only permeable cation. This procedure has been used in cells that lack voltage-activated calcium channels.[35] Furthermore, it may be of advantage when Fura-2 and endogenous Ca^{2+} buffers are not uniformly distributed throughout the cell. Small but significant differences between these two calibration procedures have been reported that appear to be most pronounced in nonspherical cells.[36]

To ensure that all Ca^{2+} ions bind to Fura-2, the concentration of Fura-2 must be high enough to overcome the endogenous Ca^{2+} buffers of the cell. Under this condition, the measurement of the change in fluorescence at a Ca^{2+}-sensitive wavelength permits quantification of the amount of Ca^{2+} entering the cell. The change in fluorescence that accompanies any Ca^{2+}-conducting membrane current is then proportional to the amount of Ca^{2+} entering the cell. The concentration of Fura-2 needed to fulfill this condition depends on the Ca^{2+} buffer capacity of the cells under study. For most types of neurons, 1 m*M* Fura-2 is sufficient. Cerebellar Purkinje neurons probably represent one exception since their Ca^{2+} buffer capacity is known to be in the range of 4000 to 5000, which is 10 to 50 times higher than that of most other excitable cells.[5] While recording at one Ca^{2+}-sensitive wavelength is sufficient for the determination of fractional Ca^{2+} currents, recording at two wavelengths makes it possible to demonstrate that the Fura-2 concentration is indeed sufficient for the cell under study. The ratio of fluorescence change to charge flow, f, can then be plotted versus the Ca^{2+}-binding capacity of Fura-2. From background-corrected fluorescence signals excited alternately at two different wavelengths (e.g., 350 and 380 nm), the intracellular concentration of Fura-2 and of free Ca^{2+} can be determined as described by Grynkiewicz *et al.*[30] and Zhou and Neher.[29] From these concentrations, the Ca^{2+}-binding capacity of Fura-2, κ_B, inside the cell can be determined. If the fura concentration chosen is high enough, f reaches a maximum f_{max} with increasing κ_B.

Figure 3 shows an example where P_f was determined for two different types of proton-activated cation channels. Whole-cell recordings and fluorescence measurements are shown on the top and on the bottom, respectively. The slowly inactivating proton-induced current response was accompanied by a decrease in the Fura-2 fluorescence excited at 380 nm (F_{380}) and therefore by an influx of Ca^{2+} (Fig. 3A). In contrast, activation of the

[35] N. Burnashev, Z. Zhou, E. Neher, and B. Sakmann, *J. Physiol.* (*Lond.*) **485,** 403 (1995).
[36] M. Rogers and J. A. Dani, *Biophys. J.* **68,** 501 (1995).

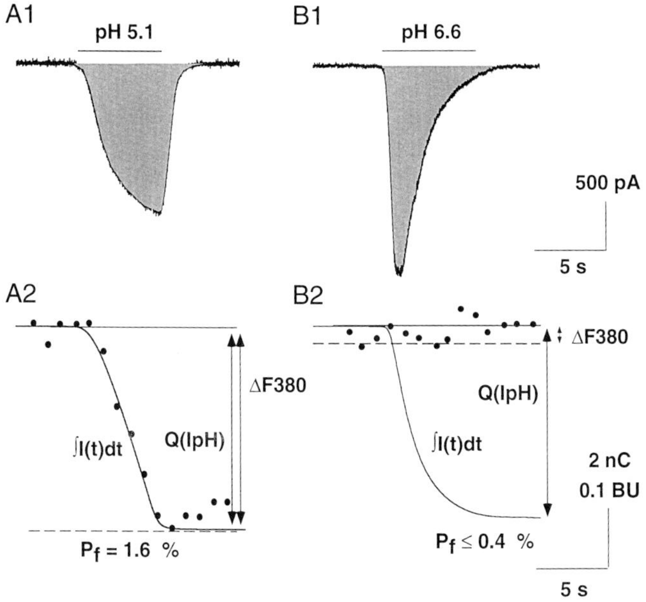

FIG. 3. The fractional contribution of Ca^{2+} to the membrane currents through two types of proton-activated cation channels was determined. (A) Activation of the slowly activating (and inactivating) current by application of pH 5.1 (A1) resulted in a decrease in the fluorescence signal excited at 380 nm, which paralleled the current integral over time [$\int I(t)dt$] (A2). Comparing the fluorescence change (ΔF_{380})/charge [Q(IpH)] ratio of this current with that of a pure Ca^{2+} current (e.g., a voltage-activated Ca^{2+} current) yields a fractional contribution of Ca^{2+} (P_f) to the slowly activating/inactivating proton-induced current of 1.6%. (B) The fast activating/inactivating proton-induced current, which was activated by application of pH 6.6 (B1), was not accompanied by a significant change in fluorescence (B2) indicating that the fractional contribution of Ca^{2+} (P_f) to this current is less than 0.4%. [Modified from Fig. 3 in H. U. Zeilhofer, M. Kress, and D. Swandulla, *J. Physiol.* (*Lond.*) **503,** 67 (1997) with permission.]

fast activating and inactivating proton-induced current had no influence on the F_{380} signal, suggesting that Ca^{2+} did not contribute significantly to this current.

The use of the patch-clamp technique for loading cells with Fura-2 provides several advantages over loading with the membrane permeable derivative Fura-2 acetoxy methylester (Fura-2AM). Cells can be loaded through a patch clamp electrode with a known concentration of Fura-2. This is important because Fura-2 inside the cell not only acts as a Ca^{2+} indicator, but also as a Ca^{2+} buffer. High concentrations of Fura-2 can therefore considerably reduce the intracellular Ca^{2+} response to any given stimulus.

A general problem of ratiometric $[Ca^{2+}]_i$ measurements is the possibility that the distribution of Fura-2 and Ca^{2+} signals may not be equal within the cell. Changes in the regional distribution of the fluorescence ratio over the cell may then not accurately represent changes in $[Ca^{2+}]_i$. Loading cells with Fura-2 via the patch pipette limits the diffusion of the charged Fura-2 molecules into hydrophilic phases of the cell and thereby reduces compartmentalization of Fura-2. It also eliminates problems arising from fluorescence of nonhydrolyzed fura-2AM molecules.

It is well known that physiologic constituents of the cytoplasm can diffuse from the cytoplasm into the patch pipette during whole-cell recordings. Processes that are influenced may include diffusional loss Ca^{2+} ions and Ca^{2+} buffers. However, resting Ca^{2+} levels of cells loaded with either Fura-2 via the patch pipette or with fura-2AM are usually in good agreement. The diffusional loss of Ca^{2+} buffers has been investigated by Neher and co-workers. They found that in half of the adrenal chromaffin cells in their study the Ca^{2+} buffer capacity decreased within 2–5 min.[29] Furthermore, whole-cell recording may disrupt intracellular messenger signaling within the cell or impair the ability of the cell to extrude Ca^{2+} or restore it into storage organelles.

Future Directions

Use of Large versus Small Cells

The most commonly used mode of patch clamping has become the whole-cell configuration largely because this technique permits voltage-clamp studies on cells too small to be clamped with microelectrodes. Slice and culture preparations of mammalian cells have effectively replaced large invertebrate cells for many studies of cellular physiology. For example, in the study of synaptic mechanisms requiring simultaneous pre- and postsynaptic measurements, important work is being published that utilizes invertebrate preparations,[37] but similar studies can now be carried out on mammalian preparations.[38] The invertebrate preparations, especially those from *Aplysia,* continue to have the advantage of a considerable background of information about individual identified neurons in learning and behavior that is important to studies of synaptic plasticity. The 10^9 synapses/mm^3 of the mammalian cortex makes it a miniaturized preparation with definite limitations to the experimentalist. Certain of the techniques discussed here require the use of large neurons where intracellular microelectrodes can

[37] J. X. Bao, E. R. Kandel, and R. D. Hawkins, *Science* **275,** 969 (1997).
[38] J. G. Borst and B. Sakmann, *Nature* **383,** 431 (1996).

be inserted with minimal damage. "Giant" invertebrate neurons still provide an important preparation to be exploited for studies where extensive measurement and control of cellular processes are desired.[4,39]

Measurement of $[Ca^{2+}]_i$ in Cell Organelles and Cell Compartments

The continuing improvement of electrophysiologic and microfluorometric methods has enabled us to look at ever smaller compartments of cells such as neuronal dendrites and even dendritic spines.[40,41] The ongoing development in fluorometric imaging techniques, especially that of confocal and two-photon laser scanning microscopy (TPLSM),[23,42] enables researchers to investigate in great detail the relation between local rises in $[Ca^{2+}]_i$ and its physiologic consequences. In a recent study, Lipp *et al.*[43] filled cardiac myocytes with DM-nitrophen and fluo-3 via the patch pipette. With the use of TPLSM, they could release Ca^{2+} from DM-nitrophen in a very small volume element (about 1–5 μm in diameter) of the cell and then follow the spread of this "artificial Ca^{2+} spark" and the resulting Ca^{2+} waves within the cytoplasm of the cell. These and similar developments will help us to investigate the interactions of local Ca^{2+} signals with their physiologic targets in confined compartments of the cells.

[39] M. Gabso, E. Neher, and M. E. Spira, *Neuron* **18,** 473 (1997).
[40] W. Müller and J. A. Connor, *Nature* **354,** 73 (1991).
[41] R. Yuste and W. Denk, *Nature* **375,** 682 (1995).
[42] K. Svoboda, W. Denk, D. Kleinfeld, and D. W. Tank, *Nature* **385,** 161 (1997).
[43] P. Lipp, C. Amstutz, C. Lüscher, and E. Niggli, *Pflügers Arch.* **433,** R78 (1997).

[22] Determining Ion Channel Permeation Properties

By TED BEGENISICH

Introduction

Considerable effort is being put into the determination of the mechanism of ion channel permeation. This effort includes many components, but a first step must be to determine which ions can permeate the pore. Additional information can be obtained by determining how many ions may simultaneously occupy the pore and by investigating the ability of small ions to block the pore. These issues are not necessarily independent of one another and, as a result, the proper interpretation of experimental results needs to consider all as part of a whole not as isolated, unrelated pieces. This article

METHODS IN ENZYMOLOGY, VOL. 293
0076-6879/98 $25.00

describes some of the methods used to measure ion channel selectivity, permeation, and blocking and some of the complexities of interpreting the resulting data.

Single-Ion Nernst Potential

If a pore is permeable only to a single ion (X) of valence z, the net current through the pore is zero at a potential (V_X) given by the Nernst (or equilibrium) potential for that ion [Eq. (1)]:

$$V_X = (RT/zF)\ \ln(a_o/a_i) \tag{1}$$

where a_o and a_i represent the external and internal activities of ion X, respectively and R, T, and F have their usual thermodynamic meanings; RT/F at 20° is approximately 25 mV. This voltage is an equilibrium property and does not depend on the theoretical framework from which it is derived. It is a function only of the ion concentrations and is independent of any impermeant ions and independent of the presence of any fixed charges on the membrane.

The activity terms are related to the ion concentrations by external and internal activity coefficients, γ_o and γ_i. Activity coefficients are used as a simple way to account for the (usually) small interactions among ions in aqueous solutions. Thus, in terms of concentrations Eq. (1) becomes:

$$V_X = (RT/zF)\ \ln(\gamma_o[X]_o/\gamma_i[X]_i) \tag{2}$$

The activity coefficients depend on the ionic strength of the solution. In normal physiologic solutions, a monovalent ion has an activity coefficient of approximately 0.85, so the ion activity is not much different from the concentration. At higher salt concentrations, the coefficients can become rather small and divalent ions, owing to their higher valence, have rather smaller coefficients than monovalent ions. However, if the internal and external solutions are of similar ionic strength, the activity coefficients in Eq. (2) could be omitted:

$$V_X = (RT/zF)\ \ln([X]_o/[X]_i) \tag{3}$$

not because they are near unity but rather because they are approximately equal. If the solutions are not sufficiently similar, then these coefficients should not be ignored. With this caution, the remaining discussion will use concentrations rather than activities.

Equation (3) suggests a test that a pore is permeable to a specific ion: measure the channel zero-current (or "reversal") potential with known internal and external concentrations of the test cation and see if this potential is equal to the Nernst potential for the ion. An accurate measurement of the reversal potential requires proper consideration of liquid junction potentials.[1]

The use of Eq. (3) as a test for ion selectivity implies knowledge of the intracellular concentration of the ion of interest. Unfortunately, in many experimental situations the internal concentration is not accurately known. This is sometimes true even for the usual whole-cell variant of the patch-clamp technique since it relies on diffusion of the pipette contents into the cell interior.[2] Thus, rather than compare the expected Nernst potential to a single measurement, the reversal potential can be measured at several external concentrations of the test ion and (often) plotted as a function of the logarithm of the external ion concentration. Equation (3) predicts a linear relationship in such a plot with a $58/z$ mV/decade slope (at 20°), and such a finding demonstrates that the pore is highly selective for the tested ion compared to other ions present.

Occasionally, a perfectly linear relationship is found when using physiologic solutions (e.g., Lucero and Pappone[3]). More often the results are in agreement with Eq. (3) only at high concentrations of the test ion, and at low concentrations the slope is often much less than predicted (e.g., Sah *et al.*[4]). Such a result suggests that the pore is not perfectly selective for the tested ion.

Ion Channel Selectivity

Few, if any, channels are permeant to only a single type of ion, and the expected dependency of pore zero-current (or reversal) potential (V_{rev}) on ion concentration is, in general, both complex and model dependent. However, if two ions (X and Y) of the same valence (z) are considered, a relatively general and simple equation can be derived (see Hille[5] for some of the details of these computations):

$$V_{rev} = \frac{RT}{zF} \ln \frac{P_X[X]_o + P_Y[Y]_o}{P_X[X]_i + P_Y[Y]_i} \tag{4}$$

where P_X and P_Y are the permeabilities of ions X and Y.

[1] E. Neher, *Methods Enzymol.* **207,** 123 (1992).
[2] J. M. Tang, J. Wang, and R. S. Eisenberg, *Methods Enzymol.* **207,** 176 (1992).
[3] M. T. Lucero and P. A. Pappone, *J. Gen. Physiol.* **94,** 451 (1989).
[4] P. Sah, A. J. Gibb, and P. W. Gage, *J. Gen. Physiol.* **92,** 264 (1988).
[5] B. Hille, *in* "Ionic Channels of Excitable Membranes." Sinauer Associates, Sunderland, Massachusetts, 2nd Ed., p. 345, 1992.

The reversal potential of Eq. (4) actually depends only on the relative permeabilities of the two ions:

$$V_{\text{rev}} = \frac{RT}{zF} \ln \frac{[\text{X}]_\text{o} + (P_\text{X}/P_\text{Y})[\text{Y}]_\text{o}}{[\text{X}]_\text{i} + (P_\text{X}/P_\text{Y}[\text{Y}]_\text{i}} \tag{5}$$

At high external concentrations of ion X, Eq. (5) predicts a linear relationship between the measured reversal potential and the logarithm of $[\text{X}]_\text{o}$. Deviations from linearity are expected at lower concentrations owing to the second term in the numerator of Eq. (5). The relative X to Y permeability can be obtained by measuring the channel reversal potential at various ion concentrations and fitting Eq. (5) to the data.

Biionic Conditions

While Eq. (5) suggests experiments that can determine the relative ion permeability of a pore, a simpler method can be used if the solution on the cytoplasmic face of the membrane can be accurately controlled. This experimental condition allows the use of a simplified form of Eq. (4) if ions of type X are the only permeant ones in the external solution and Y ions are the only permeant ions in the solution on the cytoplasmic side of the pore:

$$V_{\text{rev}} = \frac{RT}{zF} \ln \frac{P_\text{X}[\text{X}]_\text{o}}{P_\text{Y}[\text{Y}]_\text{i}} \tag{6}$$

Equation (6) can be rewritten in a form that allows computation of the relative ion permeabilities:

$$P_\text{X}/P_\text{Y} = ([\text{Y}]_\text{i}/[\text{X}]_\text{o}) \exp(zV_{\text{rev}}F/RT) \tag{7}$$

Thus, a measurement of the current reversal potential with knowledge of the ion concentrations in these biionic conditions allows the determination of the relative ion permeabilties. Chandler and Meves[6] used this technique to show that the squid axon Na^+ channel is permeable to several other cations include Li^+, K^+, Rb^+, and Cs^+.

Internal Ions Not Known But Constant

Hille[7] showed how relative ion permeabilities can be determined even if the internal medium cannot be controlled but is constant. In this technique, the channel reversal potential is measured twice: once with only ion X in the external soluton ($V_{\text{rev,X}}$) and again with only ion Y in the external solution ($V_{\text{rev,Y}}$). Under these conditions, the relative permeability can be

[6] W. K. Chandler and H. Meves, *J. Physiol.* (*Lond.*) **180,** 788 (1965).

[7] B. Hille, *J. Gen. Physiol.* **58,** 599 (1971).

computed [using a form of Eq. (4)] from the difference between these two measurements:

$$P_X/P_Y = ([X]_o/[Y]_o) \exp[zF(V_{rev,X} - V_{rev,Y})/RT] \qquad (8)$$

Using this technique, Hille[8] showed that the selectivity of the Na^+ channel in myelinated nerve is similar to that of the squid axon Na^+ channel and extended the list of permeant ions to include several organic cations. Tsushima *et al.*[9] have used this same technique to investigate the role of certain amino acids in determining the selectivity of the rat skeletal muscle isoform of the voltage-gated Na^+ channel.

It is important to note that Eqs. (4)–(8) were derived for ions of the same valence and for a pore that obeys independence.[10] Independence means that the movement of one ion is independent of the presence of any others. Ion permeation through most channels is rather complex and seems to have properties consistent with the interaction of the ions with each other while they are in the pore. Consequently, interpretations of data through the use of these equations need to done carefully although if the recording conditions are specified, the computed permeability ratios can be useful empirical parameters.

Classification of Ion Pores

One-Ion Pores

In one-ion pores, the presence of a single ion in the permeation pathway prevents entry (perhaps through electrostatic repulsion) of any other ion. Hille[10] has shown that Eq. (7) applies to one-ion pores even if independence is not obeyed. The permeabilities in Eq. (7) may not be constant for such a pore but, rather, may be functions of membrane potential but not functions of ion concentration. However, in many situations the voltage dependence may be small or nonexistent. Consequently, Eq. (7) is very useful for determining relative selectivities even for one-ion pores.

The current through one-ion pores can be rather sensitive to the type and concentrations of the ions present. Consequently, ion selectivity is best judged not by the relative current carried by different types of ions but rather from measurements of the reversal potential. Ion currents are also affected by many pore-blocking ions (see later discussion) but, providing

[8] B. Hille, *J. Gen. Physiol.* **59,** 647 (1972).

[9] R. G. Tsushima, R. A. Li, and P. H. Backx, *J. Gen. Physiol.* **109,** 463 (1997).

[10] B. Hille, *in* "Membranes—A Series of Advances, Vol. 4: Lipid Bilayers and Biological Membranes: Dynamic Properties" (G. Eisenman, ed.), p. 255. Dekker, New York, 1975.

they are impermeant, they will not contribute to the measured reversal potential, and Eq. (7) can still be used.

Multi-Ion Pores

Pores can also be simultaneously occupied by more than a single ion. Not surprisingly, the permeation properties of such multi-ion pores are more complex than those of one-ion pores. Nevertheless, Hille and Schwarz[11] have shown that Eq. (7) may apply even to these more complicated situations but that the permeability ratio may be a function of ion concentration. Consequently, a finding of concentration-dependent permeabilities in Eq. (7) indicates that the pore under study has multi-ion characteristics. It may be difficult to separate the effects of voltage and concentration because changing ion concentrations will necessarily shift the reversal potential to a new value. Sometimes, however, it is possible to separate these two effects. For example, squid axon Na^+ channel selectivities are concentration but not (significantly) voltage dependent.[12]

Pore-Blocking Studies

Some aspects of the structure of ion channel pores can be obtained from studies of small ions that block the movement of permeant ions. As described by Woodhull,[13] the dissociation constant for a blocking ion, of valence z, that binds to a site within the membrane electric field is given as

$$K_d(V_m) = K_d(0) \exp(-\delta z V_m F/RT) \tag{9}$$

where $K_d(0)$ is the zero-voltage dissociation constant and δ is the location of the binding site as a fraction (measured from the inside in this example) of the membrane voltage. Then, the fraction of current (channels) blocked, $f(V_m)$, will not only be a function of the concentration of blocking ions, [B], but also a function of voltage:

$$f(V_m) = \frac{[B]}{[B] + K_d(V_m)} \tag{10}$$

A determination of the fraction of current blocked as a function of membrane potential will, through the application of Eqs. (9) and (10), yield a value for the electrical distance to the blocking site. In one of the best

[11] B. Hille and W. Schwarz, *J. Gen. Physiol.* **72,** 409 (1978).

[12] M. D. Cahalan and T. Begenisich, *J. Gen. Physiol.* **68,** 111 (1976); T. Begenisich and M. D. Cahalan, *J. Physiol. (Lond.)* **307,** 217 (1980).

[13] A. Woodhull, *J. Gen. Physiol.* **61,** 687 (1974).

studies of this type, Miller[14] used monovalent and bisquaternary ammonium compounds to determine the spatial dependence of the electric potential in part of the pore of a K^+ channel from sarcoplasmic reticulum.

Occupancy of a pore by multiple ions significantly complicates the use of Eq. (10) for interpreting the results of blocking studies. A good example of the difficulty of obtaining the fractional field location of a blocking site can be found in the work of Adelman and French.[15] A literal application of Eq. (10) to Cs^+ block of K^+ channels reveals that the blocking site reached by external Cs^+ is more than 100% across the membrane. This nonsensical result can be a consequence of a multi-ion pore[16] and, indeed, the data of Adelman and French can be reproduced by a specific multi-ion pore model.[17] For block to occur in a multi-ion pore, the permeant ion must vacate the blocking ion binding site. Consequently, the voltage dependence of block will reflect not only the voltage-dependent binding of the blocking ion to this site but also the voltage-dependent evacuation of the pore by permeant ions.[11]

Determining Pore Occupancy

Mathematical descriptions of selectivity and block of multi-ion pores require the specification of the number and location within the membrane electrical field of both permeant and blocking ions. Doing this accurately for any ion channel certainly approaches the limits of current methodological capabilities. However, certain types of unidirectional flux measurements can provide lower estimates for the maximum number of ions that may simultaneously occupy the pore. Hodgkin and Keynes[18] measured the unidirectional influx and efflux of K^+ through the voltage-gated, delayed rectifier K^+ channels in cuttlefish axons and compared their results with the prediction made by Ussing[19] for independent movement of a single type of ion:

$$m_e/m_i = [K]_i/[K]_o \exp(V_m F/RT) \tag{11}$$

where m_i and m_e are the unidirectional K^+ influx and efflux, respectively; or, in terms of the K^+ equilibrium potential, V_K:

$$m_e/m_i = \exp[(V_m - V_K)F/RT] \tag{12}$$

[14] C. Miller, *J. Gen. Physiol.* **79,** 869 (1982).

[15] W. J. Adelman, Jr., and R. J. French, *J. Physiol.* (*Lond.*) **276,** 13 (1978).

[16] B. Hille and W. Schwarz, *Brain Res. Bull.* **4,** 159 (1979).

[17] T. Begenisich and C. Smith, *in* "Current Topics in Membranes and Transport," Vol. 22. Academic Press, New York, 1984.

[18] A. L. Hodgkin and R. D. Keynes, *J. Physiol.* (*Lond.*) **128,** 61 (1955).

[19] H. H. Ussing, *Acta Physiol. Scand.* **19,** 43 (1949).

Hodgkin and Keynes[18] found that this equation provided a poor description of their data but a good fit was obtained if the right-hand side of Eq. (11) was raised to a power, n':

$$m_e m_i = \exp[n'(V_m - V_K)F/RT] \tag{13}$$

which, to best describe their data, required a value near 2.5. To understand the meaning of the n' parameter, Hodgkin and Keynes[18] investigated several theoretical descriptions of ion permeation and showed that an expression like Eq. (13) is expected if K^+ channels simultaneously contain several ions. Subsequently, a variety of theoretical approaches[11,20–23] have all led to the same conclusion: the value of n' in Eq. (13) can approach, but not exceed, the number of ions simultaneously occupying the pore. The 2.5 value of Hodgkin and Keynes[18] means there can be at least three K^+ ions in the cuttlefish channel.

Measurements of unidirectional ion fluxes[24] have been done with cloned K^+ channels and the results suggest that *Shaker* K^+ channels, one member of the voltage-gated family, can be simultaneously occupied by at least 4 K^+ ions.[25] Neyton and Miller[26] used quite different techniques including an investigation of the K^+ sensitivity of Ba^{2+} block and demonstrated the existence of four ion binding sites in the pore of Ca^{2+}-activated K^+ channels.

Conclusion

Many ion channels have complex permeation properties best understood in terms of simultaneous pore occupancy by several ions. The interaction of ions within the channel pore complicates the interpretation of certain types of experiments and it is useful to consider a variety of approaches with special attention to the multi-ion nature of channel pores. The pore in different types of K^+ channels appears to allow occupancy by at least four ions and it remains a challenge to find where, within the pore, the ions are located and to determine how so much electrostatic charge can be confined to such a relatively small space.

[20] S. Bek and E. Jakobsson, *Biophys. J.* **66,** 1028 (1994).
[21] M. F. Schumaker and R. MacKinnon, *Biophys. J.* **58,** 975 (1990).
[22] H.-H. Kohler and K. Heckmann, *J. Theor. Biol.* **79,** 381 (1979).
[23] D. G. Levitt, *Ann. Rev. Biophys. Biophys. Chem.* **15,** 29 (1986).
[24] P. Stampe and T. Begenisich, *Methods Enzymol.* **293** [30], 1998 (this volume).
[25] P. Stampe and T. Begenisich, *J. Gen. Physiol.* **107,** 449 (1996).
[26] J. Neyton and C. Miller, *J. Gen. Physiol.* **92,** 569 (1988).

[23] Voltage-Dependent Ion Channels: Analysis of Nonideal Macroscopic Current Data

By RÜDIGER STEFFAN, CHRISTIAN HENNESTHAL, and STEFAN H. HEINEMANN

Introduction

Voltage-gated ion channels comprise a large family of proteins that can alter their conductance in response to changes in the membrane electric field. They are of chief importance for many cellular processes, in particular for the rapid electrical signaling of neuronal and muscle cells. Owing to the rapid development of molecular biological methods, various protein families have been identified that, when expressed in host cells, give rise to voltage-dependent current characteristics. Most notable is the superfamily of voltage-gated ion channels (for a review, see Catterall[1]). These channels are selective for K^+, Na^+, or Ca^{2+} and they share a common pattern of charged amino acid residues in the putative fourth transmembrane segment in each of the four subunits or internal protein repeats. This S4 segment is thought to harbor part of the voltage sensor, which "measures" the electric field across the membrane and thereby initiates conformational protein rearrangements that ultimately lead to channel opening or inactivation.

Detailed functional assay of such ion channels is only feasible under conditions when the voltage-dependent responses of the ion channels can be measured precisely; that is, the transmembrane potential must be clamped in a predictable way and the resulting current should be measured with sufficient time and current resolution. The ultimate goal of performing voltage-clamp experiments is often to derive a quantitative description of the ion channel properties. Such a description usually comprises a set of parameters (equilibrium or rate constants) associated with a specific model. Therefore, an electrophysiologic experiment with voltage-gated ion channels involves various steps that can be repeated in an iterative fashion at various levels of sophistication so as to refine the model or simply the model parameters (Fig. 1).

The purpose of the following sections is to summarize methods and strategies that proved useful in improving both the sophistication and the efficiency of voltage-clamp experiments. Aspects to be considered during

[1] W. A. Catterall, *Science* **242,** 50 (1988).

0076-6879/98 $25.00

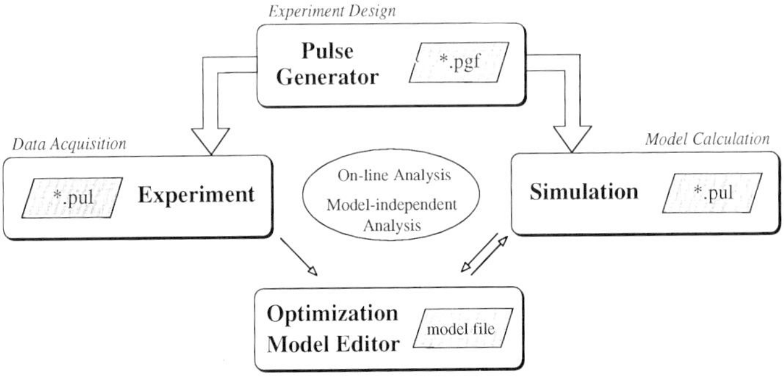

FIG. 1. Schematic diagram illustrating the use of a pulse generator for both data acquisition and kinetic simulation. For model optimization and the model editor see text.

the recording of voltage-clamp data (e.g., hardware considerations, space clamp performance, series resistance compensation) are found elsewhere in this volume. We concentrate here on the analysis of data that are strongly affected by various experimental limitations and that require special software tools. We discuss how ion channel kinetics are simulated using rate models and how these models are fit to experimental data sets. In particular, we stress how such an analysis task is facilitated when corresponding software tools are integrated into a modular program package that also supports data acquisition. The following sections address several questions which arise particularly when data of nonideal voltage-clamp experiments are to be analyzed. Among them are questions on leak correction as well as bandwidth and rise time limitations. These topics are discussed using two-electrode voltage-clamp (TEV) ionic current and gating current data from potassium channels expressed in *Xenopus* oocytes. We chose these examples because they are heavily affected by limitations requiring very careful analyses. At the end we discuss examples of how large kinetic models can be fit to heterogeneous data sets and how data simulation is used to test results arising from nonstationary noise analyses.

From Measurements to Model-Dependent Analyses

In this section we briefly summarize how various levels of sophistication are applied to analyze voltage-clamp data for voltage-dependent ion channels.

On-Line and Off-Line Analysis

Electrophysiologic experiments are often strongly interactive processes. The experimenter has to make decisions rather quickly during a running experiment, for example, one has to change the bath solution or the pulse protocols depending on the results of the ongoing experiment. Therefore, electrophysiologic software has to provide versatile on-line analysis procedures that allow for rapid and faithful decisions about the acquired data. These on-line analysis results are usually simple measures taken from predefined sections of the raw data (e.g., mean, minimum, maximum of the current as a function of the time or voltage) or simple derivative values of these (e.g., ratios of two measured values). In some cases one can even implement some kind of fits to the raw data in order to reduce the noise level and take a parameter of this fit function as the analysis result (e.g., fitting a third-order polynomial function to the peak current and calculating the maximum/minimum). However, in all of these cases the major criterion is flexibility and speed as precision can be achieved in a more time-consuming off-line analysis.

Model-Independent Analysis

As already mentioned for on-line analysis methods, the first step in analysis is usually to determine very simple values that are in some way characteristic of the measured data (e.g., relationship of peak current versus voltage; time to reach the 90% current level). This approach is completely model-independent and, therefore, rather objective. The next step after these simple "objective measurements" would be to use simple mathematical functions to fit sections of the data traces. These functions (e.g., exponentials) can be used to parameterize the data, that is, to derive a small set of values that unambiguously characterizes the raw data. Strictly, this approach is not model independent, because a mathematical function already implies some model. However, at the stage where the derived parameters are not directly associated with functional parameters of the ion channel under investigation, this approach can still be referred to as model-independent characterization or parameterization of ionic currents or ion channel behavior. This approach may sound trivial but it is very important to do this before sophisticated model building is attempted, because the derived parameters may give strong hints as to the underlying molecular mechanisms, which then can be translated into more sophisticated kinetic schemes. As an example we may consider the activation kinetics of a voltage-gated ion channel on a step depolarization. When these are fit by the simple equation

$$I(t) = I_1(\infty)(1 - e^{-t/\tau_1}) + I_2(\infty)(1 - e^{-t/\tau_2})^n \quad (1)$$

an exponent of $n = 4$ is indicative of the contribution of a Hodgkin–Huxley gating mechanism,[2] that is, the estimated exponent n may yield a good

[2] A. L. Hodgkin and A. F. Huxley, *J. Physiol. (Lond.)* **117,** 500 (1952).

prediction of the number of independent gating units involved. Such an analysis approach was applied by Terlau *et al.*[3] showing that rat-*eag* (*ether-à-go-go*) potassium channels undergo a biphasic activation where the slow phase $[I_2(t)]$ depends on extracellular Mg^{2+} and n approaches 4 at high Mg^{2+} concentrations. These results suggested that Mg^{2+} can interact with each of the four channel subunits.

Very important criteria for this part of the analysis are the flexibility and speed, because one usually has to test many alternative analysis approaches in order to get a reasonable overview of possible underlying models. Therefore, we would like to stress that the importance of the speed and the user-friendliness of the software used for this model-independent analysis cannot be overestimated.

Modeling of Ion Channel Kinetics

Kinetic models play an important role in the analysis of voltage-dependent ion channels in that they may provide further insight into the molecular mechanisms underlying ion channel function. Thus, after a preanalysis of the data, yielding rough estimates of the number of exponential components, apparent rate constants, equilibrium potentials, and, from the voltage dependencies of these components, apparent gating charges, one is seeking model descriptions that characterize the channel's behavior under a wide range of experimental conditions (e.g., voltages). For example, after a preanalysis describing the activation kinetics of rat-*eag* channels with Eq. (1), Terlau *et al.*[3] extended the activation model to a serial link of two Hodgkin–Huxley-like activation mechanisms, where the first, slow one depends on Mg^{2+} and the second rapid process describes the usual voltage-dependent activation of potassium channels. The computation of the kinetic behavior of such models requires the definition of conformational states (including their permeation characteristics) and of transitions rates between these states (including the charge transfers, i.e., voltage dependencies), and finally the numerical solution of this system, which is briefly summarized here.

We start by describing the channel kinetics using reaction schemes with N_S discrete states and with voltage-dependent reaction rates q, which, according to Eyring's rate theory, assume instantaneous transitions[4]:

$$\text{state } i \underset{q_{ji}}{\overset{q_{ij}}{\rightleftharpoons}} \text{state } j$$

[3] H. Terlau, J. Ludwig, R. Steffan, O. Pongs, W. Stühmer, and S. H. Heinemann, *Eur. J. Physiol.* **432,** 301 (1996).

[4] R. Fitzhugh, *J. Cell. Comp. Physiol.* **66,** 111 (1965).

$$q_{ij} = q_{ij}^{0} e^{\delta_{ij} z_{ij} e_0 V / kT} \tag{2}$$

$$q_{ji} = q_{ji}^{0} e^{-(1-\delta_{ij}) z_{ij} e_0 V / kT} \tag{3}$$

where δ_{ij} is the relative position of the activated state in the electric field relative to states i and j, z_{ij} is the effective valence translocated during the transition $i \rightarrow j$, e_0 is the unitary electronic charge, k is Boltzmann's constant, and T is the absolute temperature. The underlying mathematical model is a stochastic process, where for the calculation of macroscopic currents the description of the overall ensemble of realizations (i.e., single-channel currents) is used. The theory of stochastic processes and in particular the application of Markov processes to the modeling of ion channel kinetics can be found in the literature.[5,6]

As a result of using computer-aided data acquisition systems, signals are represented in digitized (i.e., time-discrete) form. However, the description of digital stochastic signals can be performed with both time-discrete and time-continuous models as outlined in the following section.

Markov Models for Ion Channel Kinetics. For a digital Markov signal, or Markov chain, the transition probabilities depend only on the immediately preceding signal value. The stochastic process is called a time-discrete Markov process and the vector of state probabilities $\mathbf{p}$ is given by

$$\mathbf{p}([n + 1]\Delta t) = \mathbf{p}(n\Delta t)\mathbf{P}(\Delta t) \tag{4}$$

where $\mathbf{P}$ is the matrix of time-discrete transition probabilities and $n\Delta t$ is an integer multiple of the sample interval. For the calculation of macroscopic currents as a response to an ideal voltage step, the description of a time-continuous Markov process is conveniently used. With Eq. (4) and the limit $\Delta t \rightarrow 0$, a set of linear first-order differential equations is derived that corresponds to the reaction equations known from enzyme kinetics:

$$\frac{d\mathbf{p}(t)}{dt} = \mathbf{p}(t)\mathbf{Q} \tag{5}$$

In Eq. (5), $\mathbf{Q}$ is the matrix of transition rates q_{ij} and q_{ji}, which depend on external parameters such as voltage and temperature [Eqs. (2) and (3)]. As a consequence of the Markov property, however, the transition rates

[5] D. R. Cox and H. D. Miller, "The Theory of Stochastic Processes," Methuen, London, 1968.

[6] D. Colquhoun and A. G. Hawkes, *Proc. Roy. Soc. London B* **199,** 231 (1977).

are rate constants and do not depend on time. From the condition that only one state can be occupied at one time, it follows that

$$q_{ii} + \sum_{j=1}^{N_S} q_{ij} = 0 \qquad j \neq i \tag{6}$$

This is used to calculate the diagonal elements of $\mathbf{Q}$. While a diffusion process can be seen and understood as a generalized Markov process, the forward and backward transitions between state i and state j of a Markov process can be treated as Poisson processes with rates q_{ij} and q_{ji}, respectively. The dwell time in state i is, therefore, exponentially distributed with mean $\bar{\tau}_i = -1/q_{ii}$.

The time courses of the macroscopic ionic $[I(t)]$ and gating currents $[I_g(t)]$ are derived as ensemble averages of the time-continuous Markov process. They are digitized for storing according to the specified sample interval Δt:

$$I(n\Delta t) = N_C \sum_{i=1}^{N_S} \hat{i}_i p_i(n\Delta t) \tag{7}$$

$$I_g(n\Delta t) = N_C \sum_{i=1}^{N_S} \sum_{j=1}^{N_S} p_i(n\Delta t) q_{ij} z_{ij} e_0 \tag{8}$$

where $\hat{i}_i$ is single-channel current level in state i and N_C is the number of identical channels. Numerical methods for the integration of Eq. (5) are outlined in the following section.

Numerical Solution. The basic method to calculate the time course of the probability function is the numerical integration of Eq. (5), where, as a first approach, Euler's method can be employed. Starting from the ordinate $\mathbf{p}_n$, the gradients at these points are used to calculate the new values $\mathbf{p}_{n+1}$, where the gradients are part of the differential equations:

$$\mathbf{p}([n+1]h) = \mathbf{p}(nh) + h\,\dot{\mathbf{p}}(nh) \tag{9}$$

The accuracy depends mainly on step size h, which is usually taken to be at least five times smaller than the sample interval. In the case of state models, the dynamic properties of the system particularly depend on the transition rates. An adjustment of the step size can, therefore, be performed by taking the minimum mean dwell time as a criterion, for example, $h = \bar{\tau}_{\min}/2$.

Analytical Solution. The set of differential equations [Eq. (5)] corresponds to an asymmetric eigenvalue problem. Given the system at equilibrium at time zero with state distribution $\mathbf{p}(0)$, the formal solution

$$\mathbf{p}(t) = \mathbf{p}(0)e^{\mathbf{Q}t} \tag{10}$$

describes the relaxation of the system into a new equilibrium.[7] The matrix exponential $e^{\mathbf{Q}t}$ can be expressed in terms of the matrix of eigenvectors $\mathbf{M}$ and the diagonal matrix of eigenvalues Λ by

$$e^{\mathbf{Q}t} = \mathbf{M}e^{\Lambda t}\mathbf{M}^{-1} \tag{11}$$

where for computing the eigenvalues and eigenvectors of the matrix $\mathbf{Q}$, the QR algorithm[8] can be used, for example. With Eqs. (10) and (11) the probability of state i as a function of time becomes

$$p_i(t) = \sum_{j=1}^{N_S} p_j(0) \sum_{k=1}^{N_S} m_{jk} m_{ki}^{-1} e^{\lambda_k t} \tag{12}$$

Equilibrium Distribution. The steady-state probabilities at equilibrium can be calculated by solving the system of linear equations:

$$\mathbf{p}_\infty \mathbf{Q} = \mathbf{0} \tag{13}$$

Alternatively, $\mathbf{p}_\infty$ can be determined by using the analytical solution [Eq. (12)]. As a result of Eq. (6), the transition matrix $\mathbf{Q}$ is singular so that one eigenvalue must equal zero; the corresponding eigenvector equals $\mathbf{p}_\infty$.

Program Environment for Analysis and Simulation

Pulse Generator and Data Storage

A pulse generator module of electrophysiologic software should provide versatile voltage protocols that can be modified rapidly. In addition, as much information as possible should be stored in the pulse generator to facilitate on-line as well as off-line analysis. In many experiments, for example, it is obvious before the start of the experiment in which segments of the expected data which type of analysis should be performed. This pulse generator information is largely independent of the acquisition hardware and can be used alternatively for data simulation. Therefore, it is advantageous to store it in a separate data file, which can then be used by various modules. This approach is applied in the *Pulse+PulseFit* acquisition software (HEKA elektronik, Lambrecht, Germany). In this package, the additional information that characterizes the hardware and further information acquired at execution time (e.g., environmental parameters, such as temperature) are stored in another file. Both files are organized in a tree structure providing very efficient access to data grouped according to their purpose

[7] M. Braun, "Differential Equations and their Application." Springer-Verlag, Berlin, 1978.

[8] W. H. Press, S. A. Teukolsky, W. T. Vetterling, and P. B. Flannery, "Numerical Recipes in C." Cambridge University Press, Massachusetts, 1994.

(e.g., data from one cell, families of pulses, segments during a stimulation protocol). This structuring facilitates an overview of the acquired data as an image of the data tree is easily displayed on the computer screen. In addition, it allows fast access to specific data types during manual or automatic data analysis because these data trees can be kept in memory throughout the experiment or the analysis. The raw data, which take up considerably more memory space, are stored on hard disk in a linear fashion and are only loaded into memory when needed.

Storage of Analysis Results

Structured data files with high-speed access as mentioned earlier require a considerable degree of predefinition (i.e. links between different data locations and their meaning). Such fixed data structures are acceptable for electrophysiologic acquisition systems because the types of data acquired do not change very much from experiment to experiment. This situation, however, is different when the storage of analysis results has to be considered, because a much larger degree of flexibility is required. To some extent, simple analysis results can be stored in the acquired data file for the sake of speed. In *Pulse+PulseFit,* for example, the on-line analysis results are stored in the tree structure thereby providing the opportunity to make quick comparisons between the analysis results of different data sections online. The storage of the results from more complicated analyses, however, requires much more data space.

Analysis File. In *PulseFit* the results of model-independent analyses (e.g., polynomials, exponentials, Gaussian functions) are stored in a separate file. This file, however, has a tree structure that is compatible with the original data file so as to maintain quick access and unambiguous links between raw data and the analysis results. The great advantage of this storage organization is that the secondary analysis (e.g., analysis of the results derived from raw data as a function of the test potential) is easily supported.

Secondary Data File. If analysis results also contain time course data (e.g., the results from nonstationary noise analysis), a second data file is created in *PulseTools* (HEKA elektronik). This output data file has exactly the same structure as the experimental data file and can, therefore, be subsequently used for data fits in *PulseFit* or data modeling in *PulseSim* (see next section).

More complicated analysis results will require deviations from this approach of fixed data structures. One possibility is to store for each analyzed data entity (e.g., current trace, family of traces) a complete description of the analysis (e.g., the fit function in one of the common computer languages) including the set of resulting parameters. This approach requires a text parser and interpreter as well as a relatively free text format. In particular, performing model-dependent data analysis using large kinetic schemes requires the storage of transition matrices in addition to a set of parameters. In the next section a model editor used in the program *PulseSim* is briefly described. This model editor stores the **Q** matrices in combination with parameters and text equations that are needed to define more complicated dependencies among the parameters.

Program for Kinetic Modeling and Simulation

The modular data structure outlined in the previous section favors the development of a program library that is not exclusively used for data acquisition or primary analysis software. Therefore, a channel modeling software module was developed for the program *PulseSim,* which can be treated in a similar way to a module for data acquisition. The close link between data acquisition and data prediction has several advantages: (1) use of common software modules; (2) consistent user interface; (3) modeling of "real" experimental situations that include limited clamp speed and causal filtering as well as nonideal leak subtraction. Kinetic modeling of this kind can be a powerful tool for electrophysiologic research (including teaching) and the development of analysis software.

The described data acquisition environment provides data structures for both, stimulation pulse protocols and storage of acquired data. The same files are also accessed and generated by the simulation program, that is, data are simulated based on a pulse generator file (*pgf*). If this *pgf* file is part of an experimental data set, the simulation program can perform model parameter optimization by applying least squares criteria for the comparison of experimental and simulated data traces. The results of a model-based experiment emulation is a data file stored exactly like experimental data. In addition to the set of data files, the specified or optimized kinetic model has to be stored in a separate *model file* (Fig. 1), which can be a text file of defined structure to be edited with any text editor. For the calculation of complex kinetic schemes all transition rates between the channel states have to be defined. Such a **Q** matrix for a 16-state model, for example, requires the definition of 240 transition rates—a task that is hard to do without introducing errors. Therefore, it is desirable for model parameters as well as the model structure to be edited in a user-friendly way.

Model Editor. High flexibility of the model editor in *PulseSim* is achieved by incorporation of a text parser that is used to define dependencies between various model entries. Using this approach, errors in specifying large models can be successfully avoided. In *PulseSim,* the model editor allows one to specify four different sections of the model file: parameters, variables, states, and transitions. Parameters are predefined entries that are used for model calculations; most of them can be determined experimentally (see later) and can be fixed during model optimization (number of channels, reversal potential, voltage clamp rise time, bandwidth, delay). Variables are defined on-line and are used to assign values to text identifiers that are parsed and interpreted. The definition of states comprises a state label, the single-channel conductance, the total charge movement with respect to state zero, and coordinates used for the graphical display of the kinetic model (see, e.g., Fig. 3). Transitions comprise information about the charge transfer, the symmetry of the energy barrier, a voltage offset, and the rate; all entries are expressed as text string formulas to be parsed. Each parameter or variable has an extra flag specifying whether or not it is to be optimized during data fits.

In addition, it has proved useful to define external variables that are directly linked to parameters in the input experimental data files. This feature allows the incorporation of varying experimental parameters into the models such that simultaneous data fits to different data sets obtained under different experimental conditions can be performed (see later).

Examples for modeling and simulation of macroscopic ionic and gating current using *PulseSim* are shown in the subsequent sections. The calculation of fluctuating current signals and the time courses of current variance and autocovariance are mandatory tools for the implementation of nonstationary noise analysis and can also be used for model optimization (see last section). An additional important feature of *PulseSim* is the consideration of the *p/n* leak correction procedures.

Leak Correction

A problem associated with most electrophysiologic recordings is the discrimination between the desired signal, for example, the current flowing through a population of ion channels, and distorting superimposed signals, generally referred to as leak. This "leak" can have "technical" problems like shunt current flowing around a not-so-tight seal or holes in the cell membrane generated by the impalement of microelectrodes. In addition, current flowing through transport systems that are not of interest for the present study are also often called "leak." Because this is a general problem in electrophysiologic recordings, data obtained for voltage-activated ion channels or transport systems are also affected by current transients arising from the charging of the membrane and the recording electrodes whenever the membrane potential is changed. Such capacitive artifacts can be largely eliminated in many amplifiers by analog compensation circuits. However, even the best compensation will often leave some artifact that has to be compensated by other means. A simple approach would be to record current, then completely block the currents of interest by pharmacologic means (if possible), then record a reference signal (i.e., a current response in the absence of the ion currents of interest), which can be used to correct for the leak and capacitive artifacts of the first record. This procedure, however, even if one is lucky enough to investigate a channel that can be blocked completely without blocking any "leak" components, has many drawbacks. One is that the capacitance of the system may change during the application of blockers. Thus, in many cases one prefers to make use of well-defined nonlinearities of the ion channel open probability. Ideally, one tries to use a voltage regime in which the ion channels of interest are not open and thus do not contribute to the current signal.

Design of Leak Templates

The design of leak pulse protocols strongly depends on the experimental situation (voltage range of channel activity, stability of the recording configuration, noise level, etc.) and on whether the leak pulses are to be used for on-line or off-line leak correction. The first case is very important for viewing corrected data during the ongoing experiment so as to be able to decide on further steps in this experiment. An automatic on-line leak correction, however, may have some deficits with respect to correction performance such that off-line methods may provide significantly better results. Thus, ideally, leak protocols are designed such that they allow both on-line correction for the running experiments and, if necessary, off-line correction for more detailed analysis or for the generation of high-quality figures.

The important parameters for the generation of leak templates are the reference potential from which the template is started (leak holding), the scaling factor (r), the number of leak pulses to be acquired and averaged (n), the time interval between test and leak pulses (leak delay), and information about alternating leak pulse polarity (r alternates between positive and negative values for every leak pulse or for every set of leak pulses when on-line signal averaging is performed).[9] Typically $n = 4$ and $r = 0.25$ ($p/4$ correction[10]), such that the noise of the corrected main trace is increased by a factor of $\sqrt{2}$.

The leak holding potential is mainly determined by the available voltage range that is free of channel activity and by the stability of the recording configuration. The number of leak pulses is determined by the noise consid-

[9] S. H. Heinemann, *in* "Single Channel Recording" (E. Neher and B. Sakmann, eds.), p. 53. Plenum Press, New York, 1995.

[10] C. M. Armstrong and F. Bezanilla, *J. Gen. Physiol.* **63,** 533 (1974).

erations and the time required for the experiment. Besides the statistical noise arising from averaging fluctuating leak records, systematic errors can be introduced owing to the limited dynamic range of the A/D–D/A (analog–digital/digital–analog) hardware. This arises primarily because the precision of the voltage protocol is limited by the minimum D/A unit. Therefore, the leak protocol may not be exactly an *r*-fold scaled version of the test protocol. Although this is not usually a serious problem (provided one is using at least a 12-bit D/A converter) it may become one in special cases, in particular when generating *p/n* records for protocols that contain ramp or sine segments, for example.

Storage of *p/n* traces requires a compromise between flexibility, speed, and disk space. Ideally, all *n* individual leak responses should be stored (and for the sake of speed also the averaged scaled leak pulse) in addition to the corrected and uncorrected test data. This storage method enables one to discard "bad" leak records in an off-line procedure. Typically, however, only the averaged leak response is stored.

The most severe limitations in acquiring leak pulses are usually the required recording time and possible instabilities due to holding the potential at very negative values, which often causes unintended instabilities. Therefore, one has to find a compromise between the time spent acquiring the *p/n* traces and the performance of the leak correction (i.e., the introduced noise). In particular, when using pulse protocols in which some segments are rather long, one can optimize the *p/n* correction by acquiring *p/n* responses only in those parts of the protocol that contain changes in potential and which are important for further analysis. One way of solving this problem is to introduce so-called conditioning segments into the pulse protocol. Conditioning segments are segments of constant voltage; they differ from "normal" constant segments in that they are not considered for the generation of *p/n* templates. Additionally, one can choose whether or not to acquire data during such segments in the main protocol. In Fig. 2 a protocol for the recording of recovery characteristics is shown, illustrating the situation in which no data are recorded during the conditioning segments. If there are data in these segments, leak correction has to consider a linear interpolation of *p/n* current before and after this segment. Figure 2 illustrates how conditioning segments can reduce the time required for the experiment and how the *p/n* correction is improved.

Linear Off-Line Leak Correction

Off-line leak correction has several advantages over on-line correction. Leak traces that are badly affected by artifacts can be eliminated before use and leak traces can be smoothed, that is, noise can largely be eliminated

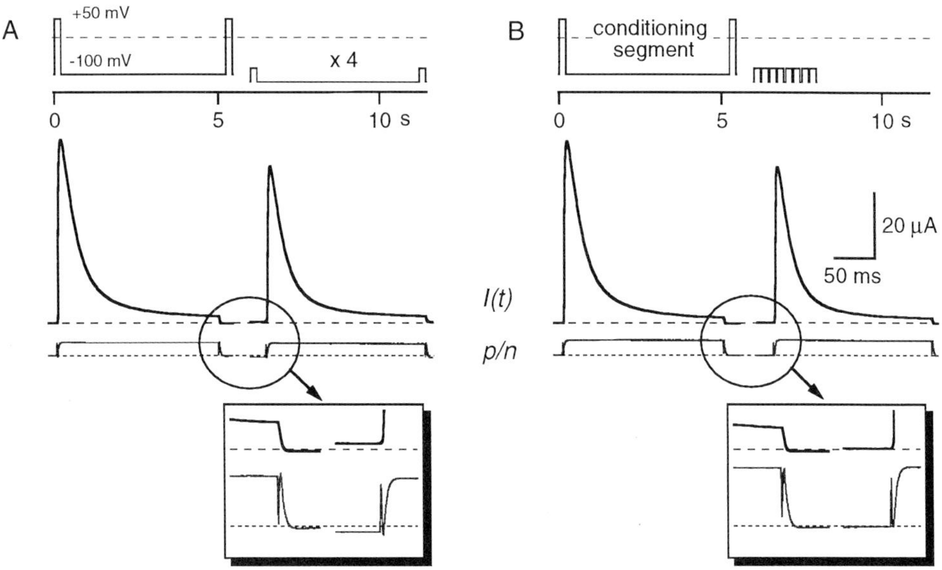

FIG. 2. Illustration of *p/n* leak correction using conditioning segments. Two-electrode voltage-clamp data of rKv1.4 potassium channels expressed in *Xenopus* oocytes were obtained after application of twin-pulse protocols in order to measure the kinetics of recovery from fast channel inactivation. *Top,* pulse protocol; *middle,* leak-corrected current data and *p/n* leak data; *bottom,* magnified data section. (A) Standard *p*/4 protocol with leak pulses following the main test pulses. The data acquired during the central part of the interpulse interval are not shown. (B) Between the two test pulses a "conditioning segment" was inserted, that is, no data are digitized in this segment and they are skipped in the *p/n* protocols. This requires a shorter time for data acquisition (in this case four times 5 sec) and a more faithful leak correction is obtained (compare bottom panels) because very long sojourns at very negative leak holding potentials, which can activate distorting endogenous current components, are eliminated.

by fitting appropriate functions to the traces. Usually, a set of linear and multiexponential functions is sufficient to describe the time course of *p/n* responses. In addition, leak traces from successive recordings can be accumulated in case that there is no significant drift in leak and/or capacitance. For symmetrical three-segment protocols in which only the central segment is varying in potential (*IV* series), leak responses from all pulses can be accumulated and used for correction after appropriate scaling. For such simple protocols, it is actually even better to generate *p/n* protocols with fixed segment amplitudes so as to optimize the signal-to-noise ratio. This requires the use of variable *r*-factors for the successive pulses of one series. This procedure is suited to both on-line and off-line correction methods.

Nonlinear Off-Line Leak Correction

All correction methods discussed thus far rely on the assumption that the leak and capacitive currents depend linearly on the voltage and that in the voltage regime where *p/n* traces are acquired no signals of interest are recorded.

The first criterion is often not valid if one considers ionic current contributions of channels and transporters that are not currently under investigation as "leak" currents. In such cases little can be done other than trying to block these components efficiently or selecting a different leak holding potential. Further nonlinearities can arise from the amplifier (e.g., saturating components when the bandwidth is too high, nonlinearity in the R_S compensation, low-quality electronic components) or from the A/D–D/A hardware. Another source is the biological membrane itself, which may undergo very small changes in electrical capacitance as a function of the potential. This phenomenon, however, which is referred to as *electrostriction,* is only a minor problem when recording in the biological membrane voltage range of approximately ±100 mV.

Under some experimental conditions it is not possible to find a leak holding potential that meets the requirements for linear leak correction. A test for nonlinearities in the range of the leak holding potential is quickly done by averaging short oppositely directed pulses from the leak holding potential. In a linear case the result should be a noisy trace without transient components. If this is not the case, one can try to characterize the residual components carefully and to describe them by a simple function (e.g., an exponential function), which is then used for nonlinear leak correction. This approach was successfully applied for the correction of gating current signals from cloned potassium channels that show nonlinear charge movement at potentials as low as −160 mV.[11,12] However, this procedure is very tedious and to some degree arbitrary because one has to make an assumption about "real" and "capacitive" signals in the *p/n* records.

As the major goal of recording voltage-dependent ionic currents or gating currents is often to derive a kinetic model that reflects the molecular mechanisms involved, at some point sooner or later in the analysis one has to decide on a model that is to be compared with the recorded data. Therefore, instead of choosing one "model" for nonlinear leak correction and one model for channel gating, one may want to try to describe both phenomena by the same model. This is done by fitting the model to the "real" linearly leak-corrected data, that is, both the test data and the *p/n*

[11] W. Stühmer, F. Conti, M. Stocker, O. Pongs, and S. H. Heinemann, *Eur. J. Physiol.* **418,** 423 (1991).

[12] S. H. Heinemann and F. Conti, W. Stühmer, *Methods Enzymol.* **207,** 353 (1992).

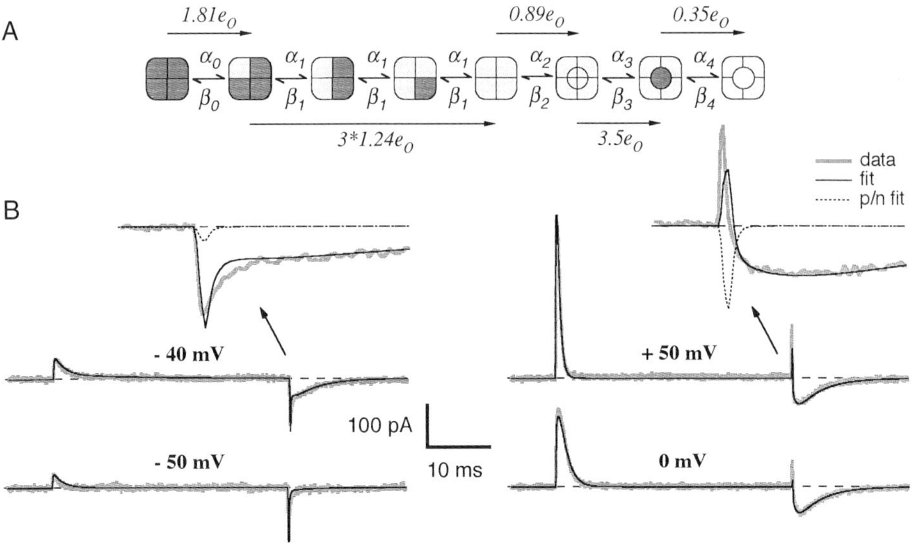

FIG. 3. Model prediction of *p/n* overcompensation. A kinetic model for *Shaker* potassium channel activation[16] with the indicated gating charge transfers (A) was fitted to a set of on-cell macropatch gating current data considering possible gating current relaxations in the *p/n* voltage range. (B) A selection of *p/n*-corrected currents is shown as gray traces. The black curves are the results of the data fit showing that the *p/n* overcompensation artifact in the Ig_{off} section is partially accounted for by the model. The insets show magnifications of the Ig_{off} data of -100 mV after steps to -40 and $+50$ mV. The dotted curves are the nonlinear *p/n* signals predicted by the model. The data traces were obtained using the double-mutant W434F · T449R of *Sh*Δ, which does not permit ionic current flow under physiologic conditions. For illustration purposes, a standard *p*/4 leak protocol with a leak holding potential of -100 mV was used; *p/n* protocols going in the negative direction (negative leak size) would result in smaller *p/n* overcompensation artifacts.

trace have to be calculated based on the model before comparison. An example of such an approach is shown in Fig. 3 where we analyzed gating currents from a *Shaker* channel mutant with a deleted N-terminal end and additional mutations in the pore region, which largely eliminate ion current conduction (*Sh*Δ W434F · T449R[13,14]). The corrected data show an interesting phenomenon at the beginning of the tail current. This overcompensation artifact is thought to be at least in part created by gating charge relaxations in the *p/n* traces (i.e., nonlinear leak signals which were subtracted in excess[11]). The figure illustrates how part of this overcompensation artifact is predicted by a model for *Sh*Δ channel gating when the *p/n* signal is

[13] E. Perozo, R. MacKinon, F. Bezanilla, and E. Stefani, *Neuron* **11,** 353 (1993).
[14] K. McCormack, W. J. Joiner, and S. H. Heinemann, *Neuron* **12,** 301 (1994).

considered in the calculations. The remaining difference is an indication that the model did not completely suffice in describing this phenomenon, which might be related to an apparent alteration of the dielectric properties of the channels after a conformational change.

Determination of Amplifier-Dependent Parameters

For the development of kinetic models of ion channel function, limitations of the voltage clamp such as the true voltage profile or distributed filters have to be considered carefully. In particular, during the optimization of model parameters based on macroscopic currents derived from two-electrode voltage clamp experiments (for methods, see Stühmer[15]), the nonideal nature of the data is obviously not negligible. In the following section we describe how the rise time of the voltage as well as low-pass filtering can be considered during model calculations. Methods for determining and verifying these amplifier-dependent parameters are devised that may help to reduce the number of free parameters during optimization.

Low-Pass Filtering

Gaussian Filter with Linear Phase. High-order Bessel filters are mostly used for low-pass filtering of voltage-clamp signals as they perform an optimal step response in the time domain. A good approximation of high-order Bessel filters is the Gaussian low-pass filter system[17] with the frequency response

$$A(f) = A_0 e^{-\ln 2/2(f/fc)^2} \tag{14}$$

where the exponent factor ln2/2 results from the condition that f_c equals the -3-dB corner frequency. Below the corner frequency, a general property of Bessel filters is that the envelope or group delay does not depend on the frequency. To approximate this property of linear phase, the phase function of the Gaussian filter is defined as

$$\Theta(f) = 2\,\pi f t_0 \tag{15}$$

where t_0 is the filter delay. The impulse response of the Gaussian filter with linear phase is obtained by Fourier back-transformation

$$h(t) = \frac{A_0}{\sigma_g \sqrt{2\pi}} e^{-(t-t_0)^2/2\sigma_g^2}; \qquad \sigma_g = \frac{\sqrt{\ln 2}}{2\pi f_c} \tag{16}$$

[15] W. Stühmer, *Methods Enzymol.* **207,** 319 (1992).

[16] E. Stefani, L. Toro, E. Perozo, and F. Bezanilla, *Biophys. J.* **66,** 1011 (1994).

[17] D. Colquohoun and F. J. Sigworth, *in* "Single Channel Recording" (E. Neher and B. Sakmann, eds.), p. 483. Plenum Press, New York, 1995.

The filter coefficients are determined by digitizing the impulse response [Eq. (16)]. The number of filter coefficients is taken to be $4\sigma_g/\Delta t$, for example, so that an approximate *finite impulse response* (FIR) filter is obtained. Therefore, considering the filter delay, the current response of the Gaussian filter is calculated by convoluting the current with the filter coefficients shifted by t_0.

Determination of Filter Parameters. Usually it is not sufficient to rely on the product specifications describing the bandwidth characteristics of an amplifier or external filter. Therefore, the filter parameters f_c and t_0 have to be determined experimentally. We illustrate this for a typical two-electrode voltage-clamp amplifier (TEC10 series, npi electronic, Germany). Instead of a real cell, a simplified model circuit, where the membrane capacitance is neglected, is used to measure the current responses to sinusoidal stimulations. Figure 4 shows the frequency characteristic of both the gain and the phase. The frequency of the low-pass filter in this amplifier is continuously tunable within three ranges, which elucidates the importance of bandwidth calibration. Estimates for the corner frequency were obtained by fitting Eq. (14) to the frequency characteristic of the gain (*A*), whereas estimates for the filter delay were obtained by fitting a straight line to the frequency characteristic of the phase (*B*). The estimated filter parameters for three front-panel settings of 2.1, 1.2, and 0.3 kHz were 2.18, 0.63, and 0.22 kHz for f_c, and 0.13, 0.45, and 1.21 ms for the filter delay t_0. This result indicates that some values deviate considerably from the corresponding specifications on the front panel of the amplifier. In addition, it is confirmed that the filter delay depends linearly on the reciprocal value of the corner frequency[18]:

$$t_0 = \frac{1}{k_0 f_c} \tag{17}$$

Fitting Eq. (17) to the determined f_c and t_0 values (Fig. 4C) yielded a factor k_0 of 3.56. For comparison, according to the theoretical group delay of a three-pole, four-pole, and eight-pole Bessel filter, the factor k_0 amounts to 3.82, 3.02, and 1.97, respectively.

Figure 4D shows the measured step responses of an ohmic model cell superimposed with the corresponding simulations using the estimated filter parameters. The simulated step responses match the measured step responses well. Deviations mainly occur during the early rising phase of the current, which is a result of the acausal properties of the Gaussian filter. The long filter delay leads to quasicausal properties of the filter so that these deviations are relatively small. However, the results can be further

[18] U. Tietze and C. Schenk, "Halbleiter-Schaltungstechnik." Springer-Verlag, Berlin, 1983.

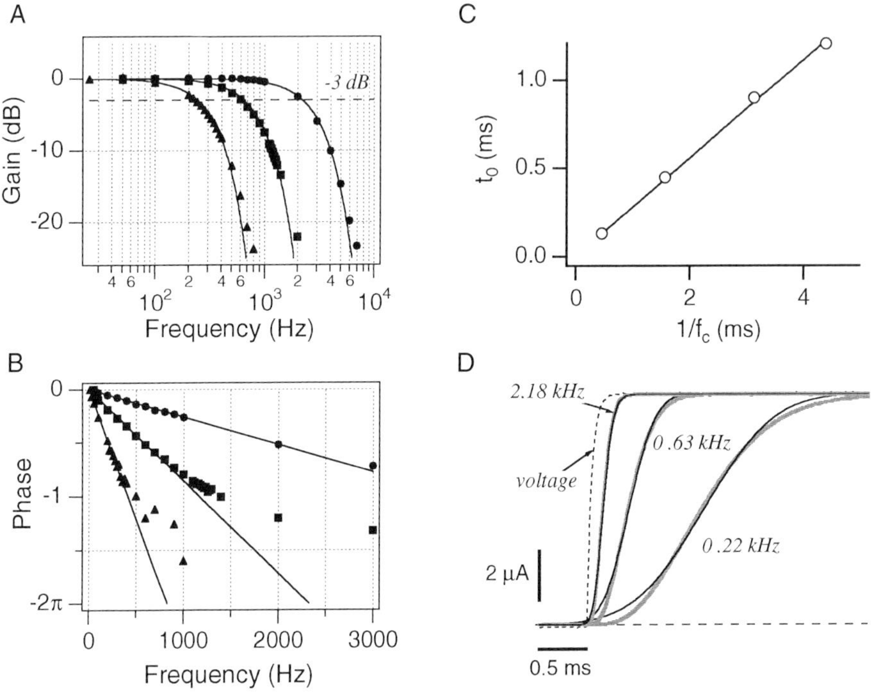

FIG. 4. Filter characteristics of a TEC10 series amplifier. (A) Gain, measured with application of sinusoidal stimulations, for three front-panel settings of the low-pass filter: circle, 2.1 kHz; square, 1.2 kHz; triangle, 0.3 kHz. (B) Phase shift introduced by the filter. For fit functions see text. (C) Filter delay as a function of $1/f_c$ with a linear fit according to Eq. (17). (D) Current responses (gray) of an ohmic model cell to voltage "steps" from 0 to +100 mV (dashed) for the indicated filter settings. The black curves are simulated step responses using the values for f_c and t_0 determined from parts (A) and (B).

improved by the design of a Gaussian filter that meets the causality criterion as described in the subsequent paragraph.

Simulation of Causal Bessel Filter. Causality of a Gaussian filter can be enforced by the elimination of filter coefficients at "negative points in time." As a first approach, this can be simply done by considering a rectangular window function, shifted by t_0, during convolution. However, the best results were obtained with a more "smoothed" transition of the window function. A commonly used window function for the design of FIR filters is the Hann window or $\cos^2$ window[19]:

[19] A. V. Oppenheim and R. W. Schafer, "Digital Signal Processing." Prentice Hall, Englewood Cliffs, New Jersey, 1975.

$$w(n) = \frac{1}{2} - \frac{1}{2}\cos\left(\frac{2\pi n}{N_w - 1}\right) \qquad 0 \le n \le N_w - 1 \qquad (18)$$

The filter coefficients for f_c = 0.63 kHz and the window functions are depicted as an example in Fig. 5A. The steepness of the window function N_w was taken to equal the number of filter coefficients at "negative time" which provided the best results. Figure 5B shows the simulated step response compared with the corresponding measurement. The simulation almost exactly fit the measured, causal step response so that the proposed method can be used to simulate the Bessel filter of a typical voltage-clamp amplifier.

The modification of the impulse response by a window function in the time domain is altering the frequency response of the filter. The altered frequency response can be calculated by convoluting the original frequency response with the Fourier transform of the window function. However, the purpose of the proposed method is to optimize the step response of the

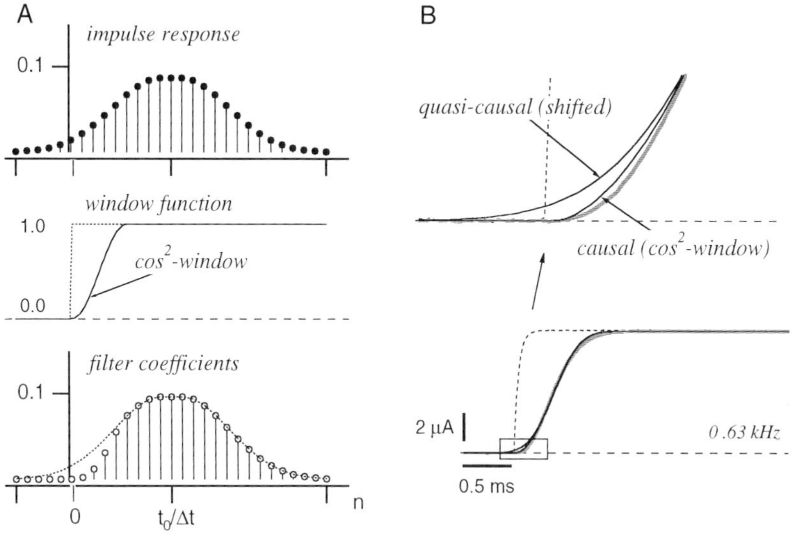

FIG. 5. Causal Gaussian filter. (A) Impulse response of an acausal Gaussian filter with filter delay t_0 (quasi-causal). Causality is enforced by multiplying the corresponding filter coefficients with a $\cos^2$ window, which eliminates the filter coefficients at negative time and removes a discontinuity at time zero. After renormalization, a new set of filter coefficients is obtained (bottom). (B) Comparison of a measured current response (gray) to the indicated voltage profile (dashed) with simulated responses using both a quasi-causal Gaussian filter and a causal filter smoothed with an additional $\cos^2$ window. The inset shows a magnified section around time zero.

digital filter in the time domain after the filter parameters f_c and t_0 were determined in the frequency domain; thus, the knowledge of the final frequency response is not mandatory for further analysis.

Rise Time Limitations

Determining Rise Time Parameters. In the case of TEV measurements, the voltage measured at the potential electrode provides a first-order approximation of the clamp rise time during data acquisition. The time course of the voltage can often be described almost exactly by a single-exponential function, where the time constant τ_r will be used as an amplifier-dependent parameter for model calculation. However, the voltage measured at the tip of the potential electrode only represents one voltage sample, which is obviously not identical to the voltage applied to all parts of the cell membrane. Distributed filters can contribute to the rise time by introducing an additional delay.

The step response of the entire system can be measured by a passive or an active cell model as outlined earlier for the determination of filter parameters. However, this approach can only roughly emulate the real situation for the rise time, so that a method has to be employed that allows the determination of both the rise time of the voltage and additional delays of the current under more realistic experimental conditions.

In Fig. 6, measured ionic current of *Sh*Δ is shown as a response to a voltage step from +30 to +40 mV, voltages at which the maximum

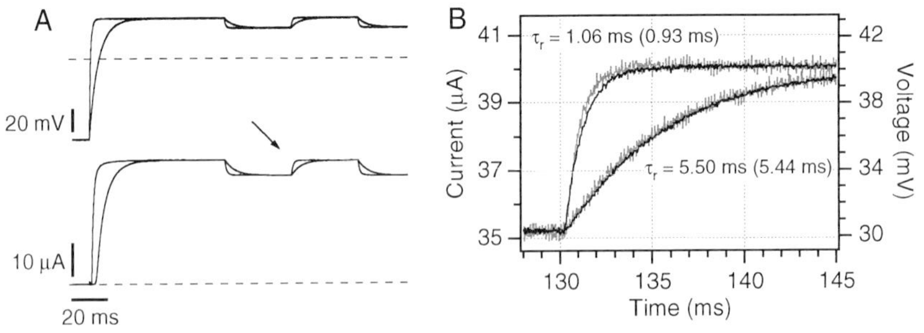

FIG. 6. Verification of rise time parameters. (A) *Sh*Δ currents (bottom) as a response to the voltage protocol shown at the top for two rise time settings of the amplifier. The arrow points to the onset of the relevant segment, which is depicted magnified in part (B). (B) Current responses (black) to voltage steps (gray) from +30 to +40 mV according to the protocol shown in part (A). The voltage is superimposed with a time delay of six sample intervals (240 μs; for details see text). The time courses of the current can be fit with a single exponential function yielding the indicated time constants. The values in parentheses are the time constants for the time courses of the voltage.

channel open probability is reached. During this voltage step the nonlinear current due to channel gating, therefore, is negligible, so that the measured current approximates the step response of the system. A superposition of the measured voltage and the current response shows that, if an additional delay is considered, the shape of the current matches the shape of the voltage. In this particular example, the delay amounts to six sample intervals (each 40 μs); the delay of one sample interval results from the A/D conversion as measured independently. Figure 6 illustrates further how this delay does not depend on the rise time. The time courses of both the current and the voltage can be well described by a single-exponential function, yielding almost the same time constants. For further details regarding the determination of delays between the theoretical and the actual voltage stimulus see Sigworth and Zhou.[20]

Considering Rise Time during Simulation. As a result of a nonideal voltage step, the transition rates are a function of time and the basic Markov property is violated. For the calculation of the probability functions $\mathbf{p}(t)$, however, numerical methods such as Euler's integration method can still be applied without modification by recalculating the $\mathbf{Q}$ matrix for each point in time. Assuming that the voltage is approximately constant within a short time interval Δt, the probability distribution for this time interval can also be calculated analytically by

$$\mathbf{p}([n + 1]\Delta t) = \mathbf{p}(n\Delta t)e^{\mathbf{Q}\Delta t} \tag{19}$$

In the case of an exponential rise of the voltage, the current response can be calculated iteratively using Eq. (19) until the voltage changes only slightly (e.g., until $5\tau_r$). In a final step, the remaining time interval can then be calculated as usual. For an arbitrary time course of the voltage, the measured voltage trace can also be used directly as a template for calculation.

Equation (19) can be seen as a combination of a numerical and an analytical method. The advantage is that it is more accurate and faster compared with a purely numerical method. Moreover, the step size of a numerical method has to be adapted according to the transition rates of the specific model. In contrast to that, for a typical exponential rising phase of the voltage ($\tau_r \approx 0.5$–2 ms), an appropriate sample interval of the data is sufficient as a model-independent value for the step size Δt in Eq. (19).

Validation of Filter and Rise Time Parameters

Simulation of Modified Amplifier Parameters. As illustrated in Figs. 4 and 5, the determined amplifier parameters faithfully describe the step

[20] F. J. Sigworth and J. Zhou, *Methods Enzymol.* **207,** 746 (1992).

responses measured with a model cell. The next step is to test the predictions under real experimental conditions. As an example we measured gating currents of *Sh*Δ W434F channels[13] with TEV methods after the application of 5 m*M* 4-aminopyridine (4-AP). It has previously been shown that 4-AP reduces the number of voltage-dependent gating transitions during activation[14] leading to symmetrical gating charge movement even at depolarizing voltages. Furthermore, if depolarization steps are performed at very low potentials, the derived gating currents can be successfully described by a simple two-state model as shown in Fig. 7. Thus, gating currents from *Sh*Δ W434F channels at negative potentials and in the presence of 4-AP may be used for internal calibration of the system responses.

The kinetic parameters of the two-state model were fitted to current responses to voltage steps from −120 to −100 mV, where the specified filter and rise time parameters were kept constant (Fig. 7A). Subsequently, gating currents measured with different rise time and filter settings were compared with the model predictions by adjusting the corresponding filter and rise time parameters for calculation (Fig. 7B). The results clearly demonstrate that this approach is a useful method for verifying the estimated filter and rise time parameters before they are used for further model calculations.

Prediction of Nonideal Voltage-Clamp Data. An additional method to verify the amplifier-dependent parameters is the comparison of patch-clamp data with data obtained with TEV methods. For this purpose, relatively

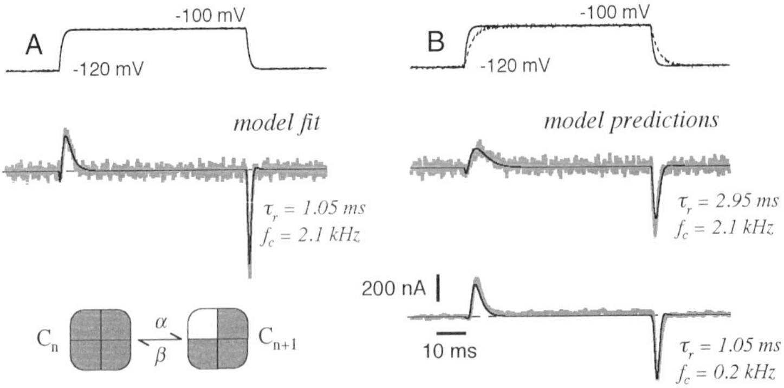

FIG. 7. (A) Gating currents of *Sh*Δ W434F channels were measured from whole oocytes in the presence of 5 m*M* 4-AP. A two-state model was optimized for the measured current trace (gray), given the independently measured voltage rise time and low-pass filter settings. The resulting model fit is superimposed (black). (B) Voltage and current records from the same oocyte as in part (A) but with different rise time and filter settings. The traces superimposed on the current records were obtained by using the model parameters of part (A) considering the modified amplifier settings.

fast gating currents prove useful for investigating merits and limits of any voltage-clamp amplifier.

Gating currents of *Sh*Δ W434F · T449R channels were measured from macropatches and are shown in Fig. 8A. Superimposed are the simulated gating currents from the optimized kinetic model shown in Fig. 3. The fixed amplifier parameters for rise time and filter were 2 μs and 5 kHz, respectively, so that these gating currents represent almost ideal voltage-clamp data. The gating currents in Fig. 8B were measured with TEV methods. The oocytes used for patch clamp and TEV were taken from the same batch in order to avoid batch-to-batch variability. The simulations superimposed on the TEV data were obtained by adjusting only the parameters for rise time, filter, and number of channels, indicating that the determined filter and rise time parameters correctly predict the TEV data.

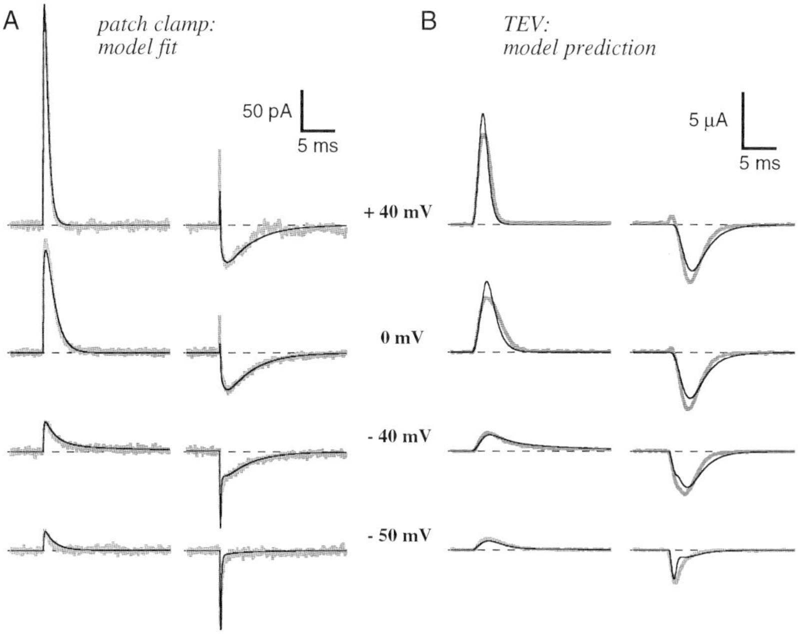

FIG. 8. (A) Gating current traces of *Sh*Δ W434F·T449R channels obtained from macropatch recordings (gray) are shown together with the superimposed results of a fit using the sequential activation model[16] as shown in Fig. 3. (B) Gating current records from the same batch of oocytes as those used for part (A) obtained with TEV methods. The superimposed black traces are model predictions based on the model parameters obtained from part (A) and by fitting the number of channels. The amplifier parameters (rise time and filter) were independently determined. Data in A and B were obtained using different *p/n* protocols.

Tools for Modeling and Simulation

Simultaneous Data Fits

Kinetic models for ion channel gating are ideally optimized if as much as possible experimental information is used during modeling. Thus, an important task for modeling software is to provide versatile tools that allow the consideration of more than one single data sweep for model optimization. In the following we outline which points have to be considered during simultaneous model fits to heterogeneous data sets and an example for application is discussed.

Given the data structure outlined, simultaneous model fits to more than one single sweep are easily implemented, because each sweep in *Pulse* files is linked to a complete description of the pulse protocol. Several individual data sweeps or families (e.g., from current–voltage protocols) can be marked in the data file. The simulation module then calculates, based on the given model, data traces within cursor-defined relevant sections and determines a normalized deviation between experimental and simulated data. This approach works well if all data are treated equally; that is, they should be recorded from identical patches or cells under identical experimental conditions and they should be weighted in the same way. Any deviation from this ideal case requires additional information to be supplied for the modeling program. If data from different cells are to be considered, the model has to use multiple variables for the number of channels, N_C, because the current magnitude may vary from cell to cell. *PulseSim* uses for that purpose a set of extra variables that is stored in the data file for each family of traces. These variables are read by the simulation module and can be used in any of the parsed equations used in the model editor. The above-mentioned problem is, therefore, solved by measuring the magnitude of current from all cells used for the data fit with a program for model-independent analysis. The magnitudes relative to the first cell are then specified in the extra parameter field such that the program has to optimize only one N_C value.

Similarly, other varying experimental conditions (e.g., temperature, concentration of a substance) are directly supplied to the simulation module. This approach was used, for example, for the fit of a kinetic model for the activation of rat-*eag* potassium channels to data obtained in various concentrations of extracellular Mg^{2+}.[3] In this case various scenarios of how extracellular Mg^{2+} could possibly affect the rates of channel activation were tested.

In Fig. 9 another example using an approach similar to that of simultaneous data fits, including all other methods described in the previous sec-

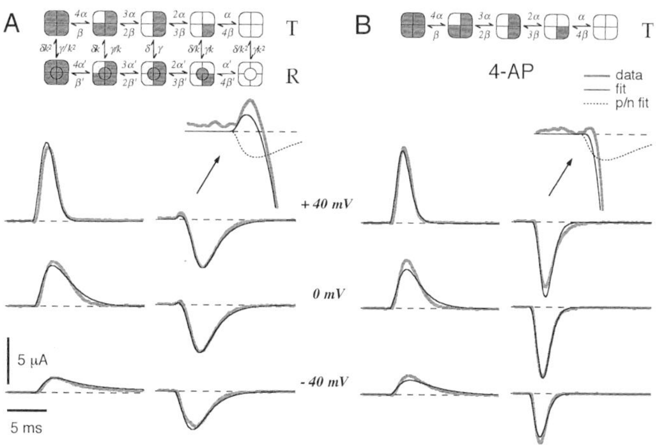

FIG. 9. Simultaneous model fit to two different data sets. Examples of *Sh*Δ W434F · T449R gating currents measured in standard solutions (A) and in 5 m*M* 4-AP (B, gray) are superimposed with fit results (black). For (A) a *tense-relaxed* (*TR*) activation model[14] was used, while the 4-AP data were fit with only the *T* section of the model using identical parameters to those in part (A). The insets show magnifications of the Ig_{off} sections on repolarization from +40 to −100 mV. The overcompensation artifact appears to be smaller in the presence of 4-AP. However, the prediced *p/n* signals are almost identical and the differences between the measured and simulated gating currents are of similar magnitude.

tions (consideration of filter, rise time, model-dependent *p/n* correction), is shown. Based on measurements of steady-state charge movements during *Sh*Δ channel activation in the presence and absence of 4-AP, McCormack *et al.*[14] developed a *tense-relaxed* (*TR*) activation model (Fig. 9A), which features two Hodgkin–Huxley-like activation pathways for the four subunits in addition to a concerted transition from the *tense* conformation to the *relaxed,* and potentially open, conformation. The *T*-to-*R* transition carries only about 10% of the total gating charge, which is efficiently suppressed by the application of 4-AP. Thus, this kinetic model should suffice to describe experimental time course data at different potentials, both in the presence and absence of 4-AP.

A selection of such *Sh*Δ W434F · T449R data traces for both conditions obtained with TEV methods from the same oocyte is shown in Figs. 9A and 9B as gray traces. The extra values in the control data sections were set to unity, while those in 4-AP data sections were set to 10^6. This extra value was then used to modify the **Q** matrix by redefining the *TR* transition ($\gamma/extra$ and $\delta \cdot extra$; see Fig. 9A) such that the TR transition is practically excluded in Fig. 9B. The black traces are the results of a simultaneous model optimization showing that both the time courses and the 4-AP effect (including the partial prediction of the *p/n* overcompensation artifact) are predicted reasonably well. However, the best fits only yielded gating charges of $1.6e_0$ (*T* and *R* transitions) and $0.9e_0$ (*TR* transition), which suggest that the TR model tends to underestimate the total amount of gating charge as measured with more direct methods.[21]

Multiple Populations of Channels

One of the most stringent assumptions for model fits is that only one population of ion channels contributes to the current signal. However, in many cases it is inevitable that additional components arising from other channel types will contaminate the signal. Moreover, the ion channels under consideration may exist in multiple forms (e.g., due to distributions or slow modes, different posttranslational modifications, partial pharmacologic modifications). In principle, such heterogeneity may always be incorporated in single kinetic models. Such models, however, quickly become very large and awkward, and model calculation is slowed down considerably. Thus, when there are no transitions between the different populations or when such transitions are slower than the time during which data were taken, it is often more appropriate to calculate multiple kinetic models (multiple **Q** matrices) in parallel. The use of global variables will then be helpful for reducing the number of free variables and for eliminating redundancy.

A data fit using two **Q** matrices was performed to describe the effect of Mg^{2+} on the activation of rat-*eag* channels.[3] While in the approach outlined earlier a graded effect of Mg^{2+} on the rate constants was modeled, in an alternative model it was assumed that the binding of Mg^{2+} to the channel alters a set of rate constants in a discrete manner, leading to two population of channels.[3]

Monte Carlo Simulations

Monte Carlo simulation is a powerful tool for the compilation of reliable statistics. In cases where the underlying process is not known Monte Carlo simulation can be performed by randomly resampling from arbitrary data according to the estimated errors of the individual data points. This approach is based on the bootstrap method,[22] where the assumption of the basic bootstrap method (resampling with replacement) is that the errors are identically and independently distributed. Alternatively, Monte Carlo simulation can also be performed based on an assumed underlying stochastic process by randomly sampling from a particular stochastic model. This

[21] N. E. Schoppa, K. McCormack, M. A. Tanouye, and F. J. Sigworth, *Science* **255,** 1712 (1992).
[22] B. Efron, *Ann. Stat.* **7,** 1 (1979).

Monte Carlo technique is frequently applied to the simulation of single-channel currents. The superposition of simulated single-channel currents enables the generation of fluctuating current records with realistic stochastic properties, which has been used for the investigation of nonstationary noise analysis.[23] A particular example, where nonstationary noise analysis is applied to "nonideal" fluctuating current data, is discussed later.

Single-Channel Simulation. The simulation of single-channel currents can be performed using the Markov property; that is, the transition probability only depends on the immediately preceding signal value. For a short time interval Δt, the time-discrete transition probability is given by

$$P_{ij}\,(\Delta t) \cong q_{ij}\,\Delta t \tag{20}$$

where q_{ij} is the transition rate (for theory, see Cox and Miller[5]). The decision of whether or not the channel moves into a new state can then be made by comparing the transition probabilities of all possible transitions with a unitary distributed random number. The advantage of this direct description of the stochastic process in discrete time is that it can be applied without knowledge of the underlying probability distribution. If Δt is sufficiently small it is valid for arbitrary input signals, that is, time courses of the voltage, recalculating the $\mathbf{Q}$ matrix for each point in time. However, the computation time as well as the number of required random numbers can be reduced by drawing from the known dwell-time distribution of a particular state. For stationary transition rates (constant voltage), the dwell times follow exponential distributions where exponentially distributed random numbers are derived by taking the negative logarithm of unitary distributed random numbers.[8] An example of a simulated single-channel current trace as a response to a voltage step is shown in Fig. 10A. Given the probability density function for waiting times τ_{ij} of a nonstationary transition $q_{ij}(t)$[24]

$$f(\tau_{ij}) = q_{ij}(\tau_{ij})e^{-\int_0^{\tau_{ij}} q_{ij}(t)\,dt} \tag{21}$$

algorithms for the generation of random numbers can also be developed for nonstationary voltages, as shown for voltage ramps.[25]

Application of Nonstationary Noise Analysis to Nonideal Data. Nonstationary noise analysis is a valuable tool for estimating single-channel parameters when excessive background noise (σ_b^2) obscures the resolution of single-channel currents (i) or when a preparation contains many active channels (N_C). The basic assumption for calculating the theoretical ensem-

[23] R. Steffan and S. H. Heinemann, *J. Neurosci. Methods* **78,** 51 (1997).

[24] A. Papoulis, "Probability, Random Variables, and Stochastic Processes." McGraw-Hill, New York, 1991.

[25] D. Sigg and F. Bezanilla, *J. Gen. Physiol.* **109,** 27 (1997).

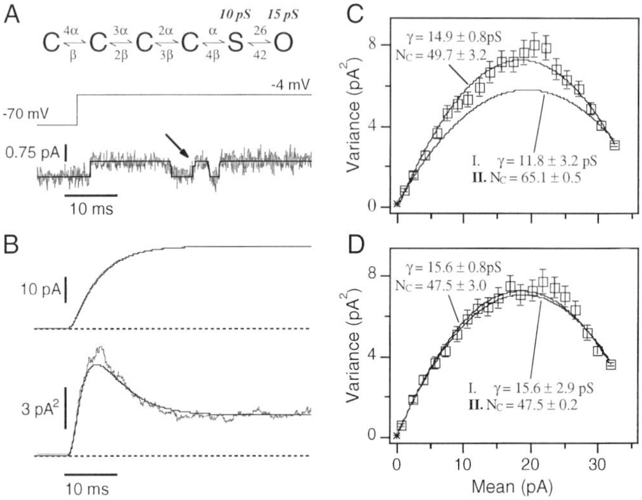

FIG. 10. (A) Simulated single-channel current data in response to a voltage step from −70 to −4 mV based on a sequential model[29] with a sublevel state preceding the main level of 15 pS. The arrow points to a current sublevel buried in the superimposed background noise of 150 fA at 3 kHz. (B) Mean current (top) and variance (bottom) as a result of nonstationary noise analysis[26,30] based on 500 fluctuating records, each generated by the superposition of 50 simulated single-channel current traces as shown in part (A). Superimposed are the analytically derived time courses of the "true" ensemble mean current and ensemble variance (smooth black curves). Mean-variance plots using 20 current bins for data grouping are shown in (C) and (D). The error bars are derived from the diagonal elements of the error-covariance matrix.[23] (C) Mean-variance plot of the data shown in part (B). Superimposed are generalized least squares fits in which i and N_C are fitted simultaneously (upper parabola) or when the fit was performed in two steps (lower parabola) yielding the indicated single-channel conductance γ and N_C (for details see text). (D) Mean-variance analysis performed as described in part (C) for data generated with the model shown in part (A), but without a sublevel state.

ble variance (σ^2) associated with the ensemble mean current (I) is that channels open only to a single conductance level[26]:

$$\sigma^2 = iI - \frac{I^2}{N_C} + \sigma_b^2 \tag{22}$$

A particular problem arises when subconductance states contribute to the mean current. When subconductance states are correlated with the activation process, as has become apparent for mutant potassium channels,[27,28] the plot of the variance versus the mean must deviate from a

[26] F. J. Sigworth, *J. Physiol.* **307,** 97 (1980).
[27] M. L. Chapman, H. M. A. VanDongen, and A. M. J. VanDongen, *Biophys. J.* **72,** 708 (1997).
[28] J. Zheng and F. J. Sigworth, *J. Gen. Physiol.* **110,** 101 (1997).

parabolic function [Eq. (22)]. Using simulated data it can be shown that subconductance states lead to sigmoidal onset of the mean-variance plot. In the following, a strategy for nonstationary noise analysis is outlined that might help to discriminate possible subconductance levels.

For the simulation of fluctuating current records, a model for potassium channel gating[29] was modified by removing inactivation and by inserting a subconductance state with 10-pS conductance preceding the main level of 15 pS (Fig. 10A). Nonstationary noise analysis was applied to 500 fluctuation records each derived by the superposition of 50 single-channel current traces. The resulting time courses of the mean current and the variance are shown in Fig. 10B. The binned mean-variance data shown in Fig. 10C can hardly be distinguished from the mean-variance data derived from the same model without a sublevel state (Fig. 10D), since the maximum occupancy probability of the sublevel state only amounts to 3.5%. The estimated single-channel parameters from a generalized least squares fit based on Eq. (22)[23] are, therefore, almost identical. However, if the fits are performed in two steps, the results are remarkably different. In a first step (Fig. 10, step I) the parameters are fitted at very low open probability ($p_{o,min}$ to $p_{o,max}$ = 1.4 to 9.4%). The resulting single-channel current is then kept constant during a second fit (Fig. 10, step II) using data up to high open probability ($p_{o,min}$ to $p_{o,max}$ = 1.4 to 85.6%). Comparison of Figs. 10C and 10D shows that the results of both fit approaches are almost identical only in the absence of a subconductance state.

Besides the mean current, the nonstationary variance and autocovariance functions contain valuable information about the underlying stochastic process.[24] Since subconductance levels are altering both the mean and the autocovariance function, it is desirable to incorporate autocovariance functions into fits of macroscopic signals. An example an of analytically derived time course of the variance[24,31] is shown in Fig. 10B. For simultaneous data fits in the time domain an appropriate weighting of current and autocovariance has to be defined for optimization.

Acknowledgments

We are grateful to K. McCormack for providing the cDNA clone for *Sh*Δ. The technical assistance of L. Kuschel, A. Rossner, and A. Grimm is appreciated as well as the valuable comments on the manuscript by A. A. Elliott and U. Faust.

[29] W. N. Zagotta and R. W. Aldrich, *J. Gen. Physiol.* **95,** 29 (1990).
[30] S. H. Heinemann and F. Conti, *Methods Enzymol.* **207,** 131 (1992).
[31] F. J. Sigworth, *Biophys. J.* **34,** 111 (1981).

[24] Signal Processing Techniques for Channel Current Analysis Based on Hidden Markov Models

By SHIN-HO CHUNG and PETER W. GAGE

Introduction

The patch-clamp technique can be used to record single-channel activity from a small patch of cell membrane. Although the noise from a small patch is much less than that from a whole-cell membrane, signals of interest are often obscured by the noise. Even if the signal frequently emerges from the noise, low-amplitude events such as small subconductance states can remain below the noise level and there may be little evidence of their presence. It is desirable, therefore, to have a method to measure and characterize not only relatively large ionic currents but also much smaller current fluctuations that are obscured by noise.

Extracting the real signal from a limited set of imperfect measurements is a problem that commonly occurs in science, and techniques have been developed to overcome this difficulty. Following digitization, a single-channel record consists of a sequence of data points. Any movement between successive data points can be due to the signal of interest, extraneous noise, or both. The challenge is to remove the noise leaving the biological signal untouched. Some of the methods that have been used to do this are linear filtering, nonlinear filtering,[1] and transition detectors.[2–6]

Hidden Markov Model

Although both linear filtering and nonlinear filtering suppress noise and nonlinear filtering produces little distortion of rapid transitions in underlying signals, neither method utilizes all of the knowledge available about the nature of the signal and interfering noise. Using such information improves the probability of reconstructing the original signal accurately. As an analogy, suppose that some printed text is difficult to read because it has been accidentally sprayed with black ink spots. To discern the true

[1] S. H. Chung and R. A. Kennedy, *J. Neurosci. Methods* **40,** 71 (1991).
[2] J. B. Patlak, *J. Gen. Physiol.* **92,** 413 (1988).
[3] S. D. Tyerman, B. R. Terry, and G. P. Findlay, *Biophys. J.* **61,** 736 (1992).
[4] A. Queyroy and J. Verdetti, *Biochim. Biophys. Acta* **1108,** 159 (1992).
[5] D. R. Laver, *J. Gen. Physiol.* **100,** 269 (1992).
[6] J. B. Patlak, *Biophys. J.* **65,** 29 (1993).

0076-6879/98 $25.00

text, it is sensible to make use of all the information available about the text and the spots: where the offending spots came from, whether the text is in English or French, the general theme of the message, etc. In attempting to reconstruct an obscured word, it is also useful to scan the text before and after the word. For example, if a word is obscured in the following sentence "I took ... bread out of the cupboard" one might be confident that the obscured four-letter word is "some." Broadly speaking, these kinds of strategies are used in the hidden Markov model (HMM) processing technique.

Signal Model

To apply a digital signal processing technique based on HMM to records of single-channel currents contaminated by noise, we first make a plausible guess about the origin of the observation sequence and then construct a signal model. It is assumed that the channel current signal can be represented as a Markov process with the following characteristics.

Discrete Time

Time is discrete, that is, it is broken up into discrete steps. It is more convenient to deal with a discrete-time rather than a continuous-time Markov process embedded in noise. Techniques for extracting continuous-time Markov processes from noise are presented elsewhere[7] but the mathematics associated with such techniques is relatively difficult and involves use of the properties of Wiener processes and Ito stochastic calculus. Because, in practice, the experimental record we deal with is digitized so that time is in discrete steps, there is no need to add unnecessary mathematical complexity by working with continuous-time processes, although it could be done.

Finite-State

For each discrete time k, the signal, s_k is at one of a finite number of states, $q_1, q_2, \ldots, q_N$. Each q_i, where $i = 1, 2, \ldots, N$, is called a *state* of the process and such a process is called an *N-state Markov chain.* In the context of channel currents, the Markov state, s_k, represents the true conductance level (or current amplitude) uncontaminated by noise at time k. The observed value, y_k, contains the signal s_k, random noise w_k, and possibly deterministic interferences p_k, such as sinusoidal interferences from electricity mains and baseline drift. Note that the meaning of the term *state* differs from that adopted in the Colquhoun–Hawkes model of channel

[7] O. Zeitouni and A. Dembo, *IEEE Trans. Inf. Theory* **34,** 890 (1988).

dynamics,[8–11] in which *state* refers to a hypothetical, not directly observable, conformation of the channel macromolecule.

First Order

The probability of the current being at a particular level (state) at time $k + 1$ depends solely on the state at time k. The transition probabilities of passing from state level q_i at time k to state level q_j at time $k + 1$, defined as

$$a_{ij} = P(s_{k+1} = q_j | s_k = q_i)$$

form a state transition probability matrix $\mathbf{A} = \{a_{ij}\}$, $i = 1, 2, \ldots, N$, $j = 1, 2, \ldots, N$. Note that $\mathbf{A}$ is an $N \times N$ matrix, with its diagonal elements denoting the probabilities of remaining in the same state at time $k + 1$ as at time k.

Homogeneous

We assume that the transition probabilities are invariant of time k. The process, in other words, is taken to be stationary. To characterize a finite-state Markov chain, we define the initial state probabilities $\pi_i = \{\pi_i\}$, where $\pi_i = P(s_1 = q_i)$. We also define the noise (known as the *probabilistic function* of the Markov chain or the *symbol probability*) as $\mathbf{B} = b_i(y_k)$.

In the special case when the noise is Gaussian,

$$b_i(y_k) = (2\pi)^{-1/2}\, \sigma^{-1} \exp[-(y_k - q_i)^2/2\sigma^2] \tag{1}$$

Specification of a signal model involves choice of the number of states N, and their amplitudes or state levels (q). Then, transition probabilities from each of N states to each of the other possible states must be chosen, giving an $N \times N$ matrix ($\mathbf{A}$). In addition, the signal model requires a prior knowledge of the variance of the noise (σ^2) and the initial probability distribution (π). We start by making initial guesses of these unknown parameters, and use the notation $\lambda = (\mathbf{q}, \mathbf{A}, \mathbf{B}, \pi)$.

Example of Signal Model

Suppose we know that a record contains four current levels but we are not sure of the exact levels nor the exact signal sequence. (Later, we

[8] D. Colquhoun and A. G. Hawkes, *Proc. R. Soc. Lond. B* **199,** 231 (1977).

[9] D. Colquhoun and A. G. Hawkes, *Proc. R. Soc. Lond. B* **211,** 205 (1981).

[10] D. Colquhoun and F. J. Sigworth, *in* "Single Channel Recordings" (B. Sakmann and E. Neher, eds.), pp. 191–263. Plenum Press, New York, 1983.

[11] D. Colquhoun and A. G. Hawkes, *Proc. R. Soc. Lond. B* **240,** 453 (1990).

discuss the situation in which a channel can assume an unknown number of conductance sublevels.) We can set up an initial model with the following assumed characteristics. First, we make a reasonable guess, and say that the baseline level is 0 pA and that there are three open states at −1, −2, and −3 pA. Second, we assume that the noise is zero-mean Gaussian, with a standard deviation of, say, 0.25 pA. Third, we provide our initial guesses of transition probabilities from one state level at time k to another state level at time $k + 1$. For a four-state Markov chain, these transition probabilities form a 4×4 transition probability matrix. In the example given here, the first entry of the first row of the matrix represents the probability that the process remained in the closed state at time $k + 1$ given that it was closed at time k, whereas the second entry of the first row represents the probability of transiting to the first open level at time $k + 1$ given that it was closed at time k. Similarly, the last entry of the last row represents the probability that the process remained in the fourth level, or the −3-pA level, at time $k + 1$ given that it was at this level at time k. Finally, we stipulate that the probability of the signal being at each one of the four levels at time $k = 1$ is 0.25.

These assumptions can be represented as:

1. State levels (**q**)		q_1	=	0 pA	
		q_2	=	−1 pA	
		q_3	=	−2 pA	
		q_4	=	−3 pA	
2. Noise characteristics	Gaussian with	$\sigma = 0.25$ pA			
3. Transition matrix		a_{11}	a_{12}	a_{13}	a_{14}
		a_{21}	a_{22}	a_{23}	a_{24}
		a_{31}	a_{32}	a_{33}	a_{34}
		a_{41}	a_{42}	a_{43}	a_{44}
4. Initial probabilities		$\pi_1 = \pi_2 = \pi_3 = \pi_4 = 0.25$			

We can also stipulate, if needed, that there is AC hum (50 or 60 Hz and its odd harmonics) embedded in the data but, for simplicity, this is not included in this example. We compactly write all these initial guesses as $\lambda^1 = (\mathbf{q}, \mathbf{A}, \mathbf{B}, \pi)$, our first signal model.

Expectation–Maximization Algorithm

The signal model is compared with the data. Essentially, the probabilities of all possible "pathways" between adjacent data points are calculated, both forward and backward, and the true current levels derived from the highest probabilities. Because the initial parameters we have supplied (e.g., the transition probability matrix and conductance levels) are only guesses,

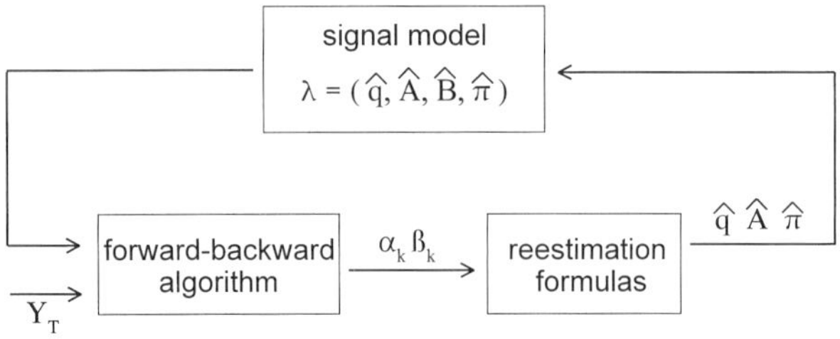

FIG. 1. A block diagram of the processing method. On the basis of the initial signal model λ, the observation sequence Y_T is processed, and the forward and backward variables, α_k and β_k, are computed for each discrete time k and each Markov state q_i. By using these variables, the parameters of the signal model are revised according to the reestimation formulas. The entire process is repeated many times.

there is going to be a mismatch between the model and data. The model is revised so that it will be more consistent with the data. Using the revised model, the observation sequence is compared again with the new model, and the model is again revised. This iterative process continues, as shown in Fig. 1.

The E Step

In this iterative process, the HMM processing technique utilizes two mathematical principles. The first is the *forward–backward procedure,* known also as the *E step* of the expectation–maximization algorithm.[12] For each data point k, the algorithm evaluates, using Bayes' rule, the forward and backward variables, α and β, defined as:

$$\alpha_k(i) = P(Y_k, s_k = q_i|\lambda), \qquad \beta_k(i) = P(\overline{Y}_k|s_k = q_i, \lambda) \tag{2}$$

where Y_k refers to the past observation sequence from $k = 1$ to k, and $\overline{Y}_k$ refers to the future observation sequence from $k + 1$ to T. In words, the forward variable $\alpha_k(i)$ is the joint probability of the past and present observation with the present signal in state q_i, given the model λ, and $\beta_k(i)$ is the probability of the future observation given that the present state is q_i and given the model λ.

The forward variable is calculated in a forward recursion and the backward variable in a backward recursion. Recursive formulas for Eq. (2) are

[12] A. P. Dempster, N. M. Laird, and D. B. Rubin, *J. R. Statist. Soc. B.* **39,** 1 (1977).

readily calculated[13] using Bayes' rule, as:

$$\alpha_k(i) = \sum_{i=1}^{N} \alpha_{k-1}(i)\, a_{ij} b_j(y_k), \qquad \alpha_1(j) = \pi_j b_j(y_1)$$

$$\beta_k(i) = \sum_{j=1}^{N} \alpha_{ij} b_j(y_{k+1})\beta_{k+1}(j), \qquad \beta_T(i) = 1$$

The M-Step

Then, using the forward and backward variables, the model is reestimated, using the Baum–Welsh reestimation formulas, known also as the *M step* of the expectation–maximization (EM) algorithm. Estimation formulas stipulate how the model parameters should be revised, given the forward and backward variables. Loosely stated, the E step of the EM algorithm expresses the expectation of the likelihood function if the parameters of the old model are replaced by a new model. In our application, this is a lengthy equation involving many variables. The M step involves finding the parameters of the new model λ^2 to maximize this expectation of the likelihood function of the "fully categorized" data.[14] Just as we do when finding the maximum of a cubic equation, we differentiate this equation with respect to each of the particular variables (with certain constraints where applicable) and obtain a new set of equations by letting the derivative equal zero. The full derivations of these reestimation formulas are given elsewhere.[15] These new sets of equations, called the *reestimation formulas,* are used to replace the parameters of the first signal model with new parameters to create the second signal model. The same segment of data is now processed with the second signal model, and then the parameters of this model are replaced, again according to the reestimation formulas, in the third signal model. The process is iterated again and again, 15 to 1000 times.

Computational Procedures

For illustration, we use the example of the signal model given earlier and follow how the first few computational steps are carried out to evaluate the forward variables. Because the model stipulates that there are four signal states (q = 0, −1, −2, −3 pA), there are four forward variables

[13] S. H. Chung, J. B. Moore, L. G. Xia, L. S. Premkumar, and P. W. Gage, *Phil. Trans. R. Soc. Lond. B* **329,** 265 (1990).

[14] D. M. Titterington, A. F. M. Smith, and V. E. Makov, "Statistical Analysis of Finite Mixture Distributions." John Wiley & Sons, New York, 1985.

[15] S. H. Chung, V. Krishnamurthy, and J. B. Moore, *Phil. Trans. R. Soc. Lond. B* **334,** 357 (1991).

[$\alpha_1(1)$, $\alpha_1(2)$, $\alpha_1(3)$, and $\alpha_1(4)$] to be calculated for time $k = 1$. Assume that the first data point happened to be -1.3 pA. The probability that the real signal at k was 0 pA but the additive Gaussian noise would have displaced this point to -1.3 pA is computed with Eq. (1), with the standard deviation of this noise taken to be 0.25 pA, as we specified. This probability is multiplied with the initial probability of the signal being at the 0-pA level, namely, 0.25, to obtain the first forward variable belonging to the first level, $\alpha_1(1)$. The forward variable at time $k = 1$ belonging to the second level, $\alpha_1(2)$, is computed by evaluating the probability that the noise would have displaced the signal from -1 to -1.3 pA and then multiplying this quantity with the initial probability of the signal being at the -1-pA level, 0.25. The remaining two forward variables, $\alpha_1(3)$ and $\alpha_1(4)$, are similarly tabulated.

When it comes to the second data point, again four forward variables, $\alpha_2(1)$, $\alpha_2(2)$, $\alpha_2(3)$, and $\alpha_2(4)$, need to be computed. Suppose that the second measured current value was -0.5 pA. If the signal at $k = 2$ was assumed to be at the 0-pA level, it would have made one of the four possible transitions from time $k = 1$; from the 0-pA level to the 0-pA level, from the -1-pA level to the 0-pA level, from the -2-pA level to the 0-pA level, and from the -3-pA level to the 0-pA level. In our signal model, the probabilities of making such transitions are stipulated in the first column of the 4×4 matrix. The probability that additive Gaussian noise would displace the current level from 0 to -0.5 pA is calculated using Eq. (1). Numerically, this probability is 0.216. In our example, $\alpha_2(1)$ is the sum of the four terms: $[\alpha_1(1) \times a_{11} \times 0.216] + [\alpha_1(2) \times a_{21} \times 0.216] + [\alpha_1(3) \times a_{31} \times 0.216] + [\alpha_1(4) \times a_{41} \times 0.216]$, as the recursive formula given in Eq. (2) states. The backward variable $\beta_2(1)$ is computed similarly, using the recursion formula given in Eq. (2).

Once the numerical values of $\alpha_2(1)$ and $\beta_2(1)$ are tabulated, we multiply them and call this quantity $\gamma_2(1)$, after suitable normalization. This is the probability of the signal being in the 0-pA level at time $k = 2$, given the observation sequence y_T and the model λ. In symbols,

$$\gamma_k(i) = P(s_k = q_i | y_T, \lambda) \tag{3}$$

and this quantity can be computed from the forward and backward variables using the formula:

$$\gamma_k(i) = [\alpha_k(i)\, \beta_k(i)] \Big/ \sum_{i=1}^{N} \alpha_k(i)\, \beta_k(i)$$

The initial parameters are reestimated using the forward and backward variables. To revise the first current level, which we specified as 0 pA in our example, we multiply the first data point y_1 by $\gamma_1(1)$, the second data point y_2 by $\gamma_2(1)$, the kth data point y_k with $\gamma_k(1)$, and the last data point y_T with $\gamma_T(1)$, and then add all these values. This total sum, normalized by dividing by the sum of all γ's belonging to the first current level, namely, $\gamma_1(1) + \gamma_2(1) + \ldots + \gamma_k(1)$, gives the reestimated current level of the first Markov state. More generally,

$$^*q_i = \sum_{i=1}^{T} \gamma_k(i)\, y_k \Big/ \sum_{k=1}^{T} \gamma_k(i)$$

where $^*q_i$ is the estimate of the ith state level. The elements in the transition probability matrix are similarly revised using the reestimation formula. The old parameters are replaced with these revised parameters, and the new signal model, now λ^2, is used and all of the computational steps are repeated. This procedure is repeated many times.

Reestimation Theorem

After each iteration, we can compute from the forward variable a numerical value that we call the *likelihood function*—that is, how likely are the model parameters given the data sequence. The closer the model parameters are to the true parameters, the higher the likelihood function. If the likelihood functions were to increase and decrease erratically with successive iterations, this procedure would have been a waste of time. The rationale behind the iterative procedure rests on the elegant reestimation theorem formulated by Baum and colleagues,[16–18] which states:

$$P(Y_T|\lambda^{n+1}) \geq P(Y_T|\lambda^n)$$

In words, the probability of the observation sequence Y_T, given the reestimated signal model, is greater than or equal to the probability of Y_T, given the previous signal model. Thus, the signal sequence estimated using the revised model is more consistent with the data than that estimated using the previous signal model. When the iterative procedure converges, then $P(Y_T|\lambda^{n+1}) = P(Y_T|\lambda^n)$, and λ^n is termed the *maximum likelihood estimate* of the HMM. This important theorem, the proof of which is based on Jensen's inequality,[16] is the core of the HMM processing scheme.

There is a choice of numerical methods for calculating the maximum likelihood estimates. One approach is the Newton–Ralphson algorithm, which, when it converges, does so quadratically and thus rapidly. The EM algorithm, on the other hand, converges linearly, and so convergence can

[16] L. E. Baum and T. Petrie, *Ann. Math. Statist.* **37,** 1554 (1966).

[17] L. E. Baum, T. Petrie, G. Soules, and N. Weiss, *Ann. Math. Statist.* **41,** 164 (1970).

[18] L. E. Baum, *Inequalities* **3,** 1 (1972).

be very slow. However, successive iterations with the Newton–Ralphson algorithm do not necessarily improve the likelihood function. In contrast, the EM algorithm is simple to implement and has the appealing property that the likelihood function is always improved after each iteration.

In the graph in Fig. 2, the likelihood function obtained from one of the data segments is plotted against successive iterations. This shows that the probability that the observed data sequence could have been generated by each successively reestimated model increases steadily and then converges by the time the process is iterated about 20 times.

Estimating the Number of States

Perhaps the most subjective part of the HMM processing method, like any parameter estimation scheme, is finding the state dimension—or the number of conductance states in our example—in hidden Markov chain processes. The error in fitting a model to a given set of data decreases with the number of free parameters in the model. Thus, it makes sense, in selecting a model from a set of models with different numbers of parameters, to penalize models having too many parameters. The question of how to

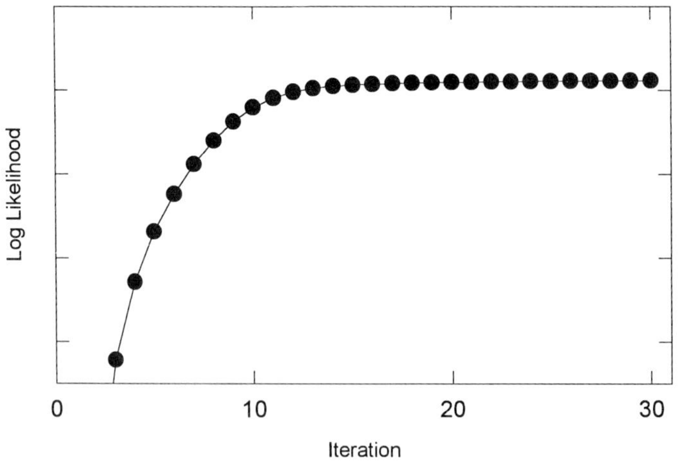

FIG. 2. Increase in the log likelihood function with successive iterations. The likelihood that the observation sequence could have been generated by each successively reestimated model increases rapidly at first and then slowly but steadily with each additional iteration. A segment of the record used to obtain the likelihood function is shown in the top panel of Fig. 3A, and the parameters of the first signal model are given in the text. In the first signal model, the diagonal and all off-diagonal elements of the transition matrix were assumed to be 0.91 and 0.03, respectively.

penalize HMM for having an excessive number of free parameters is an area of current research (see, for example, Ref. 19) and one proposed criterion for model-order selection is the compensated likelihood approach.[20]

In practice, however, it is relatively easy to identify the number of states present in the underlying Markov chains. One of several ways of doing this is by constructing the most likely amplitude histogram, as shown in Fig. 3 (also see Fig. 4B). Here we assume that the signal can be represented as a Markov chain with a large number of equally spaced states, say, 100 states, and assume that the actual state levels are fixed. We construct an amplitude histogram from the estimated signal sequence obtained under this assumption, as shown here. These three histograms were derived from the three sets of fictitious channel currents exhibited in the first panel. The maximum likelihood histograms, indicated as bars, clearly show four prominent peaks, the baseline at the right-hand side, and three open current levels. These peaks are evenly spaced at 0.38 pA. In contrast, no meaningful information can be gleaned from the amplitude histogram of the original record, indicated here as a superimposed continuous curve.

Alternatively, we can appeal to the principle of parsimony in deciding the number of conductance states. We measure the goodness of fit by evaluating the likelihood of the model, and weigh this against what is to be gained by increasing the number of parameters, which generally increases the likelihood. Thus, we process the same data segment under the assumption that the underlying signal has a different number of conductance states. If a plot of log likelihood versus model order (number of states) shows a "knee" for a certain model order (Fig. 4A, asterisk), we would prefer this model to one of higher order. This approach has been used to determine the number of conductance substates in channel currents activated by γ-aminobutyric acid (GABA).[21]

Extension of Signal Model

Under the simple assumption that the signal sequence can be represented by a discrete-time, first-order, finite-state, homogeneous Markov chain, the HMM processing technique provides the following information. First, it identifies the number of conductance states in the true signal that is hidden in the noise. Second, it identifies the state levels of the signal.

[19] D. S. Poskitt and S. H. Chung, *Adv. Appl. Probab.* **28,** 405 (1996).

[20] L. Finesso, "Consistent estimation of the order of Markov and hidden Markov chains." PhD Dissertation, University of Maryland (1990).

[21] P. W. Gage and S. H. Chung, *Proc. R. Soc. Lond. B* **255,** 167 (1994).

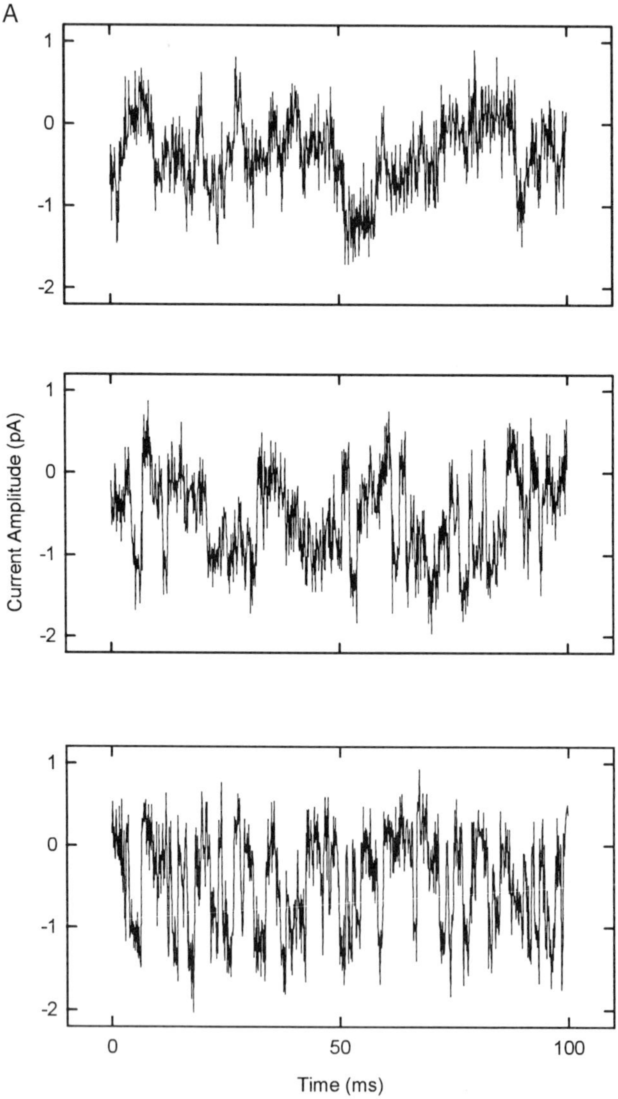

FIG. 3. Segments of synthesized channel currents and estimation of their current levels. (A) The traces show three first-order, two-state, Markov chains with $a_{11} = 0.99$ and $a_{22} = 0.98$ opening and closing independently of each other (top) or in partial synchrony (middle and bottom). The coupling coefficients of the signals contained in the three segments from top down were 0, 0.05 and 0.15, respectively. The open current level of each binary chain was 0.375 pA. To the signal sequences, white Gaussian noise (standard deviation 0.25 pA) and AC hum (50, 150, and 250 Hz) were added: the amplitude of each component was 0.25 pA. Once buried in the random and deterministic interferences, the original signal se-

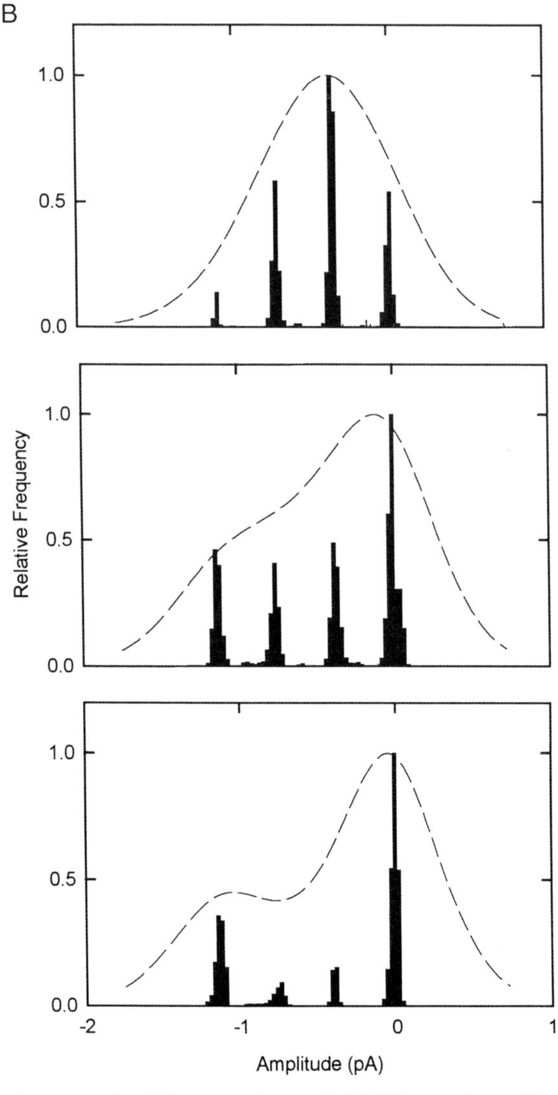

quences and their current levels became obscured. (B) The maximum likelihood estimates of the signal sequences embedded in the record and AC hum contaminating the signal were obtained using the iteration scheme shown in Fig. 1. The amplitude histograms obtained from the estimated signal sequences, corresponding to the three segments of the current traces displayed in part (A), are shown as dark bars. The peaks of the histograms are evenly spaced at about 0.38 pA. In contrast, no meaningful information can be gleaned from the all-points amplitude histograms of the original records, indicated here as continuous lines superimposed on the bar graphs. (C) Using the fact that there are four Markov states separated by 0.38

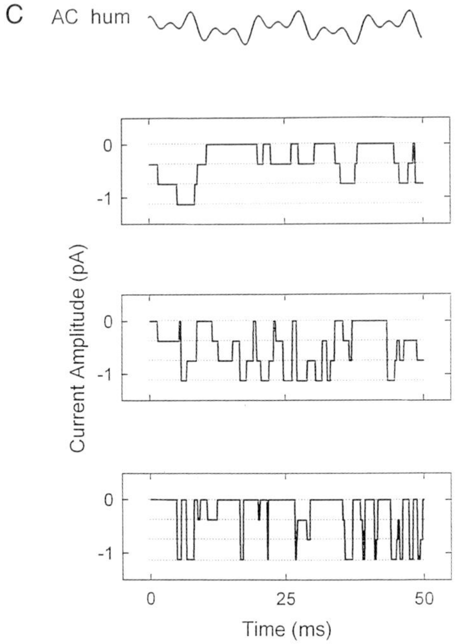

pA, the most likely signal was extracted from the currents in Fig. 3A together with the underlying hum (top trace). The three signal sequences differ in the degree of coupling between three channels. In the first example, the three binary channels were uncoupled; they opened and closed independently of each other. In the second example, they were weakly coupled in that they opened and closed not totally independent of each other but in partial synchrony. In the third example, they were strongly coupled. When one channel opened (or closed), there was a tendency for the other two channels to open (or close) in synchrony.

Third, it gives the maximum likely estimate of the signal sequence—what the signal must have been like before it became obscured by noise. Finally, it gives the stochastic matrix of the signal from which all the necessary kinetic parameters can be obtained. These include the mean open and closed times and interval histograms. If there is more than one channel (more than one Markov chain), the structure of this stochastic matrix enables us to deduce whether these chains are totally independent, or whether they show some form of dependency.[22,23]

What if the real signal deviates from the underlying assumptions? Typical biological signals may not be best approximated by a first-order Markov process, especially if the number of events sampled is small. Moreover, if

[22] S. H. Chung and R. A. Kennedy, *Math. Biosci.* **133,** 111 (1996).

[23] R. A. Kennedy and S. H. Chung, *Intl. J. Adap. Control Signal Proc.* **10,** 623 (1996).

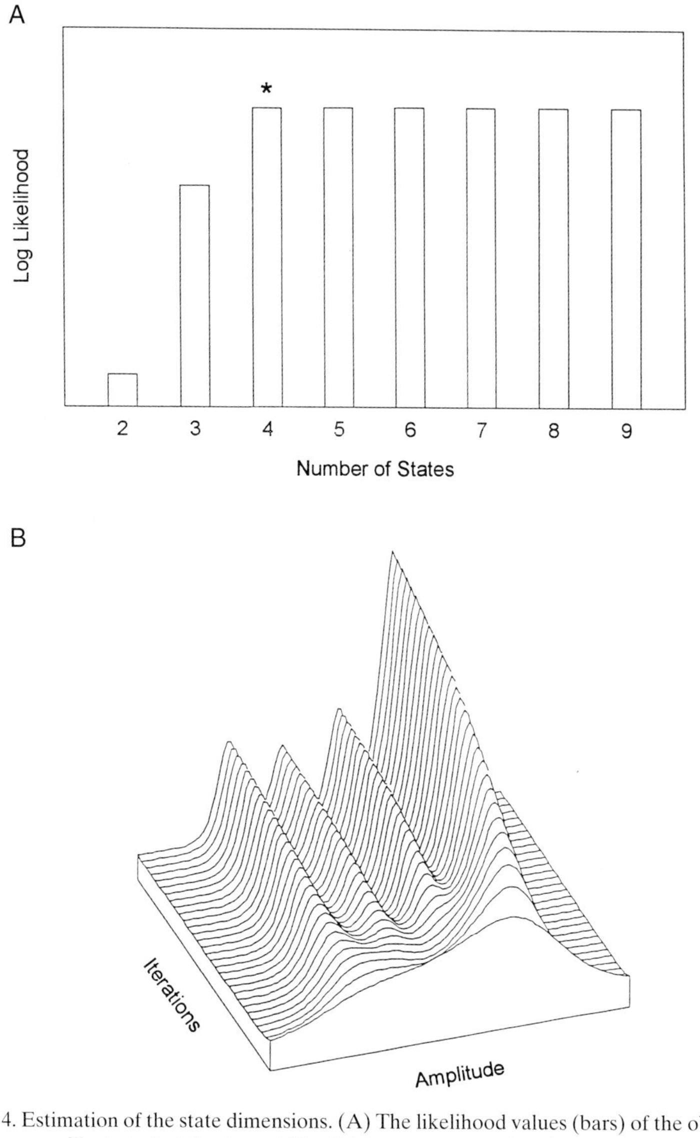

FIG. 4. Estimation of the state dimensions. (A) The likelihood values (bars) of the observation sequence, illustrated at the top of Fig. 3A, were computed under the assumption that the signal was a two-state, three-state, and up to a nine-state Markov chain. The likelihood value first increased as the number of states was increased from two to four, and then remained unchanged with increasing model order. Appealing to the principle of parsimony in determining the model order, we conclude that a four-state model (marked with an asterisk) fits the data. (B) The maximum likelihood amplitude histogram was constructed to determine the model order. We assumed that the signal can be represented as a Markov chain with 100 states. With successive iterations of the data under this assumption, the location of the peaks become more and more distinct. The large peak on the right-hand side is the baseline and three additional peaks on the left correspond to the open current levels.

the underlying conformational changes of a channel macromolecule can be represented by the kinetic scheme: $C_1 \rightleftharpoons C_2 \rightleftharpoons O$, where C_1 and C_2 are two distinct closed conformational states and O is the open conformational state and the transitions between states are governed by a first-order Markov process, the observed two-state signal sequence would not be a Markov process. (Technically, such a process can be construed as a switched Markov chain.) It has been demonstrated previously that departures from the underlying assumptions do not seriously degrade the performance of the processing method.[13,15] Even when the signal embedded in noise deviates drastically from the first-order Markov assumption and the additive noise is correlated, the technique extracts the signal and estimates its statistics with an acceptable degree of accuracy.

In the same framework of HMM techniques and the EM algorithm, the signal model can be further modified and extended to make the signal processing schemes far more versatile than those described here. For example, a Markov process, after entering one of its states, can be allowed to decay back in time exponentially (or otherwise) to the original state. Such a signal process can be formulated as an augmented homogeneous HMM problem, and a scheme for estimating this stochastic process can be devised. Also, it is not difficult to extend the signal model to a higher order Markov or a semi-Markov process and devise a scheme for estimating this process. A mathematical description of this extension is described elsewhere.[24] In the algorithm that is currently available, it is assumed that the amplitude and fundamental frequency of AC hum that is added to the recordings remain unchanged within the short data segment being analyzed. In reality, the amplitude of the harmonic wave may wax and wane slowly. It is again easy to incorporate these variables into the processing scheme at the expense of added computational cost.

Since the HMM processing techniques were introduced for analyzing single-channel currents,[13,15,25] several groups have refined and adapted the algorithms so that the method can be applied to practical patch-clamp recordings when the assumptions underlying the idealized signal model no longer hold. For example, one of the assumptions underlying the processing method is that noise is white and uncorrelated with the signal. If this assumption is violated, the algorithm performs poorly, as pointed out originally.[13] In reality, noise superimposed on single-channel currents is not white but colored. Its spectral power, instead of being flat, rises steeply at high frequencies. Moreover, the standard deviation of noise increases when

[24] V. Krishnamurthy, J. B. Moore, and S. H. Chung, *Signal Proc.* **24,** 177 (1991).

[25] D. R. Fredkin and J. A. Rice, *Proc. R. Soc. Lond. B* **249,** 125 (1992).

the channel is in the open state.[26] This issue was addressed by use of an autoregressive noise model to represent temporal correlation in the background noise contained in patch-clamp recordings and then the EM algorithms for such correlated and additive state-dependent excess noise were reformulated.[27,28] It was demonstrated with simulations that the performance of the algorithm was markedly improved when the background noise was modeled realistically.

On many occasions, recordings obtained from a membrane patch contain more than one channel and currents from several channels are superimposed. It is possible to extract, again using a modified HMM algorithm, the transition probability matrices and current amplitudes for each individual channel from such multichannel patches.[15,29] Finally, a method for handling the situation in which a Markovian signal embedded in noise is passed through a low-pass filter has also been addressed.[30]

Model Evaluation

One additional application of the HMM processing method may be very useful. The technique can be used to evaluate different types of models. Many schemes have been proposed to explain ion channel kinetics, including aggregated Markov models,[8] fractal models,[31–34] semi-Markov models encompassing fractal models,[35,36] hidden Markov models,[25,37] and coupled Markov models.[22,23] With so many feasible models, it would be useful to have an objective analytical tool to rank them and to reject any that are implausible. The HMM technique can be used for this. Not only will it estimate (in an optimal sense) the model parameters based on the observation sequence, but it will also discriminate between a set of alternative

[26] J. D. Becker, J. Hohnerkamp, J. Hirsch, U. Frobe, E. Schlatter, and R. Greger, *Eur. J. Physiol.* **426,** 328 (1997).

[27] L. Venkataramanan, J. L. Walsh, R. Kuc, and F. J. Sigworth, *IEEE Sig. Processing* (1998). In press.

[28] L. Venkataramanan, R. Kuc, and F. J. Sigworth, *IEEE Sig. Processing* (1998). In press.

[29] S. Klein, J. Timmer, and J. Honerkamp, *Biometrics* **53,** 43 (1997).

[30] V. Krishnamurthy and L. B. White, *Proc. IFAC Intl. Conf. Adapt. Syst. Cont. Signal Proc.* 633 (1992).

[31] L. S. Liebovitch, J. Fischbarg, and J. P. Koniarek, *Math. Biosci.* **78,** 203 (1986).

[32] S. J. Korn and R. Horn, *Biophys. J.* **54,** 871 (1988).

[33] L. S. Liebovitch, J. Fischbarg, J. P. Koniarek, I. Todorova, and M. Wang, *Biochim. Biophys. Acta* **896,** 173 (1987).

[34] L. S. Liebovitch and J. M. Sullivan, *Biophys. J.* **52,** 979 (1987).

[35] F. Ball and M. S. P. Sansom, *Proc. R. Soc. Lond. B* **236,** 385 (1989).

[36] F. Ball, G. F. Yeo, R. K. Milne, R. O. Edeson, R. W. Madsen, and M. S. P. Sansom, *Biophys. J.* **64,** 357 (1993).

[37] D. R. Fredkin and J. A. Rice, *Biometrics* **48,** 427 (1992).

models by assessing their plausibility.[38] Traditionally, identification of membrane channel models has focused on techniques involving the numerical fitting of the interval histograms or the power spectrum of the noisy channel data. These approaches are often not robust when the data are noisy and are limited in the amount of information they can yield about the process parameters. Such approaches only work when the relevant time constants (i.e., mean durations) differ sufficiently. Some other approaches that have been tried include maximum likelihood estimation,[35] which suffers an exponential complexity problem, and correlation functions,[33] which have only been applied to identify binary Markov chains.

Hidden Markov modeling is a computationally efficient maximum likelihood estimator that is robust in the presence of white noise and applicable to processes with many states. It may well prove useful for evaluating the likelihood of models proposed for ion channels. As an illustration, let us suppose that we observed in a coin-tossing experiment a sequence (HHHTHH). We construct two alternative signal models, the first being that the coin is unbiased and the second being that it is bent in such a way that a head would turn up 90% of time, $P(\mathrm{H}) = 0.9$. By comparing the likelihood values, we can state that this observation sequence is more consistent with the second signal model than with the first. As we try out various pairs of signal models, we discover that there is a unique $P(\mathrm{H})$ that gives the largest likelihood value, L_{T}, and we call this the *maximum likelihood estimate* of $P(\mathrm{H})$, 0.067 in this example. The principle of selecting the most likely signal model as illustrated with the simple coin-tossing example can readily be generalized to channel currents. A more complete discussion of the model selection problem is available.[38] In a similar vein, a scheme for testing whether a recorded sequence of channel currents can be described adequately as Markov chain has been proposed.[39] This statistical test relies on the fact that the asymptotic distributions of the log-likelihood will be distributed normally if the signal embedded in noise is a homogeneous, first-order Markov chain. Using simulated signal sequences modified from a homogeneous, first-order Markov chain systematically to render them non-Markovian, they demonstrated that their test procedure can detect when the signal deviates appreciably from the assumed model.

Computational Cost

Modern fast desktop computers have made the HMM technique completely accessible to all researchers. The number of computational steps

[38] G. W. Pulford, J. C. Gallant, R. A. Kennedy, and S. H. Chung, *Biometrics* **37,** 39 (1995).
[39] J. Timmer and S. Klein, *Phys. Rev. E* **55,** 3306 (1977).

involved in one forward–backward process is N^2T, where N is the number of Markov states and T is the number of data points. Typically, we analyze a 100,000-point record using about 10 allowed states, and the process is iterated about 100 times, resulting in computational steps of approximately 10^9. A modern, inexpensive, desktop computer can perform about 200 million computations per second, and mainframe supercomputers that can carry out 5×10^9 floating-point operations per second (5 gigaflops) are now readily available to researchers. Thus, the real time involved in processing such a record, once the codes are optimized, can be of the order of minutes.

[25] Investigating Single-Channel Gating Mechanisms through Analysis of Two-Dimensional Dwell-Time Distributions

By BRAD S. ROTHBERG and KARL L. MAGLEBY

Introduction

The use of single-channel recording techniques[1] and the application of stochastic theory[2] have led to great advances in our understanding of ion channel gating mechanisms. The gating of many types of ion channels can be described as a Markov process containing a discrete number of open and closed states, with the rate constants for transitions among the states remaining constant in time for constant conditions.[3–11]

To develop a Markov model that can describe the gating of a channel, it is necessary to determine how many open and closed states there are, and how these states are connected to one another to form the activation

[1] O. P. Hamill, A. Marty, E. Neher, B. Sakmann, and F. J. Sigworth, *Pfluegers Arch. Eur. J. Physiol.* **391,** 85 (1980).

[2] D. Colquhoun and A. G. Hawkes, *Proc. R. Soc. Lond. B* **211,** 205 (1981).

[3] R. Horn and K. Lange, *Biophys. J.* **43,** 207 (1983).

[4] R. W. Aldrich, D. P. Corey, and C. F. Stevens, *Nature* **306,** 436 (1983).

[5] S. M. Sine, T. Claudio, and F. J. Sigworth, *J. Gen. Physiol.* **96,** 395 (1990).

[6] O. B. McManus and K. L. Magleby, *J. Gen. Physiol.* **94,** 1037 (1989).

[7] O. B. McManus and K. L. Magleby, *J. Physiol.* (*Lond.*) **443,** 739 (1991).

[8] A. Auerbach, *J. Physiol.* (*Lond.*) **461,** 339 (1993).

[9] F. Bezanilla, E. Perozo, and E. Stefani, *Biophys. J.* **66,** 1011 (1994).

[10] W. N. Zagotta, T. Hoshi, and R. W. Aldrich, *J. Gen. Physiol.* **103,** 321 (1994).

[11] D. Colquhoun and A. G. Hawkes, *in* "Single-Channel Recording" (B. Sakmann and E. Neher, eds.), p. 397. Plenum, New York, 1995.

0076-6879/98 $25.00

and deactivation pathways. Estimates of the numbers of open and closed states in such models can be obtained by fitting sums of exponentials to the distributions of open and closed dwell times from steady-state single-channel data, where the number of significant exponential components required to describe the distributions gives an estimate of the minimum number of open and closed states.[2,3,11,12] It has proven more difficult to determine the connections among the various states because the potential number of different kinetic schemes, each with different connections, is large for most channels.

Information that can be used to distinguish among kinetic gating models with different connections among the states is contained in the correlations between adjacent open and closed dwell times.[13] A number of maximum likelihood methods are available that take the correlation information contained in the sequence of interval durations into account while fitting rate constants for kinetic models.[3,14–17] These methods share the advantage of using the kinetic information contained in the single-channel data to its fullest extent, and can also readily accommodate data containing multiple subconductance levels as well as data from multiple channels. Nonetheless, it is often useful to extract information directly from the single-channel kinetic data in the absence of the constraints of a specific model, for the purposes of gaining insight into the gating mechanism and extending or refining proposed models.

We have recently presented a method for estimating the time constants and volumes of two-dimensional (2-D) components of adjacent open–closed dwell times, by directly fitting sums of 2-D exponentials to open–closed adjacent interval distributions from steady-state single-channel data.[18] Fitting with 2-D exponentials affords increased resolution for estimating the minimum numbers of open and closed states required for kinetic models over the analysis of one-dimensional (1-D) open and closed distributions considered separately. In addition, parameters obtained through this method can be used to calculate component dependencies, which give the fractions of open–closed pairs in each 2-D component that are in excess (or deficit) of the number expected from independent (random) pairing of

[12] O. B. McManus and K. L. Magleby, *J. Physiol.* (*Lond.*) **402,** 79 (1988).

[13] D. R. Fredkin, M. Montal, and J. A. Rice, *in* "Proceedings of the Berkeley Conference in Honor of Jerzy Neyman and Jack Kiefer" (L. M. LeCam and R. A. Olshen, eds.), p. 269. Wadsworth, Belmont, California, 1985.

[14] F. G. Ball and M. S. Sansom, *Proc. R. Soc. Lond. B* **236,** 385 (1989).

[15] S. H. Chung, V. Krishnamurthy, and J. B. Moore, *Phil. Trans. R. Soc. Lond. B* **334,** 357 (1991).

[16] D. R. Fredkin and J. A. Rice, *Proc. R. Soc. Lond. B* **249,** 125 (1992).

[17] F. Qin, A. Auerbach, and F. Sachs, *Biophys. J.* **70,** 264 (1996).

[18] B. S. Rothberg, R. A. Bello, and K. L. Magleby, *Biophys. J.* **72,** 2524 (1997).

open and closed intervals. Component dependencies can be used as an indicator of whether the kinetic model should have one or multiple independent pathways between the open and closed states, and can suggest potential connections among open and closed states of various lifetimes to aid in the development of simple multistate kinetic models.

The maximum likelihood method for fitting of sums of 2-D exponentials will not be reiterated here, because it has already been presented in detail.[18] The purpose of this article is to explore additional uses and interpretations of 2-D kinetic analysis of single-channel data. We first summarize the theory underlying 2-D analysis and present methods for the display of 2-D data and interpretation of dependencies. We then present a method for quantifying irreversible transitions in single-channel gating using the 2-D distributions. Finally, we address limitations of 2-D analysis by giving examples of how the interpretation of component dependencies may lead to incorrect assumptions about the connections between states. Despite its limitations, 2-D analysis can provide a useful means to extract kinetic information from single-channel data and to test channel gating mechanisms.

Theory Underlying Two-Dimensional Analysis of Adjacent Open and Closed Interval Durations

For those unfamiliar with 2-D distributions and their component composition, it is suggested that Rothberg *et al.*[18] be read to provide the necessary background before proceeding. For discrete state models in which the rate constants for transitions among the states remain constant in time (Markov gating), the 1-D distributions of all open $f_O(t)$ and all closed $f_C(t)$ dwell times are described by the sums of exponential components, with the number of components given by the number of states:

$$f_O(t) = \sum_{i=1}^{N_O} \alpha_i \tau_i^{-1} \exp(-t/\tau_i) \tag{1}$$

$$f_C(t) = \sum_{j=1}^{N_C} \beta_j \tau_j^{-1} \exp(-t/\tau_j) \tag{2}$$

where N_O and N_C are the number of open and closed states, α_i and β_j are the areas, and τ_i and τ_j are the time constants of the components of the 1-D open and closed distributions, respectively.[2,11] The areas of all open components in Eq. (1) and of all closed components in Eq. (2) each sum to 1.0.

The 2-D distribution (joint density) of adjacent interval pairs described by each open dwell time and the following closed dwell time, $f_{OC}(t_O, t_C)$, is given by:

$$f_{OC}(t_O, t_C) = \sum_{i=1}^{N_O} \sum_{j=1}^{N_C} V_{ij} \tau_i^{-1} \tau_j^{-1} \exp(-t_O/\tau_i) \exp(-t_c/\tau_j) \tag{3}$$

where N_O and N_C are the number of open and closed states, V_{ij} is the volume of each 2-D component (fraction of total open–closed interval pairs), t_O and t_C are the open and closed times, and τ_i and τ_j are the time constants of the open and closed exponential functions.[13] For Markov gating, the time constants that underlie the 2-D dwell-time distributions are identical to the time constants underlying the 1-D dwell-time distributions.[13] The number of underlying 2-D components that sum to form the 2-D distribution is given by $N_O \times N_C$, and the volumes of the $N_O \times N_C$ underlying components sum to 1.0. An expression similar to Eq. (3) gives the 2-D distribution of closed intervals followed by open intervals:

$$f_{CO}(t_C, t_O) = \sum_{i=1}^{N_C} \sum_{j=1}^{N_O} V_{ij} \tau_i^{-1} \tau_j^{-1} \exp(-t_C/\tau_i) \exp(-t_O/\tau_j) \tag{4}$$

If the data are consistent with microscopic reversibility (detailed balance), then $f_{OC}(t_O, t_C) = f_{CO}(t_C, t_O)$.[13,19,20] In a later section we suggest the use of 2-D analysis as a means to quantify irreversibility in single-channel gating.

Obtaining Two-Dimensional Dwell-Time Distributions and Their Underlying Two-Dimensional Components

The 2-D distributions displayed and discussed in this chapter were all obtained from Schemes 1–6 using **Q** matrix calculations.[13]

For experimental data, the 2-D dwell-time distributions would be obtained by measuring and plotting the interval durations in the single-channel data. The time constant and volumes of the 2-D components underlying the 2-D dwell-time distributions would be obtained by fitting the 2-D distribution with sums of 2-D exponential components, as outlined in the Appendix of Rothberg *et al.*[18]

Two-Dimensional Dwell-Time Distributions

Figure 1A shows the 2-D dwell-time distribution (histogram) of adjacent open and closed interval pairs predicted by Scheme 1 with the rate constants

[19] I. Z. Steinberg, *J. Theo. Bio.* **124,** 71 (1987).
[20] L. Song and K. L. Magleby, *Biophys. J.* **67,** 91 (1994).

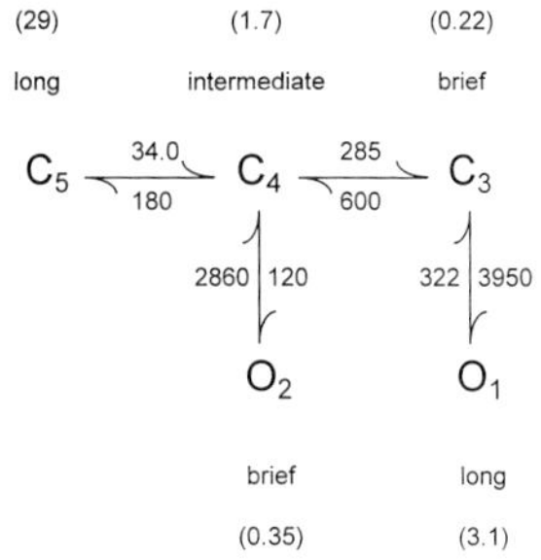

SCHEME 1

shown.[21] The durations of the open and closed intervals making up the pairs are binned and plotted on logarithmic scales on the x and y axes that define the horizontal plane of the plot, and the vertical axis shows the number of pairs per bin, which are plotted using a square root transformation[22,23] that yields peaks at the time constants of the major components. Thus, for example, Fig. 1A shows a major component of long openings adjacent to brief closings in the single-channel data.

Dependency Plots

To extract further information on the correlations between interval durations, the data in Fig. 1A were used to construct a dependency plot, shown in Fig. 1B, which displays the fraction of open–closed pairs in excess (or deficit) of the number expected if openings and closings paired independently (randomly). Details for constructing such plots have been presented previously.[24] The heavy black line on the surface indicates the zero level. The dependency plot in Fig. 1B shows a deficit of brief open intervals adjacent to brief closed intervals, and an excess of brief open intervals adjacent to long closed intervals. Such features if observed in experimental data may suggest that kinetic models for the gating should not contain dominant pathways between the states (or groups of states) which give rise to brief openings and the states (or groups of states) which give rise to brief closings. Such features if observed in experimental data would also suggest that there should be dominant transition pathways between the states (or group of states) which give rise to brief open intervals and the

[21] K. L. Magleby and B. S. Pallotta, *J. Physiol.* (*Lond.*) **344,** 585 (1983).
[22] F. J. Sigworth and S. M. Sine, *Biophys. J.* **52,** 1047 (1987).
[23] K. L. Magleby and D. S. Weiss, *Proc. R. Soc. Lond. B* **241,** 220 (1990).
[24] K. L. Magleby and L. Song, *Proc. R. Soc. Lond. B* **249,** 133 (1992).

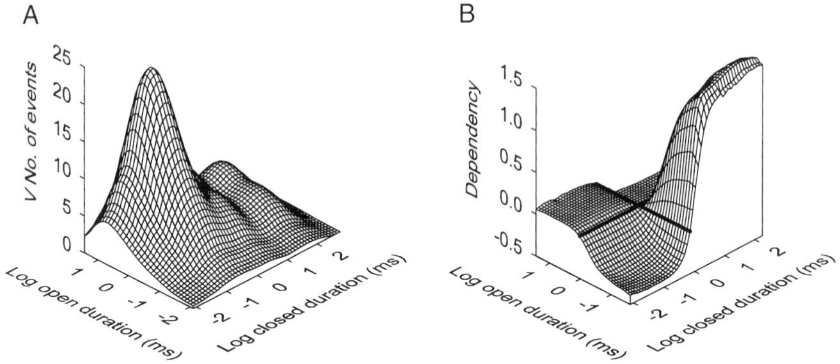

FIG. 1. (A) Two-dimensional dwell-time distribution of adjacent open and closed intervals for Scheme 1. The durations of the adjacent open and closed intervals forming a pair are plotted as the log of the durations on the horizontal plane (x and y axes). The distribution was scaled using the Sigworth and Sine[22] square root transformation as described in Magleby and Weiss[23] for a hypothetical analysis of 100,000 intervals. The 2-D components describing the distribution are given in Table I. Equation (3) relates the components to the distribution. (B) The dependency plot calculated directly from the 2-D distribution plotted in (A)$_1$ using methods described in Magleby and Song.[24] The data in (A) and (B) are plotted at a resolution of 10 bins per log unit. The plots were generated by the SURFER plotting program (Golden Software, Golden, CO) using inverse distance interpolation.

states (or groups of states) which give rise to long closed intervals.[18,24] These qualitative observations are consistent with the connections among the open and closed states in Scheme 1.

Component Dependencies

To obtain a quantitative measure of the excess or deficit of interval pairs in each underlying 2-D component over that expected if open and closed intervals paired independently, component dependencies were calculated (see below) for Scheme 1 from the data in Fig. 1A and are plotted in Fig. 2. The time constants of the open (O_1' and O_2') and closed (C_3', C_4', and C_5') components that pair to form the 2-D components are indicated in Table I along with their volumes (V_{obs}). Also included in Table I are the volumes that would be expected if the pair of intervals in the open and closed components was independent (V_{ind}), and the calculated component dependency (*cDep*).

The component dependencies for Fig. 2 and Table I were calculated from Scheme 1. For experimental data the 2-D distribution would be fit with sums of 2-D exponential components in order to estimate the observed volume of each component [Eqs. (3) or (4); see later discussion]. Next, the

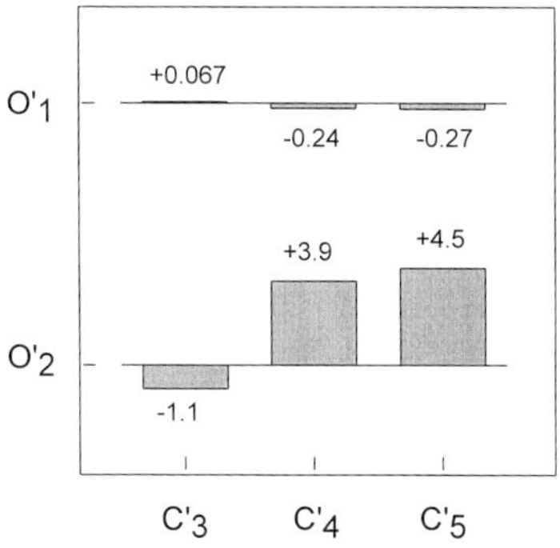

FIG. 2. Component dependency plot for Scheme 1. Component dependency values are from Table I (*cDep*). The number above or below a bar indicates its value.

expected volume of each 2-D component for independent pairing would be calculated. Because the observed volumes of the individual 2-D components underlying the 2-D distributions sum to 1.0, they may be viewed as probabilities. Thus, for independent pairing of open and closed intervals, the probability of observing an open interval from open component O_i' adjacent to a closed interval from closed component C_j' is the product of the probabilities of observing the individual intervals separately:

$$V_{\text{ind}}(O_i'C_j') = \alpha_i \times \beta_j \tag{5}$$

where $V_{\text{ind}}(O_i'C_j')$ is the expected volume of the 2-D component $O_i'C_j'$ for independent pairing of open and closed intervals, and α_i and β_j are the

TABLE I
TWO-DIMENSIONAL KINETIC PARAMETERS FROM SCHEME 1[a]

τ_{open} (ms)	τ_{closed} (ms)	V_{obs}	V_{ind}	*cDep*
O_1' 3.1	C_3' 0.22	0.80	0.75	0.067
	C_4' 1.8	0.099	0.13	−0.24
	C_5' 45	0.045	0.062	−0.27
O_2' 0.35	C_3' 0.22	−0.0036	0.048	1.1
	C_4' 1.8	0.042	0.0085	3.9
	C_5' 45	0.022	0.0040	4.5

[a] τ_{open}, open time constant; τ_{closed}, closed time constant; V_{obs}, component volume observed; V_{ind}, component volume expected for independent pairing; *cDep*, component dependency. τ_{open}, τ_{closed}, and V_{obs} were calculated with rates from Scheme 1 using **Q** matrix methods. V_{ind} was calculated with Eq. (5). *cDep* was calculated with Eq. (6).

areas of the 1-D open and closed components O'_i and C'_j, respectively. These areas can be determined from the 2-D distributions by summing the volumes corresponding to a given open component (for α_i) or closed component (for β_j).

To determine the fractional excess or deficit of interval pairs for each component, the component dependency, $cDep(O'_iC'_j)$, is calculated with

$$cDep(O'_iC'_j) = \frac{V_{obs}(O'_iC'_j) - V_{ind}(O'_iC'_j)}{V_{ind}(O'_iC'_j)} \tag{6}$$

where $V_{obs}(O'_iC'_j)$ is the observed volume of each 2-D component, and $V_{ind}(O'_iC'_j)$ is the expected volume if the open and closed intervals pair independently. Thus, component dependencies of $+0.5$ or -0.5 would indicate a 50% excess or 50% deficit, respectively, of observed interval pairs when compared to that expected for independent pairing of open and closed intervals. A component dependency of -1.0 would indicate that no interval pairs were found in a component, and a component dependency < -1.0 (supernegative) would indicate a component with negative volume.

The value of $V_{obs}(O'_iC'_j)$ can be determined using pairs consisting of (1) the open intervals and their following closed intervals, specifically denoted V_{OC}; (2) the closed intervals and their following open intervals, specifically denoted V_{CO}; or (3) both of the above types of adjacent pairs combined into a single distribution. As mentioned in the Theory section, if the data are consistent with microscopic reversibility, then $f_{OC}(t_O, t_C) = f_{CO}(t_C, t_O)$, and consequently $V_{OC} = V_{CO}$. Thus, combining the open–closed (O–C) and closed–open (C–O) pairs into a single distribution can serve as a computationally efficient approach for data consistent with microscopic reversibility.

Extracting Useful Information from Component Dependencies

Component dependencies give quantitative information on the correlations between the durations of adjacent intervals. Thus, interpretations of these quantities can be used to extract useful information on the underlying gating mechanism.

Component dependencies can be used primarily to indicate whether the gating mechanism underlying the observed single-channel data contains multiple independent pathways between different open and closed states. If component dependencies in a given distribution deviate significantly from zero, this would indicate that the data were produced by a gating mechanism containing multiple independent pathways between different open and closed states.[11,13,18,24] The component dependencies generated by Scheme 1, for example, deviate from zero (Fig. 2 and Table I), ranging from −1.1 [$cDep(O_2'C_3')$] to 4.5 [$cDep(O_2'C_5')$]. Consistent with the hypothesis that the data were produced by a gating mechanism containing multiple independent pathways between different open and closed states, Scheme 1 contains two such independent pathways; one pathway is between states O_1 and C_3, and the other is between states O_2 and C_4.

From Scheme 1 and Table I, it can be seen that the time constants of the open components are the same as the lifetimes of the open states, since the open states in Scheme 1 are not connected to one another to form a *compound* state. For the closed states in Scheme 1, which do form a compound state, the time constants of the components differ from the lifetimes of the individual states due to transitions among the closed states. Thus for Scheme 1, the observed open components O_1' and O_2' arise from states O_1 and O_2, respectively. In contrast, the closed component C_3' arises mainly from the closed state C_3, while the closed component C_4' arises from the combined sojourns to the closed state C_4 and the compound closed state C_3–C_4, and component C_5' arises from the compound closed states C_4–C_5 and C_3–C_4–C_5.

Although Scheme 1 is known to be the gating mechanism for the component dependency plot in Fig. 1, if the component dependencies in this plot were obtained from experimental data, they could be interpreted to suggest the following requirements for a kinetic model:

1. Since the component dependencies deviate from zero, indicating that the open and closed intervals do not pair randomly, the gating mechanism must contain more than one independent pathway between the open and closed states.

2. The component dependency plot for Scheme 1 shows an excess (+0.067) of long open intervals adjacent to brief closed intervals ($O_1'C_3'$). This would suggest that the state (or group of states) that gives rise to the long open intervals may be directly connected to the state (or group of states) that gives rise to the brief closed intervals. This is consistent with the direct connection between O_1 and C_3 in Scheme 1.

3. The supernegative deficit (−1.1) of brief open intervals adjacent to brief closed intervals ($O_2'C_3'$) suggests that the pathway between the state (or group of states) that gives rise to brief openings (O_2') and the state (or group of states) that gives rise to brief closings (C_3') should include open or closed states of longer durations, consistent with the position of state C_4 in Scheme 1. This would result in no observed brief openings adjacent to brief closings, since the gating must pass through a longer duration state (C_4) to connect the open and closed states of brief lifetime.[18]

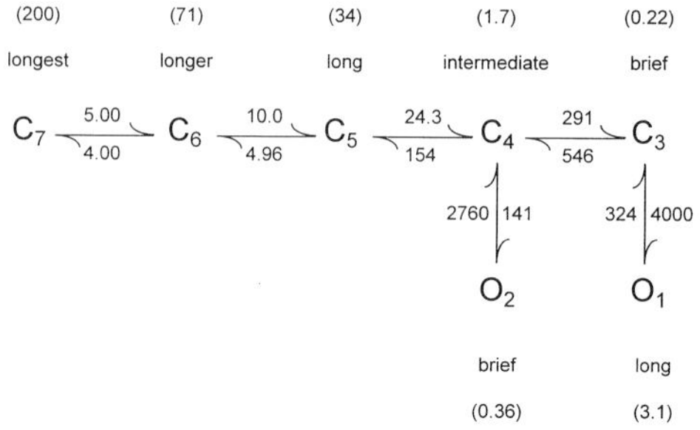

SCHEME 2

4. The positive component dependencies of +3.9 and +4.5 for brief open intervals adjacent to intermediate and long closed intervals ($O_2'C_4'$ and $O_2'C_5'$, respectively) suggests that the state (or group of states) that gives rise to brief openings should be connected directly to the state (or groups of states) that gives rise to the intermediate and long closed components, consistent with the pathway O_2–C_4–C_5 in Scheme 1.

Note that this is only one set of interpretations for the component dependencies predicted by Scheme 1, which is simplified because of the relatively simple kinetic model. For this particular model the correct interpretation of the component dependencies is intuitive. Despite these encouraging results, however, such simple interpretations would not always be possible for more complicated models. It should also be noted that while any given kinetic scheme generates a unique component dependency plot, a given component dependency plot does not necessarily define a unique model.

Similar Component Dependencies Suggest Extended Compound States

This section investigates the relation between the component dependencies and components arising from connected (compound) closed states, using the hypothetical mechanism shown in Scheme 2. Scheme 2 is similar to Scheme 1 except that two additional closed states, C_6 and C_7, have been added so that there are three connected states beyond the gateway state[11] C_4. Such compound states beyond a gateway state will be referred to as *extended compound states.* With two open and five closed states, the 1-D distributions of open and closed intervals were described by the sum of two open and five closed exponentials, respectively, giving ten individual 2-D components, as shown in Table II.

TABLE II
TWO-DIMENSIONAL KINETIC PARAMETERS FROM SCHEME 2[a]

τ_{open} (ms)		τ_{closed} (ms)		V_{obs}	V_{ind}	*cDep*
O_1'	3.1	C_3'	0.22	0.81	0.75	0.080
		C_4'	1.8	0.098	0.14	−0.30
		C_5'	38	0.017	0.024	−0.29
		C_6'	83	0.012	0.017	−0.29
		C_7'	370	0.0053	0.0077	−0.31
O_2'	0.36	C_3'	0.22	−0.0039	0.050	−1.1
		C_4'	1.8	0.047	0.0090	4.2
		C_5'	38	0.0091	0.0016	4.7
		C_6'	83	0.0066	0.0012	4.5
		C_7'	370	0.0029	0.0005	4.8

[a] τ_{open}, open time constant; τ_{closed}, closed time constant; V_{obs}, component volume observed; V_{ind}, component volume expected for independent pairing; *cDep*, component dependency. τ_{open}, τ_{closed}, and V_{obs} were calculated with rates from Scheme 2 using **Q** matrix methods. V_{ind} was calculated with Eq. (5). *cDep* was calculated with Eq. (6).

Figure 3 shows a component dependency plot of the data predicted by Scheme 2. Intervals from the three longest closed components adjacent to intervals from the brief open component, $O_2'C_5'$, $O_2'C_6'$, and $O_2'C_7'$, were all in excess, and each of these components had almost identical component dependencies of around 4.7. Intervals from the same three longest closed

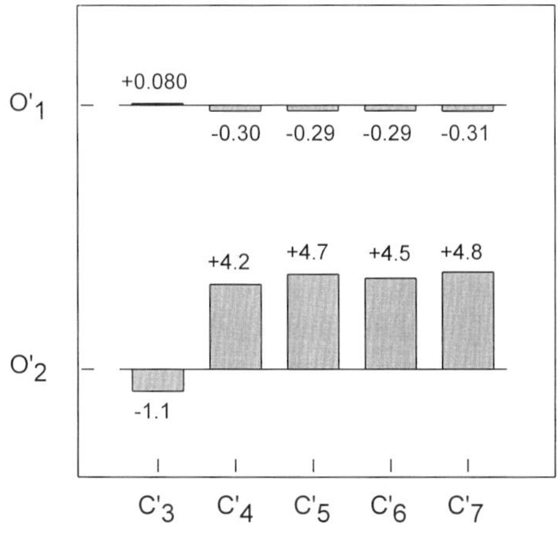

FIG. 3. Component dependency plot for Scheme 2. Component dependency values are from Table II (*cDep*). The number above or below a bar indicates its value.

components adjacent to intervals from the long open component, $O_1'C_5'$, $O_1'C_6'$, and $O_1'C_7'$, were all in deficit, and each had almost identical component dependencies of around −0.3. Thus, for Scheme 2, dependencies associated with components arising mainly from extended compound states beyond the gateway state C_4 had almost identical component dependencies for a given open component.

As further tests of the conclusion that components associated with extended compound states have essentially identical component dependencies, the rate constants $k_{6,7}$ and $k_{6,5}$ in Scheme 2 were changed to 12 and 19.96 sec^{-1}, respectively. This changed the time constants of the three longest closed components for Scheme 2 from 38, 83, and 370 ms to 31, 66, and 3600 ms, respectively, with essentially no change in the component dependencies associated with the component arising mainly from the three longer closed states (not shown). Similarly, increasing the rate constant $k_{7,6}$ in Scheme 2 to 100 sec^{-1} changed the time constants for the three longest closed components to 34, 1200, and 8.9 ms, with essentially no change in the component dependencies associated with the 34- and 1200-ms components, and only a 2% decrease in the component dependencies associated with the 8.9-ms component. It is not entirely clear from where the 8.9-ms component arises, but if it is associated with the closed state C_7, this state would not be reached without transitions through longer duration closed states C_4 and C_5, which may account for why the volume of component C_7' was vanishingly small at about 10^{-6}. Thus, the 8.9-ms component would not be detected experimentally.

These findings suggest that any detectable 2-D components associated with two or more extended compound closed states (connected closed states beyond the last gateway state) can have essentially the same dependencies for any given open component. However, there are exceptions to this rule. It will be shown in a later section that this association of extended compound states with similar component dependencies is no longer valid if the closed states have similar lifetimes.

Interchanging open and closed states in the preceding examples would lead to a similar conclusion for extended compound open states.

Using Analysis of Two-Dimensional Dwell-Time Distributions to Test for Microscopic Reversibility in Single-Channel Gating

The gating of ion channels is often described by kinetic schemes that assume reversible (time-symmetrical) transitions among open and closed states, so that the number of transitions between connected states is the same in either direction. However, since the recording of ionic currents requires nonequilibrium experimental conditions (such as voltage or concentration gradients) to produce a net flux of permeating ions through the channel pore, the coupling of the net flux of ions to the gating could provide a source of energy to drive time-asymmetrical gating. Such nonequilibrium gating has been shown for the chloride channel from *Torpedo,* where asymetry in the transitions between inactivated and conducting states was coupled to the transmembrane electrochemical gradient.[25]

Two-dimensional analysis can provide a means to detect such time asymmetry in single-channel gating. Song and Magleby[20] used 2-D dwell-time distributions to test for microscopic reversibility in the gating of large conductance calcium-activated potassium channels, by comparing the 2-D dwell-time distribution constructed from pairs of each open interval and the following closed interval, called the open–closed, O–C, or forward distribution, to the 2-D dwell-time distribution constructed from pairs of each open interval and the preceeding closed interval, called the closed–open, C–O, or backward distribution. If the O–C distribution is the same as the C–O distribution, then the gating is time reversible.[19]

A more quantitative measure of reversibility in gating can be extracted from 2-D distributions by fitting the O–C and C–O distributions with sums of 2-D exponential components, to obtain the volumes of the various 2-D components. Assuming Markov gating, the time constants underlying the O–C distribution will be identical to those underlying the C–O distribution, although if the gating is time asymmetrical, then the corresponding component volumes will differ. Once the component volumes for the O–C and C–O distributions have been estimated, the volume expected for reversible gating can be determined as the mean of the corresponding volumes from the O–C and C–O distributions:

$$V_{\text{mean}}(\text{O}_i'\text{C}_j') = \frac{V_{\text{OC}}(\text{O}_i'\text{C}_j') + V_{\text{CO}}(\text{C}_j'\text{O}_i')}{2} \tag{7}$$

From these quantities a component irreversibility factor, C_{IRR}, can then be calculated:

$$C_{\text{IRR}}(\text{O}_i'\text{C}_j') = \frac{V_{\text{OC}}(\text{O}_i'\text{C}_j') - V_{\text{mean}}(\text{O}_i'\text{C}_j')}{V_{\text{mean}}(\text{O}_i'\text{C}_j')} \tag{8}$$

where $V_{\text{OC}}(\text{O}_i'\text{C}_j')$ is the volume of component $\text{O}_i'\text{C}_j'$ from the forward (O–C) distribution, and $V_{\text{mean}}(\text{O}_i'\text{C}_j')$ is determined from Eq. (7). Thus if $V_{\text{OC}}(\text{O}_i'\text{C}_j') = V_{\text{CO}}(\text{C}_j'\text{O}_i')$, then C_{IRR} is equal to 0 (no irreversibility).

[25] E. A. Richard and C. Miller, *Nature* **247,** 1208 (1990).

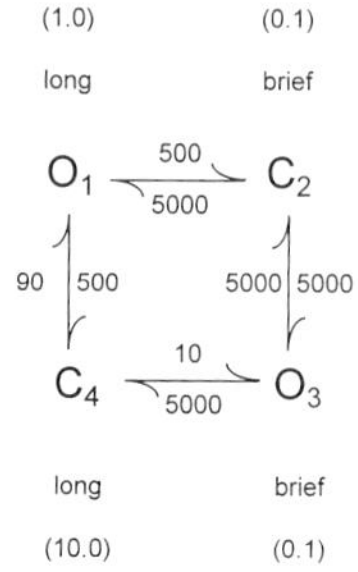

SCHEME 3

Scheme 3 is a model in which the states are connected in a loop, with two open states and two closed states in an alternating arrangement, and with rates selected so that there would be a nine-fold net excess of transitions in the clockwise direction (O_1–C_2–O_3–C_4–O_1) over transitions in the counterclockwise direction around the loop (O_1–C_4–O_3–C_2–O_1). This model has been used previously to investigate partially irreversible gating (Scheme 2 from Song and Magleby[20]). Scheme 3 should generate more long open intervals followed by brief closed intervals than brief closed intervals followed by long open intervals [$V_{OC}(O_1'C_2') > V_{CO}(C_2'O_1')$]. It should also generate more brief open intervals followed by long closed intervals than long closed intervals followed by brief open intervals [$V_{OC}(O_3'C_4') > V_{CO}(C_4'O_3')$]. Table III shows the time constants and volumes of the O–C and C–O distributions, determined from **Q**-matrix calculations.[13] As men-

TABLE III
TWO-DIMENSIONAL KINETIC PARAMETERS FROM SCHEME 3[a]

τ_{open} (ms)		τ_{closed} (ms)		V_{OC}	V_{CO}	C_{IRR}
O_1'	1.0	C_2'	0.10	0.35	0.25	0.17
		C_4'	10	0.35	0.45	−0.13
O_3'	0.10	C_2'	0.10	0.15	0.25	−0.25
		C_4'	10	0.15	0.05	0.50

[a] τ_{open}, open time constant; τ_{closed}, closed time constant; V_{OC}, component volume, O–C distribution; V_{CO}, component volume, C–O distribution; C_{IRR}, component irreversibility factor. τ_{open}, τ_{closed}, V_{OC}, and V_{CO} were calculated with rates from Scheme 3 using **Q** matrix methods. C_{IRR} was calculated with Eqs. (7) and (8).

tioned previously, the time constants are the same for both the O–C and C–O distribution. In order to quantify the degree of irreversible gating, C_{IRR} was calculated from Eq. (8) and presented in Table III. The calculations indicate a 17% excess of long open intervals followed by brief closed intervals, and a 50% excess of brief open intervals followed by long closed intervals. Because these quantities can be determined from experimental data through fitting of the 2-D dwell-time distribution with sums of 2-D exponential components, the determination of C_{IRR} could serve as a quantative test for microscopic reversibility in single-channel gating. Values of C_{IRR} significantly different from zero would indicate that the gating violates microscopic reversibility.

Interestingly, the component dependencies for the O–C distribution from Scheme 3 are all zero, which could mistakenly indicate a gating mechanism with a single gateway state.[18] The reason for this is that once the channel is in state O_1, the probabilities of making transitions to C_2 or C_4 for Scheme 3 are identical, and once the channel is in O_3, the probabilities of making transitions to C_2 or C_4 are identical. However, further analysis could show that the component dependencies for the C–O distribution, or for the combined O–C and C–O distributions, are not zero, suggesting more than one gateway state.

These observations do suggest, however, that a model with two gateway states that obeys microscopic reversibility might generate component dependencies of zero for appropriately selected rate constants. To explore this possibility, $k_{4,1}$ and $k_{4,3}$ in Scheme 3 were both set to 50 sec^{-1} to ensure microscopic reversibility. When this was done, all the component dependencies were zero, even though the scheme has two gateway states. Thus, while an observation of component dependencies greater than zero suggests more than one gateway state, this example shows that an observation of component dependencies of zero does not necessarily indicate only one gateway state.

Limitations of Two-Dimensional Dwell-Time Analysis

The kinetic schemes presented in the previous sections were selected to facilitate the explanation of the interpretation of the dependency plots. In these schemes, the time constants of the components were qualitatively similar to the lifetimes of the states that mainly generated them. In addition, the lifetimes of the extended closed states were generally longer than the lifetimes of the closed states, which had direct connections to open states. Under these restrictions, the dependency plots can have simple and straightforward interpretations. Interestingly, kinetic schemes that have been devel-

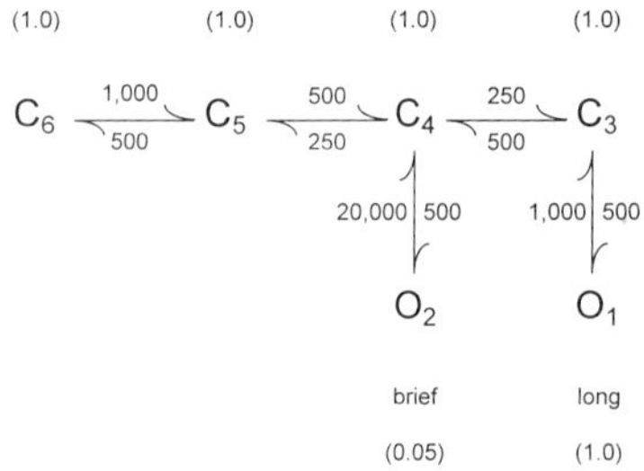

SCHEME 4

oped for several ion channels have these same properties.[7,26–28] In presenting these previous examples, the components were numbered in accord with the corresponding states of similar lifetimes for the sake of clarity, but this type of numbering would not be necessary to interpret the plots properly.

In contrast to the preceding examples, the ability to interpret dependency plots easily is compromised when all of the states in a chain forming a compound state have similar lifetimes. Scheme 4 presents an extreme case for a model containing two open states and four closed states, with rate constants selected so that the lifetimes of all four closed states are identical at 1.0 ms. The theoretical 2-D components and component dependencies are shown in Table IV. The closed time constants of 0.55, 0.76, 1.4, and 5.2 ms showed no simple relation with the lifetimes of the closed states or combinations of transitions through the states. Thus, it would be difficult to interpret the theoretical results for component dependencies in terms of mechanism. This possibly worst-case model also shows that extended compound states do not always generate similar component dependencies. Nonetheless, it is clear that the dependencies from this model deviate substantially from zero, suggesting that the correct scheme should contain more than one gateway state.

Interestingly, if data were obtained from a channel that gated like Scheme 4, it is unlikely that components C_3' and C_4' in Table IV would be resolved by maximum likelihood fitting, because of their small volumes and because they have time constants similar to C_5', which has much more volume. In this case, it is likely that a four-state model (two-open, two-closed) would be deduced from experimental data, with a long open state connected to a brief closed state, and a brief open state connected to a

[26] D. Colquhoun and B. Sakmann, *J. Physiol.* (*Lond.*) **369,** 501 (1985).
[27] D. S. Weiss and K. L. Magleby, *J. Neurosci.* **9,** 1314 (1989).
[28] D. S. Weiss and K. L. Magleby, *J. Physiol.* (*Lond.*) **453,** 279 (1992).

TABLE IV

TWO-DIMENSIONAL KINETIC PARAMETERS FROM SCHEME 4[a]

τ_{open} (ms)	τ_{closed} (ms)	V_{obs}	V_{ind}	*cDep*
O_1' 1.0	C_3' 0.55	−0.0054	0.0041	−2.3
	C_4' 0.76	−0.014	0.0011	−14
	C_5' 1.4	0.26	0.19	0.37
	C_6' 5.2	0.098	0.14	−0.30
O_2' 0.05	C_3' 0.55	0.018	0.0081	1.2
	C_4' 0.76	0.018	0.0022	7.2
	C_5' 1.4	0.32	0.38	−0.16
	C_6' 5.2	0.32	0.28	0.14

[a] τ_{open}, open time constant; τ_{closed}, closed time constant; V_{obs}, component volume observed; V_{ind}, component volume expected for independent pairing; *cDep*, component dependency. τ_{open}, τ_{closed}, and V_{obs} were calculated with rates from Scheme 4 using **Q** matrix methods. V_{ind} was calculated with Eq. (5). *cDep* was calculated with Eq. (6).

long closed state. Thus, although incorrect in detail, the deduced model would form a reasonable starting point from which to study the gating mechanism.

We next explored why Schemes 1 and 2 could be predicted from the component dependency plots and Scheme 4 could not. Because the lifetimes of the closed states C_3 and C_4 differed by nearly tenfold in both Schemes 1 and 2, we tested the consequences of this difference by changing the rate constants to change the lifetimes of states C_4, C_5, and C_6 in Scheme 4 to keep them identical at 10 ms, while the lifetime of C_3 was kept at 1 ms. These changes in the rate constants predicted a different 2-D distribution, and the predicted component dependencies from the new distribution could be interpreted to give the properties of the correct scheme. In addition,

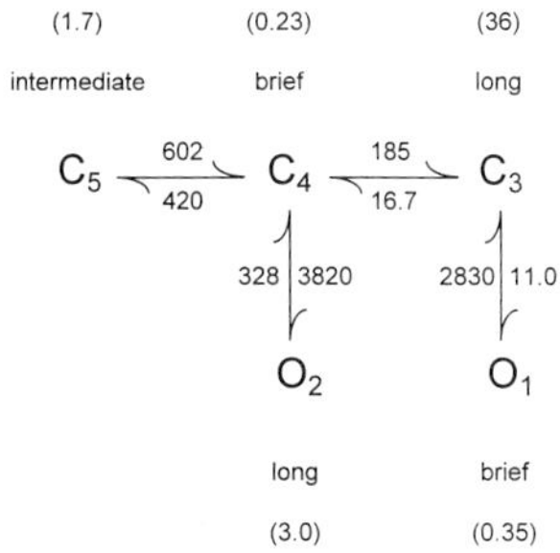

SCHEME 5

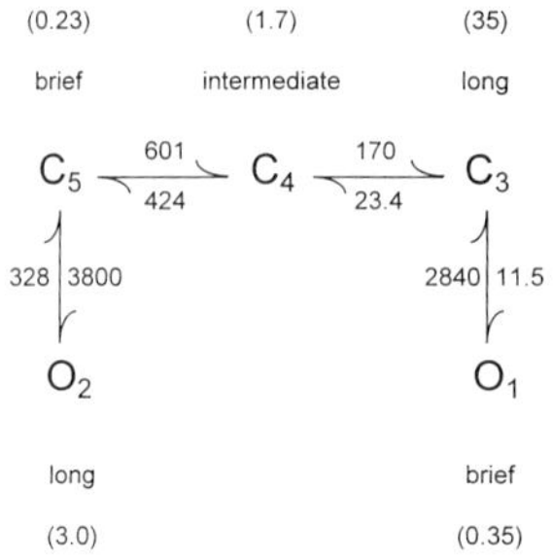

SCHEME 6

the component dependencies of the longer closed components (mainly associated with the two extended compound closed states) were similar, as in Scheme 2. Examination of the effects of changing the closed state lifetimes on the component dependencies suggested the component dependencies could be interpreted if the mean lifetimes of C_3 and C_4 differed by a factor of about 3 or more.

Experimental observations from several different ion channels suggest that brief closed intervals are adjacent to long open intervals.[27,29] It will be of interest to apply the interpretations of component dependencies, as described here, to develop and test kinetic models for the gating of these channels.

Different Gating Mechanisms Giving Rise to Similar Distributions

A key advantage of the use of 2-D dwell-time distributions of adjacent open and closed intervals is that they can be used to extract information on the correlations between the durations of adjacent intervals. This information cannot be obtained from 1-D distributions of open and closed intervals considered separately. However, while the correlation information is often necessary to identify models, this information alone is often insufficient to identify a unique model. This problem of model identification[13,14,30] is demonstrated by Schemes 5 and 6, with the rate constants shown.

Each of these two models contains two open states and three closed states. Although the states are connected differently and the transitions among the states are described by different rate constants, these two models

[29] O. B. McManus, A. L. Blatz, and K. L. Magleby, *Nature* **317,** 625 (1985).
[30] R. J. Bauer, B. F. Bowman, and J. L. Kenyon, *Biophys. J.* **52,** 961 (1987).

TABLE V
TWO-DIMENSIONAL KINETIC PARAMETERS FROM SCHEME 5[a]

τ_{open} (ms)		τ_{closed} (ms)		V_{obs}	V_{ind}	*cDep*
O_1'	3.0	C_3'	0.22	0.810	0.785	0.032
		C_4'	1.9	0.113	0.110	0.027
		C_5'	37	0.046	0.075	−0.39
O_2'	0.35	C_3'	0.22	−0.0001	0.025	−1.0
		C_4'	1.9	−0.0001	0.0035	−1.0
		C_5'	37	0.031	0.0024	12

[a] τ_{open}, open time constant; τ_{closed}, closed time constant; V_{obs}, component volume observed; V_{ind}, component volume expected for independent pairing; *cDep*, component dependency. τ_{open}, τ_{closed}, and V_{obs} were calculated with rates from Scheme 5 using **Q** matrix methods. V_{ind} was calculated with Eq. (5). *cDep* was calculated with Eq. (6).

produce nearly identical 2-D distributions and component dependencies, as can be seen by comparing Tables V and VI. Since these two models generate essentially the same distributions, this example illustrates that 2-D analysis would be unable to distinguish among these models using single-channel data obtained under one set of experimental conditions.

The problem of identification can often be overcome by obtaining single-channel data at various levels of activation by voltage, ligands, or other stimuli, and employing simultaneous (global) fitting of multiple data

TABLE VI
TWO-DIMENSIONAL KINETIC PARAMETERS FROM SCHEME 6[a]

τ_{open} (ms)		τ_{closed} (ms)		V_{obs}	V_{ind}	*cDep*
O_1'	3.0	C_3'	0.22	0.811	0.787	0.030
		C_4'	1.9	0.112	0.108	0.037
		C_5'	37	0.046	0.075	−0.39
O_2'	0.35	C_3'	0.22	0.0000	0.025	−1.0
		C_4'	1.9	−0.0011	0.0034	−1.3
		C_5'	37	0.031	0.0023	12

[a] τ_{open}, open time constant; τ_{closed}, closed time constant; V_{obs}, component volume observed; V_{ind}, component volume expected for independent pairing; *cDep*, component dependency. τ_{open}, τ_{closed}, and V_{obs} were calculated with rates from Scheme 6 using **Q** matrix methods. V_{ind} was calculated with Eq. (5). *cDep* was calculated with Eq. (6).

sets to a single kinetic scheme.[14,31,32] The simultaneous fitting approach has previously been applied with success using 1-D distributions.[7,28] Simultaneous fitting using 2-D distributions can further constrain models, and thus could serve as a powerful technique to improve the ability to identify the correct model.

In theory, 3-D (i.e., open–closed–open) or higher order dwell-time distributions, or the full sequence of dwell times, can contain more kinetic information than can be extracted from the 2-D distributions. Thus, the use of full-likelihood approaches, which use the full sequence of dwell times[3,13,14,17] should facilitate model identification. The 2-D analysis method is ideally suited for very large data sets where the full-likelihood analysis may be impractically slow. The 2-D component dependency plots can also be used to test critically models developed by full-likelihood analysis, as suggested next.

Using Component Dependency Plots to Test Mechanisms Critically

Any insight gained into mechanisms from component dependency plots will be useful, but perhaps the best use of component dependency plots will be to evaluate critically proposed gating mechanisms. The 2-D components of adjacent intervals and the component dependency plots derived from them contain essentially all of the useful kinetic information from a single-channel current record for channels without appreciable subconductance levels. As a critical test of proposed mechanisms, the component dependencies calculated from the most likely proposed kinetic schemes could be compared to those determined directly from the experimental 2-D dwell-time distributions by fitting sums of 2-D exponential components. If the model describes the component dependencies within the expected stochastic variation, then it is unlikely that a better model will be found. If the model cannot describe all of the component dependencies, then additional models will need to be investigated.

Acknowledgments

We thank R. A. Bello for helpful discussions. This work was supported in part by grants from the National Institutes of Health (AR32805, NS30584, NS007044) and the Muscular Dystrophy Association.

[31] R. Horn and C. A. Vandenberg, *J. Gen. Physiol.* **84,** 505 (1984).

[32] O. B. McManus and K. L. Magleby, *Biophysical J.* **49,** 171a (1986).

Section IV

Expression Systems

[26] Expression of Ligand-Gated Ion Channels Using Semliki Forest Virus and Baculovirus

By KATHRYN RADFORD and GARY BUELL

Introduction

Recombinant protein synthesis from baculovirus (BV) and Semliki Forest virus (SFV) vectors has been exploited to study the biochemistry, pharmacology, and structure of ligand-gated ion channels. These viral systems allow high infection efficiencies and result in expression levels that exceed those normally obtained by DNA transfection. Here, we briefly review the current status of these systems for both large- and small-scale expression of ligand-gated ion channels and provide relevant protocols. We have used these techniques to study adenosine triphosphate (ATP)-gated ion channels (P2X receptors), but there are many reports of similar methods with other receptors and channels. Table I lists some of the applications of BV or SFV expression that have been published for ligand-gated ion channels.

Baculovirus and Semliki Forest Virus as Expression Vectors

The success of baculovirus as an expression vector is based on the replacement of nonessential viral genes that are controlled by strong promoters, by exogenous genes.[1] Recombinant virus is used to infect insect cells, the natural host, and the foreign gene products are expressed at times and in amounts that depend on the promoter and infection parameters. Ligand-gated ion channels, expressed with this system, have shown ligand binding activities ranging from 18 to 71 pmol/mg[2,3] and have been purified to 0.3–3 μg/mg of total cellular protein.[4–6]

SFV offers an alternative system for the high-level expression of ligand-gated ion channels in a broader range of cells that includes mammalian.[7] Although fewer examples of ligand-gated ion channels have been expressed

[1] G. E. Smith, M. D. Summers, and M. J. Fraser, *Mol. Cell Biol.* **3,** 2156 (1983).
[2] K. Lundstrom, A. Michel, H. Blasey, A. R. Bernard, R. Hovius, H. Vogel, and A. Surprenant, *J. Recept. Signal Tr. R.* **17,** 115 (1997).
[3] J. Morr, N. Rundstrom, H. Betz, D. Langosch, and B. Schmitt, *FEBS Lett.* **368,** 495 (1995).
[4] S. Sydow, A. I. E. Kopke, T. Blank, and J. Spiess, *Mol. Brain Res.* **41,** 228 (1996).
[5] T. Green, K. A. Stauffer, and S. C. R. Lummis, *J. Biol. Chem.* **270,** 6056 (1995).
[6] M. Cascio, N. E. Schoppa, R. L. Grodzicki, F. J. Sigworth, and R. O. Fox, *J. Biol. Chem.* **268,** 22135 (1993).
[7] P. Liljeström and H. Garoff, *Bio/Technol.* **9,** 1356 (1991).

0076-6879/98 $25.00

TABLE I
LIGAND-GATED ION CHANNELS EXPRESSED BY SEMLIKI FOREST VIRUS AND BACULOVIRUS SYSTEMS[a]

Channel family	Receptor subtypes expressed	Virus	Application	Refs.
Nicotinic	Rat nAChrα-4, β-2	BV	Binding studies	41
	Bovine nAChrα	BV	Purification	42
	Chick nAChrα	BV	Fluorescence microscopy, binding studies	15
5-HT$_3$	Mouse 5-HT$_3$	BV	Large-scale expression, purification, electron microscopy	5
	Mouse 5-HT$_3$	SFV	Large-scale expression	2
	Mouse 5-HT$_3$	SFV	Electrophysiology	43
Glycine	Glycine α-1	BV	Purification	3
	Glycine α-1	BV	Purification	6
GABA	GABA(A)α-1, β-1	BV	Mutagenisis, electrophysiology	44
	GABA(A)	BV	Subunit composition, infection parameters	29
	GABA(A)α-1,2,3,5; β-1,2,3; γ-2	BV	Coexpression, binding studies	38
	GABA(A)α-1, β-1	BV	Infection parameters, electrophysiology	45
	GABA(A)β-1	BV	Stable cell line	13
	GABA(A)α-1, β-1	BV	Fluorescence studies, electron microscopy	46
AMPA	AMPAα-1, α-2	BV	Expression	47
	AMPAα-2	BV	Glycosylation, immunofluorescence, binding	48
	AMPA α-1	BV	Glycosylation, binding	26
	GluRD	BV	Purification	49
	GluRB, GLuRD	BV	Binding, electrophysiology	50
Kainate	GluR6	BV	Palmitoylation, electrophysiology	51
	GluR6	BV	Cell line comparison	52
	GluR6	BV	Glycosylation, transmembrane topology	53
	GLuR6	BV	Binding, glycosylation	54
NMDA	NMDAR1	BV	Glycosylation, assembly, purification, binding site	4
	NMDA zeta 1	BV	Binding site	55
	NMDA zeta 1	BV	Glycosylation, assembly, binding immunofluorescence	25
	NR1, NR2A, NR2C	BV	Glycosylation	40
P2X	P2X$_2$, P2X$_3$	BV	Binding, electrophysiology, assembly	19
	P2X$_1$	SFV	Binding	8
	P2X$_2$, P2X$_1$	SFV	Binding	9
	P2X$_2$	SFV	Binding, electrophysiology	2

[a] Members of all ligand-gated ion channel families are included. The references were chosen as examples of different applications and are not exhaustive, in particular for GABA receptors that have been extensively studied with the baculovirus system.

with SFV, similar protein levels have been observed.[2,8,9] The SFV genome is a single-stranded RNA molecule that functions directly as a mRNA. SFV expression vectors consist of modified SFV cDNAs which lack regions that encode viral structural proteins and they serve as templates for the *in vitro* synthesis of recombinant RNA. Unlike baculovirus vectors, the missing viral genes are essential for assembly, and helper RNA, which encodes the structural proteins, must be cotransfected with the recombinant SFV vector RNA. Vector RNA alone is contained in infectious virions because helper RNA lacks the sequence needed for packaging into nucleocapsids. Thus, in contrast with BV, recombinant SFV cannot replicate during infection.

Due to the broad host range of SFV, an additional mutation has been introduced so that only conditionally infectious virus particles are produced. This mutation in the helper RNA prevents natural proteolytic processing of the viral spike protein.[10] The resulting noninfectious virus stocks must be activated by cleavage with chymotrypsin for host cell infection. Following infection with activated recombinant virus, the amounts of ligand-gated ion channels are comparable to those obtained with BV vectors but differ in their expression kinetics. A comparison of the steps involved in recombinant protein expression by BV and SFV systems is outlined in Fig. 1.

Choice of Promoters

Currently, one promoter is used with the SFV expression system. Foreign DNA is inserted under the control of the viral promoter for the 26S subgenomic mRNA that in wild-type virus drives transcription of structural proteins. In contrast, baculovirus vectors that contain a variety of promoters exist and the choice of the promoter depends on the objective. Ion channels have been mostly expressed with vectors that exploit the very late polyhedrin promoter in order to obtain maximal yields for binding or structural studies. Also active during the very late phase of viral infection, the p10 protein promoter is not competitive with the polyhedrin promoter. They can be used together to construct multiple promoter vectors, capable of simultaneously expressing two proteins.[11]

As opposed to the very late polyhedrin promoter, use of a late promoter such as the basic protein promoter may be of interest for ion channel

[8] A. D. Michel, K. Lundstrom, G. N. Buell, A. Surprenant, S. Valera, and P. P. Humphrey, *Br. J. Pharmacol.* **117,** 1254 (1996).

[9] A. D. Michel, K. Lundstrom, G. N. Buell, A. Surprenant, S. Valera, and P. P. Humphrey, *Br. J. Pharmacol.* **118,** 1806 (1996).

[10] M. Lobigs and H. Garoff, *J. Virol.* **64,** 1233 (1990).

[11] U. Weyer, and R. D. Possee, *J. Gen. Virol.* **72,** 2967 (1991).

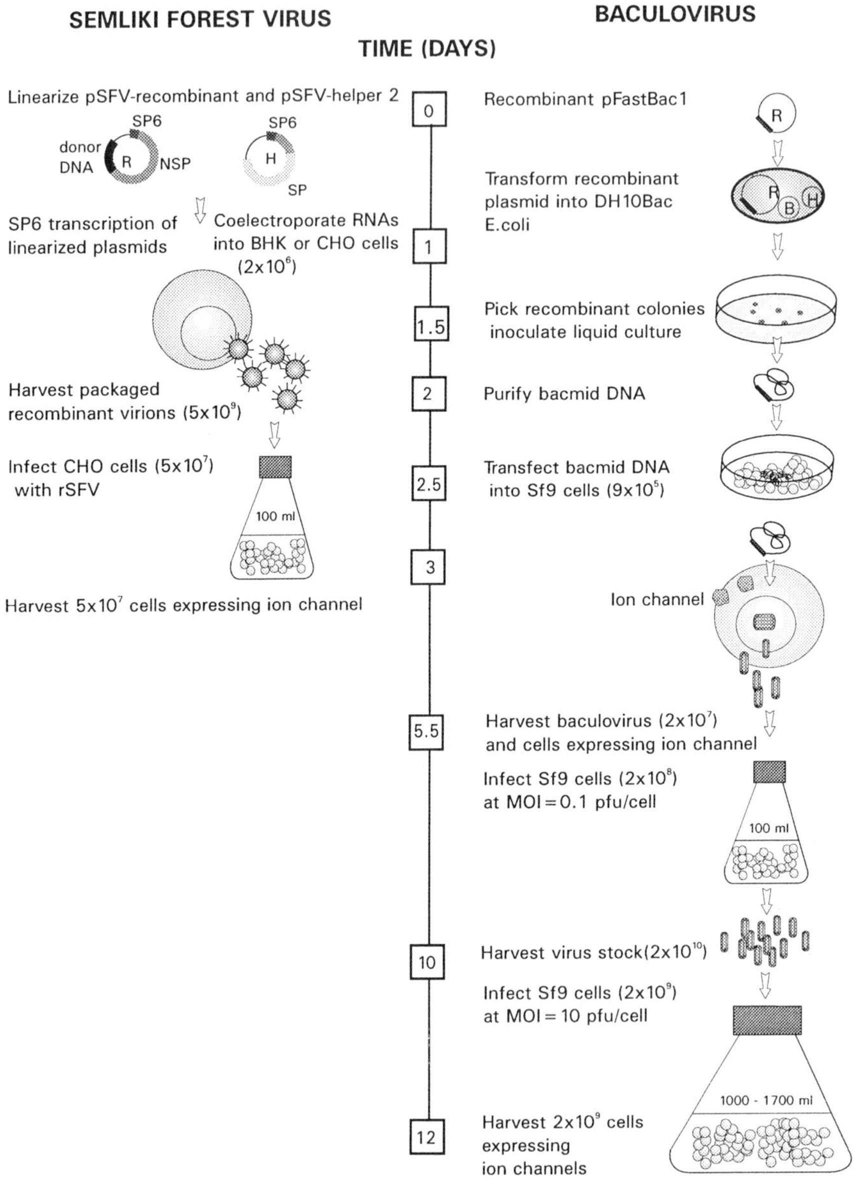

FIG. 1. Schematic representation and time scale of virus construction and expression of ion channels by semliki forest virus and baculovirus systems. (Left) Recombinant SFV plasmid (R) containing foreign DNA and helper SFV plasmid (H) are linearized with *Spe*I. RNAs transcribed by SP6 RNA polymerase are cotransfected into BHK or CHO cells and packaged virions are used to infect CHO cells in suspension to produce recombinant ion channels.

characterization.[12] The basic promoter allows maximum protein expression prior to the onset of the lytic period (30 hr postinfection) while the host cell membrane integrity is maintained. This facilitates electrophysiologic recordings due to the absence of the large holding currents that are observed after 50 hr postinfection when very late promoters generate maximum expression. The amount of foreign protein, made with the late promoter, is less than with very late promoters but is still higher than the quantities from stable cell lines. Generally, earlier promoters of baculovirus may allow better assembly, localization, and function of proteins since the host cell processing pathways remain uncompromised. Both late and very late promoter vectors are commercially available (Pharmingen, San Diego, CA).

Stable insect cell lines also exploit weaker early promoters in the absence of viral replication and efficient membrane targeting of ligand-gated ion channel receptors has been reported.[13–15] The chick nicotinic acetylcholine receptor cDNA was integrated randomly into an insect cell genome under the control of the immediate early gene promoter IE1, and efficient membrane targeting was demonstrated by fluorescence microscopy and ligand binding.[15] Stable cell lines remain an attractive alternative if maximal expression is not essential.

Incorporation of Protein Tags

Characterization and purification of proteins including ion channels is facilitated by the use of epitope tags, fused to the recombinant polypeptide. Commercially available BV vectors encoding fusion tags such as glutathione transferase (GST) or multiple histidine residues (His_6) generate proteins that can be subsequently purified on glutathione or Ni-NTA agarose beads,

[12] M. J. Paine, D. Gilham, G. C. Roberts, and C. R. Wolf, *Arch. Biochem. Biophys.* **328,** 143 (1996).

[13] K. A. Joyce, A. E. Atkinson, I. Bermudez, D. J. Beadle, and L. A. King, *FEBS Lett.* **335,** 61 (1993).

[14] F. Shotkoski, H. G. Zhang, M. B. Jackson, and R. H. French-Constant, *FEBS Lett.* **380,** 257 (1996).

[15] A. E. Atkinson, J. Henderson, C. R. Hawes, and L. A. King, *Cytotechnology* **19,** 37 (1996).

(Right) Recombinant BV plasmid (R) is transformed into DH10Bac cells that contain a helper plasmid (H) and bacmid (B). Transposition functions provided by the helper plasmid allow transfer between the recombinant BV plasmid and the target bacmid, which is capable of replicating in *Escherichia coli* and transfecting insect cells to produce infectious baculovirus. Baculovirus stock is amplified by one to two successive infections of insect cell cultures at low multiplicity. For ion channel expression insect cells are infected with baculovirus at 10 pfu per cell and cells are harvested after 50 hr.

respectively.[16,17] Other epitope tag sequences may be incorporated directly into the cDNA. Human m_2 muscarinic acetylcholine receptors, expressed in insect cells and tagged at the carboxy terminus with histidine residues have been purified with Ni^{2+} or Co^{2+} immobilized gel.[18] Their function was shown to be unaffected by the histidine tag. 5-hydroxytryptamine (5-HT_3) and P2X ion channel receptors have been successfully expressed in both SFV and baculovirus with His_6 or epitopes specific for monoclonal antibodies.[5,2,19] Whole-cell patch-clamp recordings for epitope-tagged P2X receptors were indistinguishable from those for the wild-type receptors. However, His_6-tagged 5-HT_3 receptors desensitized 50 times more slowly than the corresponding wild-type receptor.[2] Alternative tag positions may have to be tested to maintain the native channel properties. Epitope tags also serve as indicators of full-length protein expression and stability. Proteolytic cleavage of the C-terminal His_6-tagged cholecystokinin receptor expressed in BV-infected insect cells has allowed determination of the optimal time of harvest of the full-length receptor.[20]

Optimization of Expression

Despite careful consideration of the vector, the efficiency of heterologous gene expression is largely dictated by the intrinsic nature of the encoded protein. Greater than 1000-fold differences in yields have been reported for BV expressed proteins but there is little evidence that the expression level for a given gene can be influenced more than 5-fold by optimizing its sequence.[21] In our laboratory comparison of P2X receptor subtypes expressed by SFV or baculovirus have largely shown the same relative ligand binding, regardless of the expression system employed (Fig. 2). Optimization of cell culture and infection parameters can, however, improve the yields of ligand-gated ion channels. The key variables are the multiplicity of infection (MOI) and the cell density at the time of infection (TOI); they control the balance of nutrient utilization between cell growth and product expression and can be manipulated to ensure maximum yields

[16] A. H. Davies and L. M. Jones, *Bio/Technol.* **11,** 933 (1991).
[17] X. S. Chen, R. Brash, and C. D. Funk, *Eur. J. Biochem.* **214,** 845 (1993).
[18] M. K. Hayashi and T. Haga, *J. Biochem.* **120,** 1232 (1996).
[19] K. M. Radford, C. Virginio, A. Surprenant, R. A. North, and E. H. Kawashima, *J. Neurosci.* **17,** 6529 (1997).
[20] G. Gimpl, J. Anders, C. Thiele, and F. Fahrenholz, *Eur. J. Biochem.* **237,** 768 (1996).
[21] D. R. O'Reilly, L. K. Miller, and V. A. Luckow, *in* "Baculovirus Expression Vectors: A Laboratory Manual." Oxford University Press, New York, 1994.

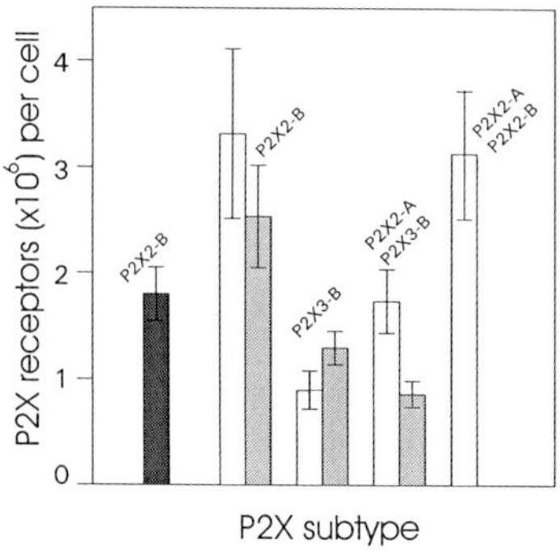

FIG. 2. Expression of $P2X_2$ and $P2X_3$ receptors with SFV or BV systems; comparison of simple and coinfection of cells and culture volume. Each bar shows the average of three determinations of the specific binding of [^{35}S]ATPγS to P2X receptors per cell. Open bars represent baculovirus expression at 150 ml and black bars at 24 liters. Hatched bars show expression with SFV in 13-ml cultures. Sf9 cell cultures were grown and infected with baculovirus in suspension cultures and CHO cell cultures were grown and infected with SFV in 75-cm^3 tissue culture flasks. Letters A and B refer to independent epitopes that were used to tag the receptors.[19]

and reproducible measurements.[22,23] The MOI is particularly important for the optimal expression of ion channels by SFV.[24] By increasing the MOI, the number of RNA molecules per cell that actively transcribes the recombinant protein is increased, resulting in more channels. In contrast baculovirus is capable of replicating within the host cell and its initial MOI has little effect on the final product yield.[22]

Glycosylation

The expected putative sites of N-linked glycosylation in P2X ion channels have been observed for BV expressed protein (K.R., unpublished results, 1996), but the extent of glycosylation varied. The expression of multiple glycosylation variants introduces sample heterogeneity that can, for example, prevent the accurate determination of the molecular weight of the fully assembled ion channel.

Glycosylation also affects the localization of ion channels but may not be essential for their function once they are fully assembled. Inhibition of glycosylation has been shown to effect the surface expression and binding

[22] K. M. Radford, C. Cavegn, M. Bertrand, A. R. Bernard, S. Reid, and P. F. Greenfield, *Cytotechnology* **24,** 73 (1997).

[23] K. M. Radford, S. Reid, and P. F. Greenfield, *Biotechnol. Bioeng.* **56,** 32 (1997).

[24] H. D. Blasey, K. Lundstrom, S. Tate, and A. R. Bernard, *Cytotechnology* **24,** 65 (1997).

of glutamate receptors[4,25,26] expressed in BV-infected insect cells. This loss of function could be due to the inability of the nonglycosylated receptors to bind transiently to molecular chaperones. The simultaneous expression of folding pathway chaperones has previously improved expression levels of secreted recombinant proteins,[27] but similar experiments have yet to be performed with ion channels.

Posttranslational modifications have been studied less extensively with SFV; however, it is expected that the broader range of host cells will allow modifications identical to those of native receptors. Short-term expression of proteins in cultured neurons has been achieved with SFV in which the expressed proteins showed appropriate intracellular transport.[28]

Receptor Assembly

Baculovirus has been used to express both heteromeric and homomeric protein complexes (Fig. 3). Successful generation of active heteromeric channels was obtained by the coinfection of insect cells with multiple baculovirus that encoded different subunit components.[29,19] Evidence of interactions between other proteins, coexpressed by baculoviruses, has been established using immunoprecipitation.[30] Although heteromeric ion channel expression has not been reported using SFV, this system has certain advantages. First, multiple viruses encoding subtype variants can be generated rapidly using the SFV system. Second, the nonreplicative nature of the recombinant virus, leading to the direct relationship between MOI and expression yield, could facilitate the study of heteromeric channel stoichiometry.

The baculovirus and SFV systems, therefore, can produce large amounts of ligand-gated ion channels for purification and in addition be used to study the assembly, interaction, and function of the component subunits.

[25] S. Kawamoto, S. Uchino, S. Hattori, K. Hamajima, M. Mishina, S. Nakajima-Iijima, and K. Okuda, *Mol. Brain Res.* **30,** 137 (1995).

[26] S. Kawamoto, S. Hattori, I. Oiji, K. Hamajima, M. Mishina, and K. Okuda, *Eur. J. Biochem.* **233,** 665 (1994).

[27] T. Hsu, J. J. Eiden, P. Bourgarel, T. Meo, and M. Betenbaugh, *Protein Expres. Purif.* **5,** 595 (1994).

[28] V. M. Olkkonen, P. Liljeström, H. Garoff, K. Simons, and C. G. Dotti, *J. Neurosci. Res.* **35,** 445 (1993).

[29] C. Hartnett, M. S. Brown, J. Yu, R. J. Primus, M. Meyyappan, G. White, V. B. Sterling, J. F. Tallman, T. V. Rambhdran, and D. W. Gallager, *Receptor Channel* **4,** 179 (1996).

[30] A. J. Barr, L. F. Brass, and D. R. Manning, *J. Biol. Chem.* **272,** 2223 (1997).

infected with	$P2X_2$-A		$P2X_2$-A and $P2X_3$-B		P2X3-B		$P2X_2$-A and $P2X_3$-B	
immunoprecipitated with	anti-A	anti-B	anti-B	anti-A	anti-B	anti-A	anti-B	anti-A
detected with	anti-A	anti-A	anti-A	anti-A	anti-B	anti-B	anti-B	anti-B

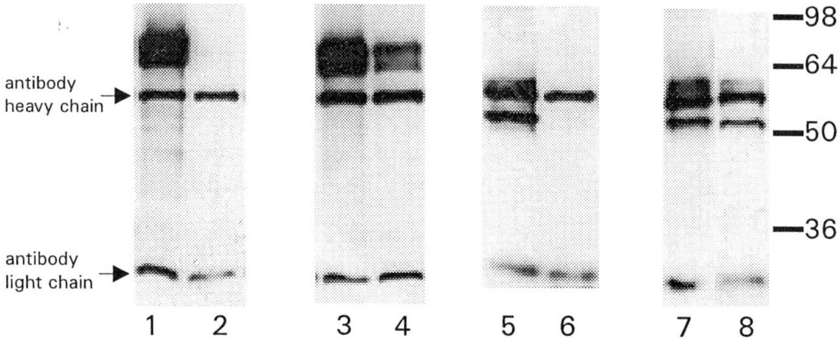

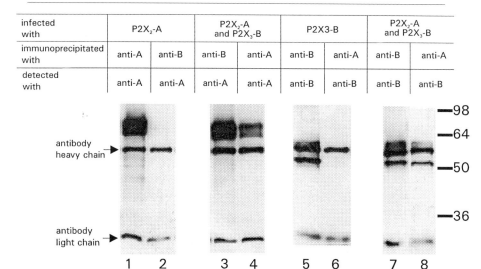

FIG. 3. Western blot analysis of $P2X_2$ and $P2X_3$ coimmunoprecipitation. Cells were infected with baculoviruses encoding a $P2X_2$ receptor tagged with an A epitope, a $P2X_3$ receptor tagged with an B epitope, or both. Cells were harvested at 52 hr, solubilized with 1% Triton X-100, and the supernatants immunoprecipitated with antiepitope peptide monoclonal antibodies as indicated. These antibodies were also used to detect the immunoprecipitated channel subunits by Western blotting. $P2X_3$ comigrated with antibody heavy chain bands. [Adapted from K. M. Radford and G. Buell (1997) Expression of ligand gated ion channels using semliki forest virus and baculovirus, *Methods Enzymol.*]

Generation of Viral DNA or RNA

Recombinant Semliki Virus Production

Two plasmids, the recombinant vector with inserted sequence and pSFV-Helper2 plasmid are linearized with *Spe*I. Recombinant RNAs are generated from these templates, using SP6 RNA polymerase, and the two *in vitro* transcripts are coelectroporated into Baby Hamster kidney (BHK) or Chinese Hamster ovary (CHO) cells. SFV structural proteins, encoded by the helper RNA, form virions that package only the recombinant vector RNA and this virus can infect other cells without subsequent replication. pSFV Helper2 should be used to prevent the low-frequency generation of replicon-competent SFV.

Protocol 1.1: In Vitro Transcription of Viral RNAs

1. 2.5 μg of pSFV1 DNA with foreign insert and pSFV-Helper2 DNA (Life Technologies, Inc., Rockville, MD) are separately digested with *Spe*I. The insert must not contain any *Spe*I restriction sites because this enzyme is used to linearize the recombinant vector plasmid. To ensure that the

template is free of ribonucleases after digestion, it is extracted with phenol and precipitated with ethanol.

2. The linearized vector and helper DNAs are used as templates for *in vitro* transcription by incubating with the following components at 37° for 1 hr in 50 μl [40 m*M* HEPES/KOH, pH 7.4, 6 m*M* magnesium acetate, 20 m*M* spermidine hydrochloride, 1 m*M* m7G(5′)ppp(5′)G, 5 m*M* dithiothreitol (DTT),] rNTP mix (1 m*M* rATP, 1 m*M* rCTP, 1 m*M* rUTP, and 0.5 m*M* rGTP), 50 U RNasin (Promega, Madison, WI), and 30 U SP6 RNA polymerase.

3. 1 μl of each transcript is analyzed by agarose (0.6%) gel electrophoresis to assess the quality and amount of RNA. The transcripts normally migrate as a single band, more slowly than the linearized plasmid.

Protocol 1.2: In Vivo Packaging of Recombinant SFV Virions. This protocol produces sufficient virus to infect 2×10^8 cells at a MOI of 100 virus/cell.

1. BKH(ATCC CCL-10 Rockville, MD) or CHO (ATCC CCL-61) cells grown in Ham's F12 medium: Iscove's (1:1) medium (Seromed, Basle, CH), 20 ml at 5×10^5 cell/ml, are centrifuged at 1000*g* for 5 min, washed, and resuspended in 1 ml phosphate-buffered saline (PBS).

2. Cells (0.8 ml) and 20 μl recombinant vector and helper RNA transcripts are mixed rapidly and electroporated with two pulses at 25 μF, 850 V, and maximum resistance (Bio-Rad Richmond, CA Gene Pulser).

3. Electroporated cells are immediately diluted 20-fold in prewarmed, complete growth medium and incubated for 36 hr at 37° in 5% (v/v) CO_2.

4. Recombinant SFV is harvested by centrifugation at 1000*g* for 5 min and can be stored at 4° for several weeks. For long-term storage the virus can be frozen in liquid nitrogen and stored at −80°; however, on thawing infective virus titers will be lower.

Compared to baculovirus, the generation of SFV stock is more rapid, but is also limited by the quantity of virus that is formed per electroporation. Repetitive generation of small-scale SPV stocks can introduce variability and necessitates the characterization of each new preparation. For these reasons we have preferred SFV for the small-scale expression of receptor subtypes and mutational analysis while baculovirus has been more useful for large-scale expression.

Recombinant Baculovirus Production

Classical construction of recombinant baculoviruses involves the cotransfection of insect cells with baculoviral DNA and a recombinant transfer plasmid that contains the foreign gene. Insertion of the gene into the baculovirus genome occurs via homologous recombination, and the resulting mixture of progeny viruses (recombinant or parent) is separated by several cycles of plaque purification. Linearized baculovirus DNA (BaculoGold DNA, Pharmingen) contains a lethal deletion in the *polh* gene, which can be complemented by recombination with the appropriate transfer plasmid so that only recombinant viruses are able to form plaques. Typically, a single plaque purification is required to ensure the absence of the parental baculovirus.

More recently, recombinant baculoviruses have been made by site-specific transposition of foreign genes into a bacmid that is propagated in *Escherichia coli* (FastBac, Life Technologies). Using this technique, recombinant viral DNA, isolated from independent colonies is free from parental nonrecombinant virus and plaque purification is not necessary. However, regardless of the system that is chosen, clonal isolation of virus by plaque purification helps to ensure homogeneous expression.

Protocol 1.3: Recombinant Virus Construction Using Linearized Baculovirus DNA

1. Seed 2 ml of Sf9 (*Spodoptera frugiperda* fall army worm ovary) cells (Invitrogen; 1×10^6 cell/ml) grown in Sf900II (Life Technologies) into 60-mm tissue culture plates and allow to attach for 1 hr at 27°.
2. Mix 0.5 μg of viral DNA (BaculoGold DNA, Pharmingen) with 5 μg of recombinant baculovirus transfer vector. Use highly purified DNA that is less than 1 month old. After a 5-min incubation at room temperature, add 1 ml of transfection buffer (25 m*M* HEPES, pH 7.1, 125 m*M* $CaCl_2$, 140 m*M* NaCl).
3. Vacuum aspirate medium from cell monolayer and add DNA/transfection buffer mixture drop by drop while rocking the plate. A fine calcium phosphate/DNA precipitate should form. Incubate plates at 27° for 5 hr.
4. Remove cotransfection mixture by vacuum aspiration. Wash monolayer gently with 3 ml of SF900II. Vacuum aspirate wash medium and add 3 ml of fresh SF900II. Incubate at 27° for 5 days.
5. Examine cells at 5 days postinfection. Infected cells are larger with a granular nucleus, 80–90% of the entire cell size. Harvest 3 ml of cotransfection supernatant and store at 4°.
6. Plaque purification is performed as for the plaque assay described later with slight modifications. Dilutions of virus from 10^{-2} to 10^{-4} are sufficient and neutral red is not required to stain plaques. Instead, 200 μg/ml X-Gal (5-bromo-4-chloro-3-indolyl-β-D-galactopyranoside) is added to the agarose/medium overlay immediately prior to use. After 5 days of incubation white recombinant viral plaques are distinguished from the small percentage (less than 1%) of the blue (*lacZ*$^+$) nonrecombinant viral plaques.

7. Pick a recombinant plaque with a sterile Pasteur pipette and transfer the entire agarose plug to 1 ml of SF900II medium. Virus is eluted from the agarose by agitation overnight at 4°. At this stage the virus can be used for large-scale amplification of virus stock.

Protocol 1.4: Recombinant Virus Construction Using pFastBAC

1. Subclone the gene of interest into the pFastBAC1 donor vector (Life Technologies). Digest vector and foreign DNA with the selected endonucleases, dephosphorylate the vector, and purify the vector and foreign insert DNA. Ligate the prepared vector and insert fragment to produce the pFastBAC-recombinant.

2. Incubate 0.1 ml of DH10B competent cells (Life Technologies) with 1 ng of the pFastBAC-recombinant on ice for 30 min. Transform by heat-shock for 45 sec at 42°. Cool 2 min on ice, add 0.9 ml rich medium, and shake at 37° for 4 hr.

3. Plate 0.1 ml of transformation mix at various dilutions (10^0, 10^{-1}, 10^{-2}) onto LB plates containing 50 μg/ml kanamycin sulfate, 10 μg/ml tetracycline, 7 μg/ml gentamicin, 100 μg/ml X-Gal, and 40 μg/ml isopropyl-thio-β-D-galactoside and incubate 24 hr at 37°.

4. Select 10 white colonies and streak a fraction of each colony onto fresh plates to verify the recombinant phenotype. Inoculate the same colonies in 2 ml LB with the antibiotics used above. Incubate at 37° with agitation for 16 hr.

5. Extract and purify DNA from 1 ml of a liquid culture, confirmed to be recombinant. A DNA extraction kit (Qiagen, Hilden, Germany) can be used and provides DNA of adequate purity for transfection. The DNA is dissolved in 40 μl of TE buffer and can be stored at −20°.

6. Sf9 cells (9×10^5) are inoculated into 2 ml of Sf900II medium containing antibiotics (50 U/ml penicillin, 50 μg/ml streptomycin), plated onto sterile 35-mm culture dishes, and allowed to attach for 1 hr at 27°. The cell condition at the time of transfection is critical and cells grown to greater than 2×10^6 cell/ml should not be used.

7. Minipreparation bacmid DNA (5 μl) in 0.1 ml of SF900II is mixed with 6 μl of CellFECTIN reagent (Life Technologies) in 0.1 ml Sf900II and incubated for 45 min at room temperature.

8. Sf9 cell monolayer is rinsed with 2 ml of Sf900II and vacuum aspirated. Sf900II medium (0.8 ml) is added to the transfection mixture and 1 ml of the diluted mixture is overlaid onto the cells. The plates are incubated for 5 hr at 27° after which the residual transfection mixture is replaced with 2 ml of Sf900II medium containing antibiotics. The transfected cells are incubated for 48 hr at 27°.

9. Supernatant is harvested for virus titration and subsequent amplification to 100 ml. At this stage it may be possible to detect expression of the ion channel proteins via immunoblot or activity assay of the residual cells.

Choice and Cultivation of Host Cells

The host insect cells most commonly used with baculovirus vectors are Sf9, Sf21, and Tn5-High Five cells (Invitrogen, San Diego, CA). These cell lines grow well in suspension in serum-free medium (SF900II for *Spodoptera frugiperda:* Sf9; Sf21 and Excell-401 for *Trichoplusia ni:* Tn5). Although neither cell line is superior for the expression of all ion channels, we have found better yields of functional P2X receptors using Sf9 than with Tn5 cells. Differences between cell lines have also been reported for other receptor families. The human muscarinic m_2 receptors have been expressed using baculovirus in several insect cell lines: Sf9, Sf21, and Tn5.[31] The expression level was slightly increased in Sf21 cells versus Sf9 cells, and Sf9 and Tn5 cells showed significant proteolysis compared to Sf21 cells. The Sf21 and Sf9 cells are smaller and have a higher surface area-to-volume ratio which could improve the expression of membrane proteins. Increased synthesis of toxic metabolites (ammonia and lactate) in Tn5 cell cultures has been reported and may decrease productivity.[32] Regardless of the cell line employed, the nutrient composition of the medium potentially imposes a limitation. Most examples of ion channels, expressed in insect cells, have used serum containing media. However, the glucose and amino acid content of the original basal medium component of serum containing formulations such as Graces medium, TNMH, or IPL41 media is very low, compared to current serum-free formulations and do not support maximal insect cell growth or optimal protein expression.[23]

Insect Cell Culture

Cultivation of large volumes of insect cells as monolayers is impractical due to the high surface area-to-volume ratio that is required for adequate oxygenation and to the disruption of the monolayer for monitoring channel expression. Since the application of pluronic polyol F-68 (Life Technologies), a shear protection agent,[33] insect cell cultures of 30–1700 ml are most efficiently cultivated in suspension. Culture volumes should not exceed 50% of the maximum capacity of Erlenmeyer flasks. For larger cultures

[31] F. Heitz, C. Nay and C. Guenet, *J. Recept. Signal Tr. R.* **17,** 305 (1997).

[32] J. D. Yang, P. Gecik, A. Collins, S. Czarnecki, H. H. Hsu, A. Lasdun, R. Sundaram, G. Muthukumar, and M. Silberklang, *Biotechnol. Bioeng.* **52,** 696 (1996).

[33] D. W. Murhammer and C. F. Goochee, *Biotechnol. Progr.* **6,** 391 (1990).

additional oxygen or air must be supplied so that dissolved oxygen tension is maintained at 50% of that in air. In stirred tank bioreactors (2–10 liters) this is achieved by a combination of sparging and stirring. Fixed-angle rather than flat-blade impellers are employed to increase the mixing without excessive shear stress. Large-scale airlift reactors (20–100 liters) can oxygenate and mix by sparging alone and the dimensions of these reactors (high height-to-surface area ratio) can minimize cell damage due to bubble bursting at the medium surface.

Due to the diversity of culture vessels and bioreactors, the following protocols describe expression only in readily available flasks. Cultures of 30–1700 ml provide sufficient material for most biochemical, pharmacologic, and preliminary structural studies.

Protocol 2.1: Routine Subculture of Insect Cells

1. Inoculate Erlenmeyer shaker flasks (250 cm^3, Schott) with 125 ml of Sf9 cells at 2×10^5 cell/ml in SF900II medium supplemented with 0.1% Pluronic F68 and rotate at 170 rpm at 27° with lids unscrewed.
2. Subculture at 2×10^5 cell/ml into prewarmed medium (27°) every 3 or 4 days while cells are in early exponential growth. Sf9 cells should have an average doubling time of 21 hr and should be replaced if a doubling time greater than 24 hr is observed routinely. Do not use cells for subculture that have exceeded 5×10^6 cell/ml without prior medium exchange. Discard cultures exceeding a density of 7×10^6 cell/ml.
3. Cell density and viability are quantified using a hemocytometer and trypan blue (0.2%) exclusion. Count samples in triplicate, where a single count includes both sides of the hemocytometer and sample is diluted so that approximately 200 cells are counted per side. The mean of triplicate estimations ±95% confidence interval can be calculated from the variance estimated by the method of Nielsen *et al.*[34]
4. Cultures can be scaled up in volume at time of subculture to 1700 ml and are typically infected for expression at a density of 1×10^6 to 5×10^6 cell/ml. Two weeks of reproducible growth kinetics should be observed before infection, especially after thawing cells from liquid nitrogen. New cell cultures from frozen parent stocks should be thawed after Sf9 cells have exceeded 50 passages.

SFV Host Cell Line

The preferred hosts for SFV are mammalian cell lines, although the virus is capable of infecting a wide variety of primary cells and most animal

[34] L. K. Nielsen, G. K. Smyth, and P. F. Greenfield, *Biotechnol. Progr.* **7,** 560 (1991).

cell lines. BHK and CHO cells are most commonly used but CHO cells have been suggested to be less efficient for SFV expression.[24] BHK cells in suspension tend to aggregate and are difficult to maintain without specialized bioreactors or nonadherent clones. We, therefore, use CHO cells for medium-scale expression (50–1700 ml) but prefer BHK for small-scale (maximum 50-ml) monolayer cultures. These mammalian cells typically reach maximum cell densities of $1–2 \times 10^6$ cell/ml. This is 5- to 10-fold less than for Sf9 cultures, but BHK and CHO cells grow faster, doubling every 15–18 hr.

CHO cell manipulations are performed in Ham's F-12 and Iscove's medium (1 : 1) supplemented with 4 m*M* glutamine and 10% fetal calf serum (FCS). Mammalian cell lines have lower oxygen requirements than insect cells, however, the oxygen demand increases dramatically after infection in both systems and CHO suspension cultures should be rotated at 120 rpm. Unlike insect cells, pH control is critical for optimal growth and ion channel expression by CHO or BHK cells and all incubations should be performed in the presence of 5% (v/v) CO_2 as sodium bicarbonate is used as a buffering component in the medium.

Protocol 2.2: Routine Subculture of CHO Cell

1. Inoculate Erlenmeyer shaker flasks (250 cm^3) (Schott) with 125 ml of CHO cells at 4×10^4 cell/ml in Ham's F-12; Iscove's (1:1) medium supplemented with 4 m*M* glutamine, 10% FCS, and 0.2% pluronic F-68 and rotate at 120 rpm at 37° in 5% CO_2 with lids unscrewed.
2. Quantify cell density and viability as explained for insect cells in Protocol 2.1.
3. Allow cells to grow and subculture at 4×10^4 cell/ml into prewarmed medium (37°) every 2 or 3 days while cells are in early exponential growth. They are ready for infection with SFV at 5×10^5 cells/ml. Methods for optimization of expression parameters are given in Protocol 4.2.

CHO cells should have an average doubling time of 16 hr and should be replaced if a doubling greater than 20 hr is observed routinely or they have exceeded 2×10^6 cell/ml during subculturing.

Amplification and Titration of Virus Stocks

The quantity of baculovirus, generated by transfection or plaque purification, is adequate for small-scale receptor expression. Virus can be amplified by repeated infections of insect cell cultures at increasing scale, but the number of serial infections should be kept to a minimum (two to three) and a low MOI employed [0.1–0.2 plaque-forming units (pfu)/cell].

Extensive serial passage of virus or high MOI will produce virus stock, dominated by defective interfering particles, which have extensive mutations and are helper virus dependent.[35] These particles interfere with replication of the normal virus and substantially diminish the titer of viral stocks.

Protocol 3.1: Amplification of Baculovirus Stocks

1. Prepare a 20-ml suspension culture of Sf9 cells in SF900II at 1×10^6 cell/ml in a 100-cm^3 flask. Cells may be centrifuged at 1000*g* for 5 min and resuspended in fresh medium to increase the potential virus titer.
2. Inoculate the culture with supernatant from bacmid transfection to give a final MOI of 0.1 pfu/cell or add 1 ml of the eluted virus prepared by plaque purification.
3. Rotate at 170 rpm at 27° for 4–5 days or until viability approaches zero as determined via trypan blue exclusion.
4. Centrifuge at 2000*g* for 30 min and store supernatant at 4° in the dark for up to 6 months. For long-term storage, virus may be frozen at −80° but is unstable at −20°. Repeated freeze–thawing will result in a large reduction of virus titer.
5. Determine the virus titer (approximately 1×10^7 pfu/ml) by one of the methods suggested later, and repeat the amplification procedure at 1700 ml scale, using a MOI of 0.1 pfu/ml. The 1700 ml stock should contain 2×10^8 pfu/ml of recombinant virus, which is sufficient for expression of the ion channel at the 10-liter scale.

Baculovirus Titration

Two forms of baculovirus are always produced that have different natural roles in insects. These are budded, infectious virus and multinucleocapsid, noninfectious virus. The multinucleocapsid form, which is normally surrounded by the protective polyhedrin coat, ensures propagation within insect populations and requires ingestion to infect, whereas the infective form permits efficient infection of adjacent cells within an insect and also insect cell cultures. Plaque and end-point dilution methods for determining baculovirus concentrations measure only infective virus and not the total concentration of virus particles, which may be 100-fold higher.

An important parameter influencing either of these assays is the condition and density of the insect cells. Uninfected controls should be included to monitor the condition of the cell monolayers. Virus is visualized in the plaque assay when a single infectious unit replicates and progeny virus

[35] M. Kool, J. W. Voncken, F. L. J. van Lier, J. Tramper, and J. M. Vlak, *Virology* **183,** 739 (1990).

infects surrounding cells that lyse and produce clear zones. The end-point dilution method is often performed when the recombinant protein is expressed with a detectable marker, such as β-galactosidase. However, coexpression of nonrelated marker proteins potentially reduces the maximum yield of the ion channel and should be avoided during its large-scale expression. For proteins without markers an alternative end-point assay is described that reliably detects the titer of virus by the degree of postinfection cell growth. End-point dilution methods are simpler to perform than plaque assays and permit large numbers of sample repeats in multiwell screening plates.

Protocol 3.2: Plaque Assay

1. Inoculate 35-mm multiwell tissue culture plates with 1.5 ml of Sf21 cell suspensions grown in TNM-FH insect medium (Pharmingen) containing 5% FCS at 8×10^5 cell/ml. Plaque definition is improved if Sf21 cells are used instead of Sf9 cells. Allow cells to attach for at least 1 hr at 27°.
2. Prepare serial dilutions of test virus (10^{-3} to 10^{-8}) in TNM-FH without serum.
3. Aspirate cell monolayers, rinse with 2 ml TNM-FH without serum, and inoculate with 0.5 ml of diluted virus. At least one well is left uninfected as a cell monolayer control. Plates are tilted immediately to ensure even distribution of virus over the monolayer. Allow virus to absorb by incubation at 27° for 2–3 hr in a humid environment. Tilt plates at least twice during the incubation period to prevent cells from drying.
4. Prepare overlay medium (2 ml per well) immediately before use by mixing sterile 3% low gelling temperature agarose (cooled to 37°), with an equal volume of prewarmed (27°) TNM-FH medium containing 10% FCS.
5. After virus adsorption vacuum aspirate monolayers and overlay 2 ml of the agarose mixture without bubbles to each well. Agarose is allowed to set for 3–4 hr at 27°.
6. Overlay agarose with 1 ml of TNM-FH medium containing 5% FCS and incubate in a humidified environment at 27° for 5–6 days.
7. Plates are stained with 1 ml neutral red (0.3% solution diluted 1:20 in PBS) and incubated for 2 hr at 27°. Excess stain is poured off and the plate is then blotted by inversion onto paper towels for 5 hr or overnight at room temperature.

Protocol 3.3: End-Point Dilution Virus Assay

This assay requires that the ion channel construct also encodes a marker protein such as β-galactosidase that can be used to visualize the viral dilution end point. The end point is the dilution of virus that would infect 50% of the cultures ($TCID_{50}$) and can be converted according to Poisson distribution to virus concentration as pfu/ml.

1. Serial dilutions of test virus (10^{-1} to 10^{-10}) are made directly into Sf9 cell suspensions (1×10^6 cell/ml) in Sf900II in a final volume of 1 ml. An uninfected control sample is prepared by mixing 0.1 ml of PBS with 0.9 ml of cells.
2. Multiple repeat aliquots (0.1 ml) of virus/cell mixtures are pipetted into 96-well plates and incubated at 27° in a humid environment. One column is loaded with the uninfected control culture.
3. 10 μl of 20 mg/ml X-Gal is added to each well at day 5 or 6 and plates are incubated 4 hr at room temperature. Blue wells are scored as positive for virus and the dilution that gives 50% infection response is calculated by linear interpolation.

Protocol 3.4: MTT End-Point Dilution Virus Assay

1. Serial dilutions of test virus are made into Sf9 cell suspensions as in Protocol 3.3, step 1, but with slightly fewer cells (8×10^5 cell/ml).
2. Aliquots of virus/cell mixtures are loaded into 96-well plates as in step 2 of Protocol 3.3.
3. After 72 hr of incubation 20 μl of MTT [3-(4,5-dimethylthiazol-2-yl)2,5-diphenyltetrazolium bromide] (7.5 mg/ml) is added to all wells and incubated for at least 6 h.
4. Plates are read spectrophotometrically at 570 nm and viral end point is determined by scoring the number of virus positive wells (those indicating a lower absorbance than the noninfected control) at each dilution. Error estimates in the virus titer can be calculated by the method of Nielsen *et al.*[36]

Semliki Forest Virus Titration

The methods just outlined for baculovirus titration rely on the single virus end point, which can be determined following virus replication. In contrast, SFV does not produce plaques and the absolute amount of infectious virus cannot be determined. The following technique is based on counting the number of infected cells and is similar to the end-point dilution method. Although this method can only estimate the titer, it can be used to compare the relative concentrations of SFV stocks and to improve reproducibility between experiments.

[36] L. K. Nielsen, G. K. Smyth, and P. F. Greenfield, *Cytotechnology* **8,** 231 (1992).

Protocol 3.5

1. Test virus is activated by addition of α-chymotrypsin (Sigma, St. Louis, MO) 200 μg/ml for 15 min at room temperature. This cleaves the viral p62 precursor into E2 and E3 membrane proteins.

2. After 15 min the protease activity is arrested by the addition of aprotinin (Sigma) to a final concentration of 400 μg/ml.

3. Serial dilutions of activated test virus (10^{-1} to 10^{-10}) in culture medium (Ham's F12 and Iscove's 1:1, 4 m*M* glutamine, 10% FCS) are used to infect CHO or BHK monolayers (1×10^4 cells per well) in 96-well plates. Multiple repeats are required for each dilution.

4. Infected plates are incubated at 37° in a humid 5% CO_2 environment for 20 hr after which β-galactosidase expressing cells are stained blue by the addition of X-Gal to 40 μg/ml. Microscopic examination of the cells in each well is used to score the percentage of cells expressing β-galactosidase at each viral dilution.

Optimization of Process Parameters

Nutrient exhaustion limits the potential yield of ion channels expressed by baculovirus and correct application of MOI and TOI is required for reproducible expression.[23] Substrate limitation arises when cultures, grown to high density, are infected at low MOI; the uninfected cells continue to divide, leaving few nutrients available for product expression. The simplest strategy is to use an MOI (10 pfu/cell) that allows synchronous infection of the population at increasing cell densities. This is also affected by virus titer since the addition of large volumes of virus inoculum, consisting of spent medium, is inhibitory. It is recommended to stay well below the maximal critical cell density of infection and at the highest MOI possible. Once the optimal cell density at time of infection has been established, the optimal time of harvest is chosen as a balance between maximum expression and minimal degradation of the receptor.

Protocol 4.1: Optimization of the Times of Infection and Harvest for Ion Channels Expressed by Baculovirus

1. Inoculate a 1700-ml shaker flask with 2×10^5 Sf9 cells/ml and allow cultures to grow under conditions described earlier.

2. At various cell densities (10^6 to 6×10^6 cell/ml) remove 100-ml aliquots from the parent culture under sterile conditions and transfer to 250-cm^3 capacity Erlenmeyer flasks.

3. Infect immediately with recombinant baculovirus at a multiplicity of 10 pfu/cell or up to 5% of the culture volume. Incubate shaker flasks as

above and sample at 30, 40, 50, 60, 70 hr postinfection (5 $\times$ 1 ml) to determine the cell density and activity.

4. Results of the activity assay are expressed on a per-cell basis. Note that total cell density will decrease late in the infection period due to virus-induced cell lysis and estimates at these times cannot be used to calculate the specific activities.

5. The optimal time of harvest for the culture is chosen as the time at which the highest total yield of active receptor is obtained. If activity assays are not available, immunoblot analysis can be used to approximate the optimal cell density at the time of infection and harvest, but active and inactive protein is not distinguished. For P2X receptors the optimal time of harvest is between 50 and 65 hpi for cultures infected at high MOI.

Similar methods can be used to optimize ion channel expression by SFV, but these conditions will vary for different cell lines and media. The MOI cannot be optimized in monolayer cultures and the results extrapolated to suspension cultures. Viruses generally adsorb more slowly in suspension than monolayer cultures. Due to the nonreplicative nature of SFV, small changes in culture infection conditions, such as MOI, may result in significant differences in total yield and expression kinetics. For this reason it is necessary to monitor the time course of protein production to determine the optimal time of harvest. The following protocol is used to determine the optimal MOI and associated optimal time of harvest for ligand-gated ion channels expressed with SFV-infected CHO cells.

Protocol 4.2: Optimization of Ion Channel Expression Parameters for SFV

1. Recombinant virus (70 ml at 10^9 virus/ml) is activated by addition of α-chymotrypsin as in Protocol 3.5.

2. CHO cells grown to a density of 5 $\times$ 10^5 cell/ml as 1-liter suspension cultures are centrifuged and resuspended at 5% of the original volume in fresh medium (pH adjusted to 6.9 with acetic acid). Volumes of 10 ml are aliquoted into five 250-ml shaker flasks ready for immediate infection.

3. Cultures are infected at various MOIs (1, 10, 50, 100, 500 virus/cell) with activated virus and incubated at 37° (120 rpm) in 5% (v/v) CO_2 for 6 hr to ensure adequate adsorption of virus. Each culture is sampled at the time of infection. Centrifuge 0.5-ml aliquots at 1000*g* for 5 min and store the pellet at −80° as a control and the supernatant at 4° to assess virus adsorption.

4. At 6 hr postinfection the cultures are centrifuged at 1000*g* for 5 min and resuspended in 100 ml fresh medium. A sample (0.5 ml) of the supernatant is stored at 4° to confirm complete virus adsorption.

5. Cultures are sampled (5×1 ml) at 6, 12, 24, 48, and 60 hr postinfection and the cell density and ion channel expression levels are determined.

We routinely use SDS–PAGE and Western blot analysis to monitor P2X expression in these cases, but electrophysiologic activity can also be used to determine the optimal MOI and its associated optimal time of harvest. Depending on the quality of the virus stock, maximum expression can occur as early as 20–30 hr postinfection with greater than 10^6 receptors per cell.

SFV and baculovirus production are thus optimized by determining the TOI, optimal time of harvest, and, since SFV is nonreplicative, its optimal MOI. Once estimates of these culture and infection parameters have been established, ion channels can be reproducibly expressed by these systems for many function and structure applications.

Applications of Ion Channels Expressed by Baculovirus and SFV

Electrophysiologic Recording

The lytic nature of the baculovirus expression system precludes the use of whole-cell recording techniques for determining optimal expression parameters. The infected host cell membrane is compromised late during infection, due to budding of progeny virions and cell lysis. Host cell membrane integrity is usually sufficient for electrophysiologic recording up to 50 hr postinfection, after which the holding currents become large and unstable. The use of alternative baculovirus gene promoters for expression of ligand-gated ion channels at earlier times than the very late polyhedrin promoter may offer an advantage for electrophysiologic recording. In general the expression levels of proteins controlled by earlier promoters such as the basic promoter are lower than those controlled by the polyhedrin promoter but the slightly earlier expression kinetics provide a better separation of the expression and lytic phases of infection.

Mammalian cells infected with SFV are more amenable to electrophysiologic applications due to the earlier onset of expression. Maximum expression of ion channels can be achieved within the first 20 hr postinfection, well before cell lysis begins. This is particularly true if a high MOI is employed (MOI $>$ 30). The following protocol for electrophysiologic recording from insect cells infected with baculovirus encoding the $P2X_2$ receptor can also be adapted for BHK cells infected with recombinant SFV by altering the culture conditions, MOI, and time of recording.

Protocol 5.1: Preparation of Insect Cells for Whole-Cell Patch Clamp

1. Prepare a 30-ml suspension culture of Sf9 cells and culture under normal conditions at a density of 10^6 cell/ml.

2. Inoculate the culture with virus at MOI of 10 pfu/cell. Synchronous expression is desirable for electrophysiologic recording so that chosen cells are representative of the entire cell population.

3. Rotate infected cultures at 170 rpm at 27° for 30 hr. Aliquot approximately 2000 cells to polylysine-treated coverslips under sterile conditions and allow to attach at 27° for 2 hr before electrophysiologic recording.

Heteromeric Receptor Expression and Detection

Baculovirus supports the assembly of both homomeric and heteromeric ion channel complexes. Pentameric 5-HT_3 receptors, expressed with baculovirus, have been observed by electron microscopy.[5] Coexpression of two subunits, resulting in coassembly and formation active heteromeric channels, has been reported for $GABA_A$ and P2X channels.[19,29,37–39] The level of expression for the $GABA_A$ receptor was found to depend on the ratio of infectious virus particles for each subunit in multivirus infections. The stoichiometry of the channel, however, was not affected and only a single type of receptor was produced.[29]

Methods for coexpression of different subunits are essentially identical to those described earlier for a single subtype except that two or more viruses are used to infect a culture simultaneously. The following protocol was used to determine whether epitope-tagged $P2X_2$ and $P2X_3$ subunits assemble to form heteromeric channels (Fig. 3).

Protocol 5.2: Coinfection and Coimmunoprecipitation of P2X Receptor Subunits

1. Insect cells at 1.5×10^6 cell/ml grown as 100-ml volumes in 250-cm^3 capacity Erlenmeyer flasks are pooled, centrifuged at 1000*g* for 5 min, and resuspended in fresh medium.

2. Resuspended cultures are inoculated at a MOI of 5 pfu/cell with baculovirus encoding $P2X_2$ (tag A), $P2X_3$ (tag B), or both receptors.

3. Cultures are shaken at 200 rpm at 27° for 50 hr and aliquots removed at approximately 0, 24, 30, and 50 hr and stored as pellets at −80°.

[37] J. F. Pregenzer, W. B. Im, D. B. Carter, and D. R. Thomsen, *Mol. Pharmacol.* **43,** 801 (1993).

[38] M. R. Witt, S. E. Westh-Hansen, P. B. Rasmussen, S. Hastrup, and M. Nielsen, *J. Neurochem.* **67,** 2141 (1996).

[39] R. J. Primus, J. Yu, J. Xu, C. Hartnett, M. Meyyappan, C. Kostas, T. V. Rambhdran, and D. W. Gallager, *J. Pharmacol. Exp. Therapeut.* **276,** 882 (1996).

4. Cells (2×10^7) harvested at 50 hr are lysed on ice in 2 ml of 20 m*M* Tris, 150 m*M* NaCl, 1 m*M* $CaCl_2$, and 1 m*M* $MgCl_2$ with 1% Triton X-100, and homogenized twice.

5. Lysates are centrifuged at 12,000*g* for 10 min at 4° and the supernatant transferred to 200 μl of a 1 : 1 mixture of washed protein A agarose/protein G agarose (Pharmacia, Piscataway, NJ) for preadsorption of background proteins. Samples are rotated 1 hr at 4° and beads centrifuged at 12000*g* for 3 min.

6. Supernatant (0.8 ml) is incubated for 1 hr at 4° with mouse monoclonal antibodies to either tag (25 μg/ml final concentration), and overnight after addition of 0.1 ml of washed protein A/protein G agarose.

7. Beads are washed three times in lysis buffer by repeated centrifugation as in step 5. Pellets are resuspended in 50 μl of sample buffer, boiled for 5 min, and 10 μl subjected to SDS–PAGE.

8. Samples immunoprecipitated with tag A antibody are detected by Western blot/ECL (enhanced chemiluminescence) format (Amersham, Amersham, UK) using the tag B antibody and anti-mouse IgG as the secondary antibody.

Glycosylation

Homogeneous expression of single-receptor subtypes is preferred for applications such as mass spectrometry or crystallography. Expression of ligand-gated ion channels by the baculovirus and SFV systems, however, typically results in the production of multiple molecular weight bands, as seen by SDS–PAGE. This heterogeneity is often the result of varying degrees of glycosylation. Inhibition of glycosylation with tunicamycin or posttranslational removal of carbohydrate moieties with glycosidase will reduce the bands to single species.[40]

[40] A. K. Kopke, I. Bonk, S. Sydow, H. Menke, and J. Spiess, *Protein Sci.* **2,** 2066 (1993).

[41] D. X. Wang and L. G. Abood, *J. Neurosci. Res.* **44,** 350 (1996).

[42] M. Iizuka and K. Fukuda, *J. Biochem.* **114,** 140 (1993).

[43] P. Werner, E. Kawashima, N. Hussy, K. Lundstrom, G. Buell, Y. Humbert, and K. A. Jones, *Mol. Brain Res.* **26,** 233 (1994).

[44] B. Binir, M. L. Tierney, M. Lim, G. B. Cox, and P. W. Gage, *Synapse* **26,** 324 (1997).

[45] B. Binir, M. L. Tierney, N. P. Pillai, G. B. Cox, and P. W. Gage, *J. Membr. Biol.* **148,** 193 (1995).

[46] B. Binir, M. L. Tierney, S. M. Howitt, G. B. Cox, and P. W. Gage, *Proc. R. Soc. Lond. B: Biol. Sci.* **250,** 307 (1992).

[47] S. Kawamoto, S. Hattori, I. Oiji, A. Ueda, J. Fukushim, K. Sakimura, M. Mishina, and K. Okuda, *Ann. N.Y. Acad. Sci.* **707,** 460 (1990).

[48] S. Kawamoto, S. Hattori, K. Sakimura, M. Mishina, and K. Okuda, *J. Neurochem.* **64,** 1258 (1995).

[49] A. Kuusinen, M. Arvola, C. Oker-Bloom, and K. Keinanen, *Eur. J. Biochem.* **233,** 720 (1995).

of a suitable promoter, the level of expression can be controlled. Expression can persist for weeks under favorable conditions. Cytotoxic effects are sometimes a problem, but usually are not severe.

Adenovirions enter the cells by a complex pathway, usually by initial attachment to an integrin,[1] followed by endocytosis into endosomes. The viral DNA is transported to the nucleus where it exists as a linear molecule of approximately 36–40 kb, with a covalently attached protein at each end.[2] This protein presumably contributes to nuclease resistance of the DNA and thus to the persistence of expression. In proliferating cells, the DNA will be diluted during cell division.

Transgenic mice, either deleted in (i.e., "knockouts") or overexpressing a gene of interest, are an attractive tool for studying effects of genes in neurobiology. For the many cases where dominant-negative mutants exist, overexpression of this mutant is a useful alternative to gene knockouts as a way to study the effect of reduced functional activity of an endogenous gene. The recombinant adenovirus system can be established in a somewhat shorter period of time than can a transgenic mouse. If an introduced gene has a lethal embryonic phenotype, the adenovirus system is much more convenient for studying the adult phenotype. On the other hand, the volume of tissue *in vivo* that can be infected with adenovirus is limited by the number of injections and volume per injection that an animal will tolerate, because the virus does not spread appreciably through the extracellular space. Therefore, for a nonsecreted molecule such as a receptor or an ion channel, the broad region of expression achieved with transgenic mice cannot be duplicated by adenovirus.

In the present article, we focus mainly on the methods and genes that have been used in our laboratory. Our studies have been carried out for hippocampal cells in culture and in acute slices. Up to the present, we have studied several potassium channels and several nitric oxide synthase constructions.

Preparation of Recombinant Adenovirus

The goal is to prepare, by homologous recombination, a virus in which the essential E1a and E1b gene products have been inactivated by removal of the sequence from base pairs (bp) 455 to 3333 and replaced by an expression cassette for the cDNA of interest.[3] The E1a and E1b genes are essential for viral replication and for all late gene expression. Human

[1] G. R. Nemerow, D. A. Cheresh, and T. J. Wickham, *Trends Cell Biol.* **4,** 52 (1994).
[2] F. L. Graham, *EMBO J.* **3,** 2917 (1984).
[3] H. van Ormondt, J. Maat, and C. P. van Beveren, *Gene* **11,** 299 (1980).

embryonic kidney 293 (HEK293) cells were originally transformed by a fragment of adenovirus including the E1 genes[4]; these cells can thus supply the gene product in *trans* for recombinant virus replication. However, the E1 deleted viruses are nonreplicating in postmitotic cells such as neurons, although they may slowly replicate in some transformed cell lines such as HeLa cells.

The general strategy we and others have used is depicted in Fig. 1. After exhaustive overdigestion of the wild-type virus with enzymes that cleave 1- to 1.3-kb from the essential left end, the longer right arm (plus any undigested full-length virus) is separated from the shorter left end by either sucrose gradient centrifugation or gel electrophoresis. The transfer plasmid, either linearized or still circular, contains an expression cassette consisting of the cytomegalovirus (CMV) immediate early promoter, the inserted cDNA, and an SV40 splice and polyadenylation sequence. This cassette is flanked by 454 nucleotides of the far left end of adenovirus and a segment spanning nucleotides 3334–6231 of the virus. The plasmids we use are depicted in Fig. 1 and are denoted pAC plasmids.[5] When cells are cotransfected with pAC plasmid and the isolated long right arm of the virus, sufficient recombination takes place so that some recombinant virus is formed.

Bett *et al.*[6] report that vectors with inserts resulting in viral DNA close to or less than a net genome size of 105% of that of the wild type, which is 35.9 kb, grow and are relatively stable.

Maps of the viruses we have used for most of our work are shown in Fig. 2. Ad5 *dl*309 has been extensively used by workers in the field.[7] When used in conjunction with transfer vectors that result in the substitution of the region from 455 to 3333 (Fig. 1) with the expression cassette, the cloning capacity is 4.7 kb. The virus Ad5 *Pac*I[8] has an E3 deletion of 2.7 kb, so the cloning capacity is increased to 7.4 kb. Ad5 RR5 and Ad5 SJS2 have the same cloning capacities as Ad5 *dl*309 and Ad5 *Pac*I, respectively. The viruses Ad5 *dl*309 and Ad5 *Pac*I contain the E1 genes; they are therefore replication competent and cytotoxic in most mammalian cells. Ad5 RR5 and Ad5 SJS2 are deleted in the E1 region. They are nonreplicating in many cells; in some transformed cells, such as HeLa, that can supply a substitute for the E1 gene, there is slow replication. Ad5 *Pac*I and Ad5

[4] F. L. Graham, J. Smiley, W. C. Russell, and R. Nairn, *J. Gen. Virol.* **36,** 59 (1977).

[5] A. M. Gomez-Foix, W. S. Coats, S. Baque, T. Alam, R. D. Gerard, and C. B. Newgard, *J. Biol. Chem.* **267,** 25129 (1992).

[6] A. J. Bett, L. Prevec, and F. L. Graham, *J. Virol.* **67,** 5911 (1993).

[7] N. Jones and T. Shenk, *Cell* **17,** 683 (1979).

[8] A. J. Bett, W. Haddara, L. Prevec, and F. L. Graham, *Proc. Natl. Acad. Sci. U.S.A.* **91,** 8802 (1994).

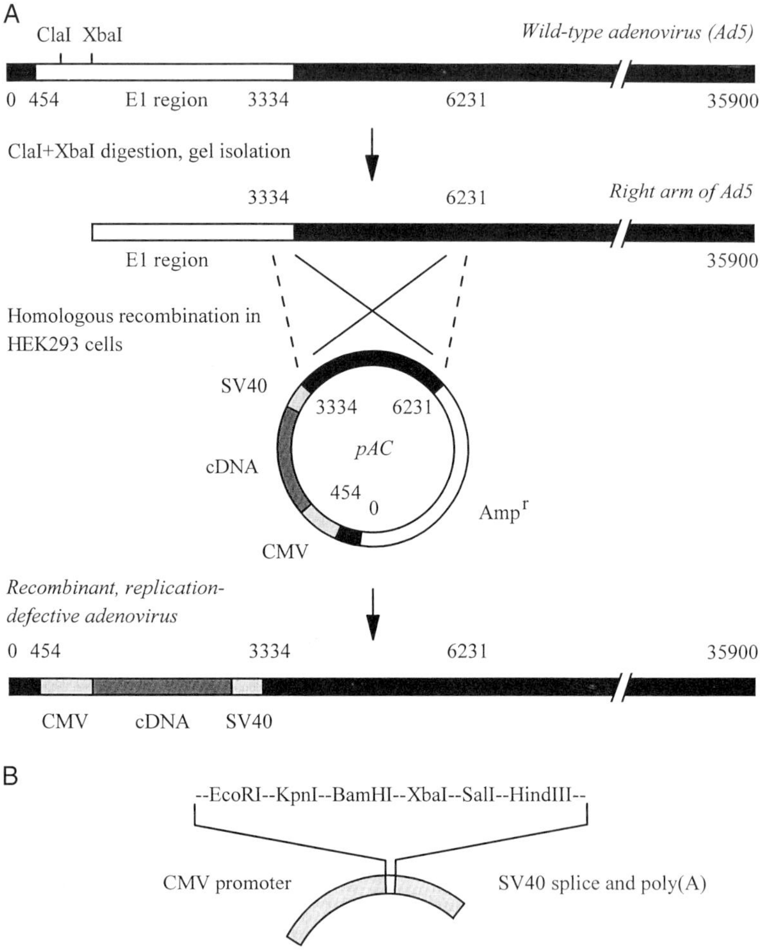

FIG. 1. Schematic representation of the construction of recombinant adenovirus. (A) The adenovirus Ad5 *dl*309 is cut with *Cla*I and *Xba*I; the resulting nonviable right arm is transfected into HEK293 cells together with the pAC adenoviral transfer vector (pACCMVpLpA). [From A. M. Gomez-Foix, W. S. Coats, S. Baque, T. Alam, R. D. Gerard, and C. B. Newgard, *J. Biol. Chem.* **267,** 25129 (1992).] The plasmid contains adenoviral sequences (nucleotides 0–454 and 3334–6231), and the cDNA of interest under the control of a CMV promoter (760 bp) and upstream of an SV40 splice and polyadenylation signal (470 bp). On homologous recombination between adenoviral nucleotides 3334–6231 of Ad5 and pAC, recombinant adenovirus is formed; it contains the necessary left-end packaging signal, but is replication-defective as E1 region genes (within nucleotides 455–3333) have been deleted in the recombination process. The recombinant virus can propagate only in HEK293 cells, which complement the factors encoded by the adenoviral E1 region [From F. L. Graham, J. Smiley, W. C. Russell, and R. Nairn, *J. Gen. Virol.* **36,** 59 (1977).] (B) Unique restriction sites in the polylinker of pAC.

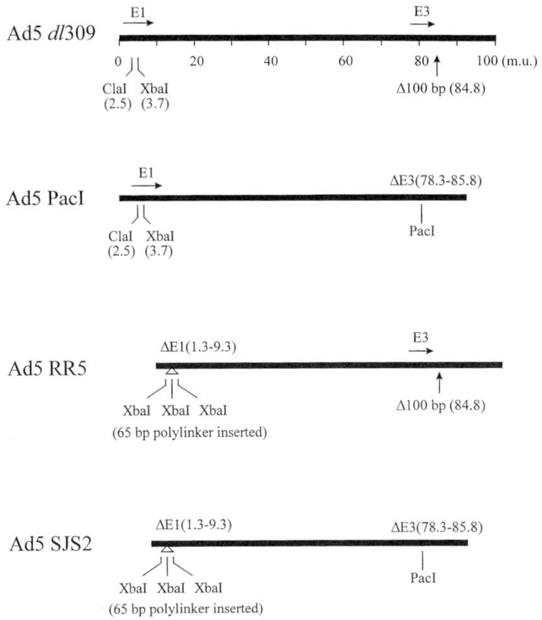

FIG. 2. Structures of adenoviruses used for recombinants. Ad5 *dl*309 is useful because it contains a unique *Xba*I site at 3.7 map units (m.u.). (One hundred map units = 35.9 kb.) The mutation at site 84.8 is due to a spontaneous deletion from 83 to 85 m.u. and partial substitution with a fragment of foreign DNA. Sequencing shows that the net effect is a loss of 104 bp [From A. J. Bett, V. Krougliak, and F. L. Graham, *Virus Res.* **39,** 75 (1995).] Ad5 *Pac*I [A. J. Bett, W. Haddara, L. Prevec, and F. L. Graham, *Proc. Natl. Acad. Sci. U.S.A.* **91,** 8802 (1994)], like Ad5 *dl*309, contains the wild-type E1 region, with an E3 deletion of 2.7 kb replaced by a unique *Pac*I site. In Ad5 RR5, the E1 region from 1.3 to 9.3 m.u. has been replaced by a polylinker, which includes three *Xba*I sites. The left arm of the virus was made by R. D. Gerard by recombination with the plasmid pACESHR. This plasmid contains the same adenovirus sequences as the pACCMV plasmids, with the CMV promoter and the SV40 splice and polyadenylation sequences replaced by the pUC19 polylinker. Ad5 SJS2 was made in our laboratory by S. J. Stary by recombination of the plasmid pACESHR with Ad5 *Pac*I. The right arm of the sequence is from Ad *dl*309.

SJS2 have a unique *Pac*I restriction site in place of the E3 region. As described later, an expression cassette may be introduced at this site by ligation. For example, we have ligated a green fluorescent protein expression cassette of length 0.9 kb into the *Pac*I site of Ad*Pac*I and of Ad5 SJS2. These viruses now have a cloning capacity of 6.5 kb in the E1 region. With such recombinants, one can determine which cells in a culture have been infected by observation of GFP fluorescence.

Techniques

Safety Considerations

Requirements from the National Institutes of Health (NIH) and Center for Disease Control (CDC) on handling adenovirus and adenoviral recombinants can be found in the current NIH *Guidelines for Recombinant DNA*. The current guidelines can be accessed on the Internet at http://www.ehs.psu.edu/nih95-1.htm. Adenoviruses are classified in risk group 2 (RG2). This includes agents that are associated with human disease that is rarely serious and for which preventive or therapeutic interventions are often available. In practical terms, adenovirus is regarded as a moderate-risk pathogen.

Recommended precautions for RG2 agents are standard microbiological practices, laboratory coats, decontamination of all infectious wastes, limited access to working areas, protective gloves, posted biohazard signs, and class I or II biological safety cabinets used for mechanical and manipulative procedures that have high aerosol potential.

Of course, any adenovirus recombinant expressing a highly toxic protein should be treated as a special risk.

Preparation of Recombinant Adenovirus

Preparation of Right Arm of Viral DNA. The separation procedure removes a portion of the 5′ end of the viral DNA extending beyond nucleotide 500. The inverted terminal repeat of length 102 nucleotides at each end, and the packaging signal, nucleotides 94–358, are essential for assembling a functional virus.

The viruses Ad5 *dl*309 and Ad5 *Pac*I both have a *Cla*I and an *Xba*I site. Ad5 RR5 and Ad5 SJS2 have three adjacent *Xba*I sites around position 450. Accordingly, these enzymes are used to release the left end.

Twenty micrograms of viral DNA (prepared as described later) is incubated with 100–200 units of restriction enzyme for 2–3 hr. The enzyme is added in two batches, 50–100 units for the first hour of incubation and then the rest. This decreases effects due to thermal degradation of the enzyme. Note that it is very important to digest exhaustively, and that minor nuclease action at the cohesive sites is not deleterious. In our laboratory we prefer to separate the long right arm from the left end by 1% agarose gel electrophoresis. We believe this procedure to be more effective as well as time saving as compared to sucrose gradient centrifugation. For the subsequent gel isolation, we use GeneClean II (Bio101, La Jolla, CA), a silica matrix-based procedure. Vortexing and vigorous pipetting should be avoided to prevent shearing of the long DNA.

Cotransfection into HEK293 Cells. HEK293 cells are maintained and propagated under standard conditions [5% humidified CO_2/95% air (v/v) in Dulbecco's modified Eagle's (DME) high-glucose medium (Irvine Scientific, Santa Ana, CA) supplemented with 10% fetal calf serum (FCS), glutamine, and antibiotics (DME-complete)]. For cotransfection, HEK293 cells are grown in 60-mm plastic dishes to 50–70% confluency. Transfection is carried out using Lipofectamine (GIBCO-BRL, Gaithersburg, MD) according to the manufacturer's instructions. Briefly, 1.5 μg plasmid DNA and 1.5 μg right-arm viral DNA (molar ratio of 3–4 to 1) are assembled in 300 μl DME, free of serum and antibiotics (DME). Fifteen μl of Lipofectamine (30 μg) reagent is diluted in 300 μl DME. The Lipofectamine solution is then added dropwise to the DNA solution.

This mixture is incubated at room temperature for 20 min, diluted to a total volume of 1.5 ml by adding 900 μl DME, and added to the HEK293 cells that have previously been rinsed once in DME. (This wash step must be done with maximal care because the HEK293 cells do not adhere tightly to the plastic.) After 3–5 hr, 3.5 ml DME containing 7% FCS but no antibiotics is added. The cells are incubated overnight and the medium is then changed to DME-complete.

About 7–14 days after transfection, a few cells start to appear round and phase bright, and will eventually die. This cytopathic effect (CPE) is a consequence of virus production; virions spread to neighboring cells, until all cells eventually become rounded. At this stage of full CPE, the cells and the medium are harvested and freeze–thawed three times to break up the cells and to release virus particles. This primary lysate is extracted with an equal volume of chloroform. The chloroform extraction dissociates virion aggregates and removes any possible microbial contamination from the lysate.

Plaque Purification. The primary viral lysate from the preceding step will usually still contain wild-type virus. Therefore, it is advisable to purify recombinant virus by isolating single plaques that presumably originate from a single infectious virus particle. In this laboratory, we usually perform two rounds of plaque purification and screen the plaques after each round by PCR (polymerase chain reaction). However, when possible, we recommend that the primary lysate be tested for functional expression prior to starting the laborious purification procedure. Usually, the primary lysate has a virus titer that is sufficiently high to achieve expression of the transgene in cells other than HEK293 (e.g., CHO) cells. If expression is weak, the proportion of wild-type virus might be high.

For plaque purification, HEK293 cells are seeded into 60-mm dishes (four dishes per sample under test) and grown to 90% confluency. Add 200 μl from a dilution series (four dilutions, 10^4- to 10^7-fold) of the primary lysate in DME-complete to the dishes and incubate for 1 hr. The dishes need to be rocked gently every 10 min. Then, add 2–3 ml DME-complete and incubate the cells for another 3–5 hr before pouring the agarose overlay. Prepare the agarose overlay by melting 1.5% (w/v) low melting point agarose (GIBCO-BRL) in water and mix it with an equal volume of 2× DME-complete (2× DME, 2× glutamine, 2× antibiotics, 10% FCS). Let the mixture cool to 37°, remove the medium from the cells, and add 5 ml of the DME–agarose mix to each 60 mm dish. Incubate the dishes at 4° for 5–10 min to let the agarose solidify before placing the dishes back into the 37° incubator.

Plaques will appear after 4–7 days as small holes (~100-μm diameter) surrounded by round cells. The plaques are marked and harvested with a sterile cotton-plugged Pasteur pipette. This material is suspended in 1 ml of DME-complete, and subjected to three freeze–thaw cycles. Each plaque is amplified in a 60-mm dish containing a confluent monolayer of HEK293 cells by adding 0.5 ml of the plaque suspension. After 3–5 days, when all cells exhibit CPE, cells and medium are harvested, freeze–thawed three times, and chloroform extracted.

Screening. Lysates derived from plaques can be screened in various ways. Expression can be tested by infecting cells, provided that an assay is available (e.g., immunostaining or functional tests such as electrophysiologic recording, enzyme assays). Alternatively, virus DNA is isolated from lysates and blotted onto nylon membranes (Southern blots), which are hybridized with a probe specific for the foreign cDNA insert or wild-type virus (e.g., the E1 region). Isolated viral DNA can also be screened by PCR with primers specific for the foreign cDNA insert or wild-type virus. In this laboratory, we routinely screen by PCR analysis.

Viral DNA is isolated by digestion of the lysate with proteinase K. Assemble 0.5 ml of lysate with 5 μl 100× TE (1 *M* Tris-HCl, 0.1 *M* EDTA, pH 7.8), 12.5 μl of 20% sodiumdodecyl sulfate (SDS) and 2.5 μl of proteinase K (20 mg/ml, Boehringer Mannheim, Germany). Incubate at 55° for 45 min. Then add 0.5 ml of phenol/chloroform/isoamyl alcohol (25:24:1), vortex thoroughly, and centrifuge in a microfuge at 14,000 rpm for 5 min. Carefully remove the upper phase and transfer it to a fresh tube. Add 0.5 ml of chloroform/isoamyl alcohol (24:1), vortex, and centrifuge as earlier for 5 min. Collect the upper phase and add 50 μl of 3 *M* sodium acetate and 1.2 ml of 95% (v/v) ethanol. Vortex and centrifuge in a microfuge at 14,000 rpm for 20 min. Wash the pellet once with ~0.5 ml 70% ethanol and let the pellet dry on the bench for 10 min. Dissolve the pellet in 40 μl of sterile water and use 1 μl for PCR analysis.

For PCR analysis, we usually use primers specific for the inserted DNA and/or the CMV promoter and the SV40 splice sequence. To check for the presence of wild-type virus (for the cases of Ad5 *dl*309 or Ad5 *Pac*I), primers specific for the E1 region are used. In this case, a faint band can usually be observed even if the recombinant virus is pure. This band presumably originates from trace amounts of DNA from HEK293 cells that contain the E1 genes. If the recombinant virus is pure, the band is about 50–100 times less intense than when wild-type viral DNA is used as template. (Different primers are used to detect Ad5 RR5 and Ad5 SJS2.)

CMV forward and SV40 splice reverse primers we use for the insert are 5′GTG′GGA′GGT′CTA′TAT′AGC′AG3′ and 5′ATC′TCT′GTA′GGT′AGT′TTG′TCC3′, respectively. The primers for the E1 region are as follows: forward primer, 5′GTG′AGT′TCC′TCA′AGA′GGC′C3′; reverse primer, 5′ACC′CTC′TTC′ATC′CTC′GTC′G3′ corresponding to nucleotides 480–498 and 976–958 of the adenovirus genome (GenBank accession no. X02996).

Plateau saturation effects of PCR can be avoided by dilution of DNA samples before PCR and/or by reducing the number of cycles of amplification so that ethidium bromide-stained gel bands of intermediate intensity are produced. In this way, a semiquantitative estimate of the amount of recombinant viral DNA and ratio of recombinant to wild-type virus can be achieved.

Large-Scale Amplification. Lysate from recombinant viral lysate or wild-type virus is used to infect HEK293 cells grown in large culture flasks (e.g., T-150 cm^2 flasks; Corning, Cambridge, MA). About 300 μl of viral lysate is sufficient to infect a 150-cm^2 flask with a confluent monolayer of cells. After 2–3 days, all the cells exhibit CPE and should then be harvested. A large fraction of the infectious particles is associated with the cells. This provides a convenient way to concentrate the virus: the cells are removed from the bottom of the flask and the cell suspension is centrifuged. The cell pellet is then resuspended in a small volume of DME-complete. One 150-cm^2 flask will produce $\sim 3 \times 10^9$ plaque-forming units (pfu) of virus. The cell suspension is freeze–thawed three times and optionally extracted with an equal volume of chloroform. Aliquots should be stored at $-80°$. The presence of cell debris enhances stability for long-term storage. If the virus was chloroform extracted, bovine serum albumin (BSA, to 0.1%) and/or glycerol (to 10%) may be added to increase the stability. We prefer glycerol alone.

Virus Purification with a CsCl Isopyknic Gradient. CsCl purification is recommended if highly pure virus is needed (e.g., for *in vivo* injections). In our experience, CsCl-purified virus has fewer toxic effects also on cultured cells and is thus suited for infection of fragile cells such as primary neuronal cultures. The virus lysate used for purification should be as concentrated as possible. Place 3 ml of 3.2 *M* CsCl (ρ = 1.4 g/cm^3) into an ultracentrifuge tube (e.g., Beckman polyallomer, 14 × 89 mm, Beckman Instruments, Fullerton CA) and overlay with 3 ml of 1.6 *M* CsCl (ρ = 1.2 g/cm^3). Carefully layer 5 ml of chloroform-extracted viral lysate on top, and centrifuge in a swinging-bucket rotor (e.g., Beckman SW41) at 37,000 rpm (~245,000*g*) for 16 hr at 4°.

The band containing the virions appears white opalescent and is located ~1.1 cm from the bottom of the tube. A faint band can be observed ~0.3 cm above the virions; it contains empty particles. Discard the top of the gradient (including the empty particle band) and then remove the band containing the virions. Dilute with an equal volume of Ad solution (137 m*M* NaCl, 10 m*M* HEPES, 5 m*M* KCl, 1 m*M* $MgCl_2$, pH 7.6). The virus is stable in CsCl for a few months at 4°.

CsCl, which is toxic to cells, can be removed by gel filtration using a 9-ml Sephadex G-25 column (Pharmacia PD-10, Piscataway, NJ). Equilibrate the column with Ad solution, apply the virus and elute with Ad solution, collecting 1.5-ml fractions. The fractions can be checked for virus content by measuring absorption at 260 nm. Alternatively, each fraction can be titered as described later. In our experience, fractions 2 and 3, or 3 and 4 contain the bulk of virus. The virus can be stored in the eluate when glycerol (to 10%) is added. The 1 m*M* Mg^{2+} in all solutions is important for stabilizing virions.

Isolation of Viral DNA. CsCl-purified virus is used to prepare DNA for transfection. The virions are digested with proteinase K, which removes the capsids. Dilute 100× TE (1 *M* Tris-HCl, 0.1 *M* EDTA, pH 7.8) to 1× in virus solution and add 20% SDS to a final concentration of 0.5% and proteinase K to a final concentration of 100 μg/ml. Incubate at 55° for 45 min. Then add an equal volume of phenol/chloroform/isoamyl alcohol (25:24:1), vortex thoroughly, and centrifuge in a microfuge at 14,000 rpm for 5 min. Carefully remove the upper phase and transfer it to a fresh tube. Add an equal volume of chloroform/isoamyl alcohol (24:1), vortex, and centrifuge as above for 5 min. Collect the upper phase and add 0.1 volume of 3 *M* sodium acetate and 2.2 volumes of 95% ethanol. Vortex and centrifuge in a microfuge at 14,000 rpm for 20 min. Wash the pellet once with 70% ethanol and let the pellet dry on the bench for 10 min. Dissolve the pellet in an appropriate volume of sterile water to a final concentration of ~1 mg/ml.

Determination of Viral Titer

PLAQUE ASSAY. The procedure described earlier for plaque purification can be used to determine the titer of a virus. Several virus dilutions (e.g., 10^7- to 10^9-fold) should be used to infect HEK293 cells in 60-mm dishes, which are then overlaid with agarose. The growing plaques are visible and can easily be counted 5 days after infection.

CELL LYSIS ASSAY. This assay may be preferable because it is less laborious and cheaper than the plaque assay. Up to three viruses can be titered simultaneously in triplicate series in a single multiwell plate. In our experience, this assay gives very reproducible results.

Pipette 50 μl of DME-complete into each well of a 96-well plate. Dilute the virus 10^5-fold, add 25 μl to the first well, and mix by pipetting. Take 25 μl from the first well, add it to the second well, and mix. Transfer 25 μl from the second to the third well, etc. Do not add diluted viral solution to the last well, and remove 25 μl from the second to last. This dilution series can conveniently be done in triplicate or even more parallel series. Add 50 μl of HEK293 cell suspension ($\sim 10^6$ cells/ml) to each well. The plate is then incubated at 37°, and cells are fed with 50 μl/well DME-complete at days 3, 6, and 9. After ~10 days, the titer can be determined by checking for CPE. It is assumed that the well with the highest dilution of virus that still exhibits CPE initially contained one infectious particle. The titer is then calculated as $3^n \times 40 \times 10^5$ pfu/ml, where n is the number of wells showing CPE.

Cloning into PacI Site of Ad5 PacI and Ad5 SJS2 by Ligation. In these viruses (Fig. 2), a 2.7-kb deletion in E3 has been replaced by a unique *Pac*I site. Thus, in addition to a gene introduced into the E1 region by recombination, a gene can be introduced by ligation into the *Pac*I site.

To prepare the viral vector, 2 μg of Ad5 *Pac*I or SJS2 DNA is digested with 20 units *Pac*I (New England Biolabs) for 2 hr. The enzyme is added in two batches, 10 units for the first hour of digestion and then the rest. On the second addition of *Pac*I, five units of shrimp alkaline phosphatase (United States Biochemical, Cleveland, OH) is added to the digestion mix to dephosphorylate the ends. The two DNA fragments in the digestion mix are purified by using GeneClean II (Bio101, La Jolla, CA) and eluted in doubly distilled H_2O. Vortexing and vigorous pipetting should be avoided to prevent shearing of the long viral DNA.

The expression cassette consisting of the gene insert with a promoter and poly(A) site needs to be flanked by two *Pac*I sites. This can be achieved by PCR or by linker addition. We prefer the former strategy, which is fast and simple. Two oligonucleotides containing the *Pac*I site plus the 5′ and 3′ sequence of the expression cassette are used as primers. There should be a 5′ extension of five or more nucleotides to ensure subsequent digestion at the *Pac*I sites. PCR is carried out with a high-fidelity PCR kit. We use the Expand High Fidelity PCR System (Boehringer Mannheim). The correct PCR product is purified by the Qiagen PCR purification kit. Two micrograms of the PCR product is digested with 20 units *Pac*I, then gel purified and eluted in doubly distilled H_2O.

The overnight ligation is carried out by mixing 2 μg DNA insert and 2 μg total of the two dephosphorylated viral vector fragments under ligation conditions at 16°. When the high concentration T4 DNA ligase (5 U/μl, Boehringer Mannheim) is used and the ligation volume is small (less than 20 μl), the ligation mix can be used directly to transfect HEK293 cells without precipitation and desalting.

The transfection and further plaque purification screening procedures are basically the same as previously described for homologous recombination. Individual plaques are screened by PCR and/or by *Pac*I digestion. For checking for multiple insertions in the *Pac*I site, a long PCR reaction using primers adjacent to the E3 region is used to measure the length of the insert. The primers for the E3 region are as follows: forward primer, 5′CCT′GAT′TCG′GGA′AGT′TTA′CCC′3′; reverse primer, 5′GTG′CTG′CTG′AAT′AAA′CTG′GAC′3′ corresponding to nucleotides 28,030–28,050 and 30,911–30,891 of the wild-type adenovirus genome (GenBank M73260).

The ligation method can also be applied for inserting genes into the *Xba*I sites of Ad5 RR5 and SJS2. For this purpose, a plasmid such as pZS2 (J. Sheng, personal communication, 1996) which contains the short right arm of adenovirus and an expression cassette with the cDNA of interest terminated by an *Xba*I site is ligated to *Xba*I-digested, dephosphorylated Ad5 RR5 or SJS2.

Toxicity Problems. We and others[9,10] have observed the death of neurons and other cells *in vitro* infected with recombinant adenoviruses. The magnitude of the effect depends on the particular cells and culture conditions, the purity of the virus, the particular gene product being expressed, and its level of expression. Expression level depends on the promoter system being used and on the multiplicity of infection. It is possible but not certain that small amounts of expression of other viral gene products in E1- and in E1 plus E3-deleted recombinants contribute to toxicity. If this is the case, the new "gutless" vectors[11] will be an improvement. In some cases, addition to the culture medium of a specific inhibitor of the gene product improves viability while permitting accumulation of the gene product. The inhibitor is withdrawn at an appropriate time for observation. Half-lives of infected cells are reported to be reduced so as to range from several days to weeks, depending on all of the factors mentioned.

For each particular system, by adjusting the level of expression, one

[9] J. Jordan, G. D. Ghadge, J. H. M. Prehn, P. T. Toth, R. P. Roos, and R. J. Miller, *Mol. Pharmacol.* **47,** 1095 (1995).

[10] H. D. Durham, H. Lochmuller, A. Jani, G. Acsadi, B. Massie, and G. Karpati, *Exp. Neurol.* **140,** 14 (1996).

[11] R. J. Parks, L. Chen, M. Anton, U. Sankar, M. A. Rudnicki, and F. L. Graham, *Proc. Natl. Acad. Sci. U.S.A.* **93,** 13565 (1996).

can usually discover a window of time during which the physiologic effects of interest can be observed before cytotoxicity becomes a factor.

Expression of Potassium Channels

To our knowledge, the only ion channel genes that have been studied by electrophysiology after adenovirus-mediated gene transfer are voltage- and G-protein-gated potassium (K^+) channels. The inactivation-removed *Drosophila Shaker B* channel has been expressed *in vitro* and *in vivo* by this method.[12] We have similarly expressed the *Drosophila Shaker H4* cDNA and several of the heteromeric G protein-gated inward rectifier K^+ channel (GIRK) cDNAs.[13]

Preliminary Transfection Tests

Because the preparation of adenovirus is a lengthy procedure, it is useful to test the ion channel gene after cloning into the pAC transfer plasmid before proceeding to make the virus. This is done using standard cotransfection procedures. For example, we use Lipofectamine transfection reagent (GIBCO-BRL) according to the directions of the manufacturer. For a 35-mm Petri dish of semiconfluent CHO cells, we use only 0.25 μg of the GIBCO-BRL (Green Lantern) GFP or 0.1 μg of the Clontech (Palo Alto, CA) pEGFP-1 as a cotransfected reporter gene, and up to 1.5 μg of the necessary plasmids. For example, for a mixture of GIRK1, GIRK2, and muscarinic type 2 acetylcholine receptor (m2AChR) we would use 0.5 μg of each. One to 2 days after transfection, 25–50% of the cells are strongly fluorescent; when these fluorescent cells are studied by whole-cell clamping, essentially 100% of them express the expected ion channel type, provided the cDNAs have not undergone mutations during preparation that make them nonfunctional. Recall that GIRKs function best as heteromers; both with plasmids and virus we see good expression with GIRK1 plus GIRK2 or GIRK1 plus GIRK4, but not with GIRK1 alone. Data illustrating typical results are given in Table I.

Because the efficiency of transient transfection into CHO cells is between 20 and 50%, the catalytic activity of nitric oxide synthase constructs and many other enzymes for which there are good assays can be tested in mass cultures. In this case the reporter gene is unnecessary, but useful to indicate that the transfection has been reasonably successful.

[12] D. C. Johns, H. B. Nuss, N. Chiamvimonvat, B. M. Ramza, E. Marban, and J. H. Lawrence, *J. Clin. Invest.* **96,** 1152 (1995).

[13] M. U. Ehrengruber, C. A. Doupnik, Y. Xu, J. Garvey, M. C. Jasek, H. A. Lester, and N. Davidson, *Proc. Natl. Acad. Sci. U.S.A.* **94,** 7070 (1997).

TABLE I
LIPOFECTAMINE-MEDIATED EXPRESSION OF POTASSIUM CHANNELS IN CHO CELLS[a]

Channel type	Peak current (nA)	Current density (pA/pF, μA/cm^2)	Channel density per μm^2
Shaker H4[b]	2.9 ± 0.2 (20)	218 ± 42 (20)	2.4
GIRK1 + 2[c]	−0.5 ± 0.1 (10)	−41 ± 10 (10)	3.1
GIRK1 + 4[c]	−1.1 ± 0.1 (39)	−126 ± 13 (39)	9.6

[a] Means ± SEM (n) assayed 1–2 days after transfection. Cells in a 35-mm dish (10–20% confluent) were transfected with 0.13–0.5 μg DNA/per pAC plasmid carrying a K^+ channel cDNA. For GIRKs, an equal amount of m2AChR in pMT2 [R. J. Kaufman, M. V. Davies, V. K. Pathak, and J. W. B. Hershey, *Mol. Cell. Biol.* **9,** 946 (1989)] was cotransfected. In addition, all cells were cotransfected with 0.25 μg Green Lantern (GIBCO-BRL) to allow for selection of successfully transfected, fluorescent cells, which were used for electrophysiologic recording.

[b] *Shaker* peak currents measured at +30 mV as described in Fig. 3 (legend); E_k was −82.9 mV. The channel density was estimated based on a specific membrane capacity of 1 μF/cm^2 [B. Hille, "Ionic Channels of Excitable Membranes." Sinauer Associates, Sunderland, Massachusetts (1992)] a single-channel conductance of 16 pS, and a channel open probability of 0.5. [From A. Karschin, B. A. Thorne, G. Thomas, and H. A. Lester, *Methods Enzymol.* **207,** 408 (19XX)].

[c] GIRK currents induced by agonists as described in Fig. 4 (legend) at −120 mV and 25 m*M* $[K^+]_o$ (E_K ~−44 mV). In the absence of agonist, some CHO cells showed relatively high basal GIRK currents of up to ~3 nA. The channel density was estimated based on a specific membrane capacity of 1 μF/cm^2 and values determined for GIRK1 + 5 channels, that is, a single-channel conductance of 40 pS and a channel open probability of 0.043. [From W. Schreibmayer, C. W. Dessauer, D. Vorobiov, A. G. Gilman, H. A. Lester, N. Davidson, and N. Dascal, *Nature* (*Lond.*) **380,** 624 (1996).]

Generally speaking, we find it very useful to test all pAC plasmids this way before making recombinant virus.

Expression of K^+ Channels by Recombinant Adenovirus

Shaker H4 and the cDNAs for GIRKs 1, 2, and 4 were cloned into the pAC vector by standard molecular biology and recombined with Ad5 *dl*309 (*Shaker;* AdH4) or Ad5 *Pac*I (AdGIRK 1, 2, and 4). The serotonin type 1A receptor ($5HT_{1A}R$) was cloned by ligation into Ad5 RR5 ($Ad5H_1AR$). In most cases, the functionality of the construct was confirmed in the plasmids by cotransfection as described earlier.

Figure 3 gives examples of *Shaker* currents induced by infection with AdH4. The insert demonstrates that the level of infection (in atrial and ventricular myocytes in this instance) increases with time postinfection up to 4 days and then levels off.

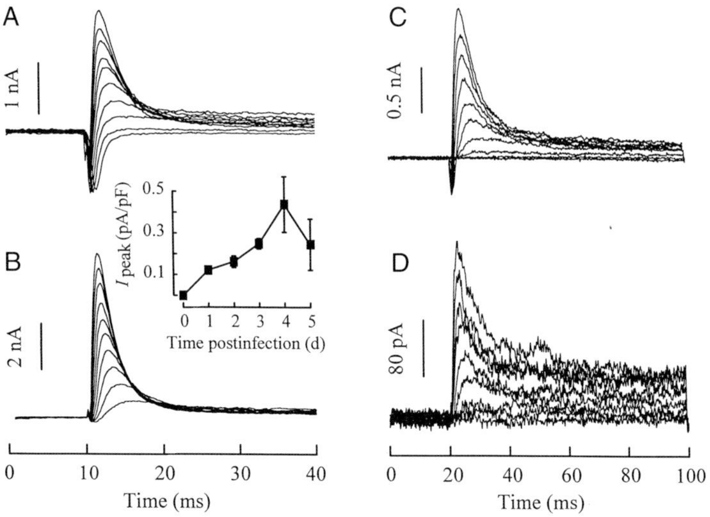

FIG. 3. Adenoviral expression of *Shaker* H4 currents in mammalian cells in primary cultures (A, B) or in mammalian cell lines (C, D). Atrial (A) and ventricular (B) myocytes from 2- to 6-day postnatal rats; CHO cells (C) and pancreatic βTC3 cells (D). Voltage-clamp experiments using a holding potential of −70 mV and leak subtraction, except for ventricular cells (−40 mV, no leak subtraction); after a 10- to 20-msec prepulse to −100 mV, the membrane potential was depolarized in 10-mV steps starting at −50 mV (−40 mV for ventricular myocytes). Currents in βTC3 cells were assayed in 10 m*M* TEA to suppress delayed outward rectifier K^+ currents. *Shaker* currents in the βTC3 cells are comparatively small because 10 m*M* TEA, which was added to block other K^+ channels, partially blocks *Shaker* channels. (Inset) Time-dependent expression of peak currents at +30 mV; means ± SEM from 3–6 myocytes are shown for each postinfection day.

Several examples for expression of GIRK currents by mixed cotransfection plus adenoviral infection of CHO cells and by infection of embryonic day 18 (E18) hippocampal neurons are shown in Fig. 4. Figure 4A shows that comparable agonist-activated GIRK currents can be induced for adenovirus-infected CHO cells either by cotransfection with m2AChR (top left) or by infection with Ad5HT_{1A}R. Table II summarizes much of these data.

With cultured hippocampal neurons, Lipofectamine-mediated transfection efficiencies are low and thus difficult to use; DOTAP-mediated transfection (Boehringer Mannheim) gives about 10% efficiency in our hands; however, the viral infection procedure gives close to 100% per cell infection frequency. G-protein stimulation of the GIRKs was via baclofen acting on endogenous $GABA_B$ receptors or by 5-HT stimulation of the 5HT_{1A}R introduced by adenovirus. Thus, these experiments show that at least three different transgenes can be effectively introduced into a single cell by adenoviral infection.

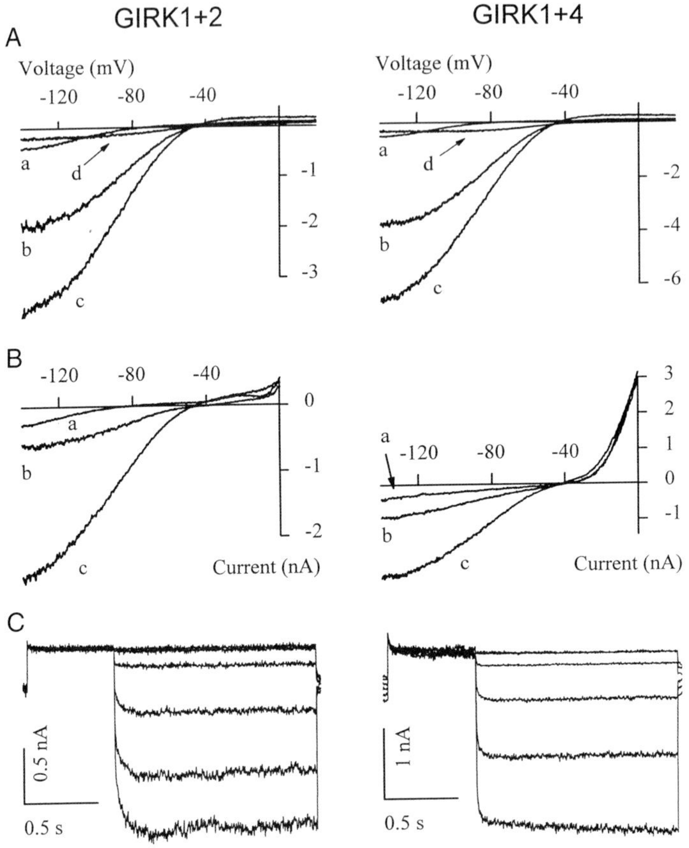

FIG. 4. Adenoviral expression of GIRK channels. GIRK currents in infected cultures of CHO cells (A) and hippocampal neurons from E18 rats (B). Cells were coinfected with AdGIRK1+2 (left; for the CHO cell, GIRK1 was cotransfected using Lipofectamine) and AdGIRK1+4 (right); CHO cells were additionally transfected with seven transmembrane-helix receptor (left, m2AChR; right, coinfection with Ad5HT_{1A}R). Currents were assessed, using 2-sec voltage ramp protocols, 1–2 days postinfection in 5.4 m*M* $[K^+]_o$ (a), and 25 m*M* $[K^+]_o$ both in the presence (c) and absence (b) of agonist (5 μM ACh and 50 μM serotonin for CHO cells, 100 μM baclofen for neurons); 0.5 m*M* Ba^{2+} blocks the basal and agonist-activated GIRK currents (in 25 m*M* $[K^+]_o$; d). (C) Time-course of agonist-activated GIRK1+2 currents in CHO cells (left) and GIRK1+4 currents in hippocampal neurons (right). After a 0.6-sec prepulse to 0 mV, the membrane potential was stepped to test potentials in 30-mV increments between −120 (bottom trace) and 0 mV (top trace). Agonist-activated currents have been isolated by subtracting traces in the absence of agonist.

TABLE II
ADENOVIRUS-MEDIATED EXPRESSION OF POTASSIUM CHANNELS IN PRIMARY AND SECONDARY CELL CULTURES[a]

Channel type and host cell	Peak current (nA)	Current density (pA/pF, $\mu A/cm^2$)	Channel density per μm^2
Shaker H4			
Atrial myocytes (rat)	5.2 ± 0.8 (14)	216 ± 37 (14)	2.4
Ventricular myocytes (rat)	4.2 ± 0.8 (9)	354 ± 96 (9)	3.9
βTC3 cells (mouse)[b]	0.9 ± 0.2 (7)	199 ± 61 (7)	2.2
CHO cells (hamster)	0.9 ± 0.3 (6)	87 ± 33 (6)	1.0
GIRK1 + 2			
Hippocampal neurons (rat)[c]	−1.7 ± 0.2 (6)	−76 ± 9(6)	55
GIRK1 + 4			
Hippocampal neurons (rat)[c]	−1.6 ± 0.5 (4)	−55 ± 17 (4)	4.0
CHO cells (hamster)	−0.6 ± 0.2 (5)	−47 ± 20 (5)	3.5
Endogenous GIRK1 + 4[d]			
Atrial myocytes (rat)	−0.7 ± 0.1 (24)	−15 ± 2 (23)	1.1

[a] Means ± SEM (*n*) of cells assayed 1–7 days postinfection (10^8–10^9 pfu/ml per virus). Atrial and ventricular myocytes were prepared from 2- to 6-day postnatal rats, and hippocampal neurons from E18 rats. Currents and channel densities were determined as described in Table I.

[b] *Shaker* H4 currents were isolated by a subtraction protocol, that is, inactivation of *Shaker* H4 currents using a holding potential of −30 mV, followed by the test pulses. [From R. J. Leonard, A. Karschin, S. Jayashree-Aiyar, N. Davidson, M. A. Tanouye, L. Thomas, G. Thomas, and H. A. Lester, *Proc. Natl. Acad. Sci. U.S.A.* **86,** 7629 (1989).]

[c] $E_K = -40.7$ mV.

[d] For comparison, GIRK1 + 4 currents activated by 5–10 μM ACh at −120 mV and 25 m*M* $[K^+]_o$ are given ($E_K = -43.5$ mV).

We note here that the cloning capacity of Ad *Pac*I is sufficiently large so that we have cloned individual expression cassettes for GIRK1 into the E1 region and GIRK4 into the E3 region (by ligation) of Ad *Pac*I and observed a comparable level of expression with a single virus.

Note that the principal scientific point of our study of GIRK-infected neurons was to analyze quantitatively the mechanisms by which GIRK activation reduces neuronal excitability.[13] The large level of expression achieved by adenovirus-mediated overexpression made this possible.

Use of Adenovirus in Acute Hippocampal Slice Physiology

The hippocampal slice has been the system of choice in the study of long-term potentiation (LTP) as a model for synaptic modification during learning and memory.[14] Slices are normally used for physiology within a

[14] T. V. P. Bliss and G. L. Collingridge, *Nature* **361,** 31 (1993).

few hours of preparation; however, in appropriately supplemented medium they appear healthy by visual examination and retain their electrophysiologic properties for at least 30 hr.

Gene transfer methods for expression of a new gene and for enhancement or inhibition of endogenous genes provide an additional tool in the study of synaptic properties of slices. The technical questions that arise in using recombinant adenovirus for this purpose include these: (1) Will there be sufficient gene expression from injected virus within the 24- to 30-hr lifetime of the slice? (2) Since virus does not spread far through the extracellular space of brain tissue, how can one achieve infection and gene expression over a volume sufficiently large for the problem being studied? Our experience so far[15] in these respects is described next.

Slices were infected by microinjection of viral stock (usually about 10^9 pfu/ml) into the extracellular space of the CA1 pyramidal cell layer with glass micropipettes with tips broken off (resistance <1 MΩ). For each of 10–20 injection positions distributed over CA1, multiple pressure injections (5–7 injections per site for 10–20 sites per slice for 100 ms at 5–10 psi) were made by advancing the electrode through the depth of the slice. The approximate volume of solution injected per site is about 0.05 μl, so that *ca.* 1 μl is needed per slice. For injection of an *Escherichia coli* LacZ reporter virus (AdLacZ), expression was detected within 8 hr after injection and was strong by 18–24 hr. As shown in Fig. 5A, fairly uniform LacZ expression was observed over a 2- $\times$ 1-mm region, which included most of the pyramidal cell layer of CA1; as shown in Fig. 5B, *lacZ* gene product was observed in the dendritic region as well as the pyramidal region. Figures 5C and 5D show that AdLacZ-infected slices show normal synaptic responses and LTP.

In this work,[15] the scientific problem that was addressed was the role of endothelial nitric oxide synthase (eNOS) in generating NO, which functions as a retrograde messenger in some forms of LTP. Endothelial NOS is myristoylated and hence membrane localized. HMA, an inhibitor of myristoylation, blocks LTP. One question then is whether this is due to specific blocking of the membrane attachment of eNOS, or blocking of membrane localization of other myristoylated proteins. A fusion protein consisting of the extracellular and transmembrane domain of the T-cell type I membrane protein CD8 fused to a nonmyristoylated mutant of eNOS was constructed and expressed in CA1, thus conferring an alternate, myristoylation-independent means of membrane association. Slices injected with this construct exhibited robust LTP, which was not abolished by the

[15] D. B. Kantor, M. Lanzrein, S. J. Stary, G. M. Sandoval, W. B. Smith, B. M. Sullivan, N. Davidson, and E. M. Schuman, *Science* **274,** 1744 (1996).

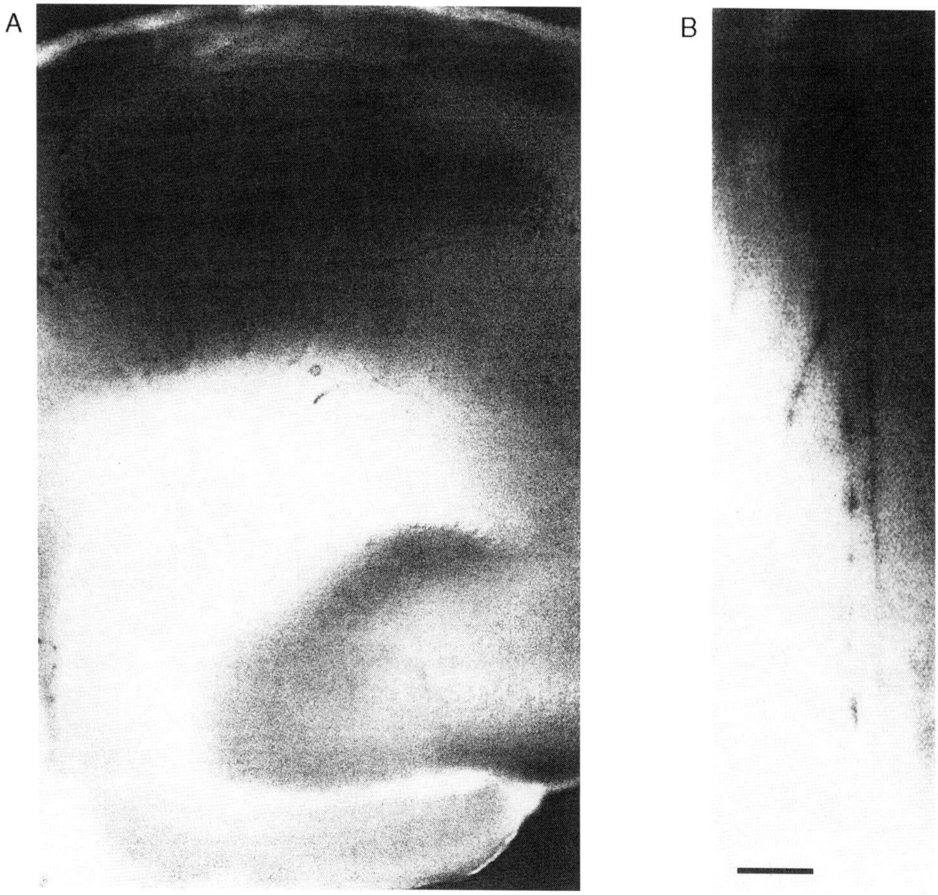

FIG. 5. Hippocampal slices infected with an AdLacZ reporter virus exhibit normal synaptic transmission and plasticity. (A) A slice infected in area CA1 by AdLacZ and stained for X-Gal 24 hr after injection. (B) X-Gal labeling of individual pyramidal neurons and their associated dendrites; scale bar: 15 μm. (C) I/O relation for saline-injected (control, solid lines) and AdLacZ-infected (dashed lines) slices. The slope of each line was calculated, and a between-group comparison indicated that they are not significantly different from one another (mean slope $\pm$SEM: AdLacZ, 0.011 $\pm$ 0.002; saline, 0.01 $\pm$ 0.001). (D) Ensemble average of LTP experiments for both control and AdLacZ-infected slices. In control slices (filled circles), the mean field EPSP slope was -0.14 ± 0.01 mV/ms before LTP and -0.27 ± 0.05 mV/ms after LTP ($P < 0.01$). In AdLacZ-infected slices (open circles), the mean field EPSP slope was -0.16 ± 0.01 mV/ms before LTP and -0.28 ± 0.03 mV/ms after LTP ($P < 0.01$). (Inset) Two representative field EPSPs from a control slice (left) and an AdLadZ-infected slice (right), taken 10 min before and 50–60 min after the LTP induction protocol; calibration bar: 0.5 mV, 10 ms. [Reprinted with permission from D. B. Kantor, M. Lanzrein, S. J. Stary, G. M. Sandoval, W. B. Smith, B. M. Sullivan, N. Davidson, and E. M. Schuman, *Science* **274,** 1744 (1996). Copyright © 1997 American Association for the Advancement of Science.]

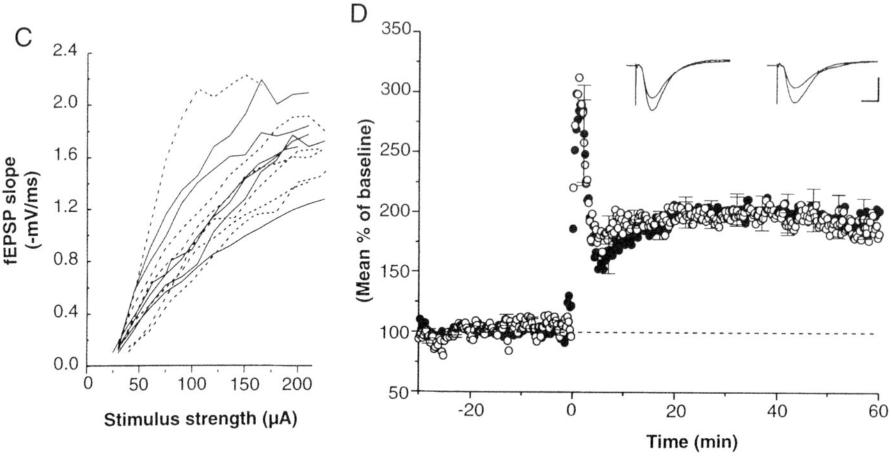

FIG. 5. (*continued*)

myristoylation inhibitor, HMA, but which was abolished by NOS inhibitors. A truncated eNOS (TeNOS) functions by heterodimerization with eNOS as a dominant-negative mutant of eNOS. Adenovirus-mediated expression of this mutant also abolished LTP without affecting normal basal synaptic transmission. Overall, these results point to eNOS as the critical myristoylated protein required for LTP induction, and suggest the importance of membrane- rather than cytosolic-localized eNOS as a critical component of the signal transduction events underlying LTP.

Korte *et al.*[16] have developed an essentially identical method of administrating adenovirus to slices and used this approach to study the role of brain derived neurotrophic factor (BDNF) in LTP. A similar procedure has also been described with vaccinia virus.[17]

Future Directions

One may anticipate many ways in which future developments will extend the usefulness of recombinant adenovirus vectors for the study of neurobiological phenomena. "Gutless" vectors with an enlarged cloning capacity and reduced immunological reaction *in vivo* and possibly reduced toxicity *in vitro* are being developed, principally for applications in gene therapy.[11] However, they will also provide enlarged cloning capacity for research

[16] M. Korte, O. Griesbeck, C. Gravel, P. Carroll, V. Staiger, H. Thoenen, and T. Bonhoeffer, *Proc. Natl. Acad. Sci. U.S.A.* **93,** 12547 (1996).

[17] D. L. Pettit, S. Perlman, and R. Malinow, *Science* **266,** 1881 (1994).

purposes and may prove to be even less cytotoxic for cells in culture as well as *in vivo*.

Present acute slice protocols are not practical for study of the presynaptic effects of genes expressed in CA3 for synaptic communication between CA3 and CA1. This is because the projections via the Schaffer collaterals from a set of CA3 cell bodies to a set of CA1 dendrites are in general not contained in a single slice; therefore, administration of adenovirus to the CA3 cell bodies of a slice will not lead to expression in the axons synapsing on CA1 dendrites of the same slice. One approach to overcome this difficulty may be the use of organotypic cultures. It is reported that in a single slice of such a culture, 76% of the CA3 pyramidal neurons are synaptically connected to CA1 neurons in the same slice.[18]

Another promising direction is the development of systems providing inducible gene expression. One such system based on the tetracycline control system described by Gossen and Bujard[19] and requiring two adenoviruses has been described by Yoshida and Hamada.[20]

Acknowledgments

We thank Sheri L. McKinney for the primary cell cultures, Dr. Paulo Kofuji for subcloning of *Shaker* H4 into pAC, Catherine Lin for construction of AdH4 and Ad5HT$_{1A}$R, and S. Jennifer Stary for advice on the adenovirus technique. We are grateful to Drs. Arnold J. Berk, Frank L. Graham, Robert D. Gerard, and Jackie Sheng for much advice. We thank Dr. Berk for Ad5 *dl*309 and AdLacZ, Dr. Graham for Ad5 PacI, Dr. Gerard for pAC, Dr. David E. Clapham for GIRK2 cDNA, Dr. John P. Adelman for GIRK4 cDNA, Dr. Brian Seed for CD8 cDNA, Dr. J. Sheng for AdRR5 and pZS2 and Dr. Simon Efrat for βTC3 cells. This work was supported by the National Institute of Mental Health, National Institute of General Medical Sciences, Human Frontier Science Program, the Swiss National Science Foundation (fellowships 81BE-40054 and 823A-042966 to M.U.E.), and the European Molecular Biology Organization (fellowship ALTF 168-1996 to M.L.).

[18] D. Debanne, N. C. Guérineau, B. H. Gähwiler, and S. M. Thompson, *J. Neurophysiol.* **73,** 1282 (1995).

[19] M. Gossen and H. Bujard, *Proc. Natl. Acad. Sci. U.S.A.* **89,** 5547 (1992).

[20] Y. Yoshida and H. Hamada, *Biochem. Biophys. Res. Commun.* **230,** 426 (1997).

[28] *In Vivo* Incorporation of Unnatural Amino Acids into Ion Channels in *Xenopus* Oocyte Expression System

By MARK W. NOWAK, JUSTIN P. GALLIVAN, SCOTT K. SILVERMAN, CESAR G. LABARCA, DENNIS A. DOUGHERTY, and HENRY A. LESTER

Introduction

We have adapted the nonsense codon suppression method[1] for the incorporation of unnatural amino acids into membrane proteins in a *Xenopus* oocyte expression system. Combining this method with electrophysiologic analysis allows us to probe structure–function relationships in ion channels and receptors in ways not possible with conventional mutagenesis.[2–4] In the absence of atomic-scale structural data for membrane proteins, these techniques can provide detailed structural information.

In the nonsense codon suppression method, *Xenopus* oocytes are coinjected with two RNAs: (1) mRNA transcribed *in vitro* from a mutated cDNA containing a TAG nonsense (stop) codon at the position of interest and (2) a suppressor tRNA containing the corresponding anticodon, CUA, and chemically acylated at the 3′ end with an amino acid. During protein synthesis, the aminoacylated suppressor tRNA directs the incorporation of the amino acid into the desired position of the protein (Fig. 1). Because the amino acid is appended synthetically, we are not limited to the natural 20, and many unnatural amino acids have been incorporated into various proteins using this method.

This article describes suppressor tRNA design and synthesis, chemical acylation of the suppressor tRNA, relevant organic synthesis methods, and optimization of mRNA and suppressor tRNA for *Xenopus* oocyte expression.

[1] C. J. Noren, S. J. Anthony-Cahill, M. C. Griffith, and P. G. Schultz, *Science* **244,** 182 (1989).

[2] M. W. Nowak, P. C. Kearney, J. R. Sampson, M. E. Saks, C. G. Labarca, S. K. Silverman, W. Zhong, J. Thorson, J. N. Abelson, N. Davidson, P. G. Schultz, D. A. Dougherty, and H. A. Lester, *Science* **268,** 439 (1995).

[3] P. C. Kearney, M. W. Nowak, W. Zhong, S. K. Silverman, H. A. Lester, and D. A. Dougherty, *Mol. Pharmacol.* **50,** 1401 (1996).

[4] P. C. Kearney, H. Zhang, W. Zhong, D. A. Dougherty, and H. A. Lester, *Neuron* **17,** 1221 (1996).

0076-6879/98 $25.00

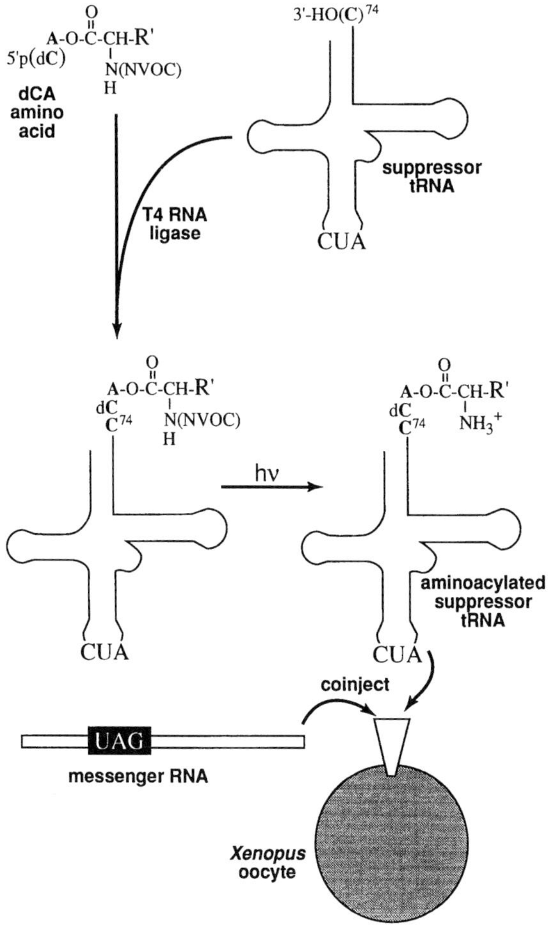

FIG. 1. Scheme for incorporating unnatural amino acids into proteins expressed in *Xenopus* oocytes. (Reprinted with permission from Saks *et al., J. Biol. Chem.* **271,** 23169 (1996).)

Materials

DNA oligonucleotides are synthesized on an Expedite DNA Synthesizer (Perceptive Biosystems, Framingham, MA.). *Fok*I, *Bsa*I, and other restriction endonucleases and T4 RNA ligase are purchased from New England Biolabs (Beverly, MA). T4 polynucleotide kinase, T4 DNA ligase, and RNase inhibitor are purchased from Boehringer Mannheim Biochemicals (Indianapolis, IN). [^{35}S]Methionine and ^{14}C-labeled protein molecular weight markers are purchased from Amersham (Arlington Heights, IL).

Inorganic pyrophosphatase is purchased from Sigma (St. Louis, MO). Stains-all is purchased from Aldrich (Milwaukee, WI). T7 RNA polymerase is either purified using the method of Grodberg and Dunn[5] from the overproducing strain *Escherichia coli* BL21 harboring the plasmid pAR1219[6] or purchased from Ambion (Austin, TX). For all buffers described, unless otherwise noted, final adjustment of pH is unnecessary.

Suppressor tRNA Design and Synthesis

Designing a suppressor tRNA presents two major challenges: (1) one must maximize efficiency of the suppressor tRNA at incorporating an unnatural amino acid at the site of interest; and (2) one must prevent editing and/or reacylation of the suppressor tRNA by endogenous tRNA synthetases. Low suppression efficiency may often be overcome by general tactics for overexpressing proteins in the oocyte, because minute signals may easily be detected. In contrast, if reacylation occurs, this must be addressed before usable data can be obtained, because the uncontrolled mixture of amino acids at the site of interest would complicate data interpretation.

Our first suppressor tRNA, tRNA-MN3, was a modification of a yeast $tRNA^{Phe}$(CUA) used previously to incorporate unnatural amino acids in an *in vitro* translation system.[1] The tRNA-MN3 was adequate with respect to efficiency and editing/reacylation in our initial study of conserved tyrosine residues in the putative agonist-binding site of the nicotinic acetylcholine receptor (nAChR).[2] However, we discovered that tRNA-MN3 did not constitute an optimal solution to the *in vivo* suppression problem.[7] By studying positions in the nAChR α subunit that were less conserved and more tolerant with regard to substitution, we found that tRNA-MN3 led to incorporation of *natural* amino acids at the mutation site along with the desired unnatural amino acid. This was probably the result of reacylation and/or editing of the chemically acylated tRNA-MN3 by synthetases endogenous to the *Xenopus* oocyte.

To reduce reacylation and to improve nonsense codon suppression efficiency, we designed a new nonsense suppressor tRNA based on $tRNA^{Gln}$(CUA) from the eukaryote *Tetrahymena thermophila.* In *T. thermophila,* the UAG codon does not signal the termination of protein synthe-

[5] J. Grodberg and J. J. Dunn, *J. Bact.* **170,** 1245 (1988).

[6] P. Davanloo, A. H. Rosenberg, J. J. Dunn, and F. W. Studier, *Proc. Natl. Acad. Sci. U.S.A.* **81,** 2035 (1984).

[7] M. E. Saks, J. R. Sampson, M. W. Nowak, P. C. Kearney, F. Du, J. N. Abelson, H. A. Lester, and D. A. Dougherty, *J. Biol. Chem.* **271,** 23169 (1996).

sis but instead codes for the amino acid glutamine. Previous *in vitro* translation experiments showed that when wheat germ[8] or rabbit reticulocyte[9] systems were supplemented with a *T. thermophila* synthetase preparation, $tRNA^{Gln}$(CUA) efficiently translated normal UAG stop codons in a variety of heterologous mRNAs.

Based on principles derived from studies of translation efficiency and tRNA recognition,[10–12] we modified *T. thermophila* $tRNA^{Gln}$(CUA) to produce a new suppressor tRNA.[7] We mutated U73 to G to reduce recognition by the oocyte's endogenous glutamine acyltransferase. The T7 RNA promoter is included directly upstream from the first nucleotide of the tRNA gene. To generate a full-length tRNA transcript, or a transcript missing the last two nucleotides (for reasons described later), *Bsa*I and *Fok*I sites are engineered into the 3′ end of the gene (Fig. 2).

T. thermophila $tRNA^{Gln}$(CUA) G73 Gene Construction

Buffers and Solutions

10× Phosphorylation Buffer

Volume	Reagent	Final concentration
7 ml	1 *M* Tris-Cl, pH 7.6	700 m*M*
0.5 ml	1 *M* Dithiothreitol (DTT)	50 m*M*
1 ml	1 *M* $MgCl_2$	100 m*M*
1.5 ml	H_2O	
10 ml		

10× Annealing Buffer

Volume	Reagent	Final concentration
2 ml	1 *M* Tris-Cl, pH 7.5	200 m*M*
5 ml	1 *M* NaCl	500 m*M*
1 ml	1 *M* $MgCl_2$	100 m*M*
2 ml	H_2O	
10 ml		

[8] C. Schull and H. Beier, *Nucleic Acids Res.* **22,** 1974 (1994).
[9] N. Hanyu, Y. Kuchino, S. Nishimura, and H. Beier, *EMBO J.* **5,** 1307 (1986).
[10] M. E. Saks, J. R. Sampson, and J. N. Abelson, *Science* **263,** 191 (1994).
[11] W. H. McClain, *J. Mol. Biol.* **234,** 257 (1993).
[12] R. Giege, J. D. Puglishi, and C. Florentz, *Prog. Nucleic Acid Res. Mol. Biol.* **45,** 129 (1993).

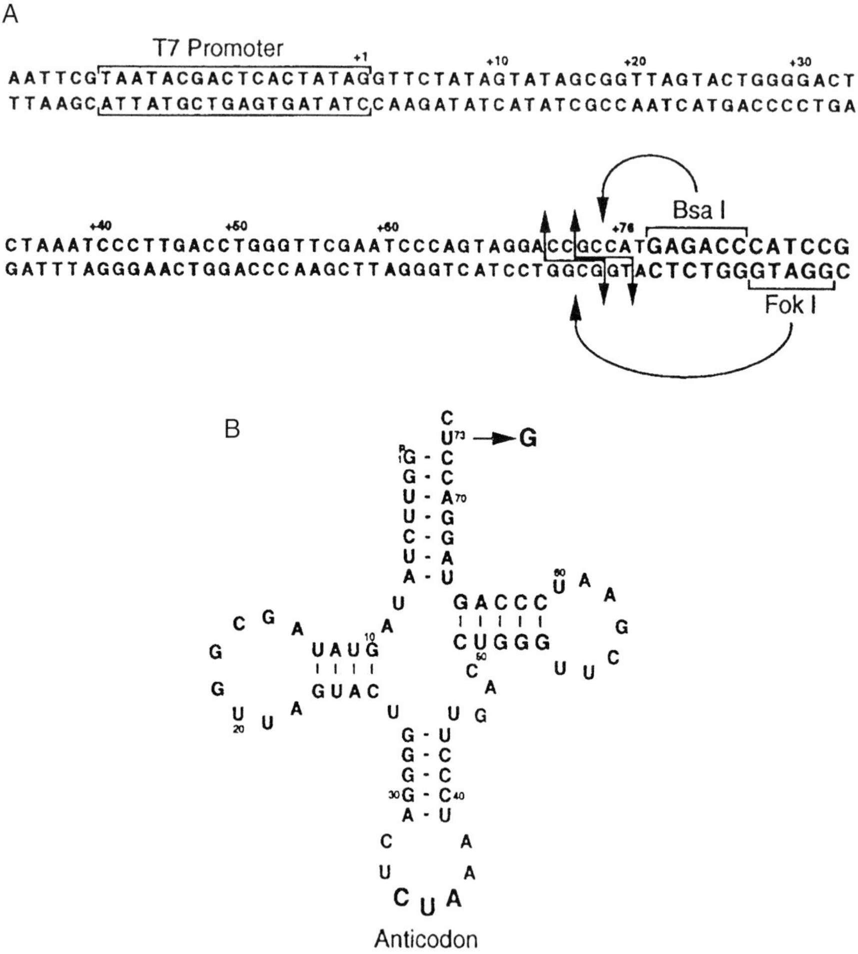

FIG. 2. Design of the engineered *T. thermophila* tRNA^{GLN}(CUA) G73 suppressor gene for *in vitro* transcription with T7 RNA polymerase. (A) The DNA sequence for the suppressor tRNA gene, the upstream T7 promoter, and the downstream runoff transcription sites are given. (B) Nucleotide sequence for the suppressor tRNA obtained from runoff transcription of *Fok*I-linearized DNA. (Reprinted with permission from Saks *et al., J. Biol. Chem.* **271,** 23169 (1996).)

10× Ligation Buffer

Volume	Reagent	Final concentration
6.6 ml	1 *M* Tris-Cl, pH 7.5	660 m*M*
0.1 ml	1 *M* DTT	10 m*M*

0.5 ml	1 *M* $MgCl_2$	50 m*M*
1.0 ml	100 m*M* ATP	10 m*M*
1.8 ml	H_2O	
10 ml		

*10× Tris-Acetate-EDTA (TAE) Buffer**

Amount	Reagent	Final concentration
48.4 g	Tris-Cl	400 m*M*
11.4 ml	Glacial acetic acid (99.7%, 2/w)	1.1% (w/w)
20 ml	0.5 *M* EDTA, pH 8.0	10 m*M*
	H_2O (bring final volume to 1000 ml)	

* Adjust pH to 8.0 with NaOH.

The gene for *T. thermophila* $tRNA^{Gln}CUA$ G73[7] is constructed from the eight overlapping synthetic DNA oligonucleotides, four sense and four antisense, shown below.

Oligonucleotide	Sequence
1	5′-AATTCGTAATACGACTCACTATAGGTTCTATAG-3′
2	3′- GCATTATGCTGAGTGATATCCAAGA -5′
3	5′- TATAGCGGTTAGTACTGGGGACTCTAAA -3′
4	3′-TATCATATCGCCAATCATGACCCCTGAG -5′
5	5′- TCCCTTGACCTGGGTTCG-3′ -3′
6	3′-ATTTAGGGAACTGGACCC -5′
7	5′- AATCCCAGTAGGACCGCCATGAGACCCATCCG -3′
8	3′-AGCTTAGGGTCATCCTGGCGGTACTCTGGGTAGGCCTAG-5′

Oligonucleotides 2 through 7 are phosphorylated as follows:

Phosphorylation Reaction

Volume	Reagent	Final concentration
5 μl	10× phosphorylation buffer	1×
5 μl	10 m*M* ATP	1 m*M*
5 μl	1 m*M* spermidine	0.1 m*M*
10 μl	10 μg/μl oligonucleotide	2 μg/μl
1 μl	T4 polynucleotide kinase (10 U/μl)	0.2 U/μl
24 μl	H_2O	
50 μl		

After incubation at 37° for 2 hr, the reaction mixture is extracted once with an equal volume of phenol (saturated with 1 *M* Tris-Cl, pH

8.0) : $CHCl_3$: isoamyl alcohol (25 : 24 : 1) and once with an equal volume of $CHCl_3$: isoamyl alcohol (24 : 1). The aqueous phase is treated with 10 volumes of *n*-butanol, vortexed for 1 min, and centrifuged at 14,000 rpm for 10 min. The pellet is washed with 70% (v/v) ethanol, dried under vacuum, and reconstituted in 20 μl H_2O. The DNA concentration is determined from UV absorption (A_{260}), assuming that an OD of 1 at 260 nm corresponds to 37 μg/ml of DNA.

Oligonucleotides are annealed to each other in pairs (P designates phosphorylated): 1 : 2P, 3P : 4P, 5P : 6P and 7P : 8. The annealing procedure is as follows:

Annealing Reaction

Volume	Reagent	Final concentration
5 μl	10× annealing buffer	1×
	2 nmol sense oligonucleotide	40 μM
	2 nmol antisense oligonucleotide	40 μM
	H_2O (bring final volume to 50 μl)	

Annealing reactions are heated at 95° for 3 min then cooled slowly to room temperature. Reactions are treated with 10 volumes of *n*-butanol, vortexed for 1 min, and centrifuged at 14,000 rpm for 10 min. The pellets are washed with 70% ethanol, dried under vacuum, and reconstituted in 25 μl H_2O. The annealed oligonucleotides are ligated as follows:

Ligation Reaction

Volume	Reagent	Final concentration
2.5 μl	10× ligation buffer	1×
5 μl	Annealed oligonucleotides 1 : 2	
5 μl	Annealed oligonucleotides 3 : 4	
5 μl	Annealed oligonucleotides 5 : 6	
5 μl	Annealed oligonucleotides 7 : 8	
1 μl	T4 DNA ligase (10 U/μl)	0.4 U/μl
1.5 μl	H_2O	
25 μl		

Ligation reactions are incubated overnight at 16°. To test if the ligation is successful, a 5-μl aliquot of the reaction is run on a 1% agarose gel with 1 μg/ml ethidium bromide in 1× TAE buffer. The 110 base pair tRNA gene is purified on a 1% low melting point agarose gel and subcloned into the *Eco*RI/*Bam*HI sites of the pUC19 vector, giving the plasmid pTHG73.

In vitro Transcription and Purification of *T. thermophila* $tRNA^{Gln}$(CUA) G73 (THG73)

Buffers and Solutions

10× Transcription Buffer

Volume	Reagent	Final concentration
4 ml	1 *M* Tris-Cl, pH 8.3	400 m*M*
1 ml	100 m*M* spermidine	10 m*M*
1 ml	5 mg/ml acetylated-BSA	0.5 mg/ml
4 ml	H_2O	
10 ml		

*10× Tris-Borate-EDTA (TBE)**

Amount	Reagent	Final concentration
108 g	Tris-Cl	890 m*M*
55 g	Sodium borate	890 m*M*
40 ml	0.5 *M* EDTA, pH 8.0	20 m*M*
	H_2O (bring final volume to 1000 ml)	

* pH should be 8.0 without further adjustment.

*8% Acrylamide: bisacrylamide (19:1)/7 M Urea in 1X TBE**

Amount	Reagent	Final concentration
7.6 g	Acrylamide	7.6%
0.4 g	Bisacrylamide	0.4%
42 g	Urea	7 *M*
10 ml	10× TBE	1×
20 μl	TEMED	
200 μl	10% ammonium persulfate	
	H_2O (bring final volume to 100 ml)	

* The TEMED and ammonium persulfate are added just prior to pouring the gel.

Bromphenol Blue/Xylene Cyanol (BPB/XC) Dye Solution (Final Volume: 10 ml)

Amount	Reagent	Final concentration
25 mg	Bromphenol blue	0.25%
25 mg	Xylene cyanol	0.25%

4.2 g	Urea	7 *M*
1 ml	10× TBE	1×
3 ml	Glycerol	30%
	H_2O (bring final volume to 10 ml)	

Stains-All Solution (0.005%)*

Volume	Reagent	Final concentration
25 ml	0.1% Stains-all (100 mg solid Stains-all in 100 ml formamide)	0.005%
225 ml	Formamide	50%
250 ml	H_2O	
500 ml		

* Stains-all is extremely light sensitive.

Template DNA for *in vitro* transcription of tRNA lacking the 3′-terminal bases C75 and A76 is prepared by linearizing pTHG73 with *Fok*I. *In vitro* transcription[13] reactions (1 ml) are set up as follows:

In vitro Transcription Reaction

Volume	Reagent	Final concentration
100 μl	10× transcription buffer	1×
50 μl	100 m*M* dithiothreitol (DTT)	5 m*M*
200 μl	Solution with 20 m*M* each of ATP, UTP, GTP, and CTP	4 m*M* each NTP
100 μl	100 m*M* GMP	10 m*M*
100 μl	200 m*M* $MgCl_2$	20 m*M*
100 μl	1.0 μg/μl linearized pTHG73	100 μg/μl
2 μl	RNase inhibitor (40 U/μl)	0.08 U/μl
4 μl	Pyrophosphatase (0.5 U/μl)	0.002 U/μl
130 μl	T7 RNA polymerase (200 U/μl)	26 U/μl
214 μl	H_2O	
1000 μl		

The transcription reaction is incubated at 42° for 3 hr. The tRNA transcripts are visualized on an analytical gel [8% acrylamide : bisacrylamide (19 : 1) in 7 *M* urea, 1× TBE; 0.4 mm thick, 40 cm long]. Before loading the samples, the gel is run at 1000 V for 30 min. Samples are prepared by adding 5 μl of BPB/XC dye solution to 5 μl of reaction followed by heating

[13] J. R. Sampson and O. C. Uhlenbeck, *Proc. Natl. Acad. Sci. U.S.A.* **85,** 1033 (1988).

at 65° for 2 min. Gels are run at 1000 V until the xylene cyanol band migrates 25–30 cm. The bromphenol blue and xylene cyanol dye bands correspond approximately to 10-mer and 55-mer oligonucleotides, respectively. The gel is stained overnight in 0.005% Stains-all solution (protected from light), destained in water with exposure to light, and dried under vacuum. We typically observe two tRNA products: the desired 74-mer and a 75-mer in a ratio of 2:1. The desired 74-mer tRNA transcript will be referred to as THG73.

THG73 is purified either (1) by polyacrylamide gel electrophoresis or (2) by using the Qiagen Total RNA kit (Chatsworth, CA) with modifications. With method 1, the THG73 can be purified away from the 75-mer, while with method 2, the products are obtained as a mixture. We have successfully used both tRNA preparations in suppression experiments. The second method provides greater amounts of material than the first method and takes less time to perform. Both methods are outlined next.

Purification of tRNA-THG73 Transcript by Polyacrylamide Electrophoresis (Method 1)

Buffers/Solutions

Tris-EDTA Buffer (TE): Prepared with Diethyl Pyrocarbonate (DEPC)-Treated H_2O

Volume	Reagent	Final concentration
1 ml	1 *M* Tris-Cl, pH 8.0	10 m*M*
0.2 ml	0.5 *M* EDTA, pH 8.0	1 m*M*
98.8 ml	H_2O	
100 ml		

*Elution Buffer (Prepared with DEPC-Treated H_2O)**

Volume	Reagent	Final concentration
20 ml	1 *M* Potassium acetate, pH 5.0	200 m*M*
2 ml	0.5 *M* EDTA, pH 8.0	10 m*M*
78 ml	H_2O	
100 ml		

* Adjust final pH to 5.0 with HCl.

tRNA Buffer (Prepared with DEPC Treated H_2O)

Volume	Reagent	Final concentration
0.4 ml	250 m*M* Sodium acetate, pH 5.0	10 m*M*
9.6 ml	H_2O	
10 ml		

For the gel-purification procedure (method 1), transcription reactions are terminated by adding 1/10 volume 250 m*M* EDTA, pH 8.0, and 1/10 volume 250 m*M* NaOAc, pH 5.0. The mixtures are extracted once with an equal volume of phenol (saturated with 300 m*M* NaOAc, pH 5.0) : $CHCl_3$: isoamyl alcohol (25 : 24 : 1). THG73 is precipitated with 2 volumes of ethanol at $-20°$, dried under vacuum, and resuspended in 500 μl TE buffer. The product is purified on a preparative gel [8% acrylamide : bisacrylamide (19 : 1) in 7 *M* urea, 1× TBE; 4 mm thick, 40 cm long]. Before loading the sample, the gel is run at 500 V for 30 min. Samples are prepared for loading by adding 500 μl of BPB/XC dye solution to the resuspended THG73 followed by heating at 65° for 3 min. No more than 500–600 μg crude THG73 is loaded onto the gel. Gels are run at 500 V until the xylene cyanol band migrates 25–30 cm. The lower half of the gel is wrapped in plastic wrap and placed on a thin-layer chromatography plate with fluorescent indicator. The band corresponding to THG73 is visualized by short-wavelength UV light and outlined with a marker. The THG73 band is cut out, transferred to a 10-ml tube, and crushed with a glass rod (the tube and rod are previously baked overnight at 180°). To elute THG73, 5 ml of elution buffer is added, and the mixture is gently agitated overnight at 4°. After centrifugation at 6000 rpm in a clinical centrifuge for 15–20 min, the supernatant is transferred to a disposable empty column (Bio-Rad Laboratories Inc., Hercules, CA) to remove any remaining acrylamide and stored at 4°. The gel is eluted again with 3 ml of elution buffer for 5 hr at 4°, and the supernatant is isolated as above. The combined supernatants are transferred to 30-ml glass centrifuge tubes. Ethanol (25 ml) is added to precipitate THG73, and the mixture is placed at $-80°$ for 24 hr and then at $-20°$ for 24 hr. THG73 is pelleted by centrifugation at 7000 rpm in a HB4 rotor at $-10°$ for 1 hr. The pellet is dried under vacuum, resuspended in 400 μl TE buffer, and transferred to a 1.7-ml Eppendorf tube. The product is precipitated a second time with 100 μl of 8 *M* NH_4OAc and 1 ml ethanol at $-20°$ overnight. THG73 is pelleted a final time by centrifugation at 15,000 rpm at 4° for 30 min, washed with cold 70% ethanol, dried under vacuum, and resuspended in 40 μl water. The RNA concentration is determined from UV absorption (A_{260}), assuming that an OD of 1 at

260 nm corresponds to 40 μg/ml of RNA. The purity is assessed by analytical PAGE as described earlier.

Purification of tRNA-THG73 Using Modification of Qiagen Total RNA Kit (Method 2)

Buffers and Solutions

Wash Buffer (Prepared in DEPC-Water)

Volume	Reagent	Final concentration
2.5 ml	1 *M* MOPS, pH 7.0	50 m*M* (21 m*M* Na^+)
6.61 ml	4 *M* NaCl	529 m*M* Na^+
7.5 ml	Ethanol	15% ethanol
33.39 ml	DEPC–water	
50 ml		550 m*M* Na^+

800 mM Na^+ Elution Buffer (Prepared in DEPC–Water)

Volume	Reagent	Final concentration
2.5 ml	1 *M* MOPS, pH 7.0	50 m*M* (21 m*M* Na^+)
9.74 ml	4 *M* NaCl	779 m*M* Na^+
7.5 ml	Ethanol	15% ethanol
30.26 ml	DEPC–water	
50 ml		800 m*M* Na^+

1000 mM Na^+ Elution Buffer (Prepared in DEPC–Water)

Volume	Reagent	Final concentration
2.5 ml	1 *M* MOPS, pH 7.0	50 m*M* (21 m*M* Na^+)
12.24 ml	4 *M* NaCl	979 m*M* Na^+
7.5 ml	Ethanol	15% ethanol
27.76 ml	DEPC–water	
50 ml		1000 m*M* Na^+

For the Qiagen kit purification procedure (method 2), the tRNA transcription reaction is mixed with 2.5 ml Qiagen buffer QRL1 containing 25 μl 2-mercaptoethanol. The solution is mixed with 22.5 ml Qiagen buffer QRV2 and applied to a Qiagen tip-500 column preequilibrated with 10 ml Qiagen buffer QRE. Wash buffer is eluted until the A_{260} of the eluent is less than 0.05 (typically 40–50 ml). THG73 is eluted with 2 × 7.5 ml of 800 m*M* NaCl elution buffer then 4 × 7.5 ml of 1000 m*M* NaCl elution buffer into six 30-ml glass centrifuge tubes (previously rinsed with ethanol and baked overnight at 180°). UV spectra are taken of a 300-μl aliquot of each 7.5-ml fraction; the fractions with high A_{260} contain the THG73 product. THG73 is precipitated by the addition of 1 volume (7.5 ml) 2-propanol to each fraction followed by centrifugation at 13,000 rpm in a JA-20 rotor (Beckman Instruments, Fullerton, CA) at 4° for 30 min. Pellets are each reconstituted in 200 μl DEPC–water, transferred to 1.7-ml Eppendorf tubes, and reprecipitated by the addition of 20 μl 3 *M* NaOAc, pH 5.0, and 600 μl cold ethanol. After storage overnight at −20°, the tRNA is repelleted by centrifugation at 14,000 rpm at 4° for 15 min. The pellets are rinsed with 100 μl cold 70% ethanol, dried under vacuum, redissolved in 100 μl DEPC–H_2O and stored at −80°. An aliquot (1 μl) is removed from each fraction and run on a small (0.75-mm-thick, <8-cm-long) analytical polyacrylamide gel (conditions as given earlier) to assess yield and purity. Typically the middle four fractions contain the bulk of the THG73 product; these are combined and the final concentration of tRNA is quantified by UV absorption (A_{260}), assuming that an OD of 1 at 260 nm unit corresponds to 40 μg/ml of RNA.

Organic Chemistry

Preparation of dCA Amino Acids

For the most part the procedures of Schultz and co-workers are followed, and the appropriate references should be consulted.[14,15]

Synthesis of dCA

This multistep sequence is carried out using the published procedure[14,15] with small modifications as described elsewhere.[3] Note that the *in vivo* nonsense suppression methodology generally operates on a much smaller scale than the *in vitro* protocol. Therefore, smaller quantities of reagents such as dCA are consumed. The total synthesis sequence is carried out on a scale to produce ~2 g of final dCA product, which is enough for a very large number of experiments. After HPLC purification, dCA can be stored indefinitely at −80° under an inert atmosphere. Alternatively, dCA can be synthesized on an automated DNA synthesizer.[16]

[14] S. A. Robertson, C. J. Noren, S. J. Anthony-Cahill, M. C. Griffin, and P. G. Schultz, *Nucleic Acids Res.* **17,** 9649 (1989).

[15] J. Ellman, D. Mendel, S. Anthonycahill, C. J. Noren, and P. G. Schultz, *Methods Enzymol.* **202,** 301 (1991).

[16] G. Turcatti, K. Nemeth, M. D. Edgerton, U. Meseth, F. Talabot, M. Peitsch, J. Knowles, H. Vogel, and A. Chollet, *J. Bio. Chem.* **271,** 19991 (1996).

Synthesis of Unnatural Amino Acids

This aspect of the protocol is highly variable, depending on the desired structure. In many cases, the unnatural amino acid is prepared by modification of a natural amino acid, while many others are commercially available. The NRC Biotechnology Research Institute Peptide/Protein Chemistry Group maintains an excellent listing of commercially available amino acids at http://aminoacid.bri.nrc.ca:1125. A representative list of amino acids (natural and unnatural) that have been successfully incorporated by the *in vivo* nonsense suppression methodology is given in Fig. 3. Note also that racemic amino acids can be used, because only L-amino acids, and not D-amino acids, are incorporated.[17]

Typically, the amino group is protected as the *o*-nitroveratryloxycarbonyl (NVOC) group, which is subsequently removed photochemically according to the previous protocols. However, for amino acids that have a photoreactive side chain an alternative must be used. We have used the 4-pentenoyl (4PO) group, a protecting group first described by Madsen *et al.*[18] Lodder *et al.*[19] have also shown that the 4PO group is compatible with the tRNA systems. We present here a representative procedure based on the unnatural amino acid (2-nitrophenyl)glycine (Npg).[20]

N-4PO-DL-(2-Nitrophenyl)glycine. The unnatural amino acid DL-(2-nitrophenylglycine) hydrochloride is prepared according to published procedures.[21,22] The amine is protected as the 4PO derivative as follows.[18,19] To a room temperature solution of (2-nitrophenyl)glycine hydrochloride (82 mg, 0.35 mmol) in H_2O : dioxane (0.75 ml : 0.5 ml) is added Na_2CO_3 (111 mg, 1.05 mmol) followed by a solution of 4-pentenoic anhydride (70.8 mg, 0.39 mmol) in dioxane (0.25 ml). After 3 hr the mixture is poured into saturated $NaHSO_4$ and extracted with CH_2Cl_2. The organic phase is dried over anhydrous Na_2SO_4 and concentrated *in vacuo*. The residual oil is purified by flash silica gel column chromatography to yield the title compound (73.2 mg, 75.2%) as a white solid. ^1H NMR (300 MHz, CD_3OD) δ 8.06 (dd, J=1.2, 8.1 Hz, 1H), 7.70 (ddd, J=1.2, 7.5, 7.5 Hz, 1H), 7.62-7.53 (m, 2H), 6.21 (s, 1H), 5.80 (m, 1H), 5.04–4.97 (m, 2H), 2.42–2.28 (m, 4H). HRMS calculated for $C_{13}H_{14}N_2O_5$ 279.0981, found 279.0992.

[17] V. W. Cornish, D. Mendel, and P. G. Schultz, *Angew. Chem. Int. Ed. Engl.* **34,** 621 (1995).

[18] R. Madsen, C. Roberts, and B. Fraser-Reid, *J. Org. Chem.* **60,** 7920 (1995).

[19] M. Lodder, S. Golvine, and S. M. Hecht, *J. Org. Chem.* **62,** 778 (1997).

[20] P. M. England, H. A. Lester, N. Davidson, and D. A. Dougherty, *Proc. Natl. Acad. Sci. U.S.A.* **94,** 11025 (1997).

[21] A. L. Davis, D. R. Smith, and T. J. McCord, *J. Med. Chem.* **16,** 1043 (1973).

[22] S. Muralidharan and J. M. Nerbonne, *J. Photochem. Photobiol. B: Biol.* **27,** 123 (1995).

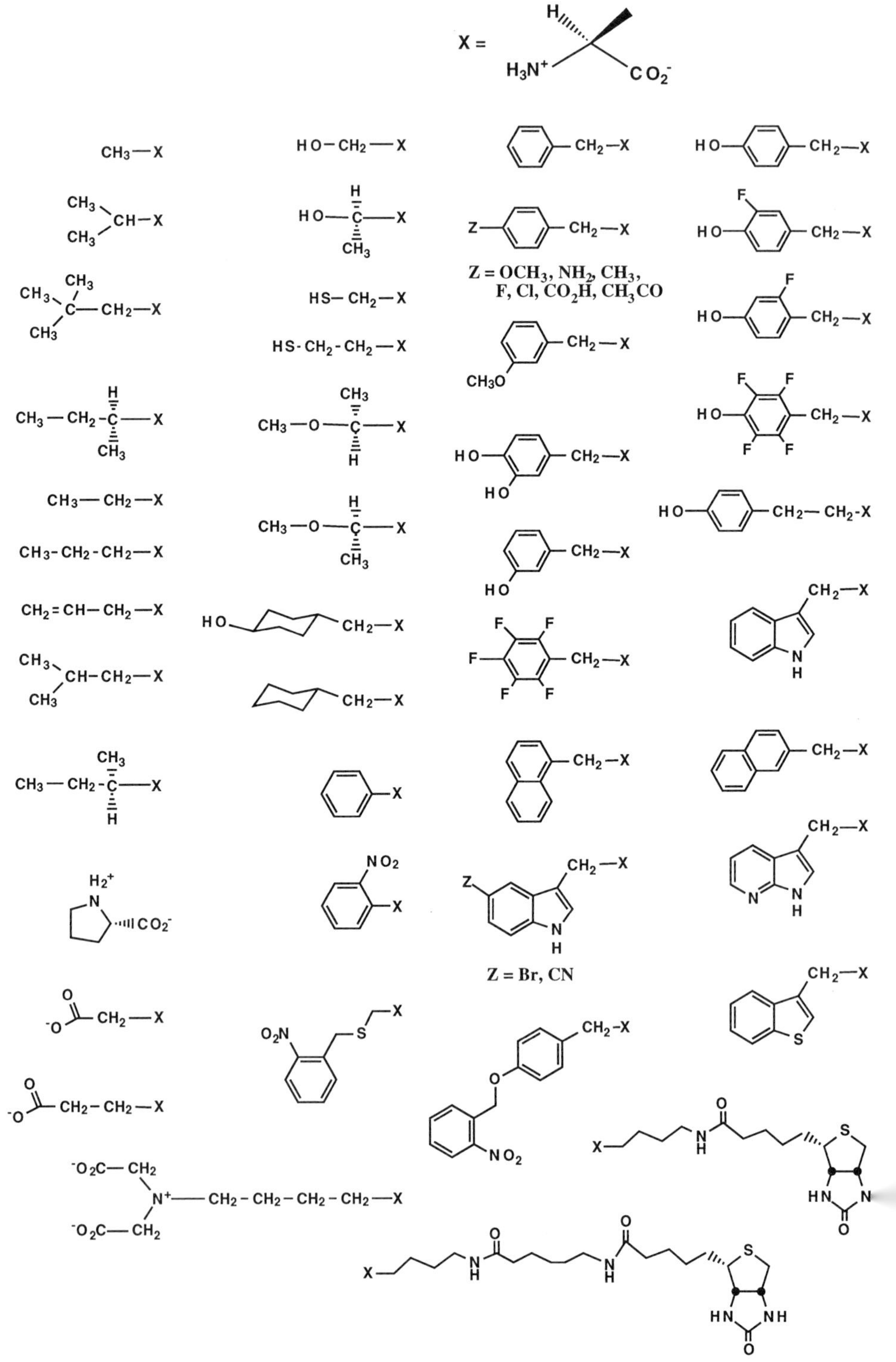
X =
Z = OCH3, NH2, CH3, F, Cl, CO2H, CH3CO
Z = Br, CN

N-4PO-DL-(2-Nitrophenyl)glycinate Cyanomethyl Ester. The acid is activated as the cyanomethyl ester using standard conditions.[14,15] To a room temperature solution of the acid (63.2 mg, 0.23 mmol) in anhydrous dimethylformamide (DMF, 1 ml) is added $N(C_2H_5)_3$ (95 μl, 0.68 mmol) followed by $ClCH_2CN$ (1 ml). After 16 hr the mixture is diluted with $(C_2H_5)_2O$ and extracted against H_2O. The organic phase is washed with saturated NaCl, dried over anhydrous Na_2SO_4, and concentrated *in vacuo*. The residual oil is purified by flash silica gel column chromatography to yield the title compound (62.6 mg, 85.8%) as a yellow solid. ^{1}H NMR (300 MHz, $CDCl_3$) δ 8.18 (dd, $J=1.2$, 8.1 Hz, 1H), 7.74–7.65 (m, 2H), 7.58 (ddd, $J=1.8$, 7.2, 8.4 Hz, 1H), 6.84 (d, $J=7.8$ Hz, 1H), 6.17 (d, $J=6.2$ Hz, 1H), 5.76 (m, 1H), 5.00 (dd, $J=1.5$, 15.6 Hz, 1H), 4.96 (dd, $J=1.5$, 9.9 Hz, 1H), 4.79 (d, $J=15.6$ Hz, 1H), 4.72 (d, $J=15.6$ Hz, 1H), 2.45–2.25 (m, 4H). HRMS calculated for $C_{16}H_{17}N_3O_5$ 317.1012, found 317.1004.

N-4PO-(2-Nitrophenyl)glycine-dCA. The dinucleotide dCA is prepared as reported by Ellman *et al.*[15] with the modifications described by Kearney *et al.*[3] The cyanomethyl ester is then coupled to dCA as follows. To a room temperature solution of dCA (tetrabutylammonium salt, 20 mg, 16.6 μmol) in anhydrous DMF (400 μl) under argon is added N-4PO-DL-(2-nitrophenyl)glycinate cyanomethyl ester (16.3 mg, 51.4 μmol). The solution is stirred for 1 hr and then quenched with 25 m*M* ammonium acetate, pH 4.5 (20 μl). The crude product is purified by reversed-phase semipreparative HPLC (Whatman, Clifton, NJ Partisil 10 ODS-3 column, 9.4 mm $\times$ 50 cm), using a gradient from 25 m*M* NH_4OAc, pH 4.5 to CH_3CN. The appropriate fractions are combined and lyophilized. The resulting solid is redissolved in 10 m*M* acetic acid/CH_3CN and lyophilized to afford 4PO-Npg-dCA (3.9 mg, 8.8%) as a pale yellow solid. ESI-MS M^- 896 (31), $[M-H]^-$ 895 (100), calculated for $C_{32}H_{36}N_{10}O_{17}P_2$ 896. The material is quantified by UV absorption ($\varepsilon_{260} \approx 37{,}000\ M^{-1}\ cm^{-1}$).

Chemical Acylation of tRNAs and Removal of Protecting Groups

The α-NH_2-protected dCA-amino acids or dCA are enzymatically coupled to the THG73 *Fok*I runoff transcripts using T4 RNA ligase to form

FIG. 3. Representative amino acids, both natural and unnatural, that have been successfully incorporated into proteins expressed in *Xenopus* oocytes by the *in vivo* nonsense suppression method. Absence from this list does not imply that a particular unnatural amino acid would not be compatible with the system.

a full-length chemically charged α-NH_2-protected aminoacyl-THG73 or a full-length but unacylated THG73-dCA.[23]

2.5× Reaction Mix

Volume	Reagent	Concentration in 2.5× mix	Final concentration in reaction
25 μl	400 m*M* HEPES, pH 7.5	100 m*M*	40 m*M*
10 μl	100 m*M* DTT	10 m*M*	4 m*M*
25 μl	200 m*M* $MgCl_2$	50 m*M*	20 m*M*
5 μl	7.5 m*M* ATP	375 μ*M*	150 μ*M*
10 μl	5 mg/ml acetylated BSA	0.5 mg/ml	0.2 mg/ml
1 μl	RNase inhibitor (40 U/μl)	0.4 U/μl	0.16 U/μl
24 μl	DEPC-H_2O		
100 μl			

Prior to ligation, 10 μl of THG73 (1 μg/μl in water) is mixed with 5 μl of 10 m*M* HEPES, pH 7.5. This tRNA/HEPES premix is heated at 95° for 3 min and allowed to cool slowly to 37°. Ligation reactions (40 μl) are set up as follows:

Ligation Reaction

Volume	Reagent	Final concentration
15 μl	tRNA-THG73/HEPES premix (see text)	0.25 μg/μl
16 μl	2.5× reaction buffer	1×
4 μl	α-NH_2-protected dCA-amino acid (3 m*M* in DMSO)*	300 μ*M*
2.4 μl	T4 RNA ligase (20 U/μl)	1.2 U/μl
2.6 μl	DEPC-H_2O	
40 μl		

* The final DMSO concentration in the ligation reaction must be 10%.

After incubation at 37° for 2 hr, DEPC–H_2O (52 μl) and 3 *M* sodium acetate, pH 5.0 (8 μl), are added and the reaction mixture is extracted once with an equal volume of phenol (saturated with 300 m*M* sodium acetate, pH 5.0) : $CHCl_3$: isoamyl alcohol (25 : 24 : 1) and once with an equal volume of $CHCl_3$: isoamyl alcohol (24 : 1) then precipitated with 2.5 volumes of cold ethanol at −20°. The mixture is centrifuged at 14,000 rpm at 4° for 15 min, and the pellet is washed with cold 70% (v/v) ethanol, dried under vacuum, and resuspended in 7 μl 1 m*M* sodium acetate, pH 5.0. The amount of α-NH_2-protected aminoacyl-THG73 is quantified by measuring A_{260},

[23] T. E. England, A. G. Bruce, and O. C. Uhlenbeck, *Methods Enzymol.* **65,** 65 (1980).

and the concentration is adjusted to 1 μg/μl with 1 m*M* sodium acetate, pH 5.0.

The ligation efficiency may be determined from analytical PAGE. The α-NH_2-protected aminoacyl-tRNA partially hydrolyzes under typical gel conditions, leading to multiple bands, so the ligated tRNA is deprotected (as described later; see section on Oocyte Injection) prior to loading. Such deprotected tRNAs immediately hydrolyze on loading. Typically, 1 μg of ligated tRNA in 10 μl BPB/XC buffer (described earlier) is loaded onto the gel, and 1 μg of unligated tRNA is run as a size standard. The ligation efficiency may be determined from the relative intensities of the bands corresponding to ligated tRNA (76 bases) and unligated tRNA (74 bases). We typically find a 30% ligation efficiency for THG73; the side products are presumably due to self-ligation.

Generation of mRNA

High-Expression Plasmid Constructs

For the nonsense codon suppression method, it is desirable to have the gene of interest in a high-expression plasmid, so that functional responses in oocytes may be observed 1–2 days after injection. Among other considerations, this minimizes the likelihood of reacylation of the suppressor tRNA.

A high-expression plasmid (designed by C. Labarca) was generated by modifying the multiple cloning region of pBluescript SK+. At the 5′ end, an alfalfa mosaic virus (AMV) sequence was inserted, and at the 3′ end a poly(A) tail was added, providing the plasmid pAMV-PA (Fig. 4). mRNA transcripts containing the AMV region bind the ribosomal complex with high affinity, leading to 30-fold increase in protein synthesis.[24] Including a 3′ poly(A) tail has been shown to increase mRNA half-life, therefore increasing the amount of protein synthesized.[25] The gene of interest was subcloned into pAMV-PA such that the AMV region is immediately 5′ of the ATG start codon of the gene (i.e., the 5′ untranslated region of the gene was completely removed). The plasmid pAMV-PA is available from C. Labarca at Caltech.

Site-Directed Mutagenesis

Any suitable site-directed mutagenesis method may be used to change the codon of interest to TAG. We have successfully used the Transformer

[24] S. A. Jobling and L. Gehrke, *Nature* **325,** 622 (1987).

[25] P. Berstein and J. Ross, *Trends. Biochem. Sci.* **14,** 373 (1989).

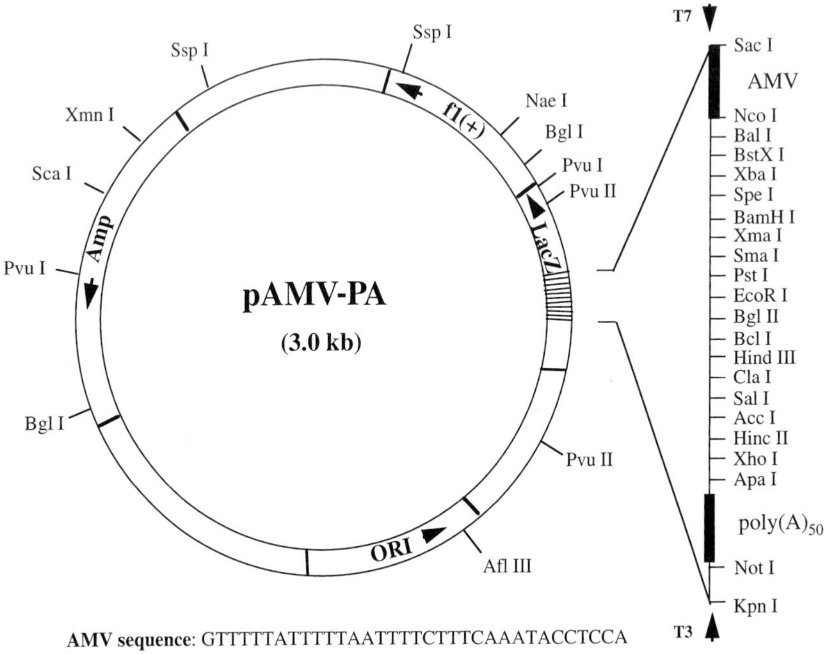

FIG. 4. pAMV-PA high-expression plasmid.

kit (Clontech, Palo Alto, CA), the Altered Sites kit (Stratagene, La Jolla, CA), and standard polymerase chain reaction (PCR) cassette mutagenesis procedures.[26] With the first two methods, a small region of the mutant plasmid (400–600 base pairs) is subcloned into the original plasmid. With all methods, the inserted DNA regions are checked by automated sequencing over the ligation sites. If the plasmid of interest naturally has TAG as its stop codon, this must be mutated to either TAA or TGA, to prevent the suppressor tRNA from inserting amino acids at the termination position.

In vitro mRNA Synthesis

The pAMV-PA plasmid constructs are linearized with *Not*I, and mRNA transcripts are generated using the mMessage mMachine T7 RNA polymerase kit (Ambion, Austin, TX). We have found that mRNA generated with this kit consistently expresses well when injected into *Xenopus* oocytes.

[26] S. N. Ho, H. D. Horton, J. K. Pullen, and L. R. Pease, *Gene* **77,** 51 (1989).

Oocytes

Oocyte Injection

Before using the α-NH_2-protected aminoacyl-tRNA for oocyte experiments, the α-NH_2 protecting group must be removed. We primarily use one of two protecting groups, either NVOC or 4PO.[18] The NVOC group is removed by irradiating 1 μl of tRNA solution for 5 min with a 1000-W Hg(Xe) arc lamp equipped with WG-335 and UG-11 filters (Schott, Duryea, PA). The filters are placed in a water-filled chamber to cool them and to remove infrared wavelengths. The 4PO group is removed by treating 0.5–1 μl of tRNA solution with an equal volume of saturated aqueous I_2 for 10 min.

Following deprotection, the tRNA is mixed with mRNA and microinjected into *Xenopus* oocytes using published methods[27] (50 nl/oocyte). The amounts of tRNA and mRNA must be optimized for each aminoacyl-tRNA and for each position within a given protein. We typically inject 5–25 ng tRNA per oocyte. Note that the endogenous level of total tRNA in an oocyte is approximately 50 ng, although any individual tRNA is present in much smaller amounts. We typically base the amount of injected UAG-mutant mRNA on the amount of analogous wild-type mRNA needed to provided a reasonable signal level. Since suppression may be only 10–15% efficient in some cases (see *In Vitro* Protein Translation section), it is often necessary to inject 10–50 times the amount of UAG mRNA relative to the wild-type mRNA to achieve comparable responses.

For heteromultimeric proteins, one must also consider the stoichiometry of the functional protein. For example, the muscle nAChR consists of four subunits, α, β, γ, and δ, in a 2 : 1 : 1 : 1 ratio. When carrying out suppression experiments in the α subunit, we typically inject the α UAG mRNA along with the wild-type β, γ, and δ mRNAs in a (2-20) : 1 : 1 : 1 ratio.

Electrophysiologic Measurements and Expression Levels

Whole-cell electrophysiologic measurements using a two-electrode voltage-clamp circuit are typically performed 18–48 hr after injection. With the nonsense codon suppression method, we routinely observe whole-cell currents on the order of microamperes from a variety of ion channels, including the muscle nAChR, the G-protein-coupled inward-rectifier K^+ (GIRK) channel, and the *Shaker* K^+ channel. In addition, we have measured single-channel currents for the nAChR using the patch-clamp technique.[4]

[27] M. W. Quick and H. A. Lester, *in* "Ion Channels of Excitable Cells" (T. Narahashi, ed.) pp. 261–279. Academic Press, San Diego, 1994.

Assessing and Optimizing Efficiency and Fidelity of Nonsense Codon Suppression in *Xenopus* Oocytes

Assessing mRNA-Dependent Background Currents

It is important to determine whether injection of the UAG mRNA without a suppressor tRNA leads to functional protein. This could occur if (1) an endogenous tRNA misreads the UAG codon or (2) an endogenous *Xenopus* subunit substitutes for the UAG mutant subunit. In some instances with the nAChR, we have observed currents when only UAG mRNA is injected, although these currents are generally small (<20 nA) and can be minimized by reducing the amount of injected mRNA.

Assessing Reacylation of THG73

The desired outcome from coinjecting oocytes with an aminoacylated suppressor tRNA and UAG mRNA is to synthesize protein incorporating the desired unnatural amino acid at the UAG position. However, if endogenous synthetases replace the unnatural amino acid on the tRNA with a natural amino acid (editing and/or reacylation), this can lead to an uncontrolled mixture of amino acids at the UAG position.

To assess the extent of reacylation, if any, oocytes are coinjected with uncharged tRNA (tRNA-dCA) and UAG mRNA. When examining a new position in a protein one *must* perform this experiment to ensure that (1) reacylation is not occurring or (2) reacylation is not a contaminating factor. For initial studies with THG73,[7] we tested for reacylation by coinjecting oocytes with tRNA-dCA and nAChR α180UAG, β, γ, and δ mRNAs. We chose position 180 since it is not highly conserved among known α subunits, and a wide range of natural amino acids are observed at this position. For these experiments, THG73 reacylation currents in response to 50–200 μM ACh were routinely <10 nA. Based on these findings, we initially thought that reacylation of the suppressor tRNA-THG73 was not occurring. However, more recent experiments at other positions within the α subunit have demonstrated reacylation currents ranging from 50 nA to as high as 1 μA. In almost all cases, this current could be reduced or eliminated by lowering the amounts of tRNA and mRNA injected.

Assessing Fidelity of Amino Acid Incorporation

When examining a new mutant, one would like to verify the fidelity of amino acid incorporation, which may be done in several ways. First, the wild-type protein should be reconstituted by incorporating the wild-type residue using the nonsense suppression method. Second, a mutant prepared by conventional site-directed mutagenesis may be compared to the same mutant obtained using nonsense suppression. We initially verified the fidelity of nonsense codon suppression in oocytes by inserting the wild-type tyrosine or mutant phenylalanine residues at positions α93, α190, and α198 of the mouse muscle nAChR.[2] Receptors obtained from nonsense codon suppression displayed indistinguishable characteristics (EC_{50} values and Hill coefficients) when compared to receptors obtained from wild-type or conventional mutant mRNAs, thus validating the method.

Additional Issues with Nonsense Suppression Method

Potential Concerns for Heteromultimeric Protein Expression

A potential complication in the oocyte expression of heteromultimeric proteins is promiscuity in a subunit–subunit assembly. For example, when one expresses the muscle nAChR in oocytes, it is possible to obtain receptors composed of only α, β, and γ subunits ("δ-less" receptors) or α, β, and δ subunits ("γ-less" receptors) simply by omitting either the δ- or γ-subunit mRNA, respectively.[28,29] Expression of δ-less or γ-less receptors could occur in suppression experiments with δ- or γ-subunit UAG mRNAs, respectively, if the suppression efficiency is poor. The δ-less or γ-less receptors could then be mistaken for receptors containing the desired unnatural amino acid. To avoid this problem, properties of the δ-less or γ-less receptors should be compared to those of receptors observed in the suppression experiment.

For many proteins like the muscle nAChR, one does not normally worry about stoichiometry variations in conventional expression experiments. But for some proteins, like those comprising the neuronal nAChRs, varying subunit compositions are a normal concern. For example, injecting oocytes with neuronal nAChR α_3 and β_4 mRNAs in mole ratios of 0.91 : 0.09, 0.5 : 0.5, and 0.09 : 0.91 leads to receptors with ACh EC_{50} values of 170, 40, and 25 μM (M. W. Nowak, unpublished results, 1997). Presumably these receptors differ in their α_3/β_4 stoichiometry. In such cases, suppression experiments with either subunit may give results difficult to distinguish from the normal stoichiometry variation.

Additional Potential Complications of the Nonsense Suppression Method

If the nonsense UAG codon is near the 3′ end of the mRNA coding region, poor suppression efficiency may yield mostly truncated protein,

[28] P. Charnet, C. Labarca, and H. A. Lester, *Biophys. J.* **59,** 527a (1991).

[29] C. Czajkowski, C. Kaufmann, and A. Karlin, *Proc. Natl. Acad. Sci. U.S.A.* **90,** 6285 (1993).

which is nevertheless functional. In this case, one could mistake the truncated protein for the protein incorporating the desired unnatural amino acid. To check for this, one may inject the UAG mRNA without suppressor tRNA to determine whether a functional protein is produced.

Related to this, truncated proteins may retain enough assembly domains to function as dominant negative mutants. In such cases, the amounts of mRNA must be optimized to control the dominant negative effect and to maximize expression. However, we note that in many cases, such as at the nAChR 9′ position in the second transmembrane domain, suppression experiments work well although dominant negative effects are observed when no suppressor tRNA is injected.[4]

What if a Functional Response is not Observed?

In some cases, a suppression experiment does not result in a functional response. This could occur because (1) the unnatural amino acid is not incorporated into the protein or (2) the unnatural amino acid is incorporated but the mutant protein is nonfunctional. To address the first possibility, an *in vitro* translational assay can be performed as described next. In the second case, an alternative protein detection method that does not require a functional response should be used. For the muscle nAChR, we have employed an α-bungarotoxin binding assay to detect expression of nonfunctional receptors. If a toxin binding assay is unavailable, a convenient epitope such as c-*myc*[30] may be inserted into an extracellular region of the protein and an antibody binding assay performed.

In vitro Translation as a Test for Suppression

Buffers and Solutions

*4× Separating Gel Buffer**

Amount	Reagent	Final concentration
18.17 g	Tris-Cl	1.5 *M*
4 ml	10% sodium dodecyl sulfate (SDS)	0.4%
	H_2O (bring final volume to 100 ml)	

* Adjust pH to 8.8.

[30] P. A. Kolodziej and R. A. Young, *Methods Enzymol.* **194,** 508 (1991).

*4× Stacking Gel Buffer**

Amount	Reagent	Final concentration
6.66 g	Tris-Cl	0.55 *M*
4 ml	10% SDS	0.4%
	H_2O (bring final volume to 100 ml)	

* Adjust pH to 8.8.

*10× Running Buffer**

Amount	Reagent	Final concentration
3.03 g	Tris-Cl	0.25 *M*
10 ml	10% SDS	1%
14.4 g	Glycine	1.92 *M*
	H_2O (bring final volume to 100 ml)	

* Adjust pH to 6.8.

Sample Buffer

Amount	Reagent	Final concentration
2.5 mg	Bromphenol blue	0.025%
2.5 ml	4× Stacking buffer	1×
2 ml	Glycerol	20%
2 ml	10% SDS	2%
0.5 ml	2-Mercaptoethanol (add just before using)	5%
3 ml	H_2O	
10 ml		

40% Acrylamide Solution (19:1)

Amount	Reagent	Final concentration
38 g	Acrylamide	38%
2 g	Bisacrylamide	2%
	H_2O (bring final volume to 100 ml)	

5% Stacking Gel

Amount	Reagent	Final concentration
6.25 ml	40% Acrylamide (19:1) solution	5%
12.5 ml	4× Stacking gel buffer	1×
10 μl	TEMED*	
100 μl	10% Ammonium persulfate*	
	H_2O (bring final volume to 50 ml)	
50 ml		

* The TEMED and ammonium persulfate are added just prior to pouring the gel.

10% Separating Gel

Amount	Reagent	Final concentration
12.5 ml	40% Acrylamide (19 : 1) solution	10%
12.5 ml	4× Separating gel buffer	1×
10 μl	TEMED*	
100 μl	10% Ammonium persulfate*	
	H_2O (bring final volume to 50 ml)	
50 ml		

* The TEMED and ammonium persulfate are added just prior to pouring the gel.

An *in vitro* translation system may be employed to test the efficiency of suppression for a specific UAG mRNA and aminoacyl-tRNA independent of protein function. This method can assess the mRNA quality, chemical ligation of THG73, and the ability of the particular unnatural amino acid to pass through the protein translational machinery. We have employed a rabbit reticulocyte lysate *in vitro* translational system (Promega, Madison, WI). Reactions are set up as follows:

In vitro Translational Reaction Premix for 10 Reactions

Volume	Reagent
70 μl	Rabbit reticulocyte lysate
2 μl	1 m*M* Amino acids (without methionine)
8 μl	[^{35}S]Methionine (1000–1200 Ci/mmol)
2 μl	RNase inhibitor (40 U/μl)

In vitro Translational Reaction

Volume	Reagent
8.2 μl	*In vitro* translational reaction premix
1 μl	1 μg/μl mRNA
1 μl	1 μg/μl Deprotected aminoacyl-tRNA

For control reactions, UAG mRNA and deprotected aminoacyl-tRNA (or tRNA-dCA) are replaced with 1 μl DEPC–water and 1 μl 1 m*M* sodium acetate, pH 5.0, respectively. After incubation at 30° for 2 hr the reactions are terminated by the addition of 10 μl of sample buffer and run on a 5% separating/10% stacking SDS–polyacrylamide gel. ^{14}C-labeled markers are run as standards. Gels are fixed for 30 min with 45 : 10 : 45 methanol : acetic acid : water, 30 min with Entensify A (DuPont, Wilmington, DE), and 30 min with Entensify B (DuPont, Wilmington, DE), dried under vacuum, and exposed to X-ray film overnight. Using this method, we estimate that the *in vitro* nonsense codon suppression efficiency of THG73 is 10–15%, although the *in vivo* efficiency may be higher or lower in any particular case.

Summary

A general method for the incorporation of unnatural amino acids into ion channels and membrane receptors using a *Xenopus* oocyte expression system has been described. A large number of unnatural amino acids have been incorporated into the nAChR, GIRK, and *Shaker* K^+ channels. Continuing efforts focus on incorporating unnatural amino acids that differ substantially from the natural amino acids, for example, residues that include fluorophores. In addition, we are addressing the feasibility of incorporating unnatural amino acids into ion channels and membrane receptors in mammalian cells.

Acknowledgments

We thank the NIH (NS-11756 and NS-34407) for support of this work.

[29] High-Level Expression and Detection of Ion Channels in *Xenopus* Oocytes

By THEODORE M. SHIH, RAYMOND D. SMITH, LIGIA TORO, and ALAN L. GOLDIN

Introduction

Since the initial demonstration by Miledi and co-workers that ion channels and neural receptors can be functionally expressed in *Xenopus* oocytes,[1–3] this system has become a standard for demonstrating that a specific cloned cDNA encodes a functional channel or receptor. Many different ion channels and receptors have been expressed in oocytes for functional analysis (reviewed in Refs. 4 and 5), and oocytes have been used for

[1] E. A. Barnard, R. Miledi, and K. Sumikawa, *Proc. R. Soc. Lond.* **215,** 241 (1982).
[2] R. Miledi, I. Parker, and K. Sumikawa, *EMBO J.* **1,** 1307 (1982).
[3] C. B. Gundersen, R. Miledi, and I. Parker, *Nature* **308,** 421 (1984).
[4] N. Dascal, *CRC Crit. Rev. Biochem.* **22,** 317 (1987).
[5] T. P. Snutch, *Trends Neurosci.* **11,** 250 (1988).

0076-6879/98 $25.00

functional cloning of receptors, as described by Lübbert *et al.*[6] and Julius *et al.*[7] RNA for injection into oocytes can be isolated from the appropriate tissue sample or cell line, or it can be synthesized *in vitro* from a cDNA clone, as previously described.[8,9] The goal of this article is to outline some of the approaches that are used for high-level expression and analysis of cloned ion channels in *Xenopus* oocytes.

Vectors for Expression in *Xenopus* Oocytes

The only essential element of a vector for RNA transcription and expression in oocytes is a promoter for a DNA-dependent RNA polymerase, such as one of the polymerases isolated from the phage T7, T3, or SP6. However, there are many instances in which it is essential to obtain high-level expression of a particular channel, such as for detection by immunofluorescence,[10,11] immunoprecipitation,[12–14] or the measurement of gating currents.[15,16] A number of plasmid vectors have been designed specifically for this purpose.

The *Xenopus laevis* β-globin gene is one of the most stably expressed mRNA transcripts. It has been shown that adding the 5′ and 3′ noncoding sequences from the β-globin gene greatly increases the expression of exogenous proteins in oocytes.[17] The pBSTA vector was designed to take advantage of these properties to improve the expression of channels in oocytes (Fig. 1A). This vector contains a T7 promoter followed by the 5′ noncoding β-globin sequences, a single *Bgl*II site for insertion of exogenous DNA, the 3′ noncoding β-globin sequences, a poly(A) tail, and a T3 promoter for the synthesis of antisense RNA. The plasmid also contains an M13 origin of replication, which can be used to synthesize single-stranded DNA for mutagenesis. There is a polylinker following the poly(A) tail so that the plasmid can be linearized before transcribing RNA. DNA encoding

[6] H. Lubbert, B. J. Hoffman, T. P. Snutch, T. VanDyke, A. J. Levine, P. R. Hartig, H. A. Lester, and N. Davidson, *Proc. Natl. Acad. Sci. U.S.A.* **84,** 4332 (1987).

[7] D. Julius, A. B. MacDermott, R. Axel, and T. M. Jessell, *Science* **241,** 558 (1988).

[8] A. L. Goldin, *Methods Cell Biol.* **36,** 487 (1991).

[9] A. L. Goldin and K. Sumikawa, *Methods Enzymol.* **207,** 279 (1992).

[10] T. M. Shih and A. L. Goldin, *J. Cell Biol.* **136,** 1037 (1997).

[11] P. Meera, M. Wallner, M. Song, and L. Toro, *Proc. Natl. Acad. Sci. U.S.A.* **94,** 14066 (1977).

[12] R. D. Smith and A. L. Goldin, *J. Neurosci.* **16,** 1965 (1996).

[13] R. D. Smith and A. L. Goldin, *J. Neurosci.* **17,** 6086 (1997).

[14] L. Santacruz-Toloza, Y. Huang, S. A. John, and D. M. Papazian, *Biochemistry* **33,** 5607 (1994).

[15] E. Stefani, L. Toro, E. Perozo, and F. Bezanilla, *Biophys. J.* **66,** 996 (1994).

[16] E. Stefani, M. Ottolia, F. Noceti, R. Olcese, M. Wallner, R. Latorre, and L. Toro, *Proc. Natl. Acad. Sci. U.S.A.* **94,** 5427 (1997).

[17] P. A. Krieg and D. A. Melton, *Nucleic Acids Res.* **12,** 7057 (1984).

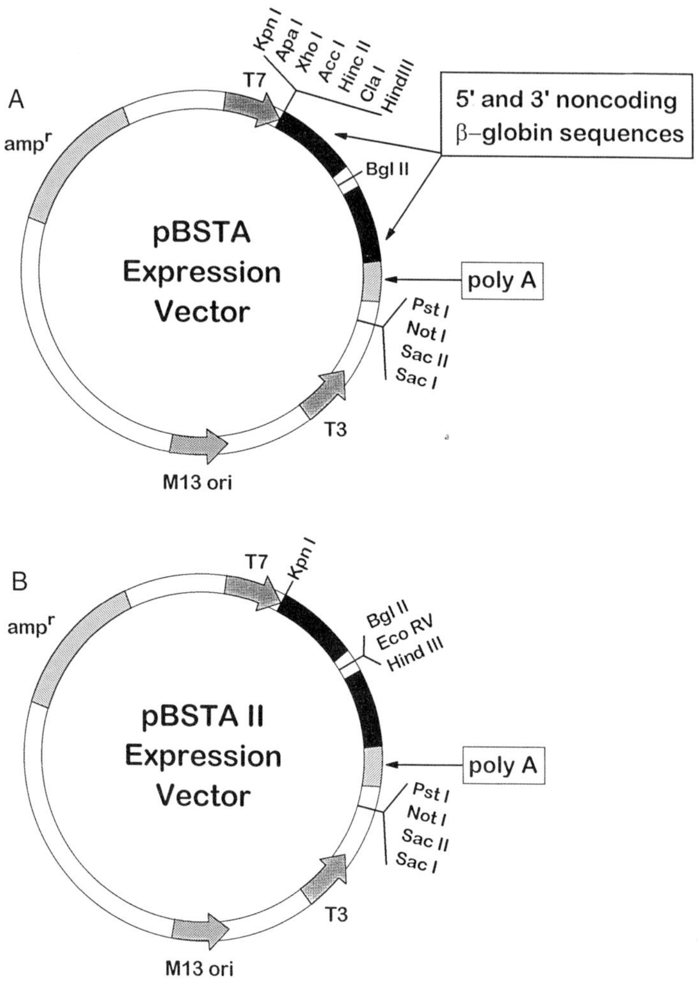

FIG. 1. Features of plasmid vectors for high-level expression in oocytes. (A) Schematic diagram of the pBSTA vector. DNA encoding the protein of interest is inserted into the *Bgl*II site between the 5′ and 3′ noncoding β-globin sequences. The DNA template is linearized with *Not*I for *in vitro* transcription with T7 DNA polymerase. The transcript contains the 5′ noncoding β-globin sequences, the coding region, the 3′ noncoding β-globin sequences, and a poly(A) tail. The M13 origin of replication facilitates the production of phagemid single-stranded DNA template for mutagenesis. (B) The pBSTAII vector was modified to include two additional restriction sites (*Eco*RV and *Hin*dIII) to facilitate the insertion of exogenous DNA.

the protein of interest can be inserted into the single *Bgl*II site by adding either *Bgl*II or *Bam*HI linkers (which have compatible ends) and cutting the linkers with *Mbo*I, which recognizes the internal four bases (GATC). The activity of *Mbo*I is blocked by *dam* methylation, which occurs in essentially all bacterial strains that are commonly used for molecular biology. Therefore, any internal *Bgl*II or *Bam*HI sites in the cDNA insert will not be digested, so that only the linker will be cut. The fragment is then ligated into the *Bgl*II site of pBSTA.

The pBSTA vector has also been modified to provide two additional restriction enzyme sites for the insertion of exogenous DNA. The pBSTAII vector (Fig. 1B) was constructed by deleting the restrictions sites upstream from the 5′ β-globin sequences and introducing two additional unique restriction sites next to the *Bgl*II site. The modified vector has a polylinker consisting of *Bgl*II, *Eco*RV, and *Hin*dIII for insertion of exogenous sequences.

To determine the extent to which pBSTA facilitated expression of ion channels in oocytes, the *Shaker* H4 cDNA was transferred from pH4U5[18] into pBSTA. This plasmid was termed pKH4T. RNA was transcribed from pH4U5 and pKH4T, injected into oocytes, and the expression levels were compared over a 4-day time period (Fig. 2). RNA transcribed from pKH4T (●) consistently expressed 4–50 times greater levels of current compared to an equivalent (▲) or larger (▼) amount of RNA transcribed from pH4U5. Currents greater than 180 μA at -20 mV were obtained after injection of 50 ng RNA, as determined by using a two-electrode voltage clamp. This level of current is suitable for biochemical, immunological, and gating current studies.

Additional manipulations can be performed to increase the level of expression, such as altering the 5′ untranslated leader of the cDNA insert. A guanine was substituted for a cytosine to construct an *Nco*I site at the initiation codon in a clone of the *Shaker* potassium channel that contains 183 bp of 5′ of untranslated region. This cysteine substitution increased the level of expression approximately 5-fold, probably by disrupting a hairpin loop in the RNA, resulting in gating currents from 0.2 to 2 μA at 10 mV using the cut-open oocyte voltage-clamp technique.[15] In the large-conductance Ca^{2+}-dependent K^+ channel (*hslo*), all of the 5′ untranslated region up to the third (M3) or fourth (M4) Kozak consensus sequence[19] was deleted. The insert was cloned into the *Nco*I and *Hind*III sites of a modified version of the pGEM9zf($-$) vector (Promega, Madison, WI)

[18] A. Kamb, J. Tseng-Crank, and M. A. Tanouye, *Neuron* **1,** 421 (1988).

[19] M. Wallner, P. Meera, M. Ottolia, G. J. Kaczorowski, R. Latorre, M. L. Garcia, E. Stefani, and L. Toro, *Receptors Chann.* **3,** 185 (1995).

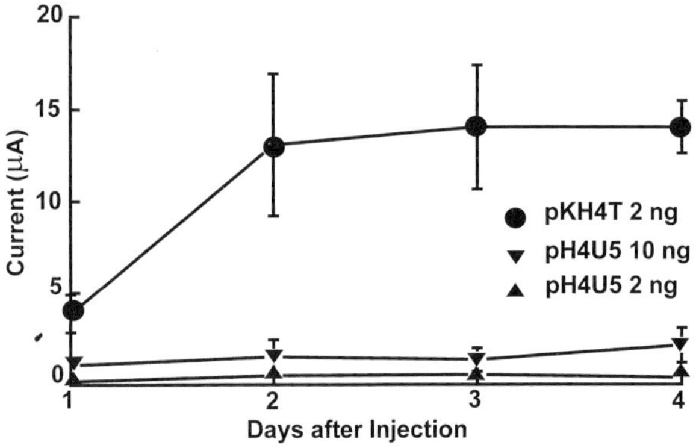

FIG. 2. Potassium currents resulting from expression of pKH4T compared to pH4U5. Stage V oocytes were injected with either 2 or 10 ng of RNA transcribed from either pKH4T, which contains the *Shaker* H4 coding region flanked by β-globin noncoding sequences and a poly(A) tail, or pH4U5, which contains only the *Shaker* H4 coding region. The oocytes were incubated at 20° in ND96 plus supplements, and potassium currents were recorded in ND96 bath solution using a two-electrode voltage clamp. Currents were elicited by depolarizations from a holding potential of −80 to 50 mV for 100 ms. Linear leak and capacitive currents were eliminated using *P*/4 subtraction. Symbols represent the average values from four oocytes for each RNA sample, and the error bars represent the standard deviations.

containing 183 bp from the 5′ untranslated region of the *Shaker* K^+ channel (as in Ref. 15) and a poly$(A)_{40}$ tail at the 3′ end.[20] These modifications resulted in a prominent increase in the level of current expressed.[16] The level of expression was increased even further (5- to 20-fold for the *Shaker* K^+ channel) by eliminating the 5′ and 3′ untranslated regions of the cDNA insert, adding a Kozak consensus sequence (GCCACC) just before the ATG start codon, and subcloning into pBSTA (D. Starace, personal communication 1997).

Preparation of RNA and Injection into Oocytes

Preparation of RNA

Plasmid DNA containing a cDNA insert is cut with a restriction enzyme to linearize the plasmid past the 3′ end of the coding region. It is preferable to use a restriction enzyme that leaves either a 5′ overhang or a blunt end, because a 3′ overhang can function as a primer for synthesis in the opposite direction, making antisense RNA that could interfere with translation of the RNA. After linearization, the DNA is extracted with phenol/chloroform/

[20] M. Wallner, L. Weigl, P. Meera, and I. Lotan, *FEBS Lett.* **336,** 535 (1993).

isoamyl alcohol, precipitated with ethanol, and resuspended in RNase-free water for use as a transcription template. The transcription reaction components can be purchased and assembled individually, as previously described.[9] However, it is more convenient and efficient to use a commercial kit. The mMessage mMachine transcription kits from Ambion (Austin, TX) are easy to use and produce large quantities of capped, functional RNA. Reactions are performed as described in the instructions to the kit, although the yield can often be increased by adding 1 μl of 20 m*M* Guanosine-5′-Triphosphate (included in the kit) to the reaction mix. The RNA can be stored at −70° for many months before injection, but thawing and refreezing the same sample multiple times will lead to degradation of the RNA.

Preparation of Oocytes

Oocytes are obtained from *Xenopus laevis* frogs shortly before injection, either the same day or 1 day before injection. Procedures for the maintenance of *Xenopus* frogs and the preparation and injection of oocytes have been described in detail previously.[21] Briefly, the frog should be anesthetized by immersion in either MS-222 (also called tricaine, Sigma, St. Louis, MO) at concentrations ranging from 0.15 to 0.35%, or in benzocaine (ethyl *p*-aminobenzoate, Sigma) at a concentration of 0.03%. The frog is anesthetized and immobilized within about 10–20 min, after which the oocytes are extracted with forceps through a small diagonal incision about 1 cm long in the abdomen. Remove as many oocytes as required by pulling out the lobes of the ovary, and place the oocytes in a 60-mm tissue culture dish containing calcium-free OR2. The incision is then closed with at least two stitches using 5-0 silk.

Follicle cells are removed either manually[22] or by treatment with collagenase at a concentration of 0.5 units/ml, which corresponds to 2 mg/ml for most lots of Collagenase B from Boehringer Mannheim (Indianapolis, IN). It is a good idea to test each lot of collagenase before large-scale use, since different lots vary in effectiveness and toxicity. For treatment, gently tease apart the oocytes while they are in calcium-free OR2 in a tissue culture dish. The oocytes are then incubated in a solution of collagenase in OR2 at 20°. Incubating the oocytes in a test tube on a rotator allows better mixing than using a tissue culture dish. After about 90 min, replace the solution with fresh collagenase, and remove the oocytes after another 30–90 min. Monitor the treatment carefully by removal of oocytes, and stop the treatment by rinsing the oocytes in OR2 several times. Increase calcium stepwise by incubating for 1 hr at 18°C with 1:4 ND96:OR2 and

[21] A. L. Goldin, *Methods Enzymol.* **207,** 266 (1992).
[22] R. Miledi and R. M. Woodward, *J. Physiol.* **416,** 601 (1989).

then for 30 min with 1 : 3 ND96 : OR2. Finally, transfer the oocytes to ND96 plus supplements for incubation, and manually select the oocytes to be used for injection. For most purposes, stage V oocytes are appropriate. These oocytes can be distinguished because they are 1000–1200 μm and the hemispheres are clearly delineated with the animal hemispheres being slightly lighter than those of the smaller, stage IV oocytes.[23]

Injection of Oocytes

RNA for injection is precipitated with ethanol at least once to remove excess salt, and is resuspended at an appropriate concentration in dilute Tris-HCl (1 m*M*, pH 6.5). The low pH of RNA solutions in water is slightly more toxic to oocytes than a buffered solution. The appropriate concentration depends on the source of the RNA, and can be as high as 10 mg/ml for a sample with low specific activity, such as total rat brain RNA. It may be helpful to use carrier RNA (such as yeast tRNA) with very low concentrations of sample RNA to prevent losses from nonspecific sticking to the glass injection needles, but this is not essential. Injections can be performed with a simple and inexpensive manual injection device that uses a Drummond 10-μl microdispenser (Oocyte Injector, Drummond, Broomall, PA) attached to a three-dimensional micromanipulator such as a Brinkmann MM-33 (Brinkmann, Westbury, NY).[24] Alternatively, a motor-driven microdispenser that is designed for oocyte injections can be used (Nanoject, Drummond), which is preferable for injecting large quantities of oocytes.

The oocytes are placed in ND96 in a 35-mm tissue culture dish under a dissecting microscope. To hold the oocytes in place during the injections, a polypropylene mesh (Spectra/Mesh PP, Fisher) can be glued to the bottom of the dish. Injection needles are made by using a pipette puller to draw out the glass bores that are used with the Drummond microdispenser. Standard bores work well with vertical pullers, but horizontal pullers require double-length bores available from Drummond to make two injection needles. After pulling, the needles are broken off at a tip diameter of 20–40 μm, as measured with a reticle under a dissecting microscope. To attach the needles to the dispenser, add about 5 mm of light mineral oil into the end of the needle opposite from the injection tip and insert the bore onto the dispenser plunger. The plunger will force the oil toward the tip of the needle, which should result in a complete seal of oil between the metal plunger and the injection needle tip.

Before drawing the RNA into the needle, the RNA solution is centrifuged in a microfuge tube for about 60 sec to pellet insoluble debris. This

[23] J. N. Dumont, *J. Morph.* **136,** 153 (1972).

[24] R. Contreras, H. Cheroutre, and W. Fiers, *Anal. Biochem.* **113,** 185 (1981).

reduces the likelihood of needle clogging. About 2 μl of RNA solution is placed on a tissue culture dish, and the solution is drawn into the Drummond microdispenser. It is important to watch the RNA being drawn into the needle through the microscope to ensure that the needle has not clogged. Once the needle is filled with RNA, it is positioned over each oocyte and gently lowered until it pierces the oocyte. As much as 100 nl can be injected into each oocyte, and it is possible to inject 20 oocytes with one sample in a few minutes. Using a new needle for each RNA sample prevents cross-contamination and decreases the risk of needle clogging.

Solutions

OR2,[25] Calcium-free
- 82.5 m*M* NaCl
- 2 m*M* KCl
- 1 m*M* $MgCl_2$
- 5 m*M* HEPES, pH 7.5 with NaOH

ND96[26]
- 96 m*M* NaCl
- 2 m*M* KCl
- 1.8 m*M* $CaCl_2$
- 1 m*M* $MgCl_2$
- 5 m*M* HEPES, pH 7.5 with NaOH

The supplements for ND96 include:
- 0.1 mg/ml Gentamicin
- 0.55 mg/ml Pyruvate
- 0.5 m*M* Theophylline

Analysis of Expression

Channels and receptors expressed in oocytes can be studied by a variety of electrophysiologic, biochemical, and physical techniques. Analysis by electrophysiologic recording is significantly more sensitive than either of the other approaches. Using a two-electrode whole-cell voltage clamp, it is possible to detect as few as 10^5 channel molecules in a single oocyte (less than 10^{-18} mol). On the other hand, biochemical techniques are generally reliable down to the level of 10^{-12} mol, although this depends strongly on the specific activity of the reagents being used and the number of oocytes examined. Therefore, it is necessary to express the channels at a much

[25] R. A. Wallace, D. W. Jared, J. N. Dumont, and M. W. Sega, *J. Exp. Zool.* **184,** 321 (1973).

[26] J. P. Leonard, J. Nargeot, T. P. Snutch, N. Davidson, and H. A. Lester, *J. Neurosci.* **7,** 875 (1987).

higher level for biochemical or physical analysis than is required for electrophysiologic recording. The techniques for electrophysiologic recording (Section III of this volume) and physical analysis[26a] are described in detail elsewhere, so they are not mentioned in this article. Instead, three specific techniques that can be used to confirm and analyze expression of ion channels are described: immunofluorescence, immunoprecipitation, and reconstitution into lipid bilayers.

Detection of Channels by Immunofluorescence

An effective means for confirming and localizing the expression of ion channels is to use channel-specific antibodies for immunofluorescent staining. There are advantages and disadvantages to consider when choosing among the different types of antibodies and the techniques for using them.

Monoclonal antibodies are highly specific and can be used to target short, defined sequences within a peptide, consisting of as few as six amino acids. In general, the best epitopes are located in large extracellular or cytoplasmic loops or one of the termini, so that they are positioned on aqueous, exposed surfaces. Developing monoclonal antibodies against short loops of peptide between transmembrane segments of proteins can be risky. These regions could normally form tight bends between the transmembrane segments, but lose their native conformation when attached to a fusion protein for immunization. Hence, the antibody may not recognize the peptide in its native conformation. In addition, the epitope being selected may be buried beneath structures that block access of the antibody. This problem is particularly important when examining sites close to the surface of the membrane, because steric hindrance from the membrane or other portions of the channel can interfere with antibody binding.

Polyclonal antibodies have the advantage that there are many different antibodies directed against large segments of the protein, so that the antibodies are more likely to recognize the native conformation of the protein. On the other hand, polyclonal antibodies are less specific, in that a positive signal may originate from any portion of the protein segment being targeted. Therefore, polyclonal antibodies are less desirable for studies that seek to define the boundaries of exposed portions of the channel. An additional problem with polyclonal rabbit antibodies is unique to their use for immunofluorescence in *Xenopus* oocytes. Nonimmunized rabbits have been shown to develop autoimmune antibodies against cytokeratin.[27,28] These antibodies cross-react with cytokeratin from many species, including *Xenopus*.

[26a] *Methods in Enzymology,* Vol. **294** (1998).
[27] W. E. Gordon, III, A. Bushnell, and K. Burridge, *Cell* **13,** 249 (1978).
[28] S. G. Ramaswamy and D. M. Jacobowitz, *Brain Res. Bull.* **25,** 193 (1990).

Therefore, using rabbit polyclonal antibodies for immunostaining can result in high background staining of cytoskeletal elements within oocytes. The background reactivity can sometimes be reduced by depleting the serum with acetone powder prepared from *Xenopus* oocytes. Noninjected oocytes are dried in acetone and then ground into a powder, which is added to the polyclonal antibody serum. The precipitate is spun down, and the depleted serum can be used for staining. Polyclonal antibodies from guinea pigs do not appear to have the same problems of cross-reactivity to cytokeratin (R. Steele, personal communication 1991).

An approach that avoids the necessity for developing antibodies is to insert an epitope tag that is recognized by a monoclonal antibody into the protein. Epitopes that have been used for this purpose include Flag, c-*myc*, hemagglutinin, and T-antigen. These epitopes are well characterized and are likely to maintain a recognizable conformation when inserted into the expressed protein. The ability to use commercially available monoclonal antibodies against the epitope reduces the risk and expense of developing antibodies against native peptides. However, since the structure of the protein is being altered by the epitope insertion, the functional activity of the altered channel needs to be examined by standard electrophysiologic techniques to assess adequately the degree of disruption to the structure of the channel. A functional channel with electrophysiologic properties equivalent to those of the unmodified channel is a strong indication that the overall structure has not been disturbed by insertion of the epitope.

A synthetic epitope that we have used extensively in the past is Flag.[10,12,13] The Flag epitope (DYKDDDDK) is a hydrophilic eight-amino-acid peptide.[29,30] The hydrophilic properties give Flag a high surface probability, so that Flag is less likely to be involved in hydrophobic interactions with other regions of the protein. Insertions of Flag into regions of the channel that are normally exposed to an aqueous environment are unlikely to disrupt the overall structure of the channel. However, insertions into hydrophobic transmembrane segments are likely to cause severe disruptions of the normal structure. These properties make a Flag good probe for mapping the structure and topology of channels. Another epitope that has been used successfully for this purpose is the human c-*myc* epitope (AEEQKLISEEDL).[11]

[29] T. P. Hopp, K. S. Prickett, V. L. Price, R. T. Libby, C. J. March, D. P. Cerretti, D. L. Urdal, and P. J. Conlon, *Bio/Technol.* **6,** 1204 (1988).

[30] K. S. Prickett, D. C. Amberg, and T. P. Hopp, *BioTechniques* **7,** 580 (1989).

Immunofluorescent Staining of Oocytes

Preparation of Oocytes

Prior to injection, manually remove any follicle cells that were not removed during the collagenase treatment. The presence of these cells can interfere with the staining reagents and can also increase the level of nonspecific background staining. Follicle cells are best viewed under high magnification with illumination from both above and below the oocyte. The follicle cell layer will show up as a rough bubbly layer surrounding the oocyte. Any red blood vessels on the surface belong to the follicle cell layer. The vitelline layer underneath will appear as a smooth glassy surface like plastic wrap, and should not be removed.

Staining for Membrane Detection

Detecting channels by immunofluorescence is facilitated by high-level expression. For *Shaker* potassium channels, a current amplitude of 15 μA or greater at -20 mV is sufficient. The overall approach is to fix the oocytes in formaldehyde to stabilize the structure, dehydrate the membrane with methanol to make the sample permeable to the labeling reagents, and then incubate the oocytes with labeling reagents. The oocytes are stable after fixing, and they can be stored in ND96 at room temperature before labeling. Labeling can be performed in 1.5-ml Eppendorf tubes laid flat on a bench top at room temperature. In this configuration, the oocytes can withstand multiple rinses without significant damage. Placing the tubes on a rocking platform during incubations is not recommended because friction from the walls of the tube can erode the surface of the oocyte. The steps involved in labeling with antibodies against the Flag epitope are as follows:

1. Rinse the oocytes in ND96 + supplements and discard unhealthy appearing cells. It is important to remove unhealthy cells at this point because it will be difficult to assess the health of the original oocytes from cryostat sections.
2. Fix the oocytes in 3.9% (v/v) formaldehyde in ND96 + supplements for 15 min at room temperature.
3. Dehydrate the oocytes in ice cold 100% methanol for 15 min.
4. Rinse the oocytes in ND96 + supplements 3×.
5. Incubate samples in 500 μl monoclonal M2 anti-FLAG antibody (1:500) in ND96 + 0.2% sodium azide overnight at room temperature. The M2 antibody concentration varies between batches, but a

1 : 500 dilution is appropriate for antibody at a stock concentration of 2 μg/μl.

6. The primary antibody step is then followed by labeling.

Labeling Protocol

An important consideration in selecting a fluorescent label for staining oocytes is the background fluorescence. Yolk proteins in *Xenopus* oocytes fluoresce at the same wavelength as fluorescein, which makes it difficult to visualize a positive signal. Rhodamine and Texas Red are not as bright as fluorescein, but using one of these labels avoids the problem of yolk protein fluorescence, so they are generally preferable for use with oocytes.

Stacking the fluorescent label by alternating between rhodamine–avidin and biotinylated anti-avidin can greatly amplify the signal. We have observed that epitopes far from the membrane surface can readily be detected without stacking the fluorescent label, but epitopes close to the surface are less accessible to antibody reagents and require stacking. For example, we have inserted the Flag epitope into 3 locations between transmembrane spanning segments S3 and S4 in the *Shaker* potassium channel (Fig. 3). The fluorescent signal obtained when the epitope was furthest away from either transmembrane segment (position 348) was significantly stronger than that obtained when the epitope was in either of the other two positions (333 or 356).

One problem with signal stacking is that it can significantly increase the background. A number of precautions can be taken to minimize this problem. First, incubate for long times using dilute reagents followed by extensive rinses between each step. Second, oocytes from some *Xenopus* frogs display high levels of background staining, and oocytes from these frogs should not be used. Finally, sheep serum is added to many of the reagents to prevent nonspecific binding of the secondary antibody. Sodium azide is also added to the reagents as a preservative to prevent microorganism contamination. The steps involved in immunofluorescent labeling are as follows:

1. Rinse samples 2× with ND96.
2. Incubate samples in 500 μl biotinylated sheep anti-mouse antibody (1 : 1000 (v/v)) in ND96 + 10% (v/v) sheep serum + 0.2% (w/v) sodium azide overnight at room temperature.
3. Rinse samples 3× with ND96 over 6 hr.
4. Incubate samples in 500 μl rhodamine–avidin (1 : 1000) in ND96 + 10% sheep serum + 0.2% sodium azide overnight at room temperature.

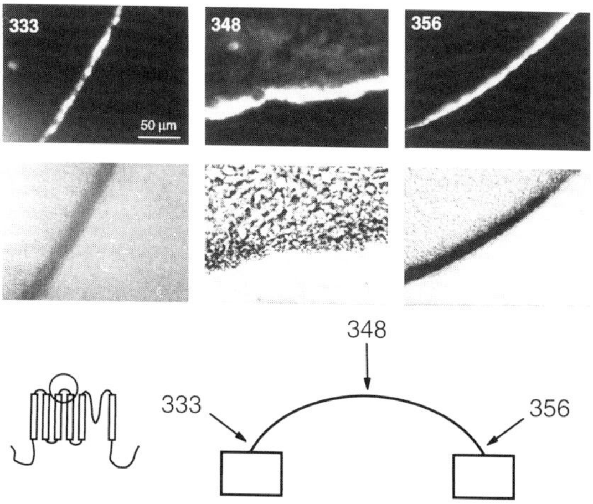

FIG. 3. Immunofluorescent labeling of the Flag epitope in a potassium channel is inhibited by steric hindrance. The Flag epitope (DYKDDDDK) was inserted into three positions (after amino acids 333, 348, and 356) in the S3–S4 loop of the *Shaker* H4 potassium channel in pKH4T. The channels were expressed in *Xenopus* oocytes, followed by immunofluorescent staining of intact oocytes to label the epitopes on the extracellular surface. The columns show the photomicrographs for each sample, with the top row containing fluorescent images and the bottom row containing phase contrast images. Although all three channels expressed comparable levels of potassium current, the intensity of the fluorescent staining at the membrane surface differed. Staining was most intense for the epitope that was furthest from the membrane (348). Adapted from Shih and Goldin.[10]

5. Rinse samples 3× with ND96 over 6 hr.
6. Incubate samples in 500 μl biotinylated anti-avidin antibody (1 : 1000 (v/v)) in ND96 + 10% sheep serum + 0.2% sodium azide overnight at room temperature.
7. Rinse samples 3× with ND96 over 6 hr.
8. Incubate samples in 500 μl rhodamine-avidin (1 : 1000) in ND96 + 10% sheep serum + 0.2% sodium azide overnight at room temperature.
9. Rinse samples 3× in ND96 over 6 hr.
10. Incubate samples in 500 μl biotinylated anti-avidin antibody (1 : 1000) in ND96 + 10% sheep serum + 0.2% sodium azide overnight at room temperature.
11. Rinse samples 3× with ND96 over 6 hr.
12. Incubate samples in 500 μl rhodamine–avidin (1 : 1000) in ND96 + 10% sheep serum + 0.2% sodium azide overnight at room temperature.
13. Rinse samples 3× with ND96 overnight.

Staining for Topology

An advantage of immunofluorescent staining is that it can be used to distinguish between epitopes on the extracellular and intracellular surfaces of the membrane. To determine localization, the staining protocol is modified by selectively targeting one face of the membrane with the primary antibody. For example, we have inserted the Flag epitope into many locations throughout the *Shaker* potassium channel, and we determined whether each of those positions was located on the extracellular or intracellular face of the membrane.[10] The results of differential staining with one intracellular (91) and one extracellular (252) epitope are shown in Fig. 4.

Extracellular epitopes are targeted by incubating whole, intact oocytes with the primary antibody, followed by rinsing, fixation, and labeling. Localizing intracellular epitopes is slightly more difficult, because it requires an appropriate method of getting the primary antibody across the membrane

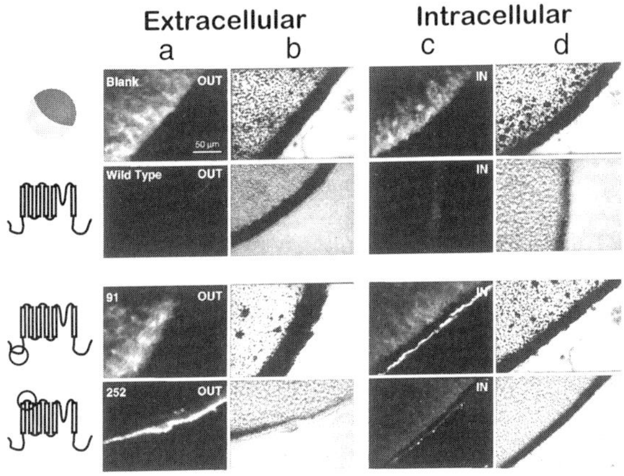

FIG. 4. Extracellular versus intracellular fluorescent staining of epitopes. The Flag epitope was inserted into either an intracellular (91, in the amino terminus) or extracellular (252, in the S1–S2 loop) region of the *Shaker* H4 potassium channel in pKH4T. Noninjected oocytes and oocytes injected with the wild-type H4 channel (which does not contain the Flag epitope) were included as negative controls. The schematic diagrams to the left of each row indicate the region of the channel containing the Flag insertion. Photomicrographs of oocytes labeled on the extracellular side of the membrane are shown in (a) fluorescence and (b) phase contrast. Photomicrographs of oocytes labeled on the intracellular side of the membrane are shown in (c) fluorescence and (d) phase contrast. Both noninjected and wild-type controls display a diffuse nonspecific background in the fluorescent images (a, c). Oocytes expressing the channel containing the amino terminal epitope (91) display a negative diffuse pattern for the extracellular stain (a) and a concentrated band of membrane staining for the intracellular stain (c). Oocytes expressing the channel containing the S1–S2 loop epitope (252) demonstrate strong fluorescence for the extracellular stain (a) and a negative diffuse pattern for the intracellular stain (c). Adapted from Shih and Goldin.[10]

barrier. Chemical methods using detergents or alkaline solutions to create holes in the membrane are not specific, because they expose both sides of the membrane to labeling.[31] They also may cause epitopes buried within the membrane to become accessible to the antibody. Injection of the antibody into the oocytes avoids these problems. Since only a tiny patch of the surface membrane is disrupted (the injection site) and the oocyte membrane reseals around the site after injection, the membrane barrier remains intact. Once antibody binding is complete, any unbound antibody must be prevented from recrossing the membrane during the subsequent labeling steps. Unbound antibody leakage is prevented by cross-linking the unbound antibody using formaldehyde fixation. The oocytes are then fixed in methanol to dehydrate the membrane and make the membrane permeable to staining reagents. The steps involved in staining to determine differential localization are as follows:

Extracellular Staining

1. In an Eppendorf tube, incubate live oocytes in 500 μl monoclonal M2 anti-FLAG antibody (1 : 500 (v/v)) in ND96 at 19°C overnight. To prevent oocyte death resulting from association with other unhealthy cells, the oocytes should be separated into small groups in individual tubes or multiwell plates. Discard oocytes that appear unhealthy.
2. Oocytes are rinsed in ND96 + supplements 2×. Discard unhealthy cells.
3. Fix the oocytes in 3.9% formaldehyde in ND96 + supplements for 15 min at room temperature.
4. Fix the oocytes in ice-cold 100% methanol for 15 min.
5. Rinse the oocytes in ND96 + supplements 3×. This step is followed by the labeling protocol discussed earlier.

Intracellular Staining

1. The monoclonal M2 anti-FLAG antibody is diluted 1:10 and 50 nl is injected into each oocyte.
2. The oocytes are incubated in dishes of ND96 overnight at 19°C. Care must be taken to avoid contaminating extracellular epitopes during this procedure. To prevent contamination, avoid injecting antibody into the dish, replace the ND96 in the injection dish between batches of injections, and do not allow the oocyte contents to ooze after injection.

[31] M. L. Jennings, *Ann. Rev. Biochem.* **58,** 999 (1989).

3. Oocytes are rinsed in ND96 + supplements 2×. Discard unhealthy cells.
4. Fix the oocytes in 3.9% formaldehyde in ND96 + supplements for 15 min at room temperature.
5. Fix the oocytes in ice-cold 100% methanol for 15 min.
6. Rinse the oocytes in ND96 + supplements 3×. This step is followed by the labeling protocol discussed earlier.

Freezing Oocytes in OCT Compound

Detecting and localizing a fluorescent signal in whole oocytes is difficult. Both epifluorescent and confocal systems focus on a plane, so that absorbance and fluorescence from nearby portions of the sample can block or obscure signals from the section being viewed. These problems can be avoided by cutting thin slices from the oocytes and mounting them onto slides, which greatly improves the resolution of the membrane at the edges of the oocyte. Cryostat sections maintain structural integrity close to the membrane, making it possible to determine localization near the surface of the oocyte. However, the yolk proteins in the interior have a grainy texture and tend to shred and fall apart. Intracellular structures, such as the golgi or transport vesicles, are not distinguishable at all in the sections, so that cryostat sections are not appropriate for studies examining protein trafficking within the cell. Confocal microscopy is more appropriate for these studies. The steps for preparing frozen sections of oocytes are as follows:

1. Transfer oocyte samples to a plastic cryomold and remove as much fluid as possible using a Pipetman. The positioning of an oocyte in a cryomold is shown schematically in Fig. 5A.
2. Cover the oocytes with a small drop of OCT (optimal cutting temperature) compound. Use a Pipetman tip to gently swirl the OCT compound around the oocytes to draw away residual fluid near the oocyte surface. This step prevents ice crystals from forming near the surface of the sample during freezing, which can cause the sections of the oocyte to fall away from the OCT slice during sectioning.
3. Fill the cryomold to the brim with OCT compound. Use the Pipetman tip to gently arrange the oocytes a few millimeters off the bottom of the cryomold.
4. Seal the top of the cryomold with tape, being careful to avoid sealing in any bubbles. It is important to get a tight, bubble-free seal, because any ethanol leaking in to the cryomold will turn the OCT into a soft jelly-like substance that makes cutting difficult.
5. Label the sample on the tape and trim the edges with scissors.

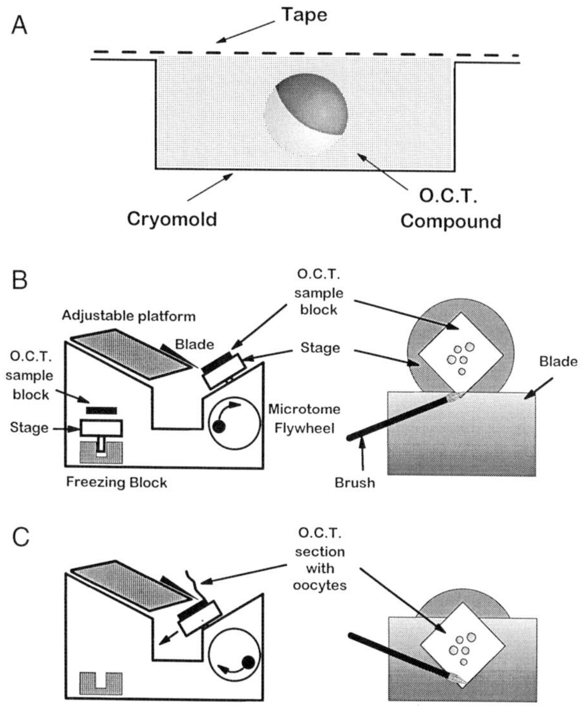

FIG. 5. Schematic diagram of a cryomold and cryostat for sectioning of oocytes. (A) Diagram of the correct positioning of an oocyte in the cryomold. (B) Diagram of the OCT sample block mounted on the cryostat stage before cutting a slice. (C) Diagram of the OCT sample block and stage after a slice has been cut.

6. Place the sample into a dry ice–ethanol bath to freeze into blocks.
7. Store the samples at −20°.

Sectioning Oocytes on Cryostat

The temperature of the OCT block is critical for obtaining good sections. The samples should be equilibrated at −20° for at least 10 hr before cutting, and care must be taken to maintain the blocks at −20° before and during cutting. Blocks colder than −22° are too brittle, causing the sections to shred during cutting. Blocks warmer than −18° are too soft to cut properly. The procedures for cutting sections in a cryostat are shown schematically in Figs. 5B and 5C, and are as follows:

1. Set the cryostat cabinet temperature to −20° and the cutting depth to 14 μm.

2. Remove the OCT block from its cryomold. Cut off any portions of the block that are soft and jelly-like from ethanol contamination. Place the stage onto the freezing block and adhere the OCT block to the stage with a drop of OCT. Mount the block flat to the surface of the stage.
3. Once the OCT has set, mount the stage on the microtome.
4. Lower the adjustable platform to bring the blade close to the sample.
5. Adjust the angle of the stage so that the face of the OCT block will cut evenly with the edge of the blade.
6. Begin rotating the flywheel, which will advance the microtome in stages toward the blade. Continue cutting into the sample until a depth is reached at which the oocytes are displayed in sufficient detail.
7. Brush the leading edge of the OCT section onto the face of the blade. Continue brushing out the leading edge and stretching out the slice as it is being cut. Take care not to contact directly the area of the slice containing samples.
8. With the leading edge being held down with a brush, and the lagging edge still attached to the OCT block, lower a BSA (bovine serum albumin) coated slide over the section. When the slide is brought close enough, the section will adhere to the surface of the slide.
9. When enough slides have been cut and mounted, the OCT sample block can be preserved by pressing it back into its mold and applying a drop of OCT compound to the cut surface to protect it from freezer burn.
10. Allow the slides to air dry overnight on a cool dry benchtop.
11. Fix the slides with a drop of acetone (in a hood).
12. Rehydrate the slides in phosphate-buffered saline (PBS) for 5 min.
13. Mount coverslips to the slides with a drop of Vectashield.
14. Seal the edges of the coverslips with nail polish and dry.
15. Store the slides at 4° overnight before viewing.

Coating Slides with BSA

Coating the slides with BSA will prevent the sections from floating off the surface of the slide. The procedure is easily performed using a removable rack to dip the slides into a staining dish.

1. Dip the slides in 70% ethanol for 10 min. Tap the slide on a paper towel several times to remove as much ethanol as possible.
2. Dip the slides in 0.2% BSA for 10 min. Set the rack on top of the staining dish and allow the solution to drip back into the dish for a couple minutes.

3. Repeat dips into 0.2% BSA 2× more (in separate staining dishes).
4. Dry the slides for 1 hr.
5. Store the slides in a cool dry place at room temperature.

Reagents

Anti-FLAG M2 monoclonal antibody, 200 μg (Eastman Kodak Company, New Haven, CT) (*Note:* Add PBS to adjust the concentration to 2 μg/μl.)
Biotinylated sheep anti-mouse antibody, 1 ml (The Binding Site, San Diego, CA)
Sheep serum 50 ml (The Binding Site)
Rhodamine–avidin DCS, 1 mg, Vector Laboratories Inc., Burlingame, CA (resuspend in 1 ml H_2O)
Biotinylated anti-avidin D, 0.5 mg, Vector Laboratories Inc. (resuspend in 1 ml H_2O)
Vectashield, 10 ml, Vector Laboratories Inc.
Tissue-Tek OCT Compound 4583 (Miles Laboratories, Elkhart, IN)
Tissue-Tek Cryomold Intermediate 4566 100/box

Preparing Oocyte Samples for Confocal Microscopy

Confocal microscopy can be used with intact oocytes to distinguish intracellular organelles. However, interference from oocyte pigments and yolk proteins can obscure the fluorescent signal. Interference from pigments can be minimized by using oocytes obtained from albino frogs. Interference from yolk proteins can be minimized by replacing the water in the oocytes with BA/BB (benzyl alcohol/benzyl benzoate), which matches the refractive index of the yolk proteins, thereby making the proteins invisible.

Fix and stain oocytes as previously described for immunofluorescence. After these steps, it is important to dehydrate the samples completely before using the BA/BB clearing agent, as any residual water in the sample will cause the oocytes to appear cloudy. The steps involved in preparing oocytes for confocal microscopy are as follows:

1. Dehydrate the oocytes completely by a series of ethanol rinses.
   a. Incubate 10 min in 20% ethanol.
   b. Incubate 10 min in 40% ethanol.
   c. Incubate 10 min in 60% ethanol.
   d. Incubate 10 min in 80% ethanol.
   e. Incubate 10 min in 100% ethanol (repeat 3×).
2. Resuspend the oocytes in BA/BB, which is benzyl alcohol/benzyl benzoate mixed in a 1:2 ratio. The solution matches the refractive

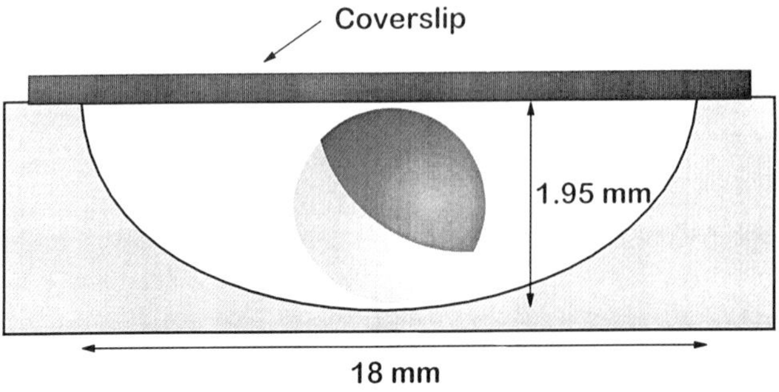

FIG. 6. Schematic diagram of an oocyte in a depression slide. Depression slides, which are also known as hanging drop slides (Fisher, Tustin, CA), have ground and polished concavities 18 mm in diameter. The depth of the concavity should be 1.50–1.95 mm deep to accommodate an oocyte. Oocytes that have been cleared with BA/BB clearing agent are mounted in the concavity with the BA/BB clearing agent. A glass coverslip is used to seal the sample. The edges of the coverslip are sealed with clear nail polish to prevent the BA/BB from leaking.

index of the oocyte yolk proteins. The oocytes should look like clear glass marbles with only the pigmented animal pole visible.

3. Mount the samples in depression slides with BA/BB, as shown in Fig. 6.
4. Cover the oocytes with glass coverslips.
5. Seal with nail polish to prevent BA/BB from leaking. The BA/BB clearing agent will make the oocytes turgid. Oocytes that are too large to fit inside the depression slide will push the coverslip up and prevent a proper seal. Avoid this problem by using slides with deep wells, not mounting too many oocytes per slide, and using smaller oocytes.

Detection of Channel Protein by Immunoprecipitation

Immunoprecipitation makes it possible to isolate exogenously expressed proteins. The procedure can be used to quantify the amount of protein synthesized, examine posttranslational protein processing such as glycosylation and phosphorylation, or detect protein–protein interactions. One problem with immunoprecipitation from oocytes is that solubilization results in isolation of both cytoplasmic and membrane proteins. Functional ion channels and receptors are all associated with the membrane, but there is frequently a large intracellular pool of molecules that either has not or cannot be inserted into the membrane.[32] Electrophysiologic recording ex-

[32] J. W. Schmidt, S. Rossie, and W. A. Catterall, *Proc. Natl. Acad. Sci. U.S.A.* **82,** 4847 (1985).

amines only the molecules that have been inserted into the membrane in a functional configuration, whereas immunoprecipitation of total oocyte proteins examines both membrane and cytoplasmic proteins. Since the properties of the cytoplasmic proteins may be different from the functional proteins inserted into the membrane, it may not be possible to correlate the results from the two types of analysis. This problem can be alleviated by isolating the oocyte membrane fraction before immunoprecipitation.

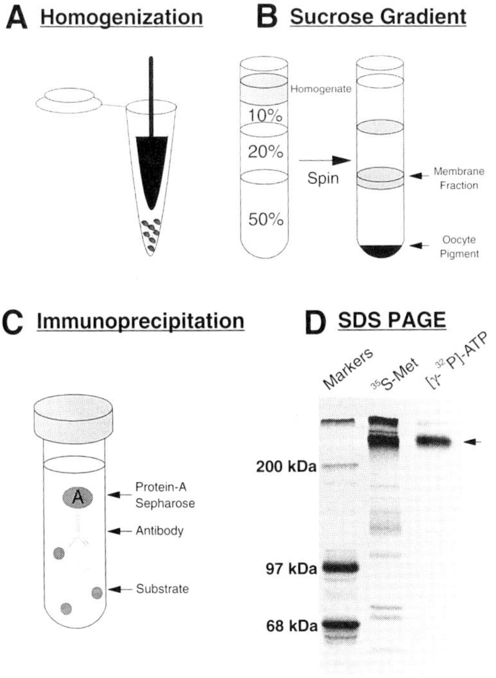

FIG. 7. Immunoprecipitation of membrane protein from oocytes. (A) Homogenization: Oocytes are homogenized using a microfuge tube dounce. (B) Sucrose Gradient: The membrane fraction is isolated on a discontinuous sucrose gradient containing steps of 10, 20, and 50% sucrose, which is centrifuged at 170,000*g* in a Beckman SW41 rotor at 4°. The oocyte homogenate in 10% sucrose solution is layered on top of the 10% sucrose layer. After centrifugation, the membrane fraction is located at the interface of the 20 and 50% layers. (C) Immunoprecipitation: The washed membrane fraction is combined with antibody and protein A-Sepharose beads for immunoprecipitation of the channel proteins. Immune complexes are isolated by spinning down the protein A-Sepharose beads and boiling in SDS–PAGE sample loading buffer to release the complexes. (D) SDS–PAGE: Sodium channel proteins expressed in *Xenopus* oocytes and labeled with either [^{35}S]methionine or [γ-^{32}P]ATP were immunoprecipitated and analyzed on a 7.5% SDS polyacrylamide gel. An autoradiograph of the dried gel is shown. The bands representing full-length sodium channels α subunit protein are indicated by the arrow.

The procedures for the isolation of sodium channel protein from membrane fractions are diagrammed in Fig. 7 and detailed next:

1. Inject each of 50 oocytes with 25 ng of *in vitro* transcribed RNA and 500 nCi [^{35}S]methionine. For maximum specific activity, the RNA pellets are resuspended directly in 10 μCi/μl [^{35}S]methionine. Alternatively, the protein can be labeled with ^{32}P by injecting 500 nCi [γ-^{32}P]ATP 20 hr prior to homogenization.

2. Incubate the oocytes for 48 hr in dishes placed in a covered tray lined with activated charcoal to prevent contamination from volatile [^{35}S]methionine. Change the incubation media 1–2 times per day, removing unhealthy oocytes.

For the following steps, work rapidly and keep all solutions at 4°:

3. Suspend the oocytes in 0.5 ml of ice-cold, buffered 10% sucrose solution.

10% Sucrose Solution

10% Sucrose
150 m*M* NaCl
5 m*M* KCl
20 m*M* HEPES, pH 7.5

Protease Inhibitors

40 μg/ml Bestatin
50 μg/ml Antipain
0.5 μg/ml Leupeptin
0.7 μg/ml Pepstatin
2 μg/ml Aprotinin
25 μg/ml 4-(Amidinophenyl)methanesulfonyl fluoride
0.1 m*M* EDTA
0.5 m*M* PMSF (phenylmethylsulfonyl fluoride, made fresh in ethanol just before use)
0.5 m*M* Dithiothreitol (DTT)

Phosphatase Inhibitors for ^{32}P Label

50 m*M* KF
5 m*M* β-Glycerophosphate

4. Homogenize the oocytes in an Eppendorf tube with a small, plastic Dounce homogenizer (Kontes, Vineland, NJ), as shown in Fig. 7A.

5. Overlay the homogenate on a 4° 10–20–50% discontinuous sucrose gradient in the same buffer without protease inhibitors. The volume is too

large to use protease inhibitors because of their cost, although phosphatase inhibitors can be used because they are significantly less expensive. A diagram of the gradient is shown in Fig. 7B.

Sucrose Gradient (bottom to top)

4.5 ml 50%
5.0 ml 20%
1.0 ml 10%
10.5 ml Total

6. Spin in a swinging bucket rotor (SW-41) at 37,000 rpm for 30 min at 4°.

7. Collect the "membrane fraction" from the 20–50% interface (a fluffy, gray band as shown in Fig. 7B) using a Pasteur pipette. This band contains the membrane fraction and secretory proteins.

8. Dilute at least 1:3 with cold water, to a final volume of 8 ml.

9. Pellet by centrifugation in an SA-600 rotor at 11,000 rpm for 10 min at 4°.

10. Solubilize the membrane proteins in 2.5 ml of solubilization buffer. Use a 7-ml glass Dounce homogenizer on ice with a type A pestle.

Solubilization Buffer

75 m*M* KCl
75 m*M* NaCl
50 m*M* Sodium phosphate, pH 7.2
2 mg/ml Soybean lipids (Avanti, Alabaster, AL)
1% Triton X-100
(Filter sterilize)
+ Protease (+ phosphatase) inhibitors, which should be added just before use

11. Eliminate insoluble material by centrifugation at 8000 rpm for 30 min at 4°.

12. Immunoprecipitation (Fig. 7C):

a. Add antibody directly to the cleared supernatant from step 11 (3 μg M2 anti-Flag monoclonal antibody). The Flag epitope had been inserted at the amino terminus of the rat brain IIA sodium channel, just after the start codon.

b. Incubate on a rotating wheel at 4° for 4 hr.

c. Add 150 μl hydrated and pH neutralized protein A-sepharose beads (Pharmacia LKB, Alameda, CA). To prepare a solution containing 40% beads, add 115 mg beads to 1 ml of 1 *M* Tris-HCl, pH 7. Also include 10 μl phenol red to monitor the pH of the solution.

d. Incubate at 4° for 2 hr.

e. Spin down the beads in a microcentrifuge.

f. Wash the beads 2× with solubilization buffer (without protease inhibitors or phosphatase inhibitors)

g. Release the immune complexes from the beads by adding SDS–PAGE sample loading buffer and boil for 3 min. Repeat 1× to ensure complete recovery of the sample, which can be stored frozen at −20° until use.

SDS–PAGE Sample Loading Buffer

20 m*M* Tris-HCl, pH 7.0
4% SDS
5 m*M* EDTA
20% Glycerol
10% 2-Mercaptoethanol
0.5 mg/ml Bromphenol blue

An autoradiograph of a gel showing rat brain sodium channels labeled with either [^{35}S]methionine or [γ-^{32}P]ATP and immunoprecipitated from the oocyte membrane fraction is shown in Fig. 7D. The band representing the α subunit of the sodium channel is indicated by the arrow, demonstrating that this procedure is sensitive enough to detect sodium channel proteins using either labeling procedure.

Isolation of Oocytes Membranes for Channel Reconstitution in Lipid Bilayers

Although ion channels can be examined by a variety of electrophysiologic techniques in *Xenopus* oocytes, there are endogenous channels in oocytes that can interfere with recordings from exogenous channels. For this reason, it is sometimes advantageous to record from channels in a lipid bilayer, which represents a more completely controlled environment. This approach is particularly useful for studying the single-channel properties of expressed ion channels. Usually, single channels are examined by patch clamping of oocytes that have been injected with a small amount of RNA, resulting in a low level of whole-cell current. However, under these conditions it is sometimes difficult to distinguish the currents through the channel of interest from those through endogenous channels. An example of this problem has recently been discussed by Krause *et al.*[33] These authors suggest that single-channel studies of expressed large conductance voltage- and calcium-sensitive potassium (MaxiK) channels in oocytes might be compromised by a scarce, endogenous MaxiK channel in the oocytes.

[33] J. D. Krause, C. D. Foster, and P. H. Reinhart, *Neuropharmacology* **35,** 1017 (1996).

There are at least two alternative experimental approaches to minimize this problem. Both approaches involve injecting larger quantities of *in vitro* transcribed RNA to decrease the ratio of endogenous/exogenous channels, and then adjusting the recording conditions to compensate for the higher levels of current. The first approach is to use extremely small pipette tips for patch clamping (tip resistance of 10–20 MΩ). However, patch excision may be difficult with high-resistance electrodes. The second procedure is to incorporate membrane vesicles in lipid bilayers. Reconstitution of ion channels in lipid bilayers has the following advantages. First, the internal and external faces of the channel are readily accessible to different agents and solutions. Second, the experiments can be carried out for long periods of time, provided the channel does not run down, because the membranes can last for hours. Third, the intrinsic channel properties are maintained.[19] For a more detailed discussion on reconstitution in bilayers, see Ref. 33a.

We have developed a method for incorporating channels synthesized in *Xenopus* oocytes into lipid bilayers. The approach is based on the procedures of Colman,[34] and it has been successfully used to characterize the single-channel properties of cloned MaxiK channels (*dslo* and *hslo*).[19,35]

Solutions

High K Buffer

600 m*M* KCl
5 m*M* PIPES, pH 6.8

Supplemented with the Following Protease Inhibitors

100 μ*M* Phenylmethylsulfonyl fluoride (PMSF)
1 μ*M* Pepstatin
1 μg/ml Aprotinin
1 μg/ml Leupeptin
1 μ*M* *p*-Aminobenzamidine

We have increased the KCl concentration of this solution from 400 m*M*[35] to 600 m*M* to improve membrane stability (see below).

Buffer A

300 m*M* Sucrose
100 m*M* KCl
5 m*M* MOPS, pH 6.8

[33a] I. Favre, Y.-M. Sun, and E. Moczydlowski, *Methods Enzymol.* **294,** [15], 1998 (this volume).
[34] A. Colman, *in* "Transcription and Translation—A Practical Approach" (B. D. Hames and S. J. Higgins, eds.), p. 271. IRL Press, Oxford, 1984.
[35] G. Perez, A. Lagrutta, J. P. Adelman, and L. Toro, *Biophys. J.* **66,** 1022 (1994).

The entire procedure is performed at 4°, and is as follows:

1. Oocytes expressing exogenous channels (30–40 oocytes) are rinsed using high K buffer supplemented with protease inhibitors and 300 m*M* sucrose, transferred to a 1-ml ground glass tissue grinder (Kontes Duall, Fisher Scientific, Tustin, CA), and manually homogenized with the same solution ($\approx$10 μl/oocyte) for about 5 min.
2. The homogenate (approximately 200 μl) is layered onto a discontinuous sucrose gradient (0.75 ml of each, 50 and 20% (w/v), in high K buffer), and centrifuged in a swinging bucket rotor (Beckman TLS55 or Sorvall RP55-S) at ~100,000*g* for 30 min.
3. The top lipid (yellowish) layer is eliminated by aspiration with vacuum, and the 20 : 50% interface (visible band) is collected and diluted 3 times with high K buffer without protease inhibitors.
4. The membranes are pelleted at ~90,000*g* for 30 min in a fixed-angle rotor (Beckman TLA100.3 or Sorvall S100-AT5), and resuspended in a final volume of 8–10 μl with a micropipette using buffer A. The membranes are divided into 2-μl aliquots, frozen in liquid nitrogen, and stored at −70°.

Using this method, proper membrane fusion and reasonably stable recordings (>1 hr) can be obtained. An important step in this procedure is to dilute the membrane-enriched sucrose interface with high K buffer, otherwise reconstitution of channel activity is poor and the bilayers are unstable. This step may enhance the separation of fatty acids and other components that can destabilize vesicles from the membrane preparation and/or destabilize bilayers.

Reconstitution into Lipid Bilayers

The membrane vesicles from oocytes expressing K_{Ca} channels are incorporated into lipid bilayers composed of phosphatidylethanolamine : phosphatidylcholine : phosphatidylserine in a ratio of 5 : 3 : 2 at 25 mg/ml in *n*-decane.[36] In most cases, multiple channels are fused into the bilayer, so the membrane fraction was diluted 2–5 times in buffer A to incorporate single channels. Currents through endogenous oocyte chloride channels can be eliminated by using methane sulfonate instead of chloride in the solutions. The membrane preparation is applied with a glass rod to the preformed bilayer from the *cis* chamber (the voltage-controlled side),

[36] L. Toro, L. Vaca, and E. Stefani, *Am. J. Physiol.* **260,** H1779 (1991).

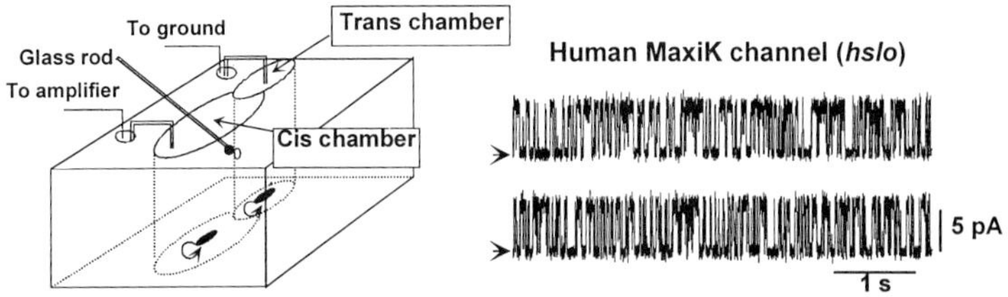

FIG. 8. Recording channel activity from lipid bilayers after reconstitution of channels expressed in *Xenopus* oocyte membranes. Both chambers are filled with 5 m*M* KCl, 10 m*M* MOPS, 100 μM $CaCl_2$, pH 7.4, before and during application of the oocyte membranes. To reconstitute channel activity, a glass rod is first inserted into the membrane preparation and then applied or "painted" on top of an existing bilayer. After "painting" the membrane sample, 3 *M* KCl is applied to the "*cis*" chamber near the bilayer to induce channel incorporation and to achieve a K gradient (e.g., 250 m*M* KCl *cis*, 5 m*M* KCl *trans*). After channel incorporation has occurred, KCl or NaCl is added to the *trans* chamber to make that chamber isosmotic with the "*cis*" chamber, which maintains the stability of the bilayer. The recordings were obtained from a single human MaxiK channel α subunit (*hslo*). The arrows indicate the closed state of the channel.

while the *trans* chamber is connected to ground (Fig. 8). Application of the sample to the preformed bilayers results in channel incorporation, and requires very small volumes (≤ 1 μl of membrane preparation per experimental session), maximizing use of the micropreparation. Single-channel currents recorded from a bilayer containing human MaxiK (*hslo*) channels are shown in Fig. 8.

Conclusion

In summary, *Xenopus* oocytes are widely used for the study of ion channels and receptors. This expression system has the advantage that channels can be expressed and analyzed by electrophysiologic, biochemical, and physical approaches. Most of the electrophysiologic recording techniques described in Section III of this volume on electrophysiology can be used for analysis of ion channels in oocytes. In addition, the system can be used effectively as an assay for the functional cloning of channels that have only been identified by their electrophysiologic properties. Finally, expression of ion channels in oocytes is an excellent system for correlating structure with function using a combination of molecular biological, biochemical, and electrophysiologic techniques.

Acknowledgments

We thank Drs. Lucia Santacruz-Toloza and Diane Papazian for the initial immunoprecipitation protocol, Dr. Jim Negrev for help with the confocal microscopy procedures, and Z. Jiang for constructing pBSTAII. Work in the authors' laboratories is supported by grants NS26729 (A.L.G.), HL54970 (L.T.) and HL47382 (L.T.) from the National Institutes of Health. A.L.G. and L.T. are Established Investigators of the American Heart Association.

[30] Unidirectional Fluxes through Ion Channels Expressed in *Xenopus* Oocytes

By PER STAMPE and TED BEGENISICH

Introduction

The measurement of unidirectional fluxes of substances across the plasma membrane of cells is a valuable tool for understanding many membrane transport processes. The movement of electrically charged substances can be followed with electrical current measurements, but the use of radiolabeled tracers is one of very few ways to assess the transport of uncharged molecules. Even though electrical current can monitor the net flux of ions, additional valuable information is contained in the unidirectional ion influx and efflux. In particular, the ratio of unidirectional fluxes through the pores of ion channels provides rather direct information on the number of ions that may simultaneously occupy the pore.[1]

The purpose of this article is to describe some methods useful for measuring the unidirectional influx and efflux of K^+ ions through channels expressed in *Xenopus* oocytes. Specifically, we have applied these techniques to the question of the number of K^+ ions that simultaneously occupy the pore of *Shaker*[2] and IRK1 K^+ channels.[3] Even though designed for the determination of unidirectional ion fluxes through ion channels, many of the methods described here could be applied to the measurement of the unidirectional fluxes of other substances.

General Methods

The methods described here were developed for voltage-gated ion channels and, for this purpose, require an ion expression system and an associ-

[1] T. Begenisich, *Methods Enzymol.* **293,** [22], 1998 (this volume).
[2] P. Stampe and T. Begenisich, *J. Gen. Physiol.* **107,** 449 (1996).
[3] P. Stampe and T. Begenisich, In Preparation.

0076-6879/98 $25.00

ated voltage-clamp system and these are described briefly. Measurement of unidirectional influx and efflux requires such different methods that these are individually described.

Oocyte Expression System

Measuring unidirectional fluxes across a cell membrane is typically accomplished by following the movement of a radiolabeled tracer. Measurement accuracy is improved if a large amount of tracer movement can be obtained. Thus, large cells with a high permeability to the substance of interest greatly facilitate these measurements. *Xenopus* oocytes are large (1000–1300 μm in diameter) cells and the techniques for expression of high levels of transport proteins in these cells are well developed[4] (Chap. 28; Shih *et al.*).

Potassium Channels

The methods described here have been employed with two types of K^+ channels: *Shaker* K^+ channels, one member of the voltage-gated family, and IRK1 channels, a member of the inward rectifier class. In either case, mRNA is made by standard methods[5] and injected into stage V or VI oocytes. Successful measurement of unidirectional fluxes may be accomplished with channel expression levels sufficient to produce 1 μA or so of current.

Voltage-Clamp Methods

Conventional two-microelectrode voltage-clamp techniques maximize the amount of tracer flux because the entire oocyte is under voltage control and the entire surface area is available for transport. However, with reasonably robust channel expression, the cut-open oocyte voltage-clamp/perfusion system could also be exploited.

Figure 1 illustrates a simplified schematic of a two-electrode clamp design suitable for flux measurements. Membrane potential (V_m) is measured (by amplifier 1) as the voltage difference between a standard intracellular glass microelectrode and an external reference electrode. The reference electrode could be a Ag|AgCl wire or pellet or a glass (or polyethylene tubing) pipette filled with, for example, 3 *M* KCl. Amplifier 2 compares the membrane potential with a command signal (the voltage-clamp steps) and the amplified difference is connected to the current-passing microelec-

[4] H. Soreq and S. Seidman, *Methods Enzymol.* **207,** 225 (1992).
[5] R. Swanson and K. Folander, *Methods. Enzymol.* **207,** 310 (1992).

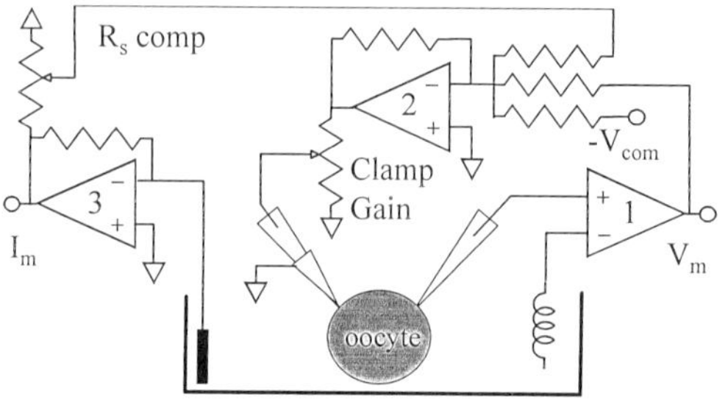

FIG. 1. Voltage clamp schematic showing simplified clamp electronics. See text for details.

trode. The operation of amplifiers 1 and 2 serves to clamp V_m to the desired voltage. Membrane current is measured with amplifier 3 in a standard current–voltage converter configuration. We use a platinized platinum foil electrode to connect this amplifier to the bath solution.

As described earlier, it is desirable to express large numbers of ion channels so as to detect unidirectional tracer fluxes more easily. This requirement places some design constraints on the voltage-clamp system. Large ion channel currents require the passing of large currents, so low-resistance electrodes should be used. We use electrodes of about 0.1–0.2 MΩ when filled with 3 *M* KCl. These have an approximate 2-μm tip diameter.

Large currents require large clamp gain in order to provide a large voltage for the current-passing electrode. Unfortunately, large voltage changes on this electrode can couple (through stray electrical capacitance) to the voltage-measuring electrode (a form of positive feedback) and cause the clamp circuit to oscillate. This capacitative coupling can be reduced by placing a grounded shield between the two electrodes. We do this by coating the current-passing electrode with a conducting, silver paint (GC Electronics, Rockford, IL) and connecting this to ground. The part of the electrode immersed in the bath solution is insulated with varnish (GC Electronics).

Large currents may also cause voltage errors through uncompensated resistance between the voltage-measuring electrodes. Such errors are minimized by using separate bath current-measuring and voltage reference electrodes (as in Fig. 1) and by placing the reference voltage electrode close to the oocyte. The magnitude of the series resistance may be on the order of a few kilohms, so "normal" currents of a few microamperes would produce errors of only a few millivolts. However, with high channel expression, currents of tens of microamperes could be generated and so series resistance errors of tens of millivolts could arise. The circuit diagram of Fig. 1 allows for compensation of series resistance errors.

Isotopes and Radioactive Counting

Because our interests are in K^+ channel permeation properties, we developed these methods to use ^{42}K as a tracer of K^+ ions. Similar techniques can be used with other radioactive tracers. ^{42}K decays with a rather short half-life of about 12 hr, which limits some types of procedures (see next section). The decay process is associated with the emission of two β particles (3.52 and 1.97 MeV) and two γ particles (1.52 and 0.31 MeV) so either scintillation counting or gamma counting can be used to detect this radionuclide. We use a model Gamma 4000 (Beckman Coulter, Inc., Fullerton, CA) gamma counter. The high energy of the ^{42}K gamma rays presents a possible problem of cross-contamination caused by samples of very high specific activity. Even in a counter with extra shielding and a 2-in. crystal (as in our Beckman model), extra spacing is needed around the very "hot" samples. A clinical type counter is therefore not suitable for use with high specific activity ^{42}K experiments.

Influx Methods

The unidirectional influx of K^+ into a *Xenopus* oocyte can be determined by counting the amount of ^{42}K that enters the egg from solution of known specific activity in a specified amount of time. To conserve isotope a low-volume (75-μl) chamber is used. The oocyte is placed in the chamber in a solution without radiotracer and voltage clamped. After allowing the egg a few minutes to recover from the electrode impalements the solution containing the ^{42}K isotope is added; three samples (5 μl each) are subsequently taken out for calculation of the specific activity of the bath solution. The actual experiment is then performed, which, for our purposes, typically consists of the repetitive application of voltage-clamp pulses to open K^+ channels and so allow K^+ movement. The chamber is then washed extensively to remove the radiotracer solution, a 5-μl sample is taken out as a control for the washing, and, finally, the egg is unclamped and removed for counting.

Specific Activity

Usable values for the specific activity of the incubating solution in influx experiments were in the range of 300 to 2500 cpm/mol $\times$ 10^9. The specific activity of the isotope was typically orders of magnitude higher on delivery, so a "true trace" amount could be used for experiments. The short half-life of ^{42}K makes it difficult to maintain the highest specific activities after 2 or 3 days but this is somewhat offset by the increased channel expression at later times.

As noted earlier, the unidirectional influx is obtained from the number of radioactive counts in the egg at the end of the experiment, converted to moles from knowledge of the solution specific activity. To obtain measurable counts within a reasonable time frame from K^+ flux into the oocyte, a high specific activity is needed in the incubating solution. Consequently, even if only a small amount of this solution is transferred with the egg, it will contaminate the influx computation. Therefore, an extensive wash of the experimental chamber is required prior to transferring the egg. The small volume of our experimental chamber allows a rather complete wash with about 5 ml and a chamber sample was routinely obtained to check for the completeness of the wash. Even with a well-designed chamber constructed of reasonably impervious material the chamber may become contaminated and may occasionally require more extensive washings between experiments.

Nonchannel Flux Correction

In addition to the usual subtraction of radioactive background counts it is important to correct the number of counts in the egg for ^{42}K that entered through other nonchannel pathways. The results of a series of control experiments[2] indicate that even after extensive washing of the experimental chamber some residual radioactivity may adhere to the oocyte. A typical value was 75 cpm in an egg incubated in a solution with a specific activity of 500 cpm/mol $\times 10^9$. Robust channel expression often allows a sufficiently high level of K^+ influx such that this correction may be insignificant.

Reflux

The exchange of isotope between the two compartments, the egg and the incubating solution, is an exponential function because as the egg gets loaded with ^{42}K some of that activity will leave again as efflux. Several measurements are needed in order to determine the time constant for this exponential function, but the oocyte can only be counted once in this type of experiment, so we estimated the flux from the initial slope of the curve. We arbitrarily chose a limit for the influx to be less than 10% (about 10 nmol) of the total potassium content of the egg for a valid approximation.

Tail Currents

A possible additional source of error specific for voltage-gated K^+ channels such as *Shaker* is the entry of isotope into the oocyte during the "tail" currents that flow at the end of a pulse before the channels close. In an experiment conducted in 25 mM KCl the Nernst equilibrium potential for K^+ is close to -40 mV and the holding potential was -80 mV, so with such a large driving force the net current of the tails is a very close approximation to the influx and can be used as a correction if necessary. Since the tail currents are of a rather short duration (usually less than 10 ms), the use of long test pulses (300 ms) can reduce the magnitude of this error to reasonable levels.[2]

Efflux Methods

Chamber

The measurement of tracer efflux requires rather different methods than those used for influx. The oocyte must first be loaded with the radioactive tracer by soaking in a solution containing the tracer. We have found that soaking in a ^{42}K solution of specific activity near 10^{12} cpm/mol K^+ for 6–12 hr is suitable for oocytes expressing K^+ channels. The loaded oocyte is extensively washed and placed in a 350-μl chamber constructed to allow a rather rapid flow of the external solution: typically at 1 ml/min driven by peristaltic pumps (LKB model 2132, LKB-Produckter AB, Bromma, Sweden) as diagrammed schematically in Fig. 2.

The inflow pump is set exactly at 1 ml/min and the outflow pump at a somewhat higher rate and it draws from the upper surface of the chamber. The outflow tubing is directed to a fraction collector (Gilson FC-80K Micro Fractionator, Gilson Medical Electronics, Inc., Middleton, WI) set for 1-min samples.

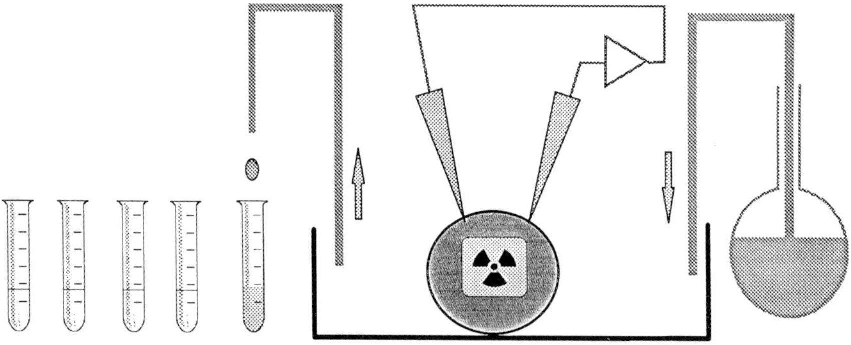

FIG. 2. Schematic diagram for efflux measurements. One channel of a peristaltic pump pulls the bath solution from a reservoir (illustrated by flask on right-hand side). Another channel of the pump draws the fluid from the top of the chamber and supplies the flow to a fraction collector (represented by the test tubes on the left-hand side).

In a typical experiment the loaded oocyte is placed in the chamber, the voltage-clamp electrodes are inserted, and the voltage clamp set to maintain a constant holding potential. The pumps and fraction collector are started and the chamber effluent collected in the tubes of the fraction collector. The time course of the tracer efflux can then be determined by counting the radioactive decay in each tube. An example of such an experiment is illustrated in Fig. 3.

The ordinate in Fig. 3 is the radioactive counts (expressed as counts per minute, cpm) in each tube. Since we typically collect 1-min samples, the abscissa is easily converted to time (in minutes). In the experiment illustrated in Fig. 3 the oocyte was injected with a modified form of the *Shaker* voltage-gated K^+ channel. At the beginning of the experiment the membrane potential was maintained at −80 mV resulting in a low, baseline efflux. During the period noted on the figure, repetitive 300-ms voltage-clamp pulses to −5 mV were applied at a frequency of 1 Hz.

The voltage-induced channel opening allowed the large K^+ efflux seen in the figure. The mixing of the effluxed ^{42}K in the chamber makes it impractical to synchronize precisely channel opening with the very beginning of a collection tube. As a consequence, the amount of radioactivity in the first tube after beginning stimulation (see Fig. 3) does not reflect an entire minute of efflux through the channels opened by the pulses. The next three tubes, however, have an approximately equal amount of radioactive

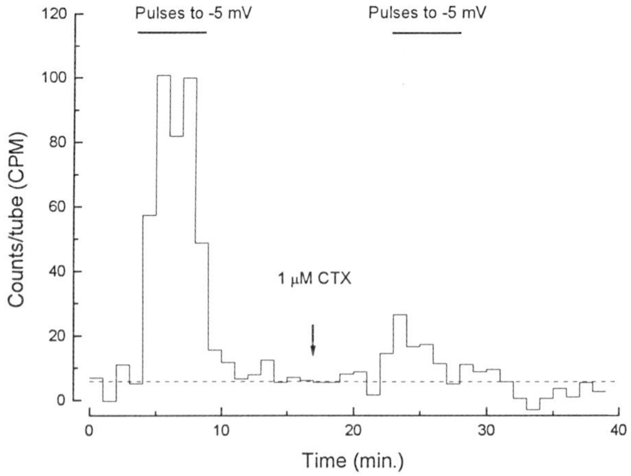

FIG. 3. Example of measurement of efflux. Ordinate: Counts per minute from each tube of the fraction collector corrected for radioactive background. Abscissa: Tube number but is equivalent to time since each tube was a 1-min sample. Repetitive voltage-clamp pulses (at 1 Hz) to −5 mV applied as indicated. Application of 1 μ*M* CTX applied where indicated. See text for additional details.

counts. Our approach to this issue is to collect all ion currents elicited by the pulses and integrate all of these to obtain the total K^+ current (net flux). Likewise, integration of the "bump" of radioactivity provides a measure of the amount of ^{42}K efflux during the pulses.

After cessation of the voltage-clamp pulses, the tracer efflux returned to the baseline level. At the time marked by the arrow in Fig. 3, 1 μM of recombinant charybdotoxin (CTX) was added to the bath solution. The voltage-clamp pulses were applied again approximately 6 min later and the resultant efflux was substantially reduced due to block of the channels by toxin.

The data of Fig. 3 provide a view of the relative amount of ^{42}K tracer efflux. It is often useful to use the tracer flux to determine the absolute level of K^+ efflux. It is necessary to know the specific activity (cpm/pmol of K^+) of the internal solution in order to make this conversion. Three general methods can be used to determine the specific activity.

If the oocytes are soaked in a known specific activity solution for a sufficiently long period of time, the internal solution will equilibrate to the same specific activity as the soaking solution. This is certainly a simple method but there is no *a priori* way to know how long a time is sufficient since the equilibration time will depend on the amount of K^+ permeability of each oocyte and this, in turn, will be a function of the amount of K^+ channel expression.

Another method allows each oocyte to serve as its own control: two voltage-clamp periods are used one of which is to a potential (near or more positive than −5 mV for an oocyte bathed in 25 mM K^+) where the net flux is essentially equal to the efflux. The total amount of current flow during the pulses and the total CPM of tracer flux are determined. The integral of the current yields the total moles of K^+ efflux and the conversion factor is the total counts per moles of K^+. A second test potential is then used to produce efflux and the calibration from the first pulse used to compute the efflux at this second potential. This method is a bit cumbersome because it always requires at least two pulses and not all oocytes are able to tolerate a complete experiment.

In the third method the amount of radioactivity in the oocyte is counted at the end of the experiment to yield the total counts per minute and the amount of K^+ in each oocyte is determined. This immediately yields the specific activity for that particular oocyte. The K^+ content of each oocyte can be obtained using the following technique: (1) extracellular potassium is removed by washing the oocyte with a K^+-free solution (e.g., 100 mM NMG/HCl, 5 mM $MgCl_2$, 10 mM HEPES, pH 7.4); (2) the egg is allowed to air dry overnight after which 1 ml of 15 mM LiCl is added to the tube and vortexed until the egg disintegrates; (3) after 4 hr the tube is spun at

14,000g for 5 min and the supernatant removed and analyzed for potassium content with a flame photometer (model IL343, Instrumentation Laboratory Inc., Lexington, MA).

Conclusion

The large size of and robust ion channel protein expression in *Xenopus* oocytes makes this system ideal for the measurement of unidirectional fluxes of substances, especially ions, through these channels. The methods described here have been used for determining unidirectional K^+ fluxes through both voltage-gated *Shaker* and inward rectifier (IRK1) channels. We described specific methods for determining either the K^+ ion influx or efflux using ^{42}K as a tracer. However, in both types of experiments the total flux is available from the measured ion current, so for any particular experiment both the unidirectional fluxes can be determined. Our particular interests are in applying these methods to aid in producing a molecular understanding of permeation and selectivity of ion channels but the techniques described here can generally be applied to many other scientific questions whose answers require unidirectional flux measurement.

Acknowledgments

We thank Jill Thompson for critically reading the manuscript and for assistance with many of the experiments that helped develop the techniques described here and, especially, the development of methods to determine the potassium content of individual oocytes. This work was supported, in part, by a grant from the National Science Foundation (IBN-9514389 to T.B.).

[31] Transient Expression of Heteromeric Ion Channels

By ALISON L. EERTMOED, YOLANDA F. VALLEJO, and WILLIAM N. GREEN

Introduction

A major advance in the study of ion channels has been the ability to express channels in foreign host cells after isolation of the channel genes. Initially, the system of choice for rapid expression of ion channel proteins was the *Xenopus* oocyte. Most of the electrophysiologic characterization

0076-6879/98 $25.00

of newly isolated ion channel subunit cDNAs has been performed using the *Xenopus* oocyte expression system. Unfortunately, the amount of ion channel protein produced using this system is relatively small. For biochemical and cell biological analysis, ion channels are generally expressed in cultured cell lines where production can be generated in enormous numbers of cells. The transfer of ion channel genes to cultured cells allows the synthesis of large amounts of these proteins, which previously could only be purified in minute quantities after much effort. Moreover, cultured cells have the additional advantage over *Xenopus* oocytes in providing an environment that is usually closer to that of the expressed ion channel's native environment.

Our objective in writing this article is *not* to provide a step-by-step description of the methods for ion channel gene transfer into cultured cell lines. Many excellent and up-to-date texts describe these methods in detail (e.g., Refs. 1 and 2). Instead, our intent is to describe problems and complexities we have encountered using these methods and to discuss changes we have made in the methods in order to counter problems and to achieve higher levels of expression. Because almost all ion channels are heteromeric proteins (see Ref. 3 for a review), we have devoted most of the results and following discussion to the problems encountered when attempting to express heteromeric ion channels.

Although cultured cell lines are widely used for the expression of ion channel proteins, the methods used are far from standard. Thus, the first problem one is faced with is sorting among the daunting number of options when considering how to express the ion channel of interest. Generally, the options are narrowed by the questions to be addressed. For example, if large quantities of the ion channel are needed for biochemical or structural analysis, a good choice may be to express the channel in Sf9 (*Spodoptera frugiperda* fall armyworm ovary) cells using a baculovirus expression vector. Or, if it is critical that the ion channel be expressed in a specific host cell type, such as a neuron, then one might choose to infect primary cultures of neurons using an adenovirus expression vector.

Usually the next problem encountered is how to introduce your ion channel subunit cDNAs effectively into the chosen cells. In our laboratory, we have been using a variety of mammalian cell lines to transiently express different nicotinic acetylcholine receptors (AChRs). We have tried several different methods to introduce the subunit cDNAs including electropora-

[1] B. Rudy and L. E. Iverson, *Methods Enzymol.* **207,** (1992).
[2] R. Ashley, "Ion Channels: A Practical Approach." IRL Press, London, 1996.
[3] W. N. Green and N. S. Millar, *Trends Neurosci.* **18,** 280 (1995).

tion,[4] retroviral infection,[5] and transfection, both calcium phosphate and lipid mediated.[6] Each method has its advantages and disadvantages. Unfortunately, with each of these methods the percentage of the cells expressing AChRs is well below 100%. At first glance, the inefficiency of transient expression might not appear to pose a significant problem except to lower the levels of expression. In fact, this seems to be the case for ion channels that are homomeric (see later discussion). However, for heteromeric ion channels such as muscle-type AChRs, which are composed of α, β, γ (or ε), and δ subunits, inefficiency of expression has additional complications. These complications appear to arise from an uneven distribution of the subunits among different populations of the expressing cells. At one extreme, one or more subunits are underexpressed or not expressed at all in some cells, which will result in the expression of different combinations of the subunits. As discussed later, the levels of AChR surface expression and efficiency of assembly are both strongly affected by the combination of subunits expressed. At the other extreme, one or more subunits are overexpressed. Overexpression of subunits can be just as deleterious as the underexpression of subunits (see below).

An alternative to transient expression that avoids problems of expression inefficiency is stable expression. A cell line stably expressing an ion channel is established by integrating the correct combination of subunit genes into the genome of a cell. This procedure involves (1) introduction into the cells of a selectable marker along with the subunit genes, (2) isolation of single colonies containing the selectable marker, (3) growth of enough cells for analysis, and (4) further screening of each single colony isolate for the correct combination of subunits. Although stable expression has a number of advantages over transient expression, it is nonetheless a lengthy process, ranging from 2 to 5 months to complete.

Because the method of stable expression involves such a considerable investment of time, we have continued to use transient expression methods in our laboratory. Transient expression is used primarily for the initial characterization of new AChR subunit constructs before the time-consuming procedure of isolating and characterizing stably transfected cell lines. Considerable effort has been expended during the last several years in our laboratory to test the reliability of these methods and to optimize AChR

[4] H. Potter, *in* "Current Protocols in Molecular Biology" (F. M. Ausubel, R. Brent, R. E. Kingston, D. D. Moore, J. G. Seidman, J. A. Smith, and K. Struhl, eds.), Vol. 1, pp. 9.3.1–9.3.6. Wiley, New York, 1995.

[5] C. Cepko, *in* "Current Protocols in Molecular Biology" (F. M. Ausubel, R. Brent, R. E. Kingston, D. D. Moore, J. G. Seidman, J. A. Smith, and K. Struhl, eds.), Vol. 1, pp. 9.10.1–9.14.3. Wiley, New York, 1995.

[6] T. Claudio, *Methods Enzymol.* **207,** 391 (1992).

expression using them. In the course of optimizing AChR transient expression, we have made changes and adjustments to the methods, many of which are described later. To further characterize the transient transfection of a heteromeric AChR, we describe how the expression of a heteromeric AChR compares to that of a homomeric AChR, and how heteromeric AChR transient expression compares with stable expression. For most of these experiments, AChR expression was monitored by measuring the total amount of cell-surface expression on a dish of transfected cells using ^{125}I-labeled α-bungarotoxin (Bgt) binding. We chose this assay because it is relatively easy to perform, but also because it is the feature of AChR expression that we most wanted to optimize for many of our experiments. However, as described later, the measurement of AChR surface levels fails to detect some of the more subtle differences between transient and stable expression of the AChR heteromer. To detect these more subtle differences, we have also assayed for intracellular and surface AChR expression using fluorescence microscopy and measured the efficiency of subunit assembly by metabolically labeling the AChR subunits.

Methods

AChR Constructs and Cell Lines

Three different AChRs are used in this study. Two of the AChRs are muscle-type AChRs from different species: (1) the *Torpedo* AChR, which is composed of α, β, γ, and δ subunits, and (2) the "adult-form" of mouse muscle-type AChR, which is composed of α, β, ε, and δ subunits. The other "AChR" is a homomeric receptor containing α_7/5HT$_3$ chimeric subunits. These subunits consist of the N-terminal half of the chicken α_7 subunit and the C-terminal half of the mouse 5HT$_3$ receptor subunit.[7] Previous studies of the α_7/5HT$_3$ chimera demonstrate that it has all the pharmacologic properties of an AChR.[7] For transient expression, all of the *Torpedo* and mouse subunit cDNAs are subcloned into the pRBG4 expression vector,[8] and the α_7/5HT$_3$ subunit cDNA is subcloned into the pMT3 vector.[9] Both expression vectors contain the same cytomegalovirus (CMV) promoter and the simian virus 40 (SV40) origin and polyadenylation sequence.

[7] J. L. Eisele, S. Bertrand, J. L. Galzi, T. A. Devillers, J. P. Changeux, and D. Bertrand, *Nature* **366,** 479 (1993).

[8] B. S. Lee, R. B. Gunn, and R. R. Kopito, *J. Biol. Chem.* **266,** 11448 (1991).

[9] A. G. Swick, M. Janicot, K. T. Cheneval, J. C. McLenithan, and M. D. Lane, *Proc. Natl. Acad. Sci. U.S.A.* **89,** 1812 (1992).

The cell lines that stably express the *Torpedo* AChR subunits,[10] the mouse AChR subunits,[11] and the $\alpha_7/5HT_3$ chimeric subunits[12] have been previously described. The human embryonic kidney line, tsA 201,[13] is used for transient transfections. All cell lines are grown at 37°, 5% (v/v) CO_2 in Dulbecco's modified Eagle's medium (DMEM, GIBCO, Grand Island, NY) plus 10% calf serum (Hyclone, Logan, UT). The *Torpedo* and mouse subunit cDNAs in the stably expressing cell lines are under the control of SV40 promoters. To enhance expression of the subunits, the medium is supplemented with 20 m*M* sodium butyrate (Baker, Phillipsburg, NJ) 36–48 hr prior to the experiment. The assembly of the *Torpedo* subunits is temperature dependent.[10] To allow assembly to occur, the temperature at which the cells are grown is dropped from 37 to 20° for the indicated times.

Transfection

For the experiments described next, cells are transfected using the calcium phosphate method. This method was chosen because of its relative ease and its low cost.

Stable Transfection. Our stable transfection protocol has been described previously in detail in another volume in this series.[6]

Transient Transfection. Transient transfection methods are also used to express the three different AChRs. The following protocol presents the details of the transient transfection methods used in our experiments.

PROTOCOL 1: CALCIUM PHOSPHATE TRANSFECTION PROCEDURE. All work involving the preparation of the cell cultures and the steps in the protocol should be performed in a sterile, laminar flow hood. *See below for advice on the choice of host cell.* Care should be taken to maintain the most sterile environment possible for the preparation of the DNA and solutions used.

1. The day before transfection, plate cells in 6-cm culture dishes at a density so that cells will be 50–70% confluent the next day.
2. In a sterile tube (polypropylene), add 250 μl of the 2× HEPES solution and the appropriate amount of your DNA. *Note:* As discussed later, the level of expression is highly dependent on the total amount of transfected DNA and, if you are transfecting more than one subunit, on the DNA ratio for the different subunits. Add sterile,

[10] T. Claudio, W. N. Green, D. S. Hartman, D. Hayden, H. L. Paulson, F. J. Sigworth, S. M. Sine, and A. Swedlund, *Science* **238,** 1688 (1987).

[11] W. N. Green and T. Claudio, *Cell* **74,** 57 (1993).

[12] F. Rangwala, R. C. Drisdel, S. Rakhilin, E. Ko, P. Atluri, A. B. Harkins, A. P. Fox, S. B. Salman, and W. N. Green, *J. Neurosci.* **17,** 8201 (1997).

[13] R. F. Margolskee, B. McHendry-Rinde, and R. Horn, *BioTechniques* **15,** 906 (1993).

deionized, distilled water to the tube to a final volume of 469.5 μl. *Note:* These are the amounts for the transfection of a single 6-cm plate and should be scaled appropriately if more plates are to be transfected.
3. Slowly add 31.5 μl of 2 *M* $CaCl_2$ *dropwise* to the solution.
4. Wait 30 min for the precipitate to form. The resulting solution should become slightly cloudy during this period.
5. Distribute the solution dropwise over the plated cells.
6. Incubate 5–6 hr at 37°, 5% CO_2, then change the medium to remove the precipitate.

Maximal expression of the mouse and α_7/5HT_3 AChRs occurs 3 days following transfection. For expression of *Torpedo* AChRs, cells are shifted from 37 to 20°C 24 hr after the transfection. Maximal expression occurs 4 days following the temperature shift.

Solutions

2× HEPES: 8.0 g of NaCl, 0.198 g of $Na_2HPO_4 \cdot 7H_2O$, and 6.5 g of HEPES. Add sterile, distilled, deionized water to a volume just under 500 ml. Adjust the pH to 7.0 and readjust the final volume to 500 ml. Aliquot and store at −20°. Aliquot currently in use is kept at 4°.

2 *M* $CaCl_2$: Mix in distilled, deionized water and sterile filter. Can be kept at room temperature.

Assays

α-Bungarotoxin Binding. The primary assay used for characterizing the expressed AChRs was binding of Bgt to the AChR. α-Bungarotoxin binds to muscle-type and α_7/5HT_3 AChRs with high affinity and it has a very long resident time. Because of this slow off rate, Bgt binding to the AChR basically serves as a method to label the receptors. Two different Bgts were used in the experiments: Bgt conjugated with ^{125}I ([^{125}I]Bgt) and Bgt conjugated with tetramethylrhodamine (TMR-Bgt). Protocol 2 presents the details of the methods used for [^{125}I]Bgt binding to the transfected cells in our experiments, and Protocol 3 presents the details of the TMR-Bgt staining.

PROTOCOL 2: CELL-SURFACE ^{125}I-BGT BINDING PROCEDURE. This protocol is designed for poorly adherent cells, such as HEK 293 and tsA 201 cells, plated on 6-cm culture plates and grown to confluency. Care must be taken since this protocol involves handling radioisotopes. All work involving [^{125}I]Bgt should be monitored with a gamma probe and, as much as possible, should be performed in a fume hood. Gloves and a lab coat should be worn. [^{125}I]Bgt-containing solutions should be aspirated into a designated vacuum flask and, along with [^{125}I]Bgt-containing dry waste, safely disposed of.

1. Aspirate media from the cell culture plates to be assayed.
2. 700 μl of the 1× phosphate-buffered saline plus EDTA solution (PBS/EDTA) is added to the cell culture plates. The cells, which are poorly adherent, are removed from the plates simply by applying a gentle fluid stream from a pipette. Cells are transferred to a 1.5-ml microfuge tube. Any remaining cells are removed from the plates by the addition of 600 μl PBS/EDTA and added to the microfuge tube.
3. Cells are pelleted in a centrifuge at 1000–2000*g* for 3–5 min. This centrifuge spin must be gentle so that the cells remain intact. The supernatant above the pellet in the tube is aspirated using a syringe needle (21-gauge 1/2).
4. The cells are washed by addition of 1 ml of PBS/EDTA gently added back to the tube. The cell pellet is gently dispersed. The cells are again pelleted, and the supernatant is aspirated.
5. A PBS/EDTA solution containing 0.1% bovine serum albumin (BSA; w/v) and 4 n*M* [^{125}I]Bgt (4 μl/ml of the [^{125}I]Bgt stock) is prepared.
6. 750 μl of this PBS/EDTA solution is added to each microfuge tube and the cell pellet is gently dispersed.
7. The microfuge tubes are incubated for 2 hr at room temperature on a rotator to keep the cells dispersed.
8. Following the incubation, the cells are pelleted, the supernatant aspirated, and the cells are washed 3–4 times in the 1× PBS/EDTA solution.
9. After the last wash, the supernatant is aspirated and the [^{125}I]Bgt bound to the cells is counted in a gamma counter.

Solutions

5× PBS/EDTA: 40.0 g of NaCl, 1 g of KCl, 7.2 g of Na_2HPO_4, 7.2 g of KH_2PO_4, and 1.2 g of EDTA. Add sterile, distilled, deionized water to a volume of just under 1 liter. Adjust to pH 7.4 and readjust the final volume to 1 liter. Store at room temperature.

[^{125}I]Bgt stock solution: Add sterile, distilled, deionized water to a 250-mCi lyophilized aliquot of [^{125}I]Bgt (NEN, Boston, MA; 140–170 cpm/fmol) to the appropriate volume in order to make a 1 μM Bgt solution (usually 2–3 ml of water). Store at 4°.

Fluorescence Microscopy. The efficiency of expression, that is, the percentage of the cells expressing AChR subunits, and the distribution of the subunits among the cells were determined using fluorescence microscopy techniques. To visualize the Bgt sites expressed on the cell surface or intracellularly we used TMR-Bgt. Subunit-specific antibodies were used to visualize the cell-to-cell distribution of the α and δ subunits.

PROTOCOL 3: STAINING PROCEDURE WITH IMMUNOFLUORESCENT TMR-BGT

1. To coat glass slides with alcian blue (Eastman Kodak, Rochester, NY), which is used to immobilize cells for staining, three to four drops of a 1% alcian blue solution (w/v) are placed on the slides. The slides are allowed to set at room temperature for 15 min, rinsed one time with distilled, deionized water, and patted dry with a Kimwipe.
2. The cells to be stained are placed on the slides in medium after removal from the culture dish (see step 2 in Protocol 2). Cells are allowed to set at room temperature for 10 min.
3. Excess medium is aspirated from the slides and three to four drops of a 2% (w/v) formaldehyde solution are placed on the slides, which are allowed to set at room temperature for 10 min.
4. Excess formaldehyde solution is aspirated and three to four drops of Tris-buffered saline (TBS) are placed on the slides and set for 10 min to quench reactive aldehydes.
5. Excess TBS is aspirated. At this point *to permeabilize cells,* TBS plus 0.1% Triton X-100 is added to the slides for 10 min and then aspirated.
6. To block nonspecific binding sites, TBS plus 2 mg/ml BSA is added to the slides for 10 min and then aspirated.
7. 300 n*M* TMR-Bgt (Molecular Probes, Eugene, OR) in TBS plus 2 mg/ml BSA is added to the slides and the slides are incubated *in the dark* for 1 hr at room temperature.
8. The slides are washed four times for 5–10 min per wash with TBS plus 2 mg/ml BSA. For additional staining with 4′,6-diamidino-2-phenylindole dihydrochloride (DAPI; Molecular Probes), DAPI is added to the TBS of the second wash at a 1 : 1000 dilution, and cells are incubated for 15 min. *Note:* Cells do not have to be permeabilized for the DAPI staining.
9. One drop of mounting medium (Vectashield from Vector Labs, Burlingame, CA) is added to each slide. A coverslip is applied over the cells, sealed with clear nail polish, and allowed to dry.
10. Slides are viewed immediately or stored at −20°.

A rabbit polyclonal anti-α-subunit antibody (Ab) and a mouse monoclonal Ab (MAb) anti-δ-subunit, MAb 88b, were used to stain cells with subunit-specific antibodies. In the protocol for the Ab staining the following two steps are substituted for step 7:

7a. To permeabilized cells, add the Abs diluted 1:100 in TBS plus 2 mg/ml BSA for 1 hr and wash two to three times for 5–10 min per wash with TBS plus 2 mg/ml BSA.

7b. The secondary antibodies are added *in the dark* for 1 hr: fluorescein goat anti-mouse IgG(H+L) conjugate at 1:500 dilution in TBS/BSA and Lissamine rhodamine (LRSC)-conjugated AffiniPure FAb goat anti-rabbit IgG(H+L) at 1:200 dilution in TBS/BSA.

Solutions

2% Paraformaldehyde/PBS: 1.24 g of paraformaldehyde (Sigma, St. Louis, MO) is added to 4.7 ml of a 0.5 *M* Na_2HPO_4 solution and 55.8 ml of distilled, deionized water. The solution is stirred and heated at 60° until the paraformaldehyde has dissolved. After, 0.36 g of NaCl is added to the solution, followed by 1.5 ml of a 0.5 *M* NaH_2PO_4 solution. If necessary, adjust the pH to 7.1 and use fresh.

TBS: 1.0 ml of a 1.0 *M* Tris solution, pH 7.5, is added to 3.0 ml of a 5.0 *M* NaCl solution and 96 ml of distilled, deionized water. If necessary, adjust the pH to 7.5 and store at 4°.

Metabolic Labeling and Immunoprecipitation. We have found that metabolic labeling of AChR subunits *in vivo* is the most sensitive assay for analyzing the subunit properties using sodium dodecyl sulfate–polyacrylamide gel electrophoresis (SDS–PAGE). Metabolic labeling has the added advantage of allowing you to pulse label and then follow or "chase" the labeled proteins through their steps of biogenesis. The pulse-chase protocol used in the results as well as the protocol for cell lysis and immunoprecipitation have been described in detail previously.[14]

Results

Dependence on Cell Line

Not surprisingly, we have found that the level of AChR expression is highly dependent on the choice of host cell. For many cell lines, expression levels are poor because of a low efficiency at taking up the DNA. This is

[14] N. S. Millar, S. J. Moss, and W. N. Green, *in* "Ion Channels: A Practical Approach" (R. Ashley, ed.). IRL Press, London, 1996.

the case for both the rat pheochromocytoma cell line, PC12, and the human neuroblastoma cell line, SH-SY5Y, where 1% or less of the cells express AChRs (S. Rakhilin and W. Green, 1997, unpublished results). Other mammalian cell lines are much more efficient at taking up AChR genes. Figure 1 shows the levels of *Torpedo* AChR cell-surface expression obtained in four different cell lines: mouse fibroblast lines, (1) NIH 3T3 and (2) L tk^- cells, and human embryonic kidney lines, (3) HEK293 and (4) tsA201 cells. For each of these cell lines, we find that the percentage of cells that take up the DNA ranges from 10 to 40% (data not shown, but see Fig. 7 for the tsA201 cells).

The five- to six-fold higher levels of expression observed for the tsA201 cells are not due to a higher expression efficiency. The tsA201 cells were established by stably transfecting HEK293 cells with the SV40 large tumor (T) antigen.[13] The large T antigen initiates replication of expression vectors containing the SV40 origin, resulting in a high copy number for these expression vectors. The higher level of expression of the tsA201 cells is, thus, the result of an amplification of the expression vector once it has entered the cell. In short, the high levels of expression we obtain in tsA201 cells are a function of both the cells' relatively high transfection efficiency and the ability of the cells to amplify vector copy number after its entry into the cells.

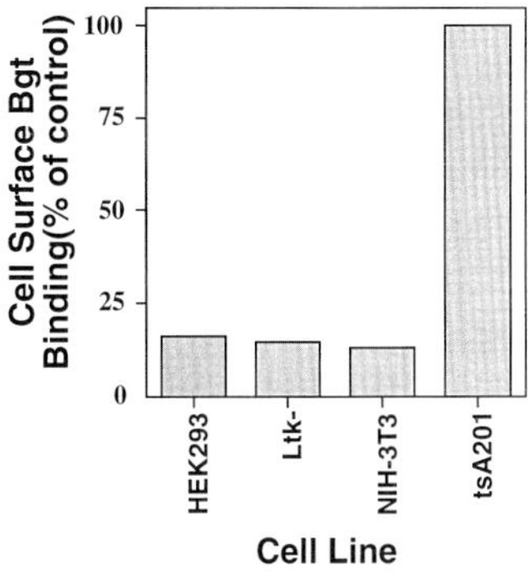

FIG. 1. Cell-surface expression of transiently transfected AChRs is dependent on the host cell line. *Torpedo* AChR subunit cDNAs were transiently transfected into either HEK293, L tk^-, NIH 3T3, or tsA201 cells. The amount of subunit cDNA transfected was 1 μg α, 0.5 μg β, 2.5 μg γ, 7.5 μg δ per 6-cm plate, and the level of AChR surface expression was determined by [^{125}I]Bgt binding to intact cells. The values represent the mean of [^{125}I]Bgt binding to two 6-cm plates. Background, determined by [^{125}I]Bgt binding to intact cells that were sham transfected with no DNA, was subtracted to obtain the data. The values are plotted as the percentage of the tsA201 value, which was 50 fmol.

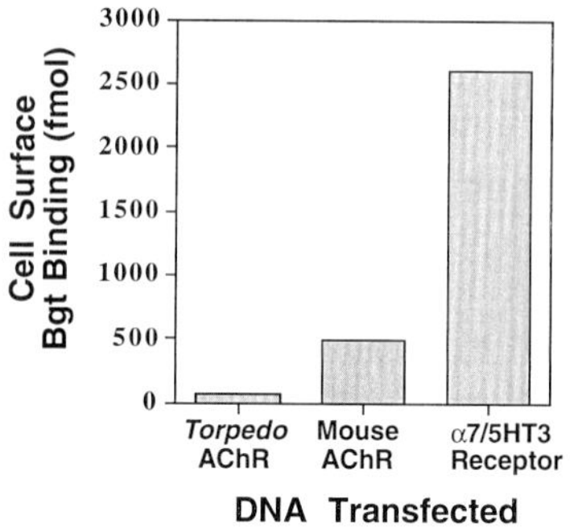

FIG. 2. Cell-surface expression of different AChRs. *Torpedo* AChR α-, β-, γ-, and δ-subunit cDNAs, mouse AChR α-, β-, ε-, and δ-subunit cDNAs, or the chimeric α_7/5HT$_3$ subunit cDNA were transiently transfected into tsA201 cells. The amount of subunit cDNA transfected was 1 μg α, 0.5 μg β, 2.5 μg γ, 7.5 μg δ per 6-cm plate for *Torpedo* AChRs; 5 μg α, 2.5 μg β, 2.5 μg ε, 2.5 μg δ per 6-cm plate for mouse AChRs, and 2.5 μg for the α_7/5HT$_3$ receptor. The level of AChR surface expression, determined as in Fig. 1, was 67 fmol for *Torpedo* AChRs, 500 fmol for mouse AChRs, and 2610 fmol for α_7/5HT$_3$ receptors.

Expression of Homomeric and Heteromeric AChRs

Different experiments were performed to compare the transient expression of heteromeric mouse $\alpha\beta\varepsilon\delta$ AChRs with that of the homomeric α_7/5HT$_3$ AChRs. One big difference between heteromeric and homomeric AChR expression was that the level of homomeric AChR expression was considerably higher than that of the heteromeric AChR. As shown in Fig. 2, cell-surface [^{125}I]Bgt binding to the α_7/5HT$_3$ AChRs was approximately five-fold higher than that to the mouse $\alpha\beta\varepsilon\delta$ AChRs, which was about five-fold higher than [^{125}I]Bgt binding to the *Torpedo* AChRs. While the transient expression of the α_7/5HT$_3$ and mouse $\alpha\beta\varepsilon\delta$ AChRs was performed under identical conditions, the conditions were different for *Torpedo* AChR expression. As described in the Methods section, cells transiently transfected with the *Torpedo* subunit cDNAs were maintained at 20° as opposed to 37° for the α_7/5HT$_3$ and mouse $\alpha\beta\varepsilon\delta$ AChRs because of the temperature dependence of *Torpedo* AChR assembly. It is likely that the lower temperature is the reason for the lower level of *Torpedo* AChR expression since at 20° most of the assembled *Torpedo* AChRs are retained in the endoplasmic reticulum.[15]

Additional differences between the transient expression of heteromeric and homomeric AChRs were apparent when we measured how varying

[15] A. F. Ross, W. N. Green, D. S. Hartman, and T. Claudio, *J. Cell Biol.* **113,** 623 (1991).

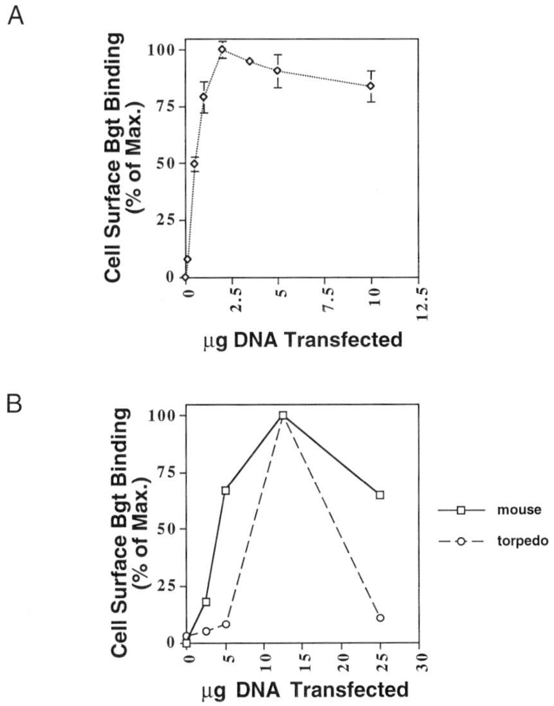

FIG. 3. Dependence of homomeric and heteromeric AChR expression on the amount of transfected DNA. (A) Cell-surface expression of homomeric α_7/5HT$_3$ receptors as a function of the amount of transfected DNA. Maximum levels of expression were obtained by transfecting 2.0 μg of DNA per 6-cm plate. The level of AChR surface expression was determined as in Fig. 1. The values represent the mean $\pm$ standard deviation for two separate experiments consisting of two to four 6-cm plates and are plotted as the percentage of maximum binding, which was 2570 fmol. (B) Cell-surface expression of heteromeric *Torpedo* and mouse AChRs as a function of the amount of transfected DNA. Maximum levels of expression for both AChRs were obtained by transfecting 12.5 μg of DNA per 6-cm plate. The level of AChR surface expression was determined as in Fig. 1. The values represent the mean for two 6-cm plates and are plotted as the percentage of maximum binding, which was 3.41 fmol for the α_7/5HT$_3$ AChR, 34 fmol for *Torpedo*, and 357 fmol for the mouse AChR.

the amount of DNA affected the level of AChR expression. Figure 3 displays AChR expression, monitored by surface [^{125}I]Bgt binding, as a function of micrograms of DNA added during transfection. The results in Fig. 3A for the homomeric α_7/5HT$_3$ AChR should be compared to those in Fig. 3B for the heteromeric mouse and *Torpedo* AChRs. First, for the homomeric AChR, expression increased linearly with increasing amounts of transfected DNA, whereas for the heteromeric AChRs expression rose sigmoidally. Second, it only took 2.0 μg of the homomeric AChR DNA for peak expression (Fig. 3A). In contrast, 12.5 μg of total subunit DNA or five-fold higher levels of DNA were needed for peak expression of both heteromeric AChRs (Fig. 3B) even though expression levels for the homomeric AChR were approximately five-fold higher than for the heteromeric AChR (Fig. 1). Finally, homomeric AChR expression plateaued with increasing amounts of transfected DNA (Fig. 3A), whereas expression of the heteromeric AChRs sharply decreased (Fig. 3B). The decrease in expression observed with large amounts of heteromeric AChR DNA is not a nonspecific effect of the increased amount of DNA. If increasing amounts of the expression vector lacking the subunit cDNA were added in the range of 10–30 μg per 6-cm plate, no decrease in AChR expression is observed (data not shown).

Complications of Heteromeric AChR Expression

In the previous section, we described how transient expression of heteromeric AChRs is different in a number of ways from that of the homomeric AChRs. To summarize, (1) heteromeric AChR expression is severalfold less than that for the homomeric AChR; (2) it requires severalfold more DNA; (3) it displays a sigmoidal rise with increasing DNA concentration as opposed to the linear rise observed for homomers; and (4) after reaching peak value, it does not plateau, but declines sharply with increasing amounts of transfected DNA. Obviously, the more complicated process of assembling an AChR consisting of four different subunits is at least part of the explanation for these differences between the expression of AChR heteromers and homomers. However, the differences may also be caused in part by the inefficiency of transient expression. As discussed in the introduction, the inefficiency of transient expression would be expected to create more problems for a heteromer than for a homomer. Problems specific to heteromeric expression are likely to be caused by the distribution of the different subunit cDNAs among the cells. An uneven distribution of the different cDNAs could cause either over- or underexpression of different subunits in different cells. In this section, we have tested how varying the amount of different subunit cDNAs affects heteromeric AChR expression.

The uneven distribution of different subunit cDNAs among a large population of cells can have several consequences that affect heteromeric AChR expression. One possibility is that one or more different subunit cDNAs fail to enter a sizable number of the cells and less than the full complement of the subunits is expressed in these cells. In Fig. 4, we have tested how different combinations of the *Torpedo* AChR α, β, γ, and δ

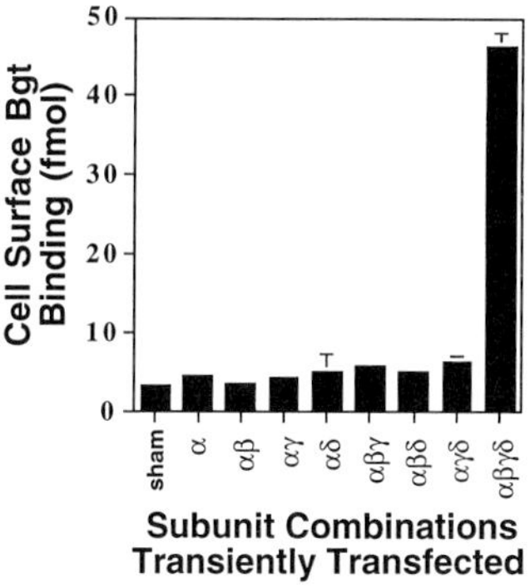

FIG. 4. Cell surface expression of different combinations of AChR subunits. The indicated combinations of *Torpedo* AChR subunit cDNAs were transiently transfected into tsA201 cells. The amount of subunit cDNA transfected was 1 μg α, 0.5 μg β, 2.5 μg γ, 7.5 μg δ per 6-cm plate. The level of AChR surface expression was determined as in Fig. 1. The values represent the mean $\pm$ standard deviation for three or four 6-cm plates.

subunits affect AChR cell-surface expression. Transient transfection of any combination of subunit cDNAs containing less than the full complement of four subunits resulted in AChR expression levels at or just above background as assayed by cell-surface [^{125}I]Bgt binding. In other experiments, we found that only the surface expression levels of the $\alpha\beta\gamma$ and $\alpha\beta\delta$ combinations were significantly above background but extremely small compared to the surface expression levels of all four subunits (data not shown). Similar results were also obtained previously for transient expression of the mouse AChR α, β, γ, and δ subunits in COS cells.[16] These findings demonstrate that if only one of the four subunit cDNAs fails to enter a cell or if relatively small amounts of one subunit enter, AChR expression on the surface of that cell is greatly diminished or eliminated. Therefore, virtually the only AChR subunit complexes that arrive on the cell surface are AChRs containing all four subunits. Note that the expression of intracellular complexes still occurs in the absence of one or more of the AChR subunits although the number and efficiency of assembly of these complexes appears to be much lower than for all four subunits. Expression of most subunit combinations containing the α subunit (i.e., $\alpha\gamma$, $\alpha\delta$, $\alpha\beta\gamma$, $\alpha\gamma\delta$, $\alpha\beta\gamma$, and $\alpha\beta\delta$) form large numbers of Bgt-binding sites even though there are few or no surface Bgt-binding sites.[11,16–18] Clearly, "quality control" mechanisms exist in the

[16] Y. Gu, J. R. Forsayeth, S. Verrall, X. M. Yu, and Z. W. Hall, *J. Cell Biol.* **114,** 799 (1991).
[17] P. Blount and J. P. Merlie, *Neuron* **3,** 349 (1989).
[18] M. S. Saedi, W. G. Conroy, and J. Lindstrom, *J. Cell Biol.* **112,** 1007 (1991).

cells that prevent the surface expression of AChRs lacking any of the four subunits.

Another outcome of an uneven distribution of the different subunit DNAs among transfected cells is that one or more subunits are overexpressed relative to the other subunits. The consequence of varying the ratio of one subunit relative to the other subunits was tested as shown in Fig. 5. In this experiment, the amount of transfected α-subunit cDNA was varied while the amount of the other three AChR subunit cDNAs was held constant. Surface expression sharply changed from zero to a peak value at 1 μg of the AChR α-subunit cDNA and then diminished to a plateaued value of half of the peak value at higher DNA concentrations. The dependence of the surface expression on the amount of α-subunit cDNA is different from its dependence on the total amount of subunit cDNA (Fig. 3B) where expression peaked at 12.5 μg and was decreased by 90% at higher DNA concentrations. The much larger decrease in surface expression observed at the higher amounts of total subunit cDNA appears to be caused by increasing the DNA concentration of subunits other than the α subunit. In previous studies, it was shown that AChR surface expression is largely blocked by increasing amounts of mouse δ-subunit cDNA relative to the other three mouse subunits.[16] We have obtained similar results for both the *Torpedo* γ- and δ-subunit cDNAs (Eertmoed and Green, 1997, unpublished results). The relatively small amount of α-subunit cDNA needed to maximize expression reflects the need for much larger amount of other subunits. The dependence on each subunit cDNA, when varied

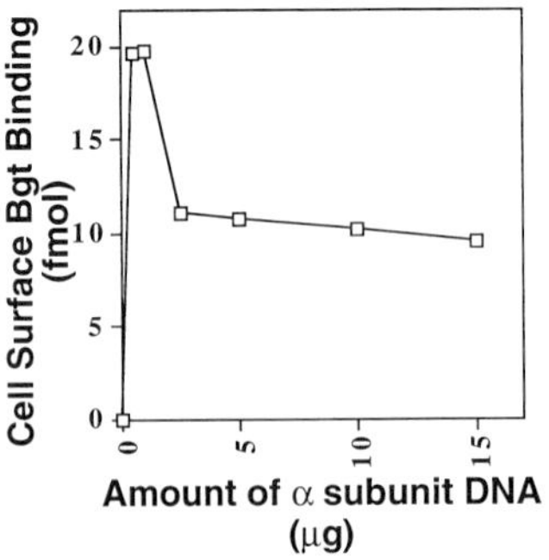

FIG. 5. Dependence of heteromeric AChR expression on varying the amount of transfected DNA for a single subunit. The indicated amount of *Torpedo* AChR α-subunit DNA was transiently transfected into tsA201 cells along with a set amount for the other three subunits (0.1 μg β, 2.5 μg γ, 7.5 μg δ per 6-cm plate). Maximum levels of expression were obtained by transfecting 1.0 μg of α-subunit DNA per 6-cm plate. The level of AChR surface expression was determined as in Fig. 1. The values were obtained from a single 6-cm plate.

relative to the other three subunits as in Fig. 5, peaked at different values. No expression above background was obtained at a subunit cDNA ratio of 2 : 1 : 1 : 1 ($\alpha:\beta:\gamma:\delta$), which is the subunit stoichiometry of the surface AChR and the subunit cDNA ratio we first tried. *Torpedo* AChR expression rose significantly above background after increasing the amount of δ-subunit cDNA relative to the other subunits, and maximum levels of *Torpedo* AChR expression were only obtained when we maintained a subunit cDNA ratio of 1 : 0.5 : 2.5 : 7.5 ($\alpha:\beta:\gamma:\delta$).

We have not determined why deviations from a subunit cDNA ratio of 1 : 0.5 : 2.5 : 7.5 ($\alpha:\beta:\gamma:\delta$) reduces AChR surface expression and, presumably, the assembly efficiency of fully assembled AChRs. The most likely explanation is suggested by the data of Fig. 4. Over- or underexpression of subunits would be expected to result in the assembly of subunit complexes containing less than the full complement of subunits. As shown in Fig. 4, the assembly of such subunit complexes would decrease both AChR surface expression and the assembly efficiency of fully assembled AChRs.

Assembly Efficiency

In the previous section, we described how transient expression of heteromeric AChRs depended on how the different subunit cDNAs are distributed among the transfected cells. High levels of heteromeric AChR transient expression were only achieved when both the full complement and the correct ratio of subunit cDNAs were transfected. Under these conditions, AChR transient expression equals and can even exceed the levels obtained with stably transfected cell lines. With comparable levels of surface expression obtained by stable and transient expression, can we assume that these two methods of expression are similar? Or could other features of AChR biogenesis differ with these two methods? To further compare stable and transient expression methods, we attempted to measure the efficiency of subunit assembly. For these experiments, the *Torpedo* AChR subunits were used because of the temperature dependence of the assembly of this AChR, which slows the rate of assembly by more than an order of magnitude.[11] To assay subunit assembly, the subunits were metabolically labeled with a 1-hr pulse of a mixture of [^{35}S]methionine and [^{35}S]cysteine, and then immunoprecipitated with the indicated antibodies. For the transiently expressed cells, each of the four subunits was successfully labeled and immunoprecipitated as shown in Fig. 6A. As assayed by the amount of [^{35}S]methionine and [^{35}S]cysteine signal, the levels of subunit synthesis obtained for the transiently transfected 6-cm cultures in Fig. 6A were considerably higher than the levels obtained for 10-cm cultures of the stably transfected cells (data not shown).

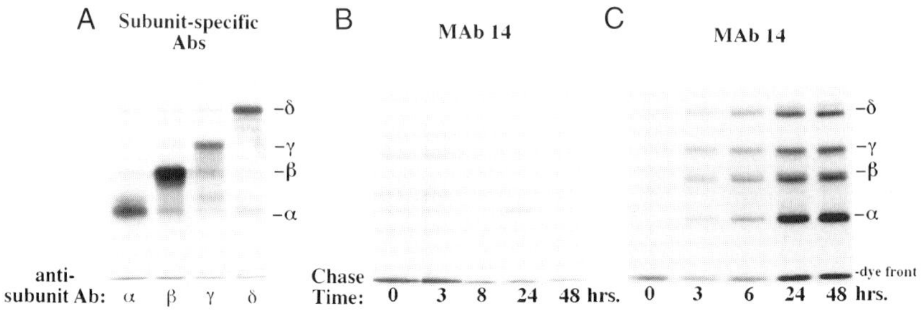

FIG. 6. AChR assembly efficiency for cells either transiently or stably expressing AChR subunits. (A) Metabolic labeling of all four AChR subunits in transiently transfected cells. The four *Torpedo* AChR subunits were transiently expressed in tsA201 cells on 10-cm plates (2 μg α, 1 μg β, 5 μg γ, 15 μg δ). One day after transfection, the cells were pulse labeled for 1 hr at 20° in methionine-cysteine-free medium supplemented with 333 μCi of an [^{35}S]methionine–[^{35}S]cysteine mixture (NEN EXPE^{35}S^{35}S). Cells were then lysed and treated with 1% SDS followed by 4% Triton X-100 to eliminate most intersubunit interactions, and the lysate from half a 10-cm plate was immunoprecipitated with the subunit-specific antibodies: polyclonal anti-α Ab, MAb 148 (anti-β), MAb 168 (anti-γ) and polyclonal anti-δ Ab. Labeled subunits were electrophoresed on 7.5% SDS–polyacrylamide gels, fixed, enhanced for 30 min and dried on a gel dryer. Shown is an autoradiogram of the gel. (B) The failure to detect metabolically labeled mature AChR complexes in transiently transfected cells. The four *Torpedo* AChR subunits were transiently expressed in tsA201 cells and metabolically labeled as in part (A) except that following the 1-hr pulse the labeled subunits were "chased" (see methods) for the indicated times and immunoprecipitated with MAb 14, which recognizes only subunit complexes that form late in the assembly process. No MAb 14-precipitable complexes were detectable, and we conclude that only a very small fraction of the total number of subunits synthesized assemble into AChRs and are transported to the cell surface. (C) Metabolically labeled mature AChR complexes in stably transfected cells. The four *Torpedo* AChR subunits were stably expressed in mouse L cells on 10-cm plates. Cells were pulse labeled for 30 min at 20° and treated as in part (B). In contrast to the transiently transfected cells in part (B), we observed metabolically labeled mature AChR complexes recognized by MAb 14 in cells stably expressing AChRs.

To follow the steps in assembly and measure assembly efficiency, the cells were "chased" after the metabolic labeling, and the labeled subunits were immunoprecipitated with MAb 14. Monoclonal Ab 14 was used because it recognizes only subunits that have been assembled during the latter stages of the assembly process.[11] The MAb 14 precipitation of labeled subunits from stably expressing cells is shown in Fig. 6C. Approximately 30% of the subunits originally labeled during the pulse are precipitated in $\alpha_2\beta\gamma\delta$ complexes 48 hr after the pulse. In contrast, we were unable to observe any labeled subunits precipitated with MAb 14 from the transiently transfected cells (Fig. 6B) even though most of the [^{125}I]Bgt-bound surface AChRs could be precipitated by this antibody (data not shown). These results demonstrate that the efficiency of AChR assembly in the transiently transfected cells is within the error of our measurements, that is, ~1% or lower. Heteromeric AChR transient expression thus differs from stable transfection in that the efficiency of assembly is at least 30-fold lower. Despite this large difference in assembly efficiency, the levels of AChR surface expression for stably and transiently transfected cells are similar. Transiently transfected cells must synthesize AChR subunits at much higher levels, and the lower assembly efficiency results in similar levels of surface AChRs.

Because transient transfection of the AChR subunits results in such a low efficiency of AChR assembly, we have been unable to carry out detailed studies of AChR assembly using transient expression methods. AChR subunits must be stably expressed in a cell line in order to follow assembly using metabolic labeling of the subunits. Our data suggest that low efficiency of assembly may occur for other heteromeric ion channels, and care should be taken when using transient expression methods to study the properties of ion channels not found on the cell surface.

Expression Efficiency and Distribution of Subunits

Why is the efficiency of heteromeric AChR assembly so low when the subunits are transiently transfected compared to when they are stably transfected? To address this question, cells transiently or stably expressing the mouse α, β, ε, and δ subunits were assayed using fluorescence microscopy techniques. The mouse subunits were chosen for study because they expressed higher levels of surface AChRs than the *Torpedo* subunits (see Fig. 2). The efficiency of expression was first tested by staining surface AChRs with TMR-Bgt. Examples of the stained cells are shown in Fig. 7, and the quantification of the results of the staining are presented in Table I. The total number of cells on the slides was visualized by DAPI staining, which stains the nuclei of the cells. As expected, almost all (95%) of the cells stably expressing AChRs were stained on the surface by TMR-Bgt. In contrast, 33% of the cells transiently expressing AChRs were stained. Based on the results of Fig. 4, where only cells transiently transfected with all four subunits expressed surface AChRs at significant levels, it appears that 33% of the transiently transfected cells expressed all four subunits.

Additional assays were performed to test whether AChR subunits are expressed in the 67% of the cells that did not show surface TMR-Bgt staining. The cells were next stained with TMR-Bgt after being permeabilized. This procedure allowed the staining of intracellular as well as the surface Bgt binding sites, and the results are shown in Fig. 8 and quantified in Table I. If α subunits in combination with only one or two other AChR subunits are expressed transiently, we would expect to see the formation

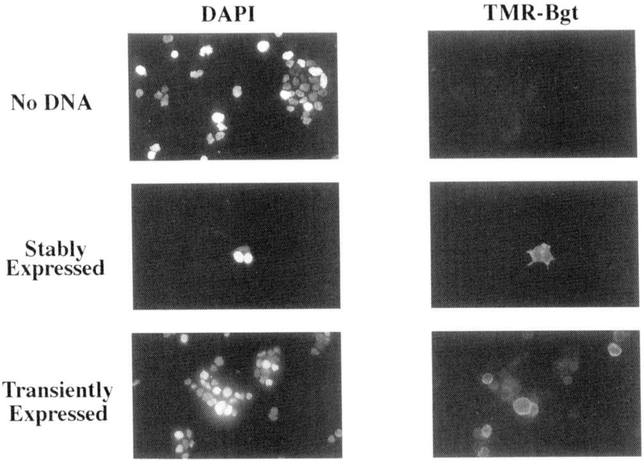

FIG. 7. Detection of cell-surface AChRs using fluorescence microscopy. Intact cells were stained with TMR-Bgt to detect cell-surface AChRs and DAPI to stain the nuclei. Displayed are cells (1) sham transfected with no DNA (top), (2) stably expressing the mouse AChR α, β, ε, and δ subunits (middle), or (3) transiently expressing the mouse AChR α, β, ε, and δ subunits (bottom). Cells were visualized and photographed, in Germany, with a Carl Zeiss Axioskop using a 40× objective and a 10× eyepiece.

of intracellular Bgt-binding AChR complexes, but no surface complexes. When the permeabilized cells were stained, the percentage of cells stained increased only to 35% compared to the 33% for the intact cells. These results indicate that AChR subunits were only expressed in the cells that express Bgt sites.

To test further whether the cells that fail to express Bgt sites were expressing AChR subunits, the permeabilized cells were stained with α- and δ-specific antibodies. These results are shown in Fig. 9 and quantified

TABLE I
TMR-Bgt STAINING OF CELLS TRANSIENTLY AND STABLY EXPRESSING $\alpha_2\beta\varepsilon\delta$ AChRs

Condition	Fixation	Total number of cells (DAPI stained)	Number stained	Stained (%)
Sham transfected	Intact	354	0	0
(no DNA)	Permeabilized	277	0	0
Stably transfected	Intact	150	142	95
	Permeabilized	154	150	97
Transiently transfected	Intact	651	213	33
	Permeabilized	541	191	35

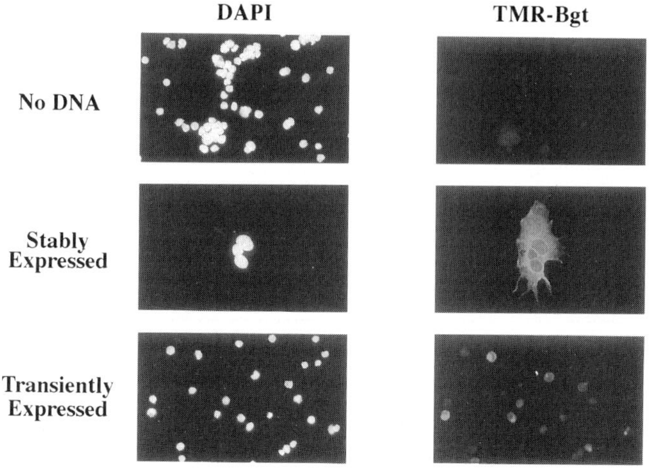

FIG. 8. Detection of cell-surface and intracellular AChRs using fluorescence microscopy. Permeabilized cells were stained and displayed as in Fig. 7.

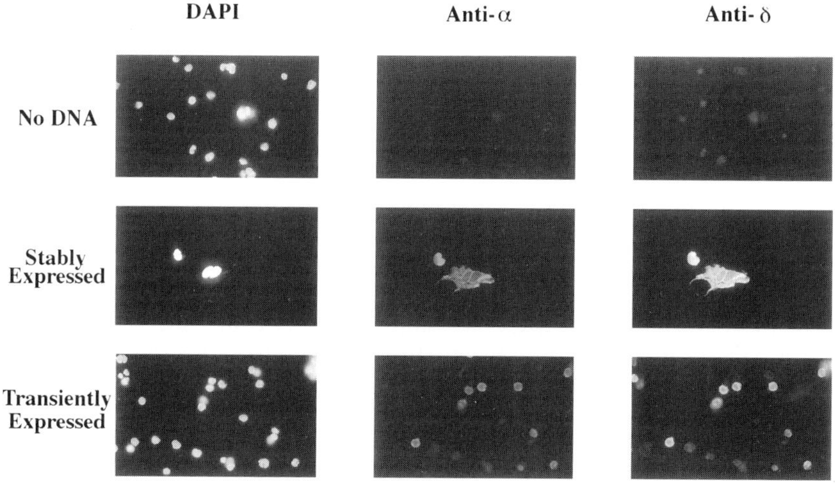

FIG. 9. Detection of AChR α and δ subunits using fluorescence microscopy. Permeabilized cells were stained with polyclonal anti-α Abs and anti-δ MAb 88b (American Type Culture Collection, Rockville, MD) followed by Texas Red and fluorescein isothiocyanate (FITC)-labeled antibodies, respectively. The cells were also stained with DAPI to stain the nuclei and displayed as in Fig. 7.

TABLE II
SUBUNIT-SPECIFIC STAINING OF CELLS TRANSIENTLY AND STABLY EXPRESSING $\alpha_2\beta\varepsilon\delta$ AChRs

Condition	Total number of cells (DAPI stained)	Number stained with α-specific Ab	Stained with α-specific Ab (%)	Number stained with α-specific Ab	Stained with α-specific Ab (%)
Stably transfected	203	194	96	202	99.5
Transiently transfected	651	151	36	158	38

in Table II. Basically, only cells with Bgt sites appeared to be stained by either the α- or δ-specific antibodies since the percentage of permeabilized cells stained with the antibodies was only 36% for the α-specific antibody and 38% for the δ-specific antibody. Based on the results using fluorescence microscopy, we conclude that the transient expression of AChR subunits is inefficient, in the range of 30–40%, but virtually all of the cells that are transfected receive all four subunit cDNAs and, therefore, express all four subunits.

If almost all of the transiently transfected cells express all four AChR subunits, the question remains: Why is the efficiency of heteromeric AChR assembly so low when the subunits are transiently transfected? There is another feature of the staining shown in Figs. 7, 8, and 9 that distinguishes transiently expressing from stably expressing cells. Although we have made no attempt to quantitate the level of staining in any of the experiments, it is clear that the staining of the transiently expressing cells is very heterogeneous while the staining of the stably expressing cells is homogeneous. For the TMR-Bgt staining of the intact and permeabilized cells, it is the cell-to-cell intensity of the signal that is heterogeneous. For the α- and δ-specific antibody staining of the permeabilized cells, the intensity of both signals also shows cell-to-cell heterogeneity. In addition, although virtually every cell that stains for the α-specific antibody also stains for the δ-specific antibody, the intensity of α-specific antibody signal often does not correlate with the intensity of the δ-specific antibody signal. These results indicate that the ratio of the four subunits varies among the transiently transfected cells.

As already discussed above, it is critical that the four subunit cDNAs are transfected in a set ratio. Deviations from this ratio cause decreases in the level of surface expression. Therefore, altogether the results suggest that cell-to-cell variations in ratio of the four subunits is the cause of low assembly efficiency in the transiently transfected cells. If variations in the subunit ratio are the cause of the low assembly efficiency, then there should be little difference in the assembly efficiency of a homomeric ion channel when either transiently or stably expressed. Although we have not measured the assembly efficiency of the homomeric AChR, a much higher efficiency would explain why the homomeric AChR expression levels are 5-fold higher than for the heteromeric AChR (Fig. 2). As we discussed previously, variations in the subunit ratio appear to result in the assembly of subunit complexes containing less than the full complement of subunits. Thus, the low efficiency of heteromeric AChR assembly obtained for transient transfection is likely to be caused by the assembly of large numbers of subunit complexes lacking one or more subunits.

Summary

Transient transfection is an excellent method for the expression and study of cell-surface, heteromeric ion channels. The cell type, the total amount of DNA, the combination of subunits and the ratio of subunit DNA are all important parameters to consider when attempting to optimize expression. A serious drawback of this method is that the efficiency of subunit assembly is very low in comparison to the efficiency of assembly for stably expressed heteromeric ion channels. The low efficiency of assembly prevents use of transient expression methods for detailed studies of heteromeric AChR assembly, and caution should be taken in the use of these methods for the study of intracellular heteromeric ion channel subunits. After the transient expression of heteromeric AChR subunits, virtually all of the expressing cells contained all four AChR subunits. However, the subunits were heterogeneously distributed among the cells, and the low efficiency of AChR assembly appears to be due to cell-to-cell variations in the ratio of the four subunits.

Acknowledgments

The authors are most grateful to Dr. T. Claudio for the cell lines stably expressing the *Torpedo* and mouse AChRs and the polyclonal anti-α and δ Abs, Dr. S. Sine for the pRBG constructs containing the *Torpedo* and mouse AChR subunit cDNAs, Dr. J.-L. Eisele for the pMT3 construct containing the α_7/5HT$_3$ cDNA, Dr. J. Lindstrom for MAbs 14, 148, and 168, and Dr. J. Kyle for the tsA201 cell line. The authors would also like to thank Christian Wanamaker for help with some of the experiments. This work was supported by grants from the National Institutes of Health and the Brain Research Foundation.

Section V

Model Simulations

[32] Molecular Modeling of Ligand-Gated Ion Channels

By MICHAEL J. SUTCLIFFE, ALLISTER H. SMEETON, Z. GALEN WO, and ROBERT E. OSWALD

Introduction

Molecular modeling is a method for mimicking the behavior of molecules and molecular systems. Simple molecular modeling studies can be performed using mechanical models. However, molecular modeling has now become synonymous with computer modeling. Computer-based molecular models are three dimensional in nature and, importantly, are interactive, allowing one to pose questions such as "what if ... ?" or "is it possible to ... ?" In the past, molecular modeling was restricted to a small cohort of people who wrote their own programs and managed their own computer systems. Computer workstations today are much more powerful than the mainframe computers of even several years ago and are relatively inexpensive. Software can be obtained readily from commercial companies, from academic laboratories, or, increasingly, via the World Wide Web (WWW). Molecular modeling produces testable hypotheses. These can be extremely useful, as in successfully predicting the structure of human immunodeficiency virus (HIV) protease.[1] They can, however, lead one astray, as in our modeling of the guanosine triphosphate (GTP) versus guanosine diphosphate (GDP) bound[2] forms of a GTP binding protein, Cdc42Hs; nevertheless, some aspects of this modeling study proved useful. Molecular modeling is particularly powerful when used as part of a multidisciplinary study in an iterative modeling/experimental verification cycle (Fig. 1). This article discusses how such techniques can be applied to ligand-gated ion channels, using efforts to model ionotropic glutamate receptors as an example.

Ligand-Gated Ion Channels

Ion channels are commonly described in terms of their ionic selectivity and gating properties. Ligand-gated ion channels are membrane-bound proteins that are activated by the binding of a ligand. This can be either a

[1] L. H. Pearl and W. R. Taylor, *Nature* **329,** 351 (1987).
[2] M. J. Sutcliffe, J. Feltham, R. A. Cerione, and R. E. Oswald, *Prot. Peptide Lett.* **1,** 84 (1994).
[3] J. L. Feltham, V. Dötsch, S. Raza, D. Manor, R. A. Cerione, M. J. Sutcliffe, G. Wagner, and R. E. Oswald, *Biochemistry* **36,** 3755 (1997).

0076-6879/98 $25.00

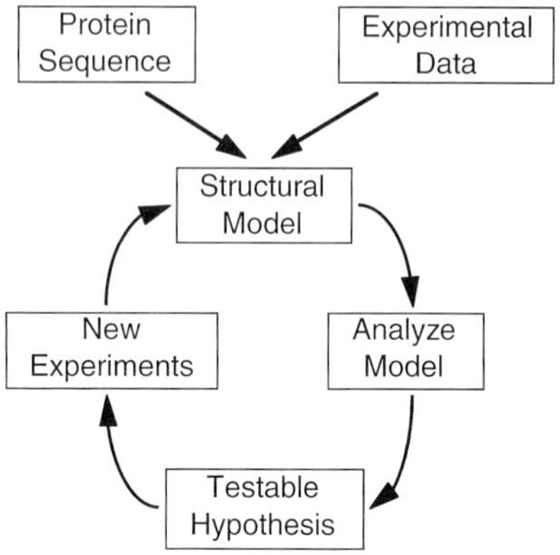

FIG. 1. Modeling/experimental verification cycle.

neurotransmitter [e.g., the nicotinic acetylcholine receptor channel (nAChR),[3a] γ-aminobutyric acid (GABA) receptor channel, and the glutamate receptor channel] or an internal ligand [e.g., calcium ions, adenosine triphosphate (ATP), and cyclic nucleotides.][3b] Ion channels that are gated by the action of neurotransmitters are involved in fast synaptic transmission in the nervous system. Glutamate receptors are the primary excitatory neurotransmitter receptors in vertebrate brain and play an important role in a wide variety of neuronal functions.[4] They are classified according to their signal transduction mechanism—metabotropic glutamate receptors (mGluRs) are linked to GTP-binding proteins and thus operate through second messengers, whereas ionotropic glutamate receptors (iGluRs) function as ligand-gated cation channels. Like other G-protein-coupled receptors, the mGluRs have seven transmembrane-spanning regions, but otherwise differ significantly from classical G-protein-coupled receptors.[5] There are three major types of iGluRs, classified according to the agonists by which they are activated: *N*-methyl-D-aspartate (NMDA) receptors (NMDAR1 to 2D), α-amino-3-hydroxy-5-methyl-4-isoxazole propionate (AMPA) receptors (GluR1 to 4), and kainate receptors (GluR5 to 7 and KA1 to 2). In addition, kainate receptors of lower molecular mass (40–50 kDa, also known as kainate-binding proteins) have been cloned from nonmammalian

[3a] M. S. P. Sansom, *Methods Enzymology* **293,** [34], 1998 (this volume).

[3b] S-P. Scott and J. C. Tanaka, *Methods in Enzymol.* **293,** [33], 1998 (this volume).

[4] D. T. Monaghan, R. J. Bridges, and C. W. Cotman, *Annu. Rev. Pharmacol. Toxicol.* **29,** 365 (1989).

[5] S. Nakanishi and M. Masu, *Annu. Rev. Biophys. Biomol. Struct.* **23,** 319 (1994).

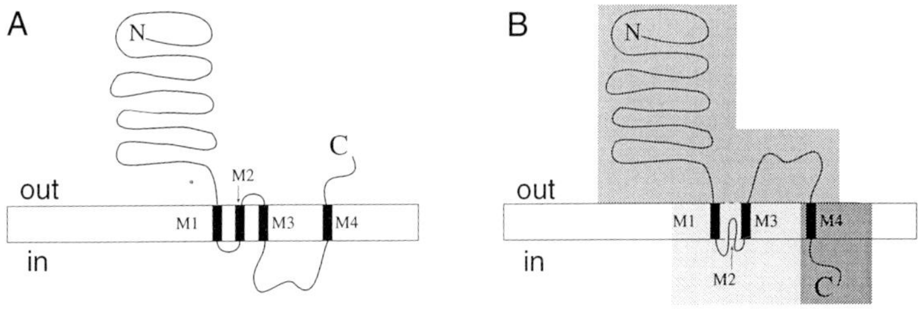

FIG. 2. Schematic representation of (A) the original four-transmembrane topology and (B) the accepted three-transmembrane topology of iGluRs. (B) The midgray region seems to have arisen from bacterial periplasmic amino acid binding proteins, the light gray region is similar to the pore-lining region of K^+ channels, and the dark gray region is a C-terminal regulatory domain of unknown origin.

vertebrates (e.g., frog,[6] chick,[7] and goldfish[8]); these exhibit considerable sequence homologies with the C-terminal portions of the 100-kDa mammalian AMPA/kainate receptors.

Transmembrane Topology

The first step in modeling ligand-gated ion channels, as with any membrane-bound protein, is to determine the transmembrane topology (i.e., which regions of the structure are extracellular, which are cytoplasmic, and which lie in the membrane itself; see Fig. 2). A knowledge of the transmembrane topology gives an indication of which regions of the sequence are likely to be close together in the three-dimensional structure—an important prerequisite in any modeling exercise. Hydropathy profiles, generated by scanning the amino acid sequence to determine where regions of consecutive hydrophobic amino acids are located, give an indication of which residues might be located in the membrane. This type of analysis is supplemented by the results from the programs MEMSAT,[9] TMAP[10] (http://www.embl-heidelberg.de/tmap/tmap_info.html), and PHDtopology[11] (http://www.embl-heidelberg.de/predictprotein/predict protein.html). These programs use information compiled from known transmembrane helical segments to predict the location of transmembrane heli-

[6] K. Wada, C. J. Dechesne, S. Shimasaki, R. G. King, K. Kusano, A. Buonanno, D. R. Hampson, C. Banner, R. J. Wenthold, and Y. Nakatani, *Nature* **342,** 684 (1989).
[7] P. Gregor, I. Mano, I. Maoz, M. McKeown, and V. Teichberg, *Nature* **342,** 689 (1989).
[8] Z. G. Wo and R. E. Oswald, *Proc. Natl. Acad. Sci. U.S.A.* **91,** 7154 (1994).
[9] D. T. Jones, W. R. Taylor, and J. M. Thornton, *Biochemistry* **33,** 3038 (1994).
[10] B. Person and P. Argos, *J. Mol. Biol.* **237,** 182 (1994).
[11] B. Rost, P. Fariselli, and R. Casadio, *Protein Sci.* **7,** 1704 (1996).

ces within the sequence. The transmembrane topology is then predicted from a consensus of the hydropathy results, the predicted transmembrane helices, and any available experimental data that help shed light on the transmembrane topology.

Identification of the correct transmembrane topology is now always clear cut, as illustrated by the iGluRs. Originally the iGluRs were thought to be structurally similar to other ligand-gated ion channels,[5] an assumption apparently supported by the hydropathy profile. This suggested that they contain four membrane-spanning regions (denoted M1 to M4; see Fig. 2A), with both the N and C termini located extracellularly. However, experimental analysis of the position of N-glycosylation sites[8,12–16] demonstrated that the transmembrane topology of iGluRs differs significantly from other ligand-gated ion channels. This showed that iGluRs consist of three membrane-spanning regions (denoted M1, M3, and M4 in Fig. 2B), with the N-terminal region located extracellularly and the C-terminal region located intracellularly. The region denoted M2, originally thought to cross the membrane, does not span the membrane but, based on its role in ion conduction, is probably inserted into the ion conduction pathway in a manner similar to the P-segment of K^+ channels.

The production of a structural model of a ligand-gated ion channel in one step would be a Herculean task. The first problem is that ligand-gated ion channels are generally composed of multiple subunits, either as homomeric or heteromeric combinations. This can introduce complexity in several ways. Interactions between subunits can potentially alter the structure of the individual subunits and this interaction can be functionally important in terms of cooperative interactions. The case of the nAChR may be even more difficult in that the ligand-binding site site seems to span two subunits.[17,18] This has been taken into consideration in models of the nAChR.[19] When possible, however, the task of producing a structural model is made much more tractable by dividing the amino acid sequence into a number of smaller "modules" based on those regions of the sequence that are predicted to be close together in the three-dimensional structure. For

[12] Z. G. Wo and R. E. Oswald, *FEBS Lett.* **368,** 230 (1995).

[13] Z. G. Wo and R. E. Oswald, *J. Biol. Chem.* **270,** 2000 (1995).

[14] F. A. Taverna, L. Y. Wang, J. F. MacDonald, and D. R. Hampson, *J. Biol. Chem.* **269,** 14159 (1994).

[15] K. W. Roche, L. A. Raymond, C. Blackstone, and R. L. Huganir, *J. Biol. Chem.* **269,** 11679 (1994).

[16] M. Hollmann, C. Maron, and S. Heinemann, *Neuron* **13,** 1331 (1994).

[17] R. E. Oswald and J. P. Changeux, *FEBS Lett.* **139,** 225 (1982).

[18] S. M. Sine and T. Claudio, *J. Biol. Chem.* **266,** 19369 (1991).

[19] I. Tsigelny, N. Sugiyama, S. M. Sine, and P. Taylor, *Biophys. J.* **73,** 52 (1997).

example, the iGluRs are composed of several evolutionarily distinct modules[20] (Fig. 2B):

1. An extracellular portion that seems to have arisen from two different classes of bacterial protein—the N-terminal half is homologous to the leucine, isoleucine, valine-binding protein (LIVBP), and the C-terminal half is homologous to the lysine, arginine, ornithine-binding protein (LAOBP).[21,22] The LAOBP-like domain in iGluRs consists of two regions of sequence, separated in sequence by a large insert that includes two of the membrane spanning regions. The LIVBP-like domain is absent from the 50-kDa kainate-binding proteins.
2. A pore-lining region that is similar in topology to the pore-lining region of K^+ channels.
3. A variable C-terminal regulatory domain of unknown origin that exhibits considerable diversity among subtypes.

An overview of our model building strategy is given in Fig. 3. To date, we have concentrated our efforts on modeling two of the domains found in iGluRs—the LAOBP-like domain and the K^+ channel-like domain. We should stress that such models represent a guess at the structure. As in any modeling study, the result is no substitute for a high-resolution experimentally derived three-dimensional structure but is extremely useful as a guide to subsequent experimental work.

Identification of Template Structure(s)

Scanning against Database of 3-D Structures

Once a structural module has been defined, a search is made for a similar protein(s) of known three-dimensional structure. This is based on the observation that a good correlation generally exists between the level of similarity in the amino acid sequence and the similarity in the three-dimensional structure. The amino acid sequence is scanned against a database containing the three-dimensional structures of proteins that have been determined experimentally. In this case, the database accessed via the SCOP[23] Web server (http://scop.mrc-lmb.cam.ac.uk/scop/aln.cgi) is used.

[20] Z. G. Wo and R. E. Oswald, *Trends Neurosci.* **18,** 161 (1995).

[21] N. Nakanishi, N. A. Schneider, and R. Axel, *Neuron* **5,** 569 (1990).

[22] P. J. O'Hara, P. O. Sheppard, H. Thøgersen, D. Venezia, B. A. Haldeman, V. McGrane, K. M. Houamed, C. Thomsen, T. L. Gilbert, and E. R. Mulvihill, *Neuron* **11,** 41 (1993).

[23] A. G. Murzin, S. E. Brenner, T. Hubbard, and C. Chothia, *J. Mol. Biol.* **247,** 536 (1995).

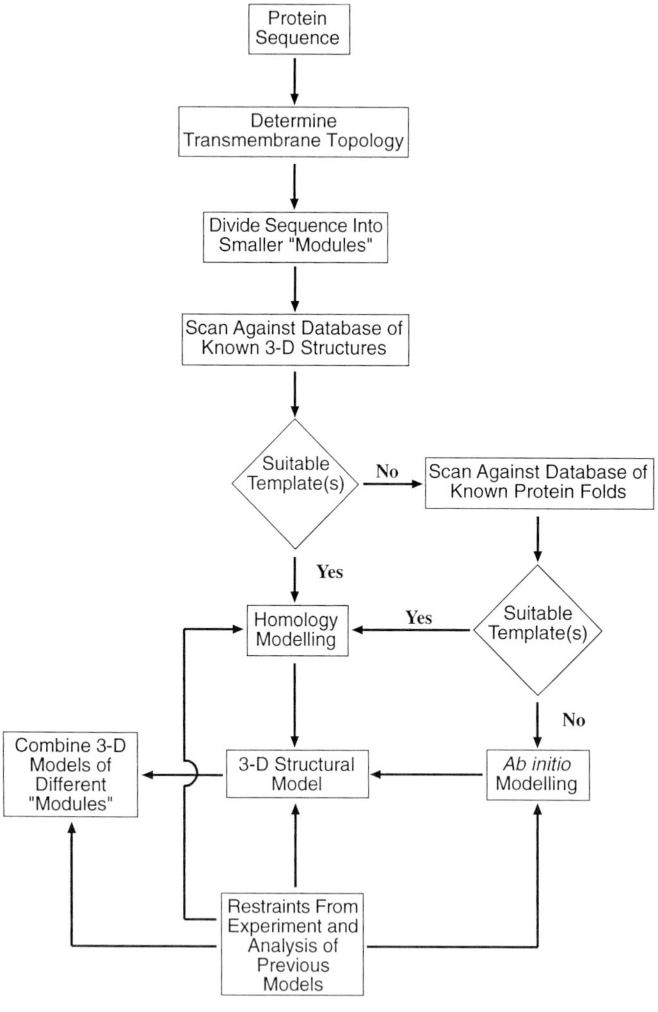

FIG. 3. Overview of model building strategy.

This computer program (based on the BLAST algorithm[24]) scans the amino acid sequence of the module being modeled against the sequence of each protein in the database in turn and returns a score for each pairwise comparison. It then ranks these scores and displays the most significant ones. This differs from a more general scan of a sequence against a database of protein

[24] S. F. Altschul, W. Gish, W. Miller, E. W. Myers, and D. J. Lipman, *J. Mol. Biol.* **215,** 403 (1990).

sequences in that only sequences for those proteins with known three-dimensional structure are included in this database.

For the iGluRs, the sequences of both the LAOBP-like domain and the K^+ channel-like domain were searched against SCOP. However, in neither case were similarities of any significance identified. An alternative approach, described below, was therefore adopted.

Scanning against Database of Protein Folds

Proteins with no obvious sequence similarity can still show remarkable similarities in their topologies (or folds), although similar folds are not identified as reliably as sequence similarities. A search is made to identify protein(s) that can favorably accommodate the amino acid sequence of the module of the ligand-gated ion channel onto their three-dimensional structure. The amino acid sequence is scanned against a database of known protein folds using two different Web servers: TOPITS[25] (http://www.embl-heidelberg.de/predictprotein/predictprotein.html) and the UCLA-DOE Structure Prediction Server[26] (http://www.mbi.ucla.edu/people/frsvr/frsvr.html). In both cases, the computer program matches the secondary structure prediction, predicted solvent accessibility, and amino acid environment of the sequence against a representative set of known protein folds. These scores are ranked and the most significant ones displayed.

Once a suitable fold is found, other proteins of known three-dimensional structure with the same fold are identified. The crystal structure of the protein identified by the fold recognition program is scanned against the structural database in the DALI[27] Web server (http://www2.ebi.ac.uk/cgi-bin/dali_html_align_pair). The computer program compares this structure with each of the known protein structures in the database in turn, assigns each of these pairwise comparisons a score, and ranks them. The most significant scores are returned.

When sequences of the LAOBP-like domain of the iGluRs were scanned against the TOPITS and UCLA-DOE servers, both returned histidine-binding protein (HBP) as the only significant hit. Thus, HBP was identified as a suitable template structure to use as the basis for modeling. The three-dimensional structure of HBP deposited in the Brookhaven Protein Data

[25] B. Rost, *in* "Third International Conference on Intelligent Systems for Molecular Biology (ISMB)" (C. Rawlings, D. Clark, R. Altman, L. Hunter, T. Lengauer, and S. Wodak, eds.), p. 314. AAAI Press, Menlo Park, California, 1995.

[26] D. Fischer and D. Eisenberg, *Protein Sci.* **5,** 947 (1996).

[27] L. Holm and C. Sander, *Proteins,* **19,** 165 (1994).

Bank[28] (PDB) is in the holo (ligand-bound) form (PDB accession number 1HSL[29]). Scanning this structure against DALI identified the holo form of lysine, arginine, ornithine-binding protein (LAOBP; 1LST[30]) as an additional template structure for modeling the holo form of the LAOBP-like domain. It also identified the apo (ligand-free) form of LAOBP (2LAO[30]) as a suitable template for modeling the apo form of the LAOBP-like domain of the iGluRs. [Additional structures that could be used as templates, but which were not available when we did the modeling, are as follows: (1) the apo form of glutamine-binding protein (QBP; 1 GGG[31]), and (2) the holo form of QBP (1WDN[31]; on hold until 5 May 1998).] In contrast, the sequences of the K^+ channel-like domain resulted in no significant hits when compared against the TOPITS and UCLA-DOE servers.

Identifying Templates by Other Means

If no suitable template structure is identified, either on the basis of sequence similarity or fold recognition, an alternative strategy is adopted. All of the available experimental data are considered in conjunction with what can be gleaned from analyzing the amino acid sequences. Approaches used include multiple sequence alignment, secondary structure prediction, hydropathy plots, and prediction of membrane-spanning helices. Aligning all of the amino acid sequences available for a particular family of proteins aids in identifying conserved features across that family. A list of the sequences in a family is derived by searching one or more of the known sequences against BLAST[24] using the "BLASTP+BEAUTY" and "WU-BLASTP+BEAUTY" (the latter allows gapped alignments, whereas the former does not) searches in the BCM Search Launcher Web server (http://dot.imgen.bcm.tmc.edu:9331/seq-search/protein-search.html), and studying those hits with the highest similarity to the query sequence supplied by the user. These amino acid sequences are then aligned by the Clustal W[32] Web server (http://dot.imgen.bcm.tmc.edu:9331/multi-align/Options/clustalw.html), and preferably at least one other program (e.g., we also use Multal[33] within the program CAMELEON; Oxford Molecular Ltd, Oxford, UK; http://www.oxmol.co.uk/). The programs return a multiple-sequence align-

[28] E. E. Abola, F. C. Bernstein, S. H. Bryant, T. F. Koetzle, and J. Weng, *in* "Crystallographic Databases—Information Content, Software Systems, Scientific Applications" (F. H. Allen, G. Bergerhoff, and R. Sievers, eds.), p. 107. Data Commission of the International Union of Crystallography, Bonn/Cambridge/Chester, 1987.

[29] N. H. Yao, S. Trakhanov, and F. A. Quicho, *Biochemistry* **33,** 4769 (1994).

[30] B. H. Oh, J. Pandit, C. H. Kang, K. Nikaido, S. Gokcen, G. F. L. Ames, and S. H. Kim, *J. Biol. Chem.* **268,** 11348 (1993).

[31] C. D. Hsiao, Y. J. Sun, J. Rose, and B. C. Wang, *J. Mol. Biol.* **262,** 225 (1996).

[32] J. D. Thompson, D. G. Higgins, and T. J. Gibson, *Nucleic Acids Res.* **22,** 4673 (1994).

[33] W. R. Taylor, *J. Mol. Evol.* **28,** 161 (1988).

ment; at least two programs are used so that a consensus alignment can be derived. Positions in the alignment of conserved amino acids are indicators of (either structurally or functionally) important amino acids in the sequences. (A good starting point is the multiple-sequence alignment of the iGluRs available via http://www.embl-heidelberg.de/srs/srsc?[PIRALN-id:FA1197].) Secondary structure prediction is performed, either on a single-sequence or a multiple-sequence alignment, using the PHDsec[34] Web server (http://www.embl-heidelberg.de/predictprotein/predictprotein.html) and the DSC[35] Web server (http://bonsai.lif.icnet.uk/bmm/dsc/dsc_form_align.html). The programs return the predicted secondary structure for each residue position in the sequence; two programs are used so that a consensus prediction can be derived. The use of hydropathy plots and prediction of transmembrane helices are discussed in the earlier section on Transmembrane Topology.

This approach was used to identify a structural template for the K^+ channel-like domain of the iGluRs. As the name implies, these were identified as having a topology similar to the central part of the K^+ channel. Because no experimentally derived three-dimensional structures of K^+ channels are available, the model structure of the inward rectifier K^+ channel IRK1[36] (also known as Kir 2.1), which was being developed by one of us (MJS), was used partly as the initial structural template for this domain. Note that, although some similarity in topology exists between the K^+ channel-like domain of the iGluRs (M1, M2 and M3) and the K^+ channels, the details of the P segments are different, as can be seen by comparing the topological studies of the NMDA-R1 and -R2C subunits[37] with topological studies of the voltage-gated K^+ channels.[38–41] Nevertheless, the region of the K^+ channels equivalent to M1, M2, and M3 in iGluRs provides a useful starting point for modeling this region.

Identifying Corresponding Residues in Template(s)

Once the structural templates are identified, the particular amino acids in the sequence of the ligand-gated ion channels that correspond to the amino acids in the template structure(s) are identified. Multiple-sequence

[34] B. Rost and C. Sander, *J. Mol. Biol.* **232,** 584 (1993).

[35] R. D. King and M. J. E. Sternberg, *Protein Sci.* **5,** 2298 (1996).

[36] P. R. Stanfield, N. W. Davies, P. A. Shelton, M. A. Sutcliffe, I. A. Khan, W. J. Brammar, and E. C. Conley, *J. Physiol.* **478,** 1 (1994).

[37] T. Kuner, L. P. Wollmuth, A. Karlin, P. H. Seeburg, and B. Sakmann, *Neuron* **17,** 343 (1996).

[38] J. Aiyar, J. P. Rizzi, G. A. Gutman, and K. G. Chandy, *J. Biol. Chem.* **271,** 31013 (1996).

[39] L. L. Kurz, R. D. Zuhlke, H. J. Zhang, and R. H. Joho, *Biophys. J.* **68,** 900 (1995).

[40] Q. Lu and C. Miller, *Science* **268,** 304 (1995).

[41] J. M. Pascual, C. C. Shieh, G. E. Kirsch, and A. M. Brown, *Neuron* **14,** 1055 (1995).

alignment (see previous section) is used for this. Because no single approach provides a "definitive" alignment, using more than one sequence alignment program allows one to derive a consensus alignment and to consider alternative alignments. This consensus alignment should take into account the predicted secondary structure (see previous section) of the ligand-gated ion channel sequences, in particular, comparing these with the secondary structural elements in the template structure(s) and, wherever possible, avoiding insertions and deletions within these secondary structural elements in the sequence alignment. Also, any experimentally derived restraints are taken into account when determining this alignment. Examples of the type of restraint used include a knowledge of those residues that need to be on the surface of the protein (e.g., glycosylated residues), those residues known to line the ligand-binding site, or those half-cystine residues known to form disulfide bridges.

For the iGluRs, this was achieved by aligning the amino acid sequences of 14 members of the iGluR superfamily with the amino acid sequences of LAOBP, HBP, and glutamine binding protein (QBP). The final consensus sequence alignment was produced manually within the program CAMELEON, subject to the constraints that (1) wherever possible, no insertion or deletion occurred within the crystallographically determined secondary structural elements of the bacterial periplasmic amino acid binding proteins, as defined in the respective "extended secondary structure" entries in the IDITIS database of protein structures (Oxford Molecular Ltd.); (2) the known N-glycosylation sites[8,12–16] corresponded to surface positions (to be glycosylated, an amino acid [in this case, an asparagine residue] position must be accessible to the incoming sugar molecule); (3) all those residues, at the time the initial modeling was performed, thought to be involved in ligand binding[42–44] were positioned in the binding site; (4) the disulfide bridge, which we showed by experiment is present,[45] could be formed; and (5) some account was taken of the secondary structure. Part of the resulting (initial) alignment is shown in Fig. 4 (see Sutcliffe *et al.*[47] for the full alignment).

Homology Modeling

Once the sequence alignment is produced, the next step is to derive a three-dimensional model using either homology modeling (this section) or

[42] A. Kuryatov, B. Laube, H. Betz, and J. Kuhse, *Neuron* **12,** 1291 (1994).
[43] F. Li, N. Owens, and T. A. Verdoorn, *Mol. Pharmacol.* **47,** 148 (1995).
[44] S. Uchino, K. Sakimura, K. Nagahari, and M. Mishina, *FEBS Lett.* **308,** 253 (1992).
[45] Z. G. Wo and R. E. Oswald, *Mol. Pharmacol.* **50,** 770 (1996).
[46] G. J. Barton, *Protein Eng.* **6,** 37 (1993).
[47] M. J. Sutcliffe, Z. G. Wo, and R. E. Oswald, *Biophys. J.* **70,** 1575 (1996).

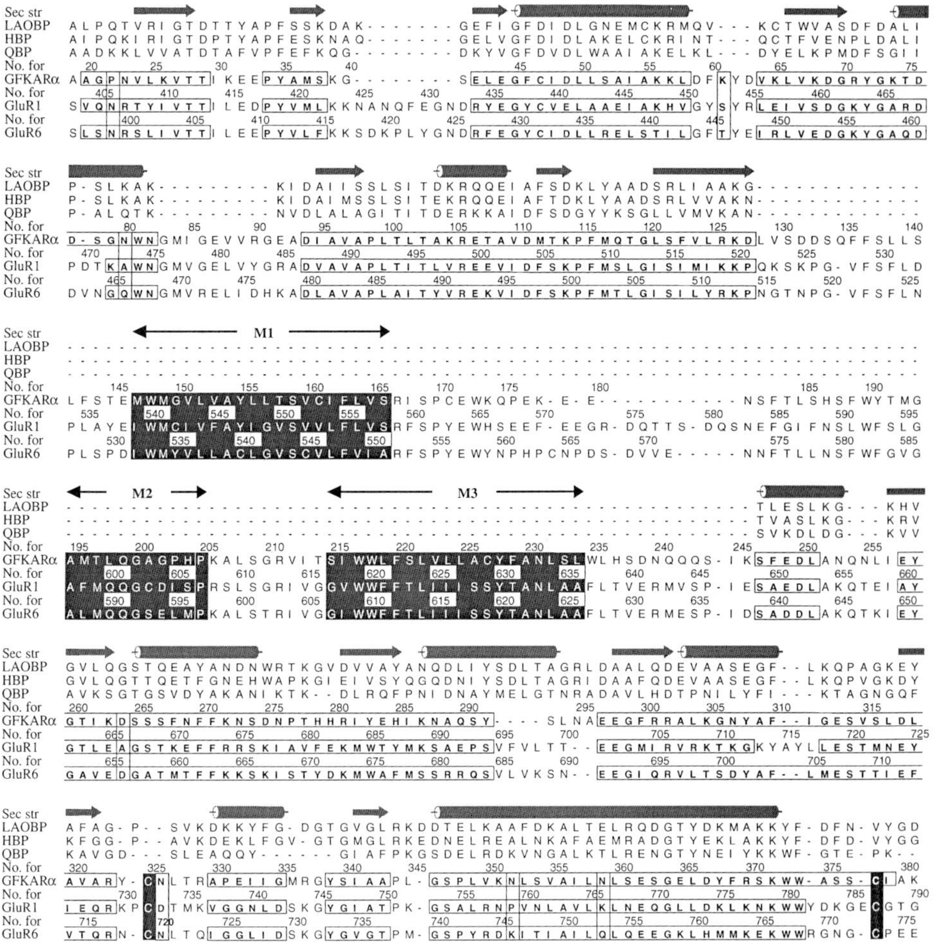

FIG. 4. The initial sequence alignment used for modeling the iGluRs (see Sutcliffe *et al.*[47] for the full alignment). The region shown includes the LAOBP-like and the K^+ channel-like domains of the iGluRs. The "framework" regions used for modeling with COMPOSER are denoted by horizontal boxes; the positions of M1, M2, and M3 are shown on a dark background; glycosylation sites are denoted by vertical boxes; and the two half-cystines are shown on a dark background. The consensus secondary structure observed in the bacterial periplasmic amino acid binding proteins is also shown. (Figure generated using ALSCRIPT.[46])

ab initio modeling (next section).[36] The results of homology modeling are critically dependent on (1) the choice of structural template(s) and (2) the sequence alignment used, emphasizing the need for care in the preceding stages. Programs for homology modeling use one of two approaches—either a fragment-based stepwise approach or a single-step approach. We have used both in our modeling of ligand-gated ion channels, and both methods produce models of equally good quality. However, the latter is now our method of choice because it enables experimentally derived restraints to be added during the modeling process, rather than in a *post hoc* fashion as with the fragment-based approach. In addition to being evaluated against the available experimental data, the "quality" of the models is evaluated during the modeling process using (1) the Verify3D[48] Web server (http://www.doe-mbi.ucla.edu/Services/Verify3D.html) to determine the compatibility between the amino acid sequence and the environment of these amino acids in the model, and (2) the Biotech Validation Suite for Protein Structures Web server (http://biotech.embl-ebi.ac.uk:8440/) to evaluate the stereochemical quality of the model.

Fragment-Based Approach

The program COMPOSER[49,50] (http://www.tripos.com/) or the SWISSMODEL[51] Web server (http://expasy.hcuge.ch/swissmod) is used for fragment-based homology modeling. They divide parts of the structural templates into three groups: the well-defined regions of the polypeptide backbone (or "structurally conserved regions"), the poorly defined regions of the polypeptide backbone (or "structurally variable regions"), and the amino acid side chains. The first stage is to build the polypeptide backbone of the structurally conserved regions of the model, which are either defined automatically by the program or entered manually by the user—this results in a disjointed set of structural fragments. The next stage is to join these with the structurally variable regions of the polypeptide backbone (often "loops" joining secondary structural elements) or, where no suitable fragment exists in the templates, to scan a database of protein structures to identify a suitable fragment. The third stage is to change the atoms in the amino acid side chains, where necessary. The final step is energy minimization of the model.

Single-Step Approach

The program MODELLER[52] (available via http://guitar.rockefeller.edu/modeller/modeller.html) produces homology models in a "single step." The individual structural features in the model (e.g., main-chain conformation and the position of hydrogen bonds) are represented by probability

[48] R. Luthy, J. U. Bowie, and D. Eisenberg, *Nature* **356,** 83 (1992).
[49] M. J. Sutcliffe, F. R. F. Hayes, and T. L. Blundell, *Protein Eng.* **1,** 385 (1987).
[50] M. J. Sutcliffe, I. Haneef, D. Carney, and T. L. Blundell, *Protein Eng.* **1,** 377 (1987).
[51] M. C. Pietsch, *Biochem. Soc. Trans.* **24,** 274 (1996).
[52] A. Sali and T. L. Blundell, *J. Mol. Biol.* **234,** 779 (1993).

distribution functions based partly on the structure of the template(s) and partly on known stereochemistry. These probabilities are used as restraints when the model is constructed. Since each feature is represented as a probability distribution function, a family of models consistent with this set of distributions, each with a different conformation, is produced. Production of a family of models, rather than a single model as in the fragment-based approach, enables the significance of different interactions and conformations in the final models to be evaluated. The "representative models" and "core residues" across the family of models are identified using the OLDERADO[53] Web server (http://neon.chem.le.ac.uk/olderado/) Also, if steric problems (particularly with bond lengths and, to a lesser extent, bond angles) occur consistently across the family of models, these are indicative of an error in the sequence alignment. The sequence alignment is therefore checked against the three-dimensional alignment of the template structures with the models using interactive molecular graphics (e.g., InsightII and Quanta, MSI, San Diego, CA; http://www.msi.com/), and the sequence alignment updated accordingly. The major advantage of this approach is that restraints derived from both experimental observations and analysis of previous models can be included in the modeling process in the form of distance restraints.

For the iGluRs, the program COMPOSER was used to produce our earlier models of the LAOBP-like domain of the apo and holo forms of GFKARα, GluR1, and GluR6—a total of six models. The structurally conserved parts of the templates were defined manually (rather than automatically; see Fig. 4). The structures produced by COMPOSER required refinement due to bad stereochemistry and because the disulfide bridge needed to be formed. This was achieved using simulated annealing and energy minimization within the program XPLOR[54] (available via http://xplor.csb.yale.edu/). The disulfide bond was formed by fixing the position of all the atoms except those in the two loops containing the two half-cystines (C325 and C378 in GFKARα) and adding a distance restraint to draw these two residues together, followed by addition of an explicit bonded term between these two residues in subsequent refinement.

Our more recent models of the LAOBP-like domain are produced using MODELLER. In these, specific protein-ligand interactions, identified from a combination of experimental results and earlier models, are included in the modeling process in the form of distance restraints. Three types of interaction are included in this way (see section on Docking of Ligands): hydrogen bonds, charge–charge interactions, and hydrophobic interactions.

[53] L. A. Kelley, S. P. Gardner, and M. J. Sutcliffe, *Protein Eng.* **10,** 737 (1997).

[54] A. T. Brünger, Xplor manual version 3.851, Yale University (1997).

The ability to add this type of distance restraint during the modeling process is the major advantage of MODELLER over COMPOSER. The two approaches performed equally well[55] in a recent "blind test" (the 1996 Critical Assessment of Structure Prediction; CASP2). This study highlighted that the accuracy of the sequence alignment is far more important than the program used to produce the model.

Ab Initio Modeling

The complete *ab initio* prediction of protein three-dimensional structures is not possible at present, and a general solution to the protein folding problem is not likely to be found in the near future. However, if some knowledge is available of (1) the secondary structural elements in the module being modeled and (2) how these secondary structural elements pack together, the modeling exercise becomes far more tractable, despite remaining highly speculative. This type of approach is illustrated later, using our modeling of iGluRs as an example.[3a]

For the iGluRs, analysis of the amino acid sequence revealed a similarity between the region M2 (see Fig. 2B) and the P segment (or H5 segment) of K^+ channels.[20] The P segment in K^+ channels was originally thought to be a β-hairpin insertion.[56] However, a revised model suggests that it is not a β-sheet conformation,[38,57] consistent with several experimental reports.[38–41] For simplicity, the P segment of the iGluRs was modeled initially as an antiparallel β barrel using XPLOR.[54] Many, but not all, of the available experimental data were consistent with this topology, suggesting subsequent adjustment was necessary. Models were built with stoichiometries of 4, 5, and 6 (the stoichiometry has not yet been definitively established) for GFKARα, GluR1, and GluR6. Twenty models (including hydrogen atoms) were generated for each structure using distance geometry, followed by simulated annealing and energy minimization. The models were restrained to be symmetrical [using the nuclear Overhauser effect (NOE) "symmetry" potential to restrain equivalent distances between backbone atoms and the NCS restraint to ensure symmetry across all atoms], appropriate distance restraints used for backbone hydrogen bonds (NH $\cdots$ O distances of 1.8–2.3 Å, and N $\cdots$ O distances of 2.5–3.3 Å) and dihedral angle restraints (ϕ, -120 to $-160°$; ψ, 115 to 155°) were applied to the β strands. Once M2 was built, the model was extended to include M1 and M3 (see Fig. 2B) by taking a single subunit of the transmembrane domain of a model of the

[55] A. C. R. Martin, M. W. MacArthur, and J. W. Thornton, *Proteins—Structure, Function, and Genetics* **Suppl. 1,** 14 (1997).

[56] S. R. Durell and H. R. Guy, *Biophys. J.* **62,** 238 (1992).

[57] H. R. Guy and S. R. Durell, *J. General Physiol. Abstr.* **17,** P7 (1994).

topologically similar inward rectifier K^+ channel IRK1[36] (also known as Kir 2.1), superposing this onto the β barrel, and then mapping the respective sequence onto M1 and M3. The model was then refined manually, using interactive molecular graphics (InsightII), to ensure that the polar residues in the α helices are oriented such that the polar residues point toward the center of the pore. The M2 region in the models was then refined interactively using SCULPT,[58] (http://www.intsim.com/) to ensure that all the residues were positioned so as to be consistent with experimental data.

An important consideration in building the models is the position of the crucial residue known to affect conductance in a number of iGluRs. This position in GluR2 and GluR6 [residue 198[600]; here and below residue numbers are given for GFKARα (see Fig. 4), with the corresponding numbering for GluR1 in square brackets; numbering begins at the start of the signal sequence] can be changed from Q to R by RNA editing[59] and is known to be involved in the blockade of NMDA receptors by Mg^{2+} (N598 of NMDA-R1[60]). Scanning cysteine mutagenesis of NMDA receptor channels supports the exposure of this residue.[37] However, this residue would not be exposed to the narrowest part of the conductance pathway in the R form, since the positive charge would impede the flow of cations through the channel. It is more likely that the Arg side chain is shielded from the conductance pathway by the protein in the R form. Our models were therefore constructed consistent with the suggestion[61] that the arginine interacts with the negatively charged residue (D or E) four residues C-terminal of it. These models suggest that this D/E residue is located in an adjacent, rather than the same, subunit to the Q/R site. In non-NMDA receptors, the Q form displays characteristic double rectifying behavior, with currents for potentials around 0 mV exhibiting a very shallow slope and weak outward currents observed only at high positive voltages. In the R form, however, the *I–V* curve is either linear or outwardly rectifying.[62] Electrostatic calculations were performed using the program DELPHI[63] (available via http://tincan.bioc.columbia.edu/delphi/) with "focusing" (a method that enhances accuracy of the calculations by using results from a low-resolution calculation as boundary values for a higher resolution

[58] M. C. Surles, J. S. Richardson, D. C. Richardson, and F. P. J. Brooks, *Protein Science* **3,** 198 (1994).

[59] B. Sommer, M. Köhler, R. Sprengel, and P. H. Seeburg, *Cell* **67,** 11 (1991).

[60] N. Burnashev, R. Schoepfer, H. Monyer, J. P. Ruppersberg, W. Günther, P. H. Seeburg, and B. Sakmann, *Science* **257,** 1415 (1992).

[61] R. Dingledine, R. I. Hume, and S. F. Heinemann, *J. Neurosci.* **12,** 4080 (1992).

[62] T. A. Verdoorn, N. Burnashev, H. Monyer, P. H. Seeburg, and B. Sakmann, *Science* **252,** 1715 (1991).

[63] A. Nicholls and B. Honig, *J. Comp. Chem.* **12,** 435 (1991).

calculation), and the results visualized using GRASP[64] (available via http://tincan.bioc.columbia.edu/Lab/grasp/). These suggest that the Q form provides a good cation-binding site, whereas the R form does not. This in turn suggests why the Q form is doubly rectifying (the D/E site is free to bind Ca^{2+}), whereas the R form is not (the predicted Ca^{2+} binding site is removed by formation of a salt bridge between the R and the D/E site; see Sutcliffe *et al.*[47] for a more detailed discussion of this point).

Additional site-directed mutagenesis data are also used for modeling M2. A number of residues in M2 of GluR3 were mutated to arginine.[61] "Killer" mutations were produced at positions equivalent to residues A195[597] and T197[599] in GFKARα. Our models are consistent with these observations in that these residues are exposed to the channel and an R in these positions would block the flow of current. Although not specifically included as constraints in our model, the exposed residues identified by scanning mutagenesis[37] are also exposed in our model. Our models also suggest why, when all Q/R sites in homomeric AMPA and kainate receptors are R, the channels conduct both anions and cations.[65] In these cases, all of the Ca^{2+}-binding sites at the D/E position are removed and, therefore, no net charge is present at this position in the channel.

In addition to M2, M1 and M3 are also likely to form part of the conductance pathway. M1 was modeled as an α helix and is predicted to form part of the ion conduction pathway. The models are consistent with the results of RNA editing of residues in M1 of GluR6.[66] This results in amino acid changes in GluR6 in positions equivalent to C150[542] and A154[546] in GFKARα. The RNA editing has no effect on Ca^{2+} permeability in the R form, whereas in the Q form the edited channel (with V, C) has a lower relative permeability than the unedited channel (I, Y[66]). The model suggests an explanation for this—a serine residue (equivalent to A201[603] in GFKARα) may be exposed in the edited form, providing a binding site for Ca^{2+}, and thus increasing the relative permeability (see Sutcliffe *et al.*[47]). M3 was also modeled as an α helix and, like M1, is predicted to form part of the ion conduction pathway. This is consistent with the suggested role of M3 in the channel blocking action of MK-801.[67] Therefore, M3 was oriented to maximize the exposure of polar groups to the pore. This suggested a possible hydrogen bond between the residue at the Q/R site in M2 and the conserved Tyr (Y228[630] in GRKARα) in M3.

[64] A. Nicholls, K. Sharp, and B. Honig, *Proteins* **11,** 281 (1991).
[65] N. Burnashev, A. Villarroel, and B. Sakmann, *J. Physiol. Lond.* **496,** 165 (1996).
[66] M. Köhler, N. Burnashev, B. Sakmann, and P. H. Seeburg, *Neuron* **10,** 491 (1993).
[67] A. V. Ferrer-Montiel, W. Sun, and M. Montal, *Proc. Natl. Acad. Sci. U.S.A.* **92,** 8021 (1995).

Stoichiometry

Subunit stoichiometry in the channel is an important consideration when modeling ligand-gated ion channels. If this is known from experiment, then there is no ambiguity. However, if this is not known, modeling is used to predict the stoichiometry. This is further complicated if the channels are heteromultimers, comprised of more than one type of subunit. The size of the pore, derived from the largest ions passing through the channel, is the guide used in modeling. The program HOLE[68] (available via http://www.cryst.bbk.ac.uk/~ubcg8ab/hole/top.html) is used to measure the pore diameter.[3a] The results are displayed using Insight II.

The subunit stoichiometry of iGluRs has not been established definitively. Sedimentation analysis and chemical cross-linking have generally been interpreted as an indication of pentameric structure.[69–72] Single-channel analysis of NMDA receptor mutants[73] also supports a pentameric structure, but evidence for a tetramer has also been reported.[74] We therefore generated a set of tetrameric, pentameric, and hexameric models to investigate systematically the different stoichiometries of the structure. Apart from the number of subunits, the main differences arise from subtleties in the packing of extracellular domains and in the size and shape of the channel. The size of the pore in the models lends support to either a pentameric stoichiometry or a hexameric stoichiometry (but a hexameric stoichiometry is not supported by any experimental data, and therefore unlikely)—the tetramer appears too small to accommodate Na^+ and Ca^{2+} permeability through the channel. (The pore size of homomeric AMPA and kainate receptors in the Q form is approximately 7.8 Å and 7.5 Å, respectively; in the R form, homomeric kainate receptors do not alter significantly; heteromeric AMPA and kainate receptors coassembled from Q and R form subunits have a pore size estimated in the range between 7.0 and 7.4 Å.[65])

Docking of Ligands

Once a model (or family of models) of the ligand-binding domain has been constructed, the ligand molecule (this is either an agonist or antago-

[68] O. S. Smart, J. M. Goodfellow, and B. A. Wallace, *Biophys. J.* **65,** 2455 (1993).

[69] C. D. Blackstone, S. J. Moss, L. J. Martin, and R. Huganir, *J. Neurochem.* **58,** 1118 (1992).

[70] N. Brose, G. P. Gasic, D. E. Vetter, J. M. Sullivan, and S. F. Heinemann, *J. Biol. Chem.* **268,** 22663 (1993).

[71] R. J. Wenthold, N. Yokotani, K. Doi, and K. Wada, *J. Biol. Chem.* **267,** 501 (1992).

[72] T. Y. Wu and Y. C. Chang, *Biochem. J.* **300,** 365 (1994).

[73] L. Premkumar, K. Erreger, and A. Auerbach, *Biophys. J.* **72,** A334 (1997).

[74] B. Laube, H. Hirai, M. Sturgess, H. Betz, and J. Kuhse, *Neuron* **18,** 493 (1997).

nist) is "docked" into its presumed binding site.[3b] The first stage is to identify where on the model protein the ligand-binding site is located. This is achieved either by analogy with the location of the ligand-binding site in the template structure(s) or, if the answer is not apparent from the template, by assuming that the largest cavity on the surface of the protein corresponds to the ligand-binding site. The largest cavity is identified using SURFNET[75] (available via http://www.biochem.ucl.ac.uk/bsm/biocomp/)—the ligand is bound in the largest cavity in over 83% of complexes analyzed.[76]

The ligand is then positioned to maximize the complementarity between its properties (e.g., shape, functional groups) and those of the binding site. To achieve this, the ligand-binding site is characterized and compared with the ligand. The simplest way is to analyze the binding site using interactive molecular graphics (e.g., InsightII or Quanta), looking in particular for hydrogen bond donors and acceptors, charged atoms, and hydrophobic patches on the protein that are complementary to the ligand. A more rigorous result is obtained by testing potential interactions in the binding site using a series of probes for each of the functional groups present on the ligand. The program GRID[77] is used to search for possible binding sites for these functional groups, and the results visualized graphically using InsightII. The ligand is then modeled manually into the contours generated by GRID for each functional group, maximizing overlap of the predicted position of each functional group with the position of ligand atoms in the modeled complex. Electrostatic interactions are also investigated using "focusing" with the program DELPHI, and visualized using GRASP (see section on *Ab Initio* Modeling). The results of these electrostatic calculations are checked to ensure compatibility with the manually modeled complex produced using the results from GRID, and the complex is remodeled if necessary.

Having identified possible protein-ligand interactions, these are then included in the homology modeling process in the form of distance restraints. For example, for a hydrogen bond the donor · · · acceptor distance (e.g., oxygen · · · nitrogen, since the models are produced without the hydrogen atoms present) is restrained to a range of 2.5–3.3 Å. The same distance restraint is used for a salt bridge. Because of the less well-defined nature of restraints between atoms thought to be involved in hydrophobic interactions, such restraints are initially entered relatively loosely, the resulting models analyzed, and the hydrophobic restraints updated accordingly.

[75] R. A. Laskowski, *J. Mol. Graph.* **13,** 323 (1995).

[76] R. A. Laskowski, N. M. Luscombe, M. B. Swindells, and J. M. Thornton, *Protein Sci.* **5,** 2438 (1996).

[77] P. J. Goodford, *J. Med. Chem.* **28,** 849 (1985).

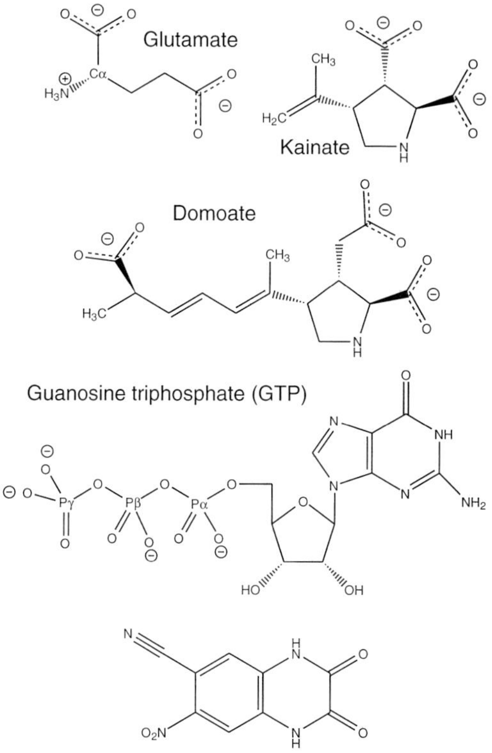

6-cyano-7-nitroquinoxaline-2,3-dione (CNQX)

FIG. 5. Chemical structures of the ligands studied—agonists glutamate, kainate, and domoate, and antagonists GTP and CNQX.

The location of the ligand-binding site in iGluRs was identified as follows. The bacterial periplasmic amino acid binding proteins contain two lobes, and it is between these two lobes that the ligand binds. By analogy with this, and consistent with both site-directed mutagenesis[42–44] and the generation of chimeric subunits,[78] ligands (agonists and antagonists) are likely to bind between these two lobes in the iGluRs. This location of the binding site in iGluRs is further substantiated by subsequent experimental results (see section on Updating the Binding Site).

The agonist glutamate (Fig. 5) was modeled into the binding site of the holo form of GFKARα, GluR1, and GluR6 using the methods described

[78] Y. Stern-Bach, B. Bettler, M. Hartley, P. O. Sheppard, P. J. O'Hara, and S. F. Heinemann, *Neuron* **13,** 1345 (1994).

earlier. In addition, the agonists kainate and domoate (Fig. 5) were modeled into the binding site of the holo form of GFKARα. The modeling positioned the three "hot spots" known to affect the binding of agonist in both non-NMDA and NMDA receptors in the binding site (residues 27[411] VTTIKE 32[416], 69[461] DGRYG 73[465], and 260[662] GTIKDS 265[667] in GFKARα), and gave insight into those amino acid residues in GFKARα likely to interact with the ligand. Our initial hypothesis, based on these models, was that E32[416], K71[463], Y72[464], R106[499], and R302[706] were involved in ligand binding (alongside other residues, to be investigated later). These suggestions formed the basis of experimental studies that partly confirmed our initial models (E32, Y72, and R106) and partly showed that some of the interactions had been modeled incorrectly (K71 and R302; see section on Updating the Binding Site). The orientation of the glutamate ligand was a concern at this stage—because of the binding site we had modeled, this was positioned "upside down" with respect to arginine and histidine ligand in LAOBP and HBP, respectively (i.e., its sidechain roughly where their C_α atoms are, and vice versa). The major reason for this was the electrostatic repulsion between the predicted position of the side chain of E32 and the sidechain carboxyl of glutamate ligand if the glutamate ligand was positioned in a similar orientation to arginine and histidine in LAOBP and HBP, respectively. Also, this "upside down" orientation is consistent with the results from solid-state nuclear magnetic resonance studies of the QBP–Gln complex.[79]

Updating Binding Site

Initial models of the ligand-binding site (or, indeed, any other part of the structure) are at best likely to be correct only in part. The extent to which they are correct becomes apparent once additional experimental data are available. This comes from the results of experiments suggested by the earlier models, and frequently includes results appearing in the literature as well—the more data used, the greater the reliability in the next round of modeling. The most useful experiments involve site-directed mutagenesis, followed by ligand-binding studies. This is explained in more detail later, illustrated by our modeling of iGluRs.

Since our initial modeling of iGluRs, three additional "hot spots" for ligand binding have been identified, giving a total of six (Fig. 6). Those that correspond to the "hot spots" included in the original modeling are denoted **S1**, **S2**, and **S4** in Fig. 6. The current sequence alignment corre-

[79] A. W. Hing, N. Tjandra, P. F. Cottam, J. Schasefer, and C. Ho, *Biochemistry* **33,** 8651 (1994).
[80] P. J. Kraulis, *J. Appl. Crystallogr.* **24,** 946 (1991).

Region	S1	S2	S3	S4
LAOBP	G T D T T Y A	T W V A S D F D A L	I S S L S I T D K R Q Q	V L Q G S T Q E
HBP	G T D P T Y A	T F V E N P L D A L	M S S L S I T E K R Q Q	V L Q G T T Q E
QBP	A T D T A F V	E L K P M D F S G I	L A G I T I T D E R K K	V K S G T G S V
GFKARα	30 I K E E P Y A 36	66 L V K D G R Y G K T 75	97 V A P L T L T A K R E T 108	261 T I K D S S S F 268
GluR1	414 I L E D P Y V 420	458 I V S D G K Y G A R 467	490 V A P L T I T L V R E E 501	663 T L E A G S T K 670
GluR6	407 I L E E P Y V 413	451 L V E D G K Y G A Q 460	483 V A P L A I T Y V R E K 494	653 A V E D G A T M 660
Change from Original GFKARα	27 V T T I K E E 33	66 L V K D G R Y G K T 75	97 V A P L T L T A K R E T 108	261 T I K D S S S F 268

Region			S5		S6		
LAOBP		w r t - - - - k g v d v	V A Y A N - - - - - - Q D L I Y	s d l t a g r l d a a -	L Q D E V A A S E	g f l k q p a g k e y a f a	
HBP		w a p - - - - k g i e i	V S Y Q G - - - - - - Q D N I Y	s d l t a g r i d a a -	F Q D E V A A S E	g f l k q p v g k d y k f g	
QBP		i k t - - - - k - - d l	R Q F P N - - - - - - I D N A Y	m e l g t n r a d a v -	L H D T P N I L Y	f i - k t a g n g q f k a v	
GFKARα	275	d n p t h h r i y e h i	K N A Q S Y S L - - N A E E G F	r r a l k g n y a f - -	I G E S V S L D L	- - - - - - - - - - - a v a	322
GluR1	677	k i a v f e k m w t y m	K S A E P S V F V L T T E E G M	i r v r k t k g k y a y	L L E S T M N E Y	- - - - - - - - - - - i e q	728
GluR6	667	k i s t y d k m w a f m	S S R R Q S V L V K S N E E G I	q r v l t s d y a f - -	L M E S T T I E F	- - - - - - - - - - - v t q	716
Change from Original GFKARα	275	d n p - - - - t h h r i	Y E H I K - - - - - - N A Q S Y	- - - - s l n a e e g -	F R R A L K G N Y	a f i g e s v s l d l a v a	322

B

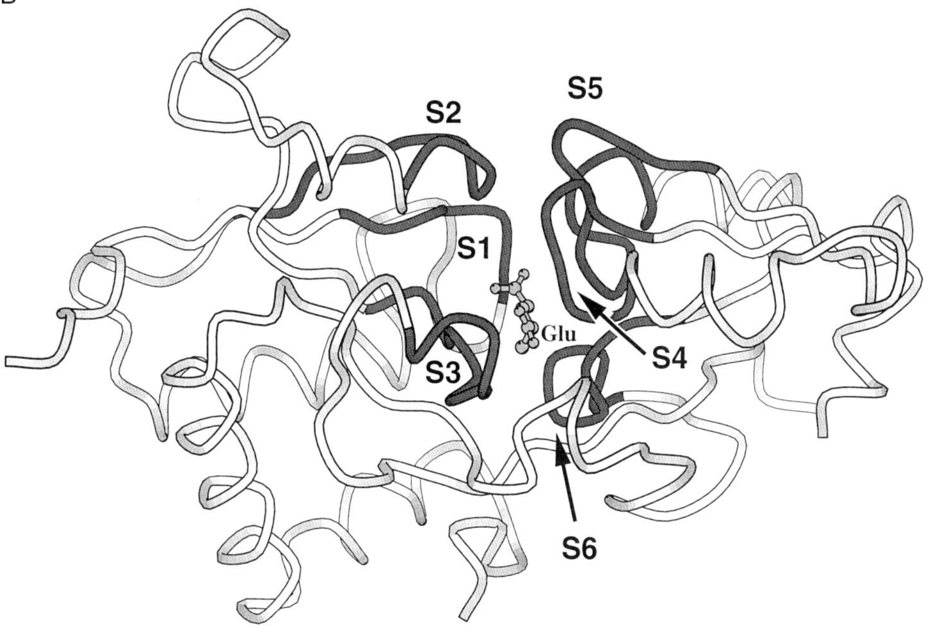

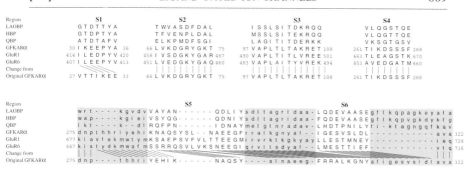

FIG. 6. (A) Modifications to original sequence alignment in regions **S1–S6**. For clarity, the sequences surrounding **S5** and **S6** are also shown (lowercase letter with a gray background). (B) Schematic representation of a model of the GFKARα–glutamate complex showing the positions of regions **S1–S6** (dark gray) in one of the LAOBP-like domains. (Figure generated using MOLSCRIPT.[80])

sponds to the original sequence alignment (Fig. 4) in two of these (**S2** and **S4**), and differs by three residues in **S1**. Of the remaining three "hot spots," the current sequence alignment corresponds to the original in one (**S3**) and differs quite significantly in the remaining two (**S5** and **S6**). Interestingly, all but one of these "hot spots" (**S5**) corresponds to regions in the ligand-binding site in the LAOBP-like crystal structures.[29–31] This further justifies the use of these crystal structures as templates. The absence of **S5** in the bacterial periplasmic amino acid binding proteins can, based on our sequence alignment, be explained in terms of the insertion (4–6 residues) in the iGluRs lengthening this region of the structure, thereby facilitating contact with the ligand. The reasons for the sequence alignments in each of these "hot spot" regions is discussed in detail next. Positions of conserved amino acids in the alignment play an important role in this realignment. All of the changes discussed leave the N-glycosylation sites accessible.

*Alignment in Region **S1***

When updating the model of the binding site, we noted an alternative alignment for the **S1** region. In the bacterial periplasmic amino acid binding proteins,[29–31] a conserved sequence motif DXX(Y/F) is observed (X is any amino acid), with D and Y or F forming part of the ligand-binding site. Inspection of the iGluR sequences reveals that the iGluRs have a conserved (E/Q)XX(Y/F) motif that is aligned with the corresponding motif in the bacterial periplasmic amino acid binding proteins if our alignment is adjusted by three residues. This also brings the Y/F site (Y35[419] in GFKARα) more completely into the binding site. Additional evidence that the E/Q site is in the binding site came from experimental work on GFKARβ[45] and an AMPA receptor.[44] Further indication that both the E/Q and Y/F sites are in the binding site came from experimental studies on ligand binding with chick kainate binding protein (cKBP[81]) and NMDA-R2B.[74] Based on this information, the sequence alignment in the region of **S1** was amended by three residue positions, as indicated in Fig. 6A. Furthermore, comparison with the crystal structures of the bacterial periplasmic amino acid binding proteins suggests that the side chain of the residue in the E/Q site in iGluRs interacts with a side-chain carboxyl group on the ligand. A direct interaction of this nature is possible for those iGluRs with a Q at this position; for those with an E at this position, the interaction (which cannot involve two negative charges) could be mediated by either (1) a water molecule, (2) a proton on the E side chain (i.e., this residue has an unusually high pKa), or (3) a cation. Possibilities (1) and (2) are seemingly discounted by studies of GFKARβ,[45] which show that an E at this position binds glutamate ligand approximately five times more strongly than does a Q. Furthermore, our preliminary experimental results suggest that glutamate ligand binding is enhanced, with either E or Q at this site,

[81] Y. Paas, M. Eisenstein, F. Medevielle, V. I. Teichberg, and A. Devillers-Thiery, *Neuron* **17,** 979 (1996).

both by Mg^{2+} and Ca^{2+}. Thus, the modeling and experimental studies suggest that a divalent cation (Mg^{2+} or Ca^{2+}) could bridge between the carboxyl side chain of glutamate ligand and the side chain at the E/Q site.

*Alignment in Region **S2***

The first indication that Y72[464] (in GFKARα) is in the binding site comes from the crystal structures of the bacterial periplasmic amino acid binding proteins,[29,30,82] where a Phe is important in ligand binding in both LAOBP and QBP (this residue is Leu in HBP). There is also a conserved Tyr in this region of the iGluRs. Consequently, this suggests that the conserved Tyr (corresponding to Y72) aligns with these Phe in the bacterial periplasmic amino acid binding proteins. The experimental studies on cKBP[81] also suggest that Y72 is important for ligand binding. Studies on NMDA-R2B[74] and GFKARα and GFKARβ[45] suggest that both R71[463] and Y72 are important for ligand binding. These results are interpreted[45] to indicate that Y72 interacts with the hydrophobic part of the ligand, while R71 forms a salt bridge with E32[416] (in **S1**), thereby interacting indirectly with the ligand-binding site. This region of the sequence has also been implicated in GTP binding.[83] The available data therefore suggest that our sequence alignment for **S2** is correct.

*Alignment in Region **S3***

Again, analysis of the crystal structures of the bacterial periplasmic amino acid binding proteins[29,30,82] is a useful starting point. These suggest that the sequence motif (S/G)X(S/T)XXXXR lines the binding site. This can be mapped onto the motif (P/G/A)X(T/A)XXXXR in the iGluRs, suggesting that, in GFKARα, P99[494], T101[496], and R106[499] lie in the ligand-binding site, and moreover that T101 could form a hydrogen bond with the α-carboxyl group of glutamate ligand, while R106 could form a salt bridge with the α-carboxyl group of glutamate ligand. P99 is also implicated in ligand binding by studies on NMDA-R2B.[74] Studies on cKBP suggest that T101 is involved in ligand binding.[81] R106 is implicated in ligand binding by studies on both NMDA-R2B[74] and AMPA receptors.[44,84] Thus, the available data suggest one initial sequence alignment in the **S3** region is correct.

[82] C. D. Hsiao, Y. Sun J., J. Rose, P. F. Cottam, C. Ho, and B. C. Wang, *J. Mol. Biol.* **240,** 87 (1994).

[83] Y. Paas, A. Devillers-Thiery, J. P. Changeux, F. Medevielle, and V. I. Teichberg, *EMBO J.* **15,** 1548 (1996).

[84] K. Keinanen, M. Arvola, A. Kuusinen, and M. Johnson, *Biochem. Soc. Trans.* **25,** 835 (1997).

Alignment in Region **S4**

The crystal structures, again a useful starting point, suggest that, in GFKARα, I262[664] and S266[668] could be important in ligand binding. Studies on NMDA-R2B[74] also suggest that I262 is important in ligand binding. S266 is also implicated in ligand binding by studies on both NMDA-R2B,[74] FKBP,[12] and cKBP.[81] Additionally, the latter study implicates S265[667] in ligand binding. Again, the available data suggest that this region of our initial sequence alignment is correct.

Alignment in Region **S5**

As mentioned earlier, the region equivalent to **S5** is not involved in ligand binding in the crystal structures, and therefore no indication about which residues are involved can be derived directly from these structures. This in turn makes sequence alignment in this region potentially unreliable, given the low level of sequence homology between the crystal structures and the iGluRs. Nevertheless, two residues have been implicated experimentally as contacting the ligand in this region. A296[698] in GFKARα, or a residue close to this (there is a two-residue deletion in GFKARα here with respect to most other iGluRs), is implicated by studies on cKBP,[81] illustrating that Y299 in cKBP is important in ligand binding. Y292[694] is implicated as being important in glycine binding from work on NMDA-R1.[85] Thus, our initial alignment for **S5** was incorrect, and has been modified as shown in Fig. 6A.

Alignment in Region **S6**

The model based on our initial alignment and orientation of the ligand suggested that R302[706] was involved in ligand binding. However, our experimental studies (L. Niu, Z. G. Wo, M. Byrnes, R. E. Oswald, unpublished results, 1997) demonstrated that this is not the case. Again, the crystal structures were useful in suggesting an improvement alignment—a conserved Asp forms a salt bridge with the α-amino group of the ligand. Analysis of the sequence alignment of the iGluRs suggests three different conserved acid residues as possibilities, corresponding to E298[702], E313[719], and D318[724] in GRKARα.

Experimental studies suggest that E313 may be the correct position.[81,84] Studies on NMDA-R2B[74] additionally suggest that S316[722] is important in ligand binding. Our initial alignment for **S6** was therefore likely to be incorrect and has been modified as shown in Fig. 6A.

[85] M. W. Wood, H. M. VanDongen, and A. M. VanDongen, *J. Biol. Chem.* **272,** 3532 (1997).
[86] A. C. Wallace, R. A. Laskowski, and J. M. Thornton, *Protein Eng.* **8,** 127 (1995).

A model of the GFKARα–glutamate complex using the updated alignment (Fig. 6A) is shown in Fig. 7.

Docking Other Ligands

6-Cyano-7-nitroquinoxaline-2,3-dione (CNQX; Fig. 5), a known potent non-NMDA receptor antagonist, was modeled into the apo form of GFKARα. The models suggest the following: (1) The aromatic ring of CNQX may pack against the aromatic ring of Y72[464]; this is supported by the observation that the mutation Y72S removes CNQX binding.[45] (2) CNQX lies against P99[492] and T101[494]; this could be consistent with the mutation in GluR1 equivalent to P99A in GFKARα. This mutation reduces CNQX binding affinity but affects glutamate and kainate binding

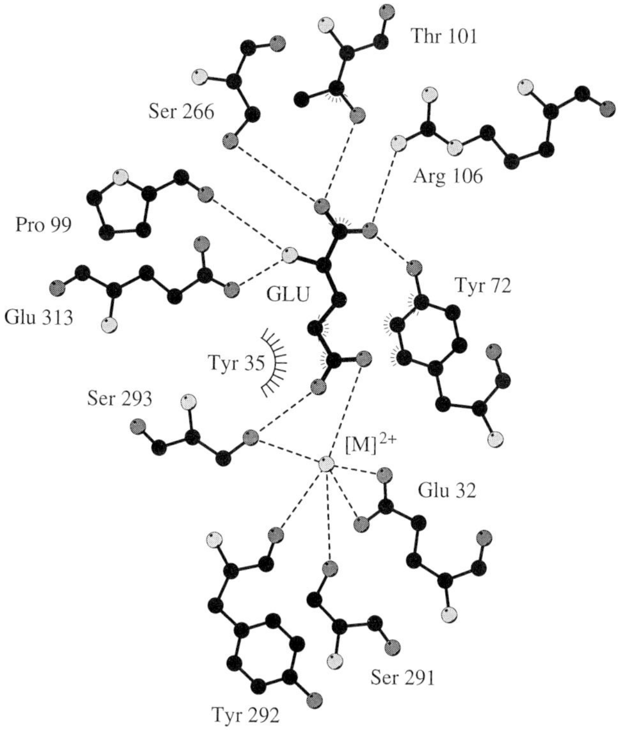

FIG. 7. Schematic representation of the glutamate-binding site in a model of the GFKARα–glutamate complex showing one of the LAOBP-like domains. Hydrogen bonds and charge–charge interactions are shown as dashed lines, hydrophobic interactions as spoked arcs, carbon atoms as black balls, oxygen atoms as dark gray balls, and nitrogen atoms as light gray balls. The divalent cation is denoted $[M]^{2+}$. (Figure generated using LIGPLOT.[86])

much less,[81] perhaps due to the rigid nature of the CNQX molecule, which could make a conformational change more difficult to accommodate. (3) CNQX is "sandwiched" between Y35[419] and Y72; this is consistent with mutagenesis studies which suggest that residues equivalent to both Y35 and Y72 in GFKARα may be involved in ligand binding (see earlier section). (4) CNQX could interact with the side chain of E32[416] via a divalent cation (Mg^{2+} or Ca^{2+}). (5) CNQX could hydrogen bond to T101 and R106. The modeled GFKARα–CNQX complex is shown in Fig. 8. Since the CNQX binding site appears to overlap significantly with the agonist binding site, but only in lobe 1, this could explain its potent antagonist behavior.

GTP (Fig. 5), a known ligand of kainate receptors,[83] was modeled into the apo and holo forms of GFKARα. The holo form had insufficient space in the binding site to accommodate GTP. It could, however, be modeled into the apo form. Our modeling of the apo form suggests that the γ

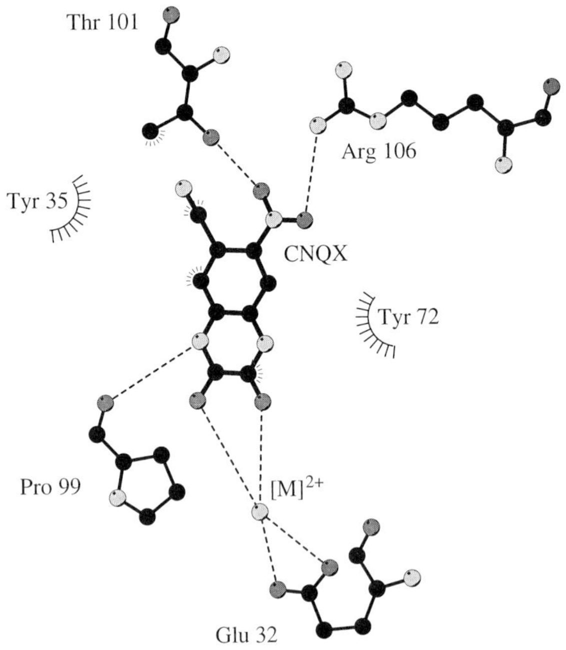

FIG. 8. Schematic representation of the CNQX ginding site in a model of the GFKARα–CNQX complex showing one of the LAOBP-like domains. Hydrogen bonds and charge–charge interactions are shown as dashed lines, hydrophobic interactions as spoked arcs, carbon atoms as black balls, oxygen atoms as dark gray balls, and nitrogen atoms as light gray balls. The divalent cation is denoted $[M]^{2+}$. Only three ligands are shown to this cation; the remaining three coordination sites could be occupied by water.

phosphate interacts with S265[667], the α phosphate with R106[499], one or more of the sugar oxygens with T101, the base is sandwiched between Y35 and Y72, and the O-6 atom of the base interacts with E32 via a divalent cation (Mg^{2+} or Ca^{2+}). Involvement of Y72 in GTP binding is consistent with mutagenesis studies on the equivalent residue in cKBP.[83] The modeled GFKARα–GTP complex is shown in Fig. 9.

Combining Different Modules

Once models have been produced for the different modules, the final stage is to combine them to create a model of the ligand-gated ion channel.

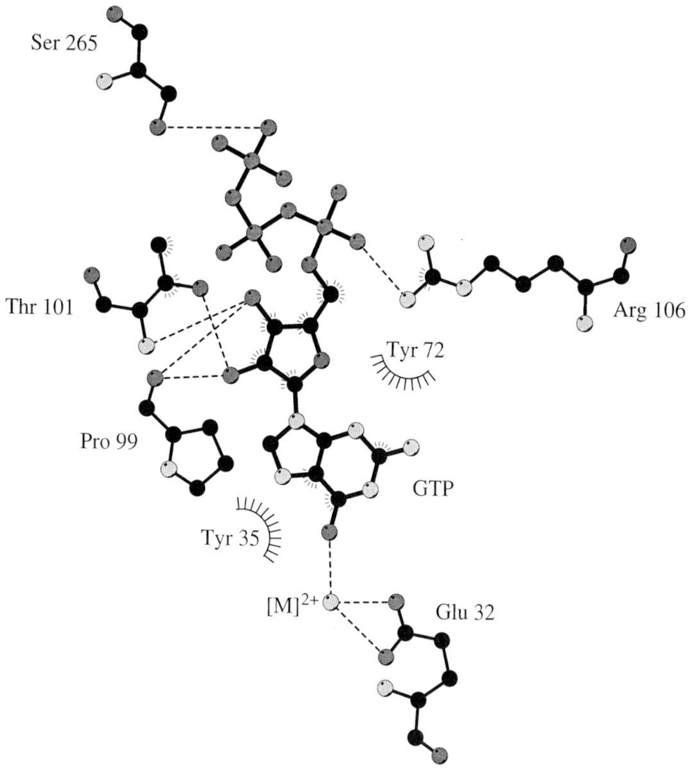

FIG. 9. Schematic representation of the GTP-binding site in a model of the GFKARα–GTP complex showing one of the LAOBP-like domains. Hydrogen bonds and charge–charge interactions are shown as dashed lines, hydrophobic interactions as spoked arcs, carbon atoms as black balls, oxygen atoms are dark gray balls, and nitrogen atoms as light gray balls. The divalent cation is denoted $[M]^{2+}$. Only four ligands are shown to this cation; the remaining two coordination sites could be occupied by water.

This is achieved by ensuring that the model is consistent with all the available data. For example, known glycosylation sites should be exposed to solvent, the ligand-binding site should be accessible to the ligand, residues known to line the ion conduction pathway should be positioned appropriately, and distances between the end of one module and the start of the next should not exceed the maximum distance corresponding to the number of amino acids between them. Methodologies for protein–protein docking (e.g., GRAMM[87]; available via http://guitar.rockefeller.edu/gramm/) can be used for this. However, regions of the protein between the modules that have been modeled are often missing, making such an approach inappropriate. The modules are combined using a "middle–out" approach. The starting point is the center of the pore (i.e., the membrane-spanning region that lines the ion conduction pathway) in the appropriate stoichiometry. Other modules are then combined with this, working outward from the center of the pore.

For the iGluRs, a module comprising M1, M2, and M3 was used as the starting point. The appropriate number of copies of the LAOBP-like domain (e.g., five for a pentamer) was then added to this using interactive molecular graphics (InsightII). The model of the LAOBP-like domain was positioned empirically with respect to both the membrane and its symmetry-related copies, so that (1) the consensus glycosylation sites were solvent accessible, (2) the agonist-binding site was accessible, (3) the distance between the end of the N-terminal of the LAOBP-like domain and M1 was in a reasonable range (approximately 17 residues were missing from our models in this region), (4) the distance between the end of M3 and the start of the C-terminal section of the LAOBP-like domain was within a reasonable range (approximately 12 residues were missing from our models in this region), and (5) the domain was as close to the pore as possible without overlapping sterically with its symmetry-related copies in both the apo and holo forms. In the resulting orientation, the long axis of the molecule was parallel to the surface of the membrane. Although not a unique solution to the positioning of this domain (due to the limited experimental data), this positioning is not inconsistent with the currently available data and is, in fact, constrained to a large extent by the experimental results. An assembled model of GFKARα is shown in Fig. 10.

Binding and Signal Transduction

Once the models of the different modules have been combined, the molecular basis of the transduction of binding energy to ion channel opening

[87] I. A. Vakser, *Biopolymers* **39,** 455 (1996).

can be investigated. The models are analyzed, paying particular attention to those residues close to the ion conduction pathway in the "channel open" (ligand-bound) and "channel closed" (ligand-free) states of the model. If particular residues appear to be involved in ion conduction, and they are conserved across the family of ligand-gated ion channels being modeled, then these are possible candidates. Note that the conformational change in the ligand-binding domain assumed to take place in the transition between the ligand-bound and the ligand-free states is likely to be propagated to other parts of the model; in particular, the membrane-spanning portion of the structure.

With our models of the iGluRs, the transition from the ligand-bound to the ligand-free states results in a change in the orientation of two sets of conserved amino acids with opposite charge. These correspond to residues E249[651] and D250[652], and K272[674] in GFKARα. In the ligand-free form, K272 (note that this is Q or S in a minority of iGluRs, making its proposed role more speculative) would be near the extracellular opening of the channel, impeding flow of cations. In the agonist-bound form, with the lobes closed, our models suggest that K272 is moved away from the opening and E249 and D250 are positioned near the opening. This ring of negative charges would attract cations to the channel in a manner analogous to the ring of negative charges in nicotinic acetylcholine receptors.[88] This proposed role of E249 and D250 is supported by studies of the NMDA-R1 receptor, which implicate the involvement of these residues in voltage-dependent spermine block.[89]

So far, the assumption has been made that the LAOBP-like domains exist in one of two states—either the ligand-free or ligand-bound states. This is a simplistic picture. In fact, there are at least two ligand-bound states—an initial channel open state and a subsequent desensitized (channel closed) state. The structural correlates remain unclear. It has been proposed[90] that the mechanism of ligand binding is akin to the venus fly trap mechanism observed in the bacterial periplasmic amino acid binding proteins, with ligand binding initially to lobe 1 to give the "channel open" state, and subsequent additional binding to lobe 2 as well to give the "desensitized" state. However, in the case of the bacterial periplasmic amino acid binding proteins, in the initial ligand-bound "open cleft" form the relative orientation of the two lobes remain unchanged (see Hsiao *et*

[88] K. Imoto, C. Busch, B. Sakmann, M. Mishina, T. Konno, J. Nakai, H. Bujo, Y. Mori, K. Fukuda, and S. Numa, *Nature* **335,** 645 (1988).

[89] K. Kashiwagi, J. Fukuchi, J. Chao, K. Igarashi, and K. Williams, *Mol. Pharmacol.* **49,** 1131 (1996).

[90] I. Mano, Y. Lamed, and V. I. Teichberg, *J. Biol. Chem.* **271,** 15299 (1996).

A

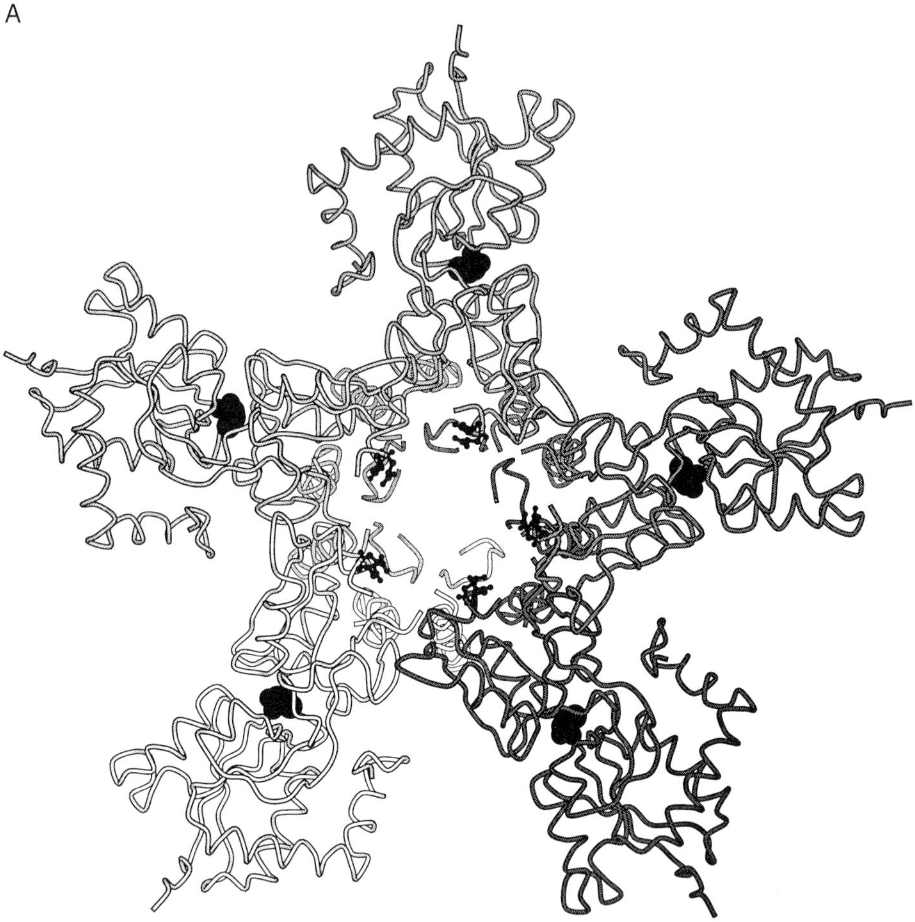

FIG. 10. Schematic representation of the pentameric structure of the GFKARα–glutamate complex. (A) View from outside membrane. (B) View along membrane. Each subunit is in a different shade of gray, glutamate ligand is in black space filling representation and the two conserved acidic amino acids E249[651] and D250[652] are in black ball and stick.

al.[82] for a discussion of the different forms). Therefore, it is difficult to envisage how ligand binding in such a manner results in a sufficiently large conformational change to bring about channel opening. This is even more difficult to understand if our proposed role of E249 and D250 (residues to lobe 2) in the channel open state is indeed correct (which it may not be)—this suggestion predicts that there is a movement in lobe 2 on ligand binding. Our models suggest the following possibility, which is speculative

B

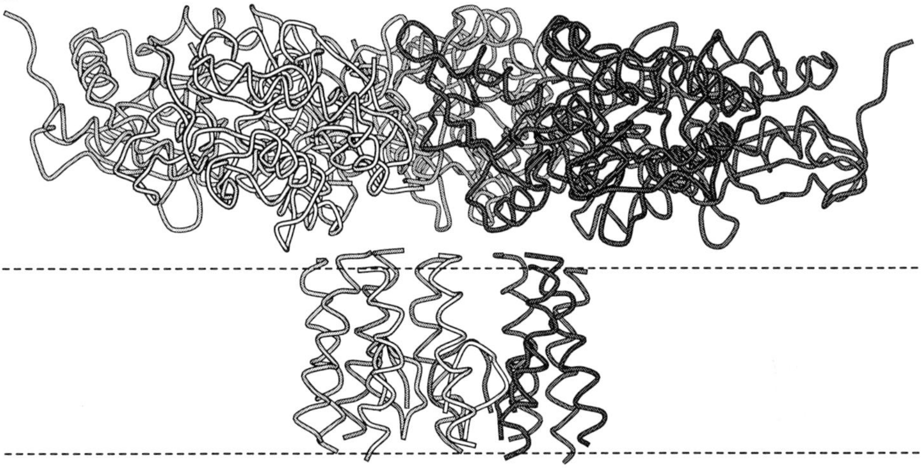

FIG. 10. (*continued*)

but not inconsistent with the experimental results[90]: Initial ligand binding causes the lobes to close, which is impeded initially by the loop we denote **S5** (Fig. 6). Loop **S5** subsequently changes conformation to allow the two lobes to come closer together. This change in the relative positions of lobes 1 and 2 results in desensitization, with the ligand now bound more tightly. The model cannot prove or disprove this notion. It merely suggests this as a potential mechanism that can be tested. The power of this approach is thus the iterative interplay between modeling and experiment.

Conclusion

Structural models of ligand-gated ion channels can be produced using molecular modeling techniques. This is achieved by dividing the channel into a smaller set of "structural modules," producing models for these, and then combining the models of the modules to produce a model of the channel. Restraints derived from experimental studies are used at all stages, and the findings from one round of modeling are used as restraints in the next. The process is best achieved using an iterative modeling/experimental verification cycle. As in any modeling exercise, the result is no substitute for a high-resolution three-dimensional structure. However, the resulting insights suggest directions for further study.

Acknowledgments

MJS is a Royal Society University Research Fellow. AHS is funded by an EPSRC studentship. This work was supported in part by a National Science Foundation grant (IBN-9309480) to REO.

[33] Use of Homology Modeling to Predict Residues Involved in Ligand Recognition

By SEAN-PATRICK SCOTT and JACQUELINE C. TANAKA

Introduction

Ion channels are typically large, multidomain membrane proteins designed to regulate ion flow across the cell membrane in response to a physiologic signal. Although we have some clues about the general architecture of ion channels, high-resolution structures of these proteins have not yet been determined.

In the absence of high-resolution structural data, it is possible to construct a three-dimensional model of a protein if a related protein of known structure is used as a template. This homology-based modeling is predicated on the assumption that the overall folds of functionally related protein regions are more highly conserved than the amino acid sequences.[1,2] This conservation may apply to functionally similar regions in seemingly distant proteins with <20% sequence identity. For example, a model of the cAMP binding domain of the regulatory subunit of cAMP-activated protein kinase (PKA)[3] was generated from the crystal structure of the cAMP binding domain of the catabolite repressor protein (CRP), an *Escherichia coli* transcription factor.[4,5] The model predictions were supported by experimental results using cAMP analogs and mutational analysis.[6–8] The crystal structure of a 1–91 deletion mutant of the regulatory subunit of PKA was deter-

[1] I. T. Weber, *Proteins Struct. Func. Genet.* **7,** 172 (1990).
[2] R. W. Harrison, D. Chatterjee, and I. T. Weber, *Proteins* **23,** 463 (1995).
[3] I. T. Weber, T. A. Steitz, J. Bubis, and S. S. Taylor, *Biochemistry* **26,** 343 (1987).
[4] D. B. McKay, I. T. Weber, and T. A. Steitz, *J. Biol. Chem.* **257,** 9518 (1982).
[5] I. T. Weber and T. A. Steitz, *J. Mol. Biol.* **198,** 311 (1987).
[6] J. B. Shabb, B. D. Buzzeo, L. Ng, and J. D. Corbin, *J. Biol. Chem.* **266,** 24320 (1991).
[7] J. B. Shabb, L. Ng, and J. D. Corbin. *J. Biol. Chem.* **265,** 16031 (1990).
[8] J. Bubis, J. J. Neitzel, L. D. Saraswat, and S. S. Taylor, *J. Biol. Chem.* **263,** 9668 (1988).

METHODS IN ENZYMOLOGY, VOL. 293
0076-6879/98 $25.00

mined.[9] The structure showed good agreement with the structure of CRP, supporting the major features of the PKA model. The crystal structure of the cAMP binding domain of CRP was also used to generate models of the cGMP binding domain of the cyclic nucleotide-activated channel (CNGC) from retina[10,11] and olfactory neurons.[12] Although the CNGC protein has not been crystallized, the available CNGC electrophysiologic data are consistent with model predictions.

In this article, we describe the methods used to generate homology-based models and demonstrate how electrophysiologic data can be used to refine the models and test the modeling predictions. The first part of this article describes the development of a model starting with the test protein sequence. We discuss how to select the best reference protein, how to model insertions and deletions in the sequence, and how to generate the ligand in the binding domain. In the next section, we discuss the energy minimization procedures including refinement of the initial models. We then illustrate the modeling process using the cyclic nucleotide binding domain of the retina cyclic nucleotide-gated channel (CNGC) β subunit. In the final section, electrophysiologic data are used to refine molecular models and generate testable hypotheses about ligand recognition.

Many useful tools for molecular modeling are available both commercially and as freeware. The Appendix includes a number of internet addresses related to this article and two related modeling chapters.[12a,12b] These addresses are intended to aid readers in locating starting points for molecular modeling and accessing specific tools used in modeling. The modeling described in this chapter was developed with tools from Biosym including InsightII for visualization, Homology for generating the initial alignment of the model, Discover for energy minimization, and Builder for generating ligand analogs. We also used several modeling programs developed by Dr. Rob Harrison including Homologuizer[13] for alignment and AMMP, Another Molecular Modeling Program,[14] for molecular mechanics calcula-

[9] Y. I. Su, W. R. G. Dostmann, F. W. Herberg, K. Durick, N. Xuong, L. Ten Eyck, S. S. Taylor, and K. I. Varughese, *Science* **269,** 807 (1995).

[10] V. D. Kumar and I. T. Weber, *Biochemistry* **31,** 4643 (1992).

[11] S.-P. Scott, R. W. Harrison, I. T. Weber, and J. C. Tanaka, *Biophys. J.* **68,** A147 (1995).

[12] S.-P. Scott, R. W. Harrison, I. T. Weber, and J. C. Tanaka, *Protein Eng.* **9,** 333 (1996).

[12a] M. S. P. Sansom, *Methods Enzymol.* **293,** [34], 1998 (this volume).

[12b] M. J. Sutcliffe, A. H. Smeeton, Z. G. Wo, and R. E. Oswald, *Methods Enzymol.* **293,** [32], 1998 (this volume).

[13] L. Menendez-Arias, I. T. Weber, J. Soss, R. W. Harrison, D. Gotte, and S. Oroszlan, *J. Biol. Chem.* **269,** 16795 (1994).

[14] R. W. Harrison, *J. Comp. Chem.* **14,** 1112 (1993).

tions. Dr. Harrison's programs are available over the Internet from the address provided in the Appendix.

Homology: Starting Point for Model Building

The initial challenge of homology modeling is the identification of a reference protein of known structure that shares a function with the test protein of interest. The crystal structure of the reference protein will provide the overall fold for the test protein domain and every effort should be made to match the secondary structure of the test protein to that of the reference protein. The reference protein should also provide the initial coordinates for ligand placement within the binding site.

Identification of Reference Proteins

A number of strategies can be employed to search for appropriate reference proteins. One approach is to use the amino acid sequence of the test binding domain to search for homologous sequences; however, this approach has several obvious weaknesses. First, the sequence-based search will identify many closely related proteins also of unknown structure. For example, using the sequence of the cGMP binding domain of the CNGC, other members of the channel family will be identified. Second, the search will not identify candidate reference proteins with low sequence identity. A better starting point for selecting a reference protein is to search for proteins of known structure that share functional characteristics with the test protein. Using the Brookhaven protein database (PDB), target proteins can be searched for using relevant keywords. To model a binding site, it is particularly helpful to search for other proteins binding to the same or similar ligands. A third approach identifies reference proteins by threading the test protein sequence onto a library of different protein folds derived from the database of protein structures.[15–17] The approaches used in the threading algorithms differ in the way that the test protein compatibility is scored relative to the fold as described in several reviews.[18,19] Some of these programs can be downloaded and others accept test sequence submissions and return candidate reference proteins. (See the Appendix for relevant starting points for these search tools.)

[15] C. A. Orengo, D. T. Jones, and J. M. Thornton, *Nature* **377,** 631 (1994).
[16] D. Fischer and D. Eisenberg, *Protein Sci.* **5,** 947 (1996).
[17] R. Sowdhamini and T. L. Blundell, *Protein Sci.* **4,** 506 (1995).
[18] D. Fischer, D. W. Rice, J. U. Bowie, and D. Eisenberg, *FASEB J.* **10,** 126 (1996).
[19] D. Jones and J. Thornton, *J. Comput. Aided Mol. Des.* **7,** 439 (1993).

Which is the best reference protein to use for constructing a model? From the initial searches, generate a list of potential reference proteins. Select the reference protein with the highest degree of functional homology. In the case of the cGMP binding domain of the CNGC, the most important functional characteristic is the structure of the ligand. Although the list of reference proteins generated from various searches gave higher scores for proteins binding GMP, GDP, and GTP, the most appropriate homologs were CRP and PKA, both of which bound cAMP. CRP was used as a reference protein for modeling the binding domain of the CNGC[10,12] rather than PKA since the structure of PKA[9] had not yet been solved. Once reference proteins have been identified, the coordinate files from the PDB can be downloaded to the computer.

Optimizing Alignment: Insertions, Deletions, and Helix Terminations

The first step in constructing the model is to align the sequence of the test protein with that of the reference protein. Although initial alignments may have been generated from the sequence searches, these alignments now have to be evaluated with respect to the secondary structure of the reference protein. Because the foundation of the model is the α carbon backbone of the reference protein, it is essential to maintain this backbone with as few changes as possible when dealing with insertions or deletions.

Insertions and deletions in the alignment should be grouped and positioned as loops between structural elements where possible. The loop regions are the most flexible and they can be manipulated with minimal perturbation to the overall protein fold. Looking at the reference protein structure, it may be obvious that some regions can tolerate insertions better than others. For example, imagine a region with three consecutive α helices. The first two helices are antiparallel, separated by a loop of five amino acids. The third helix, separated from the second by three residues, makes a 45° turn with respect to the second helix. Suppose the test protein exhibited a high sequence identity with the reference protein in the first and third helix but a low identity in the second helix. In addition, suppose the test protein had an insertion of five residues. One possibility is that the insertion could form an extra turn in one of the helices since one turn contains approximately four amino acids. The overall fold could be maintained if an extra helical turn or an addition to the loop were placed between the first and second helix since this loop does not define the relative position of these helices in the way that the loop between helix 2 and 3 does. For this reason, it is important to avoid altering the three-residue loop because doing so will likely change the relative position of the second and third helices. In general, loop regions can better accommodate inserts than highly ordered secondary structural elements. If your alignment has multiple insertions within secondary structural elements, you might want to consider a different reference protein or a different alignment before proceeding.

It is also important to consider the local protein environment when placing insertions. For example, if the alignment places a charged residue in the interior of a hydrophobic pocket with no partner to balance the charge, then a different strategy should be considered. The initial alignment is the most important step in constructing a model and later results might indicate that the initial alignment, or parts of the alignment, must be reconsidered.

At this point, we emphasize the importance of collaborating with crystallographers during model building. Because most electrophysiologists do not have the experienced eyes of a protein crystallographer when viewing protein structures, it is helpful to consult with a crystallographer about the initial sequence alignment. Often they can quickly identify potential trouble spots within the protein core that will cause problems at later stages of model building. If trouble spots are apparent, alternative strategies should be considered at this early stage of model building.

Location and Placement of Conserved and Variable Amino Acids

Once the alignment is completed, the test sequence is mapped to the coordinates of the reference protein in preparation for the initial minimization. The goal here is to retain the coordinates of the reference protein for all conserved atoms of the test protein in order to preserve the reference protein fold. Because the sequence identity may be low, the actual number of residues with all atomic coordinates initially assigned will also be low. Next, the backbone atoms of the test protein are assigned coordinates for all nonconserved residues. Because it is important to retain as many of the reference protein coordinates as possible, the coordinates for each conserved atom in the nonconserved side chains should also be retained. For example, when changing a valine to a leucine, retain the β carbon coordinates of the reference protein. By retaining the coordinates of as many of the side-chain reference protein atoms as possible, the overall side-chain packing will be preserved. At this point, the varied side-chain atoms and all atoms in the insertion regions will be the only atoms not assigned initial coordinates.

Initial Construction of Loop Regions

Those residues not yet defined by the reference protein structure lie in loop regions, outside the well-defined secondary structural elements. The importance of modeling loops may be best appreciated if one considers that loop regions are often the most variable regions between similar proteins, and the variability encodes functionally significant information.

There are several ways to generate coordinates for loop regions of a model. For larger loops of 12 amino acids or more, the sequence of the loop region can be used to search a loop database. The loop coordinates are then inserted into the model at the appropriate location. One can also generate a series of energy minimized loop structures with fixed end coordinates using programs such as Discover (BioSym) or AMMP. This approach can be used for any size of loop and is probably the best approach for medium-sized loops of about 7 amino acids. This method should be used if a loop database is not available for searching for large loops. Finally, the loop can be generated during the energy minimization of the model. This approach is best for small loops of about three amino acids, since they will be highly constrained by the defined secondary structural elements already present.

If candidate loops are generated, a loop must be selected for insertion into the model. When choosing a loop several attributes should be considered. First, the amino acids in the loop must be in a favorable environment with respect to the overall fold of the protein. Hydrophobic residues should face other nonpolar regions and charged residues should face the solvent or form salt bridges. Second, the loop should conform to the overall structure of the binding domain and show good packing with respect to other loops as well as the protein core. Larger loops may have secondary structure or extend existing secondary structural elements. Larger loops may also form an interior hydrophobic region within the loop.

Third, the loop must be consistent with available functional data such as ligand studies, cross-linking experiments, or mutant cycling with peptide blockers. The functional data set spatial constraints that provide important criteria that should be used to evaluate loop position and suggest which loops interact with the ligand. For example, cross-linking data with ligand analogs may provide distance constraints from the ligand to specific residues in the binding site. Conformationally constrained ligands of known structure are good steric probes of the binding domain. While the ligand may change conformation in the bound state going from syn to anti, for example, the experimentally verified ligand structure provides important distance constraints on the model.

During generation of the bovine retina CNGC binding domain model, the loops were formed during the minimization of the model. An illustration of the variation in loop formation is shown by comparing the original model of the bovine retina CNGC binding domain[10] and a more recent model.[12] The deletion in the CNGC between the $\beta 4$ and $\beta 5$ strands of CRP are similar in both models but the position of the insertion in the channel alignment before the $\beta 4$ strand differs. In the more recent model, this loop wraps closer to the protein core, whereas in the earlier model the loop was positioned away from the core. Although this example may be subtle, it is often necessary to model much larger loops with few clues as to how to position them.

Generating Ligands within Binding Domain

The coordinates of the reference ligand in the binding domain should be used to generate the test ligand. If the test ligand differs from that bound to the reference protein, it is better to modify the reference ligand coordinates rather than generate the ligand coordinates *de novo* or import the coordinates from another structure. The reason for using the reference coordinates is to retain the relative position of the ligand within the binding domain.

If the reference ligand structure must be modified to generate the test ligand, the test ligand can be energy minimized under two different conditions. First, the ligand coordinates can be extracted from the binding domain, modified, energy minimized, and then placed back into the binding domain. Alternatively, the ligand can be modified and minimized within the binding domain. It is important to recognize the different possible outcomes of each of these approaches. Minimizing the ligand structure in the absence of the protein coordinates may allow the ligand to achieve a structure that would not be easily achieved within the constraints of the binding domain. If, on the other hand, the ligand is energy minimized within the binding domain, a less energetically favorable ligand conformation may be achieved in order to accommodate the ligand within the constraints of the binding domain. This issue takes on greater significance as the modifications to the reference ligand increase.

Once the desired ligand structure is generated, it might be necessary to position side chains manually in order to accommodate the ligand substituents. If the changes are minor, they should be corrected during the minimization. Sometimes, however, even a slight movement will select a lowest energy path, which places a side chain in an unfavorable conformation. If this happens, note the side-chain position and examine it following the initial minimization.

Water and pH Considerations

Water often plays a direct role in ligand binding and interacts with the protein surface as well as the pore region of ion channels.[12a] The final consideration in constructing a model deals with the placement of water in the model. Two general approaches are commonly used to account for water in the models. The first approach defines each water molecule within the proposed structure and the second approach assumes that water is present by adjusting the dielectric constant to reflect the local protein environment.

The advantage of the discreet method is that water may be directly involved in the ligand binding. For example, water may be used to extend the reach of polar side chains interacting with the ligand.[20] A major disadvantage of using discrete waters in modeling is the increased computation cost due to the increased number of atoms. Interpreting the significance or meaning of discrete water molecules in a model structure is also problematic. For example, InsightII (BioSym) has a soak function that assigns water molecules within a specified volume around the protein but water may not penetrate the binding pocket. Further, the soak function surrounds the model with water despite the fact that certain regions may be in contact with other protein domains or the lipid bilayer. Clearly this is an unrealistic situation.

The other way to model water is to set the dielectric of the "media" in which the calculations are performed. The advantage is that the computation cost is decreased but the disadvantage is that a water molecule may be involved in the binding, forming an important point of contact for the ligand. Perhaps the best approach is to set the dielectric for the model and add discrete waters within the binding domain if they appear to be involved with ligand binding. If water molecules are placed in the binding site, they may have to be constrained during minimization, especially if they are in an area of the model that is open to the outside environment to keep them from moving away.

A final consideration before the initial minimization involves setting the pH for the model. The pH will affect the charges on the amino acid side chains and may affect protonation of substituents on the ligand. Most models will be built with physiologic pH values set near pH 7.2 but you may want to explore other pH values when predicting ligand–protein interactions.

Energy Minimization Procedures

When all the atoms within the model have been assigned initial coordinates, you can proceed to the initial energy minimization. To initiate the minimization algorithms, each atom in the structure must be assigned a force field, which will define the characteristics of the atom including the

[20] J. I. Yeh, H. Biemann, G. G. Prive, J. Pandit, D. E. Koshland, and S. Kim, *J. Mol. Biol.* **262,** 186 (1996).

charge, bond characteristics, and torsion angles as discussed elsewhere.[12a] The force fields for all atoms in the naturally occurring amino acids are available in any standard modeling package but force fields for ligand atoms may have to be explicitly defined. It is important to emphasize that a given atom may have different force fields depending on the type of bond it forms. The proper force field must be selected for each atom before attempting to minimize the model. A number of programs are available for model building and a detailed comparison of the programs and their implementation is beyond the scope of this discussion. One should realize that the program used to carry out the minimizations will define the algorithms and force fields available and it is possible to generate different models. The overall folds would be comparable but details of the interactions may change.

Initial Energy Minimization of Models

During the initial minimization, all backbone atoms should be inactivated or constrained to preserve the overall fold of the reference protein. Further, all amino acids and ligand atoms that are identical to those in the reference protein should also be restrained or inactivated. The reason for constraining so much of the model initially is to force the unknown structure to abide by the template folds of the reference protein. In adhering to this strategy, the initial model will have the general architecture of the reference protein and will provide the template for all future refinements.

Evaluating Initial Models

The easiest way to evaluate the initial model is to examine the energy terms after the initial minimization. The nonbonded energy includes the electrostatic and van der Waals terms. If the van der Waals energy is high (see the following example) following the initial minimization, it is likely that several atoms are too close, as can happen if the side chains are compressed in a particular region. The easiest way to alleviate this problem is to adjust the overlapping side chains manually to a more favorable packing arrangement. In general, one should attempt to reduce this energy term as low as possible during the minimizations. If the electrostatic energy term is high, there may be an unsatisfied charge within the protein core. The covalent energy terms of bond, angle, and dihedral angle only become large if there is deviation from the typical geometry of the amino acids found in proteins. When these energies are high it suggests either a packing problem or problems with a deletion area where the remaining amino acids cannot easily close the structure.

Because the backbone and many of the side chains were constrained during the first minimization, the new atoms were forced to pack between the stationary atoms. These constraints elevate the energy considerably and as the constraints are released, the energy should drop. Before releasing the constraints, however, the model should be adjusted taking particular care with regions that were not well defined in the starting model.

Adjusting Initial Models

Guided by the energy readouts and visual inspection, the next step is to adjust the initial model. At this point there are several types of problems you want to identify including (1) obvious distortions in amino acid or ligand conformation, (2) unsatisfied charges in the protein core, and (3) ligand–side chain interactions that look possible but are not present in the model.

The easiest distortions to localize and the most common distortions are found in the conjugated rings of Tyr, Phe, and Trp. These distortions usually occur when the residue replaces a smaller one in the protein core. If the ligand contains rings, distortions may also occur in these rings. Distortions in other amino acids are harder to spot and will usually be detected in regions that are densely packed or regions that do not follow the secondary structure of the reference protein. Distortions can often be visualized using a bump check, which identifies noncovalently bonded atoms that are too close. This check will also show atoms predicted to interact via hydrogen bonds and these should obviously not be considered to have packing problems.

The next check is to determine that the interior of the protein contains no unpaired charges. This task can be simplified by color coding the amino acids according to their polarity and examining regions of charge. In general, charged side chains should interact with the nearest oppositely charged residue. If there is no oppositely charged amino acid nearby, look for a conjugated ring, a polar or charged region of the ligand, or a polar amino acid. A combination of several of these interactions may be necessary to shield a charge fully.

Interactions between the side chains and the ligand should also be examined. The best starting point is to examine the interactions between the ligand and the binding domain in the reference protein structure. To be sure you identify all interactions between the ligand and binding domains, use the original description of the reference protein crystal structure. This is important because programs such as InsightII often do not define all of the hydrogen bonds described by the crystallographer. In principle, all bonds between conserved amino acids and the ligand are retained since these positions were restrained during the initial minimization. The problem now is to define interactions that might be unique in your protein. To accomplish this, we typically generate a 5-Å shell around the ligand and look for possible interactions. You can test for possible hydrogen bonding by manually manipulating candidate side chains and measuring the distance between the putative donor and acceptor atoms. Once a favorable interaction is identified, leave the side chain in that conformation for the next energy minimization.

In addition to the more traditional hydrogen bonding pairs, a number of variants involving conjugated ring interactions should also be mentioned. Although benzene rings are considered nonpolar, the π electrons in the aromatic rings are localized above and below the face of the ring, giving a partial negative charge to the face and a partial positive charge to the hydrogens at the edge of the ring.[21,22] Amino acids with aromatic rings can therefore act as weak hydrogen bond acceptors and donors. Rings within the ligand may also form ring–ring and ring–side-chain interactions with neighboring residues.

Once areas have been identified within the initial model that should be adjusted, adjust the amino acid side chains manually or perform a computer search of the conformational space to find low-energy states. In densely packed regions of the protein, the computer search is often the best method. At this stage, it is once again important to consider experimental evidence. Often information is available from family members or mutants, cross-linking data, cysteine scanning mutagenesis, or ligand analog studies that suggest specific ligand interactions within the binding domain. To the extent that such data exist, make every effort to satisfy these criteria with the model.

Subsequent Minimizations

Following manual adjustments to the initial model, a second energy minimization is performed to reposition the amino acids to favorable energy states. Once again, the α carbon backbone should be restrained. During this minimization the heavy atoms that were positioned to allow hydrogen bonding should be restrained in an attempt to retain the favorable hydrogen bonding position, defined as having a distance of $\leq$3.5 Å between the donor and acceptor and an optimum angle of 180°.

Final Minimization to Release All Constraints

The third minimization is different from previous minimizations in that all restraints are released including the hydrogen bonding distances, the

[21] S. K. Burley and G. A. Petsko, *Science* **229,** 23 (1985).

[22] J. B. O. Mitchell, C. L. Nandi, I. K. McDonald, and J. M. Thornton, *J. Mol. Biol.* **239,** 313 (1994).

backbone atom positions, and the ligand. The position of the backbone atoms of the conserved amino acids should vary little in the final model from the original structure. There may be a packing problem that was not resolved in previous steps if the backbone varies more than 1.5 Å. If the amino acids that were previously restrained in order to allow them to interact with the ligand vary by a more than 1 Å, these ligand interactions are unlikely. Further adjustments may be required to optimize the ligand positioning if the amino acids positioned for hydrogen bonding no longer predict a hydrogen bond. At this point, you can return to the second minimization, and release the restraints selectively. It is possible that some restraints will be required during the final minimization especially since the model is limited to a single domain of a multidomain or multisubunit assembly. The restraints may suggest that interactions with other parts of the protein or solvent interactions are required to stabilize the optimal conformation.

Model Building Example: β Subunit of Human Retina CNGC

To illustrate the molecular modeling steps, we describe the initial steps of generating a molecular model using the β subunit of the human retina CNGC[23] as the test protein and the coordinates of the cAMP binding domain of CRP as the reference protein.[5] The human retina β subunit is a 240-kDa protein, approximately twice the size of the α subunit.[23] When the β subunit is expressed in a heterologous system, no cGMP-activated currents are recorded; however, when the β subunit is coexpressed with the α subunit, native-like properties are conferred on the CNGC currents.[24–26]

The β-subunit binding domain sequence was aligned to the CRP sequence using the α-subunit alignment previously generated.[10] The only alignment changes in the β subunit, compared to the α subunit, were a Pro-1128 insertion following the β4 strand and a three residue deletion before Gly-1153. The PDB file containing the CRP coordinates (3GAP) was then modified. All waters in the CRP file were deleted as well as residues 1–10 and residues beyond 128. All coordinates of the second subunit and any deleted residues, compared to CRP, were deleted also. Next, the coordinates for all conserved residues were assigned to the β subunit along with the coordinates for cAMP. The backbone atom coordi-

[23] H. G. Korschen, M. Illing, R. Seifert, F. Sesti, A. Williams, S. Gotzes, C. Colville, F. Muller, A. Dose, M. Godde, L. Molday, U. B. Kaupp, and R. S. Molday, *Neuron* **15,** 627 (1995).

[24] R. S. Molday and Y. Hsu, *Behav. Brain Sci.* **18,** 441 (1995).

[25] W. N. Zagotta and S. A. Siegelbaum, *Ann. Rev. Neurosci.* **19,** 235 (1996).

[26] K. Yau and T. Chen, *in* "CRC Handbook Sensory Physiology" (R. A. North, ed.), p. 307. CRC Press, Ann Arbor, Michigan, 1995.

TABLE I
ENERGY CHANGES DURING MINIMIZATIONS OF BINDING DOMAIN WITH syn-cGMP

Energy type (kcal/mol)	Initial model (Fig. 2A)	Before Val adjustment (Fig. 2B)	After adjustments	Bovine retina CNGC (Fig. 3)	Bovine olfactory CNGC (Fig. 5)
Bond	203	145	213	208	206
Angle	2078	1621	645	515	541
Torsion	515	452	477	340	374
Hybrid	1028	1242	46	29	36
Nonbonded	1645	791	−72	−115	−673
Total	5470	4254	1311	976	485

nates from CRP were retained for nonconserved residues as well as for any other shared atoms in the side chains. Two residues were inserted in a loop region between β strands 6 and 7 and assigned coordinates of 0,0,0 equivalent to the origin of the grid. All other atoms without assigned CRP coordinates were also assigned to 0,0,0 coordinates. Using InsightII, the anti conformation of cAMP was manually adjusted to the syn position. Finally, the additional atoms in cGMP, as opposed to cAMP, were also assigned 0,0,0 coordinates.

The model was minimized in three discrete steps using AMMP. The first step was a geometric minimization[27] using only geometrical considerations, including atomic bonding distances, to give atoms a rough starting position near their covalently bonded neighbors or near the area of the insertion. During this minimization, only those atoms starting at the 0,0,0 position were free to move; all other atoms were inactivated. The second step was an energy minimization using the energy terms of angle, bond, hybrid, nonbonding, and torsion. Again, only those atoms originating at 0,0,0 were allowed to move. The final energy minimization of this round again used all of the energy terms. During this minimization, all of the side chains were allowed to move as well as the backbone atoms of two residues on either side of the insertion and deletion region. The relaxation of these constraints allowed for better packing in the insertion and deletion areas.

Table I shows the energy terms of the initial model. The very high nonbonded, angle, and hybrid energies suggest either packing problems due to overlapping atoms or atoms in unlikely conformations or unfavorable environments. One problem area in the initial model is shown in Fig. 1A. In this region, the labeled Phe is clearly distorted due to dense packing, which may not be apparent because only a few of the neighboring residues

[27] P. N. Brown and Y. Saad, *J. Am. Sci. Stat. Comput.* **11,** 450 (1990).

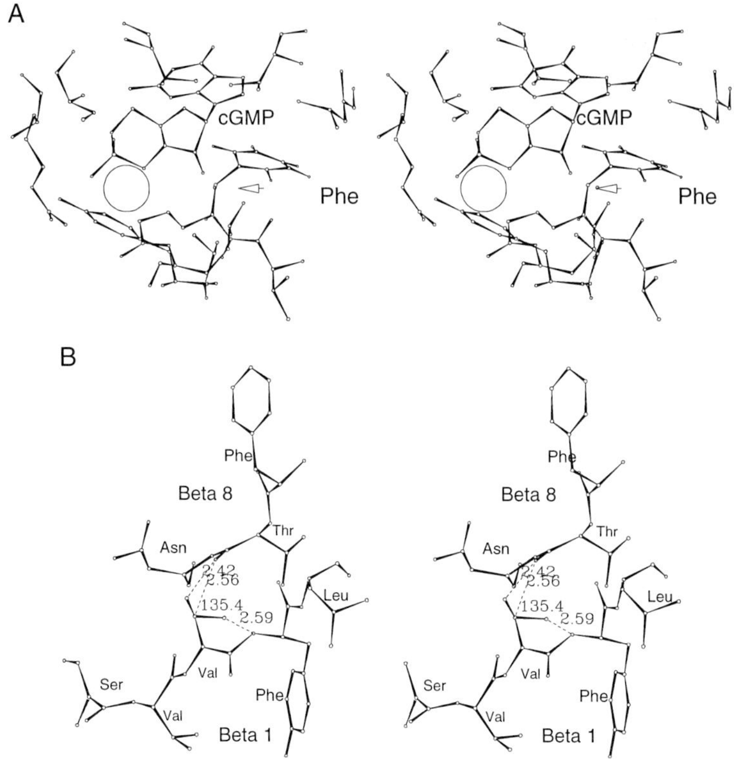

FIG. 1. Unrefined models of the cyclic nucleotide-binding domain of the β subunit of human retina CNGC. Selected residues within a 5-Å radius of the residue indicate some of the packing problems in this area. (A) Stereo view of a Phe ring that is distorted due to compressed packing. Rotation about the torsion angle between the α and β carbons from the present position to the open area represented by the circle will relieve the problem. (B) Stereo view of a compressed Val that also has packing problems. In this situation, the Val is compressed between two β strands because the backbone was inactive during the previous minimization. This packing problem can be relieved by removing backbone constraints and allowing the Val to push the β1 strand away from the β8 strand as discussed in the text.

are shown in this view. To alleviate the packing strain in this region, the distorted ring was moved manually by changing the torsion angle between the α and β carbon of the Phe in the direction indicated by the arrow to the more open area indicated by the circle. Two other rings were similarly adjusted. In addition to the ring adjustments, a lysine was manually adjusted to counterbalance an unsatisfied glutamic acid charge.

Following these adjustments, an energy minimization was performed utilizing all the energy terms and all but the backbone atoms were free to move. The energies following this minimization are also given in Table I (Before Val adjustment column). The nonbonded energy term decreased by half, in response to moving the lysine, and partially relieved the packing problems. A modest change in the angle energy was likely due to the improved ring structures. The large value of the nonbonded energy suggested additional packing problems, however. The problem region was located by using the bump check as described earlier. Figure 1B shows the region of conflict where the heavy atoms of Val were too close the backbone atoms in β strands 1 and 8. Furthermore, the ~135° angle between the singly bonded carbon atoms, which should be ~109°, suggested that the Val side chain contributed to the high angle energy term. To overcome the packing problem, a brute-force approach was used since rotation around the torsion angle between the α and β carbons would not have alleviated the problem. Because AMMP has the capability to energy minimize using selected energy terms, only the hybrid energy was used for the next round of minimizations. The model was then minimized using all energy terms. During these two minimizations, all the atoms, including those in the backbone, were allowed to move. The result is shown in Table I (After adjustments column). These energy terms are in the expected range for models of this size although they would likely continue to drop with further refinement to approximate those of the refined retina and olfactory CNGC models, also given in Table I.[28]

The secondary structure of the final model with the bound ligand is shown in Fig. 2A. The binding site is formed by an eight-stranded β barrel and three α helices. The entire cyclic nucleotide ligand is surrounded by the binding domain except for the N-2 position of cGMP. The amino acid side chains surrounding the ligand will be further refined in future iterations of adjustment and minimization but the backbone will move only slightly if at all. An overlay of the CRP α carbon backbone and that of the β-subunit model are shown in Fig. 2B. The RMS deviation of 0.616 Å corresponds to a good agreement with the reference protein. Importantly, no apparent disturbances are seen in the backbone structures in the areas of deletion (1), insertion (2), and Val packing problems (3).

Relating Structural Models to Electrophysiologic Data

Insights about ligand-binding interactions with ion channels must often be derived from electrophysiologic data because of the difficulty of measuring the ligand binding directly. It is important to recognize that electrophysiologic data in the form of dose–response curves will reflect contributions

[28] S.-P. Scott and J. C. Tanaka, *Biophys. J.* **72,** 242 (1997).

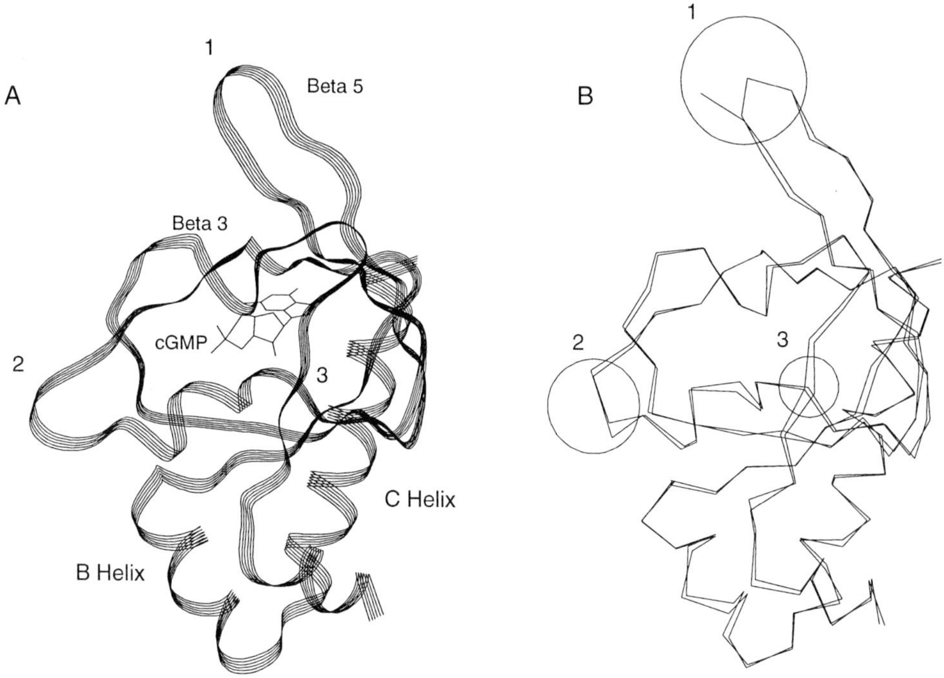

FIG. 2. The secondary structure and α carbon traces of the unrefined β-subunit model. (A) cGMP is shown in the binding site of the β subunit. The secondary structural elements are defined by the initial CRP alignment. (B) An overlay of the α carbon traces of the model and the reference protein CRP. The circles represent the (1) location of a two-residue deletion in CRP, (2) location of a two-residue insertion in the test protein, and (3) location of the Val region shown in Fig. 1B.

from channel gating processes in addition to ligand binding. The CNGC provides an illustration of the challenges of deriving insights about ligand binding from electrophysiologic data. This channel is thought to be a tetrameric complex[29] with each subunit able to bind one cGMP molecule. The dose–response relation based on macroscopic currents recorded from excised patches is cooperative, suggesting intersubunit communication within the channel. While we know that the cGMP binding domain is located in the C-terminal region of the channel, an N-terminal region is directly involved in coupling ligand binding with channel opening.[25] The interpretation of data from neuronal patches becomes even more complicated because

[29] D. T. Liu, G. R. Tibbs, and S. A. Siegelbaum, *Neuron* **16,** 983 (1996).

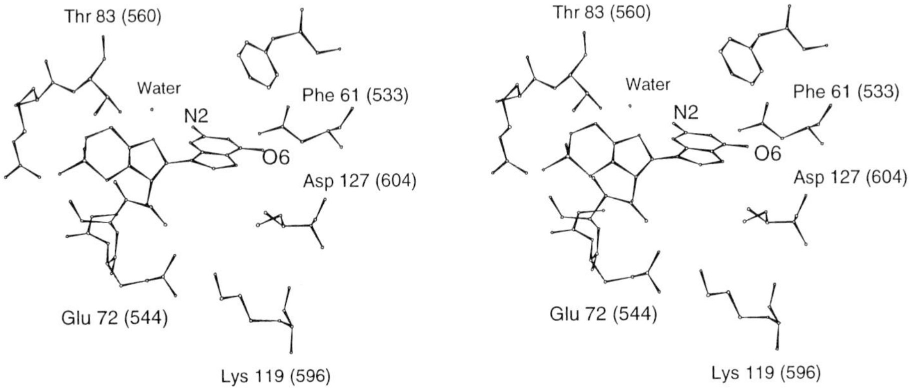

FIG. 3. Stereo closeup view of the refined model of bovine retina CNGC binding domain with *syn*-cGMP. CRP numbering is used with the channel sequences in parentheses. The predicted interactions are described in detail elsewhere.[12]

these channels are likely heteromeric in their subunit composition expressing both α and β subunits.[30]

Example: Refined Model of the Bovine Retina α-Subunit CNGC Binding Domain

A close-up view of cGMP bound to the α subunit of the bovine retina CNGC is shown in Fig. 3. The water molecule is predicted to interact with both the OH of Thr-83 and the N-2 of cGMP. This interaction has been suggested as the structural basis for the ~30-fold preference of the retina channel for cGMP over cAMP since cAMP cannot interact with Thr-83 because it lacks the C2 amino group. A mutation of Thr-83 to alanine resulted in a large increase in the $K_{0.5}$ for cGMP.[31] The initial model of the bovine retina CNGC placed cGMP in the syn conformation to maintain this predicted interaction.[10]

There is one internal salt bridge between Glu-72 on $\beta 6$ and Lys-119 on the $C\alpha$ helix. Three interactions are predicted between the purine ring and the protein. These interactions, described in detail elsewhere,[12] include the Thr-83 hydrogen bond, a hydrogen bond between N-1 of cGMP and OD1 in the charged carboxyl group of Asp-127 on the $C\alpha$ helix. The third interaction is a weaker aromatic–aromatic interaction between Phe-61 and the purine ring. In addition to the interactions with the purine, six interac-

[30] C. A. Colville and R. S. Molday, *J. Biol. Chem.* **271,** 32968 (1996).

[31] W. Altenhofen, J. Ludwig, E. W. Eismann, W. Bonigk, and U. B. Kaupp, *Proc. Natl. Acad. Sci. U.S.A.* **88,** 9868 (1991).

tions are predicted between the sugar and the CNGC binding domain. Because these interactions are invariant between cAMP and cGMP they do not contribute to the ligand specificity of the channels. Molecular models of other ion channels are presented elsewhere in this volume.[12a,12b]

Use of Ligand Analog Data

Analysis of electrophysiologic data and modeling go hand in hand once the initial models are generated. At this point, small changes in the ligand or in specific residues can easily be introduced and the energy minimization repeated. Often a series of models is needed to compare predictions with data from ligand analog or mutant studies. From the model, it is usually straightforward to predict which residues will interact with the ligand. It is more difficult to predict the effects of mutations in noninteracting residues because these effects may be long-range, affecting the overall fold of the domain.

We used a series of cGMP analogs, shown in Fig. 4, to probe the steric and electrostatic constraints of the retina CNGC binding domain.[32,33] Modifications in the ribofuranose rings were generally not tolerated suggesting that the protein–ligand interactions in this region are required. Since the only differences between cGMP and cAMP are in the purine ring, contacts with the purine ring must account for the electrophysiologic characteristics of these two ligands.

The $K_{0.5}$ values for macroscopic current activation and the maximal current produced by the analog are summarized in Table II. These data suggest a number of important features about the purine–protein contacts in the retina CNGC. First, the most useful ligands were those that did not bind to the channel because these analogs suggested contacts that are required for ligand activation. Because all of the inactive analogs were altered at either N-1 or C-6 positions, the predicted contacts with the $\beta 5$ strand or the Cα helix must be critical for binding. Other alterations in the N-1, C-2, and C-6 positions on the purine increased the $K_{0.5}$, suggesting less favorable interactions with the binding site. In contrast, alterations in the C-8 position were well tolerated and even reduced the $K_{0.5}$ in some cases, suggesting a more favorable energetic interaction compared to the parent ligand.

By examining the analog structures, one can readily appreciate why cGMP analogs are useful steric probes of the binding domain. All of these ligands have rigid ring systems that rotate about the glycosidic linkage giving syn, as shown, or anti conformations. Aside from this rotation, their

[32] J. C. Tanaka, J. F. Eccleston, and R. E. Furman, *Biochemistry* **28,** 2776 (1989).
[33] S.-P. Scott and J. C. Tanaka, *Biochemistry* **34,** 2338 (1995).

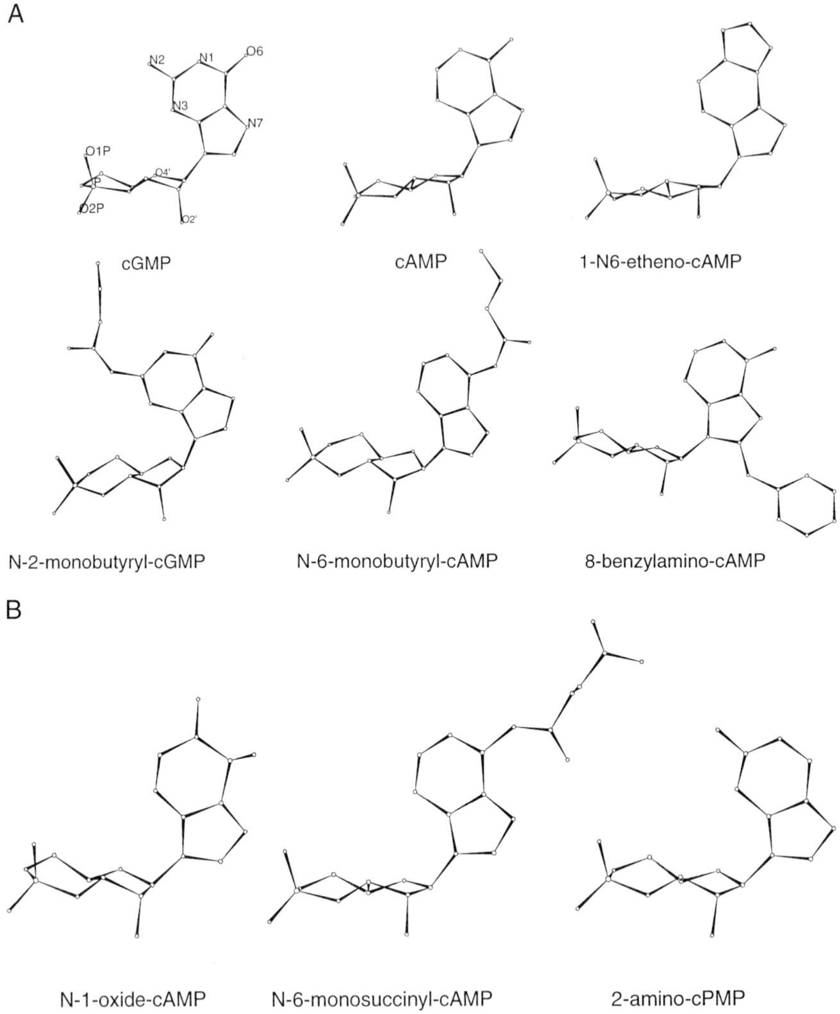

FIG. 4. Molecular models of (A) active cyclic nucleotide analogs and (B) inactive analogs. Note that all inactive analogs have alterations at either N-1 or C-2 positions, suggesting that this region of the ligand mediates an essential interaction with the protein.[33]

structures are inflexible. By generating a series of molecular models we were able to show that the CNGC binding domain could sterically accommodate all cGMP analogs known to bind to the channel. Examination of the models with the inactive ligands in the binding domain suggested that essential contacts were required between the N-1 and C-6 region of the

TABLE II
PURINE ANALOG ACTIVITIES FROM PHOTORECEPTOR CURRENT MEASUREMENTS[a]

Ligand	Purine substituent	$K_{0.5}$ (μM)	I_{max} (%)
cGMP	N^1-H; C^2-NH_2; $C^6=O$; C^8-H	13.7	100
N^2-Monobutyryl-cGMP	$C^2-NH-CO-C_3H_7$	313	45
6-Thio-cGMP	$C^6=S$	81	
8-Bromo-cGMP	C^8-Br	16	95
8-Fluorescein-cGMP	C^8-fluorescein	0.85	100
cAMP	C^2-H; C^6-NH_2; C^8-H	617	42
N^1-Oxide-cAMP	N^1-O	ND	
1,N^6-Etheno-cAMP	$N^1-CH=CH-NH-C^6$	844	20
N^6-Monobutyryl-cAMP	$C^6-NH-CO-C_3H_7$	NS	20
N^6-Monosuccinyl-	$C^6-NH-CO-C_2H_4-CO_2^-$	ND	
8-Amino-cAMP	C^8-NH_2	NS	50
8-Hydroxy-cAMP	C^8-OH	841	79
8-Bromo-cAMP	C^8-Br	724	16
8-Benzylamino-cAMP	$C^8-NH-\varnothing$	542	48
8-Methylamino-cAMP	$C^8-NH-CH_3$	356	53
8-Azido-cAMP	$C^8-N=N^+=N^-$	103	67
2-Amino-cPMP	C^2-NH_2; C^6-H; C^8-H	ND	
cIMP	N^1-H; C^2-H; $C^6=O$; C^8-H	1200	78

[a] ND, No detectable binding; NS, not saturable; $\varnothing$, benzene ring. From refs. 32 and 33.

purine and the $\beta 5$ strand of the binding domain. The other contacts between the purine and the protein side chains apparently contribute to the ligand interaction but are not required.

When modeling a ligand analog, a number of considerations need to be kept in mind. First, small ligands will not drastically change conformation or binding characteristics during minimization. For example, cis versus trans or syn versus anti conformations tend not to change during energy minimization so it is necessary to begin the energy minimization with the ligand in the desired conformation. To explore various ligand conformations, multiple models should be generated with the ligand initially placed in the desired conformation. A second consideration when modeling the ligand is that rigid substituents are easier to model than flexible ones since the number of possible conformations to explore during the minimization is less than for long, flexible substituents. For example, the addition of a ring to cAMP as in 1-N^6-etheno-cAMP is easier to model than the addition of the fluorescein group to cGMP as in 8-[[(fluorescein-5-ylcarbamoyl)] methyl]thio]guanosine 3′,5′-monophosphate (8-Fl-cGMP) because the addition of the etheno ring results in a highly constrained structure as opposed to the highly flexible hydrocarbon linker in the fluorescein substituent. It may also be necessary to define a ligand analog in a stepwise manner, minimizing after each modification and examining the models generated. For example, we had difficulty generating a model with 8-Fl-cGMP in the binding domain with a single minimization. By adding the fluorescein substituent in steps, we were able to generate the large analog within the binding domain.

Comparing Data from Related Channels

Additional clues about ligand recognition can be derived by generating models of closely related channels. The CNGC family includes channels from olfactory neurons. In these neurons, cAMP is the cytosolic transmitter, whereas cGMP conveys the signal in phototransduction.[25,26,34] Olfactory CNGCs are activated at low micromolar concentrations of both cGMP and cAMP, whereas retina channels have $K_{0.5}$ concentrations of $\sim 30\ \mu M$ for cGMP and 700–1200 μM for cAMP activation. For both olfactory and retina neurons, variability in the $K_{0.5}$ concentrations measured *in situ* might reflect both the presence of multiple subunit types having somewhat different nucleotide-binding properties or cytosolic regulation of the channel properties. In heterologous expression systems where the channels are assembled from a single subunit type, the bovine retina and bovine olfactory CNGCs show a 30- to 50-fold preference for cGMP over cAMP while catfish olfactory CNGCs are activated by equal concentrations of the nucleotides.[25,26,34]

The bovine olfactory CNGC binding domain is 85% identical in sequence to the binding domain of the bovine retina CNGC. To analyze the structural basis of nucleotide discrimination within this channel family, we constructed molecular models of the olfactory channels from the coordinates of cAMP bound to CRP.[12] A closeup view of a model of the bovine olfactory CNGC is shown in Fig. 5 with cGMP in the binding site. Three interactions were predicted between cGMP and the binding domain in this channel. The first is a hydrogen bond between Thr-83 and the N-2 group of cGMP, mediated through the water molecule. This interaction is similar to that predicted for the retina CNGC in Fig. 3. The second interaction is an electrostatic interaction similar to that of Asp-127 in the bovine retina CNGC binding site except that Glu replaces the Asp. The Glu-127, unlike Asp, is long enough to form hydrogen bonds with N-1 and N-2 of cGMP simultaneously. The third interaction with Tyr-61 is different from the bovine retina binding site, which has a Phe in this position. The addition of the hydroxyl group provides a second polar atom, which can interact

[34] T. Nakamura and G. H. Gold, *Nature* **325,** 442 (1987).

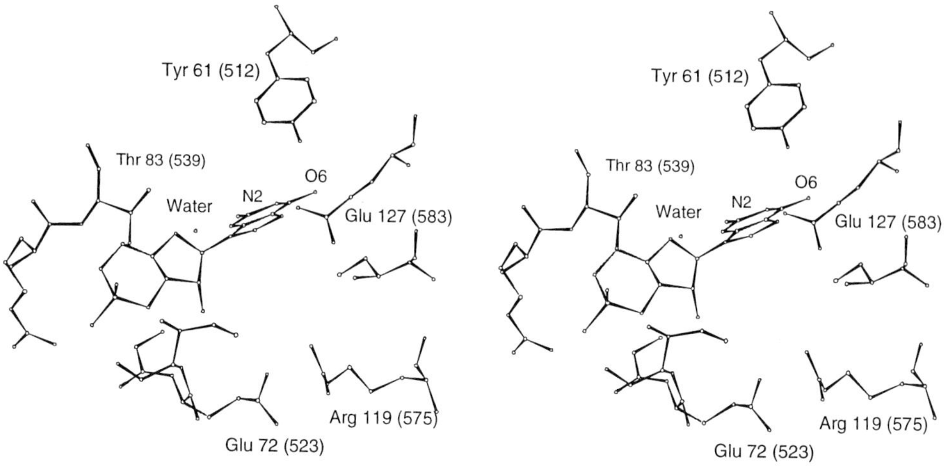

FIG. 5. Stereo closeup view of the refined model of bovine olfactory CNGC binding domain with syn-cGMP. This model is described in detail elsewhere.[12]

with the purine ring of cGMP. The OH of Tyr-61 forms a hydrogen bond with OE2 of Glu-127. The OH of Tyr-61 likely interacts weakly with the hydrogen-rich area near N-1 and N-2, becoming the fourth point of a tetrahedron formed by the atoms Tyr-61 OH, Glu-127 OE2 and cGMP N-1 and N-2. Further, the conjugated ring of Tyr-61 still acts as a hydrogen donor to the relatively polar O6 of cGMP. The two weak interactions, the OH group and ring system of Tyr-61, are additive and we considered them to be equivalent to a strong interaction. The interactions with the ribofuranose are equivalent in the bovine retina and olfactory CNGCs. The bovine olfactory binding site, therefore, has three strong interactions predicted between the purine ring of cGMP and the binding site, suggesting a tighter association with cGMP compared with that of the bovine retina binding site, which has two strong and one weak interaction.[12]

Based on a series of models, we predicted that residues in positions 61, 83, 119, and 127 (using the CRP sequence in order to compare among channels) act in concert to bind the ligand and determine the nucleotide specificity.[12] Of these residues, only Thr-83 is conserved among members of the CNGC family. In our models, a single hydrogen bond was predicted between the adenine of cAMP and the Asp-127 in the bovine retina binding domain in contrast to the guanine, which is predicted to interact with Thr-83 and Asp-127. In both the olfactory and retina channels, the aromatic residue in position 61 on the β5 strand was predicted to form weak electrostatic interactions with C-6 substituents.[12] The importance of these contacts was experimentally supported by the inactive analogs, all of which vary at N-1 or C-6 positions.

Although our models were consistent with all of the experimental data available, we were left with one open issue related to the syn versus anti conformation of the bound ligand. For all CNGC models generated, both cAMP and cGMP were sterically accommodated in either syn or anti conformation. Based on the number of strong and weak interactions predicted between the ligand and protein, the models provided useful insights about why one conformation might be selected over another; however, the issue has not been resolved with currently available tools. This point reminds us that some details of the ligand interactions will only be resolved with high-resolution structure determinations.

Limitations of Molecular Modeling

The most important thing to recognize about a model is that it is not a structure. This point is particularly relevant for models of a ligand-binding domain where the test protein consists of multiple domains and, possibly, multiple subunits. In this case, the model does not include interdomain or intersubunit interactions, which convey essential information to other regions of the protein matrix. With our models of the CNGC, for example, there are no intersubunit contacts despite the importance of these contacts in the reference protein. In CRP, the C-6 amino group of cAMP forms two hydrogen bonds, one with each Cα helix of the two subunits[5] and the interdomain contact is essential for DNA binding. At present, little is known about intersubunit contacts in the CNGC but a growing body of evidence indicates the importance of interdomain contacts in regulating channel activation.[24–26]

Testing Model Predictions Using Mutational Analysis

Modeling predictions can often be tested using either ligand analogs or mutational analysis. Our models of the retina and olfactory CNGC binding domains predict that residue 119 on the Cα helix plays an important role in coordinating the ligand interaction. This prediction was examined using mutational analysis where Lys-119 of the bovine retina channel was replaced with Arg, which occupies the corresponding position in the bovine olfactory channel. Because the olfactory channel is activated with a lower concentration of cGMP than the retina channel, we predicted that the arginine would decrease the $K_{0.5}$ for cGMP.

Figure 6 shows the effect of this mutation on the $K_{0.5}$ values for cGMP-activated macroscopic currents in excised patches. cDNA for the mutated bovine retina α-subunit CNGC was transiently transfected into tsA201 cells

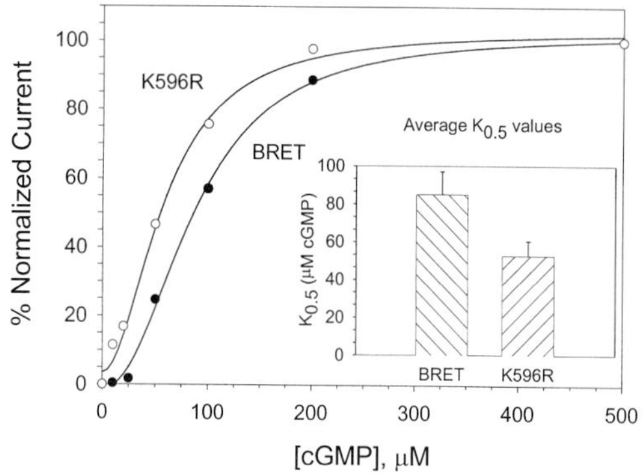

FIG. 6. cGMP activation for wild-type and Lys-119 (596) to Arg mutant bovine retina channels transiently expressed in eukaryote cells. The cGMP dose–response relations at 60 mV are shown for a single representative patch from wild-type and mutant CNGCs in the main panel. The $K_{0.5}$ values were 90 and 50 μM, respectively. The inset shows averaged data from several patches with values of 85 (n = 3) and 52 μM (n = 7), respectively. The currents were normalized to the current at 500 μM cGMP for each patch.[28]

to express the channel. The $K_{0.5}$ values decreased from 85 μM in wild type to 52 μM for the mutant[28] supporting the prediction that residue 119 is important in coordinating the ligand interaction in olfactory CNGCs. This residue would not have been predicted to interact with the ligand from the sequence alignment alone.

Additional mutagenesis experiments are under way to better define the coordination of a number of residues within the binding domain. The forthcoming data will enable us to test a number of the modeling predictions in an effort to refine the models and better understand nucleotide discrimination in different family members.

Conclusion

Protein domains are semi-independent functional units as well as autonomous structural units of folding.[15,35] Examination of the available protein structures has led to the prediction that the number of unique structural motifs in proteins is limited to about a thousand.[15,36] These basic folding motifs can therefore be achieved with multiple amino acid sequences, suggesting that the folded structure is encrypted within the amino acid sequence of the protein.[37] Although the encryption lexicon is currently obscure, the

[35] J. S. Richardson, *Methods Enzymol.* **115,** 341 (1985).
[36] C. Chotia, *Nature* **357,** 543 (1992).
[37] C. B. Anfinsen, *Science* **181,** 223 (1973).

large database of protein structures has been used to derive clues about how sequence relates to structure. Fortunately, this information has been coupled to powerful computational tools to develop searching algorithms (see Appendix) used to predict structural motifs for proteins of undetermined structure based only on their sequence.

Ion channels are membrane proteins with multiple domains of unknown structure. The nearly universal access to the Internet provides any scientist the opportunity to use molecular modeling tools at his or her own desktop. The virtual links with structural biology tools can be used to identify candidate reference proteins and construct homology models of selected domains of the channel proteins. These models will generate testable predictions about the architecture of the channel protein as we await the determination of high-resolution structures of these proteins.

Appendix

Location of Useful Starting Points for Protein Structure and Alignment Tools

1. A number of structure-based tools can be accessed from this base including CATH, a classification of protein domains found in the protein structure database:
http://www.biochem.ucl.ac.uk/bsm/biocomp/index.html
2. Protein data bank, Brookhaven National Laboratory:
http://www.pdb.bnl.gov
3. RASMOL, a free program for molecular structure display:
http://www.umass.edu/microbio/rasmol/rasintro.htm
4. OWL is a nonredundant composite of four publicly available primary sources: SWISS-PROT, PIR (1–3), GenBank (translation), and NRL-3D. SWISS-PROT is the highest priority source, all others being compared against it to eliminate identical and trivially different sequences:
http://www.biochem.ucl.ac.uk/bsm/dbbrowser/OWL/OWL.html
5. Pedro's biomolecular research tools includes molecular biology search and analysis tools, databases, and journals:
http://www.public.iastate.edu/~pedro/research_tools.html
6. Links to many resources in bioinformatics with emphasis on protein structure analysis and prediction:
http://geoff.biop.ox.ac.uk
7. Computational molecular biology at NIH:
http://molbio.info.nih.gov/molbio
8. Multiple resources including a novel prediction-based threading method used to identify remote homologs with 0–25% sequence identity:

http://www.embl-heidelberg.de/Services/index.html
http://www.mbi.ucla.edu/people/frsvr/frsvr.html
http://www.cryst.bbk.ac.uk/~ubcg09a/tlb_progs.html
http://www.globin.bio.warwick.ac.uk/~jones/threader.html

9. Another molecular modeling program: AMMP. Molecular mechanics and dynamics program written by Dr. Rob Harrison and used in models of the CNGC described in this chapter:
http://asterix.jci.tju.edu/

Sites with Software Useful for Molecular Modeling

1. Prediction of transmembrane helices:
TMAP:
http://www.embl.heidelberg.de/tmap/tmap_info.html
PHDtopology:
http://www.embl-heidelberg.de/predictprotein/predictprotein.html

2. Scanning sequences of proteins for which the three-dimensional structure is known:
SCOP:
http://scop.mrc-lmb.cam.ac.uk/scop/aln.cgi

3. Scanning against a database of protein folds:
TOPITS:
http://www.embl-heidelberg.de/predictprotein/predictprotein.html
UCLA-DOE Structure Prediction Server:
http://www.mbi.ucla.edu/people/frsvr/frsvr.html

4. Structure comparison:
DALI:
http://www2.ebi.ac.uk/cgi-bin/dali_html_align_pair

5. Sequence comparison:
BCM Search Launcher Web server:
http://dot.imgen.bcm.tmc.edu:9331/seq-search/protein-search.html
Clustal W:
http://dot.imgen.bcm.tmc.edu:9331/multi-align/Options/clustalw.html
CAMELEON:
http://www.oxmol.co.uk/prods/cameleon/

6. Secondary structure prediction:
PHDsec:
http://www.embl-heidelberg.de/predictprotein/predictprotein.html
DSC Web Server:
http://bonsai.lif.icnet.uk/bmm/dsc/dsc_form_align.html

7. Verification of models:
Verify3D:
http://www.doe-mbi.ucla.edu/Services/Verify3D.html

Biotech Validation Suite for Protein Structures:
http://biotech.embl-ebi.ac.uk:8400/

8. Homology modeling:

COMPOSER:
http://www.tripos.com/products/tri224p1.html#composer/

SWISSMODEL:
http://expasy.hcuge.ch/swissmod/

MODELLER:
http://guitar.rockefeller.edu/modeller/modeller.html

9. Selection of representative structures:

OLDERADO:
http://neon.chem.le.ac.uk/olderado/

10. Simulated annealing and energy minimization:

XPLOR:
http://xplor.csb.yale.edu/

11. *Ab initio* modeling:

SCULPT:
http://www.intsim.com/

12. Electrostatics:

DELPHI:
http://tincan.bioc.columbia.edu/delphi/

GRASP:
http://tincan.bioc.columbia.edu/Lab/grasp/

13. Pore diameter:

HOLE:
http://www.cryst.bbk.ac.uk/~ubcg8ab/hole/top.html

14. Docking of ligands:

SURFNET:
http://www.biochem.ucl.ac.uk/bsm/biocomp/

GRAMM:
http://guitar.rockefeller.edu/gramm/

Websites for Authors

1. Home page for Sansom laboratory:
http://indigo1.biop.ox.ac.uk/home.html

2. Homepage for Oswald laboratory:
http://web.vet.cornell.edu/public/pharmacology/Robert_Oswald/frame_Robert_Oswald.html/

3. Homepage for Tanaka laboratory:
http://athens.dental.upenn.edu/tanaka/index.html

Acknowledgments

We would like to thank our collaborators, Drs. Irene Weber and Rob Harrison, for assistance and guidance in our modeling efforts and Joseph Jez for comments on the manuscript. Support for the modeling research was provided from the National Eye Institute with EY06640 to JCT and EY07035, a training grant, for S-P.S.

[34] Ion Channels: Molecular Modeling and Simulation Studies

By MARK S. P. SANSOM

Introduction

Overview

The aim of this article is to describe the use of restrained molecular dynamics (MD) simulations in modeling ion channels. In particular, the technique of simulated annealing with restrained MD for modeling the pore domains of complex channel proteins and channels of formed by peptides and simple channel proteins is described in detail. Once generated, such models may form the basis of more extended MD simulations, both to refine the models and to aid our understanding of the interactions of channels with water and permeant ions. The article concludes with a discussion of continuum electrostatics calculations based on comparisons of channel models with electrophysiologic measurements.

Need for Models and Simulations

One might ask *why* it is useful to model and simulate ion channels? In principle, structures may be determined experimentally, and channel function has been investigated in detail via electrophysiologic and associated methods. However, progress in experimental structure determination for channel proteins by X-ray diffraction is rather slow (as for membrane proteins in general), structures derived by electron microscopy do not often achieve near atomic resolution, and spectroscopic methods [e.g., solid state nuclear magnetic resonance (NMR)] as yet have yielded only limited structural data. Modeling per se can play a dual role in advancing our understanding of channels. In the absence of definitive structural data, modeling may be used to predict possible structures for channels that provide a conceptual framework on which to base further experimental and theoretical studies. If incomplete structural data are available [which can range from a small

0076-6879/98 $25.00

number of distance restraints from solid-state NMR studies to a low-resolution description of the orientation of the transmembrane (TM) secondary structure elements from electron microscopy] then SA/MD may be used to extend the effective resolution of a model by enabling integration of structural data with softer information from various other sources, e.g., site-directed mutagenesis. Computational approaches allow objective and automated comparison of alternative models for a given channel with the available experimental data. Thus, as the database of information from experimental sources expands, models can be improved and updated in a semiautomatic fashion.

Once generated, one can use a molecular model of a channel to predict physiologic properties such as the single-channel conductance and ion selectivity. Such predictions are closely related to the use of structure-based simulations to understand the physics of ion permeation through a channel. Thus, modeling and simulation studies enable one to move from a molecular to an atomic level physiology of ion channels.

In the following sections the examples are taken mainly from work in the author's laboratory. This should not lead one to ignore the many interesting computational studies of channels undertaken by a number of other research groups. In particular, I do not discuss the many simulation and theoretical studies that have been made of the simple model channel gramicidin[1,1a] because these have been discussed in detail in two excellent reviews.[2,3]

Molecular Modeling

Overview

Modeling of an ion channel can be carried out on a number of different levels, depending on the questions being examined and on the experimental data available. For example, if primary interest is in the molecular basis of the agonist pharmacology of a receptor channel, then modeling techniques based on identification of homologies with known protein folds[4,5] may be used to generate models of agonist-binding sites. If the aim of modeling is

[1] G. A. Woolley and B. A. Wallace, *J. Membr. Biol.* **129,** 109 (1992).
[1a] O. S. Andersen, C. Nielsen, A. M. Maer, J. A. Lundbaek, M. Goulian, and R. E. Koeppe II, *Methods Enzymol.* **294,** [10], (1998).
[2] M. B. Partenskii and P. C. Jordan, *Quart. Rev. Biophys.* **25,** 477 (1992).
[3] B. Roux and M. Karplus, *Ann. Rev. Biophys. Biomol. Struct.* **23,** 731 (1994).
[4] J. E. Gready, S. Ranganathan, P. R. Schofield, Y. Matsuo, and K. Nishikawa, *Prot. Sci.* **6,** 52 (1997).
[5] I. Tsigelny, N. Sugiyama, S. M. Sine, and P. Taylor, *Biophys. J.* **73,** 52 (1997).

to provide a conceptual framework for the design of further experimental (e.g., mutagenesis) studies, then interactive graphics and related techniques can be used to produce models of an entire channel protein.[6–8,8a] However, it is likely that such models will be somewhat under-determined by the available data; that is, alternative models could be constructed that would provide a different, but equally satisfactory conceptual framework.

The following discussion focuses on a more limited (but perhaps more realistic) role for channel modeling, namely, the pore-forming domains of channels. There are several reasons for restricting attention to the pore per se. First, the nature of single-channel electrophysiologic techniques means experimental measurements are more directly related to channel conductance and selectivity than to some other aspects of channel function. Second, because ion permeation is a relatively simple process, it is more amenable to simulation studies than are, for example, channel gating mechanisms, thus enabling rigorous testing of pore domain models against experimental data. Third, more extensive experimental data are available concerning the pore domains of ion channels than for other regions of such proteins. Such data take the form of extensive mutagenesis studies identifying pore-lining residues within ion channel sequences, and/or direct structural data on the nature and orientation of secondary structure elements within a pore domain. Note that implicit in the decision to model the pore-forming domain of a channel is the assumption that the transbilayer topology of the channel protein has been determined. Methods for prediction of transbilayer topologies on the basis of sequence data are well established[9] and are discussed in the chapter by Oswald and Sutcliffe in this book.[8a]

Molecular Dynamics and Simulated Annealing

Molecular dynamics is a computational technique that allows one to simulate the motions of biomolecules. A more recent extension of it use has been to aid in determination of protein structures, either in refinement of structures against X-ray data or in determination of protein structures on the basis of internuclear distance restraint data from NMR studies.[10]

[6] S. R. Durell and H. R. Guy, *Biophys. J.* **62,** 238 (1992).
[7] S. R. Durell and H. R. Guy, *Neuropharmacology* **35,** 761 (1996).
[8] M. J. Sutcliffe, Z. G. Wo, and R. E. Oswald, *Biophys. J.* **70,** 1575 (1996).
[8a] M. J. Sutcliffe, A. H. Smeeton, Z. G. Wo, and R. E. Oswald, *Methods Enzymol.* **293,** [32], 1998 (this volume).
[9] D. T. Jones, W. R. Taylor, and J. M. Thornton, *Biochem.* **33,** 3038 (1994).
[10] A. T. Brünger and M. Karplus, *Acc. Chem. Res.* **24,** 54 (1991).

Indeed, the employment of restrained MD simulations in modeling studies is directly comparable to their use in NMR structure determination.

Molecular dynamics is based on numerical simulation of the Newtonian motions of a collection of particles. The MD method per se is not reviewed in detail because several excellent accounts are already available.[11–13] However, a brief description of the principles of MD is required in order to understand the discussion of its uses in molecular modeling of channels.

MD simulations of proteins require an *empirical energy function.* The protein molecule is described as a collection of atoms, to each of which is assigned a mass, a partial atomic charge, and a van der Waals radius. The energy function describes the potential energy of a protein molecule as a function of the coordinates of its constituent atoms. Such a function can be divided into terms representing the covalent and the nonbonded (i.e., noncovalent interactions) between atoms:

$$E = E_{\mathrm{COVALENT}} + E_{\mathrm{NONBONDED}} \tag{1}$$

The covalent energy term is subdivided into terms corresponding to bond lengths, bond angles, and dihedral angles:

$$E_{\mathrm{COVALENT}} = E_{\mathrm{BOND}} + E_{\mathrm{ANGLE}} + E_{\mathrm{DIHEDRAL}} \tag{2}$$

The covalent energy component of the potential energy function is calculated using simple classical approximations to describe the change in energy as, for example, a bond length is distorted from its ideal value. Thus:

$$E_{\mathrm{BOND}} = \sum_{\mathrm{bonds}} \frac{1}{2} K_{\mathrm{B}}(b - b_0)^2 \tag{3}$$

where b is the instantaneous bond length, b_0 is the ideal (equilibrium) bond length, and K_{B} is the force constant associated with stretching a given bond. Similar terms are used for E_{ANGLE} and E_{DIHEDRAL}. Thus, each term is defined by an ideal value of a length or angle, and by a force constant. The ideal values can be obtained by analysis of crystal structures of, for example, amino acids, and the force constants from analysis of infrared and related spectroscopic data. The nonbonded energy term is made up of a van der Waals and an electrostatic component:

$$E_{\mathrm{NONBONDED}} = E_{\mathrm{VDW}} + E_{\mathrm{ELEC}} \tag{4}$$

[11] M. P. Allen and D. J. Tildesley, "Computer Simulation of Liquids." Oxford University Press, Oxford, 1987.

[12] C. L. Brooks, M. Karplus and B. M. Pettitt, "Proteins: A Theoretical Perspective of Dynamics, Structure and Thermodynamics." Wiley, New York, 1988.

[13] J. A. McCammon and S. C. Harvey, "Dynamics of Proteins and Nucleic Acids." Cambridge University Press, Cambridge, 1987.

The van der Waals component is generally calculated using a standard 6–12 function:

$$E_{\mathrm{VDW}} = \left(\frac{A}{r^{12}} - \frac{B}{r^{6}}\right) \tag{5}$$

where A and B are constants for the pair of atom types involved in the interaction, and r is the distance between them. The electrostatic energy is calculated using a Coulombic expression:

$$E_{\mathrm{ELEC}} = \frac{q_i q_j}{4\pi\varepsilon_0 \varepsilon r^2} \tag{6}$$

where q_i and q_j are the partial charges on the two atoms, r is the distance between them, and ε is the dielectric constant of the medium between them. Note that calculation of these two energy terms is quite computationally costly, because it involves summation over all pairs of atoms in a molecule. However, the calculation is often speeded via truncation of the calculation so that only atom pairs within, say, 14 Å of one another are considered. Various methods are available to smooth the truncation,[14] but there is some concern about the consequences of this when applied to electrostatic interactions.[15] As discussed later, it is also possible to incorporate terms representing experimentally derived restraints into a potential energy function.

From a potential energy function it is possible to simulate the *dynamic* behavior of a protein molecule by numerical solution of Newton's second equation of motion for all the atoms of a protein moving simultaneously:

$$\mathbf{F}_i = m_i \mathbf{a}_i \tag{7}$$

where $\mathbf{F}_i$ is the force acting on atom i (which has mass m_i,) yielding an acceleration of $\mathbf{a}_i$. Numerical integration of this equation yields the change in velocity and position of the atom during a short time step δt. The force acting on atom i is the consequence of its interactions with all the other atoms of the molecule and is obtained from the derivative of the potential energy function with respect to the position, $\mathbf{r}_i$, of the atom:

$$\mathbf{F}_i = -\frac{\partial \mathrm{E}}{\partial \mathbf{r}_i} \tag{8}$$

[14] B. R. Brooks, R. E. Bruccoleri, B. D. Olafson, D. J. States, S. Swaminathan, and M. Karplus, *J. Comp. Chem.* **4,** 187 (1983).

[15] R. J. Loncharich and B. R. Brooks, *Proteins: Struct. Func. Genet.* **6,** 32 (1989).

Note that there will be one such equation for every atom of the channel molecule. Starting from an initial set of coordinates for all the atoms of a molecule, numerical solution of these equations yields a new set of coordinates following the elapse of a short period of time (typically $\delta t = 10^{-15}$ s = 1 fs). Repeating this process until a total simulation time of, say, 10^{-9} s (= 1 ns) or more has elapsed yields a MD trajectory.

A factor of crucial importance in an MD simulation is the temperature. This determines the average amplitude of the atomic motions of the atoms. The temperature is included in a simulation via initial velocities assigned to the atoms. Atomic velocities are assigned random values, drawn from a Maxwellian distribution, such that the mean velocity is equivalent to a given temperature, for example, 300 K (room temperature). At 300 K the motions of protein atoms are restricted to the vicinity of the initial structure, and so only *local* changes in conformation may occur. However, if simulations are run at higher temperatures (e.g., 600 K or above), then larger scale atomic motions occur, thus allowing a protein to undergo larger scale conformational changes. This is exploited in simulated annealing (see later discussion) to enable exploration of a larger range of conformational variations during generation of a molecular model.

In molecular modeling, as in experimental structure determination and refinement, the empirical energy function (which embodies our chemical knowledge of the conformational properties of a protein) is supplemented by an additional energy term that represents experimentally derived restraints:

$$E = E_{\mathrm{COVALENT}} + E_{\mathrm{NONBONDED}} + E_{\mathrm{RESTRAINT}} \tag{9}$$

The restraint energy, $E_{\mathrm{RESTRAINT}}$, may take a number of forms. However, in the current application it is generally composed of a collection of distance restraints. These restraints may either be between pairs of atoms within a model structure or between corresponding atoms in the model structure and a target structure, where the latter represents, for example, a low-resolution structure. For example, a distance restraint between atom i and atom j may take the form:

$$E_{\mathrm{DIST}} = K(d_{ij} - d_{\mathrm{TARGET}})^2 \tag{10}$$

where d is the distance between the two restrained atoms (or between the centers of gravity of two groups of atoms) and d_{TARGET} is the target distance for the restraint. The scale factor K is chosen to give a balance between experimentally derived restraints and the standard energy function. Thus, if some data on which restraints are based are better than others, different scale factors for different classes of restraint may be employed. Restraints may be used for two purposes: (1) to encode experimental data (e.g., to

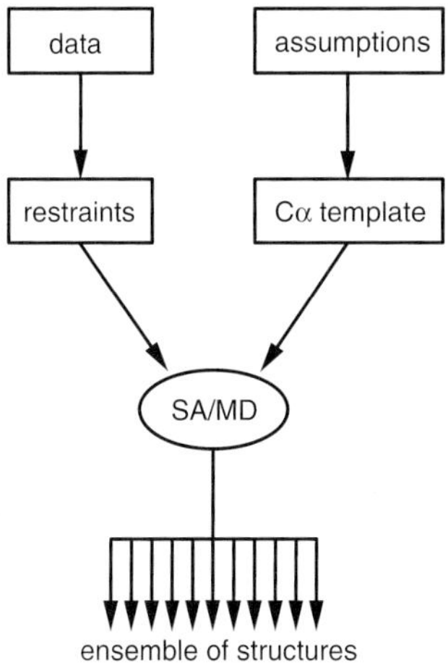

FIG. 1. Flow diagram of SA/MD. The inputs are from two sources: distance and positional restraints derived from experimental data; and a Cα template, which reflects underlying assumptions concerning, for example, the secondary structure of the pore-forming transmembrane segments. SA/MD is used to generate an ensemble of structures, all of which embody these restraints and assumptions. These models can then be analyzed in terms of their steric and electrostatic properties in order to compare them with experimental channel data.

ensure that pore-lining side chains point toward the central lumen of a channel, or to maintain the position of a TM α helix close to that in a low-resolution structure) and/or (2) to mimic the effects of the omitted components of the channel protein and its bilayer environment [e.g., to maintain a polypeptide chain in a given secondary structure (e.g., as an α helix); to hold a pair of α helices adjacent to one another]. Which restraints to employ for a given channel depends on the nature of the experimental data available, and the decision over whether to model just the pore domain or the entire channel protein.

Having defined a potential energy function and restraint terms to be added to it, it is necessary to consider how MD simulations may be used to search protein conformations compatible with the restraints. A method that has been used with some success is to employ a simulated annealing protocol,[16,17] yielding a technique known as simulated annealing via restrained molecular dynamics (SA/MD; see Fig. 1). This procedure is based on methods used for NMR structure determination[18] and has been success-

[16] S. Kirkpatrick, C. D. Gelatt and M. P. Vecchi, *Science* **220,** 671 (1983).
[17] K.-C. Chou and L. Carlacci, *Prot. Eng.* **4,** 661 (1991).
[18] M. Nilges, G. M. Clore and A. M. Gronenborn, *FEBS Lett.* **239,** 129 (1988).

fully exploited in prediction of helix packing within the GCN4 leucine zipper helix dimer.[19,20]

The key to SA/MD is to vary the temperature of the simulation in order to control the sampling of different conformations. By starting a simulation at a high temperature and then progressively decreasing the temperature until 300 K is reached, large changes in conformation are possible at the start of the simulation, while toward the end of the simulation only much smaller changes may occur. Application of SA/MD to channel models has been described in detail in several papers.[21,22] The starting point of stage 1 of SA/MD is a Cα template, which embodies underlying assumptions concerning the nature of the pore (e.g., a bundle of exactly parallel idealized α helices[21] or an idealized β barrel[22]). Remaining backbone and side-chain atoms are superimposed on the Cα atoms of the corresponding residues. These atoms "explode" from the Cα atoms, the positions of which remain fixed throughout stage 1. Annealing starts at 1000 K, during which weights for bond lengths and bond angles, and subsequently for planarity and chirality, are gradually increased. A repulsive van der Waals term is slowly introduced after an initial delay. Once the scale factors of these components of the empirical energy function reach their final values, the system is cooled from 1000 to 300 K, in steps of 10 K and 0.5 ps. During this cooling the van der Waals radii are reduced to 80% of their standard values in order to enable atoms to "pass by" one another. Electrostatic terms are *not* included during stage 1. Typically five structures are generated for each Cα template, corresponding to multiple runs of the process with different random assignments of initial velocities.

Structures from stage 1 are each subjected to, say, five molecular dynamics run (stage 2), resulting in an ensemble of $5 \times 5 = 25$ final structures from a single Cα template. Initial velocities are assigned corresponding to 500 K. Harmonic restraints are imposed on Cα atoms at the beginning of stage 2, and are gradually relaxed as the temperature is reduced from 500 to 300 K. Distance restraints are also introduced at this point. Also during stage 2 electrostatic interactions are introduced into the potential energy function. On reaching 300 K, a 5-ps burst of constant temperature dynamics is performed, followed by 1000 steps of conjugate gradient energy minimization.

[19] M. Nilges and A. T. Brünger, *Prot. Eng.* **4,** 649 (1991).

[20] M. Nilges and A. T. Brünger, *Proteins: Struct. Func. Genet.* **15,** 133 (1993).

[21] I. D. Kerr, R. Sankararamakrishnan, O. S. Smart, and M. S. P. Sansom, *Biophys. J.* **67,** 1501 (1994).

[22] M. S. P. Sansom and I. D. Kerr, *Biophys. J.* **69,** 1334 (1995).

An important practical consideration is how to implement the preceding methods. For SA/MD the program Xplor,[23] developed to implement MD simulations for X-ray and NMR determination of protein structures, is flexible and easy to use. A number of other well-established MD programs (e.g., Charmm,[14] Amber,[24] Gromos[25,26]) can also be used. Charmm employs a potential energy function similar to that of Xplor, and although less suited to SA/MD per se, offers more powerful facilities for subsequent simulations based on SA/MD-generated models.

Although this discussion has focused on SA/MD as employed in the author's laboratory, there are a number of other computational methods available. Related MD-based methods have been used to model TM helix packing in phospholamban channels[27] and to model the Kv channel pore domain.[28] Interactive graphics modeling has been applied to a wide variety of channel proteins.[6,7,29–34] Homology modeling techniques have been employed to generate models of neurotransmitter receptors and their agonist-binding domains.[4,5,8] Monte Carlo methods, treating TM helices as rigid bodies and packing them to satisfy experimentally derived distance restraints, have proved to be successful in modeling bacteriorhodopsin and G-protein-coupled receptors[35] and offer considerable potential for modeling ion channel molecules,[36] particularly as more mutagenesis data become available.

[23] A. T. Brünger, "X-PLOR Version 3.1. A System for X-Ray Crystallography and NMR." Yale University Press, New Haven, Connecticut, 1992.

[24] S. J. Weiner, P. A. Kollman, D. A. Case, U. C. Singh, C. Ghio, G. Alagona, S. Profeta, and P. Weiner, *J. Am. Chem. Soc.* **106,** 765 (1984).

[25] H. J. C. Berendsen, J. P. M. Postma, W. F. van Gunsteren, A. DiNola, and J. R. Haak, *J. Chem. Phys.* **81,** 3684 (1984).

[26] J. Hermans, H. J. C. Berendsen, W. F. van Gunsteren, and J. P. M. Postma, *Biopolymers* **23,** 1513 (1984).

[27] P. D. Adams, I. T. Arkin, D. M. Engelman, and A. T. Brünger, *Nature Struct. Biol.* **2,** 154 (1995).

[28] P. K. Yang, C. Y. Lee, and M. J. Hwang, *Biophys. J.* **72,** 2479 (1997).

[29] H. R. Guy and F. Hucho, *Trends Neurosci.* **10,** 318 (1987).

[30] H. R. Guy and G. Raghunathan, *in* "Computer-Assisted Modeling of Receptor-Ligand Interactions: Theoretical Aspects and Applications to Drug Design," p. 231. Liss, New York, 1989.

[31] H. R. Guy and F. Conti, *Trends Neurosci.* **13,** 201 (1990).

[32] S. R. Durell, H. R. Guy, N. Arispe, E. Rojas, and H. B. Pollard, *Biophys. J.* **67,** 2137 (1994).

[33] H. R. Guy and S. R. Durell, *Biophys. J.* **68,** A243 (1995).

[34] S. R. Durrell, Y. Hao, and H. R. Guy, *J. Struct. Biol.* (in press) (1998).

[35] P. Herzyk and R. E. Hubbard, *Biophys. J.* **69,** 2419 (1995).

[36] P. Herzyk and R. E. Hubbard, *Biophys. J.* **74,** 1203 (1998).

Low-Resolution Structure Availability

Perhaps the most direct use of molecular modeling is when a low-resolution (e.g., 7 Å) structure is available for a channel protein, enabling computational techniques such as SA/MD to be used to extend the effective resolution via generation of atomic resolution models of the channel which combine structural and mutagenesis data. Examples of this situation include the nicotinic acetylcholine receptor (nAChR), for which a 9-Å resolution structure is available,[37] and aquaporin (a water channel)[37a] for which two structures are available at about 7-Å resolution.[38,39] In both cases structural data were obtained by electron microscopy.

Low-resolution structural data provide powerful restraints for molecular modeling. This can be illustrated with reference to the pore domain of open conformation of the nAChR[40,41] (Fig. 2). From the electron microscope images[37] of the TM region it is evident that the central pore is lined by a bundle of five α helices. Sequence, chemical labeling, and mutagenesis studies[42–44] revealed that the pore lining is formed primarily by the M2 TM segment of the amino acid sequence of each of the five homologous subunits that make up the channel. More detailed mutagenesis studies identified those residues within M2 that contributed side chains to the lining of the pore. By SA/MD it was possible to combine all of this information in an atomic resolution model of the transbilayer pore of a nAChR.

The nature of the restraints used in SA/MD modeling of the nAChR pore is illustrated in Fig. 2A. From the electron microscope images each pore-lining helix was seen to be kinked. Thus the pore could be represented as two line segments for each helix, meeting at the position of the kink. These target line segments are shown as thick gray lines in Fig. 2A. A Cα template embodied the deduction from nonstructural data that the M2 segments formed TM helices. Thus, five exactly parallel, idealized helices were generated as the Cα template. The model shown in Fig. 2 is for a chick neuronal nAChR called α7, which is somewhat simpler to model in

[37] N. Unwin, *Nature* **373,** 37 (1995).

[37a] P. Agre, J. C. Mathai, B. L. Smith, and G. M. Preston, *Methods Enzymol.* **294,** [29] (1998).

[38] T. Walz, T. Hirai, K. Murata, J. B. Heymann, B. L. Smith, P. Agre, and A. Engel, *Nature* **387,** 624 (1997).

[39] A. Cheng, A. N. van Hoek, M. Yeager, A. S. Verkman, and A. K. Mitra, *Nature* **387,** 627 (1997).

[40] R. M. Stroud, M. P. McCarthy, and M. Shuster, *Biochemistry* **29,** 11009 (1990).

[41] N. Unwin, *Neuron* **3,** 665 (1989).

[42] J. P. Changeux, J. I. Galzi, A. Devillers-Thiéry, and D. Bertrand, *Quart. Rev. Biophys.* **25,** 395 (1992).

[43] H. Lester, *Ann. Rev. Biophys. Biomol. Struct.* **21,** 267 (1992).

[44] F. Hucho, V. I. Tsetlin, and J. Machold, *Eur. J. Biochem.* **239,** 539 (1996).

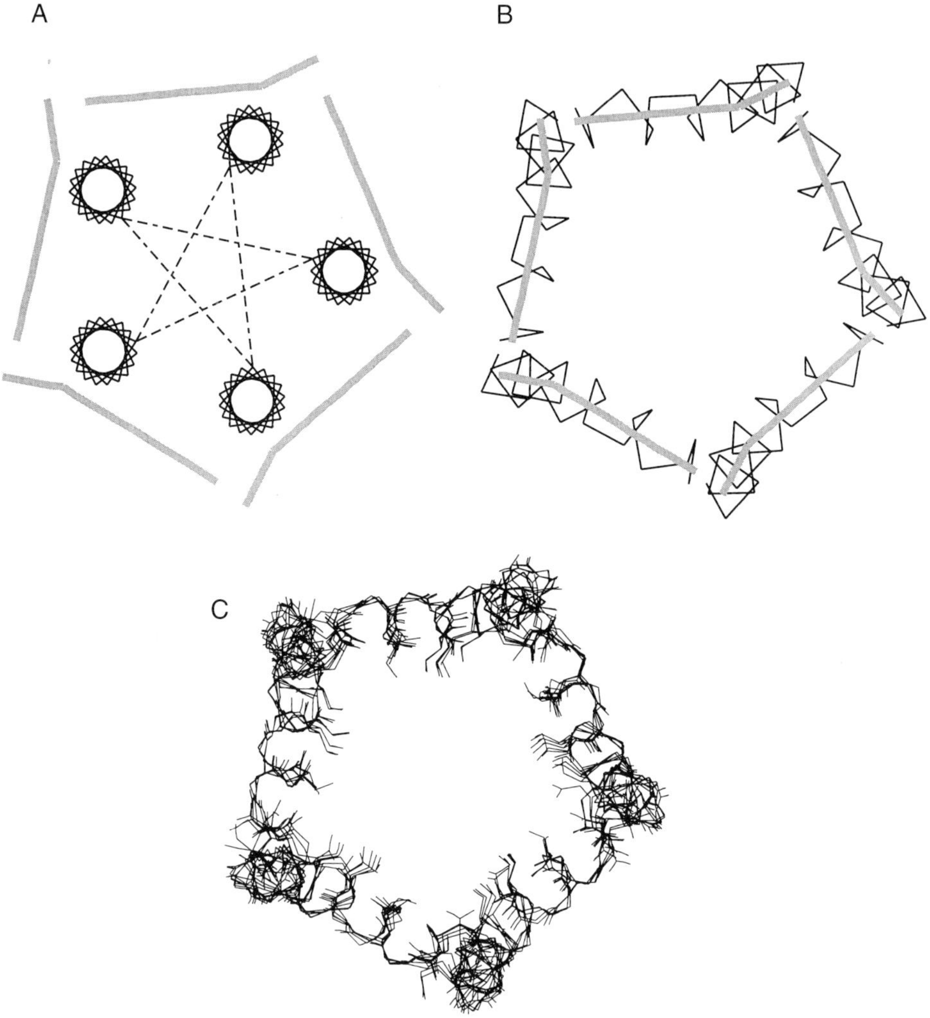

FIG. 2. The pore domain of the nAChR, in its open conformation, as modeled by SA/MD. (A) Target restraints derived from the electron microscopic images (thick gray lines); distance restraints which ensure that those residues identified by site-directed mutagenesis as pore-lining point toward the center of the pore (broken lines); and the idealized Cα template which embodies the assumption that the M2 segments are α helical in conformation (solid lines). (B) Cα trace of a single structure from the SA/MD-generated ensemble, superimposed on the target restraints. (C) Five structures (from an ensemble of 25) superimposed. The backbone atoms plus the pore-lining polar side chains (Ser and Thr), which are the subject to the distance restraints in part (A), are shown. All other side chains are omitted for clarity. In each diagram the view is down the pore axis.

that it is homopentameric, i.e., it has subunit stoichiometry α_5. The sequence of M2 for the α_7 subunit is

Glu-Lys-Ile-*Ser*-Leu-Gly-Ile-*Thr*-Val-Leu-Leu-*Ser*-Leu-Thr-Val-Phe-Met-Leu-Leu-Val-Ala-Glu

where the italic underlined residues are serine and threonine residues, which mutagenesis studies have demonstrated to form the narrowest region of the transbilayer pore.[45] This information was incorporated in SA/MD as interhelix distance restraints (shown as dotted lines in Fig. 2A) between the Cβ atoms of equivalent residues on opposite sides of the pore. Such restraints directed the key pore-lining side chains toward the center of the pore. The M2 sequence was aligned vertically (i.e., threaded) onto the Cα template by ensuring that the key serine and threonine residues corresponded to the narrowest region of the pore, and that the M2 sequence was aligned with the electron scattering data. Finally, intrahelix distance restraints (between O_i and H_{i+4}) were used to retain the M2 segments in an α-helical conformation in stage 2 of the simulation.

The results of the SA/MD run are shown in Figs. 2B and 2C. In Fig. 2B a single structure from the ensemble is shown as a Cα trace superimposed on the target coordinates derived from the electron microscopy data. It is evident that a bundle of kinked α helices has been automatically generated that fits well to the experimentally derived restraints. Ramachandran plot analysis confirms that the backbone torsion angles in the vicinity of the kinks in M2 remain within the allowed α-helical region. However, the restraints were not sufficient to define a unique structure. In Fig. 2C five structures from the same ensemble are superimposed on one another. Although it can be seen that there is good agreement between the backbone conformations of the different structures, there are variations in side-chain conformation. When analyzing the properties of the pore for this model it is important to remain aware of the variations in pore properties, both geometric and electrostatic, between different members of the same ensemble.[46,47]

This modeling technique can be readily applied to channel and other membrane proteins for which low-resolution structural data are available. It is important to consider the errors inherent in such calculations. These are as follows: (1) errors in identification of extent of the TM helices, (2) errors in restraints and sensitivity of models to restraints, and (3) bias toward the initial model. Errors in the extent of the M2 helices lining the

[45] A. Villarroel, S. Herlitze, M. Koenen, and B. Sakmann, *Proc. R. Soc. Lond. B* **243,** 69 (1991).
[46] M. S. P. Sansom, R. Sankararamakrishnan, and I. D. Kerr, *Nature Struct. Biol.* **2,** 624 (1995).
[47] R. Sankararamakrishnan, C. Adcock, and M. S. P. Sansom, *Biophys. J.* **71,** 1659 (1996).

pore are a little difficult to address. The M2 helix sequence listed earlier is consistent with most site-directed mutagenesis data and with the extent of the pore-lining rods observed in the electron microscope images.[37] However, it is difficult to specify exact start and end points of TM helices to better than a couple of residues. This is true for prediction of the location of TM helices in membrane proteins in general.[9] Because the target restraints were derived from a low-resolution structure (ca. 9 Å[37]) they may contain errors concerning the conformation of the helix kink and the overall orientation of the helices within the bundle. This is difficult to assess because the magnitude of such errors is uncertain. Errors in side-chain distance restraints and also the sensitivity of the simulation to the initial $C\alpha$ template have been tested by rotating the M2 helices in the $C\alpha$ template by $\pm 30°$ about their long axis, thus moving the pore-lining side chains somewhat away from the lumen of the pore. Analysis of the subsequent SA/MD models suggested that this had less effect on average pore radius and electrostatic potential profiles than the variation within a given ensemble of models. This confirms the importance of having an *ensemble* of models when trying to evaluate the most significant source of errors in predicted channel properties.

Structure of Constituent Monomer Available

Molecular modeling can also be combined with structural data for a number of channel-forming peptides (CFPs).[48] These are small (ca. 20 residue), amphipathic α-helical peptides that self-assemble in lipid bilayers to form ion channels. Such model systems, although evidently simpler than ion channel proteins, may provide valuable insights into the mechanisms of ion permeation. For several such peptides, although the structure of the channel assembly within the bilayer is unknown, structures (either from X-ray diffracton and/or NMR) are available for isolated peptide molecules in crystals or in nonaqueous solution. Thus, modeling techniques can be used to combine such structures with less direct evidence, from electrophysiologic and mutation studies, to generate highly plausible models of ion-conducting peptide assemblies.

The most intensively studied CFP is alamethicin (Alm), a member of a large family of peptaibols. Alm is a 20-residue peptide whose conformation, interactions with lipid bilayers, and channel-forming properties are well characterized.[1,49] Alamethicin exists as two naturally occurring variants, the R_i30 form.

[48] M. S. P. Sansom, I. D. Kerr, and I. R. Mellor, *Eur. Biophys. J.* **20,** 229 (1991).
[49] M. S. P. Sansom, *Quart. Rev. Biophys.* **26,** 365 (1993).

Ac-Aib-Pro-Aib-Ala-Aib-Ala-*Gln*[7]-Aib-Val-Aib-Gly-Leu-Aib-*Pro*[14]-Val-Aib-Aib-*Glu*[18]-Gln-Phl

and the R_i50 form in which Glu-18 is replaced by Gln-18. The N terminus is blocked by an acetyl group and the C terminus is a phenylalaninol (Phl); that is, the terminal —CO_2H of phenylalanine is replaced by —CH_2OH. The high content of a strongly helix-promoting amino acid, α-aminoisobutyric acid (Aib) ensures that Alm adopts a largely α-helical conformation. The Pro-14 introduces a kink into the center of the helix.

Channel formation by Alm is strongly voltage dependent. Once formed, the channels switch rapidly between multiple conductance levels. It is thought that the voltage-dependent step of channel formation corresponds to voltage-induced insertion of Alm into the bilayer. The resultant inserted helices self-assemble to form parallel bundles of Alm molecules surrounding a central transbilayer pore. The number of helices per bundle varies, resulting in Alm channels of different conductances. Within a burst of Alm channel activity, switching between adjacent conductance levels is due to addition/loss of Alm helices to/from a helix bundle on an approximately 10-ms time scale.

Ionic current measurements from single Alm channels enable direct determination of ion permeation properties. Such experiments support the helix bundle model, and provide restraints for molecular modeling studies of Alm channels. The helix bundle model is supported by the pattern of successive conductance levels within a single burst of multilevel openings of an alamethicin channel. Increasing the number of helices in a bundle (N) increases the radius of the central pore and hence increases its conductance.

In addition to electrophysiologic studies of channels formed in planar lipid bilayers by Alm there is a considerable body of structural data. Both X-ray[50] and NMR studies[51] indicate that Alm forms an extended α helix, kinked in its center via a proline-induced break in the H-bonding pattern, and NMR amide exchange data demonstrate that Alm is largely α helical when dissolved in methanol and retains this conformation when it interacts with lipid bilayers.[52,53] Recent simulation studies of Alm/bilayer interactions[54] suggest that a small change in kink angle may occur on insertion of the Alm helix into a bilayer, but that its overall shape is retained.

A number of models of Alm helix bundles have been proposed, starting

[50] R. O. Fox and F. M. Richards, *Nature* **300,** 325 (1982).

[51] G. Esposito, J. A. Carver, J. Boyd, and I. D. Campbell, *Biochemistry* **26,** 1043 (1987).

[52] C. E. Dempsey, *J. Am. Chem. Soc.* **117,** 7526 (1995).

[53] C. E. Dempsey and L. J. Handcock, *Biophys. J.* **70,** 1777 (1996).

[54] P. Biggin, J. Breed, H. S. Son, and M. S. P. Sansom, *Biophys. J.* **72,** 627 (1997).

with that of Fox and Richards[50] based on their X-ray structure for the Alm helix. The use of SA/MD has enabled more detailed and rigorous computational studies of Alm helix bundle models. The initial Cα templates used as input to SA/MD embodied a number of assumptions concerning the structure of channels formed by Alm. The first was that the constituent monomers of a pore-forming bundle adopted an α-helical conformation similar to that observed in the X-ray and solution NMR studies. This is justified by spectroscopic data indicating that Alm retains its α-helical conformation when interacting with lipid bilayers. The second assumption was that Alm molecules formed a transmembrane bundle of approximately parallel (rather than antiparallel) helices. Helix bundle formation is supported by, for example, in-plane neuron scattering data.[55] A *parallel* orientation for the helices of the bundle is supported by the pronounced asymmetry of Alm current–voltage curves and by the demonstration that channels formed by covalently linked parallel dimers of Alm helices resemble those of the parent Alm channels in their conductance values.[56]

From single-channel measurements, it is evident that Alm can form channels with a range of different pore sizes, likely to correspond to values of N helices/bundle. Analysis of single-channel conductance data in terms of equivalent resistances of electrolyte-filled pores modeled as simple cylinders suggested that bundles of between $N = 4$ and $N \geq 8$ helices were formed.[49] Analysis of the concentration and voltage dependence of the macroscopic conductance of Alm-containing bilayers yielded estimates of between $N = 4$ and $N = 11$ helices per bundle, depending on, for example, the lipid employed.[57] Thus Alm channels have been modeled as containing from $N = 4$ to $N = 8$ helices per bundle.

Other assumptions implicit in the Cα templates are that the hydrophilic surface of the Alm helix (defined by residue Gln-7) is directed toward the interior of the pore, and that the kinked helices are packed such that their N-terminal segments are in close contact with one another while their C-terminal segments are splayed out to form a wider C-terminal mouth to the pore. Both assumptions are reasonable on energetic grounds. The hydrophilic face of a helix will prefer to face the aqueous lumen of a pore rather than the surrounding hydrophobic fatty acyl chains. The C termini of the helices carry a bulky and negatively charged Glu-18 residue, and so

[55] K. He, S. J. Ludtke, H. W. Huang, and D. L. Worcester, *Biochemistry* **34,** 15614 (1995).

[56] S. You, S. Peng, L. Lien, J. Breed, M. S. P. Sansom, and G. A. Woolley, *Biochemistry* **35,** 6225 (1996).

[57] J. E. Hall, I. Vodyanoy, T. M. Balasubramanian, and G. R. Marshall, *Biophys. J.* **45,** 233 (1984).

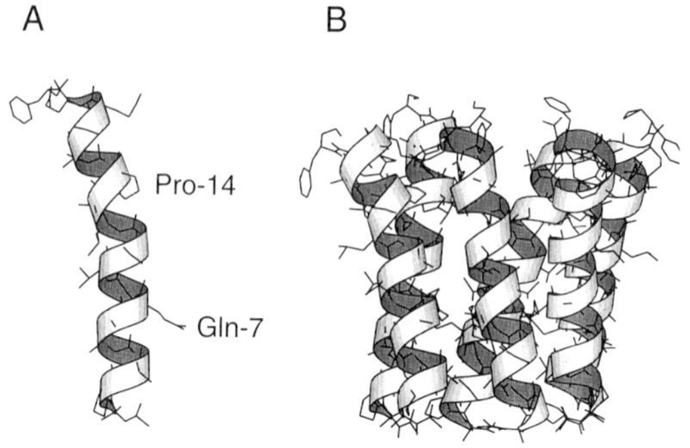

FIG. 3. A model of an Alm channel, as generated by SA/MD. (A) X-ray crystallographic structure of an isolated Alm helix. The C terminus of the helix is uppermost. (B) Hexameric bundle of Alm helices, generated by SA/MD. The view is perpendicular to the pore axis, with the C termini of the helices uppermost.

both steric and electrostatic considerations suggest that they will tend to repel one another, whereas the N-terminal helix segments do not carry a net charge and so are not expected to resist close packing.

On the basis of these assumptions, models of channels containing between $N = 4$ and 8 helices per bundle were generated by SA/MD simulations. Cα templates of $N = 4$ to $N = 8$ bundles were constructed by taking the Cα coordinates of an isolated Alm helix (monomer C) from the X-ray structure (Fig. 3A), and placing copies of this at the apices of symmetrical N-gons, with an N-terminal helix-to-helix separation of 10 Å. Helices were oriented such that Gln-7 was directed toward the center of the pore. Restraints employed during these simulations included intrahelix restraints (between backbone NH and CO groups) to maintain the Alm molecules in an α-helical conformation; and interhelix restraints (between the geometrical centers of the N-terminal segments of adjacent helices in a bundle) to hold together the helix bundle. Selected structures from the resultant ensemble were subsequently refined by 60-ps MD simulations with water molecules present within and at the mouths of the central pore (see later discussion). A model of an $N = 6$ bundle is shown in Fig. 3. The helix bundles were found to be stabilized by networks of H bonds between intrapore water molecules and Gln-7 side chains. This was confirmed by parallel mutagenesis[58] and computational[59] studies of Alm analogs in which Gln-7 was substituted by Asn or Ser, revealing a correspondence between

[58] G. Molle, J. Y. Dugast, G. Spach, and H. Duclohier, *Biophys. J.* **70,** 1669 (1996).

[59] J. Breed, I. D. Kerr, G. Molle, H. Duclohier, and M. S. P. Sansom, *Biochim. Biophys. Acta* **1330,** 103 (1997).

changes in experimental channel lifetimes and the degree of bundle stabilization by H-bond formation. Predictions of channel conductance[60] and *I–V* curves (see later discussion) based on such models were in good agreement with experimental data.

Overall, this study of using SA/MD to combine hard and soft structural data for Alm channels suggests that a pore model may be generated which predicts open channel properties with a good degree of accuracy. This supports the idea of the use of SA/MD to model channels for which the overall channel architecture is uncertain but for which the structures of the constituent TM segments are known. This applies not only to other CFPs (e.g., mellitin,[61] δ-toxin[62,63]), but also to a number of simple channel proteins, for which solid-state NMR[64] and other spectroscopic techniques may provide data on the conformation of TM segments that self-assemble to form a pore. Of course, in generating such models one must remember that the conformation of, for example, an α helix may change subtly between the isolated monomeric peptide and the pre-forming assembly within a bilayer. However, both experimental[65] and simulation[54] studies on Alm suggest that such changes in conformation are likely to be relatively small, involving mainly changes in the helix kink angle about the central proline. Indeed, it is possible that *dynamic* fluctuations in helix kink angle may occur during the lifetime of an open channel, a situation that is well suited to investigation via prolonged MD simulations of Alm channel models embedded in lipid bilayers (see later discussion).

Secondary Structure of TM Segment Known

For other CFPs and for a number of simple ion channels complete three-dimensional structures for the channel-forming TM segments are not known, but secondary structural data are available that can be used as the basis of model building. Such information may indicate that the TM segments adopt an α-helical conformation, and may also provide data on the orientation of such helices relative to the bilayer normal. Secondary structure data can be obtained from a number of spectroscopic methods, such

[60] O. S. Smart, J. Breed, G. R. Smith, and M. S. P. Sansom, *Biophys. J.* **72,** 1109 (1997).
[61] C. E. Dempsey, *Biochim. Biophys. Acta* **1031,** 143 (1990).
[62] I. D. Kerr, J. Dufourcq, J. A. Rice, D. R. Fredkin, and M. S. P. Sansom, *Biochim. Biophys. Acta* **1236,** 219 (1995).
[63] I. D. Kerr, D. G. Doak, R. Sankararamakrishnan, J. Breed, and M. S. P. Sansom, *Prot. Eng.* **9,** 161 (1996).
[64] S. O. Smith, K. Ascheim, and M. Groesbeek, *Quart. Rev. Biophys.* **4,** 395 (1996).
[65] N. Gibbs, R. B. Sessions, P. B. Williams, and C. E. Dempsey, *Biophys. J.* **72,** 2490 (1997).

as circular dichroism (CD),[66] infrared (FTIR),[67] and solid-state NMR.[64] This situation is increasingly common for a number of simple ion channel proteins. These are proteins in which the individual polypeptide chain is relatively short (ca. 100 residues) and contains a single TM segment. Application of spectroscopic techniques to either the intact protein or to synthetic peptide corresponding to the isolated TM segment may often reveal the latter to adopt a largely α-helical conformation. Channel formation is thought to occur via oligomerization of the protein to form a parallel bundle of TM α helices, in a manner analogous to that discussed earlier for Alm. Examples of such proteins include the M2 protein from influenza A,[68] the Vpu protein from human immunodeficiency virus type-1(HIV-1),[69–71] and phospholamban from mammalian sarcoplasmic reticulum.[72]

When only secondary structure data for the TM segment are available, how can SA/MD be used to generate plausible channel models which can then be tested by further experiments? The first stage is to use SA/MD to generate models of isolated TM segments. This is illustrated with two examples, one from a simple channel protein, and one from a synthetic CFP. The M2 protein from influenza A forms low pH-activated channels that are permeable to protons.[8a] The channels are blocked by the anti-influenza drug amantadine. This channel has been studied in detail, via protein chemistry,[68] characterization of the properties of amantadine-resistant mutants,[73,74] and electrophysiologic analysis of M2 channels.[75,76] The position of the TM segment within the sequence of the protein can be predicted using, for example, the program MEMSAT[9]:

Leu^{26}-*Val*-Ile-Ala-*Ala*-*Ser*-Ile-Ile-*Gly*-Ile-Leu-*His*-*Phe*-Ile-Leu-*Trp*-Ile-Leu^{43}

[66] N. A. Swords and B. A. Wallace, *Biochem. J.* **289,** 215 (1993).

[67] P. I. Haris and D. Chapman, *TIBS* **17,** 328 (1992).

[68] L. J. Holsinger, D. Nichani, L. H. Pinto, and R. A. Lamb, *J. Virol.* **68,** 1551 (1994).

[69] G. D. Ewart, T. Sutherland, P. W. Gage, and G. B. Cox, *J. Virol.* **70,** 7108 (1996).

[70] U. Schubert, A. V. Ferrer-Montiel, M. Oblatt-Montal, P. Henklein, K. Strebel, and M. Montal, *FEBS Lett.* **398,** 12 (1996).

[71] A. Grice, I. D. Kerr, and M. S. P. Sansom, *FEBS Lett.* **405,** 299 (1997).

[72] I. T. Arkin, P. D. Adams, A. T. Brünger, S. O. Smith, and D. M. Engelman, *Ann. Rev. Biophys. Biomol. Struct.* **26,** 157 (1997).

[73] A. J. Hay, A. J. Wolstenholme, J. J. Skehel, and M. H. Smith, *EMBO J.* **4,** 3021 (1985).

[74] L. H. Pinto and R. A. Lamb, *Trends Microbiol.* **3,** 271 (1995).

[75] K. Shimbo, D. L. Brassard, R. A. Lamb, and L. H. Pinto, *Biophys. J.* **70,** 1335 (1996).

[76] I. V. Chizhmakov, F. M. Geraghty, D. C. Ogden, A. Hayhurst, M. Antoniou, and A. J. Hay, *J. Physiol.* **494,** 329 (1996).

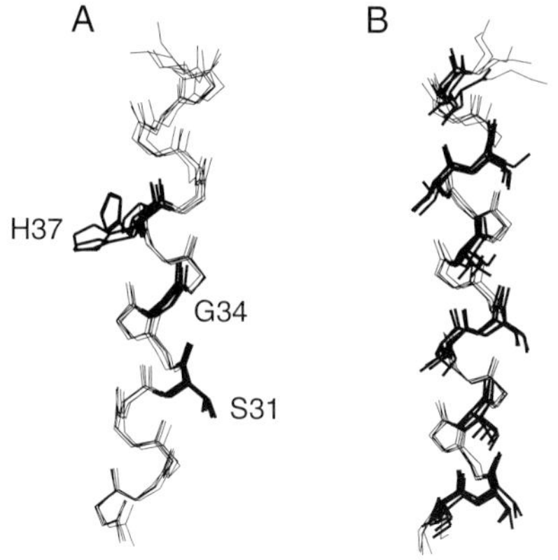

FIG. 4. Restraints determined from sequence data. (A) SA/MD-generated structures for the TM helix of the influenza M2 pore, with the pore-lining residues (S31, G34, and H37) identified by mutations shown as bold lines. (B) SA/MD-generated structure of the LS peptide with the pore-lining Ser side chains highlighted. In both cases the C terminus of the helix is uppermost, and five structures from an ensemble of 25 are shown.

Circular dichroism (CD)[77] and solid-state NMR[78] studies of synthetic peptides derived from the M2 sequence conclusively demonstrate that the TM segment adopts an α-helical conformation. Residues at positions 27, 30, 31, 34, 37, 38, and 41 (underscored) within the sequence have been identified as defining the face of the M2 helix, which is critical for channel activity and/or amantadine sensitivity.[68] To integrate this information an ensemble of 25 structures of an isolated M2 helix was generated by SA/MD. Five members of the ensemble of M2 helices are shown superimposed in Fig. 4A. Examination of these models reveals that the key residues (27, 30, 31, 34, 37, 38, and 41) lie on one face of the helix, defining a near vertical band at an angle of ca. 5° to the helix axis. This suggested that the M2 helix might adopt a tilted orientation within the helix bundle, an assumption that appears to be in agreement with solid-state NMR data.[79]

These assumptions concerning the structure of the M2 pore were embodied in a Cα template in the same manner as before. The pore is modeled as a bundle of four helices[80] because the experimental data indicate that

[77] K. C. Duff, S. M. Kelly, N. C. Price, and J. P. Bradshaw, *FEBS Lett.* **311,** 256 (1992).
[78] F. A. Kovacs, M. T. Brenneman, J. Quine, and T. A. Cross, *Biophys. J.* **72,** A399 (1997).
[79] F. A. Kovacs and T. A. Cross, *Biophys. J.* **73,** 2511 (1997).
[80] M. S. P. Sansom, I. D. Kerr, G. R. Smith, and H. S. Son, *Virology* **233,** 163 (1997).

the M2 protein is tetrameric.[81–83] The helix bundle is approximately parallel, that is, the N termini of the helices form one mouth of the pore and the C termini form the other mouth. Within the bundle, the constituent helices are rotated about their long axes such that their channel-lining faces (defined earlier) are directed toward the center of the pore. Furthermore, on the basis of the position of the surface of the M2 helix of the key side chains, the individual helices are tilted 5° relative to the pore (z) axis. Tilting of the helices ensures that all of the key residues (27, 30, 31, 34, 37, 38, and 41) are directed toward the lumen of the pore. Thus the Cα template embodies experimentally based assumptions concerning the nature of the M2 pore. SA/MD from this Cα template was used to generate helix bundle models, which consisted of a left-handed supercoil surrounding a central pore. Residue H37 has been implicated in the mechanism of low pH activation of the channel. Models generated with H37 in a fully deprotonated state exhibited a pore occluded by a ring of H37 side chains oriented toward the lumen of the pore. In models with H37 in a fully protonated state the pore was open, as the H37 side chains adopted a more interfacial location. Extended MD simulations with water molecules within and at the mouths of the pores (see later section) supported this distinction between the H37-deprotonated and H37-protonated models, and suggested a mechanism for activation of M2 by low pH in which the H37-deprotonated model corresponded to the closed form of the channel, whereas the H37-protonated model corresponded to the open form. This provides a clear example of how molecular modeling can be used not only to integrate existing experimental data but also to formulate a testable hypothesis concerning the mechanism of action of a channel.

A second case study where only secondary structural data were available concerns the synthetic LS peptide of DeGrado and colleagues, which has been shown to form cation-selective ion channels in lipid bilayers.[84–86] This sequence of this peptide

Leu-*Ser*-*Ser*-Leu-Leu-*Ser*-Leu-Leu-*Ser*-*Ser*-Leu-Leu-*Ser*-Leu-Leu-*Ser*-*Ser*-Leu-Leu-*Ser*-Leu

[81] R. J. Sugrue and A. J. Hay, *Virology* **180,** 617 (1991).

[82] L. J. Holsinger and R. A. Lamb, *Virology* **183,** 32 (1991).

[83] T. Sakaguchi, Q. A. Tu, L. A. Pinto, and R. A. Lamb, *Proc. Natl. Acad. Sci. USA* **94,** 5000 (1997).

[84] J. D. Lear, Z. R. Wasserman, and W. F. DeGrado, *Science* **240,** 1177 (1988).

[85] K. S. Åkerfeldt, J. D. Lear, Z. R. Wasserman, L. A. Chung, and W. F. DeGrado, *Acc. Chem. Res.* **26,** 191 (1993).

[86] J. D. Lear, Z. R. Wasserman, and W. F. DeGrado, *in* "Membrane Protein Structure: Experimental Approaches" (S. H. White, ed.). Oxford University Press, Oxford, 1994.

was designed to form an amphipathetic α helix. The secondary structure was confirmed by CD spectroscopy. Models of the LS peptide helix generated by SA/MD (Fig. 4B) revealed that the Ser side chains form a predominantly polar face to the helix. Thus, Cα templates of $N = 4, 5$ and $N = 6$ helices/bundle were constructed, with the Ser side chains directed toward the center of the pore. The helices in the resultant SA/MD-generated models formed left-handed supercoils, with their pores lined by the Oγ groups of the serine side chains.[87] Analysis of the H-bonding patterns of the three ensembles of LS helix bundles ($N = 4$, 5, and 6) revealed 1 or 2 interhelix H bonds for each helix of a bundle. This echoes the situation for Alm (discussed previously) where the bundle was also stabilized by interhelix H bonds. The mean radius of the $N = 5$ pore (2.7 Å) was such that passage of monovalent cations in a partially hydrated state could occur. The minima in the pore radius profile (see later discussion) corresponded to Oγ atoms of residues S2, S6, S9, S13, S16, and S20. Thus, the close interactions of serine Oγ atoms and permeant ion would be expected to confer cation selectivity on the resultant channel, as is indeed observed experimentally.[84] The SA/MD-generated models were subsequently refined by MD simulations in the presence of water molecules[87] (see later section). Comparison of predictions of ionic conductances (see later section) with experimental values provided support for the validity of these models.

Overall these studies suggest that even in the absence of definitive structural evidence for the constituent monomers of a pore assembly, by combining secondary structural data with mutagenesis and/or helix amphipathicity data, SA/MD may be used to generate plausible models of transbilayer pores. Such models can be used to design further experiments. Also, they may be used to predict open channel properties, which may then be compared with those observed experimentally.

This is an appropriate stage at which to note that an approach related to SA/MD has been used to generate a model of the transbilayer pore formed by phospholamban. This model was based on exhaustive searching, using a method similar to SA/MD, of possible modes of helix packing, in order to identify the mode that yielded the most stable helix bundle.[27] The subsequent model has been tested by a number of spectroscopic techniques[88] and is in good agreement with the experimental data. This model has been refined by prolonged MD simulations with water molecules within and at either end of the pore (see later discussion), and the resultant

[87] P. Mitton and M. S. P. Sansom, *Eur. Biophys. J.* **25,** 139 (1996).

[88] I. T. Arkin, M. Rothman, C. F. C. Ludlam, S. Aimoto, D. M. Engelman, K. J. Rothschild, and S. O. Smith, *J. Mol. Biol.* **248,** 824 (1995).

models used to compare predicted channel properties with those observed experimentally.[89]

The main problems with the approach discussed in this section are similar to those for the previous examples, namely, the quality of the restraints. In the case of influenza M2 these depend on whether mutagenesis studies correctly identify pore-lining side chains. Although objections may be raised to overinterpretation of such data, the model does seem to be consistent with experiment. In the case of the LS peptide, the restraints on helix orientation are difficult to contest. If a channel is formed in a lipid bilayer by a highly amphipathic α-helical peptide, then on energetic grounds alone it is very difficult to argue for a helix orientation other than one in which the polar side chains are directed inward toward a central water-filled pore and the apolar side chains are directed outward to interact with the surrounding lipid molecules.

No Structural Data

Modeling of channels becomes more difficult when no structural data are available, in which case any model has to be based entirely on indirect (e.g., mutagenesis) data. In particular, if even secondary structural data are unavailable then the problem is severely under-determined and it is evident that many models may exist that are compatible with the experimental evidence. In this situation SA/MD can be used to generate a number of alternative models and to attempt to discriminate between such alternatives in terms of their goodness-of-fit to the data. At the time of writing, this situation applies to the pore domain of voltage-gated potassium (Kv) channels. Because Kv channels have been the subject of several modeling studies,[7,28,34,90–92] they are an appropriate example with which to illustrate this use of modeling.

Kv channels are integral membrane proteins present in the membranes of many cells, controlling their membrane potential. They are homotetrameric proteins with four subunits surrounding a central pore. At present there is no direct structural information regarding the pore, but a large body of mutagenesis and physiologic data has allowed identification of those regions of the sequence which contribute to the pore-forming domain.[93–96]

[89] M. S. P. Sansom, G. R. Smith, O. S. Smart, and S. O. Smith, *Biophys. Chem.* **69,** 269 (1997).
[90] S. Bogusz, A. Boxer, and D. D. Busath, *Prot. Eng.* **5,** 285 (1992).
[91] J. C. Bradley and W. G. Richards, *Prot. Eng.* **7,** 859 (1994).
[92] I. D. Kerr and M. S. P. Sansom, *Biophys. J.* **73,** 581 (1997).
[93] C. Miller, *Science* 252 (1991).
[94] O. Pongs, *J. Membr. Biol.* **136,** 1 (1993).
[95] A. M. Brown, *Annu. Rev. Biophys. Biomol. Struct.* **22,** 173 (1993).
[96] Q. Lü and C. Miller, *Science* **268,** 304 (1995).

Modeling techniques have been used to explore the extent to which such data may be interpreted in terms of possible channel structures.

Each subunit of Kv consists of six putative membrane-spanning helices (S1 to S6) with an additional 21-residue region between S5 and S6. This region, termed H5, has been demonstrated to be the primary determinant of ion selectivity and block by agents such as tetraethylammonium and charybdotoxin. Mutagenesis evidence indicates that the H5 region crosses the membrane twice, adopting a hairpin-like topology. Thus the inner lining of the pore of Kv channels can be modeled as a homotetrameric assembly of H5 segments. The H5 region has been subjected to intensive site-directed mutagenesis, with every residue the subject of at least one mutagenesis study. These data have been collated data and interpreted as a set of topological restraints that any valid model should satisfy. Thus, using the sequence of the *Shaker* A[97] Kv channel the restraints are

Pro(1′)-*Asp*-Ala-Phe-*Trp*-*Trp*-Ala-Val-*Val*-*Thr*-*Met*-*Thr*-Thr-*Val*-Gly-
Tyr-Gly-*Asp*-*Met*-*Thr*-Pro(21′)

where residues Pro-1′ and Pro-21′ correspond to Pro-430 and Pro-450, respectively, in the full-length protein, and where the topological restraints are indicated as single underline, extracellular mouth of the pore; double underline, intracellular mouth of the pore; *italics*, pore-lining side chain. These restraints can be employed in modeling the pore, using SA/MD and related techniques.[92] Two classes of models have been generated, namely, eight-stranded β barrels and eight-staved α-helical bundles.

β-Barrel Models of Pore. The sequence of the H5 region has been threaded onto idealized β hairpins (see Fig. 5A) in order to score their agreement with the topological restraints. β Hairpins differ in the number of residues in the β turn and the position of the turn within the H5 sequence. Candidate threads were identified with either two, three, or four residues in the β turn. For each of the threads eight-stranded β barrels were generated differing in the orientation of the segments around the pore (either clockwise or anticlockwise) and in the shear number (S) of the barrel.[22] Three values of S are likely for an eight-stranded β barrel, namely, $S = 8$, $S = 10$, and $S = 12$. Together, this gave 42 possible configurations of β barrels to be explored as models of the Kv-H5 pore, each configuration (which consists of a specific thread, orientation, and shear number) being defined by a unique combination of distance restraints and dihedral angle restraints. Thus, the thread of the H5 sequence determines the intrahairpin hydrogen-

[97] B. L. Tempel, D. M. Papazian, T. L. Schwarz, Y. N. Jan, and L. Y. Jan, *Science* **237,** 770 (1987).

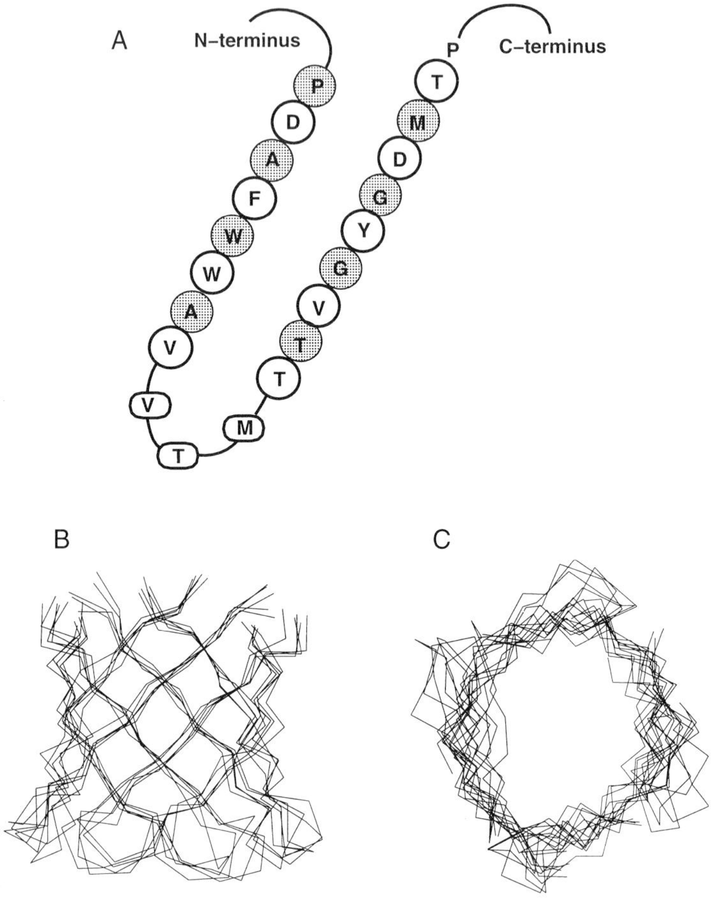

FIG. 5. Modeling the H5 pore domain of the *Shaker* Kv channel. (A) H5 sequence threaded onto a β-hairpin template (S10_839_LC) with the pore-lining residues represented as open circles, while shaded circles represent non-pore-lining residues. (B) and (C) A β-barrel model for the H5 pore, shown as superimposed Cα traces for five structures from each ensemble. In (B) the view is perpendicular to the pore axis; in (C) it is down the pore axis.

bond pattern and the pattern of the peptide backbone dihedral angle restraints, while S and the orientation determine the interhairpin hydrogen bond pattern. Ensembles of structures for each configuration were generated using distance geometry[23,98] to generate Cα templates followed by a modified SA/MD procedure to generate full structural models.

[98] I. D. Kuntz, J. F. Thomason, and C. M. Oshiro, *Methods Enzymol.* **177,** 159 (1989).

α-Helix Bundle Models of Pore. As for the β barrels a number of parameters needed to be explored in α-helix bundle models. These included (1) the position of the turn within H5, (2) the number of residues within the turn, and (3) the overall orientation of the eight-helix bundle (clockwise or anticlockwise). Together this yielded 24 alternative configurations. For each configuration there were two further geometrical parameters: the orientation of the helices by rotation about their long axis (ζ), and the crossing angle of each helix relative to its nearest neighbors in the bundle (ω). For each configuration optimal values of these angles were determined by a grid-search method using the mutagenesis data to score combinations of (ζ, ω). This defined 24 possible Cα templates, from each of which an ensemble of structures was generated by SA/MD.

Ranking Ensembles of H5 Models. Overall, in excess of 1300 models for the tetrameric Kv-H5 pore were generated. It was therefore essential to identify which ensembles were worth exploring further as models of the Kv-H5 pore. The different ensembles were ranked by a combination of two properties of the models: agreement with mutagenesis data and predicted pore conductance (see later discussion). α-Helix bundle models had predicted conductance values of ca. 20–55 pS, in general somewhat higher than the experimentally determined value of 19 pS for *Shaker.*[99,100] β-Barrel modes showed generally better agreement with the mutagenesis data, but their predicted conductances (ca. 2–8 pS) were lower. However, none of the models showed perfect agreement with the mutagenesis data. A combination of the two ranking properties enabled identification of a number of models that showed reasonable agreement with mutagenesis data, and had predicted conductance values close to that observed experimentally. Ten such models were refined further by MD simulations including water molecules in the lumen and at the mouths of the pore (see below). An example of a β-barrel model of the Kv H5 pore is shown in Fig. 5BC as superimposed Cα traces for five members of the SA/MD generated ensemble. It can be seen that there is substantial variation in the conformation of the loop regions between different members of the same ensemble. This is to be expected given the relatively soft nature of the restraints used in the generation of these models.

It is important to consider the role of modeling in this example. In the cases discussed earlier, the Cα template has been derived from experimental structural data, for example, the secondary structure of the pore lining elements. In contrast, for Kv channels, SA/MD was initiated from a number

[99] L. Heginbotham and R. MacKinnon, *Biophys. J.* **65,** 2089 (1993).

[100] M. Tagliatela, M. S. Champagne, J. A. Drewe, and A. M. Brown, *J. Biol. Chem.* **269,** 13867 (1994).

of different Cα templates, because no structural data were available to favor one or another of them. Thus, it is unlikely that in this example SA/MD will yield a definitive model for the Kv-H5 pore domain. However, in cases such as this modeling may be used as a tool to integrate all of the available experimental data and to use such data in an objective fashion to generate alternative possible models of a pore domain to, be used, in turn, as the basis of further experiments.

A number of other techniques have been used to model the Kv-H5 pore domain. These include homology modeling approaches based on existing structures for eight-stranded β barrels[90,91] and interactive molecular graphics approaches.[7,34,101] The latter approach has been used by Guy and colleagues to suggest that the H5 segments of Kv adopted a mixed conformation in which the N-terminal half of H5 is α helical, whereas the C-terminal half has a rather irregular, extended conformation. A more recent study[28] used SA/MD to generate an ensemble of models based on the secondary structure of the Guy model. Overall, it seems that modeling alone is unlikely to yield a definitive model of the Kv pore until experimental structural data[101a] become available.

Uses of Models

Generating a model of an ion channel is evidently not an end in itself. Models generated by, for example, SA/MD can be used as three-dimensional diagrams to facilitate display of key residues and to aid design of further mutagenesis experiments. However, these models are preliminary and require further refinement followed by rigorous testing against experimental data if they are to inform our understanding of the molecular basis of channel function. In the following section ways are discussed in which extended MD simulations can be used to refine such models, and in which they may be employed to predict observable physiologic properties of ion channels.

Molecular Dynamics Simulations

Levels and Approaches

A direct computational uses of channel models is to employ them as the starting point for more extended MD simulations. Such simulations provide information on the dynamic behavior of a channel over system on a short time scale (ca. 1 ns), which can be used subsequently as input

[101] G. M. Lipkind, D. A. Hanck, and H. A. Fozzard, *Proc. Natl. Acad. Sci. U.S.A.* **92,** 9215 (1995).
[101a] D. A. Doyle, R. A. Pfuetzner, A. Kuo, and R. MacKinnon, *Science,* **280,** 69 (1998).

into methods designed to simulate longer time scale events, such as ion permeation. Note that in this context a single-channel current of 1.5 pA is equivalent to a net movement of ca. 10^7 ions s^{-1} through a channel, corresponding to a mean dwell time of an ion in a channel of ca. 0.1 $\mu s \equiv$ 100 ns. Thus MD simulations based on SA/MD-generated channel models may provide a first step in the process of prediction of the physiologic properties of a channel.

It is important to remember the quality of models on which such MD simulations are based. They may vary from high-resolution X-ray structures through "hard" models based on incomplete structural data to "soft" models for which no direct structural data are available. If working with a soft model, then the results of the MD simulations are of necessity somewhat speculative. However, for all classes of model, MD simulations enable refinement of a model and may reveal interesting properties that are not explicit in the model per se.

In the following sections MD simulations at increasing levels of complexity are discussed, starting with *in vacuo* simulations and then progressively adding to the simulation system the other components (water, ions, lipid bilayer) that interact to constitute a channel and its environment *in vivo.*

In Vacuo Molecular Dynamics

This requires only brief discussion. The MD simulations at the heart of the SA/MD procedure are carried out *in vacuo.* As such they are important in enabling exploration of stereochemically feasible conformations of channel-lining TM segments. However, further, more extended *in vacuo* MD simulations based on the resultant models are of limited value, because they do not include water molecules or ions. Furthermore, the information they do provide on, for example, flexibility of channel-lining side chains is already represented by the variability within the SA/MD-generated ensemble. Although *in vacuo* MD simulations can be used to relax a model generated by homology or interactive modeling, given current computational power, it is preferable that refinement of initial models be carried out via MD simulations in the presence of explicit water molecules within and at either mouth of the pore.

Molecular Dynamics with Water Molecules

The aims of MD simulations of channel models with water molecules present within and either mouth of the pore are twofold: to refine a model by allowing, for example, polar side chains to relax dynamically in the presence of explicit solvation; and to enable analysis of the structural and dynamic properties of water within a pore. The latter is important because a number of studies (discussed later) have shown that water within channels may have distinct properties from water in its bulk state, and the properties of intrapore water are likely to be crucial to the permeation properties of a channel.

An SA/MD-generated model of a pore may be solvated by immersing it is a preequilibrated box of water molecules. For example, in a study of water within simplified pore models[102] water molecules were added to the SA/MD-generated model by superimposing a preequilibrated cylinder of water molecules and removing all water molecules closer than 1.6 Å to protein atoms plus those in the TM region that were outside the helix bundle. Solvated model pores were energy minimized prior to MD simulations, using Charmm.[14] A four-stage energy minimization was performed: (1) 1000 cycles of adopted basis Newton–Raphson (ABNR) minimization with the protein atoms fixed, (2) 1000 cycles of ABNR with the protein backbone atoms restrained, (3) 1000 cycles of ABNR with weak restraints on the protein Cα atoms only, and (4) 1000 cycles of ABNR with no positional restraints.

An important consideration in such MD simulations is that of which water model to use. There are a number of different water models that vary in their complexity and the accuracy with which they reproduce experimental data on water structure and dynamics.[103] In the author's laboratory, the widely used TIP3P model,[104] which is included as part of the Charmm parameter set, has been employed. Other water models that have been used in simulations of channels, such as SPC/E,[26] were used in simulation studies of a model of the pore domain of voltage-gated Na^+ channels.[105] All of these models have limitations in their ability to predict the properties of bulk water. However, new water models for use in biomolecular MD simulations are being developed,[106] and their use in channel simulations needs to be explored.

Once a model has been solvated, and the positions of the waters adjusted by energy minimization, a more prolonged MD simulation can be undertaken. An important consideration in such a simulation is to what extent restraints should be included to mimic the effects of the missing components

[102] J. Breed, R. Sankararamakrishnan, I. D. Kerr, and M. S. P. Sansom, *Biophys. J.* **70,** 1643 (1996).

[103] V. Daggett and M. Levitt, *Annu. Rev. Biophys. Biomol. Struct.* **22,** 353 (1993).

[104] W. L. Jorgensen, J. Chandresekhar, J. D. Madura, R. W. Impey, and M. L. Klein, *J. Chem. Phys.* **79,** 926 (1983).

[105] C. Singh, R. Sankararamakrishnan, S. Subramanian, and E. Jakobsson, *Biophys. J.* **71,** 2276 (1996).

[106] M. Levitt, M. Hirshberg, R. Sharon, K. E. Laidig, and V. Daggett, *J. Phys. Chem. B* **101,** 5051 (1997).

of the simulation system (lipid bilayer, remainder of protein, bulk solvent). The balance to be achieved is to restrain the model sufficiently that it does not fall apart during the simulations, but not to restrain it so tightly such that motions of pore-lining residue (and hence intrapore water motions) are unjustifiably restricted. Several approaches are available. The least intrusive is to include interhelix distance restraints to maintain the integrity of a helix bundle, while otherwise allowing pore atoms to move freely. Rather more restrictive is to apply harmonic positional restraints to Cα atoms, thus maintaining the overall backbone geometry of the SA/MD-generated model, while allowing movement of pore-linking side chains. In addition to such restraints on the pore, it may be advisable to apply weak restraints to prevent evaporation of water molecules from the mouth of the pore model. During energy minimization and MD simulations, restraints may be applied to water molecules to prevent their evaporation from either mouth of the pore. For example, this may be achieved using a restraining potential such as:

$$E = -\frac{F}{2}\exp\left(-\frac{\Delta}{\lambda}\right) \qquad \text{if } \Delta > 0 \tag{11}$$

or

$$E = -F\left[1 - \frac{1}{2}\exp\left(+\frac{\Delta}{\lambda}\right)\right] \qquad \text{if } \Delta \leq 0 \tag{12}$$

where F is the force constant of the restraint, λ is a smoothing constant, and Δ is $r - r_{\mathrm{WALL}}$ (i.e., the distance of a water atom from the restraining wall). Note that Eq. (11) applies to water atoms *outside* the wall and Eq. (12) applies to water atoms *inside* the wall. It is usually appropriate to choose a cylindrical restraining wall. Note that this restraining potential only comes into play for waters at the mouths of the pore as the radius of the cylinder (r_{WALL}) is set to be greater than the radius of the van der Waals surface of the pore lining. Furthermore, water molecules more than a distance λ away from the wall experience only a small restraining potential.

Although it is difficult to generalize concerning the results of model refinement by such MD simulations in the presence of water, experience with a number of channels[63,87,89,107,108] suggests that although major changes in the structure of the models do not occur, significant changes in pore

[107] J. Breed, P. C. Biggin, I. D. Kerr, O. S. Smart, and M. S. P. Sansom, *Biochim. Biophys. Acta* **1325,** 235 (1997).

[108] K. R. Ranatunga, I. D. Kerr, C. Adcock, G. R. Smith, and M. S. P. Sansom, *Biochim. Biophys. Acta* **1370,** 1 (1998).

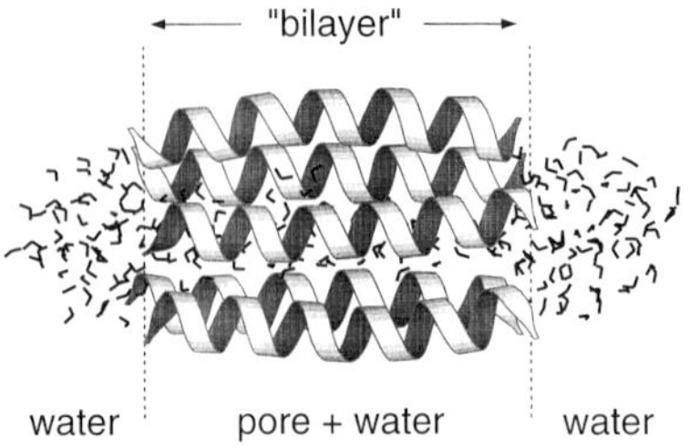

FIG. 6. Channel model plus water, illustrated using a pentameric bundle of TM helices from the NB protein of influenza B. The TM helices are shown as ribbons and the water molecules as black ⟨ shapes. The two dotted lines indicate the extent of the empirical bilayer potential employed during MD simulations using this model.

geometry may take place as a result of relaxation of pore-lining side chains and of penetration of water molecules into narrow, previously occluded regions of a pore. In particular, the latter effect may lead to an increase in the predicted conductance (see later discussion) for a given pore model following refinement and to the opening of apparently occluded pores.

As a case study of the analysis of water structure and dynamics within a model of a transbilayer pore, let us consider simulations on parallel bundle of TM helices from the NB protein of influenza B,[109] a putative channel protein[110–112]

Gly[19]-*Ser*-Ile-Ile-Ile-*Thr*-Ile-Cys-Val-*Ser*-Leu-Ile-Val-Ile-Leu-Ile-Val-Phe-Gly-Cys-Ile-Ala[40]

Helix bundles with $N = 5$ helices/bundle were generated by SA/MD. Predicted conductances (see later section) based on these preliminary models closely matched those measured experimentally for recombinant NB protein incorporated into lipid bilayers.[110] The helices within these bundles were oriented such that the polar Ser and Thr residues (*italic* above) were oriented toward the lumen of the channel. The SA/MD-generated model was solvated with 203 water molecules, and subjected to 100 ps of MD at 300 K. Although longer simulations (ca. 500 ps) have been performed, it seems that the dynamic and structural properties of water within such pores may be captured within relatively short (100-ps) simulations. During these simulations, weak interhelix distance restraints were employed to ensure the integrity of the helix bundle, and an empirical bilayer potential (see later section) was applied to the side chains of the TM helices (see Fig. 6).

[109] R. Bull, G. R. Smith, and M. S. P. Sansom, in preparation (1997).
[110] N. A. Sunstrom, L. S. Prekumar, A. Prekumar, G. Ewart, G. B. Cox, and P. W. Gage, *J. Membr. Biol.* **150,** 127 (1996).
[111] R. A. Lamb and L. H. Pinto, *Virology* **229,** 1 (1997).
[112] T. Betakova, M. V. Nermut, and A. J. Hay, *J. Gen. Virol.* **77,** 2689 (1996).

The structure and dynamics of water within the model NB pore resemble those observed in comparable simulations of a number of other models of channels formed by bundles of approximately parallel α helices. Translational motion of water molecules within and at either mouth of a pore may be characterized in terms of the self-diffusion coefficient of each water molecule, D, which is obtained by evaluation of a molecule's mean square displacement as a function of time:

$$\langle r(t)^2 \rangle = \langle (\mathbf{r}(t) - \mathbf{r}(0))^2 \rangle \tag{13}$$

using the relationship:

$$\lim_{t \to \infty} \langle r(t)^2 \rangle = 6Dt + \text{constant} \tag{14}$$

to fit the $\langle r(t)^2 \rangle$ data for t = 3–6 ps. If one then plots the value of D for each water molecule as a function of the position of the water molecules along the pore (z) axis (Fig. 7A) it is evident that those water molecules within the pore have a mean self-diffusion coefficient (D = 0.06 Å^{2}ps^{-1}) that is significantly lower than either that for water molecules in the two caps at either mouth of the pore (D = 0.25 Å^{2}ps^{-1}) or for TIP3P water in bulk simulations (D = 0.32 Å^{2}ps^{-1}).

A similar analysis of rotational motion of waters within and outside of the pore can be made, by calculation of rotational reorientation rates. A rotational correlation function can be defined in terms of the angle, $\theta(t)$, made by the dipole of a water molecule at time 0 and the dipole of the same water molecule at time t:

$$C_1(t) = \langle \cos(\theta(t)) \rangle \tag{15}$$

This was fitted (for times from 1 to 20 ps) as a monoexponential decay with time constant τ. Thus, the reciprocal of the time constant (i.e., $1/\tau$) provides a rotational reorientation rate. The reorientation rate for waters within the pore ($1/\tau$ = 0.02 ps^{-1}) is substantially less than the rates for water at either mouth of the pore ($1/\tau$ = 0.12 ps^{-1}) and for water in bulk simulations ($1/\tau$ = 0.30 ps^{-1}). Overall, such analyses demonstrate that within the relatively narrow (minimum radius ca. 2.2 Å) pore of the N = 5 NB helix bundle, the mobility of water molecules is reduced by almost an order of magnitude relative to in bulk water. Similar conclusions have been reached for simulations on intrapore water for a number of different ion channels.[3,87,102,105,113–116]

[113] S. W. Chiu, E. Jakobsson, S. Subramanian, and J. A. McCammon, *Biophys. J.* **60,** 273 (1991).
[114] M. E. Green and J. Lewis, *Biophys. J.* **59,** 419 (1991).
[115] M. Sancho, M. B. Partenskii, V. Dorman, and P. C. Jordan, *Biophys. J.* **68,** 427 (1995).
[116] M. Engels, D. Bashford, and M. R. Ghadiri, *J. Am. Chem. Soc.* **117,** 9151 (1995).

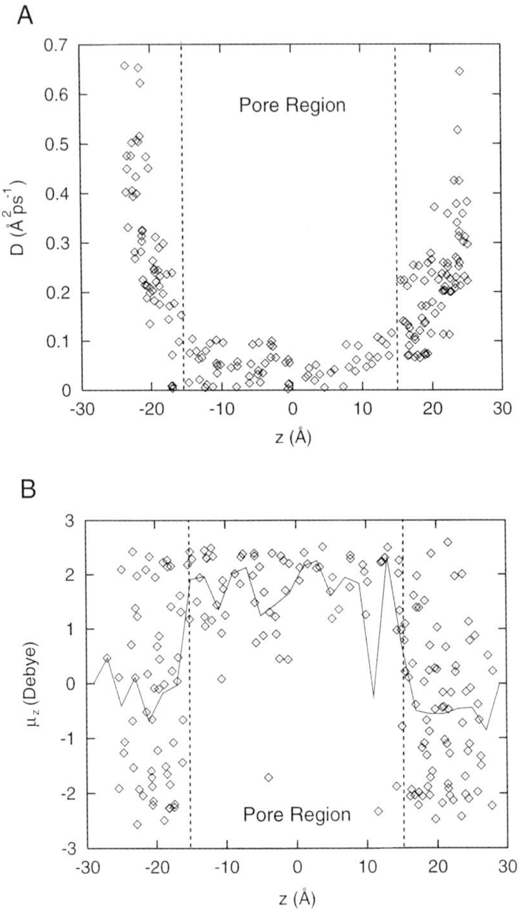

FIG. 7. Water dynamics and structure within the influenza B NB channel model (see Fig. 6). (A) Water self-diffusion coefficient (D) as a function of distance along the pore (z) axis. (B) Water dipole moments projected onto the pore axis as a function of position of the water molecule along that axis. In both diagrams the N-terminal mouth of the pore is at z ca. −15 Å and the C-terminal mouth of the pore is at z ca. +15 Å, as indicated by the vertical broken lines.

In addition to analyzing the dynamics of water molecules within the NB pore, one can examine the orientation of pore waters relative to the pore (z) axis in terms of the projection onto the z axis of the dipole moment of each water, μ_z. For an ideal TIP3P water molecule with its dipole exactly parallel to the z axis, μ_z = 2.35 Debye. Form a plot of μ_z versus z for waters in and around the model NB pore (Fig. 7B) it can be seen that while the waters are randomly oriented (as expected) at either mouth of the pore, within the pore per se the waters are on average oriented such that their O atoms point toward the N-terminal mouth of the pore. This orientation is brought about by alignment of the water dipoles antiparallel to the α-helix dipoles. Again, this is similar to the situation observed for a number of channels formed by parallel helix bundles, and indicates a further way in which pore water differs significantly from bulk water.

A consequence of alignment of water molecules by the helix dipole electrostatic field within model pores is that it perturbs the dielectric behavior (see below) of intrapore water. This also can be analyzed by MD simulations. For example, MD simulations on a simplified model pore (a parallel bundle of five Ala-20 α helices) in the presence of an external transpore electrostatic field (i.e., a voltage drop along the pore axis) have been used to estimate the resultant polarization (due to reorientation) of the intrapore water, and hence to determine local dielectric behavior within the pore.[117] In this way it was shown that the local dielectric constant of water within a pore formed by a parallel bundle of helices is reduced from $\varepsilon = 78$ to ε ca. 30. This is relevant in the context of Poisson–Boltzmann calculations of electrostatic potential energy profiles along the axis of a channel (see below). In principle, such simulations in the presence of an external transpore electrostatic field might be applied to any model of a pore.

Molecular Dynamics with Water Plus Ions

Although simulations of water within ion channels provide valuable information for use in, for example, continuum calculations on channel conductance and selectivity properties, MD simulations can also provide direct information on the energetics of an ion as it passes through a channel. In particular, such simulations can reveal the extent to which a given ion is dehydrated when it passes through a channel, and which residues of the channel lining substitute for the water molecules have been displaced.

Extensive literature exists on MD and related simulations of ions passing through the gramicidin pore, which has been well reviewed elsewhere[3] and is not discussed here. For smaller systems such as gramicidin extensive simulations have been performed that enable estimation of a free-energy profile of an ion as it moves along a pore. For more complex channels, such as those discussed in this article, simulations of channel/ion/water interactions have been generally restricted to establishing potential energy profiles for ions as moved along a pore. To date, such simulations have

[117] M. S. P. Sansom, G. R. Smith, C. Adcock, and P. C. Biggin, *Biophys. J.* **73,** 2404 (1997).

been limited to a single ion at a time in a channel and so have not addressed the complexities of ion–ion interactions within a channel environment. Estimations of ion potential energy profiles have been carried out for a number of channel models, including a model of the pore domain of the voltage-gated Na^+ channel.[105] In the following discussion, the example of a β-barrel model for the Kv-H5 pore domain of a potassium channel is discussed.[108]

The system under investigation is shown in Fig. 8A. The pore model is

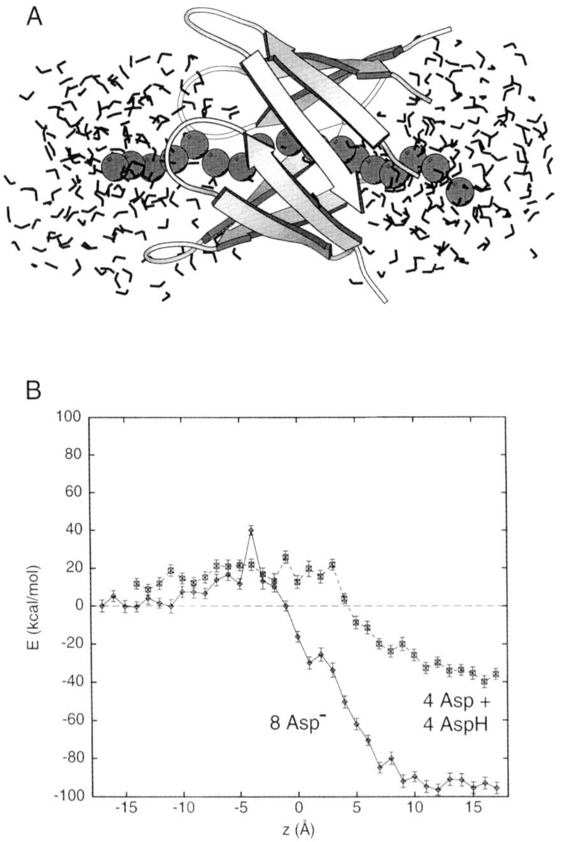

FIG. 8. Channel model plus water plus ion, illustrated using a β-barrel model of the *Shaker* Kv-H5 pore (see Fig. 5). (A) Cartoon of the channel + water molecules (black ⟨ shapes) + K^+ ions (gray spheres). The ion is shown at successive positions along the channel, although in any one simulation only a single ion is present. (B) Interaction energy profiles for translation of a K^+ ion along the solvated model, showing the potential energy of interaction of the ion with (pore + water) as a function of distance along the pore (z) axis for the $8Asp^-$ (solid line) and $4Asp^-$ + 4AspH (broken line) ionization states.

one of several models for the Kv-H5 domain discussed earlier, which yields reasonable agreement with published mutagenesis data.[92] In this particular SA/MD-generated model (S10_839_LC in the nomenclature of Ref. 92; see Fig. 5A) the Kv-H5 domain is modeled as an antiparallel eight-stranded β barrel with shear number $S = 10$. As described earlier, each H5 loop forms a β hairpin, folded such that the two ends of H5 (i.e., residues 1′ and 20′) form the extracellular mouth of the pore whereas residues 9′ to 11′ form the intracellular-facing mouth. An *in vacuo* SA/MD-generated initial model (Fig. 5BC) was refined by 250 ps of MD in the presence of 350 water molecules within and at either mouth of the pore. K^+ ion potential energy profiles were evaluated by a series of short MD simulations in which a single K^+ ion was translated along the length of the model Kv-H5 pore. The starting point for these simulations was the refined structure from the previous 250-ps MD simulation. The K^+ ion was placed at successive positions along the z axis, separated by 1-Å steps. For each position of the ion the nearest water molecule was removed, the system was energy minimized (3000 steps of ABNR), and then a short (15-ps) burst of MD was performed. The energetics of ion/protein/water interactions were averaged over the final 9-ps period for each position of the ion. These were evaluated as:

$$\Delta E_{\mathrm{ION/(PORE+WATER)}} = E_{\mathrm{PORE+WATER+ION}} - E_{\mathrm{PORE+WATER}} - E_{\mathrm{ION}} \quad (16)$$

The ion/(pore+water) interaction energies were normalized by subtraction of the mean potential energy of the corresponding ion simulated in a box of 3375 water molecules. The resultant potential energy profiles are shown in Fig. 8B. Each set of simulations was performed twice, once with all eight Asp side chains (i.e., four Asp-2′ and four Asp-18′) ionized and once with only the four Asp-2′ side chains ionized. This reflects a complication of such simulations, namely, the assumed protonation state of ionizable side chains lining such pores. Poisson-Boltzmann-based calculations of pK as (see later section and Ref. 118) for the two rings of Asp side chains guarding the extracellular mouth of the pore suggested that, in the absence of an ion not all of the Asp side chains would be fully ionized. The assumed ionization state of these side chains clearly has a major effect on the potential energy profile, with the well depth varying between ca. −100 and −40 kcal/mol depending on whether or not all of the Asp-3 residues are ionized.

Such potential energy data may be analyzed in more detail in order to yield information on the roles of water and of individual pore-lining side chains in stabilizing an ion as it passes through a channel. In the case of

[118] A. Karshikoff, V. Spassov, S. W. Cowan, R. Ladenstein, and T. Schirmer, *J. Mol. Biol.* **240,** 372 (1994).

the Kv-H5 channel model when the overall interaction potential profile of the ion is broken down into its constituent terms it appears that a deep well in the ion/pore interaction energy profile is mirrored by a barrier in the ion/water interaction energy profile. This suggests that as the ion enters the narrowest region of the pore, it is (partially) dehydrated but that favorable interactions with the pore-lining side chains outweigh the energetic costs of such desolvation. Investigation of the contribution of the individual side chains to the well in the ion/pore interaction profile reveals that the major contributions are from the ring of ionized Asp side chains. The Tyr-16′ side-chain ring also interacts favorably with the K^+ ion as it passes through the pore in its desolvated state. This correlates with experimental evidence in favor of this Tyr residue playing a role in governing the ion selectivity of Kv channels.[119]

A further complication, which has not yet been fully addressed in simulation studies of channel models, is that the protonation state of rings of ionizable side chains may be altered by the presence of a permeant ion in proximity to the side chain. Furthermore, the protonation state may depend on the nature of the ion. Clearly, more sophisticated simulation studies will be required to fully address such important questions concerning pore/water/ion interactions in order for MD simulations to yield a complete picture of the physicochemical basis of ion permeation and selectivity.

In addition to potential energy profiles, MD simulation studies of pore/water/ion systems may provide parameters for use in continuum calculations on ion permeation and selectivity. For example, Poisson–Nernst–Planck (PNP) calculations of single-channel I–V relationships (discussed later) require, as one of their input parameters, an estimate of the diffusion coefficient of an ion *within the pore.* MD simulations of pore/water/ion behavior in a simplified model[22,102] of a pore formed by an eight-stranded β barrel of sequence Ala_{10} followed by analysis of ion self-diffusion coefficients in a manner similar to that described earlier for water yielded an estimate of D_{Na^+} of 0.06 $Å^2ps^{-1}$, that is, about one-third of the corresponding value in bulk solution (Smith and Sansom, unpublished result). This illustrates how MD simulations of pore/water/ion may be used to characterize the microscopic, short time scale behavior of a system, yielding estimates of parameters that may subsequently be employed to increase the accuracy of continuum calculations of longer time scale events.

Including the Bilayer

So far an important component has been omitted from the simulations of ion channels, namely, the surrounding lipid bilayer. For simulations that

[119] L. Heginbotham, Z. Lu, T. Abramson, and R. MacKinnon, *Biophys. J.* **66,** 1061 (1994).

address only the central pore-lining domain of a more complex channel protein, it would be inappropriate to include a bilayer, because the outer layers of protein are absent from the model. However, for models of complete channels, for example, CFPs or simple channel proteins, inclusion of the bilayer is desirable. A decision must be made concerning the level of complexity one wishes to use to represent a lipid bilayer, which in turn dictates the computational power required for the simulations. It is relatively inexpensive to add a term to a potential energy function that provides an empirical bilayer potential. Such a term may be based on residue-by-residue hydrophobicities[120] in order to mimic the embedding of the helix or helix bundle within a membrane. This approach has been used in MD simulations of single transmembrane helices[54,121] and of bacteriorhodopsin.[122] The bilayer core is modeled as a hydrophobic continuum spanning from $z = -d$ to $z = +d$ (where the z axis is the bilayer normal and $2d$ is the bilayer thickness). A hydrophobicity index, H_i, is assigned to the side chain of each residue i. The potential energy of a residue is a function of the z coordinate of the geometric center of its side chain $[f(z_i)]$, such that within the bilayer $E_{\mathrm{BIL}}(z_i) = H_i$, whereas outside the bilayer $E_{\mathrm{BIL}}(z_i) = 0$. At the water/bilayer interface a simple smoothing function is employed.[120] Thus, the overall the peptide/bilayer interaction energy is given by:

$$E_{\mathrm{BIL}} = \sum_{\text{residues } i} H_i f(z_i) \tag{17}$$

where summation is over the residues of the helix. A number of different hydrophobicity scales (H_i) are available.[123] In our studies[54,120] we have used a scale taken from that used in Monte Carlo simulations of the interaction of simplified models of α-helical peptides with a bilayer.[124,125]

Inclusion of such a term will restrain the constituent helices of a pore model to remain in a bilayer-spanning orientation. However, if more detailed analyses of channel/lipid interactions or more detailed calculations of ion permeation taking into account possible long-range electrostatic interactions between permeant ion and lipid headgroups are required then it will be necessary to include the lipid molecules explicitly in the simulations. This leads to a considerable increase in computational cost, because the number of atoms in the system will increase from ca. 1500 to ca.

[120] P. C. Biggin and M. S. P. Sansom, *Biophys. Chem.* **60,** 99 (1996).
[121] O. Edholm and F. Jähnig, *Biophys. Chem.* **30,** 279 (1988).
[122] F. Jähnig and O. Edholm, *J. Mol. Biol.* **226,** 837 (1992).
[123] J. L. Cornette, K. B. Cease, H. Margalit, J. L. Spouge, J. A. Berzovsky, and C. LeLisi, *J. Mol. Biol.* **195,** 659 (1987).
[124] M. Milik and J. Skolnick, *Proteins: Struct. Func. Genet.* **15,** 10 (1993).
[125] M. Milik and J. Skolnick, *Biophys. J.* **69,** 1382 (1995).

15,000. However, MD simulations of lipid bilayers[126] and of bilayer/peptide interactions[127] are now well established, and such methods have been applied to gramicidin.[128] One may therefore envisage that such simulations will soon be applied to some of the types of model channel discussed in this article.

Limitations

The main limitation of MD simulations on ion channels is the short time scale (ca. 1 ns) currently available for realistic computational resources. This will improve as processor speeds continue to increase. In this context it is useful to remember that for a single-channel current of 1.0 pA the mean dwell time of an ion within a channel is ca. 160 ns. A further major limitation of MD simulations of channels lies in the potential energy functions used. Current potential functions used in biomolecular simulations do not take into account, for example, electronic polarizability, and yet it is likely that such effects may play an important part in ion/channel/water interactions. More complex potential energy functions[129] take into account such effects, but are more computationally demanding. Thus, the usefulness and accuracy of MD simulations of channels may increase in parallel with increases in computational power.

Continuum Calculations

It is helpful to define first what is meant by "continuum" calculations on ion channels, and to explain why they provide a valuable extension to the MD simulation approach. Continuum calculations are those that do not treat a channel/ion/water system as a collection of individual atoms, but rather as dielectric continua, whose boundaries are defined by the van der Waals surfaces of protein and water and containing fixed charges corresponding to the positions of charged and polar atoms in the atomic model of the system. Such an approximation enables classical electrostatic theory to be applied to the behavior of ions and channels, and allows one to address physiologically relevant problems such as selectivity and conductance. In particular, this approach enables one to predict single-channel *I–V* curves, which may be compared directly with the results of

[126] K. M. Merz and B. Roux, "Biological Membranes: A Molecular Perspective from Computation and Experiment." Birkhäuser, Boston, 1996.

[127] L. Shen, D. Bassolino, and T. Stouch, *Biophys. J.* **73,** 3 (1997).

[128] T. B. Woolf and B. Roux, *Proteins: Struct. Func. Genet.* **24,** 92 (1996).

[129] S. L. Price and J. M. Goodfellow, *in* "Computer Modelling of Biomolecular Processes" (J. M. Goodfellow and D. S. Moss, eds.), p. 85. Ellis Horwood, New York, 1992.

patch-clamp recording. Such comparisons between prediction and experiment enable testing and hence further refinement of channel models. Furthermore, such calculations may provide additional insights into the molecular basis of channel function.

Several excellent discussions of the applications of classical electrostatic theory to ion channels are available.[2,130,131] In the following sections an outline is provided of a relatively simple approach used in the author's laboratory that has yielded useful results.

Electrostatics

The basis of these calculations is to obtain a numerical solution of the Poisson–Boltzmann (PB) equation:

$$\nabla \cdot \varepsilon(\mathbf{r})\nabla\Phi(\mathbf{r}) - \kappa'^2 \sinh[\Phi(\mathbf{r})] = -4\pi\rho(\mathbf{r}) \tag{18}$$

where $\Phi(\mathbf{r})$, $\varepsilon(\mathbf{r})$, and $\rho(\mathbf{r})$ are the electrostatic potential, dielectric constant, and charge density, respectively, at position $\mathbf{r}$. The Debye–Hückel parameter, κ, is derived from the Debye length, and hence from the ionic strength.[132] Solution of this equation yields an estimate of the electrostatic potential (Φ) around the channel and, in particular, along the pore axis. The PB equation treats the channel model plus surroundings as a collection of dielectric continua bathed in an electrolyte solution of a given ionic strength. Thus the channel model is treated as a region of low dielectric ($\varepsilon_{\text{PROTEIN}}$ in Fig. 9) in which are embedded atomic partial charges at their atomic coordinates determined by, for example, the pore-lining side chains. The surrounding lipid bilayer can be treated as a low dielectric slab ($\varepsilon_{\text{BILAYER}}$ in Fig. 9). In most applications the pore and bilayer are treated as having the same dielectric, say, $\varepsilon_{\text{PROTEIN}} = \varepsilon_{\text{BILAYER}} = 4$. The surrounding solvent is treated as having a high dielectric. This can either be set to the same (bulk) value ($\varepsilon = 78$) for all solvent regions or the solvent within the pore can be assigned a lower dielectric (e.g., $\varepsilon_{\text{PROTEIN}} = 30$) in accordance with the MD studies of the dielectric behavior of pore water (discussed earlier).

Having set up the system it remains to solve the PB equation in order to obtain an estimate of the electrostatic potential along, say, the axis of the pore. This can be solved numerically, for which purposes a number of standard programs are available (such as UHBD[133] or DelPhi[134]) that have

[130] D. Chen, J. Lear, and B. Eisenberg, *Biophys. J.* **72,** 97 (1997).
[131] A. Syganow and E. von Kitzing, *J. Phys. Chem.* **99,** 12030 (1995).
[132] K. A. Sharp and B. Honig, *Annu. Rev. Biophys. Chem.* **19,** 301 (1990).
[133] M. E. Davis, J. D. Madura, B. A. Luty, and J. A. McCammon, *Comput. Phys. Comm.* **62,** 187 (1991).
[134] M. K. Gilson, K. A. Sharp, and B. H. Honig, *J. Comp. Chem.* **9,** 327 (1988).

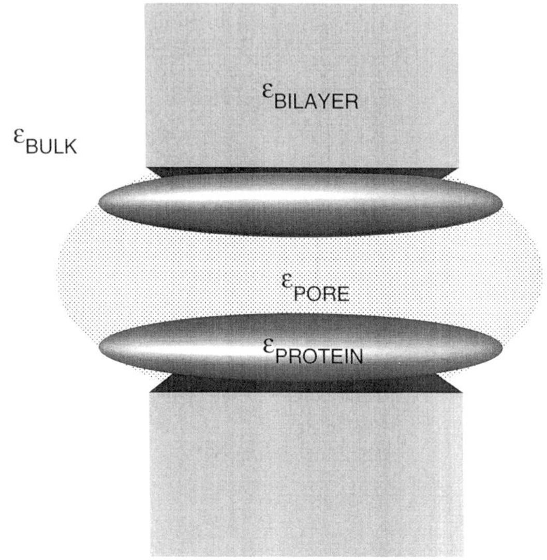

FIG. 9. Schematic diagram of the setup for PB calculations on channel electrostatics. The bilayer, protein, bulk water, and pore water regions may all be, in principle, assigned different dielectric constants.

been widely applied to proteins and other biomolecular systems[132,135] including models of lipid bilayers.[136] A number of more technical issues have to be addressed, such as: what is the most appropriate set of partial atomic charges to employ[137]; what is the best way to define the molecular surface that divides the low dielectric protein from the high dielectric solvent[138,139]; and what value to use of the Stern radius,[134] which defines the region immediately adjacent to the protein surface, which is assumed to be unoccupied by ions. Once these have been decided on and the PB equation solved, the results can be displayed as electrostatic potential profiles, $\Phi(z)$, along the center of the pore. Such profiles can be displayed in units of millivolts or, perhaps more usefully, in units of kcal/mol by multiplying the potential by $a + 1e$ probe charge [thus yielding profiles of $F\Phi(z)$, where F is Faraday's constant]. Note that this is simply a convenient change of unit, and does not directly estimate the energy of an ion (which would itself perturb the PB-calculated electrostatic field) at different positions along the pore. From

[135] M. Zacharias, B. A. Luty, M. E. Davis, and J. A. McCammon, *Biophys. J.* **63,** 1280 (1992).
[136] R. M. Peitzsch, M. Eisenberg, K. A. Sharp, and S. McLaughlin, *Biophys. J.* **68,** 729 (1995).
[137] D. Sitkoff, K. A. Sharp, and B. Honig, *J. Phys. Chem.* **98,** 1978 (1994).
[138] J. Warwicker, *J. Mol. Biol.* **236,** 887 (1994).
[139] Y. N. Vorobjev and J. Hermans, *Biophys. J.* **73,** 722 (1997).

such potential profiles one can attempt to predict, say, the anion versus cation selectivity of a channel. For example, application of this method to the nAChR M2 pore domain model discussed earlier showed that changes in the $\Phi(z)$ profile when a mutant channel was modeled were in good agreement with the change in selectivity of the pore from cationic to anionic which had been observed experimentally.[47]

Conductance Prediction

A continuum approach can also be taken to the prediction of single-channel conductance. The origin of this approach lies in a simple calculation of the electrical conductance of a small cylinder, of channel dimensions, when filled with electrolyte of a given ionic resistivity.[140] The resistance of the pore (ignoring the effects of access resistance at either mouth) is given by:

$$R = \frac{\rho l}{\pi r^2} \tag{19}$$

where ρ is the electrical resistivity of the electrolyte solution, l is the length of the pore, and r is its radius. This approach can be readily extended to channel models in which the pore is not a simple right cylinder. The first step is to estimate the pore radius profile for the model. This can be achieved using the program HOLE,[60,141] which yields the pore radius as a function of position along the pore axis, $r(z)$, by fitting a series of spheres within the lumen of the pore. Once $r(z)$ is known, then the electrical resistance of the pore can be obtained by summation of the Ohmic resistance of successive thin cylindrical sections of radius $r(z)$ and thickness Δz:

$$R_{\text{PORE}} = \sum_a^b \left(\frac{\rho}{\pi[r(z)]^2} \right) \Delta z \tag{20}$$

where the summation is from mouth a to mouth b of the pore. The reciprocal of the pore resistance gives an upper limit to the pore conductance, ($G_{\text{MAX}} = \text{R}_{\text{PORE}}^{-1}$. Initial application of this method to porins, for which both X-ray structures and accurate conductance estimates were available, suggested that this method overestimates single-channel conductance by a factor of about fivefold.[22] Thus, for a channel model a better prediction of channel conductance might be obtained by scaling G_{MAX} by a factor of 0.2: $G_{\text{PRED}} = 0.2G_{\text{MAX}}$. This was confirmed to be the case in a systematic study

[140] B. Hille, "Ionic Channels of Excitable Membranes," 2nd ed. Sinauer Associates, Sunderland, Massachusetts, 1992.

[141] O. S. Smart, J. M. Goodfellow, and B. A. Wallace, *Biophys. J.* **65,** 2455 (1993).

of predicted and experimental conductances for a number of channels for which either experimental structures or plausible models were available.[60]

I should stress that this approach has certain theoretical limitations. However, it is simpler to implement (and much quicker to run, taking just a few minutes of CPU time on a modest workstation) than more rigorous methods based on free-energy profiles from extensive MD simulations.[142] Thus, this continuum method can be used to test a large number of alternative models of a channel against experimental conductance data in order to determine the more likely models.[92] The major limitation of the method is the empirical scale factor in $G_{\text{PRED}} = 0.2G_{\text{MAX}}$. It seems likely that this arises from treating the resistivity of the electrolyte within the pore, ρ, as the same as that of the corresponding electrolyte in its bulk state. The resistivity of an electrolyte solution will be determined by the diffusion coefficients of its constituent ions. As seen earlier, these are reduced within a pore relative to their bulk values. Thus, this relatively simple approach to prediction of single-channel conductance can be placed on a firmer theoretical basis by using the results of MD simulations to eliminate the empirical scale factor. This is an area where further methodological developments are likely to occur in the near future.

I–V Curves

Single-channel current–voltage (*I–V*) curves can be calculated using Poisson–Nernst–Planck (PNP) theory. Detailed treatments of this theory[130] have been applied to channels formed by designed CFPs.[143] A simplified version of PNP theory is as follows. The current per unit cross-sectional area of the channel is calculated from the PNP equation:

$$\frac{I}{A} = C\left[\left(-q_c F\beta_c \frac{\exp(q_c F\Delta\Phi/RT) - 1}{\int_0^1 \frac{\exp[q_c F\Phi_c(z)]}{RTD_c}\,dz}\right) + \left(-q_a F\beta_a \frac{\exp(q_a F\Delta\Phi/RT) - 1}{\int_0^1 \frac{\exp[q_a F\Phi_a(z)]}{RTD_a}\,dz}\right)\right] \tag{21}$$

where I is single channel current; A, cross-sectional area of pore (see later discussion); q, charge on ion (and where the subscripts c and a indicate the cation and the anion); β, ionic partition coefficient (bulk to pore); D, ion

[142] B. Roux, *in* "Computer Modelling in Molecular Biology" (J. Goodfellow, ed.), p. 133. VCH, Weinheim, 1995.

[143] P. K. Kienker, W. F. DeGrado, and J. D. Lear, *Proc. Natl. Acad. Sci. U.S.A.* **91,** 4859 (1994).

diffusion coefficient; $\Delta\Phi$, external (i.e., applied) potential difference across the pore; $\Phi(z)$, electrostatic potential along the pore as a function of distance z along the pore axis; and l, length of the pore. The following simplifying assumptions are made: (1) $\beta_c = \beta_a = 1$; (2) $D_c = D_a$ = value of ion diffusion coefficient in bulk solution (but see earlier comments); and (3) the pore length l corresponds to the distance over which the pore exerts an effect on the electrostatic profile. The one-dimensional electrostatic potential profile along the length of the channel can be considered to be the sum of two components:

$$\Phi(z) = \Phi_{\mathrm{PORE}}(z) + \Phi_{\mathrm{APPLIED}}(z) \tag{22}$$

Thus the total potential profile is the sum of that due to the charge distribution within the pore [$\Phi_{\mathrm{PORE}}(z)$] and a linear term representing the transbilayer potential difference [$\Phi_{\mathrm{PORE}}(z)$]. The equivalent cross-sectional area (A) of the channel may be obtained from the pore radius profile (see earlier comments).

Case Study

The continuum approach is illustrated via its application to a derivative of alamethicin, namely Alm-BAPHDA.[56] In this derivative pairs of Alm helices are joined in a parallel orientation via flexible covalent linkers between their C termini. This leads to formation of stable, long lifetime (several hunderd millisecond) channels, which has enabled single-channel conductance measurements over a wide range of voltages.[144] SA/MD has been used to model pores formed by, say, three such pairs of helices (i.e., a six-helix bundle) and it is on such models (Fig. 10A) that the calculations have been based.

Figure 10B is a potential profile for an Alm-BAPHDA $N = 6$ model, calculated via numerical solution of the PB equation using UHBD. A 1-Å grid spacing and a 58 × 58 × 58 grid were used, with an ionic strength of 100 m*M*, and a Stern radius of 2 Å. The pore model was embedded in a low dielectric slab to mimic the presence of a lipid bilayer, and the dielectrics used were $\varepsilon_{\mathrm{PROTEIN}} = \varepsilon_{\mathrm{BILAYER}} = 4$ and $\varepsilon_{\mathrm{PORE}} = \varepsilon_{\mathrm{BULK}} = 78$. The resultant potential profile contains a barrier (for a + 1e charge) of height about RT and a well of depth about $2RT$. This is consistent with the weak cation selectivity of Alm and its analogs.

The pore radius profile for the Alm-BAPHDA $N = 6$ model is shown in Fig. 10C. the narrowest region of the pore is of radius ca. 2 Å, and is in

[144] G. A. Woolley, P. C. Biggin, A. Schultz, L. Lien, D. C. J. Jaikaran, J. Breed, K. Crowhurst, and M. S. P. Sansom, *Biophys. J.* **73,** 770 (1997).

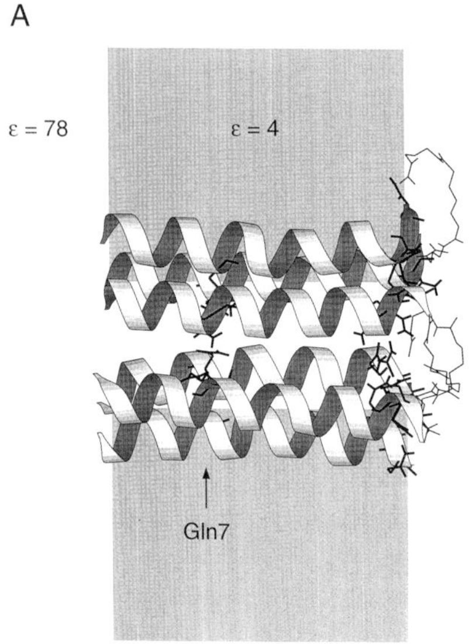

FIG. 10. Calculation of channel electrostatics and conductance properties, illustrated using a hexameric pore formed by a covalently linked dimer of alamethicin, Alm-BAPHDA. (A) Pore model embedded in a low dielectric slab corresponding to the lipid bilayer. The N-terminal mouth of the pore is on the left-hand side. The six helices are shown as ribbons. The Gln-7, Gln-18, and Gln-19 side chains are shown as bold lines; the C-terminal BAPHDA linker as thin lines. (B) Electrostatic potential energy profile along the length of the pore, calculated for 1.0 *M* ionic strength. (C) Pore radius profile along the Alm-BAPHDA pore, calculated using HOLE. (D) Single-channel *I–V* curves for 0.3, 1.0, and 3.0 *M* ionic strength calculated using the PNP equation (Adcock and Sansom, unpublished data).

the vicinity of the ring of Gln-7 side chains. Subsequent experimental studies[145] have confirmed that substitution of a smaller (Asn) side chain at this position results in an increased conductance. The model was also used as the basis of calculation of single-channel *I–V* curves as a function of the ionic strength of the electrolyte (Fig. 10D).[144] There are two main features of these curves: (1) at lower ionic strengths (e.g., 0.3 *M*) they predict a marked rectification; that is, the slope conductance of the channel is much higher for positive than for negative transbilayer potentials (where the sign of the potential refers to that of the side of the bilayer to which the peptide was added); and (2) the degree of rectification is considerably diminished as the ionic strength is increased. Both of the features are in good agreement with the corresponding experimental measurements. Furthermore, the ab-

[145] D. C. J. Jaikaran, P. C. Biggin, H. Wenschuh, M. S. P. Sansom, and G. A. Woolley, *Biochem.* **36,** 13873 (1997).

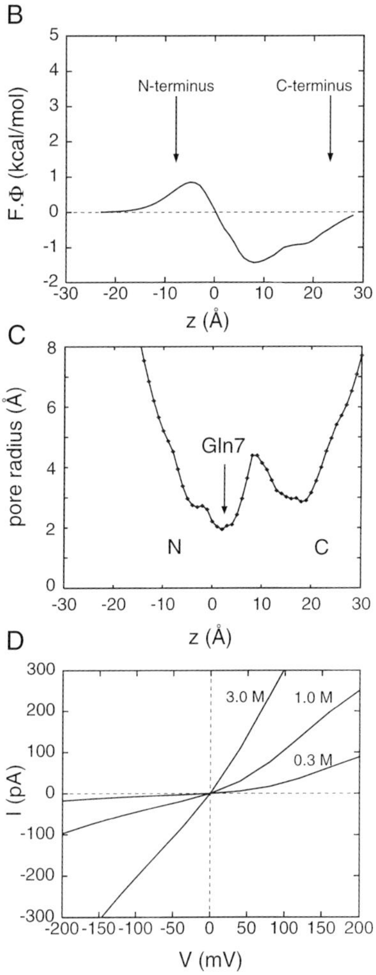

FIG. 10. (*continued*)

solute values of the single-channel currents are also in good agreement with those measured experimentally. For example, for the major conductance level of the channel at 1.0 *M* KCl, the experimental single-channel current is 100 pA at 100 mV, whereas the calculated current is 110 pA.[144] Of course, the absolute value of the single-channel conductance predicted is dependent on the cross-sectional area assumed for the pore. This was calculated on the basis of the G_{PRED} prediction of the single-channel conductance. This, in turn, suggests a way in which G_{PRED} and PNP calculations of ion channel conductance properties may in the future be combined in an integrated approach to predict single-channel I–V curves from structural models. In the context of Alm-BAPHDA the remarkably good agreement between experimental and predicted single-channel properties leads to a degree of confidence in the channel model that will justify more elaborate, and computationally demanding, simulation studies.

Problems

There remain a number of problems, both practical and theoretical, with the continuum approach to the prediction of channel properties. A number of these were alluded to earlier, residing in the difference between macroscopic (i.e., bulk) properties of water and electrolyte solutions (e.g., ε; D_{ION}) and the corresponding properties for water and ions confined within a pore of molecular dimensions. MD simulations, possibly based on simplified model systems, may provide suitable values of such parameters. It is an ongoing area of research to investigate whether these parameters can be used to improve continuum calculations or whether the main limitations are inherent in this approach per se and not in its input parameters.

A second problem, also mentioned earlier, occurs when multiple ionizable residues are in proximity to one another within or at the mouth of a pore. This will almost certainly lead to perturbation of the pK_a values of such residues so that their values *in situ* differ from those for the corresponding side chains in isolation. This problem can be approached via the PB equation approach and has been widely investigated for globular proteins and for bacteriorhodopsin.[146] For ion channels these methods have been applied with some success to the ionizable side chains that form the narrowest region of the pore in bacterial porins.[118] Titration curves are calculated for all ionizable residues within the model via a two-stage procedure for each such residue. Firstly the *intrinsic* pK_a is calculated, by assuming that all residues of the protein other than the one in question were in their electrically neutral state. Secondly, the effect of each residue's interaction with all other ionizable groups in the protein is determined for all possible combinations of ionization state, thus enabling calculation of their absolute pK_a values *in situ*. Although the pK_a values estimated by this approach are only approximate estimates they may indicate sidechains existing in a protonation state different from that which they would adopt in at neutral pH when in isolation.

[146] D. Bashford and K. Gerwert, *J. Mol. Biol.* **224,** 473 (1992).

Conclusion

An approach to modeling the pore domains of ion channels based on restrained molecular dynamics simulations has been described. The emphasis has been on ways in which restraints may be employed to embody different types of experimental structural data. The quality of the resultant models is largely dependent on the nature of such data. To arrive at a single plausible model for a channel domain, it is important to have at least secondary structure data for the pore-lining elements.

The use of extended molecular dynamics simulations both to refine initial models and to explore the dynamic and functional implications of such models has been described. Simulations of pores filled with explicitly modeled water molecules are now routinely possible with modest computational facilities. Advances in both software and hardware mean that, at least for channel-forming peptides and simple channel proteins, atomistic simulations of complete channel systems (i.e., channel + bilayer + water + ions) will become increasingly important.

In addition to MD simulations, calculations based on continuum approximations play an important role in relating channel models to physiologic function. In particular, continuum electrostatic calculations enable prediction of single-channel current–voltage relationships. The relative ease with which such calculations can be performed enables large numbers of possible models to be screened by comparison of predictions with experimental observations.

The future of modeling and simulation studies of ion channels lies in two directions. First, better experimental structural data will result in better models, which in turn will enable more accurate simulations. The second is that the continued advances in computation will allow increasingly realistic simulations of ion channel function. Thus, ion channels may be among the first membrane proteins for which a genuinely atomic description of their physiologic roles will be possible.

WWW Sites

Many of the programs described in this chapter can be obtained from WWW sites. These are listed in the Appendix to the article by Scott and Tanaka elsewhere in this volume.[147]

Acknowledgments

Work in the author's laboratory is supported by grants from the Wellcome Trust. My thanks to Charlotte Adcock, Ian Kerr and Graham Smith for their comments on this manuscript.

[147] S.-P. Scott and J. C. Tanaka, *Methods Enzymol.* **293,** [33], 1998 (this volume).

[35] Computer Simulations and Modeling of Ion Channels

By MICHAEL E. GREEN

Introduction

Ion channels have been studied by a large variety of experimental techniques during the past half century. The structure and functions of channels are becoming better understood; however, molecular level understanding of the steps in selectivity and gating, as well as relation between the water, the ions, and the protein, are still in need of further study. Because it is difficult to determine experimentally the behavior of individual protein molecules, in spite of the ability to study the physiologic properties of single channels, computer modeling by simulation of all or part of the system has become an increasingly valuable tool in understanding the channels.

In a simulation, one attempts to start with the properties of the molecules that compose the system, and, by following the effects of the molecules on each other as they move, understand the thermodynamics, often the dynamics, and other properties of the system. Other properties may include dielectric constant, and structural parameters such as orientation of the molecules, or their arrangement with respect to boundaries. In addition to intermolecular forces, boundary conditions and external fields may be important.

In recent years it has become possible to model the behavior of water and ions in pores defined by either rigid walls, walls to which water has been attached, or even models of proteins. The present state of the art does not quite allow the complete modeling of the protein with the water, a protein with more than 200 amino acids, and more than 50 water molecules in its pore, plus an ion or two, is beyond the reach of the largest computers available in 1997. However, considerable progress has begun in that direction with related systems, especially photosynthetic reaction center; that protein has water containing clefts, although it lacks a true channel.[1] In addition, gramicidin, a simple channel composed of just 30 amino acids and with a pore that can hold just seven or eight water molecules, can be modeled in increasing detail.[2,3] Woolf and Roux have carried out an all

[1] C. R. D. Lancaster, H. Michel, B. Honig, and M. R. Gunner, *Biophys. J.* **70,** 2469 (1996).
[2] B. Roux, *Biophys. J.* **71,** 3177 (1996).
[3] D. E. Sagnella, K. Laasonen, and M. L. Klein, *Biophys. J.* **71,** 1172 (1996).

0076-6879/98 $25.00

atom simulation of gramicidin, with a section of lipid membrane, and water caps.[4] Channels have large electric fields generated by charged amino acids, reaching 10^9 V m^{-1}, and voltage-gated channels generally have moderately large fields, often two orders of magnitude smaller, across their pore, when the membrane is polarized.

Channels have been modeled in the literature in several ways, neglecting different parts of the channel. In some work, the water is treated as a dielectric continuum, assigned a dielectric constant of 78 or 80, as appropriate near 25°, thereby avoiding the problem of treating the individual water molecules explicitly. In other cases, the protein is replaced by a hydrophobic wall, and the water is treated explicitly. Various intermediate levels of complexity have been attempted. However, an all atom simulation of a single transmembrane (TM) segment has been carried out by Shen *et al.*[5] on a polyalanine helix in a lipid membrane, surrounded by water; although one TM segment is not a channel, it points the way to more complete simulations of channels.

Some principles remain the same for any type of simulation, including the necessity for obtaining intermolecular potentials for all molecules included in the simulation. The general procedures also remain the same. Water, being small, allows a more complete description of its potential than does protein. In part also, the choice of the part of the system to simulate is determined by the present state of knowledge: only approximate structures are known for the normal K^+ and Na^+ channels. However, a number of simulations of water in pores have been done. One simulation has been done for a model of a sodium channel,[6] but the question of how realistic the model is remains for further study.

Therefore, although we understand simulations of water in a channel model that may be somewhat too hydrophobic to be rather limited, they are still very informative, easier to follow, and simulations of larger systems will follow the same procedures when increasing computer power and improved protein models become available. Although this is only part of what needs to be simulated, much of this article is devoted to this aspect of the problem. Modeling of the protein itself has been undertaken by several groups, for example, Guy and co-workers.[7,8] Other references are

[4] T. B. Woolf and B. Roux, *Biophys. J.* **72,** 1930 (1997).

[5] L. Shen, D. Bassolino, and T. Stouch, *Biophys. J.* **73,** 3 (1997).

[6] C. Singh, R. Sankararamakrishnan, S. Subramanian, and E. Jakobsson, *Biophys. J.* **71,** 2276 (1996).

[7] S. R. Durell and H. R. Guy, *Neuropharmacology* **35,** 761 (1996).

[8] H. R. Guy and S. R. Durell, *in* "Ion Channels and Genetic Diseases" (D. Dawson, ed.), pp. 1–16. Rockefeller University Press, New York, 1995.

given by Kerr and Sansom,[9] who modeled only the pore; they have done a number of simulations on other models (see later section).

Modeling Water in Pores

To describe the water, it is necessary to use a statistical description. The water has many possible orientations, and these cannot all be separately computed. If an attempt were made to analyze the hydrogen bonds of each of 30 or so water molecules, the number of combinations would be too large for any currently available computer. An attempt to minimize energy would not succeed because the channel is not at zero temperature, and the essential features of the pore water depend on thermal fluctuations. A completely static picture would also not allow ions to move through the pore. Therefore, one must sample many configurations, all compatible with thermal equilibrium, and average suitably. The average orientation of the water molecules, the variation in density with position, rates of diffusion (to a limited extent, because time-dependent phenomena present additional difficulties), and other properties can be obtained from simulations. It is also possible to put ions in the model, and generate the appropriate electric fields to investigate possible binding sites and forces on the ion, thus obtaining the relevant properties.

Ion channels are unique in that they are a small pore, with a dielectric boundary, which contains charges itself and in which ions are present. The consequent fields are central to understanding the channels, because they may exceed 10^9 V m^{-1}. The membrane polarization that keeps voltage-gated channels closed is equivalent to an average field of the order of only 10^7 V m^{-1} across the entire membrane, or perhaps somewhat higher locally, as the drop across the membrane need not be linear. Therefore, the calculation of the field is a critical element in the determination of the properties of the channel as a whole.

Another major difference with bulk simulations is the presence of a pore wall, which requires that the boundary conditions of the simulation be set for a finite system. In contrast, in bulk simulations, it is typical to use what are called *periodic boundary conditions*, in which the simulation volume is repeated at each boundary of the simulation cell. We describe these in the section on boundary conditions.

Intermolecular potentials are needed to describe the interactions of the molecules with each other, with ions, and with the walls. For water, a number of such potentials are available, with parameters derived in part

[9] I. D. Kerr and M. S. P. Sansom, *Biophys. J.* **73,** 581 (1997).

from *ab initio* calculations, but principally characterized by adjustments to obtain agreement with thermodynamic properties of bulk water. Because no potentials are available that have been parameterized specifically for water in confined spaces, there is no choice but to use the standard potentials; these are parameterized for uniform surroundings. Water in pores might still not behave in a manner substantially different from bulk water, in terms of the interactions of the molecules with other molecules, if the molecules do not "feel" the environment to be very different. Models are available that take into account the polarizability of the water; this has the advantage of being somewhat more accurate in high electric fields, but the disadvantage of requiring substantially more computer time. Ions, for which the largest interaction term is their charge, also have established van der Waals parameters, and their polarizabilities may be included. If explicit protein is to be included, then potentials are needed for the atoms of the protein too. Potentials are available for atom groups that can be used for this purpose.

Fundamental Statistical Ideas

Definition of Ergodicity

To get thermodynamic averages for a system, it is necessary to average over a large sample of the configurations the system could assume at equilibrium. There is usually little difference in energy between one configuration and another. For example, if a water molecule rotates, the difference in energy may well be less than $k_B T$ (k_B is the Boltzmann constant; T, temperature in kelvins), the thermal energy, making both configurations about equally likely. One way to get an adequate sample is by going through many such configurations, sampling, one hopes, enough to provide a good estimate of the thermodynamic average. There must be a method of weighting the sampling according to thermodynamic equilibrium probabilities in order to get an accurate average, or generating them with Boltzmann probability (see section on Monte Carlo simulations). Alternatively, one can take a single long trajectory as a function of time through the various configurations assumed by a system, and hope that one has sampled the possible configurations adequately. A theorem exists which states that these two possibilities are equivalent, the *ergodic theorem*. Although still not entirely proven in all instances, and of uncertain applicability in a finite system/finite time, we will assume that it holds well enough to allow us to choose either method for sampling a system.

Definition of Ensembles

Consider a harmonic oscillator, with energy $E = 1/2\, kx^2 + p^2/2m$, where the first term is the potential energy, the second the kinetic energy: k, spring constant; x, displacement from equilibrium position; p, momentum; m, mass. We can consider a large number of such systems, with energy between E and $E + dE$ (see Fig. 1). Figure 1 shows the region in which all possible configurations of the one-dimensional harmonic oscillator with energy between E and $E + dE$ can be found. A simulation seeking to define the properties of the one-dimensional oscillator would have to sample a reasonable proportion of the shaded area. If, for example, the average position were sought, position left and right of the zero would have to be sampled with about equal frequency, at roughly equal distances from the origin. In this sense, ensemble averages are averages over the possible arrangements of a system in position and momentum. The ensemble itself is, however, a conceptual entity, composed of an indefinitely large number of copies of the system, all defined by the same properties. Three particular types of ensemble are most generally used. The microcanonical ensemble is defined by the number of particles, N, the volume V, and the energy E. Figure 1 is an example of a microcanonical ensemble (in other words the energy is considered constant if all systems composing the ensemble have energy in a narrow range dE, not that all systems have identical energy). The canonical ensemble is defined by N, V, and the temperature T. This is more convenient than the microcanonical ensemble if one is interested in thermal equilibrium; temperature is maintained by exchange of heat with a thermal reservoir. The grand canonical ensemble is appropriate for an open system, in which particles as well as heat may be exchanged with the surroundings; in this, the constant quantities are μ (chemical potential),

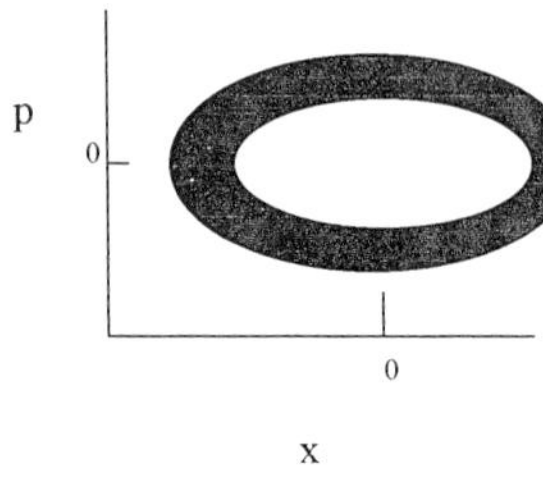

FIG. 1. Ensemble of one-dimensional harmonic oscillators with energy between E and $E + dE$. The inner ellipse is at energy E, the outer ellipse at $E + dE$.

V, and T. It is also possible to do N, p, T (p = pressure) simulations. We will discuss primarily N, V, T simulations.

Types of Simulation

Any useful simulation will be of a many-particle system. There are two main types of simulation, Monte Carlo (MC) and molecular dynamics (MD). One must define the ensemble in which one is working in any simulation. However, the two types of simulation differ in how they use the ensemble, and they require ergodicity to be equivalent. MC simulations attempt to sample a large number of configurations, moving randomly from one to another nearby configuration, and repeating until enough of the entire coordinate space has been sampled; one hopes that the sample is adequate to give a good average (but checking is necessary).

MD simulations track a system, following Newton's second law, $F = ma$, as the equation of motion, albeit in a much more elaborate fashion. Here, the hope is that the system passes through a large enough sample of configurations that the averages are valid. The equations cannot actually be solved, but particles can be moved according to the force, hence the acceleration; by keeping track of the trajectory and the velocity, one step at a time, the coordinate space is sampled. We consider the consequences of inadequate sampling later. Each step must be so short that there have been no significant motions of the other particles; otherwise, the force would have changed during the step, and the motion calculated for the particle being considered would be incorrect. For the force to be constant, the time step must be so short that even the highest frequency motions of the particles will have hardly changed position. It is therefore necessary to pick a time of approximately 1 fs (10^{-15} s) per step. This is a very short time (e.g., light moves 0.3 μm in 1 fs). Therefore, 10^6 steps produce a track 1 ns long, which is about the upper limit for practical simulations for most moderate size systems in 1997; presumably more powerful computers will be developed in coming years that will make it possible for the technique to be extended to biologically interesting times. At present, it is not possible to produce trajectories for systems as large as an ion channel, which continue for times as long as gating, generally in excess of 1 μs.

However, there is a type of technique that randomizes boundary conditions or else forces, stochastic dynamics (SD), which is capable of going to longer times. It is somewhat difficult to avoid errors in complex systems; however, SD has been used successfully, and one such case is examined later. We only mention some other techniques for going to longer times.

General references for dynamics simulations as applied to biological macromolecules include McCammon and Harvey[10] and Brooks *et al.*[11]

Potential Energy

Summary of Forces

A complete potential energy description of the system includes intramolecular and intermolecular energy terms. The former include vibrations, with a term for each normal mode, bending about bond angles, and torsion about dihedral angles. All are high-frequency motions, especially the normal mode vibrations. It is to accommodate these motions that MD simulations must take such short time steps. For each quadratic term in the energy, a constant, analogous to a spring constant, is needed (for torsions, usually a trigonometric, not quadratic, dependence appears, e.g., cos θ/a, where θ is the torsion angle and a a constant related to the symmetry of the displacement). We will spend little time on the intramolecular terms, which are similar to those in bulk simulations. Water molecules are frequently treated as rigid bodies in bulk simulations, and C–H groups, such as methyl, may be combined into "extended atoms." The groups comprising the protein in ion channels can be parameterized by comparison with other known substances; the properties of the protein are not well known independently.

Intermolecular Forces

1) Water

For either MC or MD simulations, it is critical that intermolecular potentials that represent the system as accurately as possible be used; one expects that the degree of realism will increase as experience accumulates and computers improve, understanding that "realism" is a relative term. Several types of forces exist, which are principally either electrostatic or van der Waals forces in origin. There are distributions of charge in chemical bonds, which lead to local dipoles, or charges on atoms, which in turn interact with each other. The van der Waals forces are also well known; for simulations, it is normal to include these forces explicitly in the potentials. Water is best to begin with, as the molecule which illustrates the main features. We do not cover all possible proposed potentials, but limit

[10] J. A. McCammon and S. C. Harvey, "Dynamics of Proteins and Nucleic Acids." Cambridge University Press, Cambridge, 1987.

[11] C. L. Brooks, M. Karplus, and B. M. Pettit, "Proteins: A Theoretical Perspective of Dynamics, Structure, and Thermodynamics." John Wiley and Sons, New York, 1988.

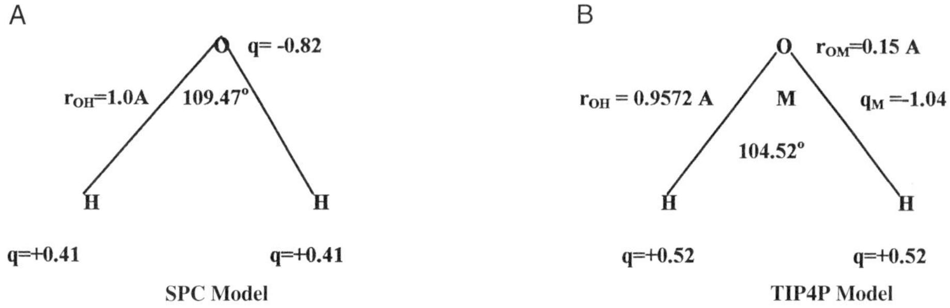

FIG. 2. Parameters of two point charge models of water. The values of q are the electronic charges on the points shown. (B) The negative charge is at point M, not on the O atom. M is 0.15 Å toward the H atoms on the angle bisector. In addition to the charges, there are van der Waals terms centered on the O atom in both models. The charges are in units of the charge on a proton.

ourselves to a few commonly used ones. We need to know how the potentials are evaluated. Two types of potentials are available; those which are polarizable and those in which the electrostatic part is represented strictly by point charges.

Point Charge Models. Several point charge models of water have been proposed. We consider only two of them: the simple point charge (SPC) model,[12] and one of the TIP models (TIP4P).[13] These are discussed in Levesque and Weis.[14] Each is characterized by a set of charges on the hydrogen and oxygen atoms (SPC), or near them (TIP4P), plus Lennard–Jones parameters. Given the polarity of the dipoles in the water molecule, the H atoms are, of course, positive, the oxygen negative. In the SPC model, the charges are distributed as shown in Fig. 2A; in TIP4P, as in Fig. 2B. The latter has four points: three for charge and the oxygen atom for the center of van der Waals forces (i.e. Lennard–Jones terms). (We note that TIP3P, a three-point model, is still often used in MD simulations, where the point with no mass found in TIP4P causes a problem, but we use TIP4P, used in MC, in this discussion.) Note that the model includes not only partial charges, but bond lengths and angles as well.

Electrostatic interaction energy among molecules can be found from $\Sigma q_i q_j / r_{ij}$, where the sum is over all pairs of interactions between charge

[12] H. J. C. Berendsen, J. P. M. Postma, W. F. van Gunsteren, and J. Hermans, *in* "Intermolecular Forces" (B. Pullman, ed.), pp. 331–342. Reidel, Dordrecht, 1981.

[13] W. L. Jorgensen, J. Chandrasekhar, J. D. Madura, R. W. Impey, and M. L. Klein, *J. Chem. Phys.* **79,** 926 (1983).

[14] D. Levesque and J. J. Weis, *in* "The Monte Carlo Method in Condensed Matter Physics," 2nd corrected ed. (K. Binder, ed.), pp. 172–177. Springer, Berlin, 1992.

sites i,j on different molecules; r_{ij} is the site-site distance. The Lennard–Jones term is $A/r_{OO}^{12} - C/r_{OO}^{6}$, where for either model oxygen–oxygen distances are used. The values A and C are similar for the two models. Again, it is necessary to sum over each pair of molecules. Therefore, the total energy of interaction can be written

$$E_{mn} = \Sigma_i \Sigma_{j>i}\, q_i q_j / r_{ij} + A/r_{OO}^{12} - C/r_{OO}^{6} \tag{1}$$

In Eq. (1), the energy is for the interaction of molecules m and n; the second sum has molecule indices $j > i$, where i is the first sum index, to avoid double counting. The O–O distances are understood to have the same molecular indices, and there are 9 terms (3×3) for the charges on each molecule–molecule interaction.

The calculation time is dependent approximately on the square of the number of interacting molecules, and thus there is a substantial advantage to cutting off the sum at a finite distance; the computation time then increases only linearly with the number of molecules. Fortunately, this is less of a problem in ion channel simulations, where the channel boundary is less than a reasonable cutoff distance away in at least two dimensions. The cutoff for the Lennard–Jones terms might be at 8 Å; $8^6 = 2.6 \times 10^5$, and the constant, C, is about 4×10^{-18} J Å^6 in both models, so that the value of the attractive term would be of the order of 10^{-23} J, less than 1% of $k_B T$; the repulsive term is much smaller at large distances (a scaling function, providing a gradual cutoff, is sometimes better). The real problem is with the Coulomb terms; these fall only as $1/r$, and are therefore long range. A variety of schemes exist for obtaining the sum at large distances, and simple cutoffs are not satisfactory. A smoothed cutoff helps; terms beyond may also be included via a reaction field, or sum term (Ewald sum). These too have disadvantages, but it would take too long to discuss them. For an ion channel, it is probably simplest to keep the complete set of interactions, as the number of molecules is not great, and the risks of using a cutoff in a system with complicated dielectric boundaries too large. This is especially necessary because of an additional complication: the interaction between charge centers is not simply dependent on distance, but depends also on the relation of position to dielectric boundary.

The quality of simulations in bulk water, using these potentials, is good, as shown in Table I. The heat capacity, C_p, is the derivative of the enthalpy, and the error with derivatives is inherently greater as they change sharply with smaller changes in the quantity of which they are the derivative. These data are from MD simulations done at 25° and 1 atm. Since this work, from the original TIP4P paper, was published, more extensive work with the same potentials has been carried out; the advent of large computers has meant more extended calculations on larger numbers of molecules. How-

TABLE I
CALCULATED AND EXPERIMENTAL PROPERTIES FOR LIQUID WATER[a]

Property	SPC	TIP4P	Experimental
d(g cm^{-3})	0.971	0.999	0.997
ΔH_{vap}(kJ mol^{-1})	45.06	44.60	43.97
C_p(J mol^{-1} K^{-1})	97.9	80.7	75.27

[a] Data from W. L. Jorgensen *et al., J. Chem. Phys.* **79,** 926 (1983).

ever, the essential points remain unchanged for these potentials. They are good for the bulk liquid, and allow a reasonably good description of its properties, including its radial distribution function (the probability of finding a molecule at a distance r from a central molecule). It is also possible to get a number of nonthermodynamic properties for the water, including properties of the hydrogen bonds. Different sets of parameters produce hydrogen bonds of slightly different length, a matter that also affects the density.

The average molecular dipole is also important. In the gas phase the dipole moment for the water molecule is 1.85 D (Debye, where 1 D = 3.336×10^{-30} C m). However, the liquid behaves as if the dipole were about 30% larger. The TIP4P dipole is approximately 2.4 D and the SPC dipole 2.3 D, about correct for the liquid, for which they were parameterized.

These models are limited in several ways. For one thing, they assume rigid molecules, thereby losing three vibrational degrees of freedom. Fortunately, the vibrations occur at relatively high energy compared with k_BT, so that the contributions of the vibrations are limited (although omitting the vibrations would be a limitation if high accuracy were needed).

Polarizable Models. The local charges in a protein produce a fairly large electric field, and this field can in turn distort the electron distribution in the molecule. The problem is more acute in a channel, as the fields are extremely high. For this reason, a model of water in which the water is polarizable would be expected to give improved results. In a polarizable potential, there is an induced addition to the dipole moment of $\mu_{ind} = \alpha\mathbf{F}$, where α is the *polarizability* of the molecule, and $\mathbf{F}$ the local electric field. Then there is an additional term in the energy of the form

$$E = -\mu_{ind}\,\mathbf{F} \tag{2}$$

In Eq. (2), one takes the scalar product of the two vectors on the right-hand side. This additional energy can be appreciable; it clearly goes as $\mathbf{F}^2$, as α is a constant. The polarizability α is of the order of the size of the molecule, 10^{-24} Å^3, or 10^{-40} C V^{-1} m^2. If $\mathbf{F}^2$ is on the order of 10^{19} V^2 m^{-2}, then the induced energy is of the order of 10^{-21} J for the molecule, or close to k_BT; a more precise calculation gives values close to k_BT for several molecules in the channel in each round of simulation. Since, in a MC calculation, the decision on accepting or rejecting a move (see later discussion) depends on quantities of this order, the polarizability does matter. For this reason we will consider one example of a polarizable potential. Several have been proposed, going back at least to the potential of Stillinger and David.[15]

The polarizable SPC (PSPC) potential was proposed by Ahlstrom *et al.*[16] as an extension of the SPC potential. It adds polarizability on the O atom of water, with $\alpha = 0.55 \times 10^{-40}$ C V^{-1} m^2. Calculating the local field is not as simple, because it includes both the external field and a dipole tensor term. The dipole tensor depends on the values of the intermolecular vectors that are being calculated. Therefore, the calculation must be made self-consistent, a substantial increase in computational effort. There are also dipole–charge interactions, and an energy term for creating the dipoles. An additional complication results from the tendency of the polarizability to draw the molecules together, leading to still higher absolute values of the energy, and forcing the molecules to come even closer together; it is necessary to put in some sort of limit on this process to avoid a polarization catastrophe. Ahlstrom *et al.*[16] added increased repulsion at short distances. Other possibilities include a hard core at an O–O distance of 2.5Å[17] or damped field at close approach. The reader is referred to Ahlstrom *et al.*[16] for details. It is in any case possible to prevent the polarization catastrophe.

The essential point of including the polarizability is the increased accuracy in the calculation, especially in high field.

In addition to the potentials listed here, a number of newer potentials have been proposed (e.g., Svishchev *et al.*[18] and Chialvo[19]). Finding an accurate representation of water is an ongoing project, especially when conditions far from 25° and 1 atm pressure are to be represented. The most significant extreme condition encountered in ion channels is the high electric field. In addition to polarization, the high field leads to an increase in

[15] F. H. Stillinger and C. W. David, *J. Chem. Phys.* **69,** 1473 (1969).

[16] P. Ahlstrom, A. Wallqvist, S. Engstrom, and B. Jonsson, *Mol. Phys.* **68,** 563 (1989).

[17] J. Lu and M. E. Green, *Prog. Colloid Polym. Sci.* **103,** 121 (1997).

[18] I. M. Svishchev, P. G. Kusalik, J. Wang, and R. J. Boyd, *J. Chem. Phys.* **105,** 4742 (1996).

[19] A. A. Chialvo, *J. Chem. Phys.* **104,** 5240 (1996).

pressure to some extent due to a phenomenon known as *electrostriction*, but the consequent increase in density is probably not enough to change the required intermolecular potentials. However, it is possible that eventually a more appropriate model of water will be used for higher accuracy simulations of water in channels.

Ions

Intermolecular potentials for ions are obviously just as important as those for water, but fortunately much simpler. We are primarily concerned with two ions, K^+ and Na^+, and sometimes Ca^{2+} or Cl^-. A simple potential can be constructed by using the charge on the ion (there is no ambiguity in position for a charge when only one atom constitutes the entire ion), van der Waals parameters, and a polarizability that may be keyed to the water model in use for the simulation. The charge is necessarily one full charge. In addition, one may observe that for consistency with the surrounding water it would be necessary to include polarizability if a polarizable model of water were used. Since the polarizability is proportional to the molar refraction, the value may be estimated by comparison to the molar refraction data. However, better results can be obtained via *ab initio* calculations.

The question of ion–water interactions has been considered in the most detail by Aqvist.[20] He gives the following ion–water interaction potential:

$$E_{\text{I-w}} = B\,\Sigma_j(q_\text{I}q_j/r_{ij}) + \Sigma_j(A_\text{I}A_j/r_{ij}^{12} - C_\text{I}C_j/r_{ij}^{6}) \tag{3}$$

where B is the constant to adjust units, and the A_j, C_j, parameters are the SPC values defined earlier, for the van der Waals terms. Then the values of A_I, C_I are the van der Waals parameters associated with the ion. The charges in the first term are the same as in the SPC model. Numerical values can be obtained from Aqvist's paper. The values are obtained by empirical adjustment of the thermodynamic values. Although *ab initio* values, obtained by direct calculation, were not as successful, they may become reliable as computer power increases, and it is possible to do more extended calculations.

Protein

Rigid Wall Representation. Channel walls are composed of protein, and must be in some way included in the calculation. The simplest solution is to evade the problem, by substituting a hard wall for the protein; the next more sophisticated solution is to substitute a dielectric medium for the

[20] J. Aqvist, *J. Phys. Chem.* **94,** 8021 (1990).

protein. At least two such simulations, using tapered pores to imitate the general structure of a Na^+ or K^+ channel, including both acetylcholine receptor channel and voltage-gated channels (at least most of which are tapered, i.e., hourglass shaped), have been carried out. This approach has been taken by Green and Lu.[17,21] In that model, the hydrophobicity of hard walls was modified by adding fixed water molecules on the channel side of the wall; these molecules, not allowed to translate, replaced a part of the hydrogen bonding ability of the protein that would have made up the channel wall. In a similar simulation, but without the fixed water molecules, Sansom *et al.*[22] used a hydrophobic wall softened by an S-shaped potential curve at the boundary, with the wall 31 kJ/mol^{-1} (5.2×10^{-20} J $molecule^{-1}$, or about 13 k_BT) high, and rising to 90% of full height over 1.15 Å. In the latter simulation, no particular dielectric properties were assigned to the wall, and only the hard wall potential was used. A MD simulation was done, while in the former simulation, a MC simulation was used. In the former, the wall was assigned a dielectric constant, and an electric field was calculated. Both concluded that with the taper, the wall aligned molecules significantly; in the Green and Lu simulation, this was true principally in the tapered region, and depended in part on the electric field. In the Sansom *et al.* simulation the outer layer of the water was oriented throughout the taper. There is little doubt that water is oriented by the geometry of the walls, and that therefore this would affect the dielectric constant in the pore. Furthermore, the orientation would affect the electric field directly, by providing a dipole moment equal to that of a number of aligned molecules. These effects are not surprising (simulations in cylindrical pores have shown some alignment in the water closest to the walls) but are important; it remains to be demonstrated experimentally that they are found in actual channels.

Explicit Representation. It is obviously better to use a complete representation of the protein, and a number of examples have been published. Some are in small channels (alamethicin, gramicidin), formed from peptides of limited size, and not close in structure to the gated channels in which we are interested. One attempt has been made[6] to simulate a sodium channel, using a model of the channel proposed by Guy and Durrell.[8] However, it is also of some interest to look at other proteins, especially if channel-like, such as photosynthetic reaction center. In this section we are concerned with the representation of the protein much more than with the results of the simulation.

The simulation of the pore lining of a model of sodium channel by

[21] M. E. Green and J. Lu, *J. Colloid Interface Sci.* **171,** 117 (1995).

[22] M. S. P. Sansom, I. D. Kerr, J. Breed, and R. Sankararamakrishnan, *Biophys. J.* **70,** 693 (1996).

Singh *et al.*[6] considered 130 amino acid residues, which line the pore in the model. Just as with water, it was necessary to have intermolecular potentials for the atoms that compose the protein. Since these were represented explicitly, there must be a distribution of charge to go with the atoms. One compromise was made in the choice of potentials; methyl, methylene, and methine groups were treated as extended atoms, and a single potential assigned to the group, rather than to each C and H atom. The water in the channel was explicitly represented also in this model, with 17- and 20-Å caps at the narrow intracellular end and the wide extracellular end respectively. In total, 933 water molecules were included, using the SPC model for water. The protein potentials were taken from a commercial package, GROMOS, developed at the University of Groningen.[23] It is one of several such, CHARMM (from Harvard University[24]) being another popular package. While it would take too long to go through the means of parameterizing a protein, it is in principle similar to what is done to parameterize water. In the Singh *et al.*[6] study, in addition to the protein and water, Na^+, K^+, and Cl^- ions were present, and the Aqvist[20] potentials mentioned earlier were used. With all the potentials in hand a MD study was carried out. Because the potentials are our immediate concern, we postpone consideration of the remainder of the study, which is one of the more interesting in the literature at the present time.

A description of the properties of a set of potentials in the form required for a simulation of a protein is given in McCammon and Harvey.[10] These include the following:

1. *An atom data file.* The properties of each type of atom must be described. This includes partial charges, van der Waals radius, and Lennard–Jones parameters. If an extended atom approach is to be used (H atoms attached to C not explicitly included) the types of atom must include such "extended atoms" as methyl groups, because these will have different parameters than the original atoms. Also, not all atoms of the same element are of the same type. For example, sp^2 and sp^3 C atoms differ in their properties, including in particular bond angles, and must be distinguished.
2. *Coordinate file.* The Cartesian coordinates of the atoms to be used explicitly in the simulation must be included in the data. If an X-ray structure is known, the coordinates may come from the Brookhaven protein database, for example. If one is starting with a model, as in the Singh *et al.*[6] calculation, the file must contain the location of the atoms according to the model.

[23] W. F. van Gunsteren, H. J. C. Berendsen, J. Hermans, W. G. B. Hol, and J. P. M. Postma, *Proc. Natl. Acad. Sci. U.S.A.* **80,** 4315 (1983).

[24] B. R. Brooks, R. E. Bruccoleri, B. D. Olafsson, D. J. States, S. Swaminathan, and M. Karplus, *J. Comput. Chem.* **4,** 187 (1983).

3. *Internal coordinate (molecular topology) file.* From this file, the Cartesian coordinates must be translated into internal coordinates, such that bond lengths and angles are correct. This refers to the list of atoms bonded to each other, the potential types to use with each, the partial charges, torsion terms, and often the information to build the starting molecule (bond lengths, angles, etc.). If the molecules are to be rigid, these internal coordinates will not change during the simulation. Also we are not generally concerned with cases in which covalent bonds break (but see Sagnella *et al.*[3]). Given three translational coordinates for a molecule, and three rotational angles, all Cartesian coordinates for atoms in the molecule would be determined if the molecule were rigid. Assuming vibrations are allowed, however, there must be displacement of the atoms from the position they would have if they were all at exactly their equilibrium position.

4. *Energy parameter file.* The Lennard–Jones energy parameters, the charges, and all other inter- and intramolecular interaction energy parameters are given here for each type of atom. If vibrations are allowed, harmonic oscillator terms for internal motions would be included. Three body interactions are not included explicitly in normal simulations, and would go with polarizability calculations, which would require self-consistency, a more elaborate computation.

In addition to the data files listed by McCammon and Harvey,[10] it may be necessary to calculate an electric field for simulations in an ion channel; this may be derived in part from the partial charges in the energy parameter file, if an all atom simulation is being carried out. However, if dielectric boundaries are involved, the field produced at one atom by another is not simply a function of the distance between the atoms, but depends explicitly on their location with respect to the boundary. We consider this matter separately later.

Once one has the data files, it is possible, in principle, to recalculate the energy in each pass through the simulation, atom pair by atom pair, as a function of distance, with the formulas built into subroutines in the program, or to have large look-up tables as a function of distance. The exact structure of a program, with some look-up tables and some distance dependent computation, has to be decided based on the nature of the computation. We will see a case later in which the electric potential and field computation depended on a preliminary computation of the field, followed by use of the field as a look-up table with interpolation for values at locations between those in the table. This approach is more feasible with ion channels than with general simulations because the volume to be covered is generally more limited. Otherwise look-up tables can quickly exceed the available memory of the largest computers. The use of virtual memory, with continual reference to disk, is slower, but may be unavoidable.

Simulations

We have already noted that two main types of simulations are of interest to us: Monte Carlo and molecular dynamics. We first describe the essential points of MC simulations, then MD. Finally, we add a few comments about stochastic dynamics.

Monte Carlo Simulations

In MC simulations, molecules are moved in turn. The change in energy is calculated, following which a decision is made as to whether to accept the change in configuration, or consider the old configuration to also be the new configuration. The point is to discover equilibrium properties. There is no explicit form of time dependence, and it has rarely been attempted to assign time equivalents to the length of a Monte Carlo simulation. It is not clear that any legitimate way to do so exists. A Monte Carlo move, as the name implies, is random.

Moving the Molecules. To move a water molecule, it is necessary to use a random number generator to get a random change in each coordinate, producing a new location for the molecule. The move may be in more than one coordinate simultaneously. If one is using a rigid water model, like SPC or TIP4P, its position is described by six coordinates: three for the position of some location in the molecule, say, the O atom, and three for rotation. It is possible to move all six simultaneously, but then the move for each should be small enough that the chance of having moved to a much higher energy position (say, within the van der Waals radius of another molecule, or into a channel wall) is small. Translations should be kept to fractions of an angstrom, rotations to a few degrees, to tens of degrees. It is critical that neither be biased, or the molecules may end up on one side of the volume.

The random number generator may be a matter of some concern, since some random number generators have subtle bias. However, the chance that this will be commensurate with the moves of the molecules, thus biasing the simulation, does not seem very large. As far as the author knows, the question has never been systematically investigated. Generally one starts with a different random number seed for each simulation, and the sequence of numbers produced by a pseudo-random number generator is very long. Therefore, different instances of otherwise identical simulations will sample different portions of phase space, and can be considered independent.

Acceptance of Moves. Generally a criterion proposed by Metropolis *et al.*[25] is used to decide whether to accept a move. If the move is to lower energy, it is accepted. If the move is to higher energy, it is tested as follows: for an increase of energy by ΔE, the move is accepted with probability $\exp(-\Delta E/k_B T)$. (Deciding whether to accept the move requires another random number.) If the move is accepted, it becomes the new configuration of that molecule. If the move is rejected, the old configuration becomes the new configuration, and is averaged in again. It is not legitimate to average only accepted moves, because this has the effect of forcing a move on each attempt. In a normal simulation, about 20–50% of the moves should be accepted. If the acceptance rate is too small, it is probable that the steps are too large.

Sampling. Monte Carlo simulations are less than perfectly efficient in sampling the phase space. If too small a portion of coordinate space is sampled, the probability of getting a good estimate of the thermodynamic equilibrium state of the system is small. This can happen if the system is trapped in a local minimum. For example, referring to Fig. 1, suppose the oscillator were always sampled in the upper half of the shaded area, where $p > 0$; some barrier prevented transition to $p < 0$ states. Then the oscillator would appear to be moving continuously to the right, an obvious error. Unfortunately, errors are not always so obvious, and incorrect sampling may lead to conclusions that are wrong but are not discarded. In systems of realistic size, there may be a huge number of minima, and some method of finding a reasonable set is required.

A conceivable alternative is to repeat the simulation several times, and average the results rather than simply extend an individual simulation, an obviously time-consuming procedure. One way around this is to use *simulated annealing.* In this procedure, the system starts at a high temperature, say, 600 K, so that the sampling can cover much more of phase space; the effective depth of any well is inversely proportional to the temperature of the system, from the point of view of a Boltzmann distribution. After the system has passed through many wells, it may be cooled in stages, in the hope that it will settle in the correct minimum, or very nearly so, at the intended simulation temperature of, say, 300 K.

If one has additional information about the system there are several tricks available to introduce this information into the simulation. One method, which has been used for biomolecules, is to introduce a coordinate representing the position of some conformation or other factor that changes the potential, such as the extent of binding a ligand. The simulation is carried

[25] N. Metropolis, A. W. Rosenbluth, M. N. Rosenbluth, A. H. Teller, and E. Teller, *J. Chem. Phys.* **21,** 1087 (1953).

out at several values of this coordinate, and the free energy examined as a function of the coordinate. Other thermodynamic properties can be obtained as well. With the coordinate variable, the potential that drives the parameter to the desired range is called the *umbrella potential* and this procedure is called *umbrella sampling*. The normal probability of the system being in some of the positions of coordinate space may be small; by adding the umbrella potential the system may be forced to spend some time in the moderately less probable regions, so that their contribution can be estimated. Then, since the umbrella potential is known, the correct contributions of the various regions of configuration space can be determined with better accuracy than would have been possible in a finite time with normal sampling. Because this is a free-energy difference technique, the umbrella potential in effect being added to the free energy, it is also related to the *potential of mean force*, defined as the derivative with respect to coordinates of the Helmholtz free energy. A method for choosing the potential has been suggested by Mezei.[26] Several other techniques for finding the free energy are also relevant (see the section on Free-Energy and Pressure Calculations).

More generally, any additional knowledge of the system, say, of the regions of coordinate space which are likely to be more strongly represented among possible configurations, can be incorporated into the simulation to ensure a better sampling of the important regions of coordinate space, by moving the system to entirely different regions of the space to start sampling from there. One example of such knowledge might be large-scale conformational changes, when it is known that more than one set of dihedral angles should be possible for a peptide, and each set must be sampled. Without the large-scale shift, the simulation might never move out of whichever local minimum it finds itself in. New methods of simulation over wide regions in phase space, or at least configuration space, are continually being proposed.

Molecular Dynamics Simulations

The basic method by which molecular dynamics simulations sample phase space is rather different than the method in MC simulations, and is often more efficient in sampling the space. MD actually tracks a system as it moves. Therefore, velocity, acceleration, and time dependence, become relevant. The equation of motion of the system is set up to follow Newton's second law, $F = ma$ (F = force, m = mass, a = acceleration). Begin by finding the force on a particle. After this the equations of motion can be

[26] M. Mezei, *J. Comput. Phys.* **68,** 237 (1987).

integrated. Some of the same considerations apply to MD and to MC. For one thing, the same intermolecular interactions, and energy calculations, apply. Therefore we need not repeat that part of the discussion.

Force Calculation. The energy is determined in the same manner as in the MC calculation. Then the force on the particle can be found from the gradient of the energy:

$$F_i = -\nabla_i E_i \tag{4}$$

where the gradient is taken with respect to particle i coordinates to obtain the force on that particle. To get useful information from the potential, it needs to be in differentiable form.

Integrating the Equations of Motion. To use the force in $F = ma$, observe that $a = d^2x/dt^2$, change in velocity $= \Delta(dx/dt) = (d^2x/dt^2)\Delta t$, and change in position is $\Delta(x) = (dx/dt)\Delta t$. However, the time step is finite, and simply projecting the system forward using the position, momentum, and acceleration particle by particle leads to serious inaccuracies. Another way of looking at this[10] is to expand the position in a Taylor series (we will write it in one dimension for simplicity):

$$\begin{aligned} x(t + \Delta t) = x(t) &+ (dx/dt)(\Delta t) + 1/2(d^2x/dt^2)(\Delta t)^2 \\ &+ 1/6(d^3x/dt^3)(\Delta t)^3 + \cdots + \end{aligned} \tag{5}$$

However, this is an infinite series, and must be truncated, which is done after the square term, the term obtained from the acceleration. A method of avoiding the error introduced by the truncation is needed. The magnitude of the error can be estimated from the fact that energy and momentum must be conserved (all the forces are internal to the system; therefore there is nothing that can change the total energy or momentum). If there are systematic errors, the total energy and momentum of the system will drift from their starting values.

There are in practice a variety of means of correcting the integration of the equation of motion. The most popular are due to Verlet,[27] to Gear,[28] and to Beeman.[29] A variant called SHAKE[30,31] has also proven very popular. Because these methods are described in the McCammon and Harvey book[10] and its references, we will only mention that these integration algorithms essentially entail more sophisticated methods of compensating for the errors introduced by simply truncating the series in Eq. (5). Observe that the

[27] L. Verlet, *Phys. Rev.* **159,** 98 (1967).

[28] C. W. Gear, "Numerical Initial Value Problems in Ordinary Differential Equations." Prentice Hall, New York, 1971.

[29] D. Beeman, *J. Comput. Phys.* **20,** 130 (1976).

[30] J. P. Ryckaert, G. Cicotti, and H. J. C. Berendsen, *J. Comput. Phys.* **23,** 327 (1977).

[31] W. F. van Gunsteren and H. J. C. Berendsen, *Molec. Phys.* **34,** 1311 (1977).

values of x, dx/dt, etc., are the values at the beginning of the time step. It would be better if the average value of these quantities during the step were used. The Verlet method, as an example, looks at the value at the midpoints of the time interval (i.e., $\Delta t/2$ before and after the Δt time step, and then "leapfrogs" the velocity from the value at $t - \Delta t/2$ to the value at $t + \Delta t/2$). The effect is to make the overall time step Δt, but take the average velocity at the midpoint of the interval rather than at the beginning. The other algorithms employ higher order corrections. SHAKE is an improvement that can be applied to an integration algorithm in that it reduces the computation time. The length of the time step is determined by the highest frequency motions of the molecule, generally the molecular vibrations. SHAKE sets constraints, typically fixed bond lengths, and takes a longer time step (therefore allowing the computation to proceed through more of space in the same computational time). At each step, the SHAKE algorithm resets the atom positions to fit the constraints within some specified relative margin (say, 10^{-4}). It is done iteratively, treating the positions sequentially, rather than exactly, to avoid solving a nonlinear matrix equation.

Sampling. While the sampling efficiency of the MD method is typically greater than that of the MC method, at least where vibrations are included, this does not eliminate the problem of finding the system in a local minimum. It is very nearly impossible to find a global energy minimum with hundreds of degrees of freedom, or more, as in a channel. Sampling a reasonable fraction is extremely important. However, the tricks that could be applied to a MC simulation are essentially the same as the tricks that could be applied to an MD simulation; details differ. Again, it helps tremendously to have some additional information concerning the possible conformations of a system.

Boundary Conditions

Very similar questions arise with MC and MD, and there is no need to discuss them separately. The problem arises because a few hundred molecules must stand in for essentially an infinite number, or at any rate a mole, in bulk simulations. Therefore, a "box" (whatever its actual geometric shape, normally determined by the symmetry of the system) equivalent to about 8 or 10 molecules on a side has to produce the same thermodynamic quantities as the macroscopic system. There are a couple of ways around this that appear to be successful. However, the most successful does not apply to ion channels. In fact, this is one way in which ion channels differ most seriously from bulk simulations, in that an "infinite" number of molecules need not be represented. Ion channels are finite in size, and must allow water molecules and ions to come into contact with the protein boundary, in whatever way the protein is represented (hard wall, softer wall, or explicit atoms).

The most common method of representing an infinite system with a small number of molecules is to use *periodic boundary conditions*.[10,32] The simulation volume ("box") is reproduced on each face by a box containing an identical set of molecules, so that the molecules near one edge see not a hard wall, but another set of molecules, which are in fact identical to those on the opposite side of the simulation box. Only one interaction is allowed with another molecule, that which is at the shortest distance, whether that distance is to the molecule in the original box or to its image in a neighboring box. Thus each molecule, whether in the center or near the edge of the simulation volume, sees an equivalent environment. Any molecule at the left boundary, say, position $(-x,y,z)$, is made a neighbor of molecules on the right boundary, near (x,y,z). This has the effect of keeping the density near any molecule, including those on the boundaries of the simulation volume, essentially the same as the density in the center. In bulk simulations, this is an adequate description, in which the hundreds of molecules in the simulation amount to only a tiny fraction of the solution being simulated. The system, infinite for practical purposes from the viewpoint of any small subvolume, needs to appear infinite in the simulation also. In effect, the boundary disappears, and the system can be translated along any axis. This is not true in a channel, unless the entire protein, including some of the boundary lipid, is included. The electric fields as well as the protein walls are not symmetric. If the model is complete enough, however, and includes enough of the boundary lipid, then the same considerations apply as to any other infinite system.

The physical environment in a channel is also not isotropic. Here, as we discussed under the section on Intermolecular Forces, the wall must be represented in some explicit fashion. For example, the boundary conditions appropriate for the channel simulation of water are the protein for at least most of the wall. However, the water at the intracellular and extracellular ends of the channel is in contact with bulk water and must be represented by either a cap of additional water molecules large enough that any edge effects would have relaxed by the edge of the functional part of the simulation volume, or else allow some sort of periodicity in one dimension, along the axis of the channel. The size of a cap would have to be at least on the order of the minimum cutoff distance of intermolecular potentials, say,

[32] K. Binder and D. Stauffer, *in* "Applications of the Monte Carlo Method in Statistical Physics" (K. Binder, ed.), pp. 1–36. Springer, Berlin, 1984.

9 Å. The techniques vary, but must take into account the specific properties of the model being used.

Methods for Longer Time Simulations: Stochastic Dynamics and Others

Stochastic dynamics attempts to speed the simulation approximately 1000-fold by averaging the high-frequency forces. Alternatively, one can allow the boundary conditions to fluctuate, but we do not consider such a case here. To simulate times longer than approximately a nanosecond, it is necessary to have a means of averaging these fast forces. Then large time steps are possible, and times up to microseconds can be simulated. The remaining forces are essentially frictional, and a random force must be added. A microsecond is long enough for an ion to pass through a channel. However, stochastic dynamics, in which averaging is applied by way of variable boundaries, is beginning to be applied to biomolecules. Wang *et al.*[33] have noted that Davis and McCammon[34] have shown that there must exist, at least in polar liquids, a solvent boundary force, in the form of a pressure on the surface atoms of the solute molecule. Wang *et al.* have extended this work to cyclosporin, a cyclic undecapeptide. A number of workers have used the *boundary element method*, which includes an explicit boundary surface to include surface pressure forces. Wang *et al.* may have carried out the first simulation to include a peptide in solution.

Another approach has been suggested by Laakkonen *et al.*[35] based on a method developed by Guarnieri.[36] This mixes MC and stochastic dynamics; the latter requires the use of the equation of motion, with a random force added:

$$m\, dv/dt = f[x(t)] + R(t) + m\gamma v \tag{6}$$

where m is the mass of the particle, dv/dt is the acceleration, and forces are $f[x(t)]$ the deterministic force, $m\gamma v$ the frictional force (γ is a friction coefficient), and $R(t)$ a random force with the property that its correlation function is

$$\langle R(t)R(t')\rangle = 2m\gamma k_B T\delta(t - t')\rangle \tag{7}$$

In other words, the average value of the product $R(t)R(t')$ is zero if the times are different; there is no correlation between different times. This is characteristic of a random force that relaxes rapidly compared with the

[33] C. X. Wang, S. Z. Wan, Z. X. Xiang, and Y. Y. Shi, *J. Phys. Chem. B* **101,** 230 (1997).

[34] M. E. Davis and J. A. McCammon, *J. Comput. Chem.* **11,** 401 (1990).

[35] L. J. Laakkonen, F. Guarnieri, J. H. Perlman, M. C. Gershengorn, and R. Osman, *Biochem.* **35,** 7651 (1996).

[36] F. Guarnieri, *J. Math. Chem.* **18,** 25 (1995).

other characteristic times of the system, and is appropriate for simulations that are to be done for a time scale long compared to the time scale of atomic motions. Parameter γ lumps fast degrees of freedom.

What is unique in the method used by Laakonen *et al.*[35] is the combination of SD simulation in the Cartesian space with MC calculations of the torsions in angular space. An MC step, with Metropolis algorithm, intervenes between dynamics steps based on a solution of Eq. (6). The velocities for the succeeding dynamics step (including the random force and frictional force) come from the preceding dynamics step, using the Verlet algorithm discussed earlier, in a form that does not require information at time $t - \Delta t/2$ (the "velocity Verlet algorithm"). The earlier (i.e., at $t - \Delta t/2$) information is unavailable here due to the MC step that intervenes between SD steps; the coordinates depend on the MC step. If the step is accepted, new coordinates are used to compute the position-dependent forces. The method is stated to converge two to three orders of magnitude faster than if "other simulations" are used. Laakonen *et al.* applied it to the TRH receptor binding pocket; the TRH receptor is a G-protein-coupled receptor (GPCR). Although GPCRs are not the primary subject of this article, they are membrane-spanning proteins and methods applicable to them can in all likelihood be applied to channels.

This is only one example of the attempt to simulate longer time intervals. A variety of techniques involving constraints, multiple time steps, integration of the equation over longer time steps with partial integrals at shorter intervals, and other methods have been tried. None is in common use yet, although the importance of the subject causes continuing effort. A recent review by Schlick *et al.*[37] covers the field, with particular reference to details of some methods not covered here.

Electric Field Calculations

A critical part of the potential in a channel simulation is the electric field from the charges on the protein. It is also not clear that all amino acids that would be ionized at physiologic pH will be ionized in the channel; a 60-mV potential shifts pK_a by one unit. Potentials of hundreds of millivolts, locally, are possible. We will consider two methods of calculating the field. One assumes the water is a continuum, to which a dielectric constant can be assigned, and a concentration of ions. This is entirely appropriate for bulk solution, and almost certainly for the larger volumes at the ends of the channels. It has also been used[1] for the fields in photosynthetic reaction center, in which local potentials as large as 1 V were found. We look at

[37] T. Schlick, E. Barth, and M. Mandziuk, *Annu. Rev. Biophys. Biomol. Struct.* **26,** 181 (1997).

that case first, because it is by far the most popular method, and uses a protein for which a complete structure is known. Second, we consider a method used by Lu and Green[17] for model simulations of a channel, in which the protein was replaced by a dielectric continuum, and a boundary element method was used to find the potential and field. In that set of simulations, the water was represented explicitly.

Nonlinear Poisson–Boltzmann Equation

The Poisson equation gives the potential for a given distribution of charges. Debye and Huckel used it to find the potential around an ion in solution from the distribution of other ions, but only if the solution was so dilute that the interaction energy was less than $k_B T$. The ionic distribution in turn is given to start by the Boltzmann distribution, since the energy of interaction of the ions with the central ion is $q_i \phi / k_B T$, where q_i is the charge on the ion, and ϕ the potential at the central ion. Since this is a form of mean field theory, with the average concentration of ions the question of interest, the concentration around the central ion is

$$c = c_0 \exp(-q_i \phi / k_B T) \tag{8}$$

The Poisson equation becomes, with this Boltzmann term (SI units, one radial dimension),

$$d^2\phi / d\mathbf{r}^2 = -1/\varepsilon \, \Sigma_i c_i = -1/\varepsilon \, \Sigma_i c_0 \exp(-q_i \phi / k_B T) \tag{9}$$

where $\mathbf{r}$ is the radial distance from the central ion, and ε is the dielectric coefficient. This is an obviously nonlinear equation, and is an instance of the nonlinear Poisson–Boltzmann (PB) equation, which Debye and Huckel linearized,[38] and then solved, by expanding the exponent to first order. This expansion is correct if other ions are far from the central ion, so that the potential is low enough for $q_i \phi \ll k_B T$. Therefore their treatment is valid as an infinite dilution limit, rather than a general result. In the nonlinear PB equation case, the expansion is replaced with a numerical solution. However, Eq. (9) is not complete, because it assumes constant ε. When there is both protein and aqueous solution, the dielectric coefficient varies from place to place in the system. If we also allow charge density to vary, one needs Eq. (9a), the most general form of the equation:

$$\nabla \cdot [\varepsilon(\mathbf{r}) \nabla \phi(\mathbf{r})] + \varepsilon(\mathbf{r}) \, \kappa^2 \sinh[\phi(\mathbf{r})] + \rho(\mathbf{r})/\varepsilon_0 = 0 \tag{9a}$$

[38] R. A. Alberty and R. J. Silbey, "Physical Chemistry," 2nd ed., pp. 230–232. John Wiley, New York, 1997; based on P. J. W. Debye and E. Huckel, *Phys. Z.* **24,** 185, 305 (1923).

where $\varepsilon(\mathbf{r})$ is the dimensionless dielectric coefficient, $\rho(\mathbf{r})$ is total charge density, **r** is now the position vector in three dimensions, and κ is the Debye–Hückel parameter, an inverse length; it depends inversely on the square root of ionic strength, the dielectric coefficient, and the temperature. With a 150 m*M* aqueous solution of 1:1 electrolyte, 25°, $\kappa \approx 8$ Å^{-1}. A number of possible methods of solving this have been proposed, of which one commonly used example is the DelPhi program.[39–41] This solves the PB equation by dividing space into small enough segments to allow a local solution. DelPhi was used by Lancaster *et al.*,[1] for example, to solve for the potential and fields in a photosynthetic reaction center molecule with a known configuraton. The isolated water molecules fixed in the protein were also represented explicitly, but in one, energy-minimized configuration. In more recent work, these water molecules are treated by Monte Carlo simulation to get the potential more accurately. Lancaster *et al.*[1] found potentials up to 1 V; more important, they found potential shifts of hundreds of millivolts near ionizable side chains, which meant that there were pK_a shifts as large as 10 units. This has a major effect on how we consider the ionization state of a protein; it is clearly too simple to assume ionization of any basic or acidic residue that would be ionized in water at physiologic pH. There is actually some experimental evidence that bears on the question of high fields. Lockhart and Kim[42] found, using Stark effect measurements, that the field at the end of a peptide due to its helical dipole was 0.43×10^9 V m^{-1}, quite a substantial value considering that it is measured at a covalently attached residue which is not enclosed in a channel or cleft, but is surrounded by water. This supports the accuracy of the calculation of fields much larger than membrane potentials in the channel.

Boundary Charge Method

It is also possible to find the field by determining the effect of a charge at any point in the volume on the potential and field at any other point. If there is a dielectric boundary, a charge anywhere in the volume, whether in the protein or the water, will induce a charge on the boundary. If all the boundary charges can be found, then the dielectric can be removed in favor of the charges induced on the boundary; the potential can then be found by Coulomb's law, summing over the real and the induced charges. To do this, it is necessary to divide the boundary into a set of elements, within which the induced charge is assumed to be constant. The size of the

[39] M. K. Gilson, K. A. Sharp, and B. Honig, *J. Comput. Chem.* **9,** 327 (1987).
[40] K. A. Sharp and B. Honig, *Ann. Rev. Biophys. Biophys. Chem.* **19,** 301 (1990).
[41] B. Honig and A. Nicholls, *Science* **268,** 1144 (1995).
[42] D. J. Lockhart and P. S. Kim, *Science* **257,** 947 (1992).

elements may be 1 Å^2, or less. One gets one linear equation for each element, leading to an $N \times N$ determinant. If this can be solved, all the induced charges are known, and the problem is, for all practical purposes, solved. The problem may be solved, with the source charges on a lattice point, and the resulting potential and field recorded for all other lattice points. For locations between lattice points, interpolation can be used to get the potential and field. This method has been used for simulations of a model channel with explicit water, and protein as a dielectric.[17,21]

Although the technique is very different from the nonlinear PB equation, potentials and fields of the same order of magnitude are found. These results turn out to be reasonable based on a rough estimate of the fields to be expected at the distances from charges found in channels. They could be avoided only by having dipoles canceling charges in a concerted fashion, something which could happen only locally. The orientation of the dipoles, of which water molecules are among the most important, can be found by MC or MD simulation, with the former probably being easier.

Other Continuum Models

Partenskii *et al.*[43] attempted to estimate the effect of the carbonyls of a gramicidin-like (in approximate geometry) model in lowering the energy barrier presented by the pore. They attempted to allow for the effect of charges in the pore by introducing a third dielectric constant for the polar part of the channel wall, and solving a nonlinear PB equation. Their work, and references therein, illustrate the difficulty of using continuum models; no really satisfactory choice of dielectric constants in a pore has yet been found for a channel model in which neither the protein nor the water is modeled explicitly.

Eventually, one would want to have a complete model of the channel protein, with all atomic positions defined. Then no dielectric problem would arise, but the complete set of charges and dipoles could be used explicitly to obtain the field. However, even this calculation would have to be made self-consistent, because the charges at one location would alter the pK values elsewhere, and the state of ionization at that location in turn would affect dipoles and pK values at the first location. So far, no calculation of this type with pK_a shifts has been attempted for a channel.

Free-Energy and Pressure Calculations

It is straightforward to get the energy, the density, and other intensive variables by direct averaging. However, free energy and other extensive

[43] M. B. Partenskii, V. Dorman, and P. C. Jordan, *Biophys. J.* **67,** 1429 (1994).

variables are not as direct, nor is there direct information on the partition function. To obtain the free energy, several special techniques have been devised.

Techniques of Free-Energy Calculation

1. *Particle insertion* (*Widom*,[44] *as discussed in Levesque et al.*[45]) If a test particle is introduced to the system, and the system has an N particle configurational energy of U_N, define $\Delta U_N = U_N - U_{N-1}$; then the chemical potential μ is given by

$$(\mu - \mu_0) = -k_B T \ln\langle \exp(-\Delta U_N/k_B T)\rangle_{N,V,T} \tag{10}$$

where μ_0 = chemical potential of an ideal gas at the same N, V, T. The brackets give the canonical average. The added particle tests the energy of its surroundings, hence the value of the free energy. There are substantial sampling difficulties with this procedure, but it has been used successfully in several instances.

2. If the free energy is known in a reference state, *thermodynamic integration*[32] can be used. Conceptually, it can be thought of as follows: Because the path by which the system moves from one state to another does not matter for differences in state variables, introduce a variable, λ, such that $0 < \lambda < 1$, where λ is the degree of advancement from the initial state, in which the free energy is known (say, at $T = 0$), to the state at which it is wanted. Then the system can be integrated along the path by increasing λ from 0 to 1, repeating the simulation with enough values of the parameter to provide an accurate integration.

There is an alternate formalism[46] giving the Helmholtz free energy, A:

$$A_1 - A_0 = -k_B T \ln\langle \exp[-\psi_1 - \psi_0)/k_B T]\rangle_0 \tag{11}$$

where A_1, A_0, are the Helmholtz free energy for values of the potential ψ before and after perturbation (the NpT ensemble would give the Gibbs free energy). If the perturbation is too large, it can proceed through several simulations with successive values of a parameter λ such that

$$\psi_\lambda = \lambda\psi_1 - (\lambda - 1)\psi_0 \tag{12}$$

where Eq. (12) helps maintain the stability of the motion. The overall free energy is then available from the integral of $d\Delta A(\lambda)/d\lambda$ over λ.

[44] B. Widom, *J. Chem. Phys.* **39,** 2808 (1963).

[45] D. Levesque, J. J. Weis, and J. P. Hansen, *in* "Applications of the Monte Carlo Method in Statistical Physics" (K. Binder, ed.), p. 37ff. Springer, Berlin, 1984.

[46] D. J. Tobias and C. L. Brooks, III, *Chem. Phys. Lett.* **142,** 472 (1987).

Another alternative, when the path would pass through energy differences large compared to k_BT, which would prevent adequate sampling, is to go through a *thermodynamic cycle*. This means that the beginning and end points of the change are connected through different, possibly hypothetical species, which can nevertheless be computed. Free energy is path independent, so the difference between initial and final state is the same. The alternate cycle may allow for the cancellation of terms difficult to compute, such as nonideality of the solutions. The method is described in McCammon and Harvey.[10]

Mezei and Beveridge[47] have given a very thorough, and very reasonable, review of the possibilities of various methods of obtaining the free energy, especially the free-energy perturbation (FEP) method, and several other methods of obtaining the free-energy difference.

Pressure

In a liquid characterized by additive pair potentials $\phi(r)$ (r = intermolecular distance), the pressure is normally obtained from the following formula[32]:

$$pV/Nk_BT = 1 - (N/6k_BT)\int_0^\infty g(r)\, d\phi(r)/dr 4\pi r^2\, dr \tag{13}$$

where $g(r)$ is the radial distribution function (the probability of finding another molecule at a distance r from the central molecule, another quantity that is typically reported as a result of the simulation). However, a channel, with its major contribution from the wall and the field, would have additional terms, and apparently no attempt to compute directly the pressure in a simulation has been attempted for a channel so far.

Examples of Ion Channel Simulations

We conclude with a few examples of ion channel simulations, illustrating the types of information that can be obtained. We begin with channels composed of antibiotics made by bacteria (gramicidin, alamethicin), and then mention a couple of more interesting voltage-gated and acetylcholine receptor examples. The gramicidin simulations are typically all-atom simulations, because the gramicidin molecule is small enough that this is feasible.

Roux and Karplus[48] simulated water with Na^+ and K^+ in an analog of a gramicidin channel, using MD, with CHARMM potentials; the model was constructed to be a periodic β helix. To understand the motion of the

[47] M. Mezei and D. L. Beveridge, *Ann. N.Y. Acad. Sci.* **482,** 1 (1986).

[48] B. Roux and M. Karplus, *Biophys. J.* **59,** 961 (1991).

ion, the potential of mean force was calculated. The energy barrier to ion motion, and the particular water configurations that appeared in the simulation to be responsible, were among the results. Sagnella *et al.*[3] have examined proton transfer, using Carr–Parinello MD, a technique that allows for breaking of covalent bonds, to facilitate consideration of proton motion. Their results indicated that carbonyl solvation was critical for proton transfer, and that water did not move with the proton, which was transferred down the "proton wire."

Breed *et al.*[49] carried out MD simulations of columns of water in several types of channels, including two types of polyalanine, alamethicin, nicotinic acetylcholine receptor M2 helix, and δ-toxin. In the narrowest channels, the self-diffusion coefficient of water was reduced by an order of magnitude compared to bulk water. Dipole–dipole interactions were strong. Sankararamakrishnan *et al.*[50] did MD simulatons of a model of the pore domain of the nicotinic acetylcholine receptor, relating the behavior of water in the channel to pore diameter and electrostatics. In addition, they carried out nonlinear PB calculations to estimate the electrostatics of the motion of a monovalent ion down the pore. From these results, they were able to make reasonable suggestions as to the configurations of the open and closed states, and the interactions of specific groups with water and with an ion in the pore.

This summary of uses of simulation obviously barely touches the literature on the subject, but does begin to suggest some of the types of information that can be obtained by simulaton of a pore, and its water and ions.

Summary of Simulation Procedures

We have seen that a simulation requires that the model include the following:

Intermolecular potentials. In a channel, these must include potentials for water, protein, and ions. It may be possible in some simulations to substitute a dielectric medium for either the water or the protein; however, a fully satisfactory model would have to include all three explicitly. Some intramolecular potentials are needed as well.

Suitable boundary conditions. These are likely to be different for a channel, in that periodic boundary conditions are not possible in two dimensions, if the protein model is less than complete. There must be some accommodation of the water to the protein, and the results will depend on the approximation made.

[49] J. Breed, R. Sankararamakrishnan, I. D. Kerr, and M. S. P. Sansom, *Biophys. J.* **70,** 1643 (1996).

[50] R. Sankararamakrishnan, C. Adcock, and M. S. P. Sansom, *Biophys. J.* **71,** 1659 (1996).

External potentials, as electric fields. These do not normally appear in bulk simulations, but are unavoidable in channels. If omitted, the model is unrealistic for a channel. The fields are large enough that intermolecular potentials which include polarization are likely to be necessary for simulations of channels.

Once one has a model, one can choose a mode of simulation. MC simulations are often simpler than MD, but MD has other advantages. Both provide thermodynamic averages and molecular distribution functions as their most important output. MD should produce time dependence, but for times too short generally to be of biological interest. Several techniques have been proposed for simulations that extend to longer times. Stochastic dynamics does this, for example, by averaging over the high-frequency motions of the molecules, replacing them with a lumped frictional coefficient; the contribution of these modes is then less well represented for other purposes. For channels, the inherently small sample size implies that errors will be somewhat larger than usual in bulk simulations; repeating simulations, or taking longer runs, may be a way of dealing with this problem.

General references for detailed descriptions of the techniques are particularly useful: for molecular dynamics, especially for biological macromolecules, McCammon and Harvey[10] and Brooks *et al.*[11]; for Monte Carlo, the works edited by Binder.[14,32,45] Harvey[51] has reviewed electrostatic calculations for macromolecules, especially the nonlinear Poisson–Boltzmann equation.

Acknowledgment

I am grateful to Dr. Mihaly Mezei for reading the manuscript and providing many helpful comments.

[51] S. C. Harvey, *Proteins: Struct. Func. Genet.* **5,** 78 (1989).

[36] Kinetic Models and Simulation: Practical Approaches and Implementation Notes

By Vladimir Avdonin and Toshinori Hoshi

Introduction: Need for Kinetic Models

Kinetic models of ion channel gating are used to *predict* behavior of channels without actually performing the measurements. Some of the parameters simulated by the models may be conventionally measurable quantities, such as whole-cell currents, or they may even be parameters that are not routinely measurable, such as gating charge movements associated with a single channel. Because of their desired abilities to *predict* the channel behavior, kinetic models of ion channel gating have several different uses. Quantitative kinetic models provide insights into the underlying physical molecular processes involved in ion channel gating. Kinetic models allow quantitative comparisons of various conceptual models of ion channel gating by providing experimentally testable predictions. Simulation using kinetic models is also necessary to verify new analysis methods. Furthermore, models implemented on computers are valuable as a teaching tool in ion channel biophysics. Simulations using ion channel kinetic models critically test our understanding of how ion channel proteins function.

Markov versus Non-Markov Models

The majority of ion channel gating models developed so far makes some common underlying assumptions; ion channels have a small to moderate number of kinetically distinguishable states separated by large energy barriers and their gating behavior has one characteristic property called the Markov property with discrete states. Detailed descriptions of the Markov property can be found in standard stochastic process textbooks. Briefly, the Markov property implies that given the present condition, the future is not dependent on the past. If a channel has one open state, knowing how the channel entered the open state does not facilitate prediction of its future behavior. Furthermore, almost all gating models assume that the rate constants among the kinetic states do not change with time. The rate constant of leaving one state does not depend on how long the channel has stayed in the state (see later discussion for alternative models). In other words, we assume that ion channels demonstrate the Markov characteristic with time homogeneity property. We could describe the Markov process

0076-6879/98 $25.00

with time homogeneity property using the one-step transition probability matrix and the initial probability vector. Using the Markov property with time homogeneity assumption, it is possible to simulate the channel kinetic behavior with these two quantities. This approach can be implemented to simulate a variety of ion channel gating properties (see later discussion).[1]

Although the vast majority of the gating models has used the Markov assumption, there are alternatives. The main "alternative" is to use the fractal theory.[2] This method was in part motivated by the finding that some proteins may have a large number of conformational states or energy minimia. In the fractal approach, the transitions between different energy minima are described by the quantity $k = At^{(1-D)}$, where k is the effective rate per unit time, A is kinetic set point, D is fractal dimension, and t is time. One interesting feature of the fractal approach is that unlike the Markov models with time homogeneity assumption, the transition rates per unit time could be dependent on time when D is not 1.

As stated earlier, almost all the gating models are Markov models with time homogeneity assumption in part because of the readily available physical interpretations. It is possible that other alternatives will be explored in the future but it is quite likely that the Markov models will remain as the mainstream models. Interesting exchanges between Markov and fractal model proponents can be found in Refs. 3–5.

Different Philosophies in Developing Kinetic Models

As with other proteins, amino acid sequences of ion channel proteins should in principle markedly constrain their kinetic behavior. In the long distant future, it may become possible to develop a kinetic model purely based on the ion channel amino acid sequence. Our current understandings of how ion channel proteins function, however, do not allow us to predict ion channel gating based on the amino acid sequences alone. As stated earlier, we develop kinetic models to gain additional insights into the principles of ion channel function. Toward this goal, there are at least two major approaches in development of channel kinetic models using the Markov property assumption. We will call them (1) statistical approach and (2) heuristic approach.

[1] D. Colquhoun and A. G. Hawkes, *in* "Single-Channel Recording" (B. Sakmann and E. Neher, eds.), pp. 589–633. Plenum, New York, 1995.

[2] L. S. Liebovitch, *Math. Biosci.* **93,** 97 (1989).

[3] L. S. Liebovitch, *Biophys. J.* **55,** 373 (1989).

[4] R. Horn and S. J. Korn, *Biophys. J.* **55,** 379 (1989).

[5] O. B. McManus, C. E. Spivak, A. L. Blatz, D. S. Weiss, and K. L. Magleby, *Biophys. J.* **55,** 383 (1989).

In the statistical approach, only those kinetic parameters that are shown to be statistically significant are included. No more and no less. The statistically simplest model is favored unless there is a statistically demonstrated clear need for including more parameters. This approach is most frequently used with single-channel data where specific statistical tests exist to determine whether to include additional model parameters. For example, if only one exponential component is statistically sufficient in the closed dwell times, as decided by a statistical test such as the likelihood ratio test, only one closed state is included in the model. One of the major advantages of this statistical approach is that there are quantitative criteria for excluding additional model parameters. However, the statistical approach does have some serious disadvantages. Purely statistical models may not always give investigators conceptual insights into the channel function. For example, if the statistical analysis of the data obtained indicates that there are three distinct kinetic states, what does it mean in terms of the protein structure and function?

The heuristic approach, on the other hand, tries to incorporate the experimentally measured information as well as some statistically nonmeasurable information into the model development. For example, if one knows that there are four identical, indistinguishable, and independent subunits in a channel protein complex, it may be useful to incorporate this information even though kinetic models with a smaller number of states may be sufficient to describe the data. In this example, if one further supposes that each subunit has two conformations (O: open and C: closed), then the model shown in Diagram 1 can be constructed.

The rate constant of transition from C_4 to C_3 is four times α because there are four subunits that can make the transition. The rate constant of transition from C_3 to C_2 is three times α because there are three subunits that can make the transition. The concepts behind the example model shown here are essentially the same as those implied by Hodgkin and

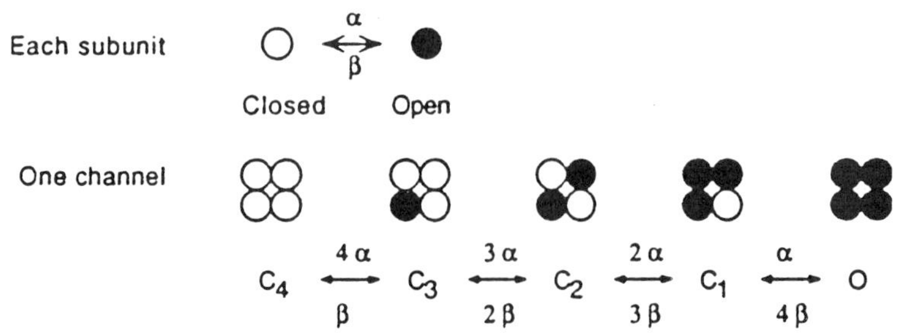

DIAGRAM 1

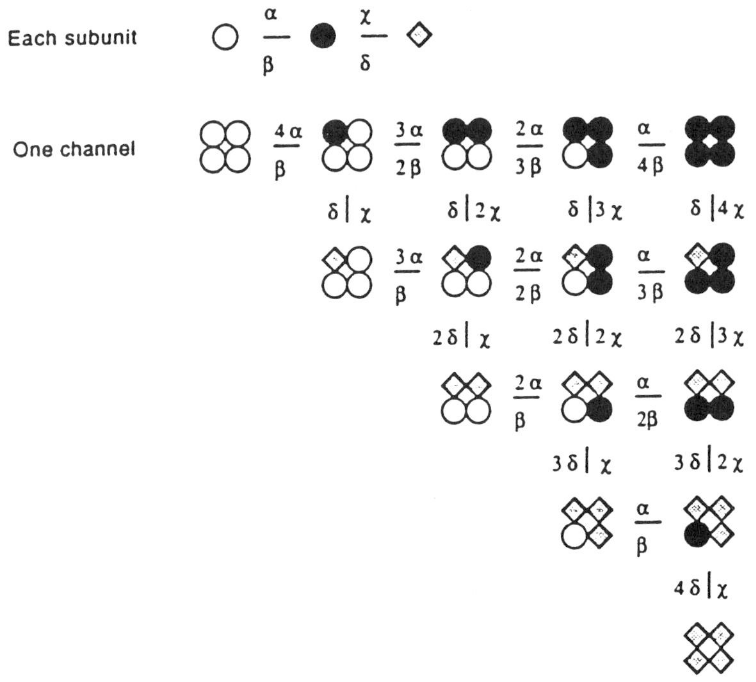

DIAGRAM 2

Huxley for the squid Na^+ and K^+ currents.[6] Similarly, if one assumes that three are four independent and identical subunits in a channel, each of which has three conformations. (C_0: $closed_0$, C_1: $closed_1$, and O: open), and that the channel is open only when all four subunits are open, the model shown in Diagram 2 is possible (see Ref. 7).

One special case of the heuristic approach is an allosteric approach to kinetic model development. This approach is largely derived from the oxygen-binding work on hemoglobin function.[8] Allosteric models of ion channel kinetics can be considered as heuristic models because many of the kinetic states in the allosteric models are not directly/statistically substantiated by the experimentally measured data. If open probability of a channel is high in the presence of some agonist, it is possible to propose a simple sequential model as shown:

[6] A. L. Hodgkin and A. F. Huxley, *J. Physiol.* **117,** 500 (1952).
[7] W. N. Zagotta, T. Hoshi, and R. W. Aldrich, *J. Gen. Physiol.* **103,** 321 (1994).
[8] J. Monod, J. Wyman, and J.-P. Changeux, *J. Mol. Biol.* **12,** 88 (1965).

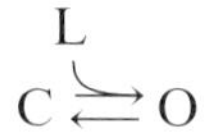

Alternatively, assuming the symmetry of protein structure–function, it is natural to propose the allosteric model shown in Diagram 3, which implies that the channel opening is promoted by this agonist (but it does not necessarily require it).

One of the major disadvantages of the heuristic approach is that in many cases those heuristic models are not the simplest models required to describe the measured data and they may contain many parameters that are not well constrained by the measured data. Because they tend to contain more free parameters than those developed using the statistical approach, simulations based on the heuristic models could be more computationally intensive. Although with some simplifying assumptions such as the channel complex having identical and independent subunits, the computational cost could be greatly reduced (see earlier discussion).

Thus, the statistical approach offers statistical simplicity with model parameters well constrained by the measured data at a possible expense of not having conceptual insights and/or of having to construct models for each set of measured data. The heuristic approach offers conceptual elegance at the possible expense of having parameters not well constrained by the measured data. Although each approach has its own advantages and disadvantages, the current trend is toward using the heuristic approach in part because of the strong desire to include the increasing amount of structural information becoming available. The model constructed can then be used as a framework to analyze the mutant channel data. Exactly which approach one adopts depends on the *scope* of the model.

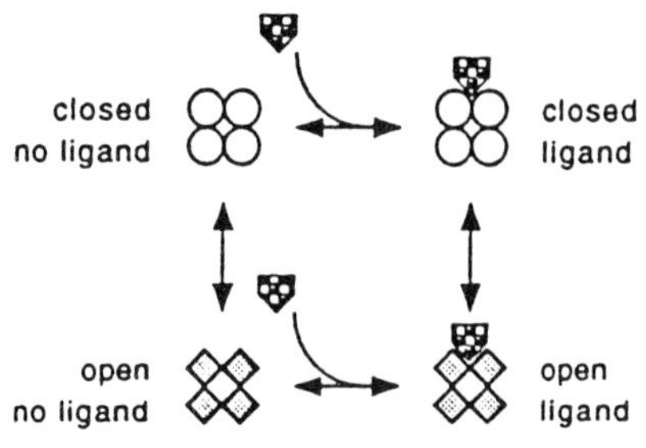

DIAGRAM 3

Scopes of Kinetic Models

One of the most important steps in channel kinetic model development is to define the *scope* of the model to be developed. In some cases, kinetic models with limited scopes are developed with specific applications in mind. For example, although we now know that the Hodgkin and Huxley models of Na^+ and K^+ currents do not accurately reproduce every aspect of the channel gating behavior, they may be perfectly fine to predict how neurons fire action potentials.

Empirical Data Required for Model Development

One of the old sayings in data analysis, "garbage in, garbage out," certainly holds true in ion channel model development. Development of ion channel gating models requires high-quality electrophysiological data. Some of the important considerations are listed here, and detailed technical information can be found elsewhere (e.g., Ref. 9).

Only One Class of Channels

As obvious as this criterion may be, this requirement that the data are to be collected from one channel type, is not easily met. Uncertainty about the number of different types of channels present in a given preparation is a serious issue that needs to be dealt with before developing kinetic models. For example, if the macroscopic current time course is described by a sum of two exponentials, this could mean that one class of channels of three states (closed and open) underlies the macroscopic current or that two channels types, each with two states, could underlie the macroscopic current.

In the single-channel recordings, the presence of only two unitary current amplitudes, zero and nonzero, is usually taken as a good sign that only one channel type is present. However, two channel types may have similar conductance states. Conversely, unequal conductance amplitude levels do not necessarily guarantee that the observed channel openings come from two channel types. It is now clear that one channel may have multiple conductance states, and their occurrences may be regulated. Although the presence of multiple conductance states may appear to be a complicating factor, especially for idealization, the multiple conductance states could be used as signatures of different kinetic states to facilitate analysis.

[9] B. Sakmann and E. Neher, "Single-Channel Recording." Plenum, New York, 1995.

In heterologous expression systems, the chance of obtaining data from one class of ion channels is probably higher than in native cells. However, most expression systems do contain "endogenous" channels that could be mistaken as heterologously expressed channels. This danger is quite high especially when the mutant channels which express at a low level are masked by endogenous currents.[10] One needs to be particularly careful when the channels under investigation form heteromultimeric channels with the endogenous channel subunits.[11]

One Class of Channels But Different Modulatory States

Even when it is reasonably certain that only one class of channels is present based on the protein primary sequence, these channels may be differentially posttranslationally modulated and the measured electrophysiological properties may represent a mixture of functionally distinct classes of channels. Many different ways to posttranslationally modify protein function have been documented (serine/threonine phosphorylation, pH regulation, cysteine oxidation, and tyrosine phosphorylation to name a few examples) and other regulatory mechanisms are also being discovered.

One extreme example is the "modal gating" behavior found in many different ion channels.[12–14] In some of these cases, it is relatively easy to identify different patterns of gating and they could be analyzed separately. Undoubtedly, however, there must be other cases where the channels with different posttranslational modification states contribute to the measured electrophysiological data, increasing the variability associated with the kinetic parameters and/or making the kinetic parameters nonstationary. The source of variability in the kinetic parameters could also be inferred by simulation (e.g., Refs. 15 and 16).

Voltage-Clamp Fidelity

Voltage clamp is conceptually a very simple and elegant way to study ion channels. However, none of the voltage-clamp implementations are perfect and one needs to minimize the artifacts and know how the voltage-clamp characteristics, such as voltage settling time, space clamp, and series resistance error, affected the measured data. For example, sigmoidal delay

[10] T. Tzounopoulos, M. Maylie, and J. P. Adelman, *Biophys. J.* **69,** 904 (1995).
[11] K. E. Hedin, N. F. Lim, and D. E. Clapham, *Neuron* **16,** 423 (1996).
[12] P. Hess, J. B. Lansman, and R. W. Tsien, *Nature* **311,** 538 (1984).
[13] A. L. Blatz and K. L. Magleby, *J. Physiol.* 141 (1986).
[14] J. B. Patlak and M. Ortiz, *J. Gen. Physiol.* **87,** 305 (1986).
[15] R. Horn, *Biophys. J.* **51,** 255 (1987).
[16] T. Hoshi, W. N. Zagotta, and R. W. Aldrich, *J. Gen. Physiol.* **103,** 249 (1994).

in the activation time course could be accounted for by assuming multiple closed states in the activation pathway and/or poor voltage-clamp settling time and poor space clamp.

Estimating Number of Required Kinetic States

Once the required high-fidelity electrophysiological data are collected, one can determine the number of kinetically distinguishable states that the channel protein may assume. Based on the macroscopic data, it may be possible to estimate the minimum number of kinetically distinguishable states that the underlying channel can enter. If only one class of channels with N states is responsible for the measured data, the time course of the macroscopic current data should in theory contain up to N-1 exponential components. In practice, however, not all the exponential components can be observed. Some of the time constants may be too close in values to be recognized as two separate components. Relative amplitudes of some components may be too small to be detected, depending on the recording condition (e.g., voltage). It is important to note that all rate constants, both forward and backward rate constants, contribute to the time constants detected in macroscopic currents unless the recording conditions are arranged such that some of them become negligible (see later discussion).

In single-channel data, both amplitude and duration information will further constrain the number of states to be included in the model. If a channel has n kinetically distinguishable open states, n exponential components in the open durations are expected. Similarly, if the channel has m kinetically distinguishable closed states, m exponential components in the closed durations are expected. To determine the number of exponential components present in a data set, one first fits the duration data with a simple negative exponential density function [$f(x) = \tau e^{-\tau x}$] using the maximum likelihood method with appropriate corrections for left and right censors (to account for the limited bandwidth and limited recording durations). With the simple negative density function, the only free parameter is the time constant. The number of exponential components is then increased by one, adding two more free parameters (one additional time constant and one relative amplitude term), and the new maximum likelihood value is recalculated. Whether to include the additional component is determined by the likelihood ratio statistic D, where $D = -2(\log LL_{n+1} - \log LL_n)$. Log LL_n refers to the natural logarithm of the maximum likelihood value estimated with n components and log LL_{n+1} refers to the natural logarithm of the maximum likelihood value estimated with n+1 components. Note that two additional free parameters are added for each additional exponential component. The distribution of D is approximated by the chi-square

TABLE I

Measured data	Exponential components	States
Macroscopic ionic current	N	$N + 1$ (combinations of closed and open states)
Closed durations	m	m closed states
Open durations	n	n open states

(χ^2) distribution. The decision to include one additional exponential component can be made by consulting the χ^2 distribution table with two degrees of freedom (for two additional parameters). Detailed procedures involved in fitting the duration data can be found elsewhere.[17]

Voltage and Concentration Jumps to Isolate Specific Rate Constants

As we gain more information into channel function, the perceived complexity of kinetic models increases. At one time, three-state models were considered to be complex. Nowadays, models with 100 states are sometimes considered. Although the actual number of free parameters in the models with a large number of states may not necessarily be very large because of some simplifying assumptions (see earlier discussion), one issue that model builders have to face is whether the rate constants are well constrained by the measured data.

Different electrophysiological measurements are suited for estimating kinetics rate constants in different parts of gating models. In general, single-channel data are most useful for obtaining information near the open states. Gating charge data are often useful to estimate the rate constant values away from the open states among the closed states.

Macroscopic currents in theory reflect contributions from all the rate constants involved although their relative contributions differ depending on the measurement conditions, such as voltage. By performing voltage jump experiments, it is possible to isolate functionally a subset of the rate constants involved in the model to better constrain the rate constant values (e.g., Ref. 18). For example, in the channels activated by depolarization, the macroscopic current time course at *very* depolarized voltages is primarily determined by the forward (opening) rate constants whereas at *very* negative voltages the deactivation time course is determined mostly by the

[17] D. Colquhoun and F. J. Sigworth, *in* "Single-Channel Recording" (B. Sakmann and E. Neher, eds.), pp. 483–587. Plenum, New York, 1995.

[18] W. N. Zagotta, T. Hoshi, J. Dittman, and R. W. Aldrich, *J. Gen. Physiol.* **103,** 279 (1994).

backward (closing) rate constants. Two large depolarizing pulses separated by a very brief interval could be used to estimate the last opening transition rate. In addition to the mean ionic current time course, ionic current noise measurements could be used to estimate the elementary charge movements (e.g., Ref. 19).

Some Formulations Useful in Simulation

The most fundamental way to simulate ion channel behavior is to simulate the time series behavior of one channel at a time. In theory, if one needs to simulate macroscopic electrophysiological properties, such as mean time course and variance properties of ionic and gating currents, this could be derived from simulation of the ensembles of channels, albeit this is not the fastest method for many applications (see later discussion). To simulate the time-series single-channel behavior, one needs to know the rate constant values in the model and the conductance values of the states involved.

The dwell time (d) for any state could be simulated by

$$d = -(1/\text{sum of all leaving rate constants}) * \ln(\text{RND})$$

where ln is the natural log operator and RND is a uniformly distributed random number. The state that the channel is going to enter next is simply described by

$$\Pr(M) = k_M/(\text{sum of all leaving rate constants})$$

where $\Pr(M)$ means the probability of the channel entering state M and k_M represents the rate constant of entering state M.

Thus, for the following three-state model,

$$0 \underset{k10}{\overset{k01}{\Leftrightarrow}} 1 \underset{k21}{\overset{k12}{\Leftrightarrow}} 2$$

once the channel enters state 1, the following sequence of events takes place:

1. The channel is to stay in state 1 for $-[1/(k_{10} + k_{12})]*\ln(\text{RND})$.
2. The probability that the channel enters state 0 is given by $k_{10}/(k_{10} + k_{12})$. The probability that the channel enters state 2 is given by $1 - [k_{10}/(k_{10} + k_{12})]$ or $k_{12}/(k_{10} + k_{12})$.
3. Thus, the channel "consults" a uniformly distributed random number generator (RND) and if the number is less than $k_{10}/(k_{10} + k_{12})$ then it enters state 0. Otherwise, it enters state 2. The dwell times in the respective states are determined as in 1.

[19] D. Sigg, E. Stefani, and F. Bezanilla, *Science* **264,** 578 (1994).

In implementing the single-channel time series simulation, it is advisable to evaluate different pseudo-random number generators. Although in many cases, the choice of pseudo-random number generation algorithms does not matter, if one is to verify new analysis protocols, it is quite conceivable that the randomness of the pseudo-random number generators becomes crucial. Different pseudo-random number generators are discussed in Ref. 20.

Comparison of Experimental Data and Model Predictions

Increased computational capabilities now allow investigators to visualize various model predictions in a semi real-time manner. One can change the model parameters and evaluate how the changes affect the predicted data. Although quantitative evaluations of channel gating models are available, we consider manual manipulations of the model parameters and visualizations of the data to be a very integral, if not the most important, part in evaluation of different gating models. These manipulations and visualizations give investigators a good "feel" about the general properties of the models being investigated.

In theory, it is trivial to compare the measured data and the data predicted/simulated by the kinetic models. One could simply superimpose both sets of data and visually inspect them. If required, more quantitative tests of goodness/badness of fits could be performed. However, in practice, the situation is much more complicated. To compare the measured data and the data simulated by the model properly, precise laboratory instrumentation characteristics must be taken into account. Ideally, the data simulated by kinetic models should be processed in the same way that the measured data are processed. Because the electrophysiological signals are usually low-pass filtered through an analog filter and digitized before analysis, it may be advisable for the simulated data to be processed in the same manner or at least the simulated data need to be "corrected" for the low-pass filtering. The effects of low-pass filtering represent an important and difficult issue, especially when evaluating single-channel data (see Ref. 17).

To discriminate more statistically among competing models, two major types of statistical tests have been used depending on the properties of the models being compared. Both of these weigh the benefits of producing better fits with more parameters and the expense of having more free parameters to help investigators develop parsimonious models (see Ref.

[20] W. H. Press, B. P. Flannery, S. A. Teukolsky, and W. T. Vetterling, "Numerical Recipes in C," pp. 1–1020. Cambridge University Press, New York, 1993.

21 for a review). If one is to test between two models, one of which is a subset of the other (i.e., nested models are being compared), the standard likelihood ratio test may be used. For example, if one is to compare the two-state model versus the linear three-state model, the likelihood ratio test is suited. The maximum likelihood values produced by the two competing models are compared using the χ^2 distribution. For nonnested models, AIC (Akaike information criterion or asymptotic information criterion) may be used to rank order the competing models. Because AIC is a rank-order test, it does not specifically say how much more likely one model is over others. More detailed discussions of model discrimination can be found in Refs. 4 and 15.

Weighing Different Measurements

Simulated data are compared with the measured electrophysiological data to examine the validity of the model. Ideally it is desirable to fit different electrophysiological parameters using a single set of parameters. For example, a model with the same set of parameters should be able to predict all of the measured parameters, such as macroscopic ionic currents, macroscopic gating currents, and single-channel currents. The fit errors in these measurements are used to evaluate different models and to optimize the model parameters. One nagging question is how to weigh the fit errors in different measurements. For example, if model A fits the macroscopic currents well but not the single-channel data, and model B fits the single-channel data but not the macroscopic current data, which model should be adopted? This is currently an unresolved issue.

Future Developments

Electrophysiological data are good but indirect footprints of conformational changes of ion channel proteins. Increasingly, standard electrophysiological data, such as macroscopic ionic and gating currents and single-channel ionic currents, are augmented by other promising measurements. For example, fluorescent measurements have been used to infer conformational changes in ion channel proteins.[22] As stated earlier, it will be necessary to weight different measurements in a statistically satisfactory manner to develop gating models.

[21] W. E. Larimore and R. K. Mehra, *Byte*, p. 167 (1985).
[22] L. M. Mannuzzu, M. M. Moronne, and E. Y. Isacoff, *Science* **271,** 213 (1996).

Implementation Notes

Software Issues

Computers have become an essential laboratory tool and their true usefulness is determined mostly by the software available. In kinetic simulation, one has two major choices: to use existing software or develop custom software. Using third-party software saves time and effort, and allows the investigator to concentrate on the scientific problems. Some commercially available programs for simulation of ion channels are *Axon Engineer for Windows*, simulation module for *Pulse/PulseFit* for Windows and Macintosh. These programs might suit the "average" need, but no software is perfect. There may be some features that are desirable but not available, whether a calculation algorithm or user interface. The alternative make-it-yourself approach might be even less acceptable, because it requires additional computer skills. Although a few notable attempts have been made to produce easy-to-use but flexible application environments using iconic protocols (e.g., LabView, Extend), that "perfect, end-all" software is yet to marketed. A fast scientific graphing/data acquisition/analysis environment with database capabilities, which can operate in the traditional command/text-based mode as well as in a graphic iconic mode, will certainly be useful in the future and it may even be commercially successful if implemented and marketed correctly.

We describe a simulation package developed in our laboratory that might offer a reasonable compromise for those who wish to try computer simulation in their work but do not want to be involved in programming issues. This simulation package is implemented in IgorPro (WaveMetrics, Lake Oswego, OR). Although this package is complete and offers the main features of a simulation system as well as a relatively simple graphical interface, the source code itself is text based, unlike some application environments which utilize more visual iconic languages.

The main advantages of our approach are based on the advantages of IgorPro itself. This program offers a rich set of tools for a wide range of electrophysiological data processing, including acquisition, processing, and preparation of publication-quality presentations and the program is very fast in displaying graphs involving many data points. We find it very convenient to have raw data, measurements, and simulated data within the same program, minimizing problems arising from interactions (or a lack thereof) among separate programs.

The source code described can be used "out of the box" in that once it is loaded in IgorPro the user can enter his or her model and run simulations of various experimental conditions. The user can readily modify this simulation package to add new features. The source code for the package is given in the Appendix and it is also available at http://hoshi-o.physiology.uiowa.edu. Those readers who do not own IGORPRO v3.x can try this package by means of a free demo that is available for downloading from http://www.wavemetrics.com, as well as from any INFOMAC archive mirror. As of this writing, IGORPRO was available only for Macintosh OS computers, but the Windows version is expected. Note that the procedure for finding eigenvalues is implemented as an external operation and currently runs only on Power Macintosh.

Theory of Operation

The program described next implements ion channel gating simulation using the Markov property with time homogeneity assumption. Each rate constant is represented by

$$k_{ij}(V,T) = K_{ij}e^{z_{ij}(eV/kT)}$$

where K_{ij} is the rate constant at 0 mV and z_{ij} is number of equivalent charges moving with the transition from state i to state j; e is elementary charge, k is Boltzmann constant, V is the transmembrane voltage, and T is absolute temperature. It is possible to modify the program if one wishes to include the higher order voltage term to account for the induced dipole effect.[23] Ligand/drug binding to a state is implemented using the pseudo-second-order reaction in the following fashion:

$$k_{ij} = K_{ij}[\mathrm{L}]$$

where [L] is the ligand concentration, and K_{ij} is the ligand-binding rate constant.

Macroscopic currents are calculated using the following standard formulation:

$$I(V, t) = N(V - V_{\mathrm{rev}})\ \Sigma G_i p_i(V, t)$$

where N is the number of channels, V_{rev} is the reversal potential for the ions permeating the channel, G_i are state conductance values, and $p_i(V,t)$ are the state occupancy probabilities calculated using the **Q**-matrix approach as described in Ref. 1.

[23] E. Neher and C. F. Stevens, *in* "The Neurosciences; Fourth Study Program (F. O. Schmitt and F. G. Worden, eds.), pp. 623–629. The M.I.T. Press, Cambridge, 1979.

Model Creation

In IGORPRO, each data sweep is called a *wave*. The waves can be multidimensional to hold matrices, as used in this simulation package. These waves are organized in a hierarchical fashion using wave folders. Use of these folders makes it possible to work with several different kinetic models concurrently, simplifying model comparisons. The name of the folder containing the model data becomes the name of the model and is present on title bars of all the windows produced by the model.

To create a model, the user first creates a folder with a descriptive name, and then selects the *Initialize model* item in the *Macros* menu. The programs invokes a dialog window to ask for the number of states in the new model. The number of states in the model can be changed later if so desired without affecting the preexisting parameters. It is also possible to clone a model by creating a copy of the folder containing the model and reinitializing the model within the new folder.

The model is represented by its **Q** matrix,[1] but the final form of this matrix corresponding to different experimental conditions is created only when simulations are performed. The user specifies the parameter values describing the model properties in four different spread-sheet-like table windows in IGORPRO: *Rate constants* window, *Voltage dependence* window, *Drug dependence* window, and *Unitary conductance* window.

Cells in the *Rate constants*, *Voltage dependence*, and *Drug dependence* tables describe different aspects of the transitions between the corresponding states of the kinetic model. The row index corresponds to the state from which the transition originates, and the column index corresponds to the destination state.

The *Rate constants* table contains the rate constant values of the model. For the transitions that do not depend on voltage or agonist, these values represent the rate constants in sec^{-1}. For the voltage-dependent transitions, the values in this table represent the rate constants at 0 mV and they are voltage dependent according to the equivalent charges entered in the *Voltage dependence* table. The agonist-dependent rate constants are calculated by multiplying the values in the *Rate constants* table by the agonist concentration during simulation. To indicate whether a particular transition is concentration dependent, the table *Drug dependence* is used. Those cells with nonzero values in this table make the corresponding transitions concentration dependent.

The *Rate constants* table uses a shortcut to simplify management of several rate constants having the same values. Only one such value needs to be specified in this table, other cells sharing this value could refer to the value by using a special reference code, constructed from row and column indices of reference cell. Each index should use exactly two digits. For example, −1001 refers to the cell in row 10, column 1; −1 (should be read as −0001) refers to cell in row 0 column 1. The fourth table has one column to specify the state unitary conductance values.

Running Simulation

Simulation is controlled within a separate sweep control window (see Fig. 1). The number of sweeps in the protocol defines the number of independent simulation runs. One sweep is divided into multiple segments with each segment being a period of simulated time having the same model parameters. Each segment is further characterized by voltage, duration, and agonist concentration. These segment values are incremented for each sweep by the value specified in the corresponding control entry.

The results of simulations are displayed in several different windows. The user has an option to display families of current traces, all or selected state occupancy traces, and several types of on-line analysis. On-line analysis is performed within the segment selected by the *Relevant segment* control. Note that the graphs are updated only if requested in the *Display current* checkbox. The *Go* button starts simulation with the currently selected parameters. The *Show Matrices* button brings up the model parameter tables.

Routines Descriptions

This simulation package uses two sets of global data. The data specifying the model are kept in the current default folder. These are three matrices; *Q_mx* for the rate constants, *Vdep* for the transition equivalent charges,

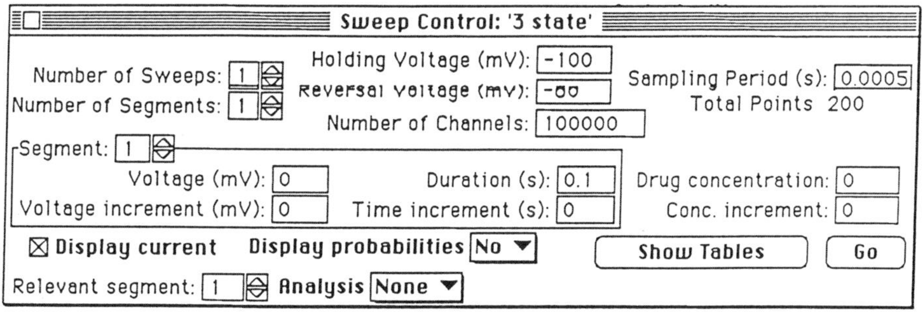

FIG. 1. Sweep control window.

Adep for denoting the agonist dependence of transitions, and one vector *Conduc* for the state conductance values. The data specifying the experimental conditions are kept in a folder *root:Packages:QModel* and they are common for all the models.

The source code is divided into three sections. The first section contains the **Q**-matrix related routines. Function *InitQ* constructs the **Q**-matrix for a given voltage and agonist concentration using the values from the three matrices specifying the model. Function *InitP0* computes the equilibrium state occupancies. These values are taken from the spectrum matrix of **Q** matrix, corresponding to the zero eigenvalues. Function *CompProb* computes the times courses of the state occupancies. It is provided by the initial probabilities, spectrum matrices of **Q** matrix, and eigenvalues. For the cases when only a subset of states is of interest, it takes parameter *idx*, which indexes the states for which probabilities is computed. The horizontal scaling of resulting wave *Pr* determines the time points to be evaluated. Function *CompCurrent* computes the ionic currents from the occupancy probabilities having nonzero state conductance values.

The main simulation code is located within the function *RunSimulation*. The simulation is performed inside two cycles, one for computing each sweep and an internal segment computation cycle. Each sweep computation starts with computation of the state occupancies for the holding voltage by calling the *InitP0* function. Then, the **Q** matrix is initialized for the first segment, and the spectral matrices and eigenvalues for this matrix are computed by a call to the *SpectrM* function. Time courses of the selected probabilities are evaluated. If the current segment is selected as the relevant segment, the on-line analysis is performed. Finally, the values of all states occupancies are computed at the end of each segment. These values are then used as the initial values for the next segment.

Function *SpectrM* computes the spectral matrices of **Q** matrix. It calls an external operation *MatrixEig* to find the eigenvalues and eigenvectors. Function *MatrixInv* finds the inverse matrix operation used in computing spectral matrices.

Conclusion

Many investigators face a hard choice between being scientists and software engineers. Even though the computational hardware has improved dramatically in recent years, much of the electrophysiological software in use today still has its roots in the days of 8-bit processors. A modern multipurpose scientific computing environment, capable of both text-based and icon-based operations, would benefit the field tremendously.

Appendix

```
#pragma  rtGlobals=1                  // Use modern global access method.

Menu  "Macros"
    "Initialize   model",   InitializeModelDialog()
End

// ===    Q-matrix routines   ===

// InitQ: Initialize Q-matrix for given parameters
// Parameters:
// Mx_name  -- String   -- name of the matrix for result
//          matrix will be created if not exists in current folder
// V        -- Variable -- Voltage in mV
// conc       -- Variable -- Ligand concentration, in units corresponding
//          to units of related rate constants in rate matrix
Function    InitQ(Mx_name,V,conc)
    string  Mx_name
    Variable  V,conc

    NVAR  Temperature  =  root:Packages:QModel:Temperature

    Wave/D Vdep = Vdep
    Wave/D Adep = Adep
    Wave/D Q_mx = Q_mx
    Variable   n=DimSize(Q_mx,0)
    Variable  i=0,  j,  k,l,sum

    MakeIfNotExistD(Mx_name,n,n)
    Wave/D Mx = $Mx_name
    Mx = Q_mx

    Mx = Mx *exp(Vdep * V / (Temperature + 273) * (1.6e-19/1.38e-23/1000))
    Mx *= (Adep[p][q]*conc + (Adep[p][q] == 0))

    i=0
    do
            j=0
            do
                    if  (Q_mx[i][j]<0)
                            l=abs(Q_mx[i][j])
                            k=trunc(l/100)
                            l=mod(l,100)
                            Mx[i][j]  =  Mx[k][l]
                    endif
                    j += 1
            while  (j<n)
            i += 1
    while  (i<n)

    i=0
    do
            j=0
            sum = 0
            do
                    sum += Mx[i][j]*(i != j)
                    j += 1
```

```
            while (j<n)
            Mx[i][i] = -sum
            i += 1
    while (i < n)

end

// InitP0: initilalize equilibrium state occupancies
// Parameters:
// P0_name  -- String   -- name of the wave for result
//          wave will be created if not exists in current folder
// V        -- Variable -- Voltage in mV
// conc     -- Variable -- Ligand concentration, in units corresponding
//          to units of related rate constants in rate matrix
Function InitP0(P0_name, V, conc)
    string P0_name
    Variable V,conc

    InitQ("Q_InitP0",V,conc)
    Wave/D Mx = Q_InitP0
    SpectrM("M_spec_P0",Mx)
    Wave/D M_spec = M_spec_P0
    Variable n=DimSize(Mx,0)
    MakeIfNotExistD(P0_name,n,0)
    Wave/D P0 = $P0_name
    P0 = M_spec[0][p][0]
    KillWaves /Z Mx, M_spec, $"M_spec_P0_val"

end

// CompProb: compute time course of selected state occupancies
//  Parameters:
//    Sp -- Wave -- Set of spectrum matricies of Q-matrix
//    EigVal -- Wave -- Vector of corresponding eigenvalues of Q-matrix
//    Pr0 -- Wave -- Vector of initial state occupancies
//    Pr -- Wave -- Resulting wave
//    idx -- Wave -- Vector with indexes of models states to compute
//  Notes:
//    Time is taken from resulting wave scaling.
Function    CompProb(Sp,EigVal,Pr0,Pr,idx)
    Wave/D Sp, EigVal, Pr0, Pr, idx

    Variable n=DimSize(Sp,0)
    Variable l = DimSize(Pr,0) // Number of points on T axis
    Variable k= numpnts(idx) // Number of states for computing
    Make /O /D /N=((n),(k)) Summ
    Variable i=0, j

    do
            Summ=  Sp[p][idx[q]][0]
            j = 1
            do
                    Summ  +=  Sp[p][idx[q]][j]*exp(EigVal[j]*pnt2x(Pr,i))
                    j += 1
            while (j < n)
            MatrixMultiply  Pr0/T, Summ
            Wave/D M_product = M_product
            Pr[i][] = M_product[q]
```

```
            i += 1
      while (i < l)
      KillWaves /Z M_product, Summ, Pr0_
 end

 // CompCurrent: compute ionic current time course
 //   Parameters:
 //     CSweep_Name -- String -- name of resulting wave
 //     Pr -- Wave -- 2D wave of state occupancies time course
 //     V -- Variable -- volateg (in mV)
 //     idx -- Wave -- 1D wave contains indexes of states for Pr wave
 Function    CompCurrent(CSweep_Name,Pr,V,idx)
      string  CSweep_Name
      wave/D Pr
      Variable  V
      wave  idx

      NVAR   NChannels  = root:Packages:QModel:NChannels
      Wave  Conduc = Conduc
      Variable  N = numpnts(Conduc)
      Variable  K = DimSize(Pr,0)
      MakeIfNotExist(CSweep_Name,K,0)
      Wave  CSweep  =  $CSweep_Name
      CopyScales  Pr,  CSweep
      SetScale  d  0,0,"A",  CSweep
      CSweep = 0
      Variable  i=0
      do
            if (Conduc[i] != 0)
                  CSweep   +=   Pr[p][idx[i]]*Conduc[i]
            endif
            i+=1
      while  (i<N)
      NVAR  Vrev  =  root:Packages:QModel:Vrev
      CSweep  *=  NChannels*(V-Vrev)/1000*1e-12

end

// RunSimulation:  main  event
Function    RunSimulation()

      DoWindow/T  SweepCtrPanel,"Sweep  Control:  "+GetDataFolder(0)
      Wave  ProbToDisplay  =  ProbToDisplay
      Wave  Conduc  =  Conduc
      Wave  SwInfo  =  root:Packages:QModel:SwInfo
      NVAR    Vhold   =  root:Packages:QModel:Holding
      NVAR   Drug   =  root:Packages:QModel:Drug
      NVAR   DrugInc   =  root:Packages:QModel:DrugInc
      NVAR   NSeg   =  root:Packages:QModel:NSeg
      NVAR   NSweeps   =  root:Packages:QModel:NSweeps
      NVAR    SweepSampling    =  root:Packages:QModel:SweepSampling
      NVAR   RelSeg   =  root:Packages:QModel:RelSeg
      NVAR    MainTauTime    =  root:Packages:QModel:MainTauTime
      NVAR   MainTauDur   =  root:Packages:QModel:MainTauDur

      ControlInfo   DispPPopup
      Variable  ShowP  =  V_value  -1
      ControlInfo  DispCBox
```

```
	Variable ShowC = V_value
	ControlInfo  AnPopup
	Variable ShowA = V_value - 1
	String AnMethod = S_value
	ControlInfo  RelAnBox
	if (V_Flag != 0)
		Variable RelAnal = V_value
	endif
	if ((!ShowC) %& (!ShowP) %& (!ShowA))
		abort "Nothing selected for display"
	endif

	Variable  N=DimSize(Q_mx,0)
	Variable i,k,l,m, NProb=0, NProbDisp=0
	Variable VSweep, Dur, SweepPts, SweepDrug, SegPts, SegV, SegDur, CPnt
	Make /D/O/N=0 ProbAndCurrIdx
	Make /D/O/N=(N) CurrentIdx, ProbIdx
	CurrentIdx = -1
	ProbIdx = -1

	i=0
	do
		if ((ShowP %& ProbToDisplay[i]) %|((ShowC %| (ShowA!=0)) %& (Conduc[i]
!= 0)))
			InsertPoints (NProb),1, ProbAndCurrIdx
			ProbAndCurrIdx[NProb]=i
			if ((ShowC %| (ShowA!=0)) %& (Conduc[i] != 0))
				CurrentIdx[i]=NProb
			endif
			if (ShowP %& ProbToDisplay[i])
				NProbDisp += 1
				ProbIdx[i] = NProb
			endif
			NProb+=1
		endif
		i+=1
	while (i < N)

	if (ShowA)
		Make/O/N=(NSweeps) AnalysisWave
		if (ShowA == 4)
			SetScale d 0,0,"s", AnalysisWave
		else
			SetScale d 0,0,"A", AnalysisWave
		endif
	endif

	Make /O/D/N=(N) AllProbIdx
	AllProbIdx = p
	String CSweepName="CurrentTrace", SwName
	Make/O/T/N=(NProbDisp) PSweepNames
	k=0
// ********** Main cycle *******************
	do
		SweepPts=TotalSweepPoints(k)
		SweepDrug = Drug + DrugInc*k
		if (ShowC)
			sprintf CSweepName, "CurrentTrace%d",k+1
```

```
                MakeIfNotExist(CSweepName,SweepPts,0)
                Wave CSweep = $CSweepName
                SetScale/P x 0,(SweepSampling),"s", CSweep
                SetScale d 0,0,"A", CSweep
        endif
        if (ShowP)
                i=0
                do
                        if (ProbToDisplay[i])
                                sprintf SwName,"ProbTrace%d_%d",i,k+1
                                MakeIfNotExist(SwName,SweepPts,0)
                                PSweepNames[i] = SwName
                                SetScale/P x 0,(SweepSampling),"s", $SwName
                        endif
                        i += 1
                while (i<N)
        endif

        InitP0("Pr0",Vhold,SweepDrug)
        Wave /D Pr0 = Pr0

// ******************** Segment Cycle **********************
        l=0
        CPnt = 0
        do
                SegV = SwInfo[l][%V]+SwInfo[l][%dV]*k
                SegDur = SwInfo[l][%T] + SwInfo[l][%dT]*k
                SegPts = SegDur/SweepSampling
                InitQ("Q_",SegV,SweepDrug)
                Wave Q_ = $"Q_"
                spectrM("Sp",Q_)
                Wave /D Sp = Sp
                Wave /D Sp_val = Sp_val
                MakeIfNotExistD("ProbSeg",(SegPts),(NProb))
                Wave/D ProbSeg = ProbSeg
                SetScale/P x 0,(SweepSampling),"s", ProbSeg

                CompProb(Sp,Sp_val,Pr0,ProbSeg,ProbAndCurrIdx)
                if (ShowC %| ((ShowA>0) %& (ShowA<4) %& ((l+1) == RelSeg)))
                        CompCurrent("SegCrnt",ProbSeg,SegV,CurrentIdx)
                        Wave SegCrnt = SegCrnt
                        if (ShowC)
                                CSweep[CPnt,(CPnt+SegPts)] = SegCrnt[p-CPnt]
                        endif
                        if ((ShowA>0) %& (ShowA<4) %& ((l+1) == RelSeg))
                                WaveStats/Q SegCrnt
                                if (ShowA == 1)
                                        i=V_max
                                else
                                        if (ShowA == 2)
                                                i=V_min
                                        else
                                                if (ShowA == 3)
                                                        i = V_avg
                                                endif
                                        endif
                                endif
                                AnalysisWave[k] = i
```

```
                        endif
                endif

                // Main tau case
                if ((ShowA == 4) %& ((l+1) == RelSeg))
                        AnalysisWave[k] = 1 /
ExpRate(SegCrnt,MainTauTime,MainTauDur)
                endif

                // Steady state case
                if ((ShowA == 5) %& ((l+1) == RelSeg))
                        InitP0("AStdStateP",SegV,SweepDrug)
                        Wave /D AStdStateP = AStdStateP
                        Redimension/N=(1,N)  AStdStateP
                        MakeIfNotExist("AStdStateIdx",N,0)
                        Wave AStdStateIdx = AStdStateIdx
                        AStdStateIdx = p
                        CompCurrent("AStdStateC",AStdStateP,SegV,AStdStateIdx)
                        Wave AStdStateC = AStdStateC
                        AnalysisWave[k] = AStdStateC[0]
                        Redimension/N=(N,1)  AStdStateP
                endif

                if (ShowP)
                        i = 0
                        do
                                if  (ProbToDisplay[i])
                                        SwName  =  PSweepNames[ProbIdx[i]]
                                        Wave ProbSegTr = $SwName
                                        ProbSegTr[CPnt,(CPnt+SegPts)]  =  ProbSeg[p-
CPnt][ProbIdx[i]]
                                endif
                                i += 1
                        while (i < N)
                endif

                MakeIfNotExistD("P0ProbSeg",2,N)
                Wave /D P0ProbSeg = P0ProbSeg
                SetScale  x  0,SegDur,"s",  P0ProbSeg
                CompProb(Sp,Sp_val,Pr0,P0ProbSeg,AllProbIdx)
                Pr0 = P0ProbSeg[1][p]

                CPnt += SegPts

                l+= 1
        while (l < NSeg)

                if (ShowP)
                        i = 0
                        do
                                if  (ProbToDisplay[i])
                                        SwName  =  PSweepNames[ProbIdx[i]]
                                        DispWave(SwName,"Prob","State
 Probability")

                                        SweepNote(SwName,k)
                                endif
                                i += 1
                        while (i < N)
```

```
				endif

			if (ShowC)
				DispWave(CSweepName,"Curr","Current")
				SweepNote(CSweepName,k)
			endif

			k += 1
		while (k < NSweeps)

		if (ShowA)
			if ((ShowA!=4) %& RelAnal)
				i = AnalysisWave[0]
				AnalysisWave /= i
				AnMethod = AnMethod + " relative"
			endif

			if (DrugInc != 0)
				SetScale/P x Drug,DrugInc,"M", AnalysisWave
			else
				if (SwInfo[RelSeg-1][%dV] != 0)
					SetScale/P x SwInfo[RelSeg-1][%V],SwInfo[RelSeg-
1][%dV],"mV", AnalysisWave
				else
					if (SwInfo[RelSeg-1][%dT] != 0)
						SetScale/P x SwInfo[RelSeg-1][%dT],SwInfo[RelSeg-
1][%dT],"s", AnalysisWave
					endif
				endif
			endif
			DispWave("AnalysisWave","Anal",AnMethod+"	analysis")
			SweepNote("AnalysisWave",-1)
		endif

		KillWaves/Z Q_, Sp, Sp_Val, Pr0, ProbSeg, P0ProbSeg
		KillWaves/Z CurrentIdx, ProbIdx, ProbAndCurrIdx
		KillWaves/Z AllProbIdx, PSweepNames, SegCrnt
		KillWaves/Z AStdStateP, AStdStateIdx, AStdStateC
	End

	// === Matrix computations ===

	// spectrM: compute spectrum matricies for Q-matrix
	//  Parameters:
	//    M_spec_name -- String -- base name for resulting matricies
	//    Q_mx -- Wave -- Q-matrix
	//  Notes:
	//    Builds 3D wave, slice of last dimension is spectrum matrix
	//     Corresponding eigenvalues are placed in a wave named M_spec_name+"_val"
	Function	SpectrM(M_spec_name,Q_mx)
		string M_spec_name
		wave/D Q_mx

		Variable n=DimSize(Q_mx,0),i
		execute "MatrixEig " + NameOfWave(Q_mx)
		Wave/D ME_val = ME_Val
		Wave/D ME_vec = ME_Vec
		Make /O/N=(n) SpSortIdx
```

```
        SpSortIdx = abs(ME_val)
        MakeIndex  SpSortIdx,SpSortIdx
        IndexSort  SpSortIdx,ME_val
        Duplicate  /O  ME_vec,tmp
        i=0
        do
                ME_vec[][i]  =  tmp[p][SpSortIdx[i]]
                i += 1
        while (i < n)

        MatrixInv("ME_vec_inv",ME_vec)
        Wave/D  ME_vec_inv  =  ME_vec_inv

        if  (!exists(M_spec_name))
                Make /D /O /N=((n),(n),(n)) $M_spec_name
        else
                Wave/D   M_spec=$M_spec_name
                if ((DimSize(M_spec,0)!=n)  %|  (DimSize(M_spec,1)!=n)  %|
(DimSize(M_spec,2)!=n))
                        Make /D /O /N=((n),(n),(n)) $M_spec_name
                endif
        endif

        Wave /D M_spec = $M_spec_name
        Make /D /O /N=(n) A_vec,B_vec
        i = 0
        do
                A_vec  =  ME_vec[p][i]
                B_vec  =  ME_vec_inv[i][p]
                MatrixMultiply    A_vec,B_vec/T
                Wave/D  M_product  =  M_product
                M_spec[][][i]  =  M_product[p][q]
                i += 1
        while  (i<n)
        KillWaves  /Z  M_product,  B_vec,  A_vec,  ME_vec_inv,  tmp,  SpSortIdx

        String  M_s_v_n  =  (M_spec_name+"_val")
        MakeIfNotExistD(M_s_v_n,n,0)
        Wave  M_spec_val  =  $M_s_v_n
        M_spec_val  =  ME_val
        KillWaves  /Z  ME_vec,  ME_val
    end

    // MatrixInv: computes inverse matrix
    //  Parameters:
    //    name -- String -- name of resulting matrix
    //           will be created or replace existing
    //    M_input -- Wave -- matrix to invert
    Function    MatrixInv(name,M_input)
        string  name
        Wave/D  M_input

        Variable    n=DimSize(M_input,0)

        MakeIfNotExistD(name,(n),(n))
        Wave/D w = $name
        w = 0
        Variable  i  =  0
```

```
    do
            w[i][i] = 1
            i += 1
    while (i<n)

    Wave/D w = $name
    MatrixSolve  SV,M_input,w
    Wave/D M_x = M_x
    Duplicate /D/O M_x,w
    KillWaves /Z M_x
end

// ===   Interface routines   ===

// InitializeModelDialog: ask for number of states in new model
Proc   InitializeModelDialog(N)
    Variable N=4
    prompt N,"Number of states:"
    PauseUpdate; Silent 1

    InitializeModel(N)
end

// InitializeModel: initialize new model in current folder
//   Parameters:
//     N -- variable -- number of states in model
Function   InitializeModel(N)
    Variable N=4

    NewDataFolder/O  root:Packages
    NewDataFolder/O  root:Packages:QModel

    MakeWithDup("Q_mx",N,N)
    MakeWithDup("Vdep",N,N)
    MakeWithDup("Adep",N,N)
    MakeWithDup("Conduc",N,0)

    Wave/D Q_mx = Q_mx
    Wave/D Vdep = Vdep
    Wave/D Adep = Adep
    Wave/D Conduc = Conduc

    ShowParameters()
    ModelGlobalVariable("root:Packages:QModel:Temperature",23)
    ModelGlobalVariable("root:Packages:QModel:Vrev",-80)
    ModelGlobalVariable("root:Packages:QModel:Holding",-100)
    ModelGlobalVariable("root:Packages:QModel:NSeg",1)
    ModelGlobalVariable("root:Packages:QModel:Drug",0)
    ModelGlobalVariable("root:Packages:QModel:CSeg",1)
    ModelGlobalVariable("root:Packages:QModel:RelSeg",1)
    ModelGlobalVariable("root:Packages:QModel:NSweeps",1)
    ModelGlobalVariable("root:Packages:QModel:SegDur",0.1)
    ModelGlobalVariable("root:Packages:QModel:SegV",0)
    ModelGlobalVariable("root:Packages:QModel:Seg_dV",0)
    ModelGlobalVariable("root:Packages:QModel:Seg_dT",0)
    ModelGlobalVariable("root:Packages:QModel:DrugInc",0)
    ModelGlobalVariable("root:Packages:QModel:SweepPts",200)
    ModelGlobalVariable("root:Packages:QModel:SweepSampling",.0005)
```

```
	ModelGlobalVariable("root:Packages:QModel:NChannels",100000)
	ModelGlobalVariable("root:Packages:QModel:MainTauTime",0.01)
	ModelGlobalVariable("root:Packages:QModel:MainTauDur",0.01)

	if (! exists("root:Packages:QModel:SwInfo"))
		Make/N=(1,4)  $"root:Packages:QModel:SwInfo"
	endif
	Wave SwInfo = root:Packages:QModel:SwInfo
	NVAR SegV = root:Packages:QModel:SegV
	NVAR SegDur = root:Packages:QModel:SegDur
	NVAR Seg_dV = root:Packages:QModel:Seg_dV
	NVAR Seg_dT = root:Packages:QModel:Seg_dT
	NVAR CSeg = root:Packages:QModel:CSeg

	SetDimLabel 1,0,V,SwInfo
	SetDimLabel 1,1,T,SwInfo
	SetDimLabel 1,2,dV,SwInfo
	SetDimLabel 1,3,dT,SwInfo
	SwInfo[CSeg-1][%V]=SegV
	SwInfo[CSeg-1][%T]=SegDur
	SwInfo[CSeg-1][%dV]=Seg_dV
	SwInfo[CSeg-1][%dT]=Seg_dT
	MakeWithDup("ProbToDisplay",N,0)
	Wave ProbToDisplay = ProbToDisplay

	if (strlen(WinList("SweepCtrPanel",";","WIN:64")) == 0)
		MakeSweepCtrPanel()
	else
		DoWindow/F SweepCtrPanel
		DoWindow/T SweepCtrPanel,"Sweep Control: "+GetDataFolder(0)
	endif
end

Function SetNSegProc(ctrlName,varNum,varStr,varName) : SetVariableControl
	String ctrlName
	Variable varNum
	String varStr
	String varName

	DoWindow/T SweepCtrPanel,"Sweep Control: "+GetDataFolder(0)
	Wave SwInfo = root:Packages:QModel:SwInfo
	Redimension/N=(VarNum,-1) SwInfo
	SetVariable SetCSeg limits={1,VarNum,1}
	SetVariable SetRelSeg limits={1,VarNum,1}
	NVAR CSeg = root:Packages:QModel:CSeg
	NVAR RelSeg = root:Packages:QModel:RelSeg
	if (VarNum<CSeg)
		CSeg = VarNum
		SetCSegProc("",VarNum,"","")
	endif
	if (VarNum<RelSeg)
		RelSeg = VarNum
	endif
	ControlUpdate DispTotP
End

Function SetCSegProc(ctrlName,varNum,varStr,varName) : SetVariableControl
	String ctrlName
```

```
    Variable  varNum
    String  varStr
    String  varName

    DoWindow/T  SweepCtrPanel,"Sweep  Control:  "+GetDataFolder(0)

    Wave  SwInfo  =  root:Packages:QModel:SwInfo
    NVAR   SegV   =  root:Packages:QModel:SegV
    NVAR   SegDur   =  root:Packages:QModel:SegDur
    NVAR   Seg_dV   =  root:Packages:QModel:Seg_dV
    NVAR   Seg_dT   =  root:Packages:QModel:Seg_dT
    SegV=SwInfo[varNum-1][%V]
    SegDur=SwInfo[varNum-1][%T]
    Seg_dV=SwInfo[varNum-1][%dV]
    Seg_dT=SwInfo[varNum-1][%dT]
End


Function  SetNSweepsProc(ctrlName,varNum,varStr,varName)  :  SetVariableControl
    String  ctrlName
    Variable  varNum
    String  varStr
    String  varName

    DoWindow/T  SweepCtrPanel,"Sweep  Control:  "+GetDataFolder(0)
    ControlUpdate  DispTotP
End

Function  SetSegVProc(ctrlName,varNum,varStr,varName)  :  SetVariableControl
    String  ctrlName
    Variable  varNum
    String  varStr
    String  varName

    DoWindow/T  SweepCtrPanel,"Sweep  Control:  "+GetDataFolder(0)
    NVAR   CSeg   =  root:Packages:QModel:CSeg
    Wave  SwInfo  =  root:Packages:QModel:SwInfo
    SwInfo[CSeg  -1][%V]  =  varNum

End

Function  SetDurProc(ctrlName,varNum,varStr,varName)  :  SetVariableControl
    String  ctrlName
    Variable  varNum
    String  varStr
    String  varName

    DoWindow/T  SweepCtrPanel,"Sweep  Control:  "+GetDataFolder(0)
    NVAR   CSeg   =  root:Packages:QModel:CSeg
    Wave  SwInfo  =  root:Packages:QModel:SwInfo
    SwInfo[CSeg  -1][%T]  =  varNum
    ControlUpdate  DispTotP

End

Function  SetVIncProc(ctrlName,varNum,varStr,varName)  :  SetVariableControl
    String  ctrlName
    Variable  varNum
```

```
    String varStr
    String varName

    DoWindow/T SweepCtrPanel,"Sweep Control: "+GetDataFolder(0)
    NVAR CSeg = root:Packages:QModel:CSeg
    Wave SwInfo = root:Packages:QModel:SwInfo
    SwInfo[CSeg -1][%dV] = varNum

End

Function SetTIncProc(ctrlName,varNum,varStr,varName) : SetVariableControl
    String ctrlName
    Variable varNum
    String varStr
    String varName

    DoWindow/T SweepCtrPanel,"Sweep Control: "+GetDataFolder(0)
    NVAR CSeg = root:Packages:QModel:CSeg
    Wave SwInfo = root:Packages:QModel:SwInfo
    SwInfo[CSeg -1][%dT] = varNum
    ControlUpdate DispTotP
End

Function SetSampProc(ctrlName,varNum,varStr,varName) : SetVariableControl
    String ctrlName
    Variable varNum
    String varStr
    String varName

    DoWindow/T SweepCtrPanel,"Sweep Control: "+GetDataFolder(0)
    ControlUpdate DispTotP
End

Function /S ProbMenuItems()

    String items="No;Yes"
    if (!exists("ProbToDisplay"))
            return items
    endif
    Wave ProbToDisplay = ProbToDisplay
    Variable N=numpnts(ProbToDisplay), i = 0
    do
            if (    ProbToDisplay[i])
                    items = items + ";\\M1!•"
            else
                    items = items + ";"
            endif
            items = items + num2istr(i)
            i += 1
    while (i<N)
    return items
end

Function DispPPopMenuProc(ctrlName,popNum,popStr) : PopupMenuControl
    String ctrlName
    Variable popNum
    String popStr
```

```
    DoWindow/T  SweepCtrPanel,"Sweep  Control:  "+GetDataFolder(0)
    Wave ProbToDisplay = ProbToDisplay
    if (popNum ==1)
        ProbToDisplay=0
    else
        if (popNum == 2)
            ProbToDisplay=1
        else
            ProbToDisplay[popNum-3]=   (!ProbToDisplay[popNum-3])
            PopupMenu  DispPPopup  mode=2
        endif
    endif

    PopupMenu  DispPPopup  value=ProbMenuItems()
End

Function  AnalMenuProc(ctrlName,popNum,popStr)  :  PopupMenuControl
    String  ctrlName
    Variable  popNum
    String  popStr

    ControlInfo  RelAnBox
    if (V_Flag != 0)
        KillControl  RelAnBox
    endif
    ControlInfo  setMTTime
    if (V_Flag != 0)
        KillControl  setMTTime
        KillControl  setMainTauDur
    endif
    if (popNum == 1)
        return 0
    endif
    if (popNum != 5)
        CheckBox  RelAnBox,pos={335,143},size={173,20},title="Relative  to  first
sweep",value=0
    endif
    if (popNum == 5)
            SetVariable     setMTTime,pos={322,144},size={89,17},title="Time:"
            SetVariable     setMTTime,limits={0,INF,0},value=
root:Packages:QModel:MainTauTime
            SetVariable     setMainTauDur,pos={412,144},size={80,17},title="Dur:"
            SetVariable     setMainTauDur,limits={0.0001,INF,0},value=
root:Packages:QModel:MainTauDur
    endif
End

Function  ShowMxButProc(ctrlName)  :  ButtonControl
    String  ctrlName

    DoWindow/T  SweepCtrPanel,"Sweep  Control:  "+GetDataFolder(0)
    ShowParameters()
End

Function  TotalSweepPoints(SweepN)
    Variable  SweepN

    if  (!exists("root:Packages:QModel:SwInfo"))
```

```
            return 0
      endif
      Wave SwInfo = root:Packages:QModel:SwInfo
      NVAR  NSeg  = root:Packages:QModel:NSeg
      NVAR  SweepSampling  = root:Packages:QModel:SweepSampling
      NVAR  NSweeps  = root:Packages:QModel:NSweeps

      if (SweepN < 0)
            SweepN = NSweeps
      endif

      Variable  i=0,dur=0
      do
            dur += SwInfo[i][%T] + SweepN*SwInfo[i][%dT]
            i += 1
      while (i < NSeg)
      return  dur/SweepSampling
end

Function  SweepNote(WaveName,  SweepN)
      string WaveName
      variable  SweepN

      Wave SwInfo = root:Packages:QModel:SwInfo
      NVAR  Vhold  = root:Packages:QModel:Holding
      NVAR Vrev = root:Packages:QModel:Vrev
      NVAR  Drug  = root:Packages:QModel:Drug
      NVAR  DrugInc  = root:Packages:QModel:DrugInc
      NVAR  RelSeg  = root:Packages:QModel:RelSeg

      Wave w = $WaveName
      string cmnt
      Variable  val
      Note/K w
      Note  w,GetDataFolder(0)
      sprintf cmnt,"Holding: %g mV; Reversal: %g mV",Vhold,Vrev
      Note  w,cmnt
      if (SweepN>=0)
            val = SwInfo[RelSeg-1][%V] + SwInfo[RelSeg-1][%dV]*SweepN
            sprintf  cmnt,"Voltage:  %g  mV",val
            Note  w,cmnt
            val = Drug + DrugInc*SweepN
            sprintf  cmnt,"Drug:  %g",val
            Note  w,cmnt
      endif
end

Function  GoButProc(ctrlName)  :  ButtonControl
      String  ctrlName
      RunSimulation()
End

// ShowParameters: brings to front windows with model parameters
Function  ShowParameters()
      Silent  1;  PauseUpdate
      if  (!exists("Conduc"))
            abort "There is no full defined model in this folder: use initialize"
      endif
```

```
	EditMx("Conduc","State	conductance")
	EditMx("Adep","Drug	dependence")
	EditMx("Vdep","Voltage	dependence")
	EditMx("Q_mx","Rate	constants")
	Arrange_Windows()
end

// EditMx: show wave in table and set window title
Function EditMx(wname, Title)
	string wname,Title

	string folder = CurrentFolder()
	Wave/D w=$wname
	CheckDisplayed/A w
	if (V_Flag == 0)
		Edit $wname
		DoWindow/C/T	$(wname+"_"+folder),Title+":"+GetDataFolder(0)
		execute "ModifyTable width=32"
	else
		string win= FindTableWithWave(w)
		DoWindow /F $win
		DoWindow/C/T	$win,Title+":"+GetDataFolder(0)
	endif
end

// DispWave: plot wave in new window or append to existing
Function DispWave(WaveName, WinName, Title)
	string WaveName, WinName, Title

	string folder = CurrentFolder()
	string TheWin = WinName+"_"+folder
	Wave/D w=$WaveName
	if (strlen(WinList(TheWin,";","WIN:1")) == 0)
		Display $WaveName
		DoWindow/C $TheWin
	else
		DoWindow /F $TheWin
		CheckDisplayed/W=$TheWin w
		if (V_Flag == 0)
			AppendToGraph w
		endif
	endif
	DoWindow/T	$TheWin,Title+":	"+GetDataFolder(0)
end

// Arrange_Windows: arrange windows for current model
Function Arrange_Windows()
	WAVE Q_mx = Q_mx
	WAVE Vdep = Vdep
	WAVE Adep = Adep
	WAVE Conduc = Conduc
	string wlist = FindTableWithWave(Q_mx) + ","
	wlist += FindTableWithWave(Vdep) + ","
	wlist += FindTableWithWave(Adep) + ","
	wlist += FindTableWithWave(Conduc)

	if (strlen(wlist) == 0)
		return 0
```

```
        endif
        execute  "TileWindows/W=(6,24,624,444)  "+wlist
    end

    // MakeSweepCtrPanel: Macro creating main control panel
    //          called only once, during initialization
    Function MakeSweepCtrPanel() : Panel
        PauseUpdate; Silent 1         | building window...

        NewPanel /W=(2,314,566,477) as "Sweep Control: "+GetDataFolder(0)
        DoWindow/C  SweepCtrPanel
        SetDrawLayer UserBack
        DrawRect  3,69,381,117
        SetVariable  SetHolding,pos={187,5},size={187,17},title="Holding  Voltage  (mV):"
        SetVariable   SetHolding,limits={-INF,INF,0},value=   root:Packages:QModel:Holding
        SetVariable  SetVrev,pos={179,25},size={195,17},title="Reversal  Voltage  (mV):"
        SetVariable   SetVrev,limits={-INF,INF,0},value=   root:Packages:QModel:Vrev
        SetVariable  SetNchan,pos={197,45},size={199,17},title="Number  of  Channels:"
        SetVariable   SetNchan,limits={1,INF,0},value=   root:Packages:QModel:NChannels
        SetVariable
SetSamp,pos={383,16},size={179,17},proc=SetSampProc,title="Sampling   Period   (s):"
        SetVariable    SetSamp,limits={1e-05,1,0},value=
root:Packages:QModel:SweepSampling
        ValDisplay   DispTotP,pos={423,33},size={138,17},title="Total   Points",format="%d"
        ValDisplay    DispTotP,frame=0,limits={0,0,0},barmisc={0,1000}
        ValDisplay  DispTotP,value=  #"TotalSweepPoints(-1)"

        SetVariable
SetNSweeps,pos={15,15},size={155,17},proc=SetNSweepsProc,title="Number   of   Sweeps:"
        SetVariable   SetNSweeps,format="%d"
        SetVariable   SetNSweeps,limits={1,20,1},value=   root:Packages:QModel:NSweeps
        SetVariable   SetNSeg,pos={2,35},size={168,17},proc=SetNSegProc,title="Number    of
Segments:"
        SetVariable     SetNSeg,format="%d",limits={1,10,1},value=
root:Packages:QModel:NSeg
        SetVariable      SetCSeg,pos={7,61},size={97,17},proc=SetCSegProc,title="Segment:"
        SetVariable   SetCSeg,limits={1,1,1},value=   root:Packages:QModel:CSeg

        SetVariable      SetSegV,pos={75,79},size={125,17},proc=SetSegVProc,title="Voltage
(mV):"
        SetVariable   SetSegV,limits={-INF,INF,0},value=   root:Packages:QModel:SegV
        SetVariable      SetVInc,pos={8,97},size={192,17},proc=SetVIncProc,title="Voltage
increment  (mV):"
        SetVariable   SetVInc,limits={-INF,INF,0},value=   root:Packages:QModel:Seg_dV
        SetVariable   SetDur,pos={259,79},size={119,17},proc=SetDurProc,title="Duration   (s):"
        SetVariable   SetDur,limits={0.0001,INF,0},value=   root:Packages:QModel:SegDur
        SetVariable      SetTInc,pos={214,97},size={164,17},proc=SetTIncProc,title="Time
increment  (s):"
        SetVariable   SetTInc,limits={0,INF,0},value=   root:Packages:QModel:Seg_dT
        SetVariable   SetDrug,pos={390,79},size={166,17},title="Drug   concentration:"
        SetVariable   SetDrug,limits={0,INF,0},value=  root:Packages:QModel:Drug
        SetVariable   SetDrugInc,pos={409,97},size={147,17},title="Conc.   increment:"
        SetVariable   SetDrugInc,limits={0,INF,0},value=   root:Packages:QModel:DrugInc

        CheckBox   DispCBox,pos={13,119},size={124,20},title="Display   current",value=1
        PopupMenu
DispPPopup,pos={147,119},size={182,19},proc=DispPPopMenuProc,title="Display
probabilities"
```

```
        PopupMenu    DispPPopup,mode=1,popvalue="No",value=    #"ProbMenuItems()"
        SetVariable   SetRelSeg,pos={5,144},size={158,17},title="Relevant    segment:"
        SetVariable   SetRelSeg,format="%d"
        SetVariable   SetRelSeg,limits={1,1,1},value=    root:Packages:QModel:RelSeg
        PopupMenu
AnPopup,pos={166,143},size={118,19},proc=AnalMenuProc,title="Analysis"
        PopupMenu      AnPopup,mode=1,popvalue="None",value=
#"\"None;Maximum;Minimum;Mean;Time    const;Steady    State\""
        Button       ShowMxBut,pos={365,121},size={134,20},proc=ShowMxButProc,title="Show
Tables"
        Button     GoBut,pos={510,121},size={50,20},proc=GoButProc,title="Go"
    EndMacro

    // ===   Utilities   ===

    // MakeIfNotExistD: create new wave or redimension existing
    Function   MakeIfNotExistD(wname,N,M)
        string wname
        Variable  N,M

        if (exists(wname))
                Wave/D  w=$wname
                if ((DimSize(w,0)==N)  %&  (DimSize(w,1)==M))
                        return 0
                endif
        endif

        Make /D/O/N=((N),(M)) $wname
    end

    // MakeIfNotExist:  create new wave or redimension existing
    Function   MakeIfNotExist(wname,N,M)
        string wname
        Variable  N,M

        if (exists(wname))
                Wave/D  w=$wname
                if ((DimSize(w,0)==N)  %&  (DimSize(w,1)==M))
                        return 0
                endif
        endif

        Make /O/N=((N),(M)) $wname
    end

    // MakeWithDup: redimension wave preserving values
    Function   MakeWithDup(wname,N,M)
        string wname
        Variable  N,M

        Variable  i,j,k
        string tmp_name = wname + "_0"
        if (exists(wname))
                Make /D /O/N=((N),(M)) $tmp_name
                Wave/D w = $wname
                Wave/D tmp = $tmp_name
                k=DimSize(w,0)
                i = min(k,N)-1
```

```
            k=DimSize(w,1)
            j = min(k,M)-1
            tmp = 0
            tmp[0,i][0,j]=w[p][q]
            Duplicate /D /O tmp,w
            KillWaves  tmp
      else
            Make /D/O/N=((N),(M)) $wname
            Wave/D w = $wname
            w = 0
      endif
end

// CurrentFolder: strip alien characters from folder name
Function /s CurrentFolder()
      string  folder=GetDataFolder(1)

      Variable idx = strlen(folder)
      Variable spc = char2num(" "), colon=char2num(":"), amp=char2num("'"), c
      folder = folder[5,idx-2]

      idx = strlen(folder)
      do
            c=char2num(folder[idx,idx])
            if (c==spc %| c==colon)
                  folder[idx,idx]="_"
            endif
            if (c==amp)
                  folder[idx,idx]="Z"
            endif
            idx -= 1
      while (idx >= 0)

      return folder
end

// FindTableWithWave: returns string with name of table window for given wave
//   Parameters:
//    w -- Wave
Function/S  FindTableWithWave(w)
      wave w

      string win=""
      variable  i=0

      do
            win=WinName(i, 2)                              // name of ith table window
            if( strlen(win) == 0 )
                  break;                                          // no more
table wndows
            endif
            CheckDisplayed/W=$win  w
            if(V_Flag)
                  break
            endif
            i += 1
      while(1)
      return win
```

```
end

Function  ExpRate(Curr,Start,Dur)
    wave Curr
    Variable  Start,Dur

    NewDataFolder/O    root:Packages:QModel:ExpRateF

    if ((Start+Dur)>rightx(Curr))
            return NAN
    endif
    String CFolder = GetDataFolder(1)
    SetDataFolder    root:Packages:QModel:ExpRateF
    Variable   V_FitOptions=4
    CurveFit/Q exp Curr(Start,(Start+Dur))
    Wave  W_coef=W_coef
    Variable res = W_coef[2]
    SetDataFolder  CFolder
    return res
end

// ModelGlobalVariable: create global variable and set its value
Function     ModelGlobalVariable(name,val)
    string name
    Variable  val

    string  comm
    if (!exists(name))
            sprintf comm,"Variable /G %s = %g",name,val
            execute  comm
    endif
end

// ModelGlobalString: create global string and set its value
Function     ModelGlobalString(name,val)
    string name
    string val

    if (strlen(val) == 0)
            val = "\"\""
    endif
    string  comm
    if (!exists(name))
            sprintf comm,"String /G %s = %s",name,val
            execute  comm
    endif
end
```

Author Index

Numbers in parentheses are footnote reference numbers and indicate that an author's work is referred to although the name is not cited in the text.

A

C

D

E

F

G

H

K

L

M

Q

R

S

T

U

V

W

Subject Index

A

B

C

D

E

F

G

H

I

K

L

M

N

P

Q

R

S

T

Y

ISBN 0-12-182194-3

90038

9 780121 821944